Cholesterol Autoxidation

Cholesterol Autoxidation

Leland L. Smith

The University of Texas Medical Branch
Galveston, Texas

PLENUM PRESS · NEW YORK AND LONDON

Library of Congress Cataloging in Publication Data

Smith, Leland L.
  Cholesterol autoxidation.

  Bibliography: p.
  Includes index.
  1. Cholesterol. 2. Oxidation. 3. Cholesterol—Metabolism. I. Title.
QP752.C5S56   599.01'92431   81-10616
ISBN 0-306-40759-0   AACR2

QP
752
.C5
S56

© 1981 Plenum Press, New York
A Division of Plenum Publishing Corporation
233 Spring Street, New York, N.Y. 10013

Printed in the United States of America

Most components of the biosphere are continuously exposed to oxygen from the atmosphere. Accordingly, the inexorable deterioration of all organic compounds by the slow attack of oxygen must occur. Despite this eventuality, a definitive treatment of oxygen-dependent decomposition of any single important natural product has not heretofore been made. The instant monograph attempts to provide a complete description of the autoxidation of one such important natural product, cholesterol, as the matter is currently understood.

The autoxidation of cholesterol in Nature has been a matter of interest to others since the close of the nineteenth century and to me for the past three decades. In this monograph I present aspects of what I have learned about cholesterol autoxidation during that interval. Because of the diffuse and troublesome nature of the subject I have selected to cite references to the literature rather fully, so that all items discussed may be properly evaluated by the interested reader. Though such extensive citation of references makes for labored reading, I hope the text will serve as a definitive treatment of the subject from which other studies may be engendered without extensive recourse to the older material.

An attempt has been made to include much related information so that a detailed awareness of the interrelationships between cholesterol autoxidation and other aspects of chemistry, metabolism, and toxicology may be had. I risk teaching more on the subject than any reader care to learn. The material utilized in the preparation of the monograph is that which I have encountered in the ordinary perusal of literature for interest, and each item cited has been personally examined by me for relevance and content. No exhaustive or retrospective search of the literature via Chemical Abstracts or similar resource has been attempted, so there may be a body of information on the subject I have not discovered.

Some familiarity with sterol chemistry and steroid nomenclature is presumed. As best I can accomodate them, systematic names for all steroids have been provided according to the IUPAC-IUB 1967 Revised Tentative Rules for

Steroid Nomenclature.   Common trivial names have also been
used in a few instances.   The usual α-, β-, and ξ-designa-
tions for nuclear substituents are used, but the Cahn-
Ingold-Prelog sequence rule nomenclature is used for desig-
nation of configuration in the side-chain.   Fisher projection equivalents (α_F- and β_F-) are given in some cases.
Not every stereochemical detail is drawn in the structural
formulas, and except where additional detailed presentation
of stereochemistry is essential, such matters as configuration of nuclear carbon atoms, side-chain substituents, etc.
are not drawn.   Thus, the 5α- and 5β-stereochemistry about
the C-5 atom is always specified, but that at other nuclear
centers is not.

     The monograph begins with a historic treatment of the
topic but then careers through the initial chemical events
of autoxidation to the more complicated subsequent reactions
that provide the many identified cholesterol autoxidation
products, now numbering approximately eighty.   Many of
these autoxidation products have been confused with genuine
cholesterol metabolites.   Furthermore, many oxidized sterols
exhibit interesting biological activities in divers assay
systems.   Thus, the role of oxidized cholesterol derivatives
in metabolism and cellular function becomes a topic of importance deserving coverage.   A final chapter deals with
the question of what constitutes pure cholesterol.

     Yet another major theme prevades throughout the mono-
graph, that of artifact versus genuine metabolite.   The
insidious nature of cholesterol autoxidation suggests that
autoxidation has intruded without recognition into many
past studies.   Artifacts of air oxidation have received
much attention.   Indeed, the question of artifact versus
genuine metabolite plagues most investigations, with the
issue being unrecognized or ignored in some instances but
also with arguments on a case by case basis for genuine
metabolic status, for artifact status, and for uncertain
status.   All too often unwarrented claims for genuine natural product status have been advanced, and there is still
an unawareness of the care with which biological material
must be handled to avoid artificial oxidations of sensitive
sterols and unsaturated lipids.

     In no other area of biochemistry does the problem of
artifact versus metabolite loom so large.   The issue has
not heretofore been given the careful scrutiny necessary

for resolution that the present monograph attempts to provide. The recorded evidence, which is considerable, on
balance supports an artifact status for a great many sterol
oxidation products found in biological materials, and the
thrust of this monograph is that all claims to genuine
metabolite status for cholesterol autoxidation products
unsupported by experimental evidence be rejected! In the
absence of specific and compelling experimental evidence
that a given autoxidation product be also derived by enzyme
action, artifact status is the better assignment. Indeed,
autoxidation products may measure the extent of uncontrolled
air oxidation that has occurred in the system.

It is hoped that this monograph will provide the basis
for understanding and acceptance of this point of view but
not as a dogmatic closure of the field for study. Rather,
these lines should serve as provocation to those who seek
to argue their discovery of a sterol autoxidation product
in a special biological setting be result of enzyme action
that they also provide supporting experimental evidence and
not resort to asservation or disputation. Recollection and
judicious application of Ockham's razor, *Frustra fit per
plura quod potest fieri per pauciora* (what can be explained
by the assumption of fewer things is vainly explained by
the assumption of more things), is appropriate.

A considerable body of new experimental evidence has
accumulated over the past decade which clarifies much of
the many prior observations of sterol oxidation and justifies its presentation at this time in this new synthesis
of ideas regarding factitious oxidation of cholesterol.
Moreover, recent discoveries of powerful biological activities of some cholesterol autoxidation products presage other
new dimensions of interest in cholesterol autoxidation
which call for the present monographic treatment.

Much of the recent unpublished experimental work contributing to the present state of understanding of cholesterol autoxidation described herein was conducted by Drs.
G.A.S. Ansari, Yong Y. Lin, and Jon I. Teng of my laboratory
and was supported financially by the Robert A. Welch Founda-
tion, Houston, Texas and by U.S. Public Health Service research grants HL-10160 and ES-00944.

CONTENTS

LIST OF TABLES

LIST OF FIGURES

Cholesterol (cholest-5-en-3β-ol) (1), the chief sterol
of mammalian tissues and obligate precursor of steroid hormones and bile salts, has received much attention to its
chemistry and biochemistry, to its role in membrane integrity
and function, and to its association with human diseases

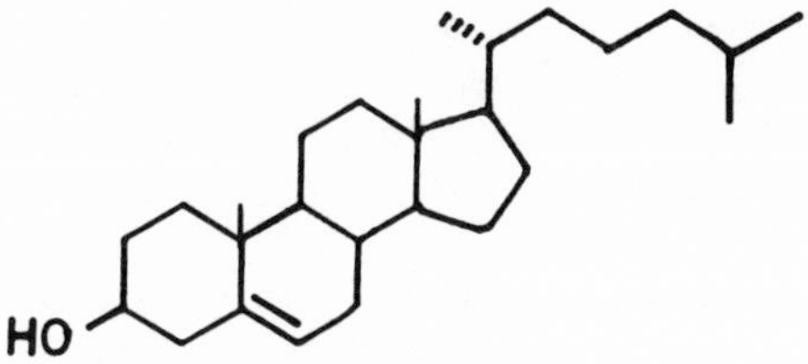

HO

1

such as atherosclerosis, gallstones, and cancer. The relationships between cholesterol and human disorders have
attracted interest for over two centuries and date from the
work of Vallisneri [2545] in 1733 on the properties of human
gallstones, from which cholesterol was ultimately isolated.
An association between cholesterol deposits and human aortal
plaque formation was well recognized by 1847 [2597], and
several other relations of cholesterol to pathological
states were summarized in 1862 [779].

Systematic study of the chemistry of cholesterol and
interest in metabolism of cholesterol began at the end of
the nineteenth century. As new techniques of experimenta-
tion and general advances in chemistry and physiology were
developed, added insight into the complexity of cholesterol
metabolism and of its importance in human medical matters
also obtained. Necessarily, detailed understanding of
sterol involvement in metabolism had to await assignment of
the correct chemical structural formula to cholesterol in
late 1932 [998, 2694], after which time a burgeoning of work
relating cholesterol to the bile acids, vitamins D, steroid
hormones, and triterpenoids has accorded us our present
extensive knowledge of this complex field of study.

As a necessary consequence of experimental investigation
of cholesterol biochemistry in these contexts, the concommitant factitious oxidation of cholesterol by the oxygen
of the air has intruded into all matters. This process of
oxidation, or autoxidation, involves a variety of reaction

types, and as encompassed herein Uri's definition of autoxidation holds: "Autoxidation is the *apparently* uncatalyzed
oxidation of a substance exposed to the oxygen of the air"
[2539]. Autoxidation being apparently uncatalyzed is thus
distinct from the two other forms of natural oxidation
combustion and respiration (metabolism). The chemistry of
cholesterol combustion has never been addressed, although
it is known that cholesterol survives the burning of cigar-
ette tobacco [924]. By contrast, the metabolism of cholesterol has received great attention, and with such interests
has come awareness of autoxidation as a problem with which
one must contend.

The primary processes of cholesterol autoxidation are
those of incorporation of both atoms of molecular oxygen
into the sterol molecule, but dehydrogenation of cholesterol to ketones also occurs as a minor primary process.
Oxidations involving acknowledged chemical oxidants other
than oxygen and dehydrogenations such as those moderated by
I_2 and O_2 [2680], chloranil [557], etc. leading to polyenes,
dimeric polyenes, and such derivatives with angular methyl
group migrations and/or skeletal rearrangements are not
considered as autoxidations but as chemical oxidations.

The press of investigation into cholesterol biochemistry has provided a fairly comprehensive understanding of
cholesterol in medical matters. A much more protracted and
less complete development of understanding of the chemistry
of cholesterol autoxidation derived in parallel. Two major
reviews by Bergström of cholesterol autoxidation [197,200]
and repeated treatment of the subject in other monographs
[198,285,361,747,748,863,1388,2037,2154,2240,2321,2372] but
not in all [737,738,1333,1459,1460,2243] evince continuing
concern for the problem, but the latest monographs on cholesterol devote but a few lines to the matter [2055,2322].
There now exists adequate new experimental evidence to provoke formulation of a unified concept of cholesterol autoxidation as it occurs in close association with cholesterol
metabolism [2290], a matter which this monograph attempts
to treat.

Moreover, broad interests in the similarities between
autoxidation and related controlled chemical oxidations of
natural products on the one part and formally similar
enzymic oxidations on the other are supported by much experimental work. Interest in such biomimetic oxygenations of

divers classes of organic compounds is growing [1586], and
we have previously examined the topic for the sterols
[2563].

STEROL REDUCTIONS

Much of the interest in cholesterol biochemistry derives from the facile esterification of the 3β-hydroxyl
group, but oxidation-reduction reactions figure prominently
in its biochemistry as well. In that cholesterol is an
olefinic alcohol its biochemistry also includes reduction
as a means of metabolism. Indeed, the reduction of cholesterol is a predominant mode of transformation in anaerobic
environments such as the gastrointestinal tract, where
enteric microflora reduce cholesterol to the stanol 5β-
cholestan-3β-ol (coprosterol)(4). In anaerobic processes
of diagenesis of sediments, the fully saturated steranes
5α-cholestane (3) and 5β-cholestane (5) have been detected
[136,839,1015], along with the olefins 5α-cholest-2-ene
[563], cholest-4-ene (9), and cholest-5-ene (7) [1950].
Moreover, recent marine sediments may also have a burden of
sterols including the 5β-stanol 4 from contamination by
human sewage [977].

Additionally the stanol 5α-cholestan-3β-ol (2) occurs
ubiquitously with cholesterol in mammalian tissues. Although biosyntheses of the stanols 2 and 4 are overall reductive in character, an initial dehydrogenation of cholesterol to cholest-5-en-3-one (6) and isomerization of 6 to
cholest-4-en-3-one (8) preceed double bond and carbonyl
group reductions. By contrast the steranes 3 and 5 of
marine sedimentary rocks may form reductively by geologic
nonenzymic or microbial enzymic processes from the stanols
2 and 4 [854] or from the C_{27}-monoolefins found in sediments [563,1950]. However, an obscure metabolism of cholesterol may occur in mammals, for the olefin 7 and cholesta-
3,5-diene (11) have been found among the hydrocarbons
of human aortal atheromatous plaques under conditions which
do not suggest artificial origins [370].

Cholesterol appears to be particularly sensitive as
regards oxidation, and the chemistry and biochemistry of
cholesterol is far more extensively dominated by oxidative
processes. Recognition of the role which oxidative processes play in the chemistry and biochemistry of cholesterol

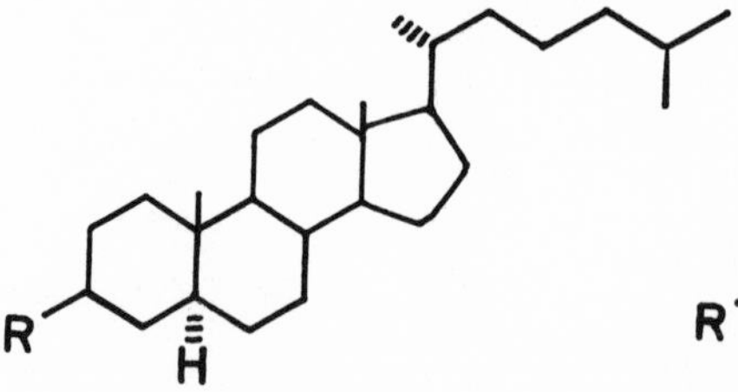

2 R = OH   4 R = OH

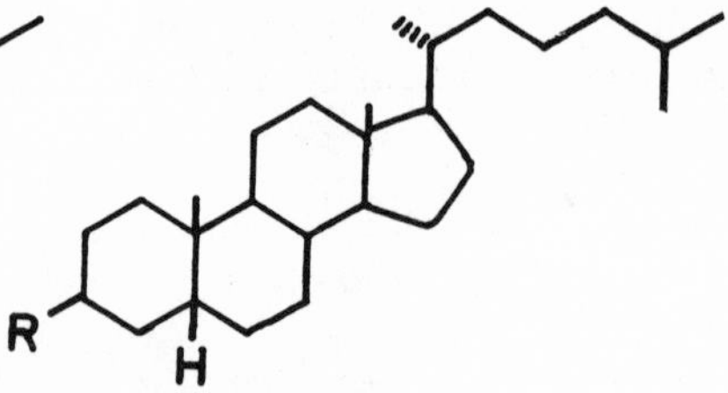

3 R = H   5 R = H

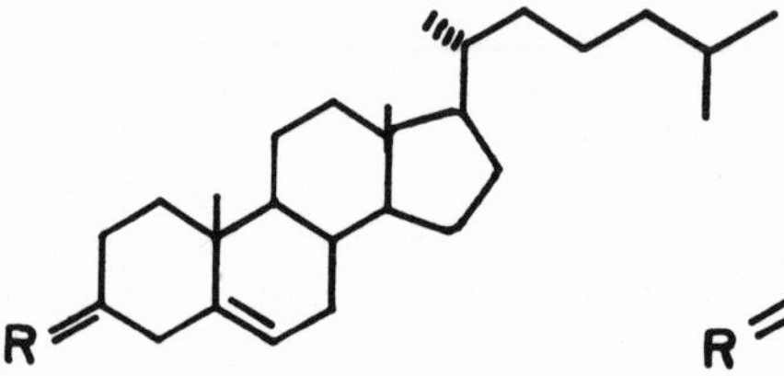

6 R = O   8 R = O

7 R = H$_2$ 9 R = H$_2$

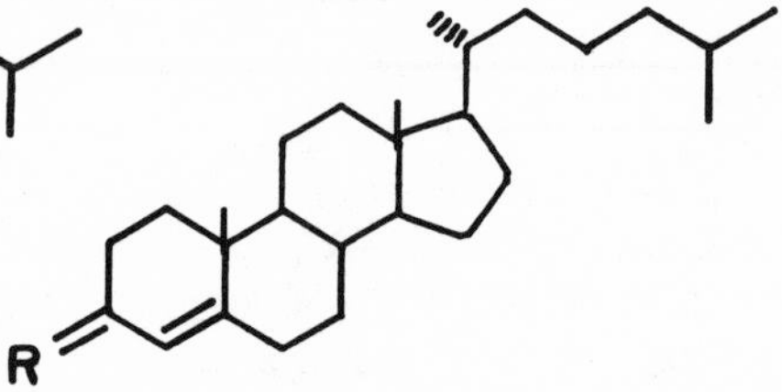

has been had for over 75 years, but a general confusion
about the differential participation of enzymic and nonenzymic processes associated with the very earliest studies
of cholesterol oxidation has persisted to the present day
with little respite.

ARTIFACTS OF MANIPULATIONS

Experimental difficulties associated with the recovery
of pure sterols from biological sources have continuously
plagued progress in these matters. Contaminations of
sterols with stopcock grease [849,1888,2621], friedelin
from cork stoppers [960,1888,1954], of phthalic acid
diesters from plasticware [1589,1952] are obvious hazards
of faulty manipulations which should no longer occur. Moreover, artifacts of decomposition of digitonin used in the
digitonin precipitation of 3β-hydroxysterols [971-973,1551]
or of Girard derivatives used for fractionation of steroid
ketones fron nonketones [1006,1888,2023] are now eliminated
by advances in chromatography which have displaced these

older procedures.

The storage or handling prior to isolation or analysis
of biological samples containing sterols is of critical
concern, as artifacts may form in the samples without specific manipulations being involved. Storage of such samples
unprotected from radiation and air or from continuing
enzyme action should not occur. Moreover, exposure of
samples to preservatives such as alcohol or formaldehyde
must be carefully avoided, as primary products of cholesterol autoxidation (cholesterol hydroperoxides) [2571]
and higher degradation products such as cholesta-3,5-dien-
7-one (10) cholesta-4,6-dien-3-one (12), and 5α-cholestane-

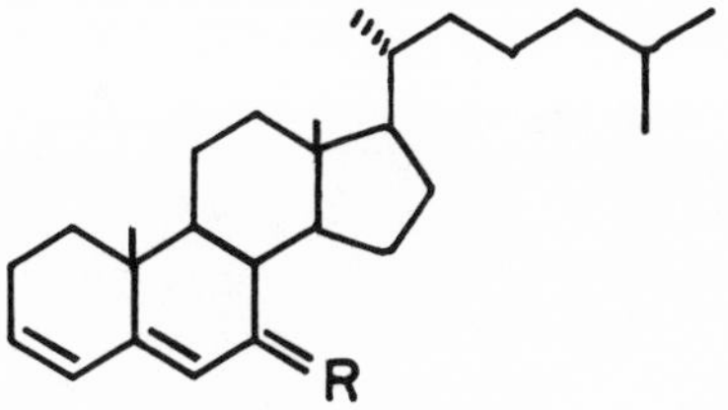

10 R = O

11 R = H$_2$

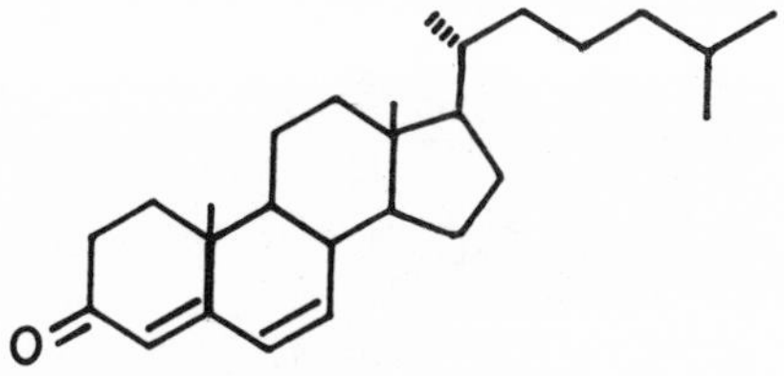

12

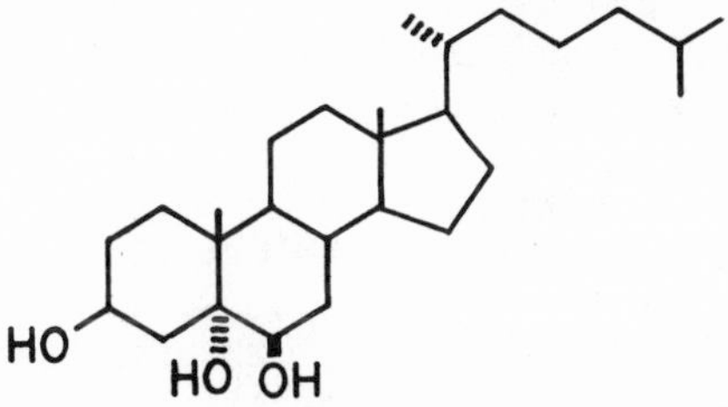

13

3β,5,6β-triol (13) have been detected in human tissues
fixed in formaldehyde [746,1964].

A variety of sterol alteration products may also be
formed as artifacts of standard procedures used for isolation of sterols from total lipids mixtures from tissues.
Cholesterol occurs in mammalian tissues as the free sterol,
as fatty acid esters, and as sulfuric acid esters, and
recovery of total cholesterol as the free sterol necessarily
requires saponification or solvolysis of these esters.

Alkaline saponification in the presence of air is characterized by the formation of low levels of the common cholesterol autoxidation products cholest-5-ene-3β,7α-diol (7α-
hydroxycholesterol) (14), cholest-5-ene-3β,7β-diol (7β-hy-
droxycholesterol) (15), 3β-hydroxycholest-5-en-7-one (7-
ketocholesterol) (16), and the 3,5-dien-7-one 10 derived
from 16. The epimeric 3β,7-diols 14 and 15 and the 7-ketone
16 constitute the most frequently reported pattern of cholesterol autoxidation products, which pattern suggests

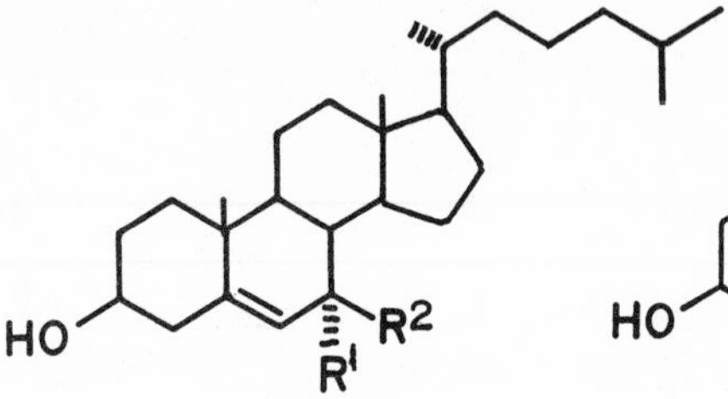

14 R^1 = OH, R^2 = H   16

15 R^1 = H, R^2 = OH

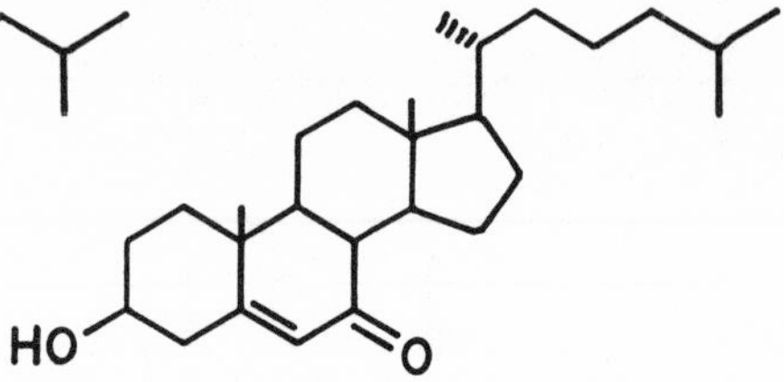

factitious oxidation of cholesterol in the sample during
collection of processing.

    Solvolysis of cholesterol and other lipid esters using
organic solvent solutions and inorganic acids is noted for
formation of artifacts not derived by autoxidation. Thus,
the methanolysis of cholesterol esters in lipid mixtures is
notorious for the formation of free cholesterol together
with the condensation products cholesterol 3-O-methyl ether
(17) and dicholesterol ether (18) as well as the elimination
product cholesta-3,5-diene (11) [683,1263,1383,1688,1715,
2199]. Anent the occurrence of cholesterol ethers in tissues,
the isolation of dicholesterol ether from bovine spinal cord
cholesterol [2252] appears to be artifactual, whereas cholesterol 3-O-hexadecyl ether (19) recovered from bovine cardiac
muscle [817] may be a new natural product of endogenous origins or an artifact.

    Sterol saponins which occur in plant tissues and sterol
glucuronides which may occur as conjugated metabolites in
urine likewise require hydrolysis for recovery of the free
sterols, and elimination reactions of such Δ^5-3β-hydroxy-
sterol glycosides yielding the corresponding 3,5-dienes as
artifacts are well known. Thus, recovery of sitosterol

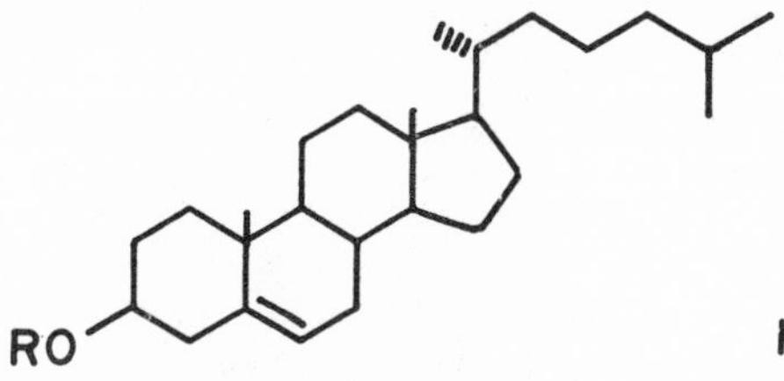

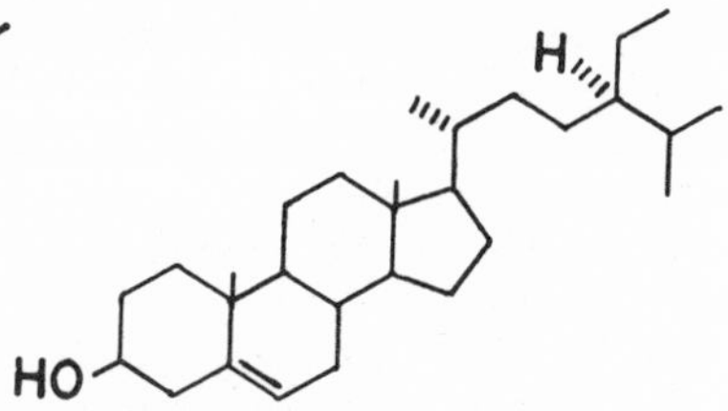

$\underline{17}$ R = CH$_3$

$\underline{18}$ R = C$_{27}$H$_{45}$

$\underline{19}$ R = C$_{16}$H$_{33}$

$\underline{20}$

(stigmast-5-en-3β-ol, 24-ethyl-(24R)-cholest-5-en-3β-ol, 24α$_F$-ethylcholest-5-en-3β-ol) ($\underline{20}$) from tall oil rosin yields stigmasta-3,5-diene as by-product [692,1452], and acid hydrolysis of (25R)-spirost-5-en-3β-ol (diosgenin) glycosides from various *Dioscorea* species regularly is accompanied by (25R)-spirosta-3,5-diene [171,305,344,1574, 1833,2752]. Moreover, the refining of edible oils is notorious for the occurrence of elimination reactions on both sitosterol and cholestrol, yielding steroidal olefins as artifacts [1260,1760,1763-1766,1970].

STEROL OXIDATIONS

Factitious alterations of cholesterol are relatively readily recognized where fairly strong chemical conditions well removed from those likely to be encountered in biological systems are implicated. However, it is the less obvious insidious alteration of cholesterol by autoxidation occurring from the moment of specimen collection through final processing under mild conditions and analysis, which transformations may not be as clearly recognized or understood, that will be treated in detail in this monograph. The omnipresent alteration of cholesterol by air oxidation may be projected as present in most studies of cholesterol, only one notable example of a preparation of pure cholesterol being stable for years in contact with air having been published [684]. Except for this unique case and for instances where protection against autoxidation is provided by antioxidants, protective colloids, enzymes, etc. one may project that where adequate analysis methods are utilized for their detection, the several oxidation products of

cholesterol will be formed and found in chemical and enzyme
systems exposed to the air.

The problem of recognition of autoxidation in studies
of tissue sterol metabolism is exacerbated by the presence
in some tissues of simple cholesterol oxidation products
that are formed by established enzyme processes acting on
cholesterol as substrate and which presumably serve some
physiological purpose. In several cases these simple cholesterol oxidation products derived enzymically are also prominent autoxidation products of cholesterol, and the problem of differentiating between metabolite and artifacts has
yet to be solved definitively. This problem occurs in both
plant and animal tissues, but the matter is of much greater
concern where the oxidized cholesterol derivatives are implicated in endogenous cholesterol metabolism, particularly
in rate-limiting reactions subject to hormonal or feedback
control.

Two prominent examples of monohydroxylated cholesterol
derivatives in these circumstances are the 3β,7α-diol (14)
and (20S)-cholest-5-ene-3β,20-diol (cholest-5-ene-3β,20α$_F$-
diol, 20α-hydroxycholesterol) (21). The 3β,7α-diol 14

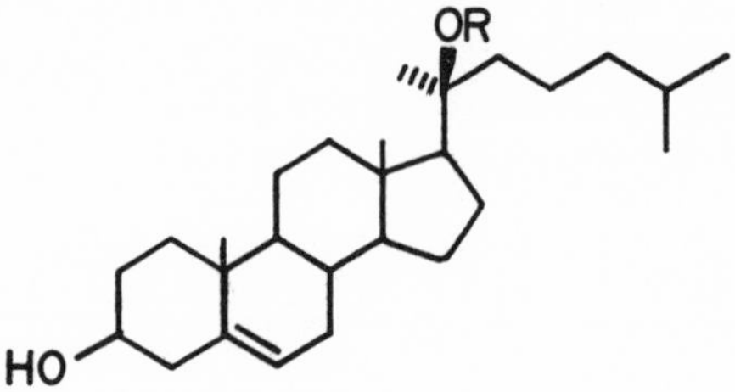

21 R = H

22 R = OH

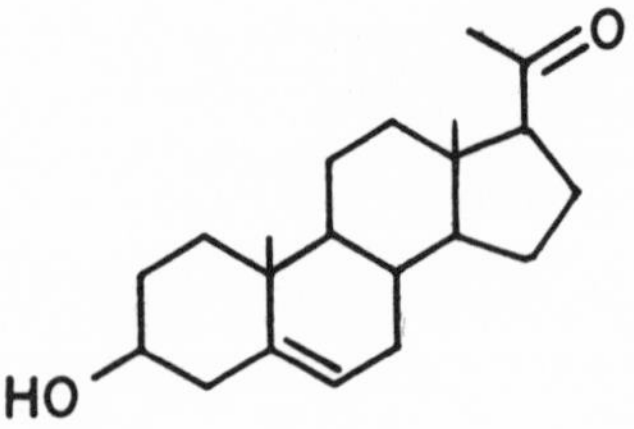

23

formed by a specific microsomal cholesterol 7α-hydroxylase
of mammalian liver appears to be the initial, committed
step in hepatic biosynthesis of bile acids from cholesterol,
the 7α-hydroxylase being under product feedback control.
However, the 3β,7α-diol 14 is also one of the most frequently
encountered, major autoxidation products of cholesterol, and
means of distinguishing simultaneous autoxidative and enzymic 7α-hydroxylation of cholesterol have occupied the atten-
tions of investigators for two decades. Similarly, the
(20S)-3β,20-diol 21 is an autoxidation product derived from

the corresponding hydroperoxide 3β-hydroxy-(20S)-cholest-5-
ene-20-hydroperoxide (22) [2576,2578] but which has also
been proposed as an initial intermediate in the biosynthesis
of pregnenolone (3β-hydroxypregn-5-en-20-one) (23) from
cholesterol in adrenal cortex mitochondria, a step considered to be under trophic hormone and product feedback control.
The (20S)-3β,20-diol 21 has been detected at very low levels
in bovine adrenal tissues [1963] and in human aortal tissue
[1427].

Other monohydroxylated cholesterol derivatives found
in tissues pose the same problems. Thus, the 3β,7β-diol
15 may be product of enzymic reduction of the 7-ketone 16
[265,1653,1655] as well as a major autoxidation product.
However, direct 7β-hydroxylation of cholesterol by rat
liver, though postulated [1616], has not been demonstrated,
the product 3β,7β-diol 15 arising via generalized lipid
peroxidation processes instead [1192,2311,2461].

The 22-hydroxycholesterol derivative (22R)-cholest-
5-ene-3β,22-diol (cholest-5-ene-3β,22β$_F$-diol)(24) [403,1670]
found in bovine adrenal [619] and plant [2333] tissues and
probably as sulfate esters in human excretions is a cholesterol metabolite, being an initial oxidized intermediate
in the scission of the sterol side chain by adrenal cortex
mitochondria. Furthermore, 3β,22ξ-diol 24 sulfate esters
found in human meconium [680], infant feces [931] and
urine [1557], and umbilical cord plasma [654] as well as a
(23ξ)-cholest-5-ene-3β,23-diol sulfate ester also found in
human meconium [680] appear to be cholesterol metabolites.
No autoxidation process has yet been discovered which gives
a 22- or a 23-hydroxylated cholesterol derivative as product.

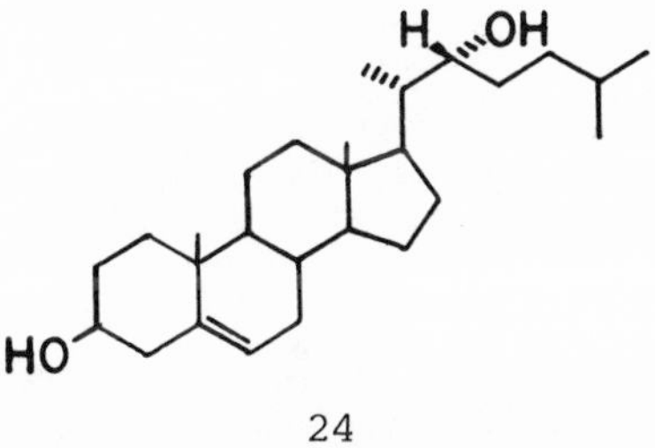

24 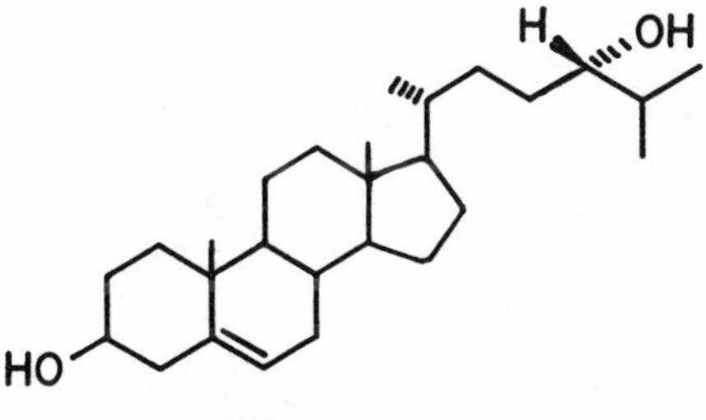   25

The four cholesterol derivatives monohydroxylated
towards the terminus of the side chain, (24S)-cholest-5-

ene-3β,24-diol (cerebrosterol, cholest-5-ene-3β,24β$_F$-diol,
24-hydroxycholesterol) (25), cholest-5-ene-3β,25-diol (25-
hydroxycholesterol) (27), and (25R)-cholest-5-ene-3β,26-
diol (26-hydroxycholesterol) (29), and (25S)-cholest-5-ene-
3β,26-diol (31) lend additional dimension to the problem.
The (24S)-3β,24-diol 25 present in mammalian brain [607,

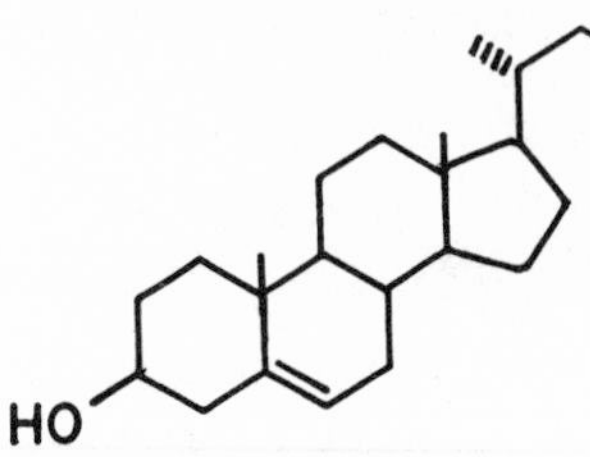

    26 R = OH   28 R = OH

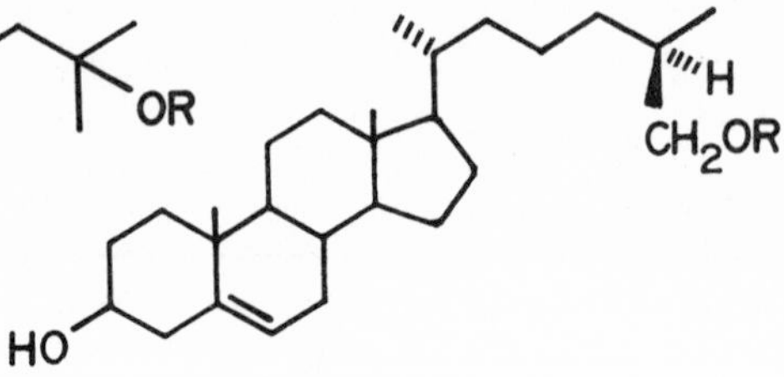

    27 R = H   29 R = H

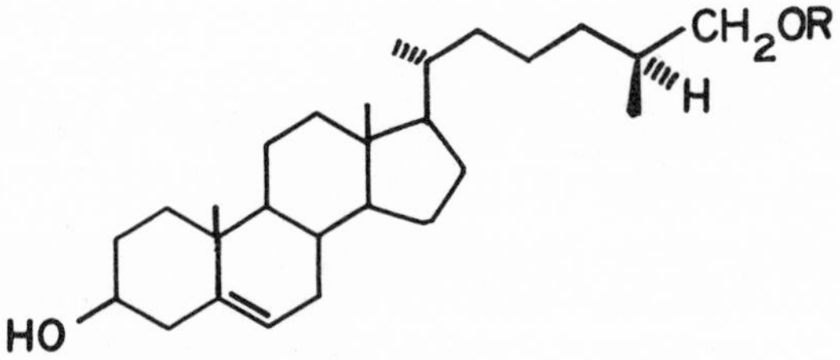

    30 R = OH

    31 R = H

613,693-695,697,1952,2154,2306,2316,2571] has a demonstrated
enzyme origin [607,1511] and a (24ξ)-cholest-5-ene-3β,24-
diol found as fatty acid esters in the human aorta [396,
2460] and as sulfate esters in human meconium and infant
feces [680,931], urine [1557], and plasma [2396] must likewise be a metabolite of cholesterol. Moreover, a direct
autoxidation pathway leading to a 24-hydroxylated cholesterol derivative from cholesterol has not been discovered
[2579].

    The 3β,25-diol 27 found in various mammalian tissues
[2567] is clearly a major secondary product of cholesterol
autoxidation derived from the hydroperoxide 3β-hydroxy-
cholest-5-ene-25-hydroperoxide (26) [2576,2578],but enzymic
origins for 27 have also been suggested [94,268,269,271,272,

797,798,927,928,1247,1390,1639,1835,1837]. Finally, (25R)-
cholest-5-ene-3β,26-diol (29) is present in human aortal
tissues [364,816,1914,2304,2315,2351,2567] and a (25ξ)-3β,
26-diol 29 and/or 31 is present in human brain [2280,2316]
and as fatty acyl esters in human aorta [362,369,874,2460]
and as sulfate esters in human meconium [680], infant feces
[931], plasma, and urine [2396]. Enzymic 26-hydroxylation
of cholesterol has been repeatedly demonstrated in murine
liver systems [94,207,260,267,268,545,797,798,927,928,930,
1247,1390,1639,1645,1834-1837], and evidence of enzymic
26-hydroxylation of cholesterol in human liver mitochondria
[269], human arterial segments [1563], and potatoes [994]
has been presented. Although the cholesterol 26-hydroxy-
lase of mouse liver has long been recognized as stereospecifically forming but one isomeric 3β,26-diol 29 or 31 [207],
hepatic biosynthesis of the (25R)-3β,26-diol 29 has recently
been demonstrated the case (cf. Chapter VII). Moreover, a
well defined autoxidation pathway for formation of both
(25R)- and (25S)-3β,26-diols 29 and 31 from cholesterol via
the corresponding 3β-hydroxy-(25R)- and (25S)-cholest-5-
ene-26-hydroperoxides 28 and 30 has been established [2559].

    Other monohydroxylated cholesterols described in the
literature, including the 1α- [150,822,1237,1673,1674,1679,
1842,2079], 1β- [1629], 2α- [451,1237,1238], 2β- [451,1674],
4α- [759], 4β- [424,2019], 11α- and 11β- [2117], 12α- [546,
664], 16β- [459,2095], 17α- [459], 19- [1213,1671], 21-[428,
2672], 22α$_F$(or 22S)- [403,1884,2516], 24α$_F$ (or 24R)-[634,693,
695,697,1363,1512,2575], and (25S)-26- [446,1915,2152,2584]
hydroxysterols, are products of chemical synthesis for
which little biological interest has been advanced. The
isomeric 20β-alcohol (20R)-cholest-5-ene-3β,20-diol (20β-
hydroxy-20-isocholesterol) (33) [1630] nominally considered
also of synthesis origin may also be derived by autoxidation, for the corresponding hydroperoxide 3β-hydroxy-(20R)-
cholest-5-ene-20-hydroperoxide (32) has been found as product of cholesterol autoxidation [2565].

    Other simple oxidized derivatives which conceivably
could be formed enzymically directly from cholesterol include the 5-en-3-one 6, the isomeric 4-en-3-one 8, and the
7-ketone 16 all implicated in autoxidation as well, and the
24-ketone 3β-hydroxycholest-5-en-24-one (24-ketocholesterol)
(34) isolated from marine plants [1338,1692,2057,2654] and
animals [135,444,592,2475] but for which a direct autoxidation pathway has been demonstrated [2579]. Other further

oxidized cholesterol derivatives such as the 3,5-dien-7-one
10 and 4,6-dien-3-one 12 isolated from mammalian tissues
[960,1888,1964,2316] are all variously implicated in the
autoxidation of cholesterol, as are also the 5,6-epoxides,
5,6α-epoxy-5α-cholestan-3β-ol (cholesterol α-oxide) (35)
and 5,6β-epoxy-5β-cholestan-3β-ol (cholesterol β-oxide) (36)
and 3β,5α,6β-triol 13. Other cholesterol derivatives oxi-

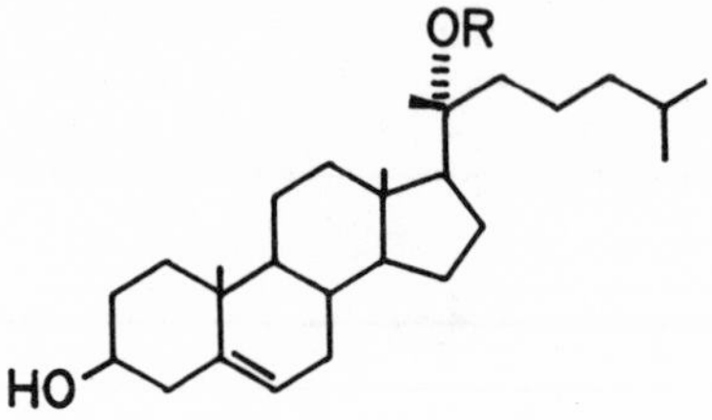

32 R = OH

33 R = H

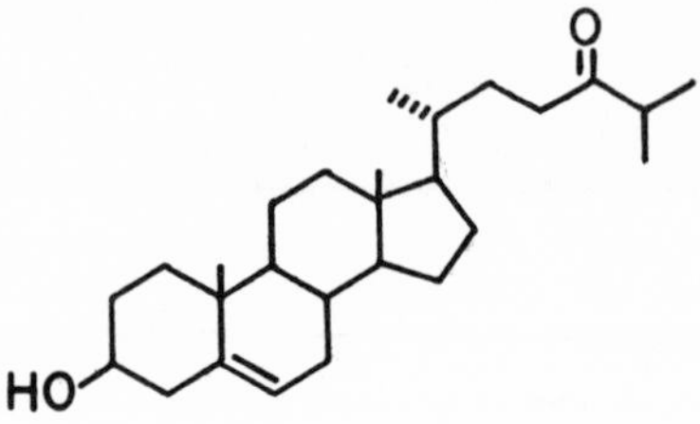

34

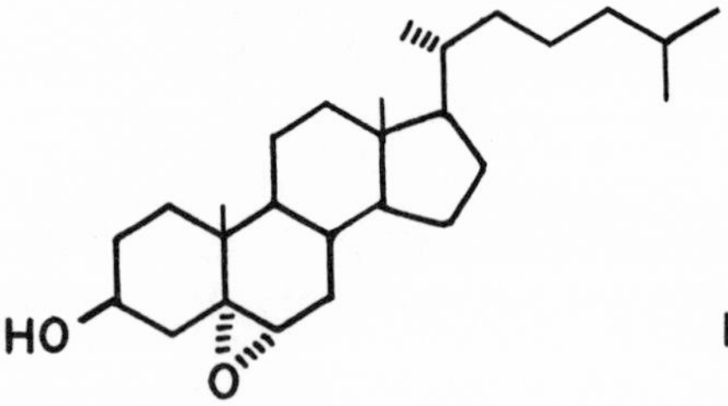

35

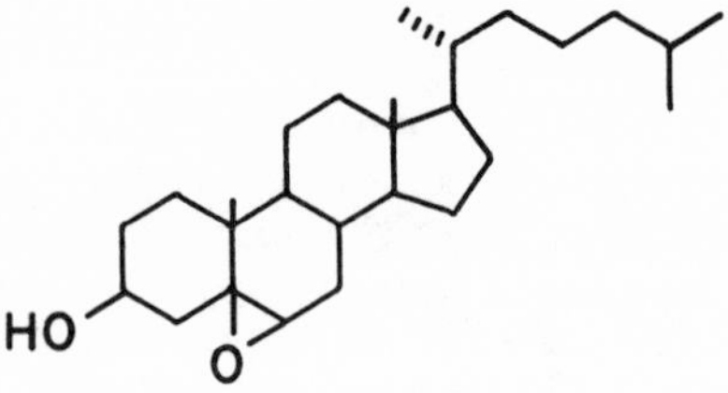

36

dized in a simple or one step fashion include several
sterol hydroperoxides whose nature as autoxidation products
is established but for which some evidence of enzyme origins
also exists.

     There are thus ample opportunities for confusing sim-
ple oxidized cholesterol derivatives of a variety of
structural types derived from autoxidation with the same or
very closely related derivatives which are tissues metabolites. It is small wonder that a satisfying description of
these oxidative processes or an adequate resolution of the
issue of ultimate origins has continuously eluded investigators.

CHAPTER II.   HISTORY OF CHOLESTEROL AUTOXIDATION

Systematic study of the chemical oxidation begun by Mauthner by 1894 [1594] was closely followed by the first published accounts by Israel Lifschütz of encounters with cholesterol autoxidation products whose nature was unrecognized at the time [561,1473]. Experimental work on cholesterol autoxidation may be traced back to Lifschütz' work of 1895 on wool fat in which the use of different color tests for cholesterol figured prominently [558-561,1473]. From that time the history of cholesterol autoxidation closely parallels the advances made in cholesterol chemistry and suggests that the early biochemistry of cholesterol derived directly from the early attempts to overcome some of the problems associated with cholesterol in which cholesterol autoxidation intruded. The history of study of cholesterol autoxidation may be conveniently divided into three separate phases: (i) an early phase dominated by the work of Lifschütz which included many random observations of oxidative transformations of cholesterol, made before the structure of cholesterol was known and ending with the assignment of the correct structure of cholesterol in 1932; (ii) an intermediate phase involving additional random observations, with isolation from tissues for the first time of identified cholesterol oxidation products, including the 3β,5α,6β-triol 13, the 7-ketone 16 and the epimeric 3β,7-diols 14 and 15; and (iii) the most recent phase dating from about 1960 in which application of effective thin-layer chromatography procedures in systematic fashion led us to the present stage of our knowledge of cholesterol autoxidation as adumbrated in this monograph.

Each of these arbitrarily defined historical phases (indeed, the whole history of the matter) is characterized by a common, unresolved question which overshadows in essence every report on the topic, namely, the question of the true origins and biological significance of oxidized cholesterol derivatives encountered in biological material. Are the oxidized cholesterol derivatives so frequently found in biological systems natural products or metabolites formed by the action of individual enzymes acting within the natural biological system or are they artifacts of manipulation formed by the acts of analysis or isolation? In each of the historical phases one after another report asseverates true natural product, biosynthesis intermediate, or metabolite status for the several oxidized cholesterol derivatives.

Other papers equally vigorously deny biological significance
to the common autoxidation products. Some reports fail to
address the issue of ultimate origins whereas yet others
pose the question but equivocate on an answer or provide no
conclusions. This pattern of uncertainty has continued to
the very recent period of 1975-1976, where major investigations on the presence of oxidized cholesterol derivatives
in human tissues [103] and in enzyme incubations [327]
attest the unusual experimental difficulties associated with
this matter.

EARLY HISTORY, 1895-1932

The historical account of cholesterol autoxidation
which may be traced back to about 1895 necessarily evolves
from the history of cholesterol itself. Recovery of cholesterol from alcoholic extracts of human gallstones was
probably achieved by Poulletier de la Salle by 1782 [779,
1845], possibly as early as 1769 [788], and later in association with de Fourcroy [787-790,1845] as well as by Conradi
in 1775 [510]. Windaus attributed isolation in 1788 of
crystalline cholesterol from gallstones to Gren of Halle,
who apparently named the sterol "cholestearin" [221,2695].
However, the history of the sterol generally dates from the
isolation of cholesterol from human gallstones in 1815 by
Chevreul [477] who named the material "cholesterine" (from
χολη, bile and στερεοσ, solid) [478]. The chemistry of
cholesterol developed slowly following Chevreul's description of the sterol, with recognition achieved of the
hydroxyl functional group by 1859 [210], olefinic double
bond by 1868 [2715], correct $C_{27}H_{46}O$ molecular formula by
1888 [1787, 1928] and monomeric nature by 1890 [3]. Discovery of the ready hydration of cholesterol [1002], of its
optical rotation [1021, 1516], and concern over matters of
saponification, melting point, crystal form, and differentiation of cholesterol from isocholesterol and phytosterol
preparations [314,315] characterized this early period
prior to initiation of systematic study of cholesterol
organic chemistry by Mauthner by 1894 [1594,2017].

The biochemistry of cholesterol evolved even more
slowly with the broad distribution of cholesterol in a
variety of human tissues being recognized by 1862 [779] and
in other biological materials such as wool fat by 1872
[2161,2162]. The fundamentals of cholesterol metabolism

had been grasped by Flint by 1862, at which time he suggested
biosynthesis of cholesterol in the brain, transport in
blood, liver absorption, and biliary secretion, catabolism
in the intestinal tract, and fecal excretion [779]. That
the article expressing these perspicacious concepts was followed by one dealing with Indian arrow wounds [220] clearly
presaged a long wait before details of cholesterol metabolism could be had.

Application to cholesterol of specific color tests such
as the sulfuric acid-chloroform test of Salkowski [2063] and
the acetic anhydride-sulfuric acid test of Liebermann [1472]
and Burchard [397] made by 1855-1890 accorded means of
detection and measurement of sterols in biological systems
for systematic experimental study of cholesterol biochemistry. Progress on these matters at the turn of the century
was very much influenced by the work of Lifschütz, whose
isolation of cholesterol from wool fat in 1895-1898 and
exploitation of color test methodology [558-561,1473,1474],
led to his early suggestions of 1908-1909 regarding the
biosynthesis of cholesterol from fatty acids [1478] and of
bile acids from cholesterol [1480,1484,1486].

One must also credit Lifschütz with the first recorded
instance of confusion about the oxidation of cholesterol by
air as an enzymic event or as an artificial result of manipulation. The suggestions that oleic acid be a precursor of
cholesterol [1478] and that cholesterol or its oxidations
product oxycholesterol be precursor of the bile acids [1480,
1484,1486] in animal tissue were derived from observations
that both oleic and cholic acids in reaction with benzoyl
peroxide in acetic acid gave upon addition of sulfuric acid
[1477] the same blue violet or blue green color as did
oxycholesterol derived from cholesterol by oxidation [1484,
1486,1487].

Although Lifschütz had applied color tests involving
acetic and sulfuric acids to detection of sterols other
than cholesterol in wool fat by 1897 [561,1473,1474], his
further studies of cholesterol oxidation were clearly stimulated by the report of Schulze and Winterstein in 1904 that
pure cholesterol deteriorated in air [2163]. Indeed, in
his first papers on cholesterol oxidation in 1907, Lifschütz
recognized the reported deterioration of cholesterol as an
instance of oxidation [1475,1476]. In his early chemical
and metabolic studies of cholesterol oxidation Lifschütz

encountered several materials which he viewed as distinct
oxidation products of cholesterol, such products including
oxycholesterol (or oxycholesterol II) formulated as a diol
bearing one atom of oxygen more than cholesterol, oxycholesterol ether (or oxycholesterol I) formulated as a condensation product of oxycholesterol which also could be derived reductively from oxycholesterol, and a third undefined
oxidation product related to oxycholesterol. The existence
of these three oxidized sterols was postulated on the basis
of different colors and spectral bands found with different
oxidized cholesterol preparations in acetic acid-sulfuric
acid [1452,1475,1476,1479], a matter adversely critized
directly by Windaus [2690].

    Lifschütz utilized three separate color tests for his
work with oxidized cholesterol. In distinction to the
Liebermann-Burchard color test which uses acetic anhydride-
sulfuric acid, Lifschütz used acetic acid-sulfuric acid
(10:1) which gave a deep blue color with oxycholesterol but
not with cholesterol [561,1473,1475,1476,1479]. A variation
of this test involved preliminary treatment of the sterol
with benzoyl peroxide in acetic acid followed by addition
of sulfuric acid, giving a characteristic blue violet or
blue green color [1477]. Yet another variation utilized
acetic and sulfuric acids containing ferric chloride and
gave a characteristic emerald green color with oxycholesterol [1479].

    These color tests were recognized by 1923 as involving
dehydration reactions of the sterols to yield polyunsaturated
steroid hydrocarbons [2665] and are today viewed as positive
for sterol enediols, dienols, and related derivatives which
form cholesta-2,4,6-triene (37) and/or cholesta-3,5,7-triene
(38) upon acid induced elimination reactions. The acetic
and sulfuric acid color test of Lifschütz is given by the

37   38

2,4,6-triene 37 [2116], by the epimeric 3β,7-diols 14 and
15, and by a variety of other steroids which can yield the
2,4,6-triene moiety upon dehydration or elimination reaction
in the acid system. Thus, a dicholesta-4,6-dienyl ether (39)
(or isomeric cholesta-3,5-dienyl ether) gives a positive
Lifschütz color test [2714], which observation is reminiscent

39

of the dioxycholesterol ether formulated earlier by
Lifschütz as a product among others in his oxycholesterol
preparations [1475,1476,1479].

The Lifschütz color test once popular in major monographs on steroids [737,738] has survived to the present
day in two separate forms. The Lifschütz color tests have
been variously used by name as test for steroid Δ^5-3β,7-
diols [77,103,114,203,242,958,1107,1128,1131,1160,1690,
2320,2540,2745-2747,2749,2750], as a general color test for
characterization of sterols and for recognition of unidentified sterols [797,1888,2172], and in continuing consideration of oxycholesterol in various preparations [242,1160,
1785,1855,1888]. Additionally, the widely used Zak colorimetric procedure for serum cholesterol determination utilizing acetic acid-sulfuric acid solutions of ferric chloride
giving a purple color with cholesterol, λ_{max} 560 nm [374,
2771], also adapted as a chromatography color test [1538],
is a direct but unacknowledged embodiment of the much older
Lifschütz test for oxycholesterol [1479].

Quantitative estimates of oxycholesterol levels in
sterol samples were attempted by Lifschütz from spectral
band absorption intensities using his various color tests
[1481,1483,1485,1495,1506] in combination with the Liebermann-Burchard color test which measured both cholesterol
and oxycholesterol [1483,1495,1498] and with the digitonin
precipitation procedure in which cholesterol was quantitatively precipitated but oxycholesterol only partially

precipitated [1485,1490,1495].

Lifschütz regarded oxycholesterol as a single chemical
compound whether it be derived chemically by the action of
potassium permanganate [1475] or benzoyl peroxide [1477,
1481,1494,1506] on cholesterol or by the process of bromination of cholesterol and debromination of cholesterol dibromide by sodium acetate [1492], or whether it be isolated as
a putative metabolite of cholesterol from a variety of animal tissues [1476,1479-1482,1486-1488,1491,1493-1496,1498,
1499,1502,1503]. However, results were highly variable and
in some cases oxycholesterol could not be detected in tissues. Thus, Lifschütz repeatedly did not detect oxycholesterol in liver [1480,1482,1490]. Lifschütz was also under
the impression that oxycholesterol esters did not respond
to the color tests but that free oxycholesterol was required
for detection. However, other investigators have reported
recovery of sterol esters from tissues which give positive
oxycholesterol color tests [338,343,1305,2012].

Oxycholesterol was formulated by Lifschütz as a monohydroxylated derivative of cholesterol as early as 1906,
with molecular formula $C_{26}H_{44}O_2$ [1475,1476,1479,1480] later
revised in view of the correct $C_{27}H_{46}O$ formulation for cholesterol [2690] to $C_{27}H_{46}O_2$ [1485,1486,1489,1492,1506].
Although directions for a standardized preparation of oxycholesterol were provided [1481,1506], no pure or crystalline preparation of oxycholesterol appears to have been
described by Lifschütz. However, chemical characterizations
included evidence for formation of a dibenzoate ester [1489],
a dibromide [1492], and a digitonide more soluble than that
of cholesterol [1485,1489,1490,1495,1506], as well as evidence that oxycholesterol was not the 5α,6α-epoxide 35
[1489,2660]. In no way could Lifschütz ever have dealt with
an oxycholesterol preparation as a single molecular species,
as subsequent chromatographic analyses of material prepared
according to his directions later established [1588,2288,
2303]. This matter was recognized as early as 1913 [2135]
and continuously thereafter by others [863,1965,2012,2015].
Moreover, methods of the period for recovery of cholesterol
from tissues and from reactions guarranteed that factitious
air oxidation of cholesterol would occur. Hot alkali saponification of sterol esters in air is characterized by cholesterol autoxidation. The use of diethyl ether not freed
from peroxides clearly effects cholesterol oxidations [496,
1067], and Lifschütz typically evaporated ether extracts

of cholesterol in air on a water bath [1491]!

On balance, evidence suggests that Lifschütz oxycholesterol was an impure mixture of sterols in which the epimeric 3β,7-diols 14 and 15 predominated. However, each
oxycholesterol preparation was unique in composition [1588,
2288,2303] and recognition or identification of any components was not achieved in this early period. Nonetheless,
a sterol later recognized as 6α-benzoyloxy-5α-cholestan-
3β-ol was obtained in 1921 from an oxycholesterol made
with benzoyl peroxide [2706], and cholest-4-ene-3β,6β-diol
(40) initially thought to be cholest-5-ene-3β,4β-diol (41)
[2018] was also recovered later from an oxycholesterol derived by the action of sodium acetate upon cholesterol
dibromide [424,2018,2019]. Many components of other oxycholesterol preparations have never been identified.

40   41

In addition to oxycholesterol and oxycholesterol ether
initially postulated by Lifschütz [1475,1476,1479] there
were several other materials which he considered to be related oxidation products, such entities as dioxycholesterol
[1495], isooxycholesterol [1492], "noncholesterol" fraction
[1486], cholesterol oxydate [1496], metacholesterol [1491,
1492,1494-1497,1499-1501,1504,1505], and oxymetacholesterol
[1495] being variously named. Of these other materials,
Lifschütz paid additional attention only to metacholesterol,
which he found with oxycholesterol in animal tissues and
which he regarded as being an isomer of cholesterol, one
also formed chemically by the action of benzoyl peroxide
on cholesterol [1492,1497,1504,1505] and upon debromination
of cholesterol dibromide [1494]. None of these other oxidized species except metacholesterol, m.p. 139-141°C, was
characterized by the common chemical criteria of the period
and what chemical and physical data exist [1497,1500] suggest that the material be impure cholesterol. This same
conclusion and subsequent rejection of metacholesterol as a

distinct sterol different from cholesterol made by Windaus
by 1920-1921 [2693,2705,2706] had not been accepted by
Lifschütz as late as 1935 [1505] or by others of the period
[579,581,1662-1664]. Although the errors of the period
anent oxycholesterol and metacholesterol should no longer
persist, the problem of impure cholesterol as a distinct
entity recurs from time to time, compare "cholesterol II"
from gallstones [2397,2398].

Reports of the instability of cholesterol to storage
in light and air dating from 1901 [1960,2163,2164] and the
clear demonstration that cholesterol heated in air yielded
a product giving the color test response of oxycholesterol
made in 1914 by Lamb, together with his caveat that drying
tissues in air be avoided [1420] provided the basis for
awareness of the problem of artifact formation from cholesterol at this early time. Nonetheless, air drying of biological tissues to be used for sterol analyses and isolations has continued to the present day.

Although Lifschütz in 1917 [1491] and others [2012,
2136] posed the question whether oxycholesterol from tissues be cholesterol metabolite or artifact generated *post
mortem* in isolation, Lifschütz regarded oxycholesterol as a
metabolite, a view which he retained despite contrary evidence to the end of his career [1499,1502,1505].

Studies of 1926-1930 coincident with new interests in
the photoactivation of sterols as antiracchitic agents suggested that Lifschütz oxycholesterol form from cholesterol
upon irradiation in air [211,2205], but other work of the
same period ascribed the effects to heat alone [1252,1666].
However, definitive studies of Blix and Löwenhielm in 1928
[296] and of Bischoff in 1930 [247], confirmed by others
[2129,2477], established beyond doubts the indispensable
role of oxygen in the formation as artifacts of oxidized
derivatives of cholesterol during manipulation of samples.

These several findings should have conclusively clari-
fied the matter, and as summed in Bischoff's statement that
oxycholesterol was without significance in the intermediary
metabolism of cholesterol [247], interest in oxycholesterol
should have ended. However, this was not to be the case,
as revival of Lifschütz color tests and the term oxycholesterol and continued confusion between artifact and metabolite status for select cholesterol oxidation products

occurred as well in later history of the subject.

The concepts and experimental procedures of Lifschütz influenced the progress of sterol biochemistry throughout this early historical period, as searches by others for oxycholesterol in animal tissues [109,247,802,1576,1657,1667, 1806,2012,2013,2128-2131,2134,2136,2476,2538] and in chemical oxidations [211,247,296,1666,2205,2477], spectral studies including comparisons between the Lifschütz test using acetic acid-sulfuric acid and the Liebermann-Burchard test using acetic anhydride-sulfuric acid [2159,2160,2476, 2538], development of new color tests for oxycholesterol [1806,2014-2016], and biological properties of oxycholesterol including its effects upon isolated frog heart, guinea pig uterus, and rat skin [780,2181] and its ability to affect multiplication in infusoria [1965] and to stabilize emulsions [2135,2136] attest.

Thus, at the end of this historical period of study of cholesterol biochemistry achieved before the structure of cholesterol was known and closed in 1932 with the assignment of the correct structures to cholesterol, ergosterol, and cholic acid, the sensitivity of cholesterol to air oxidation, the artifact nature of oxycholesterol as autoxidation product and not metabolite, and the importance of careful control of isolation procedures for sterols were well established items.

INTERMEDIATE HISTORY, 1933-1960

The intermediate period of development of cholesterol chemistry was marked by use of correct structure assignments for the first time and by an ever increasing understanding of the fundamental organic chemistry of cholesterol. The chemical relationships among the sterols, bile salts, steroid hormones, and triterpenoids were elucidated during this period, and the generic term "steroids" was proposed by Callow in 1936 [433] for all these related materials. Utilization of effective chromatographic methods for purification and improvements in instrumental methods for analysis, together with applications of spectral and optical rotation data to structure elucidation made these years very productive ones for the study of steroids. Very substantial advances were made in the study of the chemistry of the steroids during this period. Indeed, most of the fundamental chemistry of the steroids was developed to the full

during this intermediate period, including commercial introduction of the vitamins D, steroid hormones, and synthetic
hormone analogs for therapeutic measures.

Study of cholesterol autoxidation during this period
was characterized by extensive but random isolations of
specific oxidation products from biological materials, by
investigations with model oxidation systems and with
radiation, and near the close of the period by the first
preparation of identified hydroperoxides of cholesterol.

The intermediate period of study of cholesterol autoxidation opened in 1933 with the first isolation from one of
Lifschütz' described oxycholesterol preparations of a pure
oxidized sterol derivative to which a modern though incor-
rect structure was assigned. As previously mentioned, the
initially assigned structure as cholest-5-ene-3β,4β-diol
(41) [2018] was later shown to be correctly cholest-4-ene-
3β,6β-diol (40) [424,2019].

Studies of the steroid hormones and vitamins D pre-
empted most interests in the mid-1930s period, but search
for yet undiscovered tissue sterols and their metabolites
was resumed at this time. Color test responses of unsaponi-
fiable matter from which cholesterol had been removed by
digitonin precipitation, using the Salkowski and Liebermann-
Burchard tests, affording suspicions that blood [348,912,
1785] and liver [795] contained sterols other than cholesterol may have influenced early systematic examination of
these tissues for their extraneous sterol content.

The first recovery of a cholesterol autoxidation product from animal tissue was reported in 1939 by Haslewood
who isolated the 3β,7β-diol 15 from bovine liver [971].
Subsequent reports of isolation of the epimeric 3β,7α-
diol 14 from mare serum [2713] and hog liver [1551] in 1940
and of the 3β,5α,6β-triol 13, from bovine liver in 1941
[972] completed the common picture of cholesterol autoxidation products including epimeric 3β,7-diols 14 and 15 and
3β,5α,6β-triol 13, the 7-ketone 16 not having been isolated
as yet. Related studies on pig tissues directly confirmed
these findings, with the 3β,7β-diol 15, the 3β,5α,6β-triol
13, the 3,5-dien-7-one 10 [1887,1888,1954] derived from the
7-ketone 16, as well as the 7-ketone 16 [1889] providing the
same pattern of cholesterol autoxidation products from these
animal tissues. A more complete listing of detection and

isolation of cholesterol autoxidation products is made in Chapter III.

The question of possible artifact status for the 3β, 7α-diol 14 raised without resolution by MacPhillamy [1551] and Wintersteiner and Bergström [2713] in 1940, by Haslewood in 1942 [974], and repeatedly thereafter by Ruzicka and Prelog [1886,1888,1889,1954] revived the same doubts as were expressed earlier by others over Lifschütz' oxycholesterol preparations. Haslewood concluded that where found the 3β,7-diols were genuine metabolites, but his subsequent questioning in 1944 [975] once more left the matter open.

A similar though different case of confusion regarding a cholesterol oxidation product is that of the 4-ene-3-ketone 8 now recognized as a proper biosynthesis intermediate in the enzymic reduction of cholesterol to either 5α- or 5β-stanol 2 or 4 respectively. Isolation of the 4-ene-3-ketone 8 from animal tissues [1886,1888,1889] with the concomitant questioning whether it be metabolite or artifact presented the very same problem at the time as did the other cholesterol autoxidation products. In that defined 3β-hydroxysterol dehydrogenases act upon cholesterol to yield the 5-ene-3-ketone 6 in turn transformed by isomerases to the 4-ene-3-ketone 8, the detection of the 4-ene-3-ketone 8 among tissue sterols in fact presents at this time an unresolved matter, in much the same as does detection of the 3β,7α-diol 14 as a genuine liver metabolite of cholesterol.

The question of other steroid 3-ketones isolated from tissues is another matter. The 4,6-diene-3-ketone 12 [960,1888,1964,2316] must be viewed as an autoxidation product or as a secondary product of attack of electronically excited (singlet) molecular oxygen($^1O_2$) upon cholesterol, as no metabolism system forming 12 has ever been described. The 5α-cholestane-3,6-dione 42 isolated from pig testis [1889] (and later from an human adrenal tumor [1274]) thought to be genuine metabolite at the time because no photooxidation pathway for its formation had been discovered is in a similar situation. There now has been demonstrated a means of formation of the 5α-3,6-dione 42 by autoxidation processes, but no enzyme metabolic pathway is known for this compound. However, isolation of cholesterol, 5β-cholestan-3-one, the 5β-stanol 4, 5β-cholestan-3α-ol, and 3,4-

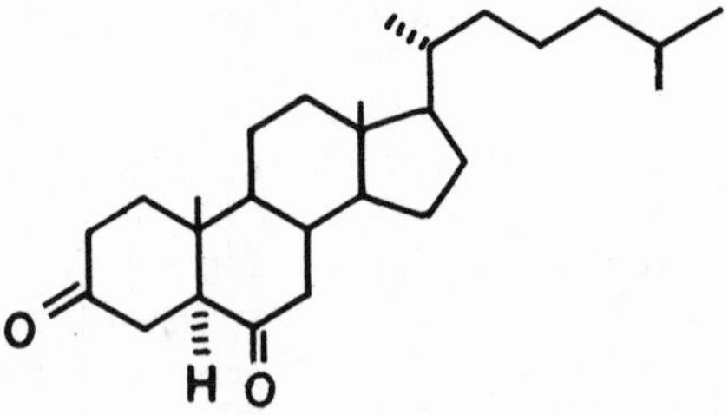

42

seco-5β-cholestane-3,4-dioic acid (43) from whale ambergris [1440] may represent genuine metabolism of cholesterol but may also represent microbial action upon the ambergris in the marine environment.

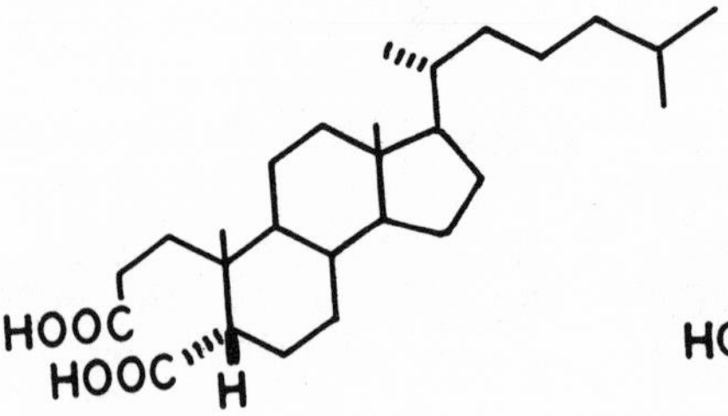

HOOC
HOOC'''H

43

HO

R O

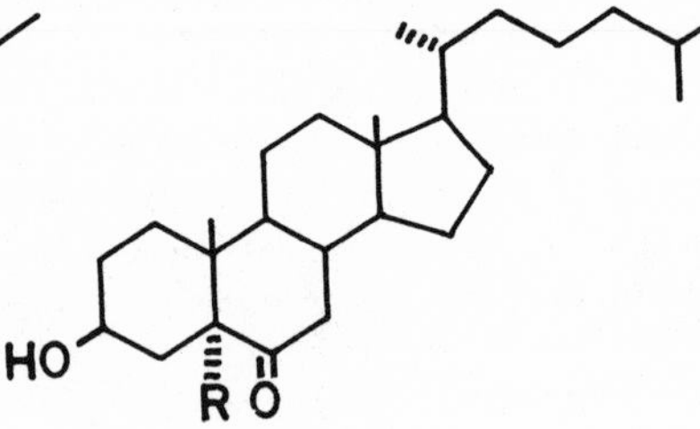

44 R = OH

45 R = H

Isolation of yet other esoteric oxidized cholesterol derivatives from tissues was also made during this period. The recovery of the 6-ketones 3β,5-dihydroxy-5α-cholestan-6-one (44) from pig liver [2172] and of 3β-hydroxy-5α-cholestan-6-one (45) from pig spleen [1888] are individual instances of isolation for each compound, but whereas an obvious autoxidation pathway exists for the 3β,5α-diol-6-ketone 44, an autoxidation pathway for the 3β-hydroxy-6-ketone 45 has not been discovered.

Isolation of cholest-4-ene-3β,6β-diol (40) from pig spleen [1888] and later from rat adrenal tissue [1894] may suggest a metabolic origin for the diol, but its origin as thermal decomposition product of a sterol hydroperoxide formed from cholesterol by attack of $^1O_2$ is also possible [1401] (cf. Chapter VII.)

Strong chemical support of studies of cholesterol autoxidation developed throughout this period also. Several cholesterol oxidation products had been described prior to assignment of their correct structure, among which were the 3,5-diene-7-ketone 10, the 7-ketone 16, the 4-ene-3-ketone 8, the 5α,6α-epoxide 35 and the 3β,5α,6β-triol 13. All the other autoxidation products of cholesterol, excepting only some cholesterol hydroperoxide derivatives, were synthesized during the intermediate history period. Examination of many chemical means of oxidation of sterols, reviewed by Fieser and Fieser in 1959 [748] attest to interests in such matters, many of which were directed towards the possible use of cholesterol as a source material for steroid hormone manufacture.

Very similar results were obtained at the same time in other oxidation studies. Although a single pattern of oxidation products suggesting common reaction pathways did not emerge from these experiments, most may be reconciled with concepts of initial formation of 7-hydroperoxides in much the same manner as for the studies using aqueous sodium stearate dispersions of cholesterol and as developed later in Chapter IV. Studies of this intermediate period utilized forcing conditions which would mitigate against recovery of early oxidation products, and the isolation approaches severely limited generalities drawn from these experiments.

Radiation-induced oxidations of cholesterol yielded slightly different product mixtures depending on the system studied. Products reported were: from solid cholesterol, the 3β,7β-diol 15 and the Δ^4-3β,6β-diol 40 [2701]; from aqueous acetic acid systems, the 3β,5α,6β-triol 13 and the 7-ketone 16 [1275,2656]; from methanol solutions, the 3β, 7β-diol 15, 7-ketone 16, and 3β,5α,6β-triol 13; and from acetone solutions, the 3β,7β-diol 15 and the isomeric 5,6-epoxides 35 and 36 [503].

Oxidations in related systems not involving radiation yielded the same kind of products. The 3β-acetate of cholesterol yielded the 7-ketone 16 3β-acetate upon oxidation with iron phthalocyanine in heated xylene [512], whereas the 3β,5α,6β-triol 13 and 7-ketone 16 3β-acetates were recovered in oxidations by the classic Fenton reagent of H_2O_2, Fe(II) salts, and acetic acid [500]. Cholesterol oxidized by the Fenton reagent also yielded the 3β,5α,6β-triol 13 [500].

A key item exploited during this intermediate period
was that of the model system for oxidation studies. The
old discovery that cholesterol was soluble in soap solutions
reported by Gobley in 1846 [879] was embodied in the aqueous
sodium stearate dispersions of Blix and Löwenhielm in 1928
[296] for their studies of cholesterol autoxidation but full
exploitation of this system was made by Bergström and
Wintersteiner during the 1940-1942 period [197,198,201,2711].
This aqueous sodium stearate dispersion system has been of
continuous interest since that time as the cholesterol dispersions may simulate the dispersed state of the sterol in
aqueous fluids of animal tissues.

Results of study of the aqueous sodium stearate dispersion model oxidation system included recognition of all
of the commonly encountered B-ring autoxidation products of
cholesterol, although identification of several oxidation
products was not accomplished until the advent of thin-layer
chromatography, as discussed in the section on more recent
istory. Thus, the epimeric 3β,7-diols 14 and 15, the 7-
ketone 16, and the 3,5-diene-7-ketone 10 therefrom derived
were all identified directly by Bergström and Wintersteiner
in the 1940-1942 period [197,198,201,2711], with the 3β,5α,
6β-triol 13 being identified by 1953 [1690,2288,2303].
This five component mixture of 10, 13, 14, 15, and 16 thus
constitutes a fundamental pattern whose nature strongly
suggests autoxidation of cholesterol as that process generating the mixture!

Identification of yet other autoxidation products in
the sodium stearate dispersions of cholesterol was not made
before the advent of thin-layer chromatography. Thus, the
isomeric cholesterol 5,6-epoxides 35 and 36 were not identified as products of the reaction until 1968-1975 [479,2297],
and the primary products of the oxidation, the epimeric
3β-hydroxycholest-5-en-7α- and 7β-hydroperoxides (cholesterol 7α- and 7β-hydroperoxides) 46 and 47 were not
identified until yet later. The presence of side-chain
oxidation products, such as the 3β,25-diol 27 found as an
autoxidation product of crystalline cholesterol have not
demonstrated in the aqueous sodium stearate dispersion
model systems.

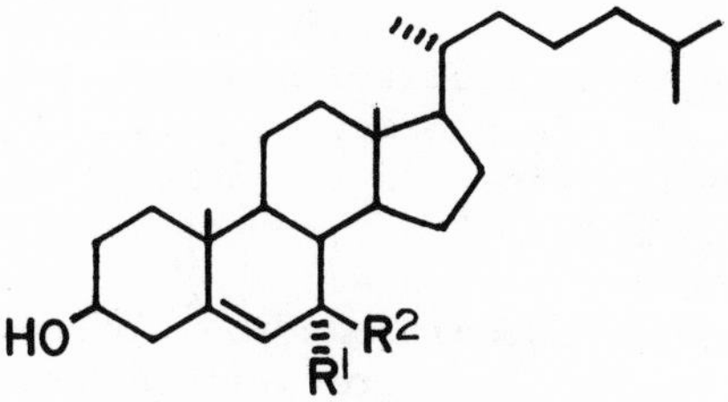

$$\underline{46} \quad R^1 = OOH, \; R^2 = H$$

$$\underline{47} \quad R^1 = H, \; R^2 = OOH$$

Although the concept of formation of sterol peroxides as initial products of cholesterol oxidation by air had been advanced by 1926 [2020] and early in the intermediate historical period as well [2349], with specific suggestion that a cholesterol 7-hydroperoxide be that derivative first formed made by Bergström and Wintersteiner by 1942 [197,198, 203], progress towards understanding of the primary events of cholesterol autoxidation could not be made in this intermediate period of study for want of adequate analytical methods and because of limitations of knowledge about the major secondary products of cholesterol autoxidation, namely the epimeric 3β,7-diols 14 and 15. At least four chemical problems, each presenting difficulties causing misunderstanding or confusion in experimental work with the 3β,7-diols, needed recognition and solution before understanding of the initial events of autoxudation could be approached or proper awareness of the role which the 3β,7α-diol 14 played in hepatic bile acid biosynthesis could be addressed.

These problems were: (i) assigment of the correct absolute stereochemistry at the C-7 carbon atom of the epimeric 3β,7-diols, (ii) development of appropriate physical methods for identification of either 3β,7-diol, (iii) recognition of the facile formation of 7-alkyl ethers by the 3β,7α-diol 14 and (iv) discovery of the ready epimerization of the 7α-alcohol 14 and 7α-hydroperoxide 46 and of the facile acid-catalyzed interconversions of the epimeric 3β,7-diols 14 and 15 and of their corresponding 7-esters and 7-alkyl ethers.

The epimeric 3β,7-diols 14 and 15 known from 1935-1936 were synthesized initially by different means, the 3β,7β-diol 15 first by chemical reduction of the 3β-acetate of the corresponding 7-ketone 16 by Windaus

[2703,2710], the epimeric 3β,7α-diol 14 second via permanganate oxidation of cholesterol hydrogen phthalate by Barr
et al. [141]. As the product of permanganate oxidation was
the second one described in the literature, it was designated by the trivial term "β-7" and that of Windaus as "α-7".

Despite clear admonitions about these trivial designations [1888] and molecular rotation arguments suggesting
revisions of the trivial nomenclature for the epimeric 5α-
cholestane-3β,7-diols (48, 49) [1868,2712] (which could be
correlated chemically with the corresponding epimeric 3β,7-
diols 14 and 15) revision of the nomenclature for the 3β,7-

48 R^1 = OH, R^2 = H

49 R^1 = H, R^2 = OH

diols 14 and 15 was not clearly proposed before 1946 [1924,
2240]. Although adoption of the correct 3β,7α- and 3β,7β-
configurational assignments for the "β-7" epimer 14 and
"α-7" epimer 15 respectively began directly by some [419],
others retained provisionally and with knowledge the older
trivial designations for the 3β,7-diol epimers and their
derivatives [214,1007,1008].

The assignment of correct absolute configuration to
the epimeric 3β,7-diols 14 and 15 by Fieser in 1949 [750],
confirmed by chemical evidence [146,1022] and molecular
rotation arguments [354,1332,2370], was further supported
by equilibration of the epimeric 3β,7-diol 3,7-diacetates
in acetic acid to yield a mixture in which the quasiequatorial 3β,7β-diol 3,7-diacetate predominated [2051,2354].
More recently, the quasiequatorial conformation of the 7β-
hydroxyl group of 15 has been demonstrated by proton nuclear
magnetic resonance spectra at 300 MHz in which the quasiaxial 7α-proton signal of the 3β,7β-diol 15 appears at 4.15
ppm as a doublet of doublets with $J_{6,7}$ 1.5 Hz, $J_{7,8}$ 8 Hz and
the quasiequatorial 7β-proton signal of the 3β,7α-diol 14
appears at 0.99 ppm as a doublet of doublets with $J_{6,7}$ 5.5 Hz,

J$_{7,8}$ 1.5 Hz [2455].

   It was thus possible to use the correct absolute configurational assignments for the epimeric 3β,7-diols 14 and
15 after the period 1947-1952, as was done as a matter of
course by some [419,753]. However, the reassignments were
questioned during the period [2094], and there were reports
in which the older trivial nomenclature persisted, without
awareness [1953] and in reviews using original authors'
designations [420,1622,1623,1625,1724-1728].

   In the absence of any reports of appropriate physical
properties, such as melting point data or optical rotations
on the 3β,7-diol or on its easily crystallized 3,7-dibenzo-
ate ester or relative chromatographic mobility data, several
reports of a 3β,7-diol 14 and 15 remain obscure. The simple
expedient of noting relative chromatographic mobility in
comparison with reference sterols suffices to establish
which 3β,7-diol 14 or 15 is at hand, the quasiaxial 7α-
alcohol 14 preceding its 7β-epimer 15 in paper [514,2154,
2288], liquid-liquid column [1690], and gas liquid [2454,
2568] partition chromatographic systems, with the reverse
order being the case in adsorption systems, the quasiequatorial 7β-alcohol 15 being more mobile than the 7α-epimer
14 on adsorption thin-layer chromatographic systems [492,
495,496,548,1107,2303] and on adsorption columns [2051].

   The ready etherification of the quasiaxial 7α-hydroxyl
group of the 3β,7α-diol 14 in acidified alcoholic solution
has also provided some confusion in the study of cholesterol
oxidations. Thus, report of the sterol "Compound A", formulated as 5α-cholest-6-ene-3β,5-diol (50) formed by allylic
isomerization of the 3β,7α-diol 14, among cholesterol autoxidation products [197,201,202] and of the putative 5β-cholest-

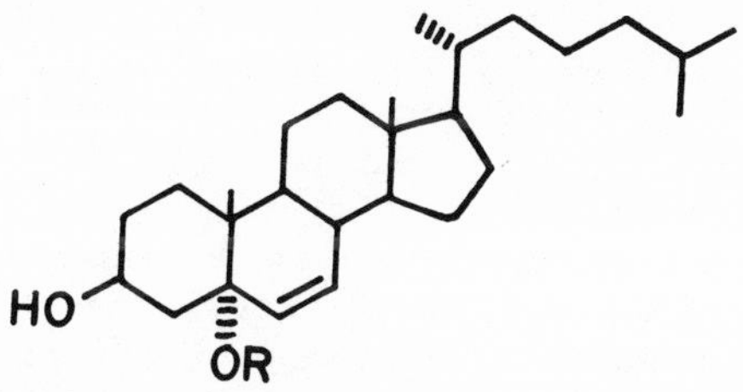

50 R = H

51 R = OH

6-ene-3β,5-diol isomer thereof isolated from pig spleen
[1888] represent cases of recovery of the 7α-ethers 7α-
ethoxycholest-5-en-3β-ol (53) and 7α-methoxycholest-5-en-
3β-ol (52) respectively as artifacts of manipulation of
sterol mixtures containing the 3β,7α-diol 14 [1006,1008].

Neither the 3β,5α-diol 50 nor its 5β-isomer 5β-cholest-
6-ene-3β,5-diol only recently synthesized [2064] have ever
been detected in tissues, as artifact or as natural product.
Moreover, their formation by isomerization of either 3β,7-
diol 14 or 15 does not occur. Rather, the 3β,5α-diol 50
is a reduction product of the 5α-hydroperoxide 51 derived
from cholesterol by the action of $^1O_2$.

The 3β,5α-diol 50, 3β,7α-diol 14, and 3β,7β-diol 15
constitute a family of enediols related to one another by
allylic isomerization and epimerization, and these dynamic
relationship add another dimension to the problem of recognition of the epimeric 3β,7-diols 14 and 15 as products of
oxidation processes whatever kind. Although the proposed
isomerization of the 3β,7α-diol 14 to the 3β,5α-diol 50
does not occur, allylic rearrangement of the 3β,5α-diol 50
to the 3β,7α-diol 14 occurs in acidified solvents and under
pyrolysis conditions. The corresponding 3β-hydroxy-5α-
cholest-6-ene-5-hydroperoxide (51) similarly isomerizes to
the 7α-hydroperoxide 46 [1544,2104,2336].

The interconversions of 3β,7-diacetate esters of the
epimeric 3β,7-diols 14 and 15 in acetic acid [2051,2354]
taken with our observations of epimerization of the 3β,7α-
diol 14 to the 3β,7β-diol 15 and of the 7α-hydroperoxide 46
to the 7β-hydroperoxide 47 [2454,2455] as well as of interconversions of the epimeric 3β,7-diols 14 and 15 in acidified
solutions [1404] and under pyrolysis conditions [2454] establish the interrelationships generally for this family of
allylic esters, alcohols, and hydroperoxides. Moreover, the
same relationship exists for the corresponding 7-alkyl
ethers, a matter suggested but not demonstrated much earlier
[1008]. The interconversion of epimeric 3β,7-diols 14 and
15 in acidified solvent solutions is exactly matched by the
interconversions of epimeric 7-methyl ethers 52 and 7β-
methoxycholest-5-en-3β-ol (54) and of epimeric 7-ethyl
esters 53 and 7β-ethoxycholest-5-en-3β-ol (55) [959,1404].
These interconversions are shown in FIGURE 1 and although
etherification of the quasiequatorial 7β-hydroxyl group of
the 3β,7β-diol 15 is not directly observed, the equilibrium

among these species 14, 15, 52, (or 53), and 54 (or 55) is such as to favor ultimate formation of the 7β-alkyl ethers 54 or 55 as preponderant products [1404].

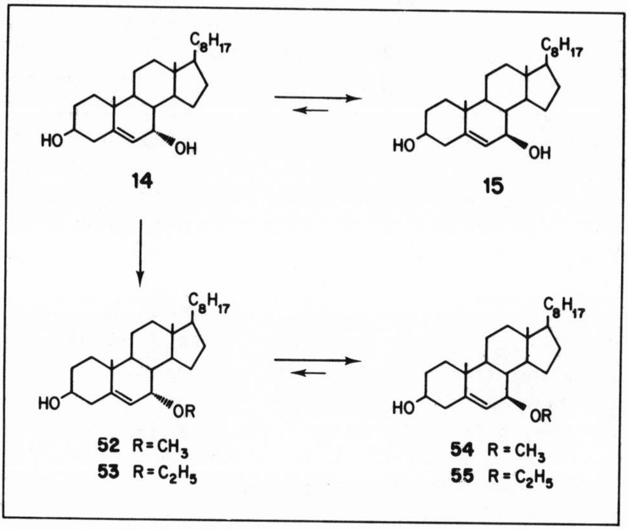

FIGURE 1. Acid-catalyzed interconversions of the epimeric cholest-5-ene-3β,7-diols 14 and 15 and of their 7-alkyl ethers.

The detection in tissue sterol samples of unidentified sterols responding to 50% aqueous sulfuric acid color test as blue colors though at low level, represents opportunity for additional confusion and difficulty in study of trace level sterols in tissues except where a recognition of this matter of artifact formation is kept in mind. The use of alcohols, particularly of methanol, for total lipid extractions with chloroform-methanol (2:1), for adsorption column separations of free sterols using diethyl ether-methanol (9:1) mixtures, for recrystallization of tissue sterols, and as solvent in alkaline saponifications of sterol esters offers many opportunities for methanol to react with any 3β,7α-diol 14 which may be present in the system, thus causing the appearance of the 7α-methyl ether 52 among products and ultimately also the epimerized 7β-methyl ether

<u>54</u> [1404].

    The same facile formation of Δ^5-7α- and -7β-methoxyl
derivatives in the C_{19}-series has also been noted [24].

    The results of Bergström and Wintersteiner in their
studies of cholesterol autoxidation attracted much continuing interest and in essence closed concern for the earlier
period of study of cholesterol autoxidation by presenting a
fairly complete concept of the process and by identifying
the most frequently encountered oxidation products. These
same studies also strengthened the general concept that
autoxidation was a troublesome event which might intrude
into studies of sterol biochemistry but not necessarily one
which must be accepted as the ultimate explanation for the
detection or isolation of any one of the common autoxidation
products from biological systems. Although it could have
been stipulated at the time that biological systems containing the now classic pattern of five sterol autoxidation pro-
ducts, including the 3β,7-diol <u>14</u> and <u>15</u>, the 7-ketones <u>16</u>
and <u>10</u>, and the 3β,5α,6β-triol <u>13</u> had clearly been subject
to uncontrolled autoxidation, this final judgement was not
made at that time, and possibly is only made now for the
first time in such absolute terms.

    New interests in discovery of other sterols as possible
companions of cholesterol, the 5α-stanol <u>2</u>, and cholesta-
5,7-dien-3β-ol (7-dehydrocholesterol, provitamin D) (<u>56</u>)
present ubiquitously in animal tissues were implemented in
this period as well. Discovery of the isomeric stenol 5α-
cholest-7-en-3β-ol (lathosterol) (<u>57</u>) by Fieser in 1950 [739]
opened a phase of discovery of other sterol biosynthesis intermediates of cholesterol. Discovery of the (24S)-3β,24-diol
<u>25</u> in human and equine brain tissue in 1953 and of an iso-
meric 3β,26-diol <u>29</u> or <u>31</u> and 3β,25-diol <u>27</u> among products of

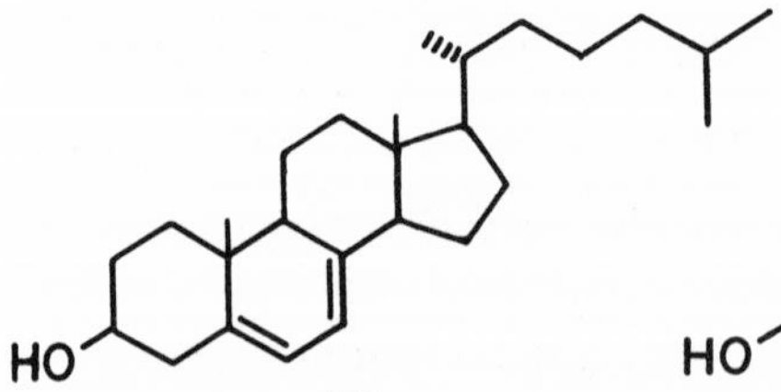

HO   <u>56</u>   HO   H <u>57</u>

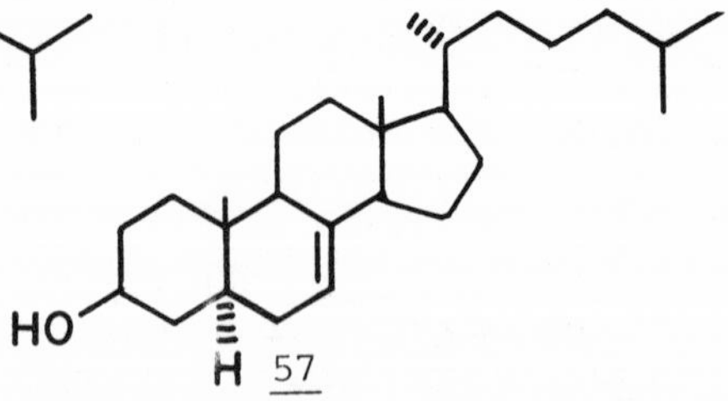

murine liver metabolism *in vitro* in 1956 led to a renascence
of general interest in cholesterol metabolites, biosynthesis
intermediates, and companion sterols.

Also at this time postulation that the 3β,7-diols 14
and/or 15 be a biosynthesis intermediate in the transformation of cholesterol to the provitamin D_3 5,7-dien-3β-ol 56
was formulated [2037,2320,2711]. Although this notion was
unsupportable, the intermediacy of the 3β,7α-diol 14 as the
initial enzymic oxidation of cholesterol in the hepatic biosynthesis of the bile acids was also conceived during this
period [543,1518]. In either case, that of vitamin D_3 or
bile acid biosynthesis, the potential relationship of the
major oxidation products 14 and 15 of cholesterol to such
vital physiological matters necessitated further studies of
these early steps in cholesterol oxidation.

Towards the close of this intermediate period of study
it became obvious that examination of sterol hydroperoxides
was possible. Indeed the facile preparation of the 5α-hydro-
peroxide 51 from cholesterol by photosensitized oxygenation
reported in 1957 [2098,2101,2102] and of the allylic isomerization of the 5α-hydroperoxide 51 to the 7α-hydroperoxide 46
shortly thereafter [1544,2104,2336] and report in 1960 of
the isolation of a C_{29}-sterol hydroperoxide 3α,22-dihydroxy-
(20S,22S,24R)-stigmast-5-ene-7α-hydroperoxide (58) from
leaves of *Aesculus hypocastanum* [767] demonstrated that past
concern for stability of such sterol hydroperoxides was
unjustified. At the same time synthesis of 6β-hydroperoxy-
cholest-4-en-3-one (59) was achieved [752,2101]. These
hydroperoxides 58 and 59 as well as the cyclic 5α,8α-perox-
ides 5,8-epidioxy-5α,8α-cholest-6-en-3β-ol (60) [2096],

58   59

5,8-epidioxy-5α,8α-ergost-6-en-3β-ol (61) [2702], 5,8-epidi-
oxy-5α,8α-ergosta-6,E-22-dien-3β-ol (62) [2700], and 5,8-
epidioxy-5α,8α-ergosta-6,9(11), E-22-triene-3β-ol (63) [2674,

2704] derived from the parent 5,7-dienes 56, ergosta-5,7-
dien-3β-ol (64), ergosterol (ergosta-5,7,E-22-triene-3β-ol),
(65), and ergost-5,7,9(11), E-22-tetraen-3β-ol (66), respectively clearly represented steroid peroxides which could
be handled with appropriate care and subjected to experimental work.

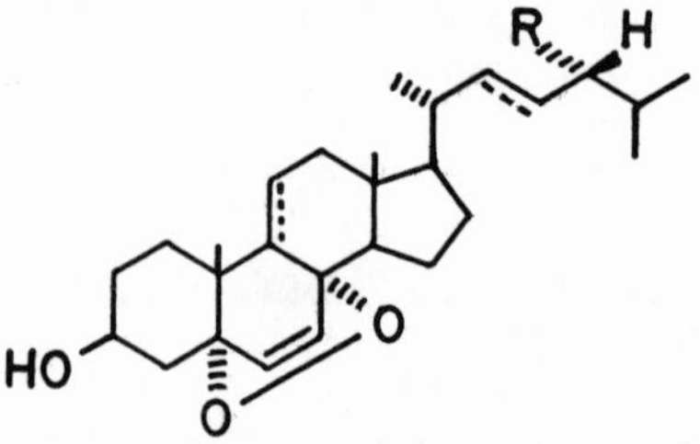

60 R = H

61 R = CH$_3$

62 R = CH$_3$,Δ^{22}

63 R = CH$_3$,Δ$^{9(11),22}$

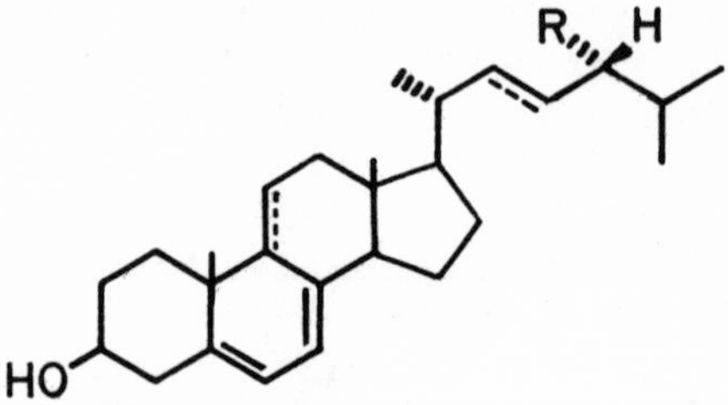

64

65 Δ^{22}

66 Δ$^{9(11),22}$

It thus became possible to look for sterol hydroperoxides as natural products, products of enzyme reactions,
and as intermediates in oxidative events, whether autoxidative or other. This matter was not settled until the very
recent period of study of sterol oxidations however.

It was during this historic period that the first proposal in 1944 for evaluation of possible physiological
activities of specific oxidized cholesterol derivatives was
made [975] and that the first evidences supporting such
activities were obtained. Such matters as cutaneous hyper-
sensitivity to lanolin traced to a crude (oxidized) cholesterol fraction [708], influences on mouse liver enzymes
by overtly oxidized cholesterol [384,1925], antiglucocorti-
costeroid activity associated with the 7-ketone 16 [1572,
1573], and induction of tumors in mice by some cholesterol
autoxidation products [230,240,241] suggested that biological responses to oxidized cholesterol derivatives might in
fact obtain broadly.

Also during this period attempts to use oxycholesterol
preparations commercially were recorded, the superior

emulsification properties of oxycholesterol for cosmetics
formulations [1179,1180,1198,1532-1534] and for embalming
fluid [1202] being noted. Also, although oxycholesterol as
a chemical entity or as a sterol metabolite had been
thoroughly discounted earlier, search for it in animal tissues continued [900].

The intermediate period of study of cholesterol autoxidation may be considered closed following three additional
events: the publication of the major text STEROIDS by
Fieser and Fieser in 1959 [748], by the appearance of the
second major review of cholesterol autoxidation by Bergström in 1961 [200], and by the introduction of thin-layer
chromatography to the study of oxidized cholesterol preparations in 1960.

RECENT HISTORY, 1960-1980

The recent historical period of study of cholesterol
autoxidation began with full awareness of the ease of oxidation of cholesterol under a variety of circumstances [200],
and the notorious instability of cholesterol towards molecular oxygen leading to a "galaxy of more polar products" was
reemphasized early in the period [986]. Although the prior
intermediate historical period had provided the suspicion
that cholesterol 7-hydroperoxides were implicated in the
oxidation processes, no adequate description nor uniform
concept of autoxidation processes had been achieved. The
recent period as defined herein is that period during which
exploitation of adequate chromatographic and instrumental
methods applied to the problem led to resolution of complex
oxidized sterol preparations and thereby to synthesis of
past and present experimental data affording our present
understanding of the matter, as unfolded in this monograph.

So long as reliance was had upon isolation methods and
approaches, only very limited progress in unraveling the
complex questions associated with autoxidized sterol mixtures was achieved, and uncertainties about the status of
given oxidized sterols in biological samples could not be
allayed. However, with the proper use of sensitive analytical means and with an analysis approach, abandoning direct
isolations, the problem yielded. The intermediate and
recent historical periods overlap one another conceptually,
for applications using an analysis approach had been

attempted from 1953, witness analyses of various biological
materials by paper [1011,1012,1953,2288,2308] and column
[574,1690] partition chromatography and of oxidized sterol
preparations by reverse phase paper partition chromatography
[1396,1579]. However, full recognition of the complexity of
the autoxidation processes had to await the application of
thin-layer chromatography after 1960 to oxidized cholesterol
preparations [1729,2553], USP cholesterol [1744], wool wax
[57], egg products [12], and plasma sterols [492,494-496],
including early preparative procedures [508].

Subsequently, general application of thin-layer chromatography to lipid peroxides and sterol hydroperoxides [1789]
was closely followed by direct applications to oxidized
cholesterol preparations in 1965-1966 [1072,1588,2302], cul-
minating in definitive demonstration of the complexity of
natural autoxidation of cholesterol in 1967 [2303]. In this
later study two dimensional thin-layer chromatography resolved
at least thirty-two oxidation products formed over twelve
years of natural air aging of a cholesterol sample which had
been purified via the dibromide procedures of Fieser [741,
744] and repeatedly recrystallized, cf. FIGURE 2.

Application of yet more sensitive procedures such as
gas chromatography has been examined by several laboratories
[492,495,496,765,2568]. The typical gas chromatographic
elution curve of FIGURE 3 attests to the complexity of
autoxidation of a pure cholesterol sample exposed to air
[2568], but the marked thermal instability of the primary
oxidation products as well as of secondary products [2454,
2559,2578,2579] restricts these procedures to specialized
uses, such as identification of thermally stable autoxidation
products [2454,2568], identification of sterol peroxides by
their characeristic pyrolysis patterns upon gas chromatography [300,905,2454] and purification of thermally stable
sterols [2569].

We have also conducted analyses of autoxidized cholesterol preparations by high performance liquid column
chromatography with microparticulate adsorption or reversed
phase systems [71,72]. Complex elution curves result, cf.
FIGURE 4, which are comparable in complexity to that disclosed by the prior chromatographic techniques. High performance liquid chromatography is readily applied to analysis for detection of specific impurities [505], for special
classes of oxidation products such as the C_{27}-3-ketones

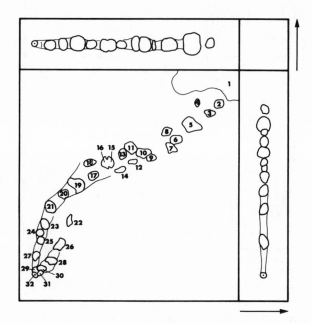

FIGURE 2. Two dimensional thin-layer
           chromatogram of once highly
           purified cholesterol aged for
           12 y. Component identities:
           Spot No. 5, cholesterol; No. 3,
           10; No. 10, 27; No. 11, 16;
           No. 15, 15; No. 16, 14; No. 18,
           44; No. 21, 13. (Reprinted
           with permission of Elsevier
           Scientific Publishing Co.,
           Amsterdam, from J. Chromatog.,
           27, 187 (1967)).

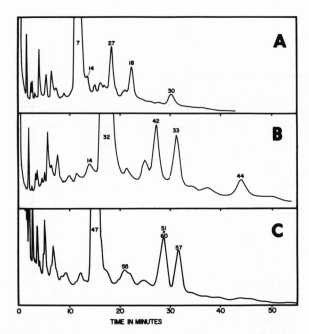

FIGURE 3. Gas chromatogram (3% SE-30) of
           autoxidized cholesterol. Curve
           A, same 12 y old sample of FIGURE
           2; Curve B, acetylated sample;
           Curve C, trimethylsilyl ether
           derivatives. Component identities:
           No. 7, cholesterol; No. 14, 10;
           No. 27, 27; No. 18, 16; No. 30, 13;
           No. 32, cholesterol 3β-acetate;
           No. 33, 16 3β-acetate; No. 42, 27
           3β-acetate; No. 44, 13 3β,6β-
           diacetate; No. 47, cholesterol 3β-
           trimethylsilyl ether; No. 51, 16 3β-
           trimethylsilyl ether; No. 55, 15 3β,
           7β-ditrimethylsilyl ether; No. 57,
           27 3β,25-ditrimethylsilyl ether;
           No. 60, 13 trimethylsilyl ether.
           (Reprinted with permission of the
           Academic Press, New York, from Anal.
           Biochem., 24, 419 (1968)).

[1105], and preparatively for pure sterol samples [1104].

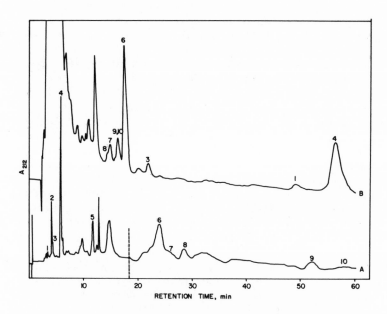

FIGURE 4.   High performance column chromatography of a naturally autoxidized
cholesterol.   Curve A, μPorasil
microparticulate adsorption column; Curve B, μBondapak C18 microparticulate reversed phase column.
Component identities: No. 1, 10;
No. 2, 8; No. 3, 108; No. 4, cholesterol; No. 5, 27; No. 6, 16; No. 7,
46; No. 8, 47; No. 9, 15; No. 10, 14,
(Reprinted with permission of
Elsevier Scientific Publishing Co.,
Amsterdam, from J. Chromatog., 175,
307 (1979)).

The recent history of cholesterol autoxidation is
characterized by yet additional random observations of
autoxidation products in biological samples. For the first
time sterol hydroperoxides were included among such oxidized

products thought to be genuine sterol metabolites. Indeed, cholesterol hydroperoxides were variously implicated in hepatic bile acid biosynthesis and steroid hormone biosynthesis by adrenal cortex mitochondria [2572-2574]. Suspicions that sterol hydroperoxides and other cholesterol autoxidation products be carcinogenic were aroused in this period, and initiation of systematic evaluation of other biological actions of cholesterol oxidation products was had.

The chemistry of cholesterol autoxidation was broadly examined by many laboratories during this recent period such that rational explanation of most observed details is now possible. A key experimental observation affording such rationalization was provided by electron spin resonance signals derived from cholesterol subjected to ionizing radiation in vacuum during the period 1959-1970 [661,899, 1004,1005,1948]. The correct analysis of these electron spin resonance signals [1004] together with the discovery of cholesterol 7β-hydroperoxide 47 as predominant product of radiation-induced air oxidation of cholesterol [2313] laid the basis for our present interpretations regarding free radical type oxidations of cholesterol. Furthermore, recognition of the possible participation of $^1O_2$ in the oxidation of cholesterol, yielding the 5α-hydroperoxide 51 as chief product as well as of hydrogen peroxide in the epoxidation of cholesterol, and of the unreactivity of superoxide radical anion towards cholesterol was achieved very recently.

Inasmuch as attempts to resolve complex cholesterol autoxidation preparations by inadequate paper and column partition chromatography systems had begun by 1953, so also have concepts of the prior intermediate period of history carried over into the modern period as well. Thus, several additional examples of period thinking and study continued to be reported well into the most recent phase of study, therefore long after the awareness of artifact formation had pervaded the literature. Accordingly, human tissue samples subjected to some sterol recovery protocols [1956] or to storage frozen for 2-9 y [103] prior to analysis are suspect. Claims of metabolite status for the common autoxidation products continue to be made as late as 1976 [327] despite the clear suggestion that in the absence of a demonstrated enzymic pathway for such formation, no such claim of natural product be advanced for the common autoxidation products. Indeed, this caveat need be recalled in

studies of other oxidized steroid derivatives of biological
material, as the number of identified cholesterol autoxida-
tion product of C_{19}-, C_{24}-, C_{26}-, and C_{27}- types now exceeds
65, with no account of unidentified very polar or very non-
polar products taken [2071,2576].

The complete recent history of the subject is in fact
the material of the remainder of this monograph, which will
be recounted without further recourse to the historical
approach.

OTHER STEROLS

Whereas the history of sterol autoxidation to this
point has been strictly limited to that of cholesterol, a
parallel but subdued accumulation of information about the
autoxidation of other sterols occurred over the same time
period. Recognition of the autoxidation of ergosterol by
Tanret in 1890 [2445] opened up the long and tortuous pathway of study of this sterol, and investigations of cholesterol autoxidation initiated 1895-1898 by Darmstaedter and
Lifschütz in fact also ushered in the study of the autoxidation of lanosterol 5α-lanosta-8,24-dien-3β-ol (67) and its
24,25-dihydro derivative 5α-lanost-8-en-3β-ol (68) as components *inter alia* of "isocholesterol" from wool fat [561,
1473,1474]. Lifschütz continued some interests in these

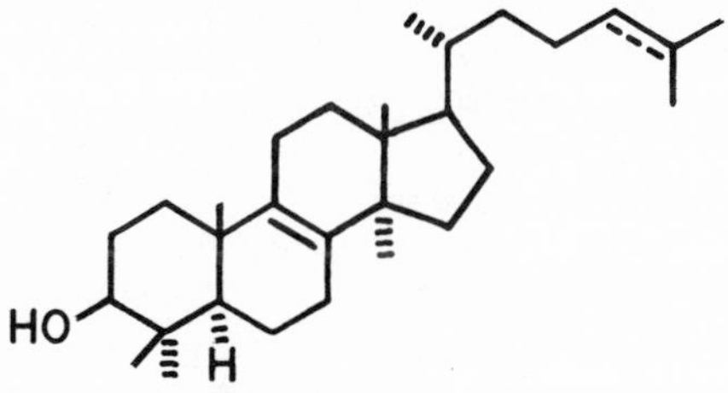

67 Δ^{24}

68

oxidized materials [1493,1498] and the inherent instability
on storage of "isocholesterol" as well as that of ergosterol
(65) and sitosterol (20) was reemphasized in 1906 [2164].

There followed preparations of oxysitosterol as an
analog of oxycholesterol, formed via catalysis with blood

solids in 1914 [1488], by action of benzoyl peroxide on
sitosterol in 1928 [2181] and later [1179], and by heating
sitosterol in air [247]. Otherwise, little interests in
the autoxidation of other common sterols and triterpenoids
developed.

However, the differential sensitivity of ergosterol
and cholesterol toward air oxidation was realized very
early, the greater instability of ergosterol being correctly
attributed by Tanret to air oxidation. The now classic
stigmata of sterol autoxidation, including development of
color and odor and lowering of melting point and optical
rotation, all accelerated by heat [2445,2446], clearly
distinguished ergosterol stability from that of cholesterol.
This greater sensitivity of ergosterol towards autoxidation
was repeatedly noted thereafter [2164], with much attention
given the matter in the late 1920s in connection with world-
wide interest in preparation of the vitamins D by photochemical activation of sterols [222,432,1398,1621,1926,2700].
Indeed, three separate processes of oxidation of ergosterol
were suggested [1621]: (i) autoxidation in the dark, inhibited by cyanide and increased by iron compounds, (ii) photochemical oxidation without sensitizer, and (iii) photochemical oxidation with sensitizer.

Discovery of the characteristic cyclic 5α,8α-peroxide
62 of ergosterol was made in 1928 [2700], with several analogous 5α,8α-peroxides 60, 61, and 63 described shortly thereafter [2096,2674,2702,2704]. In parallel with developments
for the Δ^5-sterol oxidation products, proper understanding
of the oxidation chemistry of ergosterol had to await assignment of the correct structure for the readily formed peroxide. However, in this case the initially assigned dioxetane
structure 5,6ξ-epidioxy-5ξ-ergosta-7,E-22-dien-3β-ol (69)
[1697,2700] was rapidly revised to the correct transannular
feature of the 5α,8α-peroxide 62 [193,194,737,2696].

HO

69

A considerable amount of chemistry of the ergosterol peroxide was developed in 1952 for the possible use of ergosterol as a starting material for commercial manufacture of the steroid hormones [286,288,289,498,499,1009]. Additional details of ergosterol peroxide biochemistry are presented later in Chapter VI. To this day, other than for cholesterol only the relatively available sterols ergosterol 62 and 5α-lanost-8-en-3β-ol 68 have received systematic attention to their autoxidation chemistry, the details of which are developed in ensuing Chapter VI.

The history of autoxidation of sterols other than cholesterol and ergosterol is best considered as initiating about 1940-1942 with the studies of Haslewood [974] and Bergström and Wintersteiner [197], with hot soap dispersions of several sterols. This history thus encompasses only about thirty-five years. Despite the paucity of detailed studies in support, the air oxidation of common Δ^5-sterols, including those with additional isolated unsaturation in the side-chain, appears to proceed by the same processes as for cholesterol and to yield the same 7-alcohol and 7-ketone products derivatives, most probably via the corresponding 7-hydroperoxides. Thus, campesterol (24α_F-methylcholest-5-en-3β-ol,24-methyl-(24S)-cholest-5-en-3β-ol) (71) and stigmasterol (stigmasta-5,E-22-dien-3β-ol) (70) were shown in 1942 to yield the 7-ketones 3β-hydroxy-24-methyl-(24S)-cholest-5-en-7-one (74) and 3β-hydroxystigmasta-5,E-22-dien-7-one (73) respectively and fucosterol (24-ethylcholesta-5,E-24(28)-dien-3β-ol) (75) probably also yielded its corresponding 7-ketone in like circumstance [197]. Furthermore,

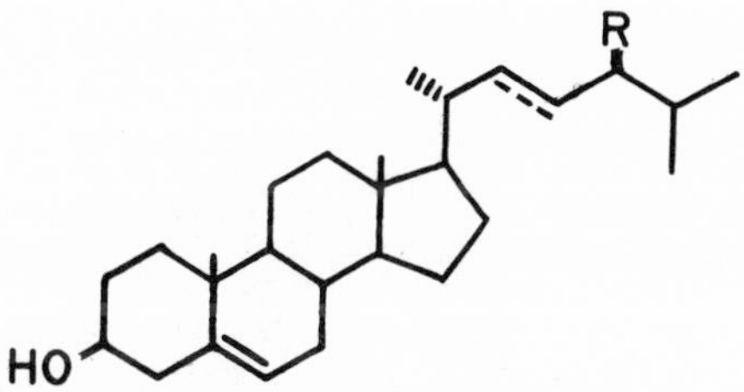

20 R = C₂H₅

70 R = C₂H₅, Δ²²

71 R = CH₃

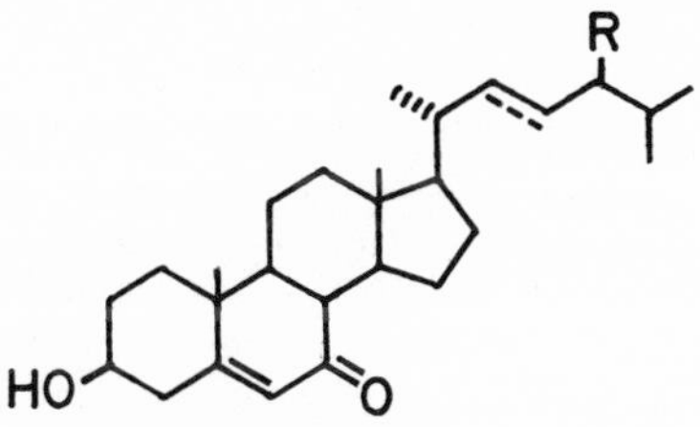

72 R = C₂H₅

73 R = C₂H₅, Δ²²

74 R = CH₃

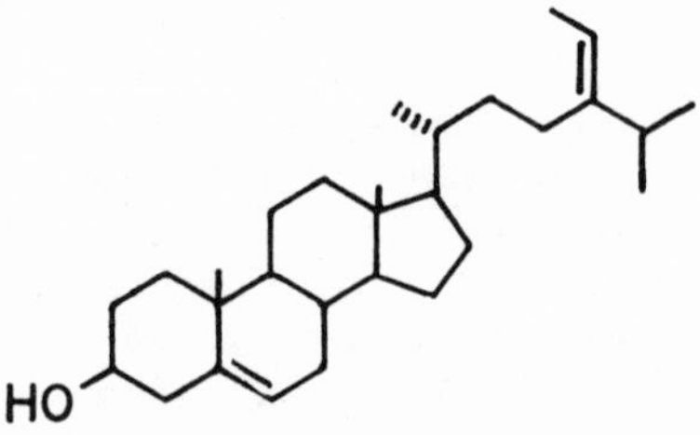

75

colorimetric analyses suggested that all three Δ^5-sterols 70, 71 and 75 and sitosterol (20) also yielded their corresponding 7-alcohol deeivatives as well. The instability of fucosterol in air has been recognized [794,2546]. The characteristic trio 3β-hydroxystigmast-5-en-7-one (72), stigmast-5-ene-3β,7α-diol (76) and stigmast-5-ene-3β,7β-diol (77) isolated from air-aged sitosterol [615] taken with the demonstrated presence of sterol hydroperoxides in autoxidized sitosterol samples [2295,2758,2759] further confirm the similarity of autoxidation of sitosterol with that of cholesterol.

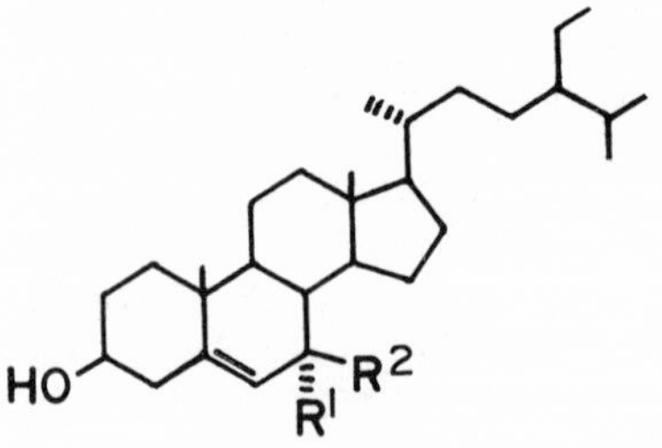

76 R^1 = OH, R^2 = H   78

77 R^1 = H, R^2 = OH

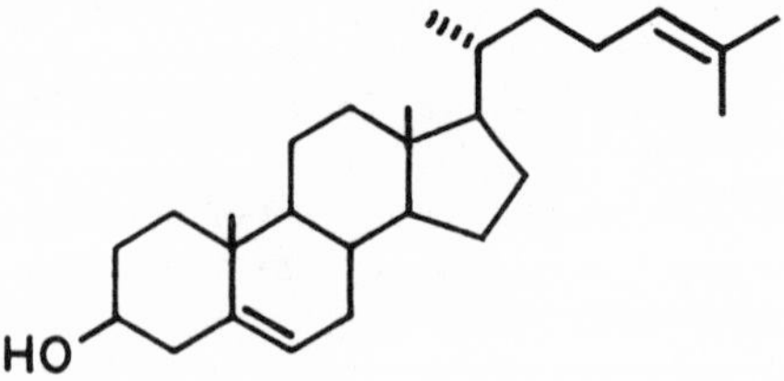

Other common sterols of this type have received but very limited attention as regards their autoxidation even though such deterioration obviously occurs. Thus, (25R)-spirost-5-en-3β-ol (diosgenin) [306] and desmosterol (cholesta-5,24-dien-3β-ol) (78), [794,2403,2480] have both been shown to decompose on storage, but studies of products formed have been limited. However, both the air oxidation of desmosterol [794] and the photosensitized oxidation of its 3β-acetate in reaction probably involving $^1O_2$ [1678] involve oxidations in the side-chain rather than in the B-ring as the case for cholesterol.

The history of study of autoxidation of other related monounsaturated sterols is of the same genre, with relatively little established by systematic investigations. Although cholest-4-en-3β-ol (79) was not oxidized in aqueous sodium stearate dispersions [197], its deterioration on storage for years is recorded [1050], and radiation-induced oxidations yielding derived 6-hydroperoxides as well as the corresponding 4-ene-3-ketone 8 have been noted [1402]. Similarly, the isomeric stenols 5α-cholest-6-en-3β-ol and 5α-cholest-7-en-3β-ol (57) are oxidized to sterol hydroperoxides in radiation induced oxidations [1402], but they have not been

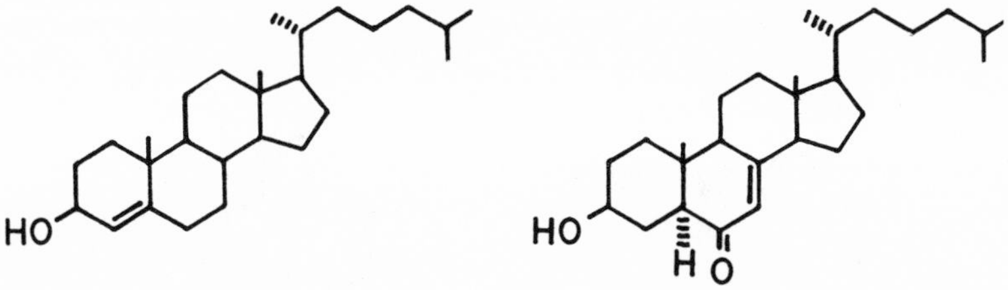

79   80

examined otherwise for their capacity for autoxidation. The 7-stenol 57 is more susceptible to SeO_2 oxidations than is cholesterol however [740]. Air oxidation of the related α-spinasterol (5α-stigmasta-7,E-22-dien-3β-ol) [750] in hot aqueous sodium stearate dispersions appears to yield conjugated ketone derivatives [197], possibly a Δ^7-6-ketone in light of formation of 3β-hydroxy-5α-cholest-Δ^7-6-6-one (80) from the 7-stenol (57) by radiation-induced air oxidation [1402]. That the autoxidation of 7-stenol 57 proceeds by hydroperoxide formation has been established by isolation of such hydroperoxides [1402] and by chromatographic demonstration of their presence in air-aged samples of 57, cf. FIGURE 5 [2295]. Likewise, other Δ^{24}-steroids are unstable to storage in contact with air, both 5α-cholesta-7,24-dien-3β-ol [2403,2480] and 5β-cholest-24-ene [2728] deteriorating in air.

Oxidation of the isomeric sterols 79 and 57 by $^1O_2$ developed as a chemical matter entirely separate from autoxidation, reports of $^1O_2$ attack on these stenols dating from 1961-1965 [665,1751,2100].

Air oxidation of 8-stenols such as lanosterol (67) and its 24,25-dihydro derivative 68 figured in the early

studies of Lifschütz as previously mentioned. Both sterols
were shown to be highly susceptible to air oxidation, with
hydroperoxide formation demonstrated by 1956 [1070,1071,1175,
1176] confirmed by systematic thin-layer chromatography over
the period 1963-1972 [357,1744,2173-2175,2295,2480]. Although autoxidation products associated with lanosterol itself have been variously isolated from oxidized wool fat,
systematic studies have uniformly used the dihydro derivative 68 as substrate, such systematic studies appearing
from 1956. The air oxidation of the 3β-acetate of 68 was
shown to yield 5α-lanosta-7,9(11)-dien-3β-ol (81) 3β-acetate

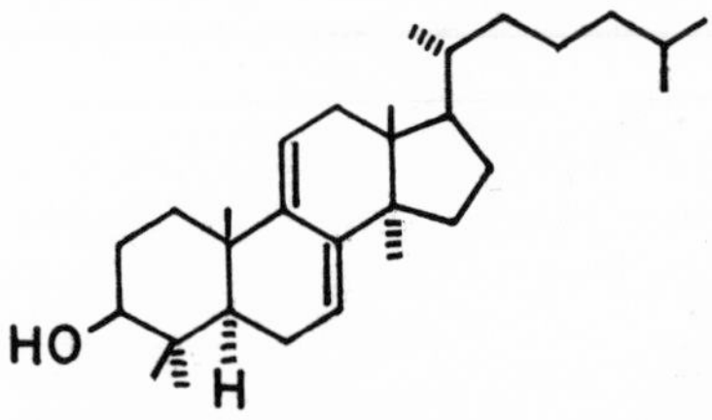

81

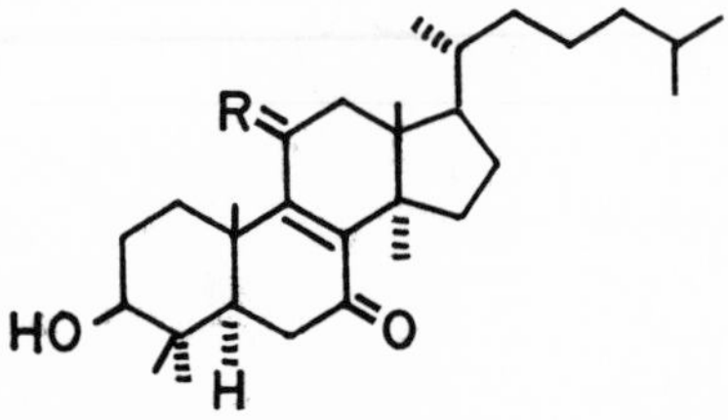

82 R = H$_2$

83 R = O

and the 3β-acetates of 3β-hydroxy-5α-lanost-8-en-7-one (82)
and 3β-hydroxy-5α-lanost-8-ene-7,11-dione (83) thought to
derive from initially formed hydroperoxides [1070,1071].
Oxidation of the 3β-acetate of 68 by $^1O_2$ likewise yielding
hydroperoxide, ketone, and other products was reported in
the same period [792,793]. The several oxidized C$_{30}$-sterols
from wool fat are catalogued later in Chapter III, and the
details of the oxidation of the 8-stenols will be presented
in Chapter VI.

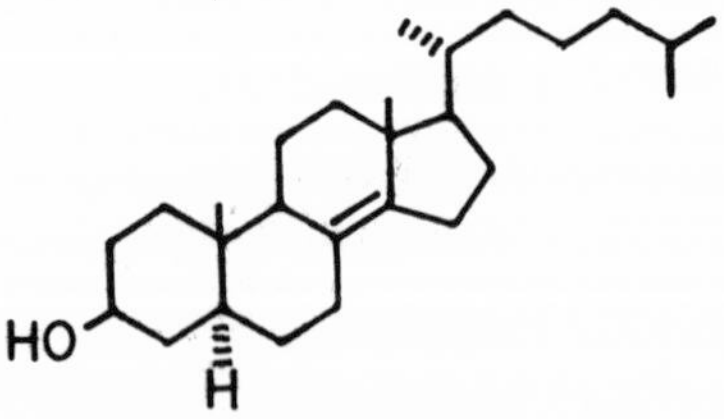

84 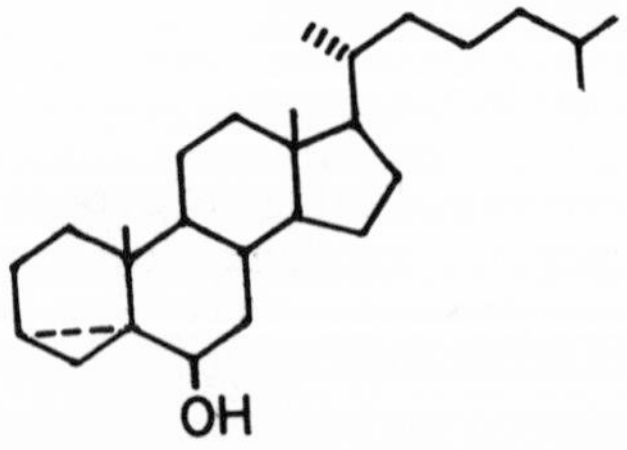   85

The general instability of other Δ^8-sterols to storage
was demonstrated in 1965 in studies showing that zymosterol
(5α-cholesta-8,24-dien-3β-ol) decomposed on storage [2480].
Likewise, stored samples of 5α-cholest-8(14)-en-3β-ol (84)
developed hydroperoxide contaminants, as did a 20 y old
sample of 3α,5-cyclo-5α-cholestan-6β-ol (ί-cholesterol) (85)
[2295]. The presence of hydroperoxide contaminants in stored
samples of sterols 56,67,68,84, and 85 as well as of the
5-stenols cholesterol, sitosterol, and stigmasterol is evinced in FIGURE 5.

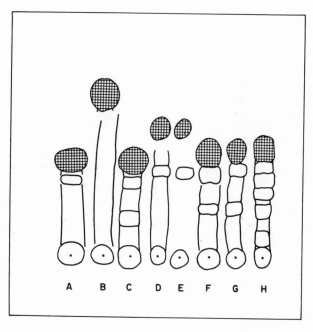

FIGURE 5. Hydroperoxides (cross-hatched spots)
           detected in air-aged sterols with N,
           N-dimethyl-p-phenylenediamine. Par-
           ent sterols: A, 8(14)-stenol 84; B,
           85; C, 7-stenol 57; D, 8-stenol 68;
           E, lanosterol; F, sitosterol; G,
           stigmasterol; H, cholesterol. (Re-
           printed with permission of Elsevier
           Scientific Publishing Co., Amsterdam,
           from J. Chromatog., 66, 101 (1972)).

CHAPTER III.   DISTRIBUTION OF
AUTOXIDATION PRODUCTS

I discuss in this chapter the distribution of cholesterol autoxidation products in Nature and also of their
natural occurrence as reported over the years by other
investigators. The term "natural occurrence" should pose
no difficulties if we retain the obvious meaning of occurrence in Nature, the term conveying no implication of
enzymic derivation. Nonetheless, there has accumulated a
substantial literature proposing that the common cholesterol
autoxidation products found in various tissues be natural
products, implication being that these be metabolites formed
by enzymic processes in the tissue of origin. It is instructive to review these many reports and the paucity of evidence
adduced for this often expressed viewpoint.

There are several oxidized cholesterol derivatives
found in tissues that are genuine enzymic metabolites
involved in established biosynthesis processes but that are
also established cholesterol autoxidation products as well.
The 3β,7α-diol __14__ is a prime example, the sterol being both
an important cholesterol metabolite leading to liver bile
acid biosynthesis but also a major autoxidation product
encountered just about everywhere that cholesterol autoxidation occurs. However, for most of the oxidized sterols
treated in this chapter evidence of enzymic origins is
lacking, but the onus of autoxidation artifact is manifest.
In general the investigators who have found cholesterol
autoxidation products in biological materials have not been
equally capable of providing experimental evidence anent
processes involved.

By the viewpoint developed in this monograph, most if
not all of the discoveries of cholesterol autoxidation products in tissues listed here appear to be in fact artifacts
of human intervention. Whereas there is suggestion that
genuine natural events such as photosensitized oxygenations
as might occur in tissues exposed to light and air and
in vivo lipid peroxidations may account for some of the
cases discussed in this chapter, to speak of natural occurrence for most of these examples is more correctly to speak
of the presence of cholesterol autoxidation products in
naturally occurring biological material subjected to a
natural environment but also to collection, handling,
storage, processing, and analysis by chemists.

It is crucial to the orientation of this chapter to
define those cholesterol autoxidation products of interest
with respect to their distribution. As disclosed later in
this monograph almost seventy oxidized steroids have been
implicated in the autoxidation of cholesterol, including
$C_{19}-$, $C_{20}-$, $C_{21}-$, $C_{22}-$, $C_{23}-$, $C_{24}-$, $C_{26}-$, and C_{27}-deriva-
tives of a variety of structural types. Thus, pregnenolone
(23) and the C_{19}-steroids 3β-hydroxyandrost-5-en-17-one
(86) and androst-5-ene-3β,17β-diol (87) are minor cholesterol autoxidation products [2549,2576]. Review here of

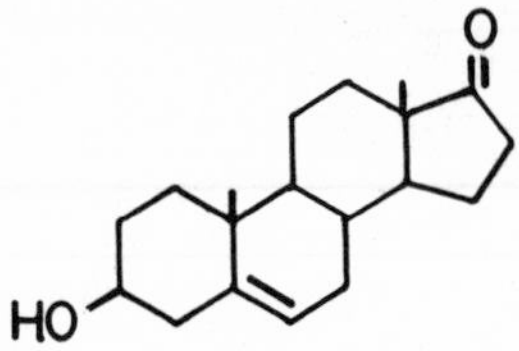

86 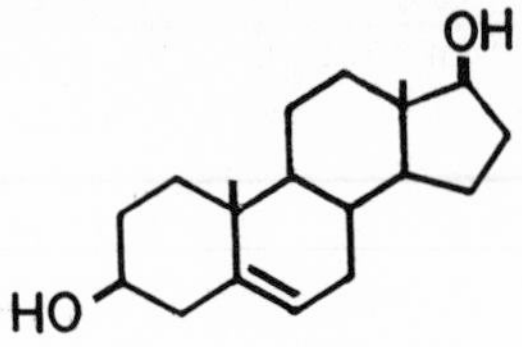   87

the natural occurrence of these steroids that are also
established enzymic metabolites involved in steroid hormone
biosynthesis would divert attention from the matter of cholesterol autoxidation and accordingly is not done. Rather,
only the observed distributions of those autoxidation
products of cholesterol that retain the full $C_{27}-$ carbon
content are given detailed treatment here. For balance,
a few classes of modified or degraded sterols with less
carbon content are also included along with highly oxidized
steroids not suspected as deriving via autoxidations.
Finally, the distribution of nonsteroidal peroxides in
Nature has been included as an item of related interest.

COMMON CHOLESTEROL AUTOXIDATION PRODUCTS

     The common cholesterol autoxidation products whose
distribution among natural sources was considered by
Bergström in 1961 [200] numbered twelve: the 3β,7-diols
14 and 15, the 7-ketone 16, the 3β,5α,6β-triol 13, enone 8,
the dienones 10 and 12, the 3β,25-diol 27, the 3β,6β- diol
40, the 6-ketones 44 and 45, and the 3,6-diketone 42. To
this group may be added the isomeric 5,6-epoxides 35 and
36 and the 7α-methyl ether 52, making the total of fifteen

autoxidation products listed in TABLE 1 and TABLE 2. These
fifteen sterol derivatives are all secondary or higher order
autoxidation products of cholesterol formed almost certainly
in the cases cited as artifacts of tissue storage and handling of the manipulations in isolations.

The most frequently encountered cholesterol autoxidation products include the epimeric 3β,7-diols 14 and 15,
the 7-ketones 10 and 16, and the 3β,5α,6β-triol 13, all
found in a variety of tissues (cf. TABLE 1). Examination
of experimental details in most cases suggests that unrecognized autoxidation account for products found. Although
claims to the contrary characterize these accounts, none of
these discoveries nor all of them together has engendered
study of the metabolic processes implied in the several
cases. Were these examples accepted as anything other than
cases of autoxidation, we could expect to have more evidence
of their metabolic origins than is apparent (cf. Chapter VII),
as is the case for the 3β,7α-diol 14.

However, cholesterol autoxidation is an insidious
process, encountered erratically in odd cases and not in
others. Except where massive autoxidation is involved and
products isolated, cholesterol autoxidation may escape
detection unless efficient chromatographic methods are used.
The more careful investigator using adequate methods will
encounter cholesterol autoxidation whereas the less attentive
may miss them. Moreover, there are cases where despite
adequate analysis means cholesterol autoxidation products
are not found. Thus, although autoxidation products 14-16
have been found in animal tissues (cf. TABLE 1) and in
foodstuffs (cf. TABLE 3) neither bovine muscle [817,2522]
nor broiled meat [2521] was found to have these sterols.
Many other reports of tissue sterols utilizing proper handling and analysis procedures also failed to note the presence
of the prominent autoxidation products 14-16.

Literature accounts of the natural occurrence of several
other recognized but less frequently encountered cholesterol
autoxidation products are summarized in TABLE 2. Here the
question of dual origins becomes more obscure, as these oxidized sterols occur at lower levels in tissue than do those
in TABLE 1 and evidence supporting enzymic origins is more
abundant for some but wholly lacking for others. Thus,
cholest-4-en-3-one (8) is an autoxidation artifact but is
also a recognized intermediate in the enzymic transformation

TABLE 1.   Distribution of the More Common Cholesterol
           Autoxidation Products

Sterol	Found In
Cholest-5-ene-3β,7α-diol (14)	Human adrenal cortex [443], arteries [960,1011,2567], brain [2154], meconium [1310,1435] Mare serum [2713] Bovine testis [1729] Pig liver [1551] Sheep wool fat [57,542] Rat feces [1953], liver [2074] Toad skin [1162] *Bombyx cum Botryte* [476]
Cholest-5-ene-3β,7β-diol (15)	Human adrenal cortex [443], arteries [1011,1012,2567], brain [2154], meconium [1310] Mare serum [2711] Bovine liver [971-973], testis [1729], fetal calf serum [1233] Pig spleen [1888] Rat feces [1953], liver [2209] Toad skin [1162] *Bombyx cum Botryte* [476]
Cholest-5-ene-3β,7ξ-diol (14 and/or 15)	Human adrenal tumor [2317], arteries [1011], brain [1873], erythrocytes [1160], sebum [735,1549,2662], serum [2319] Bovine brain [1952] Pig adrenals, blood, fat [1065] Dog blood, brain, liver [1722] Rabbit blood, heart, kidney, lung, [1874] *Glossostelma carsoni* roots [1923]
3β-Hydroxycholest-5-en-7-one (16)	Human adipose tissue [2304], aorta [364,1807,2132,2567], brain [844,2316], erythrocytes [1160], heart, kidney, liver, lung [2304], meconium [1310], muscle [2304], pancreas [2304], sebum [735], spleen [2304], urine [2153] Baboon adrenal vein blood [888], aorta [2304]

(continued)

TABLE 1   (continued)

Sterol	Found In
	Rhesus monkey adrenal vein blood [888]
	Horse aorta [2304]
	Bovine aorta [2304], brain [2023], testis [1729,1888]
	Pig adrenals [1065],testis [1889], fat [1065,2683], blood [1065], liver [39,2169]
	Sheep wool fat [542,1178,1631-1633]
	Cat aorta [2304]
	Rat adrenal [1894], aorta [2304], feces [1953], liver [2074]
	Chicken ovary, plasma [820]
	White Carneau pigeon aorta [2304]
	Sponge [592]
Cholesta-3,5-dien-7-one (10)	Human aorta and atheromas [493, 960,1242,1964,2567], brain [2316], erythrocytes [1160], serum [493,915]
	Bovine liver [1249], milk fat [776]
	Horse liver [431]
	Pig adrenal, blood,and fat [1065], spleen [1888], testis [1887,2050]
	Sheep wool fat [542,726,1069, 1178,1631,1632]
	Rat feces [1953], tissues [1243]
	Sponge [592]
	Yeast [1554]
5α-Cholestane-3β,5,6β-triol (13)	Human arteries [960,1011,1012, 2567], brain [746], gallstones [746], feces [1911,1913]
	Bovine liver [972]
	Pig testis [1887,2050]
	Sheep wool fat [542,1633]
	Rat feces [1910,1911,1953]
	Sponge [592]

TABLE 2.   Distribution of Less Common Cholesterol
           Autoxidation Products

Sterol	Found In
Cholest-4-en-3-one (8)	Human brain [2316], lymph [299], plasma [914] Bovine anterior hypophysis [1886] milk fat [776] Pig adrenals, blood, fat [1065], testis [1888,1889] Rat feces [2023] Sponge [592]
Cholesta-4,6-dien-3-one (12)	Human aorta and atheromas [960, 1964], plasma [914] Pig spleen [1888]
Cholest-5-ene-3β,25-diol (27)	Human adipose tissue [2304], arteries [1011,2567], brain [2316], heart, liver, lung, muscle, pancreas, spleen [2304] Baboon aorta [2304] Horse aorta [2304] Bovine adrenal cortex [443], aorta [2304] Cat aorta [2304] Rat aorta [2304], liver mitochondria [1615] White Carneau pigeon aorta [2304] Sponge [592]
Cholest-5-ene-3β,25-diol (27) and/or cholest-5-ene-3β,26-diol (29)	Rat liver mitochondria [1390] Mouse liver mitochondria [797, 798]
Cholest-5-ene-3β,25-diol (27) and/or cholest-5-ene-3β,24-diol (26)	Human arteries [1011] Rat feces [1953]
5,6α-Epoxy-5α-cholestan-3β-ol (35)	Human plasma [914]
5,6α-Epoxy-5α-cholestan-3β-ol (35) and/or 5,6β-epoxy-5β-cholestan-3β-ol (36)	Human plasma [1400] Hairless mouse liver [279] Rat lung [2192,2193]
3β,5-Dihydroxy-5α-cholestan-6-one (44)	Pig liver [2172]
5α-Cholestane-3,6-dione (42)	Human adrenal tumor [1274]

(continued)

TABLE 2 (continued)

Sterol	Found In
3β-Hydroxy-5α-cholestan-6-one (45)	Pig testis [1889]
	Pig spleen [1888]
	Spleen [486]
Cholest-4-ene-3β,6β-diol (40)	Pig spleen [1888]
	Rat adrenal [1894]
7α-Methoxycholest-5-en-3β-ol (52)	Human aorta [1404]
	Pig spleen [1888]

of cholesterol to the stanols 2 and 4. Likewise, the 3β,25-diol 27 is an established autoxidation product but one for which suggestion of enzymic origins have been made (cf. Chapter VII).

The 3β,5α,6β-triol 13 so broadly found in tissues is an autoxidation product derived via the 5,6-epoxides 35 and 36, and no other process for its formation other than via the 5,6-epoxides has been suggested. It is highly unlikely that the triol 13 be a product of separate 5α- and 6β-hydroxylations of a putative precursor such as the 5α-stanol 2, for example. However, enzymic 5α,6α-epoxidation and enzymic hydration of the 5α,6α-epoxide 35 have been demonstrated in selected systems, and thus both enzymic and autoxidative pathways for 35 and 13 exist. The isomeric 5β,6β-epoxide 36 has only an autoxidation origin but may also be formed in lipid peroxidizing systems (cf. Chapter VII).

The recorded presence of the 3β,5α,6β-triol 13 in tissues is thus easily rationalized as a hydration product of the 5,6-epoxides, but the 5,6-epoxides in tissues listed in TABLE 2 remain uncertain as to origins.

The remaining less frequently encountered autoxidation products of TABLE 2 are more properly pure autoxidation products despite claim to the contrary. Thus, the 4,6-dien-3-one 12 is a cholesterol autoxidation product derived from the 3β,7-diols 14 and/or 15 but is also a secondary product of the oxidation of cholesterol or of the enone 6 by $^1O_2$ (cf. Chapters IV and V). No enzymic processes for the biosynthesis of dienone 12 have ever been proposed, let alone supported by experimental work. The 4,6-diene-3-ketone moiety is not a feature of many naturally occurring steroids, but several such derivatives have been isolated from plant and

animal sources. Thus, secretions of the prothoracic protective gland system of several water beetles contain *inter
alia* 20-hydroxy-(20R)-pregna-4,6-dien-3-one from *Cybister
lateralimarginalis* [2108], *Cybister limbatus* [2263], and
Dytiscus marginalis [2106,2108]; 21-hydroxypregna-4,6-diene-
3,20-dione from a Mexican *Cybister* sp [2107] and *C. limbatus*
[2263]; and 12β-hydroxypregna-4,6-diene-3,20-dione from a
Mexican *Cybister* sp [2107], *Cybister limbatus* and others
[453] but also from the plant *Nerium odorum* [2753].

     The 3β,5α-dihydroxy-6-ketone 44 is clearly an artifact
formed from the artifact 3β,5α,6β-triol 13 in the one case
of its isolation by oxidation by elemental bromine used
during isolation procedures [2172].

     Less obviously, the 5α-3,6-diketone 42 is also a thermal
degradation product of 6β-hydroxycholest-4-en-3-one (88) and
of 6β-hydroperoxycholest-4-en-3-one (59) and therefore has a
demonstrated autoxidation pathway for its derivation. The
5α-3,6-diketone feature is not otherwise recognized among
steroid metabolites. Likewise the 3β,6β-diol 40 has no
demonstrated status as a cholesterol metabolite but is a
thermal decomposition product of 3β-hydroxycholest-4-ene-
6β-hydroperoxide (89) formed by attack of $^{1}O_2$ on cholesterol
[1401].

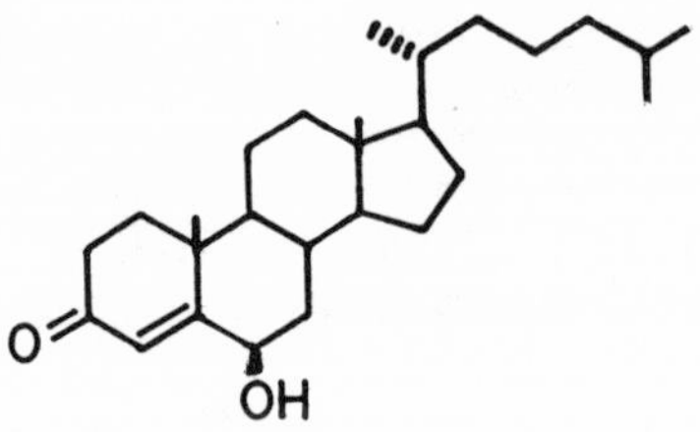

           88 R = H   89 R = OH

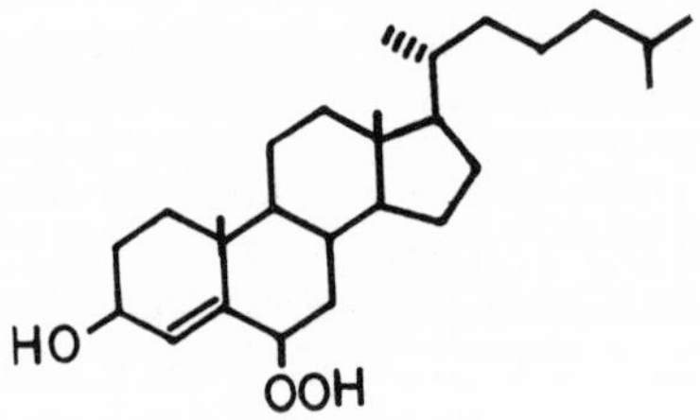

           59 R = OH   40 R = H

     The 3β-hydroxy-6-ketone 45 found in pig spleen [1888]
is yet another matter, for I am unaware of any autoxidation
process which generates this sterol. However, there is
evidence that the 6-ketone 45 be a cholesterol metabolite
in spleen [486] and in *Bombyx mori* prothoracic gland

[2060] incubations.

The 7-methyl ether 52 listed in TABLE 2 is obviously an artifact, indeed an artifact of an artifact.

In addition to these fifteen acknowledged cholesterol autoxidation products listed in TABLE 1 and TABLE 2 there are several other established cholesterol autoxidation products which have been found in tissues under circumstances which allow them to be considered in this same light. Thus, the (20S)-3β,20-diol 21 is found in bovine adrenal tissue at the level of 37 ng/g [1963] and in the human aorta [1427], but whether 21 is metabolite or artifact cannot be decided. The ease with which cholesterol is autoxidized via the (20S)-20-hydroperoxide 22 to 21 [2576, 2578] is offset by many reports that 21 be an enzymically derived intermediate in the biosynthesis of the 20-ketone 23 from cholesterol. In that the (20S)-3β,20-diol 21 also is found as fatty acid esters in bovine adrenal tissue at the level of 70 ng/g [1963] an argument may be put forth that 21 obviously is involved in enzymic metabolism. However, the facile autoxidation of the 3β-acetate of cholesterol to the corresponding (20S)-20-hydroperoxide 22 3β-acetate [2455] clearly indicates alternative processes to explain the presence of 21 fatty acid esters in tissues. At the very low levels implicated no satisfying conclusion can be drawn.

The matter of the presence of other side-chain hydroxylated cholesterol derivatives in select tissues will be discussed elsewhere in this monograph.

Almost all of the observations of the incidence of the common cholesterol autoxidation products of TABLE 1 and TABLE 2 deal with vertebrate tissues, and the very few reports involving these sterols and other life forms deserve special mention here. The recent discovery of the epimeric 3β,7-diols 14 and 15 together with ergosterol 5α,8α-peroxide 62 (and other 7-hydroxylated C_{28}- and C_{29}-sterols) in the Chinese drug *Bombyx cum Botryte* (Jiān cān), a preparation of the silk worm *Bombyx mori* infected with the fungus *Botrytis bassiana* Bals *(Beauveria bassiana* Bals) [476] is a striking example! Isolation of the 7-ketone 10 from yeast [1554] is another example.

This paucity of results in invertebrates, plants, and

microbial samples probably reflects the relatively low
levels of cholesterol in these organisms as well as the
early lack of chemical interest in these agents. In the
more recent expansion of work into these other forms of
life using modern methods and where adequate care has been
exercised these artifacts have generally not been demonstrated!

In that pronounced biological activities have been
associated with many of these common cholesterol autoxidation products (cf. Chapter VIII), the presence of such
sterols in foodstuffs becomes of special interest. In
TABLE 3 the distribution of cholesterol autoxidation products in natural and processed food materials is summarized.
Only very recent work has been conducted with adequate
modern analysis methods and with awareness of the insidious
nature of sterol autoxidation. Accordingly, these data
cannot be regarded as definitive as to whether sterol
autoxidation products are present significantly in our
foods.

A few non-oxidative processing artifacts, chiefly products of eliminations such as sterenes, steradienes, and
steratrines, are also included in TABLE 3.

In view of the highly suspicious nature of most of the
observations dealing with the several common cholesterol
autoxidation products of TABLE 1 and TABLE 2 one is tempted
to formulate the thesis that other oxidized sterols bearing
the same structural features as derivatives $\underline{10},\underline{12}$-$\underline{16},\underline{27},\underline{35}$,
amd $\underline{36}$, etc., thus Δ^5-3β,7-diol, Δ^5-7-ketone, $\Delta^{3,5}$-7-ketone,
3β,5α,6β-triol, 25-alcohol, and 5,6-epoxide features, are
prima facie artifacts of autoxidation. This concept will
be tested over the following sections of this chapter.

There are obvious major exceptions to this viewpoint
such that it cannot be accepted without reservation. However, in the cases where these suspect features are metabolic products, considerable work has been done to establish
true metabolite status. Thus, the 25-hydroxylated derivatives 9,10-secocholesta-Z-5,E-7,10(19)-triene-3β,25-diol
($\underline{91}$) and 9,10-secocholesta-Z-5,E-7,10(19)-triene-1α,3β,25-
triol ($\underline{92}$) are metabolites of cholecalciferol (vitamin D$_3$,
9,10-secocholesta-Z-5,E-7,10(19)-trien-3β-ol) ($\underline{90}$). However,
the 3β,25-diol $\underline{91}$ also appears to be formed nonenzymically
from cholecalciferol [271,272]. The 25-hydroxylated bile

TABLE 3. Oxidized Sterols in Foodstuffs

Food	Sterols Found	References
Egg Products:		
Egg yolk	5α-Cholestane-3β,5,6β-triol (13)	[746]
	Cholesta-3,5-dien-7-one (10)	[1850]
Egg dough*	Cholest-5-ene-3β,7-diols (14,15)	[12]
	Cholesterol hydroperoxides (46,47)	[12]
Spray-dried egg*	Cholest-5-ene-3β,7-diols (14,15)	[481]
	3β-Hydroxycholest-5-en-7-one (16)	[481]
	5,6β-Epoxy-5β-cholestan-3β-ol (36)	[481]
	5α-Cholestane-3β,5,6β-triol (13)	[481]
Heat-dried egg	5,6α-Epoxy-5α-cholestan-3β,ol (35)	[2507-2509]
	5,6β-Epoxy-5β-cholestan-3β-ol (36)	[2507-2509]
Dried egg mix	5,6-Epoxycholestan-3β-ols (35,36)	[1619]
	3β-Hydroxycholest-5-en-7-one (16)	[1619]
	5α-Cholestane-3β,5,6β-triol (13)	[1619]
Milk Products:		
Anhydrous milk fat	Cholest-4-en-3-one (8)	[776]
	Cholesta-3,5-dien-7-one (10)	[776]
Nonfat dry milk	Cholest-4-en-3-one (8)	[776]
	Cholesta-3,5-dien-7-one (10)	[776]
	Campest-2-ene**	[775]
	Stigmast-2-ene**	[775]
Butterfat	5α-Cholestan-3-one (365)	[1818]
Butter oil	Cholesta-3,5-diene (11)**	[1970]

(continued)

TABLE 3 (continued)

Food	Sterol Found	References
Other Products:		
Pork fat	3β-Hydroxycholest-5-en-7-one ($\underline{16}$)	[2683]
Brewer's yeast	Ergosterol 5α,8α-peroxide ($\underline{62}$)	[826]
	Cerevisterol ($\underline{154}$)	[826]
	5α,8α-Ergosta-6,E-22-diene-3β,5,8-triol ($\underline{222}$)	[826]
Baker's yeast	Cholesta-3,5-dien-7-one ($\underline{10}$)	[1554]
Beef*	Cholestatriene**	[2544]
	Cholesta-3,5-dien-7-one ($\underline{10}$)	[2544]
Edible oils	Stigmast-5-ene-3β,7-diols ($\underline{76,77}$)	[1260,1761,1764]
	3β-Hydroxystigmast-5-en-7-one ($\underline{72}$)	[1761,1764]
	Stigmasta-3,5-dien-7-one ($\underline{143}$)	[1761]
	steroid hydrocarbons**	[1760,1763,1765,1766]

* Foodstuffs variously irradiated
** Elimination (not oxidation) products

alcohols 5β-cholestane-3α,7α,12α,22ξ,25-pentaol (93), 5β-cholestane-3α,7α,12α,23ξ,25-pentaol (94), 5β-cholestane-3α,7α,12α,24,25-pentaols (95), and 5β-cholestane-3α,7α,12α, 25,26-pentaol (96) [502,2209] all are proper metabolites, no autoxidation process for their derivation being described.

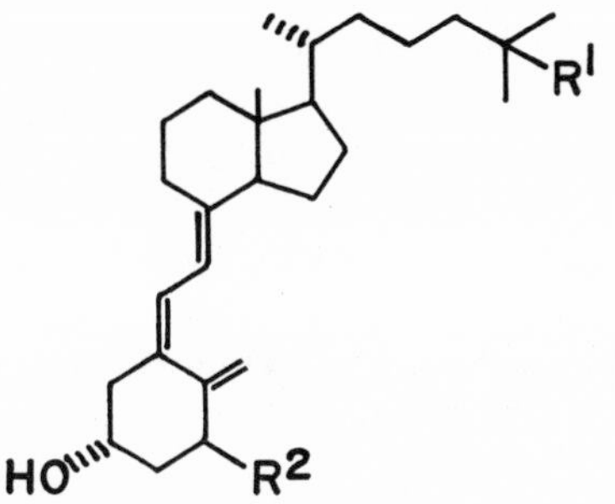

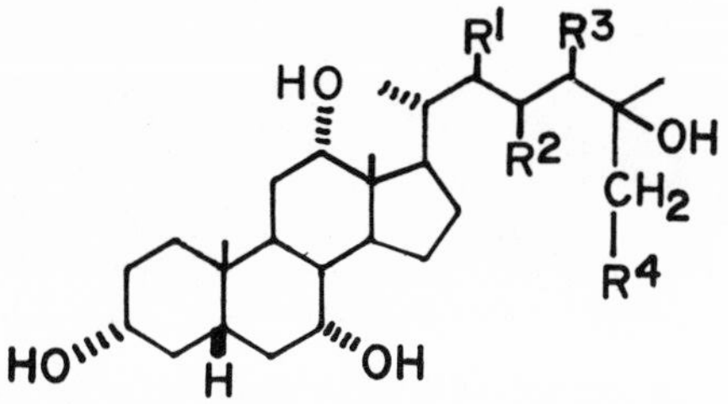

90 $R^1 = R^2 = H$	93 $R^1 = OH$, $R^2 = R^3 = R^4 = H$
91 $R^1 = OH$, $R^2 = H$	94 $R^2 = OH$, $R^1 = R^3 = R^4 = H$
92 $R^1 = R^2 = OH$	95 $R^3 = OH$, $R^1 = R^2 = R^4 = H$
	96 $R^4 = OH$, $R^1 = R^2 = R^3 = H$

A few Δ^5-3β,7-diols and Δ^5-7-ketones other than 14-16 have been isolated from biological sources, some of which will be mentioned in later sections of this chapter. Not all such 7-oxygenated-Δ^5-sterols can be regarded as arti- facts, for the Δ^5-7-ketone feature is present in several sterol sex hormones governing reproduction in the aquatic fungus *Achlya bisexualis*. These interesting sterols in- clude antheridiol (24-ethyl-3β,22,23-trihydroxy-7-oxo-(20S, 22S,23R)-cholesta-5,24(28)-dien-29-oic acid lactone (29→23)) (97) [97], 24-ethyl-3β,22-dihydroxy-7-oxo-(20S, 22R)-cholesta-5,24(28)-dien-29-oic acid lactone (29→22) [916], and several oogoniols which are fatty acyl esters of 24-ethyl-3β,11α,15β,29-tetrahydroxy-(20R,24ξ)-cholest- 5-en-7-one (98) [1608,1610] and are not only metabolically derived Δ^5-7-ketones but biologically active ones!

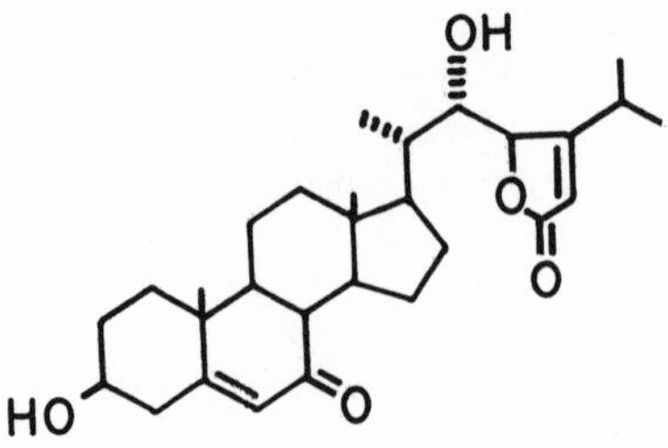

97 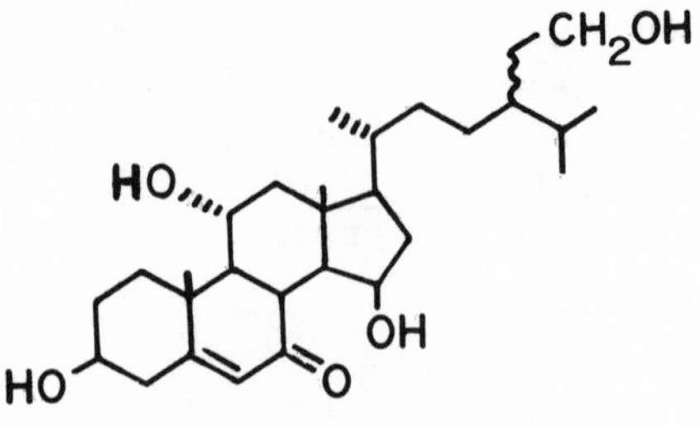   98

Were these or other similar oxidized sterols recovered
from tissue or enzyme incubations subjected to air-drying,
lengthly aeration, or storage for indefinite periods of
time under poorly controlled circumstances in the same
fashion as has been so often the case for the cholesterol
derivatives 10,13-16,27, etc., then metabolite status for
these kinds of sterols would be equally questionable.

MORE HIGHLY OXIDIZED PRODUCTS

The presence of more highly oxidized derivatives of
cholesterol in biological samples may be noted here. The
cholenic acid 3β-hydroxychol-5-enic acid (99) formed by
autoxidation processes has been found in human newborn
meconium as a sulfate ester conjugated with glycine or taur-
ine [118], in human urine and serum from patients with
cholestasis [117,119,590,1557,2396] or with hepatomas and
in human amniotic fluid [589,591,1829,1830]. The acid 99
has also been found in the brain of guinea pigs afflicted
with an experimental allergic encephalomyelitis [1713,1714].
In none of these cases is autoxidation particulary suspect.
Rather endogenous metabolic processes appear to be involved.

However, the lower homologs 3β-hydroxy-23-norchol-5-enic
acid (100) and 3β-hydroxy-22,23-bisnorchol-5-enic acid
(101) have been isolated from commercial supplies of sitosterol (20) under circumstances which suggest that the
noracids be an autoxidation products [2404,2405].

Although the formation of acidic autoxidation products
in air-aged samples of cholesterol [2576] and lanolin [1069]
is indicated by preliminary extractions of autoxidized

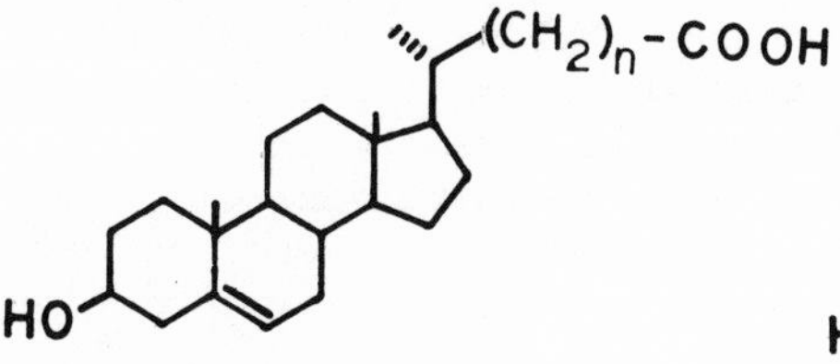

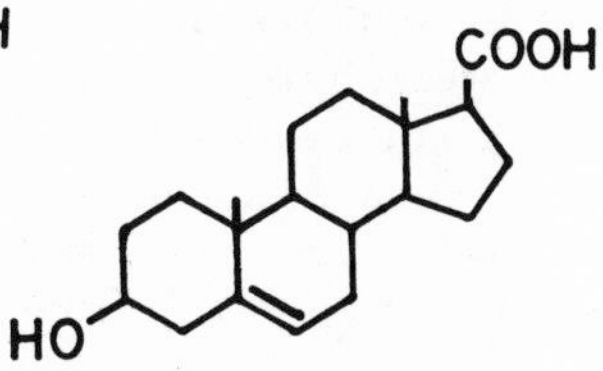

99 n = 2   102

100 n = 1

101 n = 0

material with base, these acid fractions have not been
examined systematically. Only very recently have we
adduced high performance liquid column chromatographic
evidence for the presenced *inter alia* of the cholenic acid
99 and the C_{22}-acid 101 and also of the etianic acid 3β-
hydroxyandrost-5-ene-17β-carboxylic acid (102) in air-aged
cholesterol.

Other carboxylic acid derivatives of cholesterol are
known, including side-chain oxidized acids and secoacids
derived by carbon-carbon bond scissions associated with the
A- and B-ring functionality. Thus, the B-secoacid 3β-hydroxy-
5-oxo-5,6-secocholestan-6-oic acid (103) is derived by
CrO_3 [2708], $KMnO_4$ [141], and ozone/H_2O_2 [524] oxidations
of the B-ring but has not been found in tissues of air-
aged cholesterol. The A-secoacid 5β-3,4-secocholestane-
3,4-dioic acid (43) has been recovered from whale ambergris
[1440], but as the sterols accompanying 43 were all reduced
stanols and no cholesterol autoxidations products were
isolated [1440], 43 must be a metabolite, perhaps of marine
microbial agents. The microbial decomposition of cholesterol
and other steroids obviously involves various degraded
secosteroid acids identified in controlled incubations but
not in tissues or in air-aged pure cholesterol.

Concern for the presence of more highly oxidized cholesterol derivatives in tissues, whether as metabolites or as
autoxidation artifacts, has not materialized. Other than
interests in carboxylic acids implicated in bile acid biosynthesis and in the aforementioned microbial degradation

of steroids and disposition of the cholenic acid 99, only
one major search for highly oxidized cholesterol derivatives
in tissues has been mounted. In this case, limitations of
methodology and misunderstanding of adequate control measures led to interesting oxidation chemistry of cholesterol
but not to new areas of cholesterol metabolism or autoxidation.

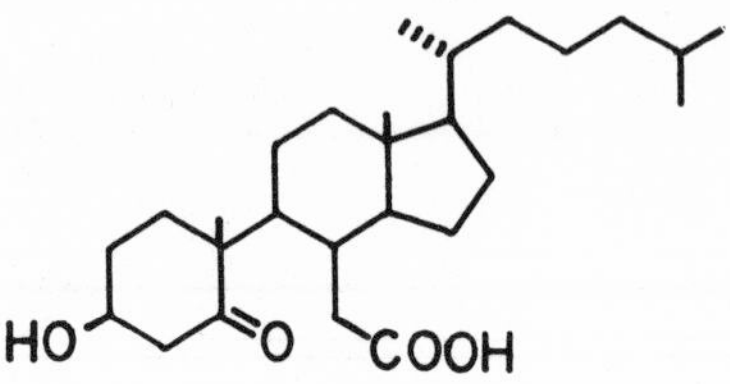

103 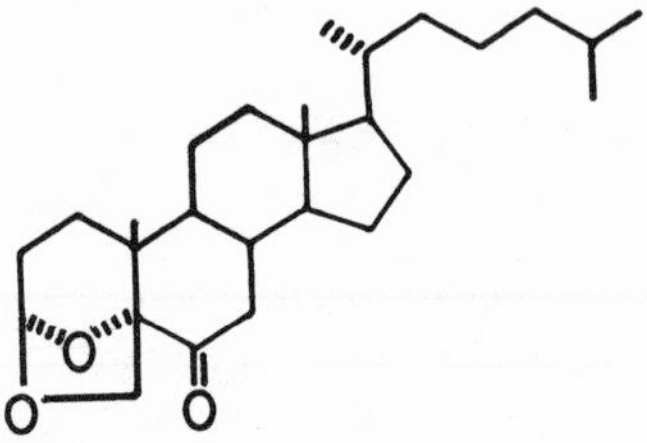   104

The search for "ketone 104" (3α,5-epoxy-5α-A-homo-4-
oxacholestan-6-one) (104) named after the notebook page
upon which it was first described [740] attracted much
interest at the time, as the compound or its hypothetical
precursor "proketone 104" was thought to be present in a
variety of tissues [740,746,754]. However, "ketone 104"
was eventually shown to be a product of cholesterol oxidation by CrO_3 used in the isolation procedures [751,754,755].
Only the repeated encounters with the obvious artifact
ketones 10 and 16 in tissues and attendant claims to natural
product status rival the effort spent on unraveling the
"ketone 104" issue. Nonetheless, work on "ketone 104" has
heuristic value in reemphasizing the absolute necessity of
detection of a companion sterol, metabolite, or autoxidation product in biological samples prior to exposure of the
sample to harsh chemical treatments.

PRODUCTS IN AIR-AGED CHOLESTEROL

As a coda to the previous remarks about the natural
distribution of the common cholesterol autoxidation products
in tissues it is fitting to mention the many minor autoxidation products which have been recovered from air-aged
cholesterol but which have not herefore been found in
tissues. In that much cholesterol is manufactured and

subjected to indefinite storage and exposure to air, such
cholesterol supplies in fact become natural sources for
the several autoxidation products in much the same manner
as summarized for products 8,10,12-16, and others in
TABLE 2.

TABLE 4 lists forty cholesterol autoxidation products
variously isolated or confidently detected in cholesterol
samples subjected merely to storage in contact with air.
These products include twenty-eight C_{27}-derivatives, two
C_{24}-derivatives, and one C_{22}-derivative, five C_{21}-deriva-
tives, one C_{20}-derivative, and three C_{19}-derivatives.
Thus, about half of the presently established autoxidation
products of cholesterol have been specifically found in
air-aged cholesterol. The remaining products have been
established by inferences from other chemical information
discussed later in Chapter V.

The yields of individual cholesterol autoxidation products recovered from air-aged cholesterol obviously depend
upon the precise sample history, and the 7-oxygenated
derivatives invariably predominate. Yields of the less
common autoxidation products range from 15 mg/g for the
3β,25-diol 27 and 1.2 mg/g for the corresponding 25-hydro-
peroxide 26 to the parts-per-million level for the (20S)-20-
hydroperoxide 22 at 120 μg/g, (24RS)-24-hydroperoxides
105/106 at 50 μg/g; pregnenolone (23) at 12 μg/g [2576,2579].

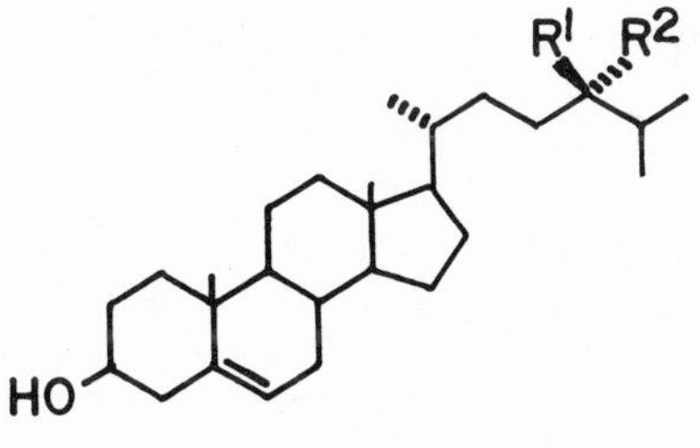

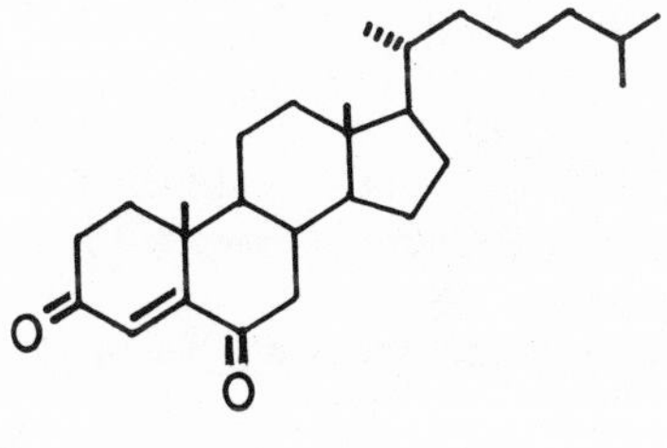

$\underline{105}$ R^1 = OOH, R^2 =H $\underline{108}$

$\underline{106}$ R^1 = H, R^2 = OOH

$\underline{107}$ R^1 = OH, R^2 = H

TABLE 4. Autoxidation Products in Air-Aged Cholesterol

Sterol	References
C_{27}-Steroids:	
Cholest-5-en-3-one (6)	[70]
Cholest-4-en-3-one (8)	[70]
Cholesta-3,5-dien-7-one (10)	[1072,2303,2576]
Cholesta-4,6-dien-3-one (12)	[72]
5α-Cholestane-3β,5,6β-triol (13)	[1072,2303,2576,2580]
Cholest-5-ene-3β,7α-diol (14)	[1072,2303,2576]
Cholest-5-ene-3β,7β-diol (15)	[1072,2303,2576]
3β-Hydroxycholest-5-en-7-one (16)	[754,1072,2303,2305,2576]
(20S)-Cholest-5-ene-3β,20-diol (21)	[2576]
3β-Hydroxy-(20S)-cholest-5-ene-20-hydroperoxide (22)	[2565,2576]
3β-Hydroxycholest-5-ene-25-hydroperoxide (26)	[2576]
Cholest-5-ene-3β,25-diol (27)	[754,2303,2576,2580]
3β-Hydroxy-(25R)-cholest-5-ene-26-hydroperoxide (28)	[2559,2560,2562]
(25R)-Cholest-5-ene-3β,26-diol (29)	[2580]
3β-Hydroxy-(25S)-cholest-5-ene-26-hydroperoxide (30)	[2559,2560,2562]
(25S)-Cholest-5-ene-3β,26-diol-(31)	[2580]
3β-Hydroxy-(20R)-cholest-5-ene-20-hydroperoxide (32)	[2565]
5,6α-Epoxy-5α-cholestan-3β-ol (35)	[74,93,765]
5,6β-Epoxy-5β-cholestan-3β-ol (36)	[74,93]
Cholest-5-ene-3β,4β-diol (41)	*
3β,5-Dihydroxy-5α-cholestan-6-one (44)	[2303,2172]
3β-Hydroxycholest-5-ene-7α-hydroxyperoxide (46)	[2576]
3β-Hydroxycholest-5-ene-7β-hydroperoxide (47)	[2455]
6β-Hydroperoxycholest-4-en-3-one (59)	[2455,2560]

(continued)

TABLE 4. (continued)

Sterol	References
6β-Hydroxycholest-4-en-3-one (88)	*
3β-Hydroxy-(24R)-cholest-5-ene-24-hydroperoxide (105)	[2579]
3β-Hydroxy-(24S)-cholest-5-ene-24-hydroperoxide (106)	[2579]
Cholest-4-ene-3,6-dione (108)	[70]

C_{24}-Steroids:

3β-Hydroxychol-5-enic acid (99)	*
Chol-5-en-3β-ol (109)	[2576]

C_{22}-Steroids:

3β-Hydroxy-22,23-biosnorchol-5-enic acid (101)	*

C_{21}-Steroids:

3β-Hydroxypregn-5-en-20-one (23)	[2576]
Pregn-5-en-3β-ol (110)	[2576]
Pregn-5-ene-3β,20α-diol (111)	[2576]
3β-Hydroxypregn-5-ene-20α-hydroperoxide (112)	[2562]
3β-Hydroxypregn-5-ene-20β-hydroperoxide (113)	[2562]

C_{20}-Steroids:

3β-Hydroxyandrost-5-ene-17β-carboxylic acid (102)	*

C_{19}-Steroids:

3β-Hydroxyandrost-5-ene-17-one (86)	[2576]
Androst-5-ene-3β,17β-diol (87)	[2576]
Androst-5-en-3β-ol (114)	[2576]

* Heretofore unpublished data

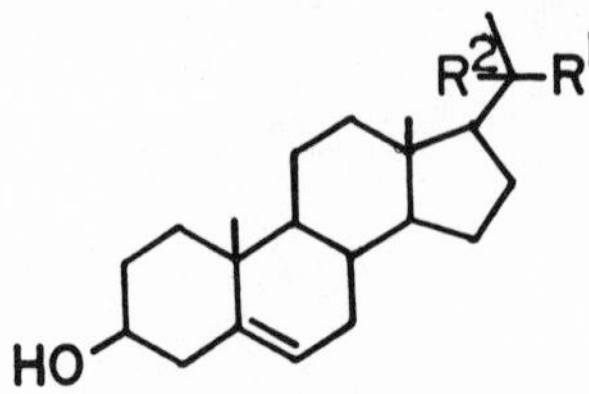

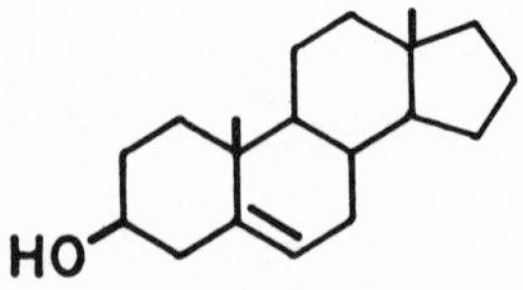

$$\underline{109} \ R^1 = CH_2CH_2CH_3, R^2 = H \qquad \underline{114}$$

$$\underline{110} \ R^1 = R^2 = H$$

$$\underline{111} \ R^1 = OH, R^2 = H$$

$$\underline{112} \ R^1 = OOH, R^2 = H$$

$$\underline{113} \ R^1 = H, R^2 = OOH$$

Our most recent examination of air-aged cholesterol have sought after yet lower yield components but range from the parts-per-million level for cholest-5-ene-3β,4β-diol (41) at 180 μg/g, 6β-hydroxycholest-4-en-3-one (88) at 87 μg/g, and the isomeric 5,6-epoxides 35 and 36 at 83 μg/g and 75 μg/g respectively [74] down to much lower levels. Thus, we have identified the dienone 12 at 40-120 ng/g [72], the C_{22}-acid 101 at 219 ng/g, the C_{24}-acid 99 at 14 ng/g, and the C_{20}-acid 102 at 2 ng/g. We presently rest our efforts at this parts-per-billion (milliard) level, but there are still many other unidentified components present in naturally autoxidized cholesterol!

Two steroids previously thought to be components of air-aged cholesterol have not been detected in our recent specific search for them. The 5α-hydroxyperoxide 51 inferring $^1O_2$ attack on cholesterol previously suggested as present in air-aged cholesterol [2576] in fact has not been detected by very careful work, nor has its decomposition product 5α-cholest-6-ene-3β,5-diol (50) been found [72]. Moreover, the presence of cholesta-1,4-dien-3-one (115) in air-aged cholesterol suggested by inconclusive chromatographic data [765,1523] has not been confirmed by our recent searches [70]. Accordingly, 50,51, and 115 may not now be regarded as natural autoxidation products

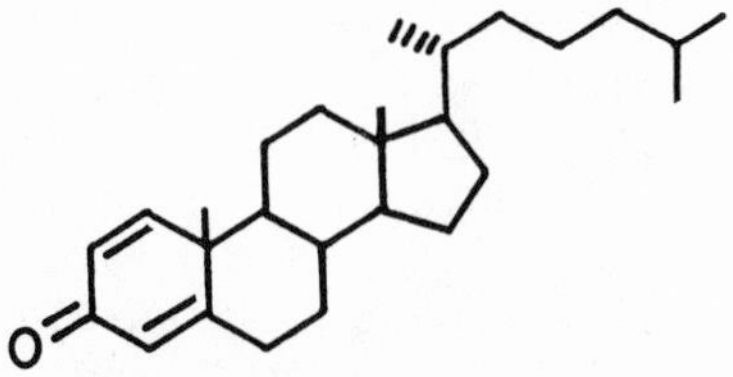

115

of cholesterol. The dienone 115 is, however, a metabolite implicated in the degradation of cholesterol by microorganisms [90,1705,1706].

OXIDIZED LANOSTEROL DERIVATIVES

Sheep wool fat or lanolin contains cholesterol, the 5α-stanol 2, lanosterol (5α-lanosta-8,24-dien-3β-ol) (67), 5α-lanost-8-en-3β-ol (dihydrolanosterol) (68), and 5α-lanosta-7,9(11),24-trien-3β-ol [355,2130] *inter alia* and is quite sensitive to air oxidation [58,1175,1176]. It is thus no surprise to find the cholesterol autoxidation products 10,14,15, and 16 listed in TABLE 1 in oxidized lanolin. The 3,5-dien-7-ketone 10 in such preparations is well recognized as deriving from the 7-ketone 16 during alkaline saponification of sterol esters [542,726,1069, 1178,1632] or during dissociation of sterol digitonides by pyridine [1631].

The sterol ester fraction of sheep wool fat is particularly complex [1844,2409], and in addition to artificial oxidations during saponification, some eliminations may occur, witness the recovery of 4-methylcholesta-2,4-diene from wool fat [2248].

No oxidation products associated with the 5α-stanol 2 have been found in wool fat, indeed in any biological material. The presence of the 5α-stanol 2 in wool fat once questioned [1631] has only recently been confirmed [355, 2130].

Ketone derivatives of lanosterol and its dihydro derivative 68 have also been found in oxidized lanolin. The 7-ketone 3β-hydroxy-5α-lanosta-8,24-dien-7-one [355,356] and

3β-hydroxy-5α-lanost-8-en-7-one (82) [726,1069,1070,1633], and the 7,11-diketones 3β-hydroxy-5α-lanost-8-ene-7,11-dione (83) [1070,1633] and 3β-hydroxy-5α-lanostane-7,11-dione [355,356] are regularly reported.

These several 7-ketones must surely derive from putative corresponding 7-hydroperoxide and 7,11-dihydroperoxide derivatives of the 8-stenol 68 as suggested at their first isolation [1070,1071] and fully demonstrated in the air oxidation of 68 3β-acetate [2173-2175] (cf. Chapter VI).

Oddly, no 7-hydroxylated lanosterol derivatives have been detected in oxidized wool fat. However, the (24R)-3β,24-diol 5α-(24R)-lanosta-8,25-diene-3β,24-diol (116) has been isolated [356]. This derivative and (24S)-cholest-5-ene-3,24-diol (cerebrosterol) (25) also isolated from wool

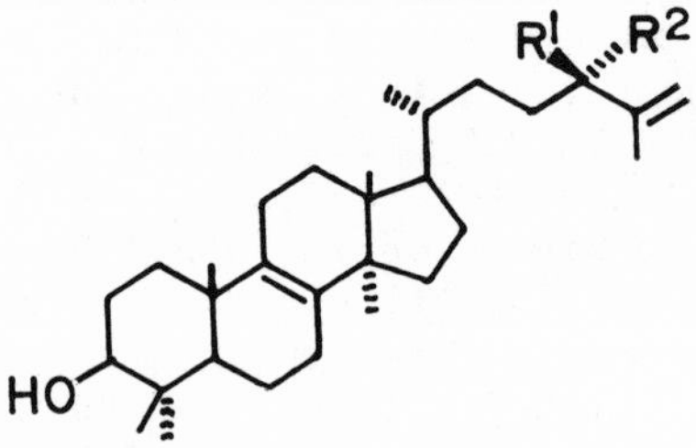

116 R^1 = OH, R^2 = H

117 R^1 = H, R^2 = OH

fat residues [754] are of obscure origins in that only the one stereoisomeric (24R)-alcohol 116 and (24S)-alcohol 25 were found where racemic (24RS)-alcohols would be expected of autoxidations.

However, a question of the stereochemical purity of the C_{30}-(24R)-3β,24-diol 116 has been raised [309], and from the recently recorded photosensitized oxidation of 68 3β-acetate (in which $^1O_2$ is implicated) whereby both C-24 epimers 116 and 5α-(24S)-lanosta-8,25-diene-3β,24-diol (117) were obtained as the respective 3β,24-diacetates [1704], it would appear that the 3β,24-diol preparation from oxidized wool fat be an artifact rather than an enzymic metabolite of lanosterol.

The two 3β,24-diols isolated from oxidized wool fat are of two separate types, the (24R)-3β,24-diol 119 (or

mixture of 116 and 117 [309]) being an allylic alcohol, the
(24S)-3β,24-diol 25 being an established metabolite of
cholesterol in brain but of uncertain origin in wool fat.
Although both cholesterol 24-hydroperoxides 105 and 106
occur in air-aged cholesterol as autoxidation products
[2579], the thermal decomposition of the 24-hydroperoxides
does not give the corresponding 3β,24-diols 25 and 107 but
the 24-ketone 34 [2579] (cf. Chapter V), but enzymic
reduction of the 24-ketone 34 or the 24-hydroperoxides 105
and 106 would afford the epimeric 3β,24-diols 25 and 107.
Borohydride reduction of the 24-ketone 34 affords the
24-epimers 25 and 107 in approximately the same ratio, only
slightly favoring the (24R)-epimer 107 [1512]. The autoxidation of cholesterol to the 24-hydroperoxides 105 and 106
however appears to favor the (24R)-epimer 107 over the (24S)-
epimer 25 by 2:1 [2579].

These items together with the absence of the 3β,24-diol
107 fron oxidized wook fat suggests that the (24S)-3β,24-
diol 25 in wool fat be a metabolite but that the case for
the lanosterol 3β,24-diol 116 be more likely of artificial
origin. A related case obtains in the cycloartenol series
discussed in a later section of this chapter.

In that wool fat may contain sterol phosphate esters
linked to carbohydrate and amino acids, saponification
adequate for hydrolysis may also yield artifacts more polar
than cholesterol, whose nature has not been examined [1174].

OTHER OXIDIZED STEROLS FROM INVERTEBRATES

As previously noted, few accounts of the distribution
of the common cholesterol autoxidation products in invertebrates are known. However, oxidized sterols bearing the
same structural features of the common cholesterol derivatives, thus Δ^5-3β,7-diol, Δ^5-7-ketone, 3β,5α,6β-triol,
25-hydroxy, etc. moieties, have become suspect as autoxidation products of putative parent sterols bearing the usual
sterol structural features.

Only few accounts of the presence of analogous 7-hydroxy-
lated Δ^5-sterols in animal tissues have issued, one being
the reported isolation of 24-methylcholesta-5,24(28)-diene-
3β,7β,19-triol (118) and its 7β-monoacetate ester 7β-acetoxy-
24-methylcholesta-5,24(28)-diene-3β,19-diol (119) from sun-

dried(!) soft coral *Litophyton viridis* [324]. Although the
7β-alcohol 118 is suspect as a possible artifact, the acety-
lated derivative 119 suggests a metabolic derivation.

118 R = H

119 R = COCH$_3$

Moreover, several sterol 3β,5α,6β-triols have been
isolated from air-dried (or wet) marine creatures, including
the C$_{27}$-sterols 5α-cholestane-1β,3β,5,6β-tetraol (120) from
soft coral *Sarcophyton glaucum* [1340] and 5α-cholestane-
3β,5,6β,9α-tetraol (121) from gorgonian *Pseudopterogorgia
elisabethae* [2115] and a variety of C$_{28}$-homologs of triol
13. Among these are a 24-methyl-5α-(24ξ)-cholestane-
3β,5,6β-triol (122), 24-methyl-5α-cholest-24(28)-ene-
3β,5,6β-triol, and their corresponding 6β-acetates from
Sinularia dissecta [325], a 24-methyl-5α-(24ξ)-cholestane-
1β,3β,5,6β-tetraol (123) from soft corals *Lobophytum pauciflorum* [2740] and *S. glaucum* [1340], 24-methyl-5α-(24ξ)-
cholestane-3β,5,6β,25-tetraol (124) and 24-methyl-5α-cholest-
24(28)-ene-1β,3,5,6β-tetraol from *L. pauciflorum* [2740], and
the 25-acetates 25-acetoxy-24-methyl-5α-(24ξ)-cholestane-
3β,5,6β-triol (125) from *L. pauciflorum* [2740] and *Sarcophyton elegans* [1658,1659], 25-acetoxy-24-methyl-5α-(24ξ)-
cholestane-1β,3β,5,6β-tetraol (126) from *S. glaucum* [1340],
and 25-acetoxy-24-methyl-5α-(24ξ)-cholestane-3β,5,6β,12β-
tetraol from *S. elegans* [1658,1659].

The 3β,5α,6β-triol feature of these several sterols is
suspect as being artifically formed. The same arguments
regarding biosynthesis applicable to the 3β,5α,6β-triol 13
apply equally to the 3β,5α,6β-triols of these marine creatures. However, metabolic processes in which such highly
oxidized sterols may participate is suggested by the isolation from sun-dried soft coral *L. pauciflorum* of the poly-
hydroxy ketone 25-acetoxy-3β,4β,5-trihydroxy-24-methyl-5β-
(24S)-cholestan-6-one [2531] containing A- and B-ring features in autoxidations.

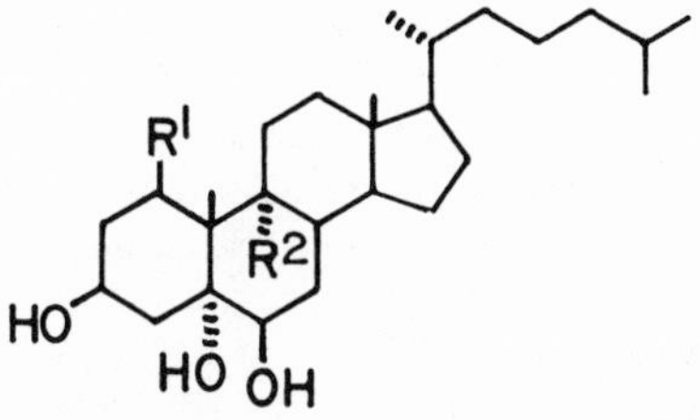

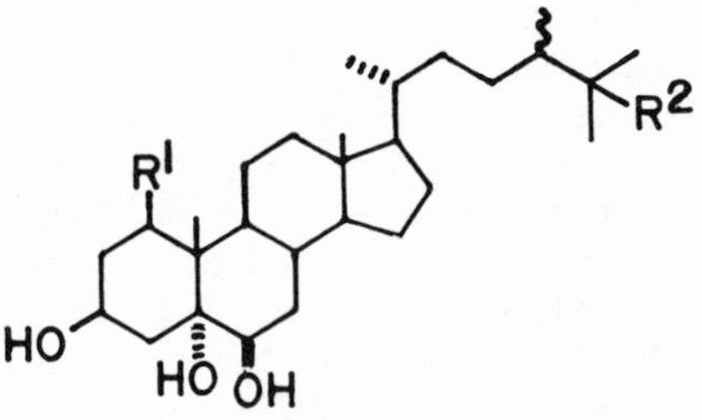

120 R^1 = OH, R^2 = H   122 R^1 = R^2 = H

121 R^1 = H, R^2 = OH   123 R^1 = OH, R^2 = H

                                   124 R^1 = H, R^2 = OH

                                   125 R^1 = H, R^2 = $OCOCH_3$

                                   126 R^1 = OH, R^2 = $OCOCH_3$

Furthermore, other metabolic processes are also implicated in the derivation of the several sterol tetraols and
pentaols mentioned. The 1β-,9α-, and 12β-hydroxylations
evident cannot be rationalized as nonenzymic oxidations
although the 25-hydroxylation may be. The main issue
raised is whether a 3β,5α,6β-triol be hydroxylated metabolically to give a tetraol metabolite or whether a putative
Δ⁵-3,X-diol parent sterol be oxidized via 5,6-epoxide
derivatives to the 3β,5,6,X-tetraol. The 6β- and 25-
acetylations inferred by the structures of several of the
isolated sterols are to be regarded as metabolic transformations of the parent alcohols.

The presence of the 3β,25-diol 24-methyl-(24ξ)-cholest-
5-ene-3β,25-diol (127) in a soft coral Nephtea sp. together
with 24-methyl-(24S)-cholest-5-en-3β-ol (campesterol) (71)
has been noted and the likely origin of the 3β,25-diol 127
by autoxidation of campesterol suggested [685]. A homologous
pair 24-ethyl-(24ξ)-cholest-5-en-3β-ol and 24-ethyl-(24ξ)-
cholest-5-ene-3β,25-diol occur in the sponge Damiriana
hawaiiana, along with cholesterol and the 3β,25-diol 27
[592], both pairs likewise suggesting autoxidations.

Besides the cases of the 25-alcohols and their esters
124-127 there are several interesting members of a class of

more highly oxidized toxic sterol glycosides found in sea
cucumbers *(Holothurioidea)* which are 25-hydroxylated [1890],
therefore are formal analogs of the recognized artifact
3β,25-diol 27. Although there is no direct reason to suspect an artifact nature for these materials now, they have
generally been isolated from dried skins of the marine
animals, and other artifacts of manipulation are manifest
for this class. The four aglycones seychellogenin (3β,20-
dihydroxy-5α-(20ξ)-lanosta-7,9(11)-dien-18-oic acid lactone
(18→20)) (128), koellikerigenin (3β,20,25-trihydroxy-5α-
(20ξ)-lanosta-7,9(11)-dien-18-oic acid lactone (18→20))
(129), ternaygenin (3β,20-dihydroxy-25-methoxy-5α-(20ξ)-lano-
sta-7,9(11)-dien-18-oic acid lactone (18→20) (130), and
praslinogenin (3β,17α,20-trihydroxy-5α-(20ξ)-lanosta-7,9(11)-
dien-18-oic acid lactone (18→20))(131) isolated from the

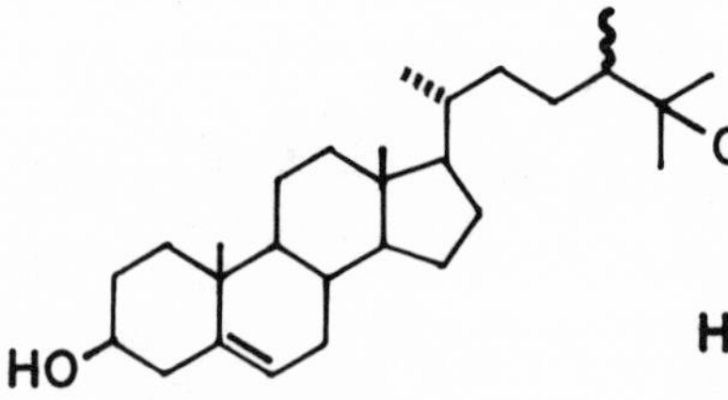

127

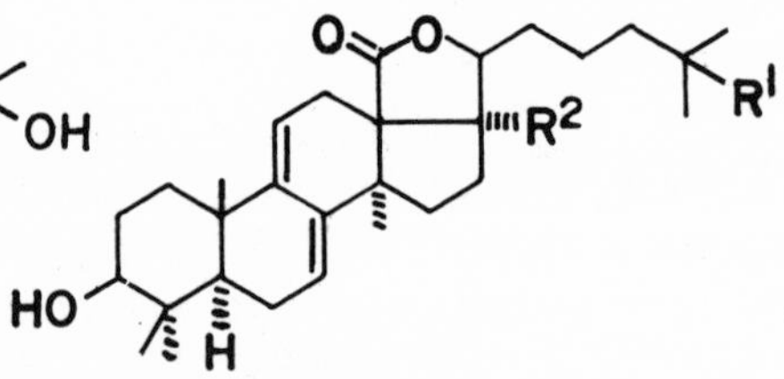

128 $R^1 = R^2 = H$

129 $R^1 = OH$, $R^2 = H$

130 $R^1 = OCH_3$, $R^2 = H$

131 $R^1 = OCH_3$, $R^2 = OH$

sea cucumber *Bohadschia koellikeri* [2001,2530] compose such
a group, the 25-0-methyl ethers 130 and 131 surely deriving
artifically during acid hydrolysis of the sterol glycosides
present in the dried skin specimens. A similar case in
related work with the sea cucumber *Stychopus japonicus*
Selenka [676,1322,1323] is recorded.

Although the 25-0-methylation of the 25-alcohol 129 by
acid methanol is observed [2001] and both parent 129 and
25-0-methyl ether product 130 are present together, another
interesting possibility for artifact formation has been
reported. The acid catalyzed addition of solvent to the

terminal methylene group of stigmasta-5,E-22,25-triene-3β-ol
(132) provides both 25-methoxystigmasta-5,E-22-dien-3β-ol
(133) and stigmasta-5,E-22-diene-3β,25-diol (134) as artificial products [1810].

Resolution of the question of the structure of the
terminal portion of the side-chain of the true sterol glycosides of these sea cucumbers might best yield in studies
of the glycosides isolated from fresh material.

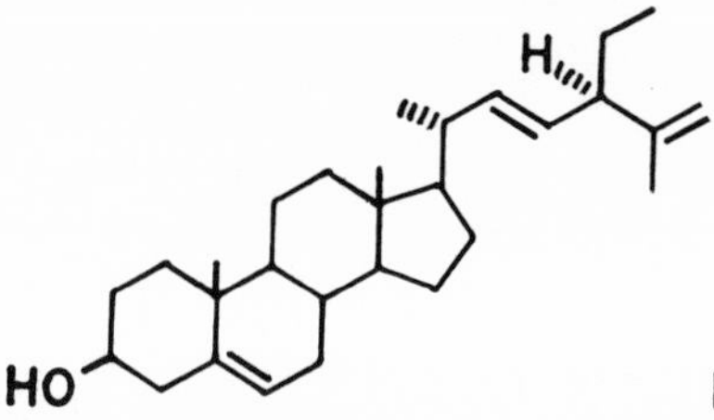

132 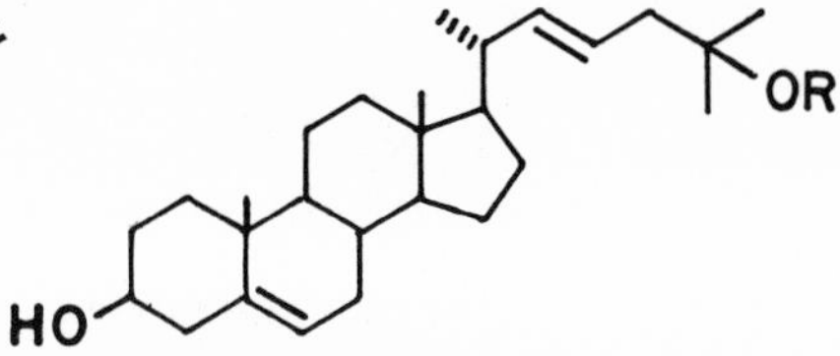   133 R = CH$_3$

134 R = H

Yet other sterols from marine invertebrates bear formal
resemblances to products of artificial oxidations of the
sterol side-chain, particularly side-chain allylic alcohols
and ketones.   The case of the C_{30}-Δ^{25}-3β,24-diols 116 and
117 of oxidized wool fat has already been mentioned.   Allylic
ketones such as 3β,9-dihydroxy-4α,24-dimethyl-5α,9β-cholest-
24(28)-en-23-one (135) isolated from the soft coral L. viridis [326] and 3β,6α-dihydroxy-5α-cholesta-9(11),24-dien-
23-one (136) isolated from the starfish Marthasterias glacialis [1550,1750,2285,2527] and Asterias amurensis [1135]
possess these suspicious features but are apparently not
artifacts.   Nonetheless, the $\Delta^{24(28)}$-23-ketone 135 was
recovered from the same preparation which yielded the suspect 3β,7β-diol 118 [324,326].

By contrast the sterols accompanying the Δ^{24}-23-ketone
136 allay doubts.   The 23-ketone 136 is accompanied by its
saturated analog 3β,6α-dihydroxy-5α-cholest-9(11)-en-23-one
(137) in M. glacialis [2285] and by the saturated alcohols
5α-(23ξ)-cholest-9(11)-ene-3β,6α,23-triol and 5α-(24ξ)-
cholestane-3β,6α,15α,24-tetraol in A. amurensis [1134,1135,
1219] (all as 6α-glycosides).

However, artifacts of isolation are recognized in these

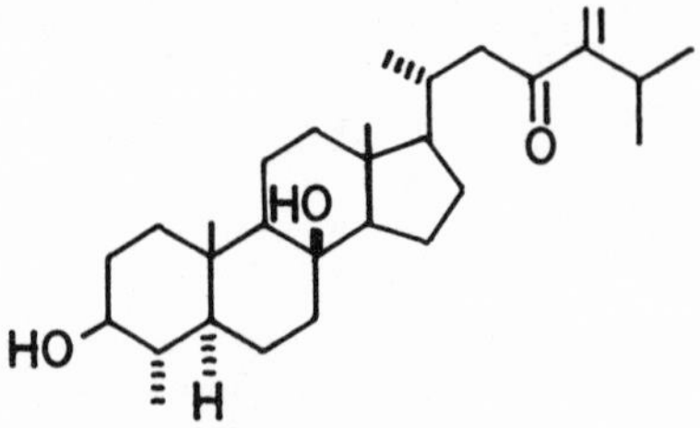

135

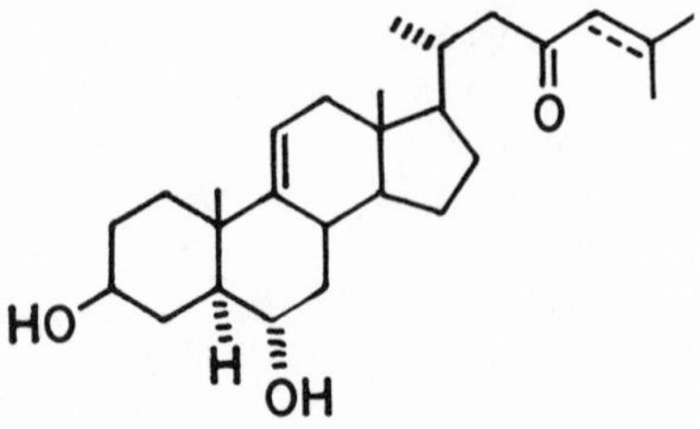

136 Δ^{24}

137

starfish sterols, both dehydration and degradation involving carbon-carbon bond scission being indicated. Thus, the 20-hydroxysterols 3β,6α,20-trihydroxy-(20ξ)-5α-cholest-9(11)-en-23-one(138) and 3β,6α-20-trihydroxy-24-methyl-5α-(20ξ,24ξ)-cholest-9(11)-en-23-one (139) are both present as conjugates in the crown of thorns starfish *Acanthaster planci*

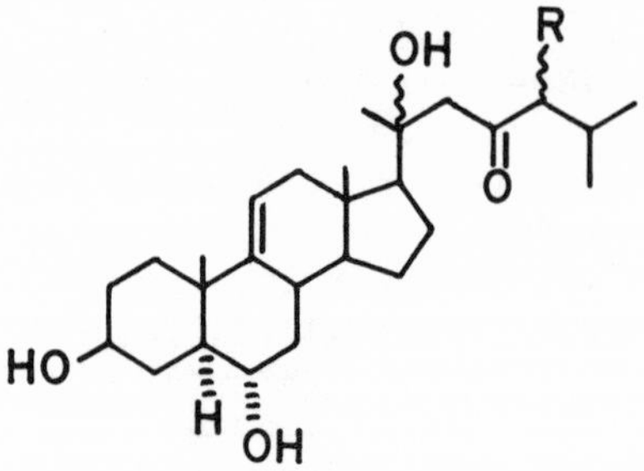

138 R = H

139 R = CH₃

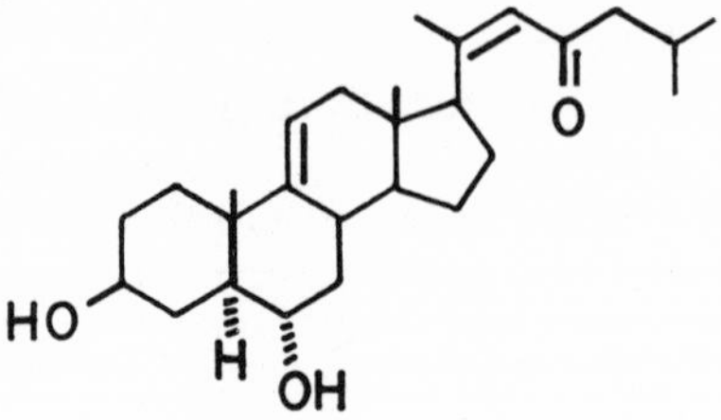

140

probably as the genuine sapogenols [1318-1321]. Acid treatment of the 20ξ-hydroxy-23-ketone 138 yielded the 20(22)-dehydro-23-ketone 3β,6α-dihydroxy-5α-cholesta-9(11), 20(22)-dien-23-one (140) also recovered along with 24-methyl-5α-(24ξ)-cholesta-9(11),20(22)-diene-3β,6α-diol and 5α-cholesta-9(11),17(20),24-triene-3β,6α-diol from *A. planci* [1318,2216,2218-2220], thus indicating that the $\Delta^{17(20)}$-and $\Delta^{20(22)}$-sterols are most likely acid dehydration artifacts [1321,2220].

   That even more severe alterations of sterols originally
likely present in these preparations may occur is evinced
by the repeated isolation of 3β,6α-dihydroxy-5α-pregn-9(11)-
en-20-one (141) from a variety of starfish, including *A.
planci* [2218,2219,2234], *A. amurensis* [1132,1133,1136,1137],
Asterias forbesia [80], *Asterias rubens* [2284], *Asterias
vulgaris* [876], and *M. glacialis* [2284]. In that the
20-ketone 141 was recovered from acid treatment of the
20-hydroxy-23-ketone 138 [1318,1321], its recovery from
acid hydrolysates of this and other related starfish sterol
glycosides is highly suspect.

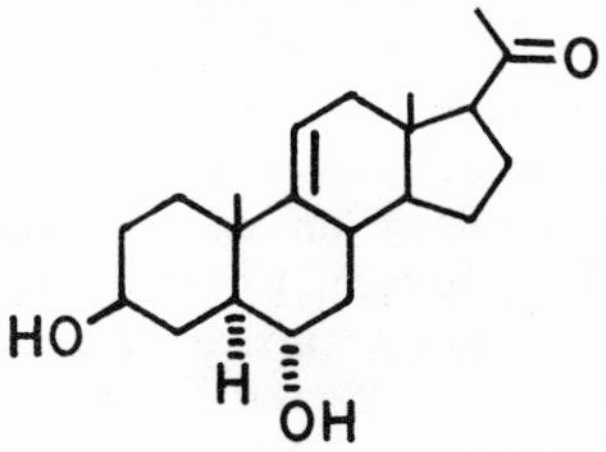

141 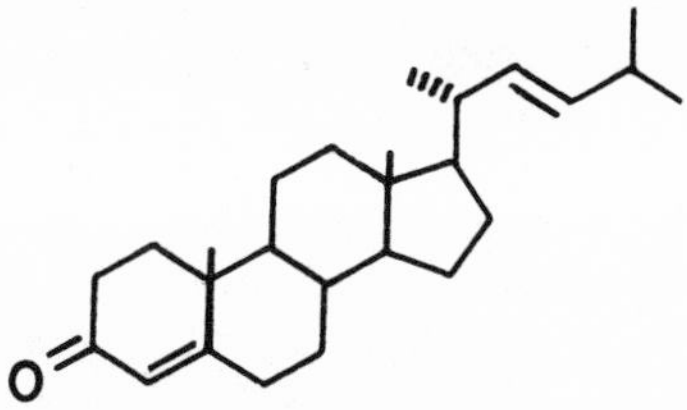   142

   Marine invertebrates also have been reported to contain Δ^4-3-ketones. The enone 8 has been found in the
sponge *D. hawaiiana* [592] and in the gorgonian *Pseudo-
plexaura porosa*, *Ps. porosa* also containing a 24-methyl-
(24ξ)-cholest-4-en-3-one and gorgost-4-en-3-one (23,24-
dimethyl-22,23-methylene-(22R,23R,24R)-cholest-4-en-3-one)
[1877]. Additionally, 24-norcholesta-4,E-22-dien-3-one
(142), cholesta-4,E-22-dien-3-one,24-methyl-(24ξ)-cholesta-
4,22-dien-3-one, 24-methylcholesta-4,24(28)-dien-3-one, and
24-ethylcholesta -4,E-24(28)dien-3-one have been found in
the sponge *Stelleta clarella* [2217].

   The special cases of sterol 5α,8α-peroxides and of
C_{19}-C_{25} sterols of short side chains found in marine
organisms are deferred to in later sections of this chapter.

                OTHER OXIDIZED STEROLS FROM PLANTS

   The presence in plant material of oxidized C_{28}-, C_{29}-,
and C_{30}-sterols which are the exact homologs of the estab-

lished cholesterol autoxidation products 10,13,14,15, and 16
may be interpreted, in the absence of other evidence to the
contrary, as representing autoxidation. As relatively few
studies have been reported on oxidized plant sterols, the
same range of distribution or natural occurrence for such
C_{28}-,C_{29}-, and C_{30}-sterol derivatives as given in TABLES 1
and 2 for cholesterol derivatives is not had. However,
numerous parallel examples exist.

Common A- and B-ring Oxidation products

The epimeric C_{29}-3β,7-diols 76 and 77 being homologs
of the 3β,7-diols 14 and 15 have not been frequently reported
in plant material, but both are found in sugar cane *Saccharum
officinarum* [605,1652] and in roots of *Glossostelma carsoni*
[1923]. The 3β,7α-diol 76 has been found in dried leaves
of the common pineapple *Ananas comosus* [10,1811,1812]. The
putative parent sitosterol (20) was present in these several
cases also. The equally suspect 3β-hydroxystigmast-5-en-7-
one (72) as homolog of 7-ketone 16 has been isolated from
pine bark [2030,2031], from the plant *Cryptocarya foveolata*
together with congener 3β-hydroxy-24-methyl-(24R)-cholest-
5-en-7-one (74) [248], and from dried powdered kidney bean
Phaseolus vulgaris roots together with 3β-hydroxystigmasta-
5,22-dien-7-one (73) [1257]. A nonenzymic origin may be
accepted in each case for these several 7-oxygenated sterols.

Moreover, stigmasta-3,5-dien-7-one (143) as homolog of
the dienone 10 has been found in aspen poplar heartwood [6],
pine bark [2030,2031], and in sugar cane wax [6,1652].
Stigmasta-3,5,E-24(28)-trien-7-one (144) has been found in
the marine brown alga *Fucus evanescens* [1139], all surely
artifacts. 5α-Stigmastene-3β,5,6β-triol (145) found in
sugar cane [605,1652] and 24-ethylcholesta-5,E-24(28)-diene-
3β,7α-diol in *F. evanescens* [1139] are further such examples.

Other C-7 oxygenated Δ^5-sterols isolated from plant
sources may be suspect as possible autoxidation artifacts
despite the paucity of evidence for ultimate conclusions.
Thus, the diosgenin 7-ketone 3β-hydroxy-(25R)-spirost-5-
en-7-one (146) and 3β-hydroxy-(25R)-spirost-5-ene-7,11-
dione (147) isolated from *Tamus edulis* Lowe [801,892] and
the C_{29}-sterol (22S)-stigmast-5-ene-3α,7α,22-triol (148)
isolated from chestnut *Aesculus hyppocastanum* and hazlenut
Corylus avellana leaves [767,2186] are particularly suspect.
In the case of the diosgenin 7-ketones, a 3β,25-dihydroxy-

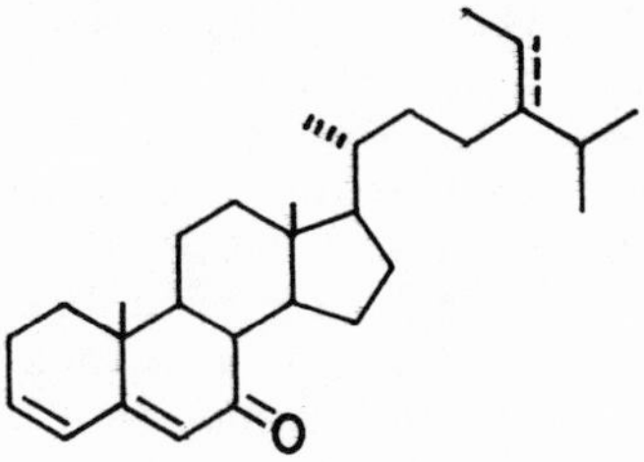

143 24R   145

144 $\Delta^{24(28)}$

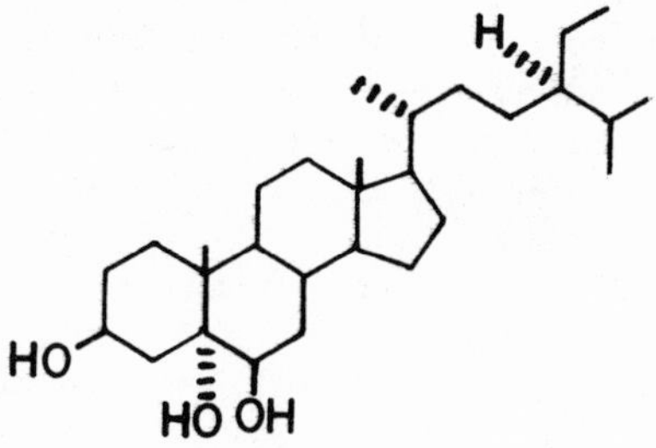

(25R)-spirost-5-en-11-one was also isolated [801,892], and
although an autoxidation pathway (via initial 25-hydroper-
oxidation) for 25-hydroxylation of sterols other than

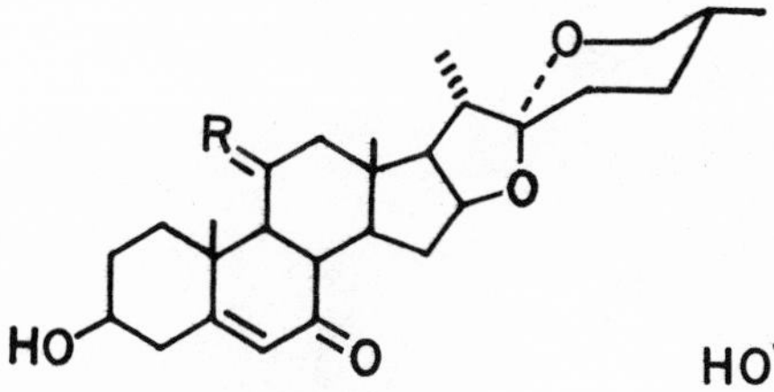

146 R = H$_2$ 148 R = H

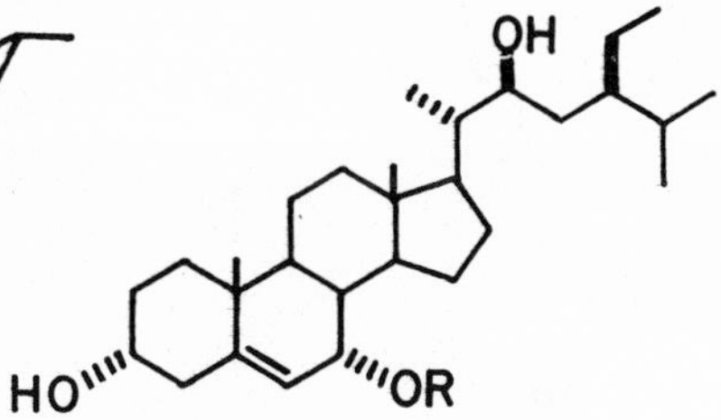

147 R = O   58 R = OH

cholesterol has not been demonstrated, the combination of
7-ketone and 25-alcohol derivatives in this isolation is
reminiscent of similar experience with the autoxidized
sterol composition of human aortal tissue, for instance
[2567]. Moreover, the $3\alpha,7\alpha,22\alpha_F$-triol 148 from chestnut
leaves was accompanied by the hydroperoxide 3α,22-dihydroxy-
(22S)-stigmast-5-ene-7α-hydroperoxide (58) [767]. Although
an artifact origin for the 7α-hydroperoxide 58 was rejected
by the investigators on the argument that no sitosterol
hydroperoxides were detected in processing of the leaves,
the case remains highly suspect.

Steroid Δ^4-3-ketones and Δ^4-3,6-diketones have also
been found in higher plants in association with the corresponding Δ^5-stenols under conditions which do not resolve
the issue of mode of origin for the ketosteroids. Stigmast-
4-en-3-one (149) has been found in pine bark [2030,2031],
various tree woods [1434], and in air-aged beech tree leaves
[1062]. Moreover, the enone 149 is formed in incubations of
Cheiranthus cheiri leaf homogenates, *Apocynum cannabinum*
callus tissue [1444], and in aerated soybean suspension
cultures [2646]. In the soybean cultures the enone 149 was
in company of ergost-4-en-3-one, stigmasta-4,22-dien-3-one,
and the 3,6-diketones ergost-4-ene-3,6-dione, stigmast-4-
ene-3,6-dione (151), and stigmasta-4,22-diene-3,6-dione
[2646].

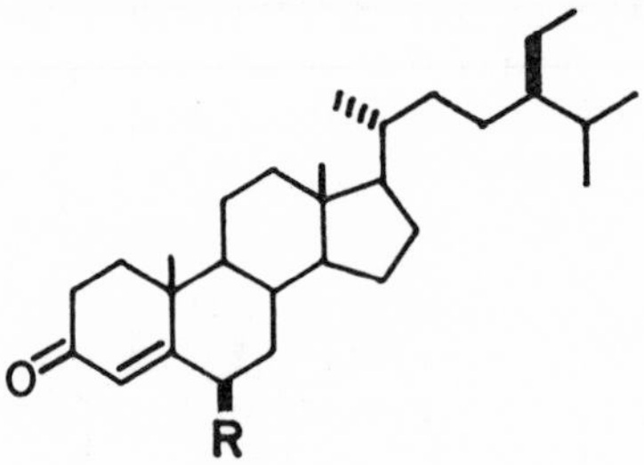

149 R = H   151

150 R = OH

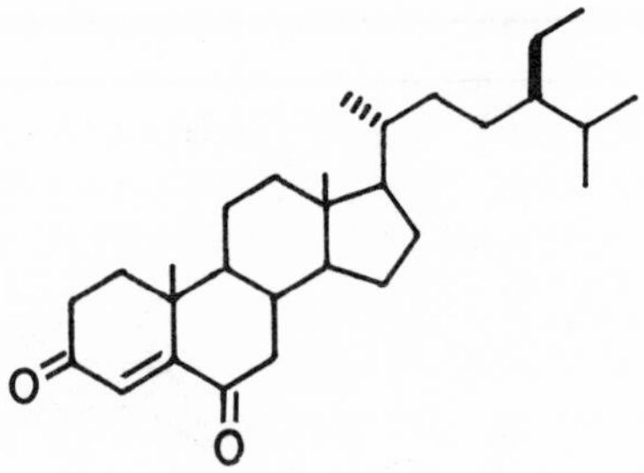

    The 3,6-diketone 151 has been isolated from commercial,
dried dwarf elder *(Sambucus ebulus)* roots [2523,2524] and
dried leaves of *Hamelia patens* and *Clitoria ternatea* [1959].
Callus tissue from *Stephania cepharantha* but not from tissues from the original plant contain the 4-en-3-one 8, a 24-
ethyl-(24ξ)-cholest-4-en-3-one, and a 24-ethyl-(24ξ)-cholesta-
4,22-dien-3-one, as well as a 24-methyl (24ξ)-cholesta-4-ene-
3,6-dione, a 24-ethyl-(24ξ)-cholest-4-ene-3,6-dione and a
24-ethyl-(24ξ)-cholesta-4,22-diene-3,6-dione [1165].

    Yet other types of steroid 3-ketones have been isolated
from plant tissues, 6β-hydroxystigmast-4-en-3-one (150)
being isolated recovered from dried dwarf elder roots [2523,
2524], and, along with 6β-hydroxystigmasta-4,22-dien-3-one,
from dried *P. vulgaris* roots [1257]. Whereas these hydroxy-
ketones appear to be artifacts, 5α-stigmast-7-en-3-one and
the corresponding alcohol 5α-stigmast-7-en-3β-ol found

together in *Coccinia indica* and related Cucurbitaceae [2389] may be metabolite and substrate respectively.

Another artifact case is that of C-7 oxidized C_{21}-Δ^5-steroids from *Metaplexis japonica* Makino roots. Here 7β-methoxy-8β,14β,17α-(20ξ)-pregn-5-ene-3β,8,12β,14,17β,20-hexaol (7β-methoxysarcostin) (152) and 7α-methoxy-8β,14β, 17α-(20ξ)-pregn-5-ene-3β,8,12β,14,20-pentaol (gagaimol 7-methyl ether) (153) appear to be artifacts of initial oxidation at C-7 followed by methyl ether formation occurring during recovery of the steroids [1647,1775].

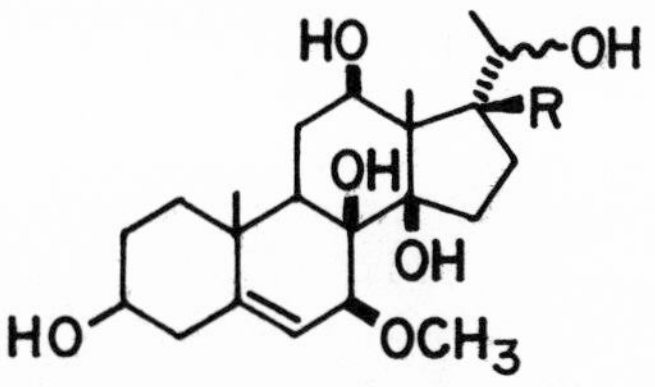

152 R = OH

153 R = H

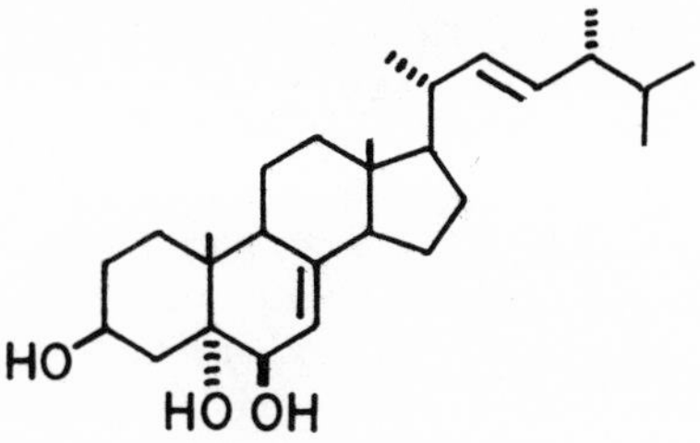

154

3β,5α,6β-Triols

5α-Stigmastane-3β,5,6β-triol (145) found in sugar cane with the 3β,7-diols 75 and 76 already mentioned represents a typical case of autoxidation of the common plant sitosterol (20). A more widely reported case involves the minor yeast sterol cerevisterol (5α-ergosta-7,E-22-diene-3β, 5,6β-triol (154) [43] variously isolated from *Saccharomyces cerevisiae* yeast [222,827,829,1068,2042,2043], from the mushroom *Amanita phalloides* [2674], from ergot [2674], and most probably in 1918 (!) from the fungus *Polyporus nigricans* [674,2042]. Cerevisterol has also been isolated together with ergosterol 5α,8α-peroxide 62 from various other fungi, including *Acremonium luzulae* [450], *Aspergillus flavus* [2543], *Cantharellus cibarius* [1358], *Fusarium oxysporum* [2346], *Gibberella fujikuroi* [160], and *Penicillium rubrum* [160].

Cerevisterol is almost certainly an artifact of autoxidation of ergosterol, a matter suggested at the time of its

initial early isolation [674] but not otherwise clearly
stipulated in subsequent reports. This proposition is well
supported although the direct experimental demonstration of
an autoxidation process yielding cerevisterol has not been
forthcoming. The isolation of cerevisterol from stored
mother liquors from commercial manufacture of ergosterol
from yeast [222,1068,2042] is suggestive, and the observation of $5\alpha,8\alpha$-peroxide 62 and cerevisterol together in
variable levels in *F. oxysporum* [2346] and in *A. flavus*
grown in light but not at all in the same culture grown in
the dark [2543] support the contention. Yet more persuasive
is the report that $5\alpha,8\alpha$-peroxide 62 and cerevisterol levels
in stored dried yeast increase during storage at the expense
ergosterol levels [823,824,827,828]! These several considerations clearly distinguish cerevisterol as an artifact of
the air oxidation of ergosterol.

Some of these suggestive points have been rationalized
in other ways. For instance, the possibility of mutations
occurring in *F. oxysporum* cultures displaying variable
levels of 62 and 154 has been advanced [2346]. Furthermore,
although cerevisterol has been viewed recently as an artifact by others [2543,2647], it has recently been listed as
a natural product from microorganisms [361]. Thus, the same
argument and counterargument for metabolite versus artifact
status for cerevisterol match those for the other sterol
autoxidation products already discussed.

Several digitanol $3\beta,5\alpha,6\beta$-triols have been isolated
from the plant *Cynanchum caudatum* MAX. Among these steroids
are $3\beta,5,6\beta,8,12\beta,14$-hexahydroxy-$5\alpha,8\beta,14\beta,17\alpha$-pregnan-20-
one (155), its 12β-ester $3\beta,5,6\beta,8,14$-pentahydroxy-12β-3',4'-
dimethylpent-2-enoyloxy-$5\alpha,8\beta,14\beta,17\alpha$-pregnan-20-one (156),
$5\alpha,8\beta,14\beta,17\alpha$-(20S)-pregnane-$3\beta,5,6\beta,8,12\beta,14,17,20$-octaol
(glycosarcostin, the most highly oxygenated steroid found
todate in Nature) (157), and its 12,20-diester 20-acetoxy-
12β-cinnamoyloxy-$5\alpha,8\beta,14\beta,17\alpha$-pregnane-$3\beta,5,6\beta,8,14,17$-
hexaol (158) [2741-2744,2752]. However, these triols occur
in company with corresponding Δ^5-digitanols $3\beta,8,12\beta,14$-
tetrahydroxy-$8\beta,14\beta,17\alpha$-pregn-5-en-20-one (lineolone) (159),
its 12β-ester $3\beta,8,14$-trihydroxy-12β-3',4'-dimethylpent-2-
enoyloxy-$8\beta,14\beta,17\alpha$-pregn-5-en-20-one (160), $8\beta,14\beta,17\alpha$-
(20S)-pregn-5-ene-$3\beta,8,12\beta,14,17,20$-hexaol (sarcostin) (161),
and its 12β-ester 12β-cinnamoyloxy-$8\beta,14\beta,17\alpha$-(20S)-pregn-
5-ene-$3\beta,8,14,17,20$-pentaol (162) [1648-1651,2235], and are

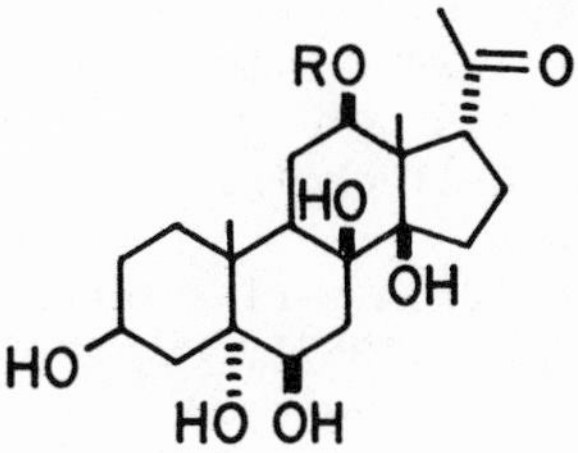

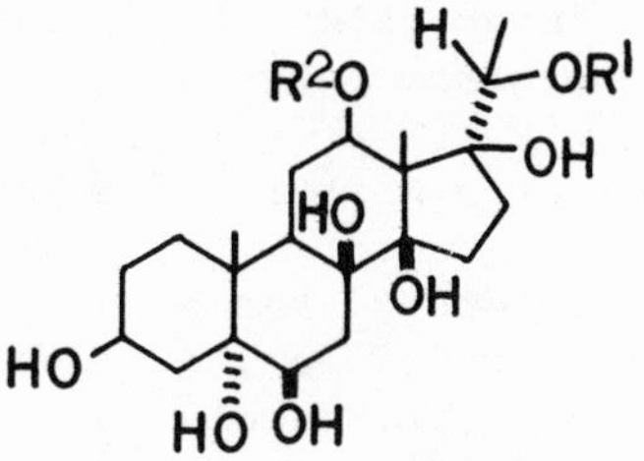

155 R = H

156 R = COCH=C(CH₃)CH(CH₃)₂

157 R¹ = R² = H

158 R¹ = COCH₃, R² = COCH=CHC₆H₅

thus suspect as autoxidation products.

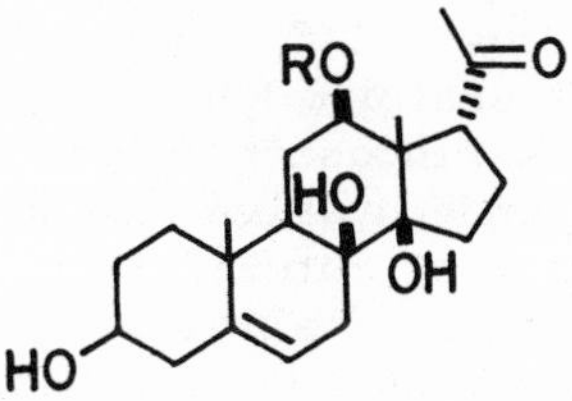

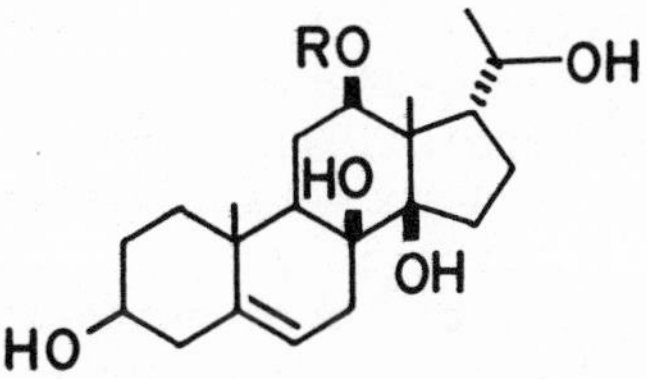

159 R = H

160 R = COCH=C(CH₃)CH(CH₃)₂

161 R = H

162 R = COCH=CHC₆H₅

Side-Chain Oxidation Products

Saringosterol. Two oxidized sterols isolated from
marine brown algae *(Phaeophyceae)* suspected as being artifacts of autoxidation are saringosterol (24ξ)-stigmasta-
5,28-diene-3β,24-diol (163) and 3β-hydroxycholest-5-en-
24-one (34). Saringosterol has been isolated from the
brown algae *Agarum cribosum* [1746], *Alaria crassifolia*
[1140], *Ascophyllum nodosum* [1338,2057], *Costaria costata*
[1140], *Cystophyllum hakodatense* [1140], *Dictyopteris divari-
cata* [1140,1141], *Fucus evanescens* [1140], *Laminaria digitata*
and *Laminaria faeroensis* [1825], *Laminaria saccharina* [2057],
Pelvetia wrightii [1140], *Sargassum confusum* [1140], *Sargas-*

sum ringgoldianum [1140,1141], and *Sargassum thunbergii*
[1140] but not from *Sargassum fluitans* [2292]. It is as
yet uncertain whether saringosterol is a single C-24
stereoisomer or a mixture of epimeric 24-alcohols.

The 24-ketone 34 has been isolated from marine brown
algae *A. cribosum* [1746], *A. nodosum* [1338,2057], *Fucus
distichus* [2654], *L. saccharina* [2057], and *Pelvetia
canaliculata* [1692]. Fucosterol (75) was isolated with the
24-ketone 34 and saringosterol in most cases, thus suggesting a possible origin for 34 and saringosterol from fucosterol. Moreover, the 24-ketone 34 has also been found in
the tunicate *Ascidia mentula* [135] and in sponges *D. hawaiiana* [444,592] and *Callyspongia diffusa* [2475], in the latter
case in company with the C_{29}-sterol allene 24-ethylcholesta-
5,24(28),28-triene-3β-ol (165) [2475].

A preliminary account states that natural air-aging of
fucosterol yields a hydroperoxide 3β-hydroxy-(24ξ)-stig-
masta-5,28-diene-24-hydroperoxide(164) transformed by base
to saringosterol [794]. The pattern of saringosterol (possibly both C-24 epimers), 24-ketone 34, and putative parent
sterol fucosterol is not that of free radical autoxidation
of cholesterol as represented by the pattern of allylic
alcohols 14 and 15, 7-ketone 16, and parent cholesterol but
is one in which double bond isomerization occurs at oxidation. Such double bond isomerization is known in free
radical autoxidations such as that of 5α-cholest-6-en-3β-ol

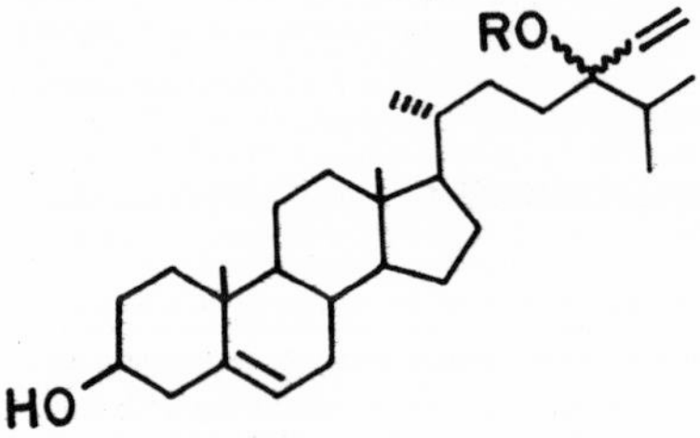

163 R = H   165

164 R = OH

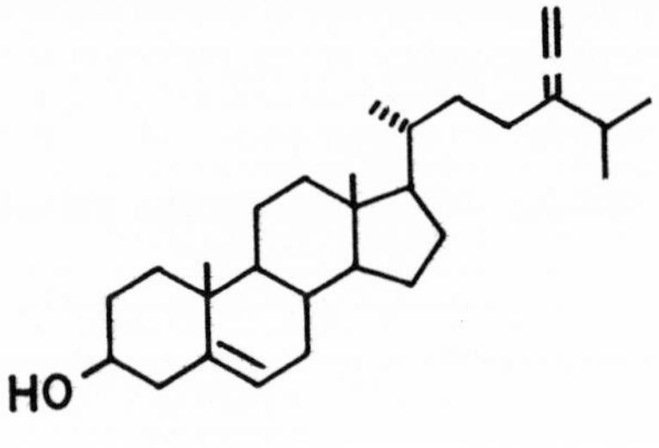

where products are Δ^5-7-hydroperoxides 46 and 47 [1402].
The autoxidation of fucosterol with associated double
bond allylic shift resembles the actions of $^1O_2$ acting

on olefins to give an allylically rearranged hydroperoxide. Subsequent thermal decomposition of the product 24-hydroperoxide 164 could reasonably be expected to yield the corresponding 24-alcohol saringosterol 163 and 24-ketone 34 as product of β-scission, thus exactly the product pattern observed.

The details of the thermal decomposition of sterol hydroperoxides to give analogous patterns of degraded products, on which this hypothesis is based, are presented in Chapter V. However, no detailed study has been made of the attack of $^1O_2$ on sterols with side-chain unsaturation where the thermal decomposition of initially formed hydroperoxides has been examined. Indeed, no studies at all on $^1O_2$ attack on fucosterol has been described! Neither has the putative hydroperoxide 164 been sought by others conducting saringosterol isolations [1140,1141,1338,1825].

Nonetheless, as saringosterol is isolated from pigmented marine algae, it appears likely that photosensitized oxygenation of fucosterol in reactions involving $^1O_2$ may occur, the algal pigments acting as sensitizers in the same manner as previously demonstrated for the photosensitized oxygenation of ergosterol to the 5α,8α-peroxide 62 involving fungal anthraquinone pigments [16,160].

Customary experimental procedures used in isolation work on brown algae involve powdered dried material [1140, 1141], commercially dried, milled material [1338], air drying at 30-40° for 48 hrs. [1692], and oven drying [1825], all procedures conducted in air with no precautions to limit autoxidations. These extended drying and handling procedures in air, as well as extractions with diethyl ether in air [1140,1141,1692], almost certainly invite autoxidations, and in the one instance where fresh material (*A. nodosum*) was examined, neither 24-alcohol 163 nor 24-ketone 34 was found. However, dried *A. nodosum* exposed to air for four weeks developed more polar sterol components recognized as the 24-alcohol 163 and 24-ketone 34 [1338]. As is frequently the case with animal tissues and cholesterol autoxidation products, the possibility that the 24-ketone 34 be an artifact in *P. canaliculata* has been discounted "in view of the great care taken to avoid uncontrolled oxidation during the various operations" [1692].

Cycloartenol Derivatives. A more difficult case of
possible artifacts obtains in the several sterols isolated
from plant sources that contain conjugated ketone or allylic alcohol features in the sterol side-chain. In most
cases the biological specimens were dried or sundried in
air (!) prior to analysis and processing, and a discussion
of these products is included here as indication of the
suspicious nature of these findings.

A series of side-chain allylic alcohol and ketone
derivatives of cycloartenol (5α,9β-9,19-cyclolanost-24-
en-3β-ol (166) isolated from Spanish moss *Tillandsia
usneoides* and other plants are of an interesting but suspicious nature. From the moss there was isolated 5α,9β-9,
19-cyclolanost-23-ene-3β,25-diol (167), 5α,9β-(24ξ)-9,19-
cyclolanost-25-ene-3β,24-diol (170), and 3β-hydroxy-5α,9β-9-
19-cyclolanost-25--en-24-one (172) [105,623,1604], whereas
only the Δ^{23}-3β,25-diol 167 was found in *Euphorbia cyparis-
sias* [2345] and *Tricholepis glaberrima* [461]. Both 167 and
a 24-epimer of 170 were isolated from *Pachysandra terminalis*
[1295]. Additionally, the 25-0-methyl ether 25-methoxy-5α,
9β-9,19-cyclolanost-23-en-3β-ol (168) was also isolated
from Spanish moss by one laboratory [623] but not by another
[105].

A biogenesis sequence for the sterols 167,168,170, and
172 has been suggested [623] and *de novo* formation of the
$\overline{\Delta^{23}}$-3β,25-diol 167 from [1-^{14}C] acetate in *Euphorbia helio-
scopia* has been suggested [1875]. Nonetheless, non-enzymic
formation of all appears equally likely and is supported by
results obtained in the related lanosterol series.
The sterol 5α-(23S)-lanosta-8,24-diene-3β,23-diol (173) has
been isolated from dried, ground peridium of an English
basidiomycete *Scleroderma aurantium* [689,690] and a 5α-(23ξ)-
lanosta-8,24-diene-3β,23-diol of different melting point,
possibly the (23R)-epimer of 173 from fresh peridium from a
Bohemian *S. aurantium* [2618].   The Δ^{24}-(23S)-23-alcohol 173
and its putative (23R)-epimer both isomerize and react in
acidified ethanol to yield 5α-lanosta-8,23-diene-3β,25-diol
(176), and its 25-0-ethyl ether 25-ethoxy-5α-lanosta-8,23-
dien-3β-ol (177), and the epimeric 23-0-ethyl ethers 23-
ethoxy-5α-(23R)-lanosta-8,24-dien-3β-ol (174) and 23-ethoxy-
5α-(23S)-lanosta-8,24-dien-3β-ol (175) [691,2618].

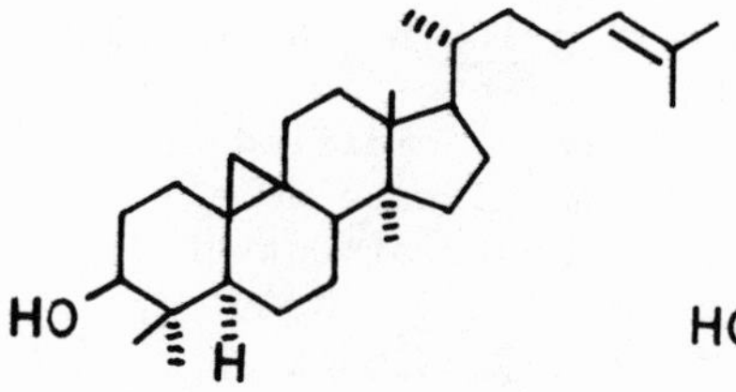

166

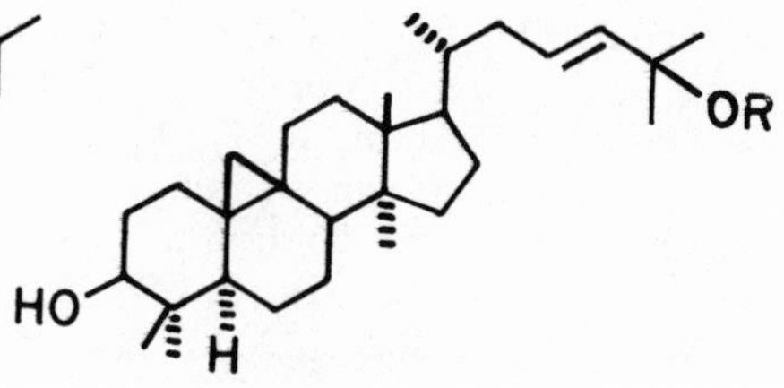

167 R = H

168 R = CH$_3$

169 R = OH

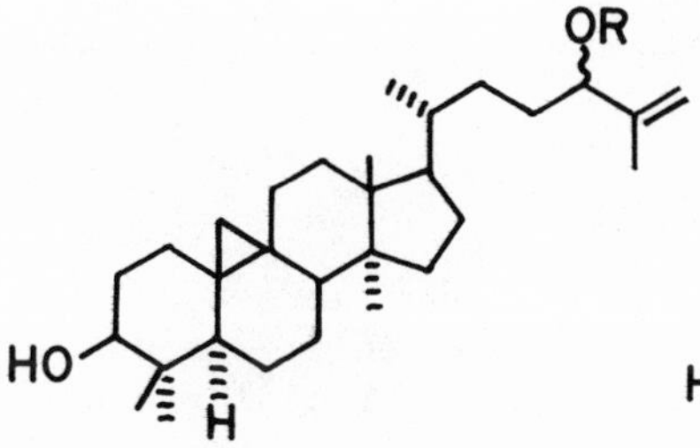

170 R = H

171 R = OH

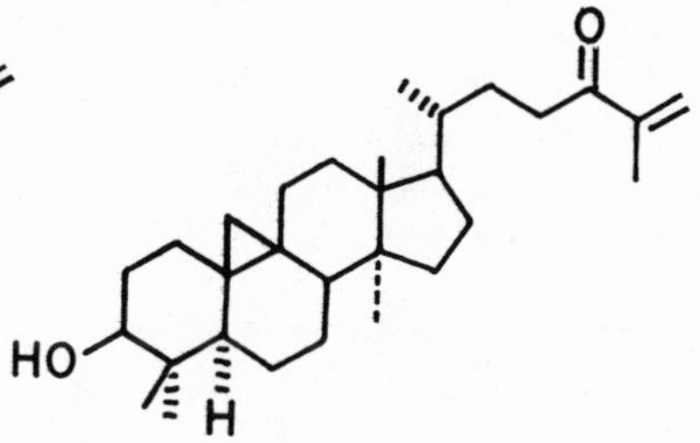

172

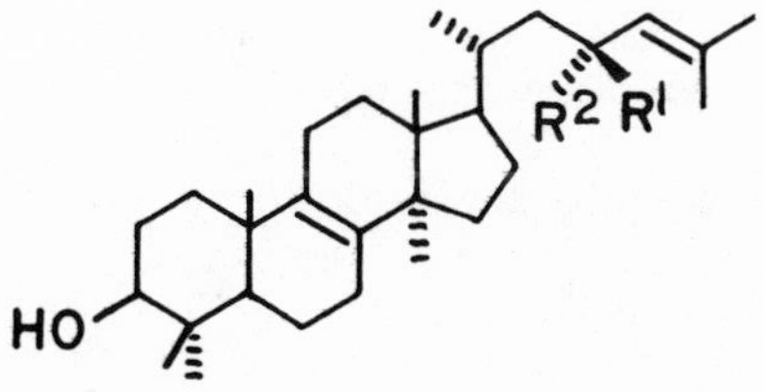

173 R^1 = OH, R^2 = H

174 R^1 = H, R^2 = OC$_2$H$_5$

175 R^1 = OC$_2$H$_5$, R^2 = H

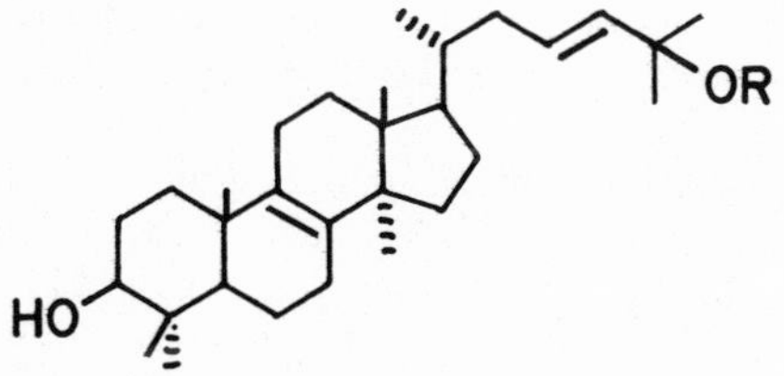

176 R = H

177 R = C$_2$H$_5$

The parallel between the 25-0-methyl ether 168 of the
cycloartenol series isolated from moss [623] and the obvious
artifact 25-0-ethyl ether 177 of the lanosterol series suggests that the cycloartenol derivative 168 be an artifact
of manipulations. Moreover, introduction of any of the
three 0-ethyl ethers 174,175 or 177 into acidified ethanol
results in formation of an equilibrium mixture of all
three [691]. Even though the same acid treatment of the
corresponding alcohols (23S)-3β,23-diol 173 and Δ^{23}-3β,25-
diol 176 favors 176 [691], the suggested anionotropic rearrangement of the Δ^{23}-3β,25-diol 176 to 173 poses a means
of formation of 173 as an artifact.

Yet, other potential for artifact formation from these
allylic side-chain allylic alcohols exists in their ready
dehydration to dienes. Acid treatment of the (23S)-3β,23-
diol 173 or of the isomeric 3β,25-diol 176 yielded 5α-lanos-
ta-8,22,24-trien-3β-ol and possibly some isomeric 5α-lanosta-
8,23,25-trien-3β-ol [691].

As the (23S)-3β,23-diol 173 appears to be an enzymic
product of lanosterol metabolism [689,690], so a putative
analogous Δ^{24}-3β,23-diol of the cycloartenol series not
heretofore postulated might likewise be an enzyme product
and parent of the Δ^{23}-3β,25-diol 167 by artificial processes. However, a Δ^{23}-25-alcohol 167 could also arise from
cycloartenol by an unrecognized oxidation involving a
putative hydroperoxide 3β-hydroxy-5α,9β-9,19-cyclolanost-
23-ene-25-hydroperoxide (169) derived by attack of $^1O_2$
generated by photosensitization provided by moss pigments
in the same manner as suggested for the oxidations leading
to saringosterol (163) and the 5α,8α-peroxide 62.

Given such an origin for the Δ^{23}-3β,25-diol 167 from
cycloartenol (166) the same process involving $^1O_2$ but
yielding the other possible allylic hydroperoxide 3β-hydroxy
5α,9β-(24ξ)-9,19-cyclolanost-25-ene-24-hydroperoxide (171)
that accounts in similar manner for the other products 170
and 172 by thermal reduction and dehydration processes
(described in Chapter V).

Just exactly these photosensitized processes acting on
three Δ^{24}-sterols desmosterol (78) 3β-acetate, lanosterol
(67) 3β-acetate, and (22R)-cholesta-5,24-diene-3β,22-diol
have been observed! The photosensitized oxygenation of 78
with subsequent reduction of the product hydroperoxides

yielded both 3β-acetoxy-(24RS)-cholesta-5,25-dien-24-ol
(178) and a 3β-acetoxycholesta-5,23-dien-25-ol (180 and/or
182) [1678,1704]. Similarly, 67 3β-acetate gave the
Δ^{23}-25-alcohol 176 3β-acetate and 5α-(24RS)-lanosta-8,25-
diene-3β,24-diol (184) 3,24-diacetate [1704]. A similar
pair of Δ^{23}-25-alcohol and Δ^{25}-24-alcohol products was
obtained from (22R)-cholesta-5,24-diene-3β,22-diol [1704],
and a tetracyclic triterpenoid glycoside bearing the Δ^{24}-
double bond gave the same type of product pattern [2737].
The selectivity of oxidative attack at the Δ^{24}-double bond
over nuclear Δ^{5}- or Δ^{8}-double bonds is apparent.

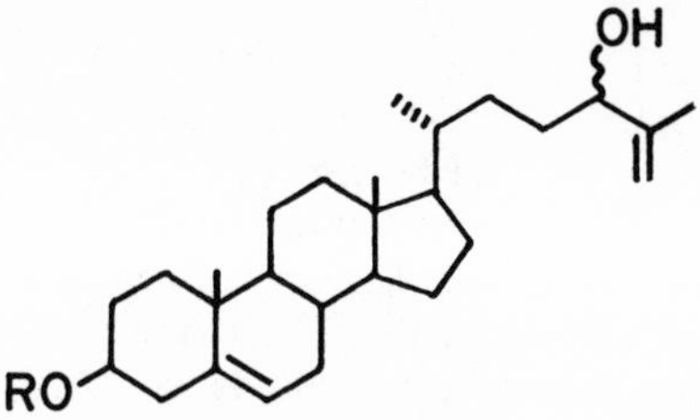

178 R = COCH₃

179 R = H

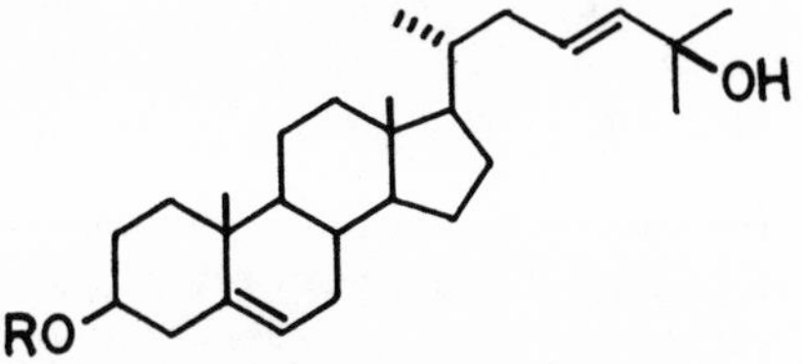

180 R = COCH₃

181 R = H

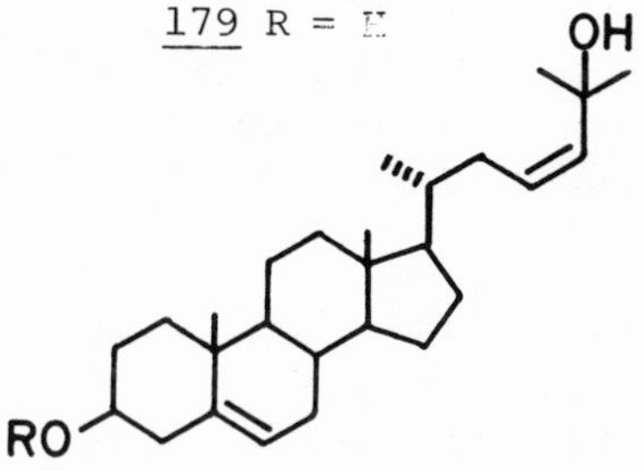

182 R = COCH₃

183 R = H

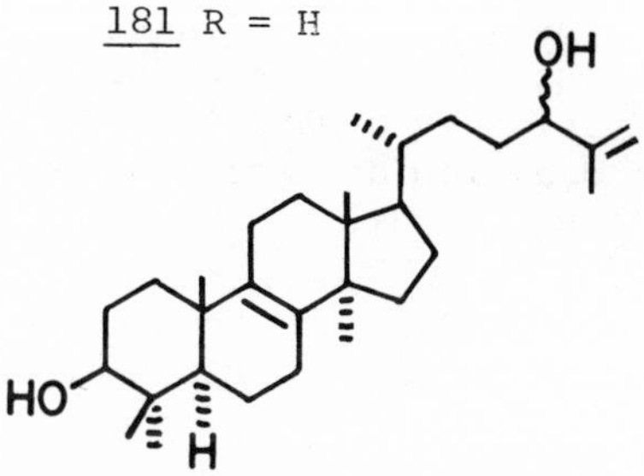

184

Moreover, the C_{27}-product dienediols cholesta-5,Z-23-
diene-3β,25-diol (183), and cholesta-5,E-23-diene-3β,25-
diol (181) have been recovered from marine red algae *Liagora
distenta* and *Scinaia furcellata* together with cholesterol
but not with desmosterol [718], and a cholesta-5,23-diene-
3β,25-diol (181 or 183) and a (24ξ)-cholesta-5,25-diene-3β,24-
diol (179) have been isolated along with desmosterol from
another red algae *Rhodymenie palmata* [1676]. Furthermore,
the Δ^{23}-3β,25-diol 181 and a Δ^{25}-3β,24ξ-diol 179 have been
recovered along with putative parent desmosterol (78) *inter*

alia from fresh red alga *Asparagopsis armata*. Moreover, in
this case the C_{27}-3β,25-diol 27 and a 24-methyl-(24ξ)-cholest-
5-ene-3β,25-diol (127) were also isolated [794].

The possibilities of artificial origins for the dienediols 179,181, and 183 in these cases is fully recognized
[718,794,1676], but the presence in plant tissues of these
and other sterol 25-alcohols as analogs of the 3β,25-diol
27 and therefore as artifacts of autoxidation has not been
often considered, and only inconclusive evidence adduced
[2388].

The Cucurbitacins. The class of highly oxidized cytotoxic sterols isolated from members of the Cucurbitaceae
based on the theoretical parent 5α-cucurbitane (5α-19(10→
9β)-*abeo*-lanostane)   [1433] contains the suspicious Δ^{23}-25-
alcohol structural feature just discussed for cycloartenol
and lanosterol derivatives. Typical of the class are
cucurbitacins B (25-acetoxy-2β,16α,20-trihydroxy-10α-(20ξ)-
cucurbita-5,23-diene-3,11,22-trione) (185), D (2β,16α,20,25-
tetrahydroxy-10α-(20ξ)-cucurbita-5,23-diene-3,11,22 -trione)
(186),and E (25-acetoxy-2,16α,20-trihydroxy-10α-(20ξ)-cu-
curbita-1,5,23-triene-3,11,22-trione) (187). Others of the
class do not retain the Δ^{23}-bond, which is subject to reduction by a NAD(P)H: cucurbitacin Δ^{23}-reductase [2092],
and a rich diversity of other structural features has been
found [1433].

Despite the formal similarity of the Δ^{23}-25-alcohols
of this group to the cycloartenol and lanosterol derivatives 167 and 176, etc. the cucurbitacins appear to be
genuine metabolic products, and no evidence supporting an
artifact nature has been adduced. Moreover, it is note-
worthy that most work on these steroids has been accomplished
using fresh fruits, leaves, or roots, the case of recovery
of cucurbitacin D 185 and 185 16α-glycoside datiscoside from
dried *Datisca glomerata* roots several years old [1406] being
an exception.

7-Dehydro-6-ketones

There is some question about the possible artifact
nature of some sterol Δ^7-6-ketones found in air-dried plant
samples. As the autoxidation of 5α-stigmasta-7,E-22-dien-
3β-ol appears to give a Δ^7-6-ketone product [197] as does
the radiation-induced oxidation of the 7-stenol 57 [1402]

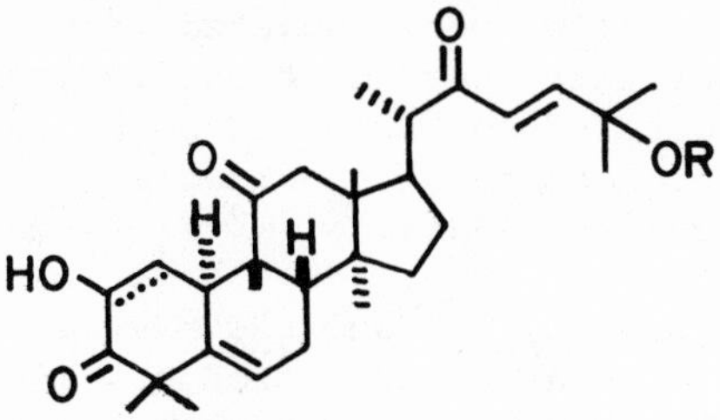

185 R = COCH₃

186 R = H

187 R = COCH₃,Δ¹

(cf. Chapter VI), this structural feature of oxidized sterols becomes suspect.

These remarks obviously do not pertain to all sterol Δ⁷-6-ketones found in Nature, for the many ecdysterols of arthropods and plants [31,1044] possess the 5β-7-ene-6-ketone feature and have certain enzymic origins. Moreover, a class of 22-hydroxylated 4α-methyl-22-ethyl homolog 3β-benzoate esters of the Δ⁷-6-ketone 80 have been isolated from dried, ground fruit of *Solanum xanthocarpum*, including 3β-benzoyloxy-24-ethyl-22-hydroxy-5α-(22R,24R)-cholest-7-en-6-one (carpesterol) (188) [177,2510] and several related (22ξ,24ξ)-analogs [1410,1411], which do not appear to be artifacts though they be recovered from air dried material. Additionally, related 11α-hydroxylated analogs of these sterol esters have been isolated from seeds of the lantern plant *Physalis franchetti*, including 3β-benzoyloxy-11α-hydroxy-24-ethyl-5α-(20ξ,24ξ)-cholest-7-en-6-one (physanol B) and 3β-benzoyloxy-11α-hydroxy-24-ethyl-5α-(24ξ)-cholesta-7,20-dien-6-one (physanol A) with its unprecedented Δ²⁰-double bond [2200], do not appear to be artifacts of B-ring autoxidations.

Less certain in origin is the set of Δ⁷-6-ketones isolated from air-dried roots of the cactus *Wilcoxia viperina* [622] and of *Peniocereus greggii* [1336]. The Δ⁷-6-ketone 80, 3β,9-dihydroxy-5α,9α-cholest-7-en-6-one (viperidone) (189) and 3β,9,14-trihydroxy-5α,9α,14α-cholest-7-en-

6-one (viperidinone (190) were isolated from both plants,
and from *P. greggii* there was additionally isolated 3β-
hydroxy-5β,14α-cholest-7-en-6-one and 3β-hydroxy-5β,14β-
cholest-7-en-6-one, both being regarded as probably artifacts of saponification [1336].

The Δ⁷-6-ketones 80,189, and 190 may be metabolites
and have been accepted as natural products [361], but isolations from fresh tissue seems necessary to remove doubts.
The possibility that the 3β,9α-diol 189 and the 3β,9α,14α-
triol 190 derive by factitious air oxidation of the putative
parent sterol 80 during isolation was recognized at the time
and shown not to be the case [622]. However, the question
of air oxidation during collection and air-drying of the
cactus roots has not been heretofore addressed.

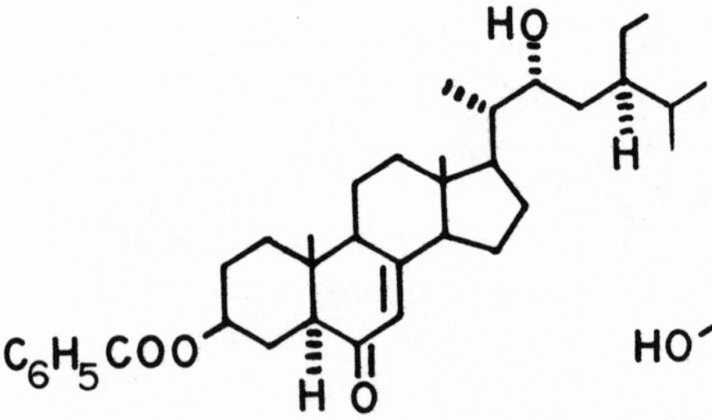

188 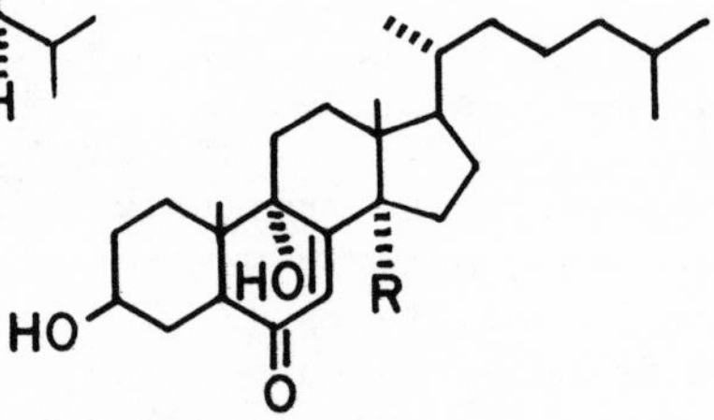   189 R = H

                              190 R = OH

OTHER SUSPICIOUS OXIDIZED STEROLS

A number of oxidized sterol derivatives have been
isolated from natural material which could be metabolites
of as yet unrecognized enzyme processes or might be equally
products of unrecognized pathways of autoxidation. Thus,
5α-cholest-7-en-3-one from butter fat [1818] may be an
artifact, as the artifact 3,5-dien-7-one 10 and enone 8
have been isolated from anhydrous milk fat and from nonfat
dried milk [776]. However, other possible artificial transformations of sterols of milk fat may occur in commercial
milk processing, as the hydrocarbons 24-methyl-5ξ-(24ξ)-
cholest-2-ene and 24-ethyl-5ξ-(24ξ)-cholest-2-ene have
been isolated from nonfat dried milk [775]. Steranes have

also been isolated from human aortal plaques, where cholest-
5-ene (7) and cholesta-3,5-diene (11) were reported [370].

Yet other oxidized sterols indicative of autoxidation
have been described in biological material. Of particular
interest is a group of C_{19}-C_{25} sterols with side-chains
shorter than the C_8 side-chain of cholesterol which have
been discovered recently in several invertebrate phyla,
including Porifera, Coelenterata, and Arthropoda. Ten
such Δ^5-3β-alcohols with side-chains ranging from 17β-H of
androst-5-en-3β-ol (114) and 17-H of androsta-5,16-dien-
3β-ol (191) to C_2-C_6 side-chains of pregn-5-en-3β-ol (110),
pregna-5,17(20)-dien-3β-ol (192), pregna-5,20-dien-3β-ol
(193), 23,24-bisnorchol-5-en-3β-ol(194), 23,24-bisnorchola-
5,20-dien-3β-ol (195), 24-norchol-5-en-3β-ol (196), chol-5-
en-3β-ol (109), and 26,27-bisnorcholest-5-en-3β-ol (197)
compose the group, with distribution among invertebrates
as listed in TABLE 5. Other unrecognized short side-chain
sterols may occur in marine life, for a C_{22}-sterol has been
reported in the clam *Tapes philipparum* [1221,2114] and a
C_{25}-sterol may be present in the sponge *Axinella polypoides*
[604].

Other modified short side-chain steroid derivatives
also have been found, the C_{21}-sterol 193 and congener
pregna-1,4,20-triene-3-one (198) being isolated from a
North Atlantic coral *Gersemia rubiformis* [1308,1309]. The
trienone 198 has also been recovered from an unidentified
air-aged Pacific soft coral, along with congener 5α-pregna-
1,20-dien-3-one [1042]. Moreover, other analogs 11α-
acetoxypregna-4,20-dien-3-one and 18-acetoxypregna-1,4,20-
trien-3-one have been isolated from gorgonian *Eunicella
cavolini* [488] and coral *Telesto riisei* [2025] respectively.

Also, there have been isolated two functionalized C_{24}-
sterols, methyl 3β-hydroxy-(20R)-chola-5,E-22-dienoate (199)
from oven-dried sea pen *Ptilosarcus gurneyi* [2554] and the
acetylene chol-5-en-23-yn-3β-ol (200) together with cholest-
5-en-23-yn-3β-ol (201) from the sponge *Calyx nicaaensis*
[2357].

Five of the short side-chain 5-stenols of TABLE 5 (109,
110,114,191,196) have established autoxidation origins, being isolated from air-aged cholesterol or having demonstrated autoxidation processes for their derivation from cholesterol [2562,2565,2576,2578,2579]. On this basis,

autoxidation origins for the class is suspected, indeed

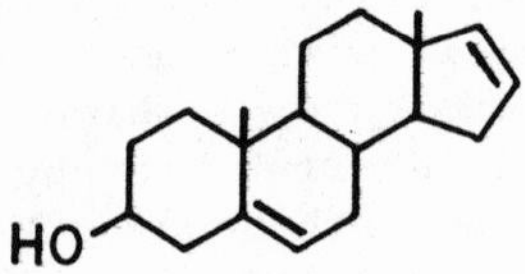

191

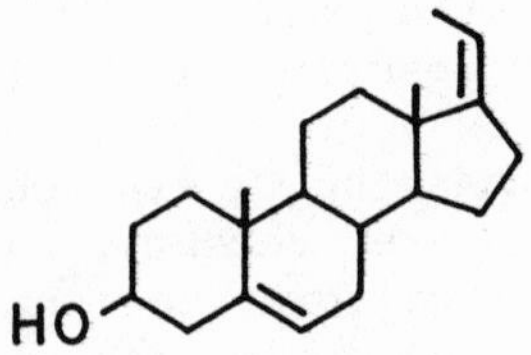

192

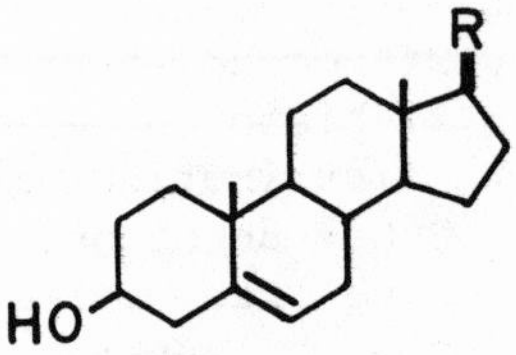

109 R = CH(CH$_3$)C$_3$H$_7$

110 R = C$_2$H$_5$

114 R = H

193 R = CH=CH$_2$

194 R = CH(CH$_3$)$_2$

195 R = C(CH$_3$)=CH$_2$

196 R = CH(CH$_3$)C$_2$H$_5$

197 R = CH(CH$_3$)C$_4$H$_9$

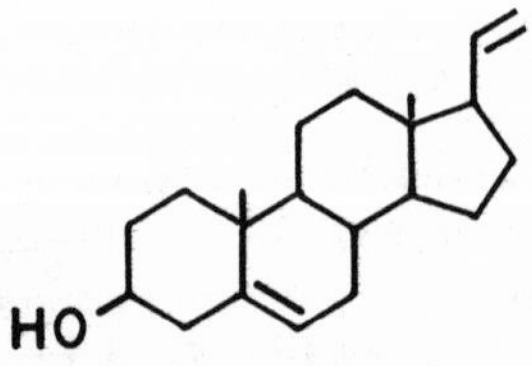

198

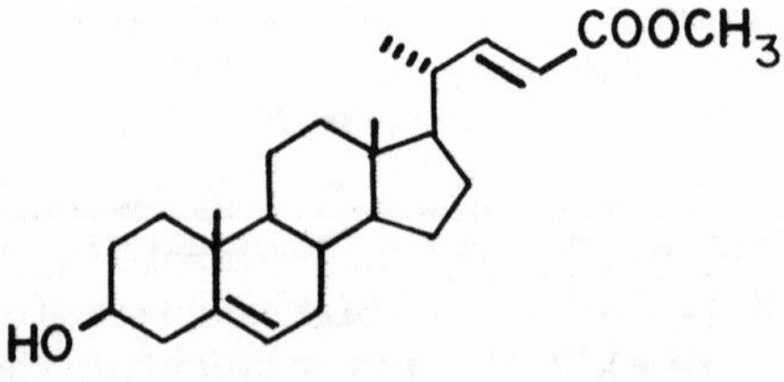

199

TABLE 5. Short Side-Chain Stenols from Invertebrates

Side Chain	Sterol	Occurrence	References
Δ^{16}-17-H	191	Sponge *Damiriana hawaiiana*	[444,592]
17β-H	114	Gorgonians *Murecia californica*, *Plexaura homomalla*, *Pseudoplexaura porosa*	[444,620,1877]
		Sponges *Callyspongia plicifera*, *D. hawaiiana*, *Luffariella* sp.	[444,592]
		Termite *Nasuititermes rippertii*	[2533]
	110	Gorgonians *P. homomalla*, *Ps. porosa*	[444,620,1877]
		Sponges *C.plicifera*, *D.hawaiiana*,*L.*sp.	[444,592]
		Termite *N. rippertii*	[2533]
$17(20)$=CHCH$_3$	192	Gorgonian *Ps.porosa*	[444,620,1877]
17β-CH=CH$_2$	193	Gorgonian *M.californica*	[444]
		Sponges *C.plicifera*,*D.hawaiiana*	[444,592]
		Coral *Gersemia rubiformis*	[1309]
17β-CH(CH$_3$)$_2$	194	Gorgonians *P.homomalla*,*Ps.porosa*	[444,620,1877]
		Sponges *C.plicifera*,*D.hawaiiana*,*L.*sp.	[444,592]
		Termite *N. rippertii*	[2533]
17β-CH(CH$_3$)=CH$_2$	195	Sponges *C.plicifera*,*D.hawaiiana*,*L.*sp.	[444,592]
17β-CH(CH$_3$)C$_2$H$_5$	196	Gorgonians *M.californica*,*Ps.porosa*	[440,620,1877]
		Sponge *D.hawaiiana*	[444,592]
17β-CH(CH$_3$)C$_3$H$_7$	109	Gorgonian *Ps.porosa*	[444,620,1877]
		Sponges *C.plicifera*,*D.hawaiiana*	[444,592]
17β-CH(CH$_3$)C$_4$H$_9$	197	Gorgonian *Ps. porosa*	[444,620,1877]
		Sponge *D.hawaiiana*	[444,592]

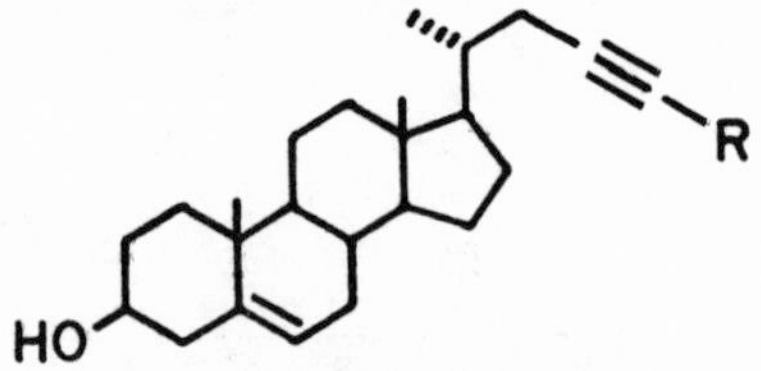

200 R = H

201 R = CH(CH$_3$)$_2$

systematic examination of artificial oxidation processes
using computer assistance has been employed to advance
the argument [592,620].

    Nonetheless, the matter is not certain, and these
short side chain sterols may represent instead unprecented
metabolic processes not heretofore recognized in animals.
For example, the C$_{19}$-C$_{22}$-sterols 114,110, and 109 found in
termites *N. rippertii* (preserved in alcohol) were accompanied by two 5ξ-dihydro derivatives 5ξ-androstan-3β-ol
and 5ξ-23,24-bisnorcholan-3β-ol [2533]. Moreover, the
C$_{22}$-sterol 194 was isolated from the gorgorian *Ps. porosa*
in esterified form [444,1877]. Both double bond reduction
and esterification bespeak of enzymic transformation and
not of autoxidation. However it is not possible to deter-
mine whether reduction or esterification follow or preceed
side chain alterations.

    Still other points bear on this issue. Another class
of six C$_{26}$-sterols with branched short side chains has
been identified in marine invertebrates and plant life.

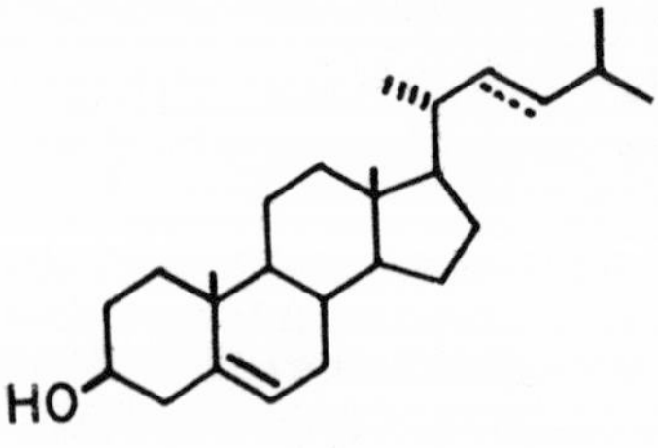

202

203 Δ^{22}

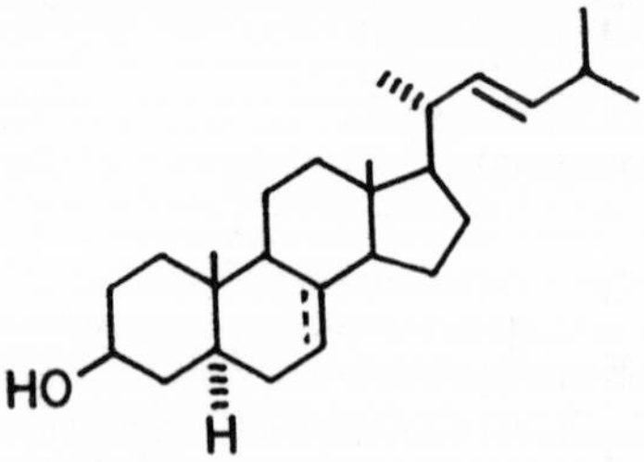

204

205 Δ^7

These sterols include 24-norcholest-5-en-3β-ol (202), 24-norcholesta-5,E-22-dien-3β-ol (203), 5α-24-norcholest-E-22-en-3β-ol (204), 5α-24-norcholesta-7,E-22-dien-3β-ol (205), 5α-24-norcholest-E-23(25)-en-3β-ol (206), 5α-24-norcholesta-7,E-23(25)-dien-3β-ol (207), and 5α-24-norcholestan-3β-ol (208). These distinctive C$_{26}$-sterols are distributed throughout nine animal phyla, with specific sterols being

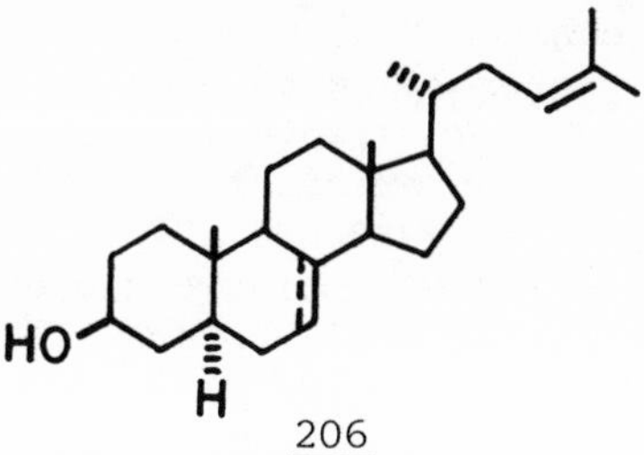

206

207 Δ^7

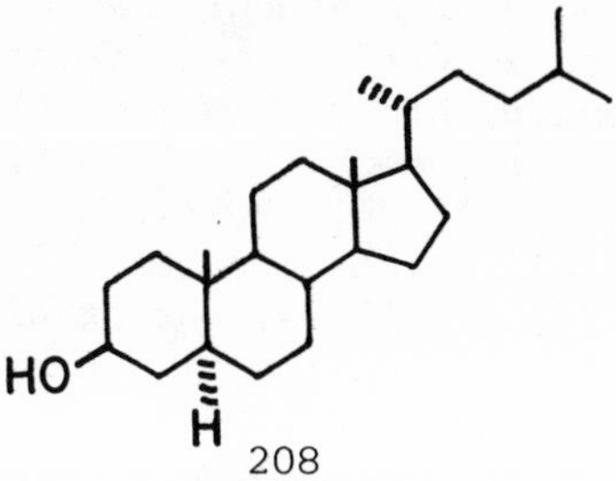

208

identified in most, as follows: Porifera (202-205,208 [130,134,594,604,657,699,700,1224,1307,2217,2611]); Coelenterata (jellyfish 203,204 [131,132,1222,1223,2761], sea anemones, 202 [733,2613,2761], gorgonians, 203 [298, 1222,2362]); Nemertinea (Cerebratulus marginatus, C$_{26}$-sterol [2606]); Brachiopoda (Terebratalia transversa, C$_{26}$-sterol [1121]); Echinodermata (starfish, 203-205 [878,1342, 1345-1347,1585,2283,2612,2617]; echinoids, 203 [878,2281, 2762]; holothurians, 203 [878,1346];ophiuroidean Ophiura albida, 203 [878]); Mollusca (scaphopod Dentalium entale, 203 [1114]; gastropods 203,205 [1114,1121,1346,2362,2466, 2470,2605]; pelecypods, 202-204 [62,63,133,1115,1118,1120-1122,1221,1223,1343,2466,2467,2469,2471,2609,2610]; opisthobranch Aplysia depilans, C$_{26}$-sterol [2614]); Annelida (polychaetes, 203 [1341,1344,2607]; oligocaetes, C$_{26}$-sterol [2608]); Arthropoda (decapod crustaceans, 203 [2760]); Chordata (tunicates, 203,204,206,207 [38,133,925,1345, 1774,2594,2615,2763]).

The presence of C$_{26}$-sterols in tissues of marine fishes is also indicated, as a norcholestadienol has been detected in chimera (Chimera phantasma) liver and sardine (Sardinops melanosticta) viscera [2419]. Moreover, the C$_{26}$-sterols have also been found as esters together with the free sterols in sea anemones [2613]. The C$_{26}$-5,22-diene 203 most frequently encountered in these marine invertebrates has also been identified in marine phyto-

plankton (chiefly *Chaetoceros* genera) [332] and in the red
alga *Rhodymenia palmata* [732,1119]. An unidentified C_{26}-
sterol has also been detected in the marine brown alga
Sargassum fluitans [2292].

    The biological origins of C_{26}-sterols in marine inver-
tebrates are uncertain. Although metabolic interconversions
and *de novo* biosynthesis may occur for C_{27}-sterols present
in marine annelids, coelenterates, echinoderms, molluscs,
and sponges, biosynthesis of the C_{26}-sterols in such cases
has not been demonstrated [878,2281,2467,2604,2605,2606,2607,
2608,2611,2613,2614,2616,2617]. A dietary origin for the
C_{26}-sterols in these animals is suspected [332,604,1223,
1341,1774,2617,2760], but origins involving symbiotic algae
living in association with the host marine creature are
also possible [444,2362,2533].

    A particularly compelling demonstration of possible
relationships has been demonstrated by analysis of the
many sterols found in isolated zooxanthellae symbionts
from the gorgonian *Briareum asbestinum*, sterols from the
whole gorgonium *B. asbestinum*, and sterols from its predator, the "flamingo tongue" gastropod *Cyphoma gibbosum* (and
also *C. gibbosum* feces). Qualitatively the same pattern of
approximately twenty sterols was found in each case, thus
suggesting dietary origins for gastropod predator sterols
possibly derived ultimately from zooxanthellae symbionts
of the prey [2362].

    Additional suspicion that marine plankton, microbe,
and plant sources contribute to the broad distribution
of these sterols is had in the detection of the C_{26}-sterols
219-221 in sea water [832,833,835] and recent marine sediments [1442].

    Although the ultimate biological origins of the C_{26}-
sterols 202-208 remain obscure, their metabolic origin
is surely the case. The class may be regarded as formal
24-methyl-26,27-bisnor-C_{27}-sterols formed by the actions
of as yet undefined methyl group transfer reactions which
add, remove, or rearrange individual methyl groups in the
terminus of the side chain [1739]. This postulated process
may involve parent 24-methyl-(24S)-C_{27}-sterols and putative
intermediate 24-methyl-(24S)-27-norsterols, as the 24-methyl-
(24S)-27-norsterols 24-methyl-(24S)-27-norcholesta-5,E-22-

dien-3β-ol (occelasterol) (209) and 24-methyl-5α-(24S)-
27-norcholest-E-22-en-3β-ol (patinosterol) occur in the
scallop *Patinopecten yessoensis* with the C_{26}-sterols
203 and 204 [1343] and 24-methyl-5α-(24S)-27-norcholesta-
7,E-22-dien-3β-ol (amuresterol) occurs in the asteroid
Asterias amurensis with the C_{26}-sterols 203 and 205
[1342].

    Alternative biosynthesis schemes for the C_{26}-sterols
202-208 involving cyclization of a modified or norsqualene
[332,809] or of synthesis from C_{21}- or C_{24}-steroids lack
experimental support. Moreover, no recognized autoxidation
process satisfactorily accounts for the C_{26}-sterols, as such
autoxidation would have to cleave both 25(26)- and 25(27)-
bonds of a parent 24-methyl-C_{27}-sterol. Such cleavages

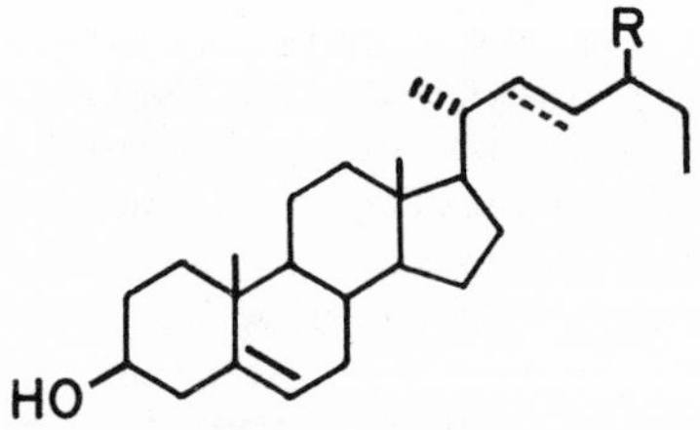

    209 R = CH_3, Δ^{22}

    210 R = H

    211 R = H, Δ^{22}

would require highly improbable autoxidations of both terminal 26- and 27-methyl groups, followed by β-scission of
the 25(26)- and 25(27)-bonds to provide the requisite
C_7-side chain of 24-norcholest-5-en-3β-ol (202).

    However, the thermal decomposition of the isomeric
cholesterol 26-hydroperoxides 28 and 30 formed by autoxidation of cholesterol gives 27-norcholest-5-en-3β-ol (210)
inter alia [2559]. Notably, the Δ^{22}-derivative of this
C_{26}-sterol 27-norcholesta-5,E-22-dien-3β-ol (211) has been
found in the gorgonian *P. homomalla* and in sponges *D.
hawaiiana* and *M. californica* (in company with the branched
side-chain C_{21}-sterols 202-207) [444,592] but most oddly
also in the urine and serum of a patient suffering congeni-
tal hyperplasia [1138]!

Yet another interesting relationship obtains in considerations of the group of short side chain sterols. In this case androsta-5,16-dien-3β-ol (<u>191</u>) has been found in the sponge *D. hawaiiana* [444,592] but also among congeners 5α-androst-16-en-3α-ol,5α-androst-16-en-3β-ol, 5α-androst-16-en-3-one, and androsta-4,16-dien-3-one in boar testis, sub-maxillary gland, and other tissues, possibly implicated as mammalian sex pheromones [906]. However, the 5,16-dienol <u>191</u> is an established cholesterol autoxidation product, being derived by thermal decomposition of the cholesterol 20-hydroperoxides <u>22</u> and <u>32</u> [2565,2578] (cf. Chapter V). Although the C_{19}-sterols derive enzymically from C_{21}-steroid precursors in mammals, the presence of <u>191</u> among sponge sterols cannot presently be rationalized.

The case for the short side chain sterols of TABLE 5 is another classic instance where uncertainty abounds regarding true origins. In our abysmal ignorance of invertebrate metabolism, the question whether these sterols be endogenous metabolites serving some purpose within the living organism or whether they be artifacts of unrecognized processes cannot now be answered, and disputation about origins seems fruitless. Although the well recognized possibilities for adventitious air oxidation of sterols generally insure that chemical processing of biological material is conducted properly, almost no attention is regularly paid to specimen collection and conservation prior to analysis. Thus, it is common practice to air dry marine creatures or to store them in alcohol for indeterminate times prior to analysis! Air drying almost certainly insures that autoxidations will ensue.

The well studied case of the sterols of the sponge *D. hawaiiana* is a paradigm, for among the forty identified sterols are parent C_{27}-,C_{28}-, and C_{29}-sterols, the C_{26}-sterol <u>202</u>, nine short side chain sterols, three sterols oxidized in the side chain, and common cholesterol autoxidation products [444,592]. The common cholesterol autoxidation products found include the enone <u>8</u>, dienone <u>10</u>, 3β,5α,6β-triol <u>13</u>, 7-ketone <u>16</u>, and 3β,25-diol <u>27</u>, these soundly establishing the intrusion of autoxudation processes into this work. The presence of a 24-ethyl-(24ξ)-cholest-5-ene-3β,25-diol and putative parent 24-ethyl-(24ξ)-cholest-5-en-3β-ol additionally supports this thesis.

Moreover, the 24-ketone <u>34</u> and 3β-hydroxy-26,27-bisnorcholesta-5,E-22-dien-24-one (<u>212</u>) found in *D. hawaiiana* [592] and 3β-hydroxy-26,27-bisnorcholest-5-en-24-one (<u>213</u>) found in the sponge *Psammaplysilla purpurea* [115] may be oxidation artifacts and not enzymic products.

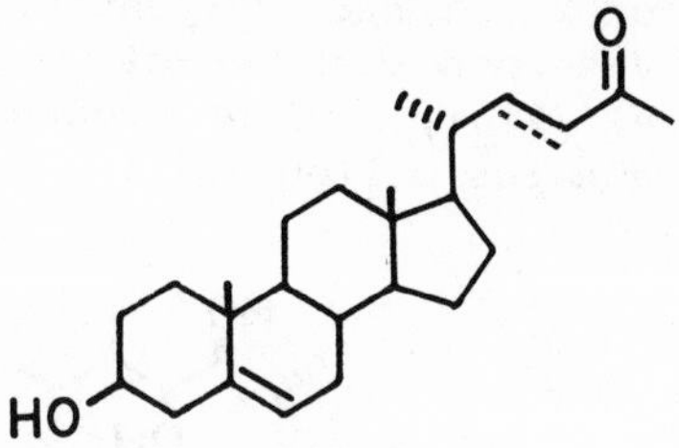

<u>212</u> Δ²²

<u>213</u>

That approximately one-third of the twenty-seven sterols found in *D. hawaiiana* suggest autoxidation casts question upon the origins of the nine short side chain sterols of TABLE 5, and oxidation occurring during specimen collection and preservation may be suspected. Even though freshly collected sponge be used, these oxidations nonetheless may occur. An alternative concept of "*in vivo* autoxidation" has been advanced [592], such formulation appearing to incorporate *in vivo* lipid peroxidations and/or photosensitized oxygenations, both of which may occur in these marine organisms or in symbiotic algae associated with them.

Indeed, by computer assisted calculation in which both free radical oxidations involving $^3O_2$ and photosensitized oxygenations involving $^1O_2$ were allowed oxidative processes acting on 27 different sterol side chains which have been found in various marine invertebrates, it was demonstrated that 34 side chain hydroperoxides could be derived from processes involving $^1O_2$ and 78 hydroperoxides from processes involving $^3O_2$! By further allowed chemical processes of hydroperoxide reductions and thermal decompositions theoretically multiple pathways were formulated for the generation of nine of the short side chain sterols <u>110</u>,<u>114</u>, and <u>191</u>-<u>197</u>, with only the C_{24}-sterol chol-5-en-3β-ol (<u>109</u>) having but one unique potential origin from cholesterol [444].

Many other more highly oxidized sterol derivatives
with interesting biological activities occur in plants,
but nonenzymic origins are of little concern. In some of
these more highly oxidized sterols besides side-chain and
nuclear hydroxylations ring scissions also occur, a recent
example being that of 2α,3α,7,22,23-pentahydroxy-24-methyl-
5α-(22R,23R,24S)-6,7-secocholestan-6-oic acid lactone
(6 → 7) (brassinolide) (214), a plant growth promoting
factor from *Brassica napus* pollen [923].

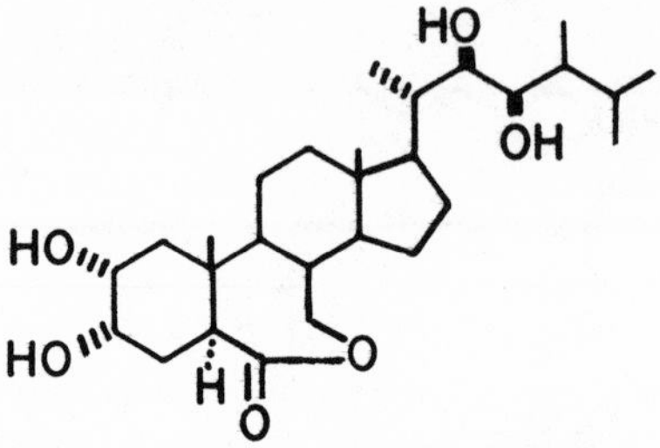

214

STEROL 5α,8α-PEROXIDES

The distribution of sterol autoxidation products
heretofore presented has been limited to oxidized derivatives as cholesterol, dihydrolanosterol (68), sitosterol
(20), etc. and to those with additional unconjugated
unsaturattion in the side-chain, such as lanosterol (67),
desmosterol (78), stigmasterol (70), and fucosterol (75).
The commonly encountered oxidation products of these
sterols are ketones and alcohols which are in reality
thermal decomposition products of initially formed sterol
hydroperoxides. In sharp distinction the autoxidation
behavior of sterol conjugated dienes such as cholesta-5,7-
dien-3β-ol (56) and ergosterol (65) is different, as the
cyclic peroxide derivatives initially formed in autoxidation are stable and readily isolated from biological
material. Secondary thermal decomposition products of
the cyclic peroxides have not been extensively detected in
Nature.

The important sterol 5,7-dienes 56 and 65 yield corresponding cyclic 5α,8α-peroxides, 5,8-epidoxy-5α,8α-
cholest-6-en-3β-ol (60) and 5,8-epidioxy-5α,8α-ergosta-

6,E-22-dien-3β-ol (62) under a variety of oxidation conditions, and both 60 and 62 have been isolated from natural
sources. The presence of the C_{27}-5α,8α-peroxide 60 in rat
liver from experimental animals treated with the sterol
Δ⁷-reductase inhibitor trans-1,4-bis-(2-cholorobenzylamino-
methyl) cyclohexane dihydrochloride (AY-9944) [650] is
considered to be artificial [649], but the peroxide 60 also
appears to be formed by NADPH-dependent microsomal lipid
peroxidation system of rat liver [1225]. The 5α,8α-perox-
ide 60 has also been found in fish liver [301] and in the
sponges Axinella cannabina [717] and Tethya aurantia [2217].

These same sponges appear to contain other 5α,8α-perox-
ides, A. cannabina containing additionally ergosterol
5α,8α-peroxide 62 and 5,8-epidioxy-5α,8α-cholesta-6,E-22-dien-
3β-ol,5,8-epidioxy-24-methyl-5α,8α-(24ξ)-cholest-6-en-3β-ol,
5,8-epidioxy-24-methyl-5α,8α-(24ξ)-cholesta-6,E-22-dien-
3β-ol, 5,8-epidioxy-24-ethyl-5α,8α-(24ξ)-cholesta-6-en-3β-ol,
and 5,8-epidioxy-24-ethyl-5α,8α-(24ξ)-cholesta-6,E-22-dien-
3β-ol [717], and T. aurantia additionally ergosterol peroxide 62 and 5,8-epidioxy-5α,8α-ergosta-6,E-22,24(28)-trien-
3β-ol [2217]. It is assumed that this latter 5α,8α-peroxide
is derived from a putative parent 24-methylcholesta-5,7,E-22,
24(28)-tetraen-3β-ol, with peroxidation occurring at the
homoannular 5,7-diene but not the side-chain 22,24(28)-diene
feature.

Although ergosterol peroxide 62 has been found in
sponges [717,2217], the sterol is preponderantly associated
with microbial systems from which it has been repeatedly
isolated. Thus, the 5α,8α-peroxide 62 has been found in
the fungi Acremonium luzulae [450], Aspergillus flavus
[2543] and Aspergillus fumigatus [2677], Cantharellus
cibarius [1358], Daedalea quercina [16,2443], Fusarium
moniliforme [2190], Fusarium oxysporum [2346], Gibberella
fujikuroi [160], Lampteromyces japonicus [679], Penicillium
rubrum [160,2666,2667], Penicillium sclerotigenum [490],
Piptoporus betulinus [16], Rhizoctonia repens [85], Trichophyton schönleini [164], and Trichophyton tonsurans [1074];
in basidiomycetes Ganoderma applanatum (Fr.) Pat. (Fomes
applanatus Gill) [2375,2402], and Scleroderma aurantium
[2618]; in lichens such as Cetraria richardsonii Hook
[2402], Dactylina arctica [2377], Hypogymnia vittata [1049],
Peltigera apothosa [2420], Peltigera dolichorrhiza [2420],
Ramalina tingitana [890], Thamnolia subuniformis [2376],
and Usnea annulata [2377]; and in yeasts including Saccharo-

myces [827,950,2666].

Moreover, the 5α,8α-peroxide 62 has been isolated as a 3β-divaricatinate ester 215 from lichen *Haematomma ventosum* [387]. This example is another one in which enzymic esterification of the oxidation product or oxidation of the parent sterol ester may occur.

One instance of the occurrence of the 5α,8α-peroxide 62 in higher plants has been recorded, that of air-dried leaves of the common pinapple *A. comosus* [1811].

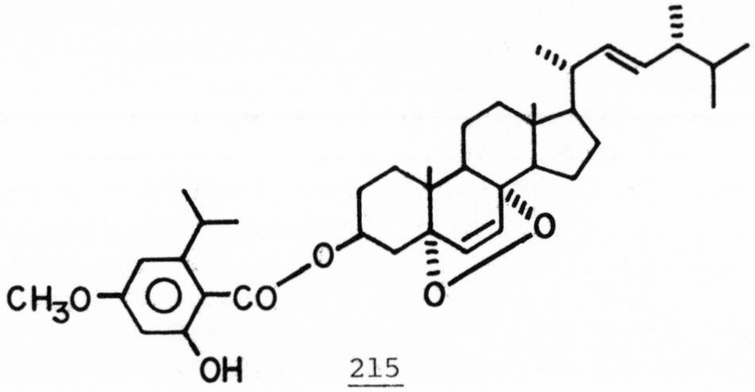

215

The sponges *A. cannabina* and *T. aurantia* together contain as many as eight 5α,8α-peroxides [717,2217], but no such diversity has been reported for the occurrence of sterol peroxides among microorganisms. Where sterol peroxides be found, ergosterol peroxide 62 occurs ubiquitously, but only the basidiomycete *S. aurantium* and the fungi *F. moniliforme*, *G. fujikuroi*, *P. rubrum*, and *R. repens* have been reported to contain additional sterol 5α,8α-peroxides. The 9(11)-dehydro peroxide 63 appears to be a constituent of *F. moniliforme*, *R. repens*, and *S. aurantia* [85,766,2190 2618]. However, it must be pointed out that the photosensitized oxygenation of ergosterol (65) yielding the 5α,8α-peroxide 62 also has been reported to give 5,7,9(11), E-22-tetraenol 66 as by-product [566], and the subsequent photosensitized oxygenation of 66 to its 5α,8α-peroxide 63 [2674,2704] then accounts for the presence of 63 in these microorganisms.

The fungi *F. moniliforme*, *G. fujikuroi*, and *P. rubrum* additionally contain the peroxyketone 5,8-epidioxy-3β-hydroxy-5α,8α-ergosta-6,9(11),E-22-trien-12-one (216)[160,2190]. The co-occurrence of the $\Delta^{9(11)}$-peroxide 63 and $\Delta^{9(11)}$-

peroxyketone 216 in these fungi provokes the speculation that the 12-ketone 216 be formed from 63 by a second oxidation. In this instance an allylic free radical autoxidation of 63 yielding putative 12-hydroperoxides 217 whose thermal dehydration yield the 12-ketone 216 would account for the presence of sterols 63,66 and 216.

Since ergosterol and its 5α,8α-peroxide 62 are also present in these fungi and the artificial formation of the $\Delta^{9,(11)}$-sterol 66 from ergosterol in photosensitized oxygenations occurs [566], overall artificial oxidation processes deriving all products 62,63,66, and 216 with ergosterol as parent sterol appears probable. The reported

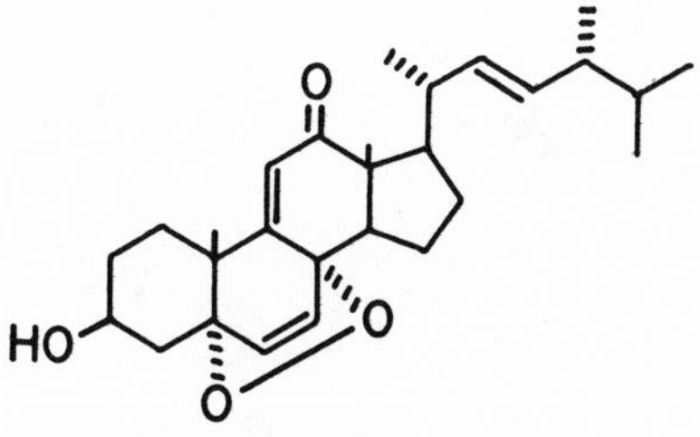

216 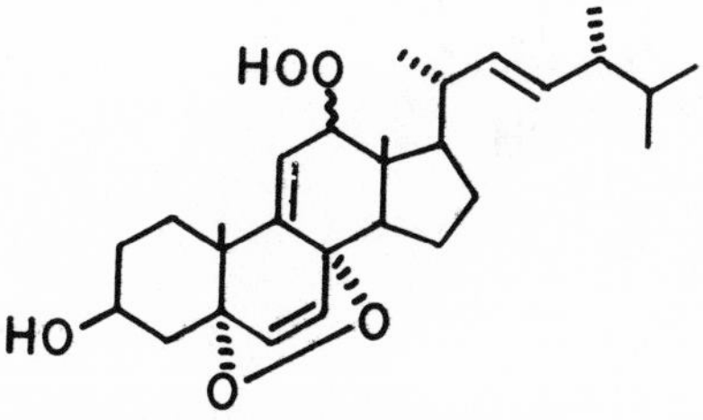   217

presence of the isomeric ergosterol peroxide 5,8-epidioxy-5β,8β-ergosta,6,E-22-dien-3β-ol in R. repens [85] is erroneous, the sterol peroxide in question in fact being the $\Delta^{9(11)}$-5α,8α-peroxide 63 [766] formed possibly by just this process. No 5β,8β-peroxides of sterol 5,7-dienes of the natural 10β-configuration have been described, but isomeric 5β,8β-peroxides of steroids of 10α-configuration are formed chemically [284,1607] (cf. Chapter VI).

We may anticipate that thermal alteration products of sterol 5α,8α-peroxides such as 60 and 62 also be present in biological materials. The recognized thermal [300,905] and alkali [1813] instability of these 5α,8α-peroxides, the formation of other unspecified ergosterol hydroperoxides [828] and of unidentified metabolites from 56 and 60 in the NADPH-dependent microsomal lipid peroxidation system of rat liver [1225], as well as formation of oxidized products from 65 not the 5α,8α-peroxide 62 [1167,2401] support such anticipation.

However, in distinction to the case of cholesterol and related 5-stenols where secondary autoxidation products but not primary hydroperoxides are found in tissues, the primary autoxidation product 5α,8α-peroxide 62 from the 5,7-diene ergosterol (65) is readily isolated from natural sources with but little in the way of secondary alteration products from 62. The most prominent secondary autoxidation product is cerevisterol (154) already discussed.

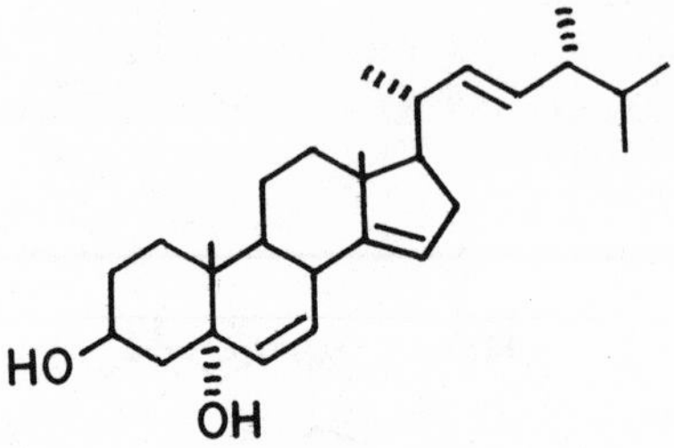

218

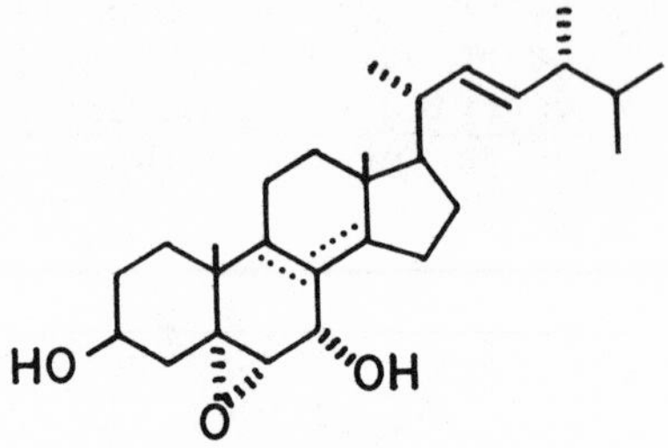

219 Δ^8

220 $\Delta^{8(14)}$

Additionally, the minor sterols 5α-ergosta-6,14,E-22-triene-3β,5-diol (218), 5,6α-epoxy-5α-ergosta-8,E-22-diene-3β,7α-diol (219), and the rearrangement product anthraergosterol derivative 1(10 → 6)-abeo-ergosta-5,7,9,14,E-22-pentaene (221) have been found in dried yeast [824,827,829]. The anthrasteroid rearrangement product 221 or an isomer thereof has also been found as an artifact in conjunction with the presence of 5α,8α-peroxide 63 in a Aplysinosis sp. sponge [593].

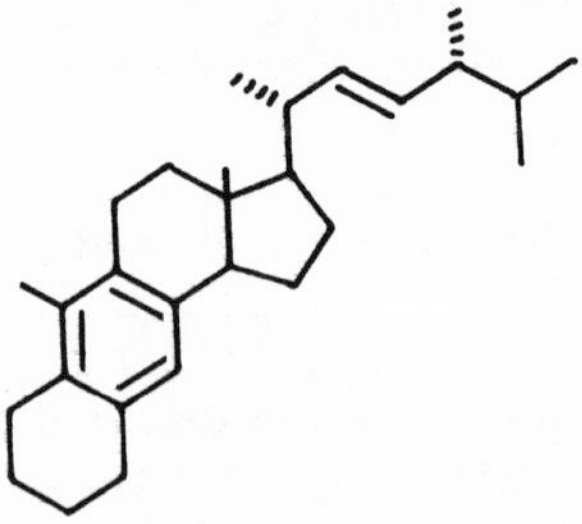

221

The epoxydiol 219 in dried yeast is one of two chief products of thermal decomposition of the 5α,8α-peroxide 62

[195], and its presence in dried yeast [824,827,829] must surely represent the thermal decomposition of ergosterol peroxide during storage. The same epoxydiol 219 as well as the isomeric epoxydiol 5,6α-epoxy-5α--ergosta-8(14),E-22-diene-3β,7α-diol (220), which also may be a thermal decomposition product of the 5α,8α-peroxide 62 [195,300,905] have been reported as metabolites of ergosterol 5α,8α-peroxide 62 by a variety of *Mycobacterium*, *Nocardia*, and *Fusarium* species of molds [1858], but the observed microbial action may in fact represent mere artifactual thermal decomposition of the 5α,8α-peroxide 62 as no control incubations were described in this report [1858].

5α,8α-Ergosta-6,E-22-diene-3β,5,8-triol (222), a chemical reduction product of the 5α,8α-peroxide 62 [641,2698, 2704], has also been isolated from *P. rubrum* [2666].

One additional steroid possibly a metabolite of ergosterol 5α,8α--peroxide 62 has been repeatedly isolated from various molds. The 3-ketone ergosta-4,6,8(14),E-22-tetraen-3-one (223) has a peculiar fluorescence aiding in its detection and has been found in mold-damaged wheat [519] and in

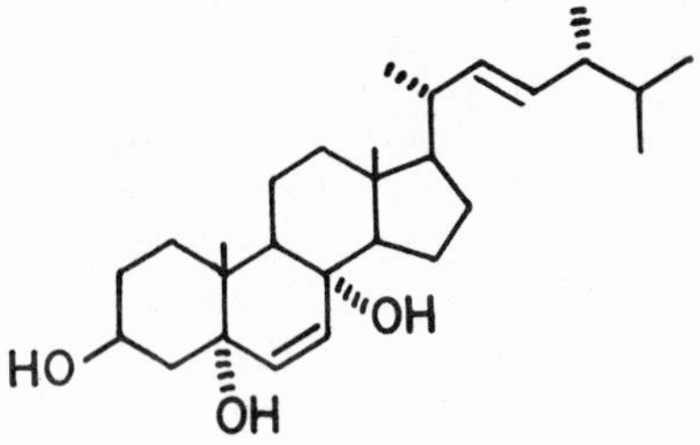

222 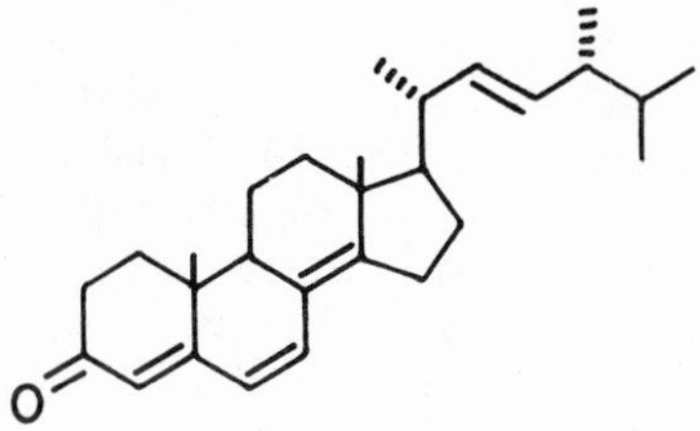   223

rice infected with *Aspergillus*, *Penicillium*, and *Fusarium* species of mold [1891] as well as in the defined organisms *Alternaria alternata* [2187], *Balansia epichloe* [1881], *Penicillium citrinum* [1891], *P. rubum* [[2666,2667], *Candida utilus* [1672], *Fomes officinalis* [2158], and the bioluminescent mushroom *Lampteromyces japonicus* [679].

As much as sterol 5α,8α-peroxides may be isolated from biological materials, there arises the inevitable question whether they be genuine metabolites or artifacts as is the case for previously mentioned autoxidation products. It is important to note that the 5α,8α-peroxide 60 has variously been proposed as a biosynthesis intermediate in the aerobic

transformation of the 7-stenol 57 to the 5,7-dien-3β-ol 56
in rat liver [1813] and also as an intermediate in the enzymic conversion of the provitamin 5,7-dien-3β-ol 56 to
cholecalciferol (90) in fish [301]. In like manner, the
ergosterol peroxide 62 has been suggested as precursor of
ergosterol in yeast [950]. In this case anaerobic enzymic
reduction of the 5α,8α-peroxide 62 to ergosterol has indeed
been demonstrated in yeast [2498]. These and other metabolic dispositions of the 5α,8α-peroxides 60 and 62 are
reviewed in Chapter VII.

Neither the occurence of 5,8-peroxides nor demonstration of their enzymic metabolism *in vitro* necessarily supports their genuine metabolite status. The question of
artifact status for the ergosterol peroxide 62 was raised
on its first isolation from *Aspergillus fumigatus* fungi in
1947 [2677], and exactly the same case of doubt obtains for
these cyclic 5α,8α-peroxides as obtains for cholesterol
autoxidation products, with asservations of true natural
products status, of uncertain dual status, and of artifact
status published.

In that the first reported synthesis of 5α,8α-peroxide
62 was by means of a photosensitized oxidation of ergosterol
[2700], it has apparently been generally or unofficially
considered of late that the 5α,8α-peroxide 62 be a product
of $^1O_2$ attack on ergosterol [85]. The ability of fungal
mycelium pigments and specific anthraquinones [16] and
biacetyl [1393-1395] to serve as photosensitizers soundly
supports an artificial photosensitized oxygenation process
for the formation of cyclic 5α,8α-peroxides, especially in
marine life subject to air drying in sunlight as is so
frequently the case. Despite these matters, the occurence
of sterol 5α,8α-peroxides in marine creatures even though
formed by photosensitized oxygenations have been viewed as
formed via a "biological process" [2217].

Indeed, the 5α,8α-peroxide 62 is formed under conditions
generating $^1O_2$ [160,766], but pure stored ergosterol samples
have low levels of the 5α,8α-peroxide 62 [649,1394], and the
same 5α,8α-peroxide 62 is clearly formed as the sole product
of oxidations of ergosterol by ground-state (triplet) molecular oxygen under a variety of circumstances, including
chemical [147-149,153,766], and enzyme [160] systems, and
it is now quite clear that the 5α,8α-peroxide may arise via
either or both pathways.

NONSTEROIDAL PEROXIDES

As with the sterols, where several Δ^6-5α,8α-peroxides and one $C_{29}\Delta^5$-7α-hydroperoxide 58 have been recovered from plantlife and marine invertebrates, so also is there record of several nonsteroidal peroxide derivatives in similar biological sources. Such peroxide derivatives are in distinction to the peroxide and hydroperoxide derivatives of polyunsaturated fatty acids that are products of various lipoxygenase actions, flavin hydroperoxides that are implicated in enzymic oxidations, and other similar peroxides that are established enzyme products. The peroxides of present concern have had no appropriate examination of possible enzyme reaction origins and accordingly carry the onus of potential artifact pending further investigations.

The most prominent of such peroxides is the well known [83,194] classic case of ascaridole (224) found as a major component of oil in chenopodium (derived from the plant *Chenopodium ambrosioides* var. *anthelminthicum* [2110]) which has anthelmintic properties. The cyclic peroxide nature of ascaridole was recognized very early [1732,1733,2626], and its synthesis by photosensitized oxygenation of α-terpinene (225) in which $^1O_2$ is implicated [2099,2105] and by oxygenations involving ground state molecular oxygen and Lewis acids [148,989] is established. The peroxide bond of ascaridole is relatively stable, surviving catalytic reduction [1808] and reduction with potassium azodicarboxylate [20].

Although early speculation that ascaridole derive by autoxidation of the 1,3-diene 225 was made [310], little

224   225

concern for ascaridole biosynthesis has been expressed. However, autoxidation of the 1,3-diene 225 does not yield ascaridole but gives low molecular weight peroxidic polymer

[310], and the biosynthesis of ascaridole has been considered
to implicate a dioxygenase speculatively [129].

Besides ascaridole, the status of which remains to be
definitively described, there are several reports of nonsteroidal peroxide derivatives in marine invertebrate and
plant sources. Beginning in 1970 a series of examples have
been recorded, as summarized in TABLE 6. These examples
include not only allylic hydroperoxides and cyclic allylic
peroxides whose nonenzymic origins be questionable but also
several other classes of compounds not so obviously suspect. In keeping with the same frustrations evident in
work with sterol autoxidation products recovered from
natural sources, none of the nonsteroidal peroxides is
viewed by its discoverers as artifact of manipulation or
as product of other adventitious nonenzymic oxidation.

Some of these peroxides exhibit biological activities,
including inhibition of root formation by cyclic peroxyketals from Eucalyptus leaves [538] and antibiotic [1043,
2365], tremorgenic [662,727,2754-2756], and ichthytoxic
[37] actions.

Verlotorin (peroxycostunolide), peroxyferolide, and
peroxyparthenolide clearly represent a class of sesquiter-
pene allylic hydroperoxides isolated from air-dried, powdered plant leaves [632,633,670,671,859]. The fifth example of the class in TABLE 6 is neoconcinndiol hydroperoxide,
found in one collection of *Laurencia* seaweed but not in
another [1075]. The cyclic peroxide baccatin found in tree
bark [2058] is a very close nortriterpene analog of cyclic
sterol peroxides such as ergosterol $5\alpha,8\alpha$-peroxide (62),
thereby suspect.

Although some of these peroxidic compounds may well
be products of endogenous metabolism within the tissues of
apparent origin, such demonstration remains to be offered.
Plausible nonenzymic oxidation processes may be posited for
the allylic peroxides, allylic hydroperoxides being formed
by free radical or photosensitized oxygenations of an appropriate parent olefin, cyclic peroxides by cycloaddition of
O_2 to an appropriate 1,3-diene precursor. The suspicious
circumstance where product peroxide and putative parent
olefin or 1,3 diene exist together in the same biological
source is sufficient comment, particularly when dried

TABLE 6. Nonsteroidal Peroxides Isolated from Biological Material

Compound Name	Structure	Biological Source	References
Allylic Hydroperoxides			
Verlotorin (Peroxycostunolide)	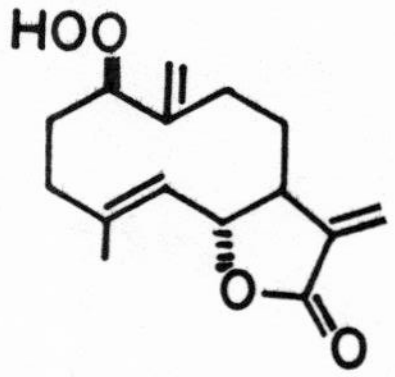	*Artemisia verlotorum* *Magnolia grandiflora*	[859] [670,671]
Peroxyferolide	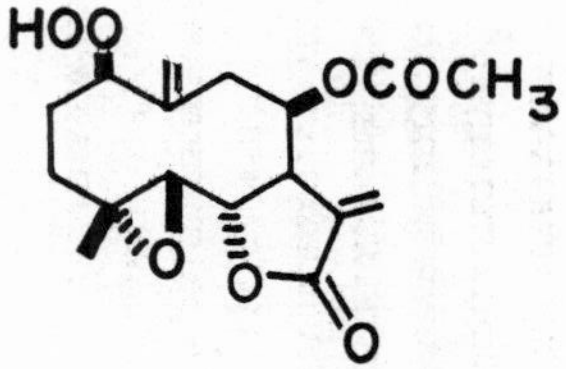	*Liriodendrum tulipifera*	[632,633]

(continued)

TABLE 6. Nonsteroidal Peroxides Isolated from Biological Material (continued)

Compound Name	Structure	Biological Source	References
Allylic Hydroperoxides			
Neoconcinndiol hydroperoxide	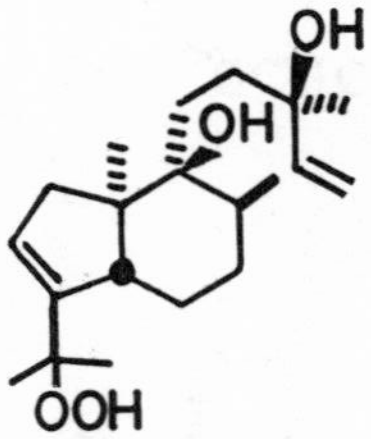	*Laurencia snyderiae* (seaweed)	[1075]
Peroxyparthenolide	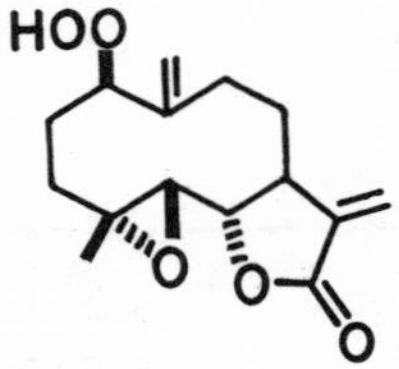	*Magnolia grandiflora*	[670,671]

(continued)

TABLE 6. Nonsteroidal Peroxides Isolated from Biological Material (continued)

Compound Name	Structure	Biological Source	References
Cyclic Allylic Peroxides			
Baccatin	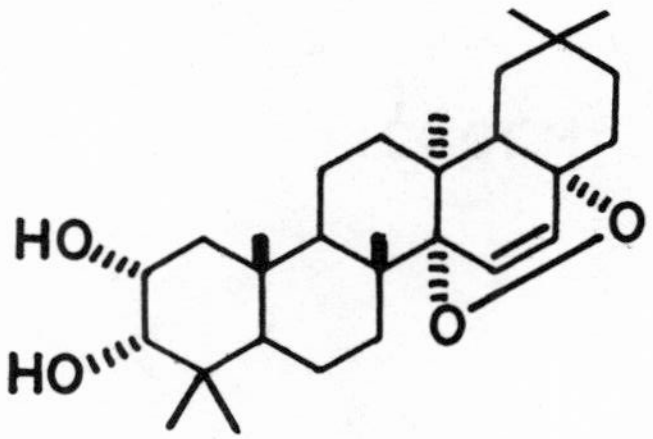	*Sapium baccatum* (bark)	[2058]
	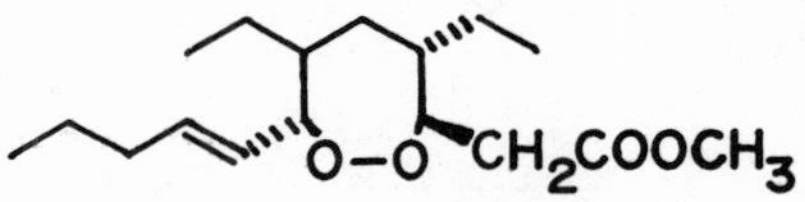	*Plakortis halichondrioides*	[2365]

(continued)

TABLE 6. Nonsteroidal Peroxides Isolated from Biological Material (continued)

Compound Name	Structure	Biological Source	References
Cyclic Allylic Peroxyketals			
	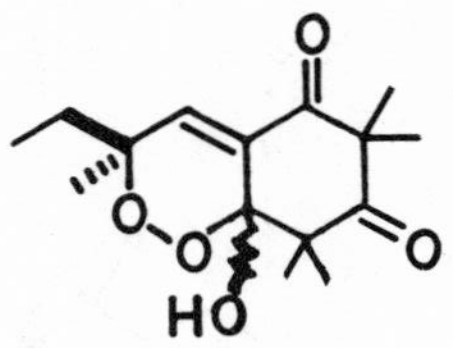	*Eucalyptus grandis* (leaves)	[538]
Chondrillin	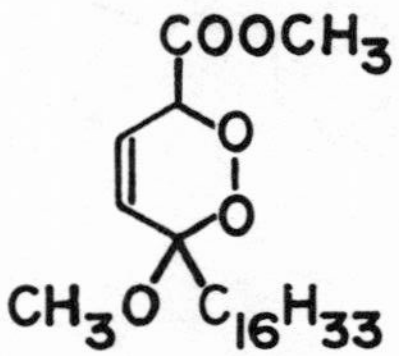	*Chondrilla* sp (sponge)	[2658]

(continued)

TABLE 6. Nonsteroidal Peroxides Isolated from Biological Material (continued)

Compound Name	Structure	Biological Source	References
Cyclic Aza Peroxides			
Fumitremorgin A (R = C₅H₉)		Aspergillus fumigatus, Aspergillus caespitosus	[662,727,2754-2756]
Verruculogen (R = H)		Penicillium verruculosum, Aspergillus caespitosus	[727]
Homoallylic Hydroperoxide			
Peroxy Y-base		Bovine liver	[297]
		Lupinus luteus	[729]

(continued)

TABLE 6. Nonsteroidal Peroxides Isolated from Biological Material (continued)

Compound Name	Structure	Biological Source	References
Cyclic Vinyl Peroxide			
Rhodophytin	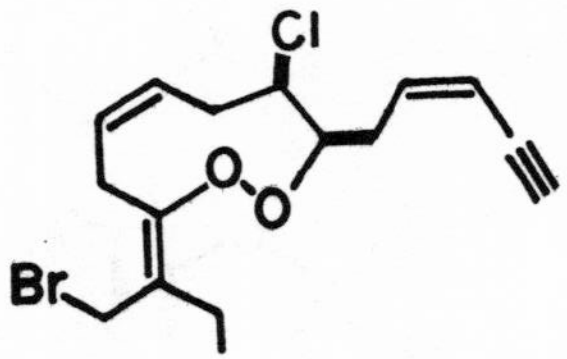	*Laurencia sp* (seaweed)	[731]
Cyclic Peroxides			
Plakortin	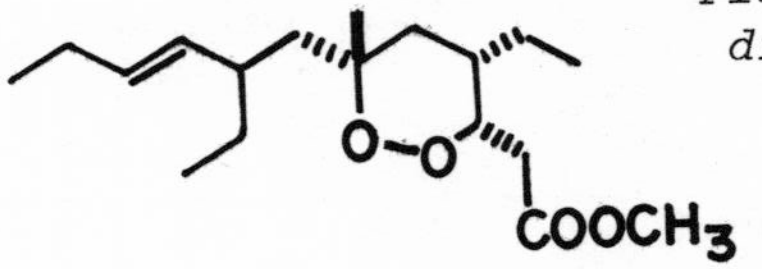	*Plakortis halichon-drioides* (sponge)	[719,1043]

(continued)

TABLE 6. Nonsteroidal Peroxides Isolated from Biological Material (continued)

Compound Name	Structure	Biological Source	References
	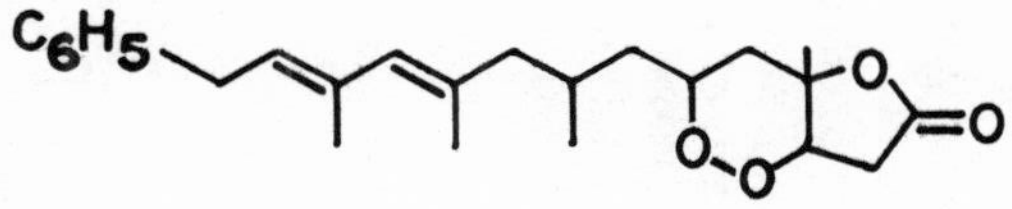	Plakortis halichondrioides	[1908]
Sigmosceptrellin-A	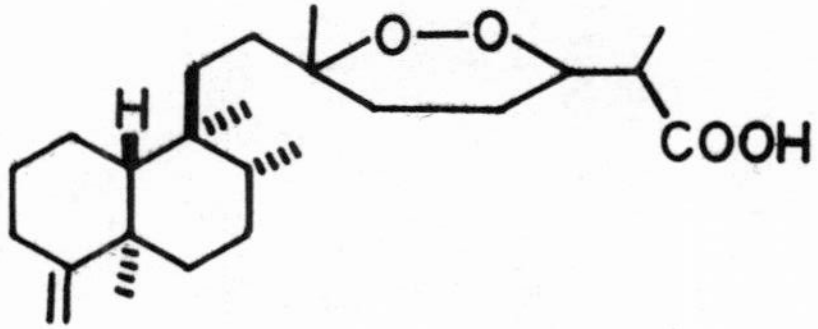	Sigmosceptrella laevis (sponge)	[37]
Muqubilin	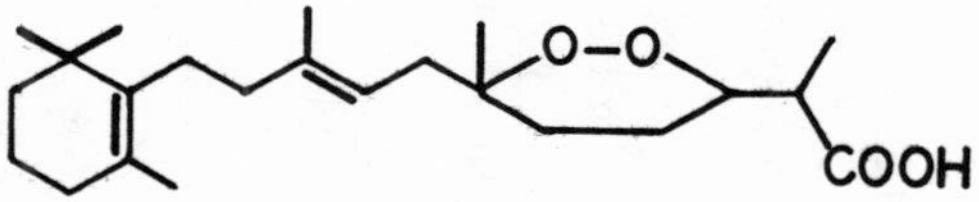	Prianos sp (sponge)	[1253]

leaves [633,670] or sun dried marine creatures [37] serve
as sources! How much more impressive would such work be
were proper handling of tissues and adequate analytical
control the case.

The cyclic allylic peroxyketals of TABLE 6 are potentially derived by cyclizations of putative γ-hydroperoxy-
α,β-unsaturated ketone parents, the 8-membered cyclic aza
peroxides fumitremorgin A and verruculogen by cyclizations
on an appropriate olefinic bond of putative parent such as
fumitremorgin B present as a congener.

However, the last three classes of compounds of
TABLE 6 lie outside our present experience for simple
rationalizations. The homoallylic hydroperoxide peroxy Y-
base recovered from bovine phenylalanine transfer RNA of
bovine liver and the plant *Lupinus luteus* [297,729], the
vinyl peroxide rhodophytin [731], and the series of cyclic
peroxides from sponges obviously represent new kinds of
peroxidic compounds about which we should learn more.

These provocative reports of naturally occurring peroxides and hydroperoxides need to be considered in light of
even more provocation, the isolation of ozonides of naturally occurring olefinic pentacyclic triterpenes from plant
material! From the plant *Quercus gilva* hop-17(21)-en-3β-ol
17,21-ozonide (226) has been isolated [1166], and from fresh
leaves of the fern *Adiantum monochlamys* Eaton and from the
Formosan plant *Oleandra wallachii* Presl adian-5-ene 5,6-
ozonide (227) was recovered [28]. Synthesis of both ozonides from the parent olefin confirmed the assigned structures. The presence of the parent olefin adian-5-ene in

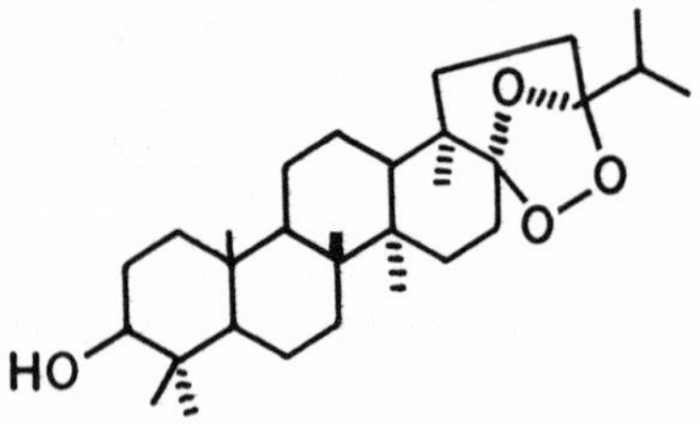

226 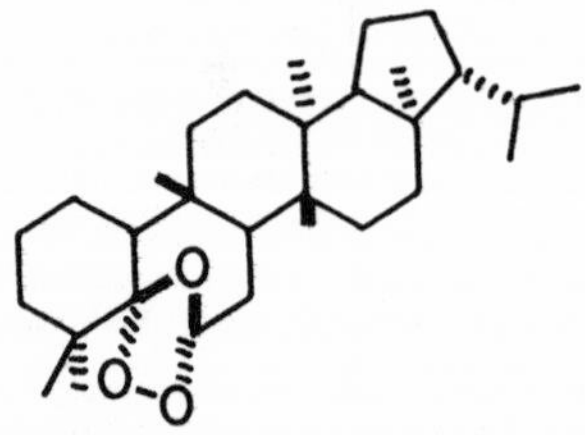   227

A. *monochlamys* leaves together with the ozonide 227 provides

a basis for the formation of the ozonide in the leaf, possibly by attack of O_3 present in the air. Alternatively,
one must postulate heretofore unprecedented oxygen insertion
reactions utilizing other oxygen species or the equally
undescribed process of biological production of O_3!

It is clear that these several studies of nonsteroidal
peroxides and ozonides do not cast light upon the cases of
sterol hydroperoxide and peroxide derivatives. Rather, all
such instances remain in an uncertain state, but one for
which almost no satisfactory evidence has been advanced for
bone fide enzymic derivation.

STEROLS IN THE BIOSPHERE

Review of the natural occurrence of sterol autoxidation
products in animal and plant tissues does not exhaust possibilities, as the remaining parts of the biosphere contain
sterols as well. Sterols found in marine and fresh waters,
soils, peat, recent sediments, petroleum, coal, and ancient
sediments may be variously exposed to air and sunlight or
thermal energy over long periods, and it is logical to
examine these materials for their possible sterol autoxidation products content.

Although these phases of the biosphere are less well
studied than are animal and plant tissues, the prevailing
evidence indicates that sterol autoxidation products do not
exist in the biosphere. Whether autoxidation does not occur
at all or whether autoxidation products are further transformed, either by microbial action or by geological influences, cannot be presently addressed. However, it is clear
that the status of the biosphere is such as to favor reductive processes and not oxidative ones.

Sterols occur in seawater at low levels [334,1189,
1190,1589,2061,2062], but the presence of cholesterol
autoxidation products in natural waters [1589,2294] or in
domestic activated sludge sewage treatment effluent [2296],
both subject to aeration and sunlight, is not indicated.
Although the autoxidation of cholesterol suspended at
300 µg/mL in seawater has been demonstrated [1589], sterol
derivatives found in natural waters appear to be of unoxidized nature. Indeed, sterols in deep marine waters *ca.*
500 y old remain unoxidized or undegraded [2062]. However,

degradation of sterols in stored fresh water has been noted
[646], and reductive processes are likely. The fecal 5β-
stanol 4 has been detected in polluted marine [977,2294]
and fresh [646,647,1163,1700,2433] waters, but a totally
reduced 5ξ-lanostane has been found in waters of San
Francisco bay [2254].

The presence of Δ^5-sterols has been demonstrated in
recent marine [106,1094,1095,1124,2168] and fresh water
[534,853-855,1096,1769,1771,1791] sediments as well as in
ancient sediments, but little evidence of the presence of
unrecognized autoxidation product of cholesterol or other
Δ^5-sterol has been reported. Rather, C_{26}-$C_{29}\Delta^5$-sterols have
been found in sediments accompanied by corresponding 5α-
stanols [853,855,1124,1769,1771,1791,2629], thus evincing
reducing environment where sterol autoxidation products
would not likely form or persist. The predominantly anaerobic character of aqueous sediments at deposit, subject to
anaerobic microbial transformations, and of the continuing
anaerobic environment during subsequent diagenesis might
reasonably account for the presence of 5α-stanols in both
marine and lacustrine sediments [853,854]. Experimental
conditions simulating the anaerobic environment of these
sediments support this formulation [791,854,1950].

However, the 5α-stanols of recent lacustriene sediments may derive in good part from phytoplankton and zoo-
plankton. A variety of C_{27}-C_{29} stenols and corresponding
5α-stanols, including cholesterol and the 5α-stanol 2 have
been found in such organisms [1769-1771]. The predominance
of 5α-stanols over isomeric 5β-stanols in recent fresh
water sediments [855] supports the contention. Furthermore,
the 5β-stanol 4 is produced from cholesterol by microbial
action in incubations of estuarine sediments and associated
seawater [2468], thus evincing a natural source for 4 in
addition to any contamination by human sewage.

Despite the overall reducing nature of natural waters
and their sediments, there are examples of oxidized steroids
being found in such sources. Recent marine ooze has yielded
up a 5α-cholanic acid (228) and 5α-27-norcholestan-26-oic
acid (229) inter alia [321]; C_{27}-C_{29} stanones have been
found in others [836]. Microbial action seems implicated
in both instances. In other recent marine sediments C_{27}-
C_{29} sterenes, steradienes, and steratrienes have been
found [834], the trienes being dehydration products of

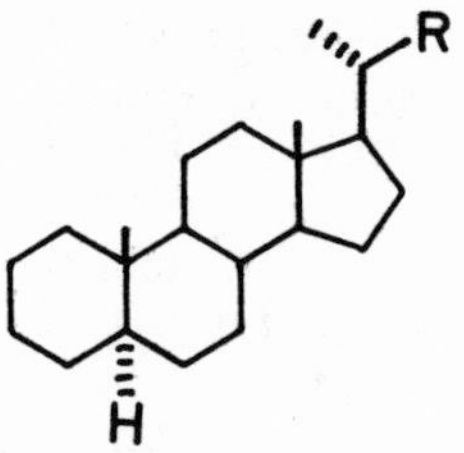

$\underline{228}$ R = CH$_2$CH$_2$COOH

$\underline{229}$ R = CH$_2$(CH$_2$)$_3$COOH

5,7-dienes but also of Δ^5-3β,7-diols. Furthermore, in Black Sea surface waters but not in deeper layers the 24-ketone $\underline{34}$ has been found [835], possibly as an autoxidation product of fucosterol or of other $\Delta^{24(28)}$-congeners.

On balance, the several steroid types found in natural bodies of water and in their sediments appear to have multiple origins, including plankton [322,1770,1771] and other life forms, sewage contaminations [977,1696], anaerobic microbial actions, and diagenesis (geological) factors.

Sterols have been found in soils [1613,2525], but no study of sterol autoxidation or of microbial action in the soil has been published. However, much work has been done on the oxidation of sterols by soil microorganisms as possible means of utilization of sterols for commercial synthesis of steroid hormones [2289]. Sterols have also been found in peat [2525] and in 2000 y old human coprolites [1509] with no evidence of autoxidation being reported.

A variety of sterols have been recovered from ancient sediments too, including Green River shale [1014,2352], Eocene Messel shale $ca.$ 50 My old [1587], Pleistocene lacustrine sediments $ca.$ 130 ky old [1013], and other sediments $ca.$ 100 ky old [366]. Interestingly, evidence suggested that the Δ^5-sterols 24-methyl-(24ξ)-cholest-5-en-3β-ol, 24-ethyl-(24ξ)-cholest-5-en-3β-ol, and 24-ethyl-(24ξ)-cholesta-5,22-dien-3β-ol might have survived in the Messel shale but were decomposed by autoxidation (!) within days of their isolation from the shale [1587]. Another unique feature of the Messel shale is that this is the only known case where evidence for the presence of oxidized steroids have been adduced. Thus, the 4α-methyl-3-ketones 4α-methyl-

5ξ-cholestan-3-one, 4α,24-dimethyl-5ξ-(24ξ)-cholestan-3-one, and 24-ethyl-4α-methyl-5ξ-(24ξ)-cholestan-3-one were preserved and isolated as such [1587].

Moreover, fully saturated steranes have been detected in Precambrian shale (ca. 2.7 Gy old) [402], in Jouy-aux-Arches bituminous shale (ca. 180 My old) [2043],Nebi Musa bituminous shale [1125], and in the Eocene Green River shale (ca. 60 My old) [56,60,136,402,839,840,1015,1016, 1695,1699]. The Jouy-aux-Arches shale additionally contained the rearranged steroids 5,14-dimethyl-5β,8α,9β 10α, 14β-(20R)-18,19-bisnorcholest -13(17)-ene (230) and its stereoisomer 5,14-dimethyl-5β,8α,9β,10α,14β-(20S)-18,19-bisnorcholest-13(17)-ene (231) [2036].Fully saturated derivatives of these rearranged steranes have also been found in marine and continental oil shale [688]. Oil shales of the Paris basin 180 My old also contain a series of C_{27}-C_{29} Ring C-aromatic rearranged steratriene derivatives [2093].

Petroleum has been shown to contain reduced sterol derivatives but sterols per se appear not to have been detected. Fully saturated steranes have been found in petroleum [182,688,1048,1695] as have also fully saturated derivatives of the rearranged sterols 230 and 231 [688], and the degraded steroid acids 5α-cholanic acid (228), 5β-cholanic acid,and the epimeric 5β-(20R)- and (20S)-22,23-bisnorcholanic acids(232)have been isolated from petroleum [2183-2185]. The reduced nucleus but oxidized side-chain of these derivatives suggest microbial interventions.

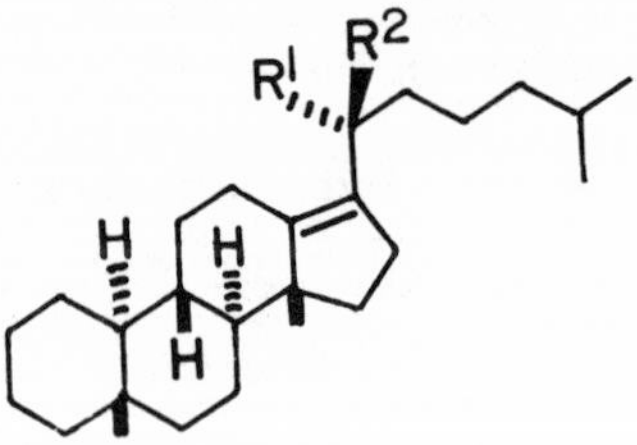

230 R^1 = CH_3,R^2 = H   232

231 R^1 = H,R^2 = CH_3

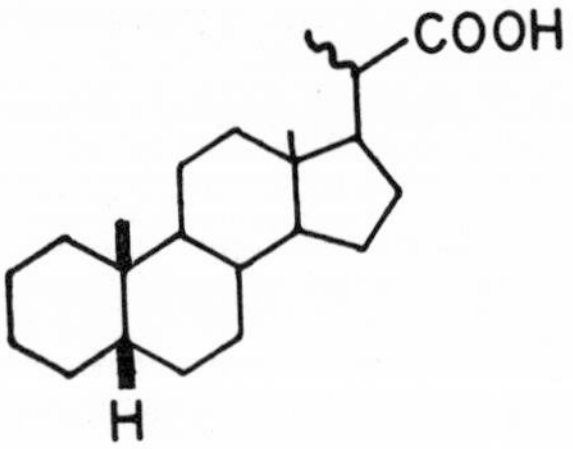

Discussions in this chapter regarding the distribution of the autoxidation products of cholesterol and of other naturally occurring sterols, including some oxidation

products of very uncertain origins, have all dealt with
the common steroid nucleus as found in these sterols.
Thus, all such findings have involved sterol derivatives
of the natural 8β,9α,10β,13β,14α-configuration.

    (20S)-Cholest-5-en-3β-ol (20-isocholesterol) [1370,
2441] and a few nuclear isomers of cholesterol such as
10α-cholest-5-en-3β-ol [658], 14β-cholest-5-en-3β-ol [54],
and 14β,17β(H)-cholest-5-en-3β-ol [53] as well as racemic
cholesterol [1197,1267] have been synthesized but are not
encountered in Nature. The sensitivity of these isomers
to autoxidation has not been examined, but some may be
metabolized [845]. However, 19-norcholest-5-en-3β-ol
(233) and several 24-alkyl-19-norhomologs have been detected in marine gorgonians Ps. porosa and P. homomalla
[1877,1878],and we may presume that autoxidation of the
19-norsterol 233 may be ultimately encountered. No hint
of the propensity of the 19-norsterol 233 towards autoxidation has been recorded [1877,1878,2590].

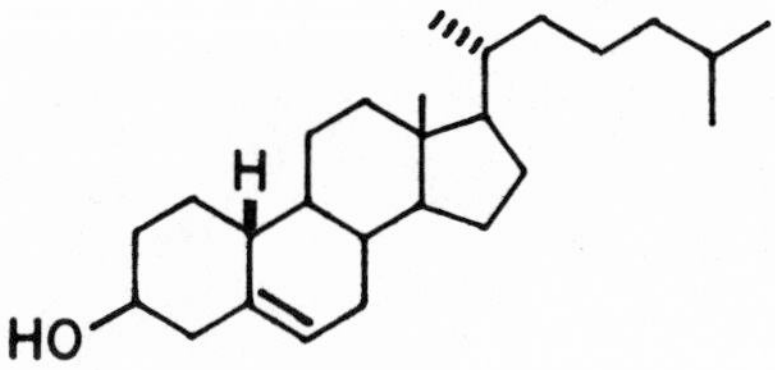

233

CHAPTER IV. INITIAL EVENTS OF AUTOXIDATION

The autoxidation of natural products remains a very complex matter only partially understood for any given class of compounds despite much study. Several recent monographs [568,983,1024,1542,1895] address the problem to different degrees, but it is clear that the natural air oxidation of organic compounds is still in need of major attention. The present chapter deals with much that is known of the autoxidation of cholesterol but only as regards the first stable products which can be isolated and identified. Very little experimental work has been done on the initiation phase of cholesterol autoxidation, where very low levels of highly excited species undergoing very rapid reactions may be the case.

In order to consider the details of cholesterol autoxidation it is important to reiterate the definition of autoxidation assumed for this work: "the *apparently* uncatalyzed oxidation of a substance exposed to the oxygen of the air" [2539]. Other definitions of autoxidation stress spontaneity of reaction under mild conditions [1157,2408], reaction at temperatures below 120°C without the intervention of a flame [568,1553], etc. These several limitations clearly distinguish autoxidations from the reactions of respiration and combustion. Two key features of Uri's definition, both of which lend subtle complexities to the topic are the apparent lack of catalysis and dependence upon molecular oxygen of the air. Accordingly, it is necessary to examine events likely to initiate autoxidations and to examine details of the chemistry of oxygen.

It is generally considered that autoxidations involve free radical species generated through one-electron transfer processes in chain reaction sequences which ultimately yield peroxide or hydroperoxide products. Subsequent reactions of these initial products may then promote more extensive autoxidations and moderate the processes through product inhibition and chain reaction terminations. Other kinds of autoxidations involving strong base or base with transition metal ions are separate processes which will be discussed in Chapter VI.

The free radical autoxidation of an organic compound RH is formulated as proceeding through three stages of initiation, propagation, and termination reactions to stable products, as summarized in Equations 1-7:

Initiation: $RH \rightarrow R\cdot + H\cdot$ Eq. 1

Propagation: $R\cdot + O_2 \rightarrow ROO\cdot$ Eq. 2

                  $ROO\cdot + RH \rightarrow ROOH + R\cdot$ Eq. 3

Termination: $2R\cdot \rightarrow RR$ Eq. 4

                    $ROO\cdot + R\cdot \rightarrow ROOR$ Eq. 5

                      $2ROO\cdot \rightarrow O_2 + ROOR$ Eq. 6

                      $2ROO\cdot \rightarrow O_2 + ROH + R = O$ Eq. 7

The initiation reaction of Eq. 1 is one which provokes keen interest. In the case of cholesterol we are fortunate in having sound experimental evidence of this initiation process in radiation-induced oxidations, to be discussed later in this chapter. Of the propagation reactions peroxy radical formation (Eq. 2) is generally thought to be very fast, and may be diffusion controlled ($k \sim 10^9$ L mol^{-1} s^{-1}) [1157]. The propagation reaction of Eq. 3 by comparison is much slower, as it involves scission of a carbon-hydrogen bond. However, Eq. 3 accounts for formation of the first stable products of autoxidation and provides means of continuation of the chain reaction sequence, and we have experimental evidence of the formation of hydroperoxides as initially formed stable products of cholesterol autoxidation [2313].

Although termination reactions for cholesterol autoxidation must occur, there is relatively little tangible evidence supporting any process of Eqs. 4-7. Kinetics data for the disproportionation of the cholesterol 7-peroxyl radical to the 3β,7-diols 14 and/or 15, 7-ketone 16, and molecular oxygen [1005] suggest the process of Eq. 7, but products 14-16 also derive from the thermal decomposition of the initially formed cholesterol 7-hydroperoxides 46 and 47 [2300] and direct proof of the liberation of molecular oxygen in these cases was not provided [1005]. Moreover, product analysis has not implicated dimeric sterene or disterene peroxide derivatives in cholesterol autoxidations as suggested in Eqs. 4 and 6 respectively, although these putative products would likely be nonpolar in

character and may have escaped recognition in prior isolation studies.

The initiation and propagation reactions of cholesterol autoxidation must be combined with subsequent thermal decomposition reactions of the initially formed products in order to provide a proper description of the overall process as observed. Accordingly, the present treatment of cholesterol autoxidation is conveniently divided
into three phases: (i) initiation events, (ii) reaction
with molecular oxygen of the air, and (iii) subsequent
transformations of initially formed products. Each of
these phases is to be considered individually, the first
events of initiation and of reaction with molecular oxygen
being covered in this chapter, the subsequent transformations following in Chapter V.

INITIATION EVENTS

Although the autoxidation of susceptible organic compounds is well recognized as occurring, the initiation of
the processes by which an otherwise unexcited organic molecule reacts with ground state molecular oxygen remains obscure. The formal initiation of free radical autoxidations
involves the homolysis of a susceptible carbon-hydrogen (or
possibly other) bond, as suggested by Eq. 1. Whereas bond
homolysis to yield two free radicals occurs relatively
readily for some organic peroxides, the process does not
occur at measurable rate for many stable organic compounds,
and some other means of inducing the formation of a free
radical in the autoxidizable molecule must exist. The
direct reaction between an organic substrate and ground
state molecular oxygen with no induction agent or catalyst
as given in Eqs. 8 does not satisfactorily account for the
initiation step, as this reaction between unexcited, ground

$$RH + O_2 \rightarrow R\cdot + HOO\cdot \qquad \text{Eq. 8a}$$

$$R\cdot + HOO\cdot \rightarrow ROOH \qquad \text{Eq. 8b}$$

$$RH + O_2 \rightarrow ROOH \qquad \text{Eq. 8c}$$

state (singlet) substrate with an even number of electrons
(all electrons paired) with ground state (triplet) molecular

oxygen ($^3O_2$) with two unpaired electrons is spin forbidden
and does not represent a predominant process.

   The presence of an agent which overcomes the spin for-
biddenness by pairing anti-bonding electron spins of molecular oxygen (yielding electronically excited singlet molecular oxygen $^1O_2$), or by removal of an electron transforming
the even-electron substrate to an odd-electron radical catalyzes the process. Such agents may be stable free radicals
themselves (nitroxides, NO, NO_2), azo compounds, peroxides,
and hydroperoxides readily subject to bond homolysis,
transition metal ions, or excited oxygen species capable of
initiating free radical processes. Other conditions which
promote free radical processes include other concommitant
chemical reactions, enzyme transformations, and radiolysis
and photolysis reactions.

   However, the formal lack of catalysis associated with
autoxidation must be reconciled with the true means of free
radical autoxidation initiation. It is here that experimental evidence is lacking, for almost no proper experiments
have been attempted to study the initiation of cholesterol
autoxidation. In fact, there are several potential agencies
by which cholesterol autoxidation may be initiated, and
until specific cases are examined only general speculations
can be provided for the case of cholesterol.

   One must consider the possibilities of trace amounts
of transition metal ions and of other transition elements,
of traces of sterol peroxides or hydroperoxides formed by
such catalysis or by other means, and of traces of other
possible free radical initiators present as adventitious
impurities in cholesterol samples, as well as the agency of
gaseous initiators present in the air utilized in the autoxidation. Finally, in a later section the initiation of cholesterol autoxidation by radiation, a process for which experimental support is had, will be given detailed treatment.

   Each of these several agencies involve one-electron
transfer processes, as is inherent in the nature of free
radicals. However, the initiation of free radical processes
may not be a simple one-step event but may involve a multi-
step sequence, such as an initial oxidation reaction not
involving free radicals to form a peroxide or hydroperoxide,
which via subsequent homolysis provides free radicals which
in turn initiate a general free radical chain reaction

involving cholesterol.

Peroxide Homolysis

It is in the homolysis of the oxygen-oxygen peroxide bond that most general treatments of autoxidation place emphasis for process initiation. Peroxide bond homolysis may occur via several means, through natural instability, thermal sensitivity, transition metal ion catalysis, etc. Uncatalyzed peroxide bond homolysis yielding alkoxyl and hydroxyl radicals, as represented in Eq. 9, has an associated peroxide bond energy of *ca*. 44-52 kcal/mole and thus is more likely than homolysis of the oxygen-hydrogen bond of a hydroperoxide (Eq. 10) with a bond energy of *ca*. 90 kcal/ mole [185,2165]. These same homolytic reactions catalyzed

$$ROOH \rightarrow RO\cdot \quad + HO\cdot \qquad \text{Eq. 9}$$

$$ROOH \rightarrow ROO\cdot \quad + H\cdot \qquad \text{Eq. 10}$$

by transition metal ions M^{n+} with two or more valence states separated by one electron such that the ion may donate or accept one electron are rendered in Eqs. 11 and 12. Transition metal ions having appropriate valance states allowing

$$M^{n+} + ROOH \rightarrow M^{(n+1)+} + RO\cdot \quad + HO^- \quad \text{Eq. 11}$$

$$M^{n+} + ROOH \rightarrow M^{(n-1)+} + ROO\cdot \quad + H^+ \quad \text{Eq. 12}$$

the ions to donate or accept two electrons may also catalyze peroxide homolysis reactions.

Given initiating peroxide bond homolysis to oxyl or peroxyl radicals, propagation reactions such as those of Eq. 3 or Eq. 13 then continue the process. There remains

$$RO\cdot + RH \rightarrow ROH + R\cdot \qquad \text{Eq. 13}$$

the question of the chemical nature of the original hydroperoxide or peroxide implicated in the specific case of cholesterol autoxidation. Here one must speculate, as there is no evidence.

Cholesterol autoxidation in tissues may be linked to initiator peroxides which are more easily formed from

polyunsaturated fatty acid esters of sterols, glycerol,
phospholipids, etc. Systematic studies of such possibilities have not been attempted. However, the sensitivity
of polyunsaturated fatty acid derivatives to air oxidation
is well known and considerably more rapid than that of
cholesterol.

The ready autoxidation of cholesterol esters of the
polyunsaturated fatty acids linoleic (octadeca-Z-9,Z-12-die-
noic), linolenic (octadeca-Z-9,Z-12,Z-15-trienoic), and
arachidonic (octadeca-Z-5,Z-8,Z-11,S-14-tetraenoic) acids
in aqueous sodium dodecyl sulfate dispersions has been
demonstrated. The much slower autoxidation of saturated and
monoolefinic fatty acid esters of cholesterol under the same
conditions contrasts markedly with the rapid reaction rate
for the polyunsaturated esters. The autoxidation of cholesterol 3β-linoleate (234) and 3β-linolenate at 85°C in
aqueous sodium dodecyl sulfate dispersions is essentially
complete within 4 hrs. whereas that of cholesterol 3β-
arachidonate is even more rapidly completed within 2 hrs.
By contrast, the monounsaturated fatty acid ester cholesterol 3β-oleate (3β-octadec-Z-9-enoate) is stable to
autoxidation during the time studied [1777,1778]. Neat
cholesterol 3β-arachidonate is readily autoxidized by the
air in light [893].

Evidence that the arachidonate and linoleate esters
of cholesterol are metabolized by rats to sterol esters in
which the polyunsaturated fatty acid moiety had been oxidized has been presented [2407]. Also, cholesterol esters
of 9-hydroxyoctadeca-10,12-dienoic and 13-hydroxyoctadeca-
9,11-dienoic acids [365,369,962], 9-hydroperoxyoctadeca-10,
12-dienoic and 13-hydroxyperoxyoctadeca-9,11-dienoic acids
[362,961], and 9- and 13-oxooctadecadienoic acids [961]
have been identified in human aortal atheromatous plaques.
The amounts of these oxidized cholesterol esters are considerable, ranging as high as *ca.* 100 μg/g lipid for hydroperoxyacyl esters [362] and 76 mg/g lipid for the hydroxyacyl esters [962] in severely atherosclerotic human aortas!

Although the origins of these oxidized cholesterol
esters are obscure, only 6% of the hydroperoxy esters found
in *post mortem* material could be artifact of manipulation.
The prospect that these hydroperoxides arise *post mortem* in
the period between death and isolation as result of failing
enzyme protection systems (peroxidases) against oxidation

has been advanced [961].

In further speculation, as shown in FIGURE 6,the postulated ready free radical oxidation of cholesterol 3β-lino-
leate (234) yields the cholesterol 3β-linoleate 11'-carbon
radical 235 whose reaction with O$_2$ putatively gives cholesterol 3β-9'-peroxyoctadeca-E-10,Z-12-dienoate (236) which
in turn gives the corresponding hydroperoxide cholesterol
3β-9'-hydroperoxyoctadeca-E-10,Z-12-dienoate (237). The
isomeric cholesterol 3β-13'-peroxyoctadeca-Z-9,E-11-dienoate
and 3β-13'-hydroperoxyoctadeca-Z-9,E-11-dienoate esters
would also be generated but are not shown in FIGURE 6 for
simplicity. Subsequent decomposition of the 9'-hydroperox-
ide 237 to the corresponding 9'-alcohol cholesterol 3β-9'-
hydroxyoctadeca-E-10,Z-12-dienoate (240) and 9'-ketone
cholesterol 3β-9'-oxooctadeca-E-10,Z-12-dienoate would then
account for the presence of hydroperoxyacyl, hydroxyacyl,
and oxoacyl esters of cholesterol found in human aortal
tissues [362,365,369,961,962].

Yet other modes of transformation of the 11'-carbon
radical 235 may occur. The simple isomerization of the
11'-carbon radical 235 to the cholesterol 3β-linoleate
7-carbon radical 238 has been suggested [1777], but intermolecular radical propagation processes are also possible.
As shown in FIGURE 6, hydrogen atom abstraction from 234
by the 11-carbon radical 235 would yield the 7-radical 238.
The putative ester 7-radical 238 would then be precursor of
3β-linoleate esters of the cholesterol 7-hydroperoxides 46
and 47 from which 3β-linoleate esters of the epimeric
3β,7-diols 14 and 15 and 7-ketone 16 could derive.

The scheme of FIGURE 6 also includes two other radical
propagation reactions potentially generating the 7-radical
238. The 9'-peroxyl radical 236 formed by reaction of the
11'-radical 235 with O$_2$ in concept is capable of abstract-
ing hydrogen from substrate 234 with formation of the stable
9'-hydroperoxide 237 and the 7-radical 238. Furthermore,
as posed in this section dealing with peroxide bond homolysis,
homolysis of the peroxide bond of the 9'-hydroperoxide 237
would provide the 9'-oxyl radical cholesterol 3β-9'-oxy-
octadeca-E-10,Z-12-dienoate (239) that could also abstract
hydrogen from substrate 234 to yield the 7-radical 238 and
the 9'-alcohol cholesterol 3β-9'-hydroxyoctadeca-E-10,Z-12-
dienoate (240). By concept, it is the 9'-oxyl radical that
should be highly active in furthering this autoxidation

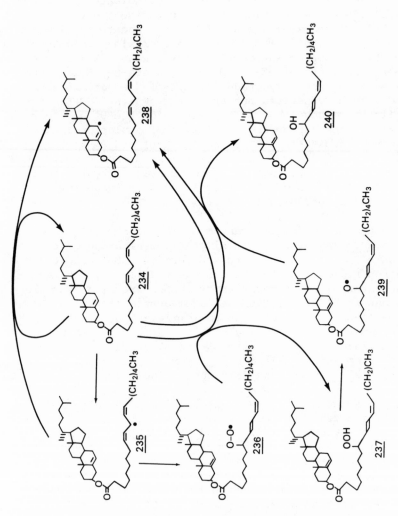

FIGURE 6. Putative cholesterol 3β-linoleate oxidations

sequence. Necessarily, companion processes involving the
13'-isomers of these 9'-oxidized derivatives may also be
postulated.

Yet other speculative means of initiating autoxidation
of cholesterol at the C-7 position via autoxidized cholesterol esters such as the 9'-hydroperoxide 239 have been
suggested [338]. As these sterol esters retain the long
flexible C_{18}-fatty acid linear carbon chain, there is a
possibility that intramolecular oxidations occur, as outlined in FIGURE 7. The indicated peroxide bond homolysis
and C-7 hydrogen atom abstraction and transfer would then
yield as intermediate the cholesteryl 3β-9'-hydroxyoctadeca-
E-10,Z-12-dienoate 7-carbon radical 240 whose subsequent
reaction with $^3O_2$ and hydrogen atom abstraction, etc. would
yield as putative products hydroxyacyl esters of the 7-hydro-
peroxides 46 and 47. Understandably, no experimental means
of demonstration of these radical processes have been
applied to the issue. Rather, these several speculations
provide bases for the initiation of cholesterol autoxidation
in tissues via the prior autoxidation of the more sensitive
polyunsaturated fatty acyl esters and for rationalizing the
demonstrated presence of oxidized cholesterol esters in
tissues [338,343,362,365,369,961,962].

These several items of evidence suggest that the polyunsaturated fatty acid esters of cholesterol be a likely
source of organic peroxides acting as initiators of cholesterol autoxidation. Moreover, the initiation of autoxidation of cholesterol in films of corn oil or of corn oil
fatty acids has been demonstrated [1776], such evidence
clearly supporting the role of unsaturated fatty acyl esters
in the initiation of cholesterol autoxidation in tissues.

Indeed, relatively few controlled observations of the
initiation of cholesterol autoxidation by catalysis provided
by any peroxides appear to have been made. The oxidation
of cholesterol by benzoyl peroxide under the conditions
described by Lifschütz [1481,1506] reexamined by us using
thin-layer chromatography [2303] is the classic example of
the induction of autoxidation by peroxides, although the
products formed have not been satisfactorily identified nor
have the multistep processes which are bound to be involved
been elucidated.

Our own studies of the reactions between cholesterol

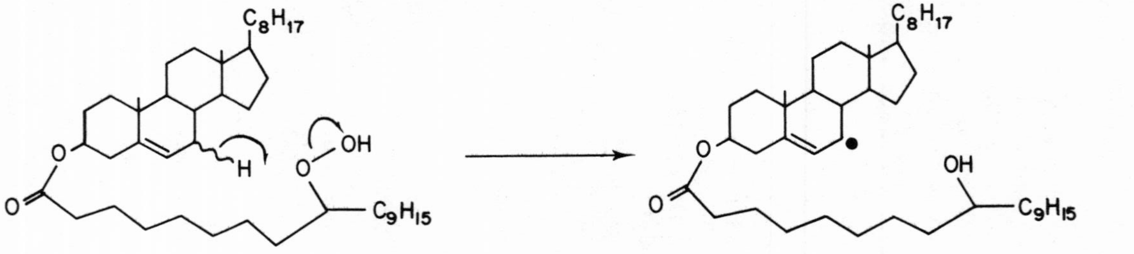

237 240

FIGURE 7. Putative scheme for intramolecular autoxidation of oxidized cholesterol fatty acyl esters.

and organic hydroperoxides in aqueous dispersions clearly
also involve formation of the common autoxidation products
14-16 *inter alia* [2297]. In those experiments involving
the epimeric 7-hydroperoxides 46 and 47 with cholesterol
the presence of the 3β,7-diols 14 and 15 and the 7-ketone
16 among products does not necessarily support an induction
of autoxidation, for the 14-16 are thermal decomposition
products of the 7-hydroperoxides 46 and 47 introduced into
the system. However, the attack of cumene hydroperoxide
on cholesterol yielding 14-16 *inter alia* must be viewed as
a instance of the induction of cholesterol autoxidation by
radicals derived from cumene hydroperoxide. Moreover, since
these experiments were conducted under N_2, one must postulate that the O_2 required for autoxidation be derived from
disproportionation of cumene peroxyl radicals putatively
generated, according to Eqs. 6 or 7. However, as no sterol
hydroperoxides were detected in these experiments, the oxidation products 14-16 may derive by other processes, for
instance by attack of hydroxyl radical (HO·) on cholesterol,
a process also recognized as yielding 13,35 and 36 as well
as 14-16, discussed later in this chapter.

    A related case obtains in our studies of the reactions
between cholesterol and H_2O_2 in aqueous dispersions. Here
also the products 13-16,35,36, *inter alia* were isolated,
and as the system was conducted under N_2, again O_2 required
must have been derived from the disproportionation of H_2O_2.
In this case the requisite epimeric 7-hydroperoxides 46 and
47 establishing autoxidation by O_2 were isolated [2299].
Other related experiments using other systems incorporating
H_2O_2 with Fe (II) salts [1303] and with Ti (III) salts [1304]
also yields the 3β,7-diols 14 and 15. The indicated autoxidations are clearly free radical induced, but whether peroxides, transition metal ions, or other active radical
species induce cholesterol autoxidation is indeterminate.

    One of the simplest yet most impressive reports of the
induction of cholesterol autoxidation by peroxides is that
of the use of diethyl ether not freed from its peroxides,
used in the isolation of cholesterol. Here a simple thin-
layer chromatogram clearly showed the several polar autoxidation products, although the individual products were not
identified [1067].

    The special case of enzyme catalyzed hydroperoxide
formation with polyunsaturated fatty acid esters, chole-

sterol, etc. will be discussed later in Chapter VII.

Whereas the autoxidation of cholesterol in tissue may be rationalized as being initiated by the catalytic effects of lipid peroxides, this explanation cannot hold for the autoxidation of pure cholesterol isolated from tissues. Rather, cholesterol itself and its congeners likely to be present must be viewed as the substrate source for any organic peroxide formation.

Cholesterol is known to be accompanied by the related sterols 5α-cholest-7-en-3β-ol (57) [739,740,746,1712] and cholesta-5,7-dien-3β-ol (56) in many tissues, and both of these sterols are more susceptible to oxidations than is cholesterol. The 5,7-diene 56 is notoriously unstable in air and is much more rapidly discolored and decomposed on storage than is cholesterol [333,1439]. The ease with which the 5,7-diene 56 reacts with $^3O_2$, though recognized, has not been the subject of systematic study, but we may presume that the differential oxidation of the 5,7-diene 56 present as a minor impurity in cholesterol isolated commercially or in individual cases of recovery from tissues may lead to many autoxidized products, some of which may yield free radical initiators by peroxide bond homolysis.

The ease with which the 5,7-diene 56 adds O₂ to give the 5α,8α-epoxide 60 is of special note, for this transformation appears to occur in Nature, cf. Chapter III, as well as under controlled laboratory conditions. However, the 5α,8α-peroxide 60 has not been examined for its ease of peroxide bond homolysis. In that no primary reaction product between molecular oxygen and the 5,7-diene 56 other than the 5α,8α-peroxide 60 has been found, we must assume that the 5α,8α-peroxide 60 is in fact the initially formed product and that subsequent transformations of the 5,7-diene 56 to the myriad of decomposition products [1439] follows from homolysis of the 5α,8α-peroxide bond and from other related one-electron transfers.

Other biosynthesis precursors linking cholesterol with lanosterol (67) do not generally occur in tissue cholesterol samples to the same extent as the congeners 56 and 57 previously mentioned. However, both lanosterol and 5α-lanost-8-en-3β-ol (68) are present in elevated amounts in some tissues, and both of these sterols are readily autoxidized, (cf. Chapter VI). The 5,24-dien-3β-ol desmo-

sterol (78) also found at low levels in cholesterol is
notoriously unstable to shelf storage in air [2403,2480]
and accordingly appears to be more susceptible to air oxidation than cholesterol. Systematic studies involving 67,
68, or 78 as initiators of cholesterol oxidation have not
been made.

Other sterols present in tissue cholesterol include
the 5α-stanol 2, the 5β-stanol 4, the (24S)-3β,24-diol 25,
the 3β,26-diol 29, as well as esters of these and of the
congeners 56,57,67,68, and 78. Any involvement of these
sterols as initiators of cholesterol autoxidation would be
purely speculative.

Even were these unsaturated sterols reasonable prospects for an initial, selective reaction with oxygen, with
subsequent peroxide bond homolysis leading the cholesterol
autoxidation, this explanation is unlikely for highly purified cholesterol freed from these congeners by purification
via the dibromide with repeated recrystallizations or
chromatography. Nonetheless, very highly purified cholesterol is as sensitive to air oxidation, perhaps more so,
than is ordinary USP cholesterol (cf. Chapter IX). Very
highly purified cholesterol, processed through the Windaus
dibromide purification with recrystallization and debrominated using Zn rapidly autoxidized within a matter of
months to a complex mixture of odorous autoxidation products [2303]. As this preparation was free of demonstrable
amounts of the congeners 56,57,67,68,78, etc. these could
not serve as initiators. In this regard it is pertinent
to recall that cholesterol purified via the dibromide but
using the Schönheimer debromination method which utilizes
NaI instead of Zn is remarkably stable to storage in contact with air for years [684].

It is possible that the initiation of autoxidation of
USP cholesterol be by one agency and of autoxidation of
highly purified cholesterol by another, such as residual
amounts of transition metal (Zn) ions discussed in the
next section. It must be obvious that by postulating
autoxidation initiation by the oxidation of a more highly
autoxidizable congener present at trace levels one still
has not explained the overall process satisfactorily, for
there remains to be derived a fundamental understanding of
the reaction between unexcited ground state substrate and
$^3O_2$ of the air.

Transition Element Catalysis

The transition elements recognized as effective catalysts exist in two or more valence states which allow the element to donate or accept a single electron, thereby aiding free radical processes. The metal ion catalyzed homolysis of organic peroxides of Eqs. 11 and 12 is the most obvious way by which metal catalysis may initiate free radical autoxidations. Whether transition metal ions may enter into reactions involving bond scissions of unoxidized substrate molecules, that is, catalyze the reactions of Eqs. 1,8,9, or 10 cannot be now answered.

No recent systematic studies of transition metal ion catalysis of cholesterol autoxidation have been reported, and much of the older information allows only generalized conclusions regarding metal ion catalysis. The transition metal most obviously implicated in these processes in tissues is iron, which is present ubiquitously in the hemoglobin of blood, in myoglobin, and in the cytochromes, etc. The close association between blood solids and autoxidized cholesterol described by Lifschütz from 1907 [1476] and thereafter [1479,1481,1495,1496,1502], including the fairly clear case of catalysis of cholesterol oxidation by dried powdered blood solids [1488] and work of others [247,1412], implicated heme Fe as catalytic agent.

Only a few observations of the effects of Fe (II) and Cu (II) salts on the autoxidation of cholesterol have been recorded. In the model system for cholesterol autoxidation involving aqueous sodium stearate dispersions the combined effects of time, temperature, pH, and transition metal ion catalysis have to be considered. This model system is characterized by marked temperature effects, with slow autoxidation at 37°C but more extensive autoxidation at 50°C and 85°C [197,2303], and by marked pH effects. Whereas pH effects in the range of 6.5-11.5 did not affect the autoxidations much [197], autoxidation did not occur over the range pH 2.0-7.0, with or without transition metal ion catalysis [2303].

Catalysis by Cu (II) ions in this model system was demonstrated by Bergström and Wintersteiner. Using specially purified materials for making the aqueous dispersions there was no autoxidation of cholesterol observed at 37° but 1-100 µM $CuCl_2$ markedly stimulated cholesterol autoxidation.

However, 10 μM FeSO$_4$, FeCl$_3$, MnCl$_2$, or NiCl$_2$ had no such
effect. Moreover, addition of NaCN, diethyldithiocarbami-
nate, ethylenediamine, or α-nitroso-β-naphthol, acting as
complexation agents for heavy metals inhibited the autoxidations. Bergstrøm viewed Cu (II) ions as necessary for
cholesterol autoxidation in the model system [197]. We
have confirmed the catalysis of CuCl$_2$ additions on the
autoxidation of cholesterol using densitometric measurements from thin-layer chromatograms [2303].

A characteristic feature of cholesterol autoxidation
in aqueous sodium stearate dispersions is the variable induction period during which no autoxidation is observed.
One presumes that during this period the necessary initiation reactions occur which lead subsequently to generalized
autoxidation of larger amounts of cholesterol. Here catalysis by Cu (II) ions does not alter the length of the induction period [197,2303] but stimulates the autoxidation rate
following the same length of time induction period as
occurs without catalysis, cf. FIGURE 8. These results are
consistent with the concept that transition metal ions catalyze peroxide bond homolysis which stimulates more extensive
cholesterol autoxidations but do not catalyze bond homolysis
in the substrate cholesterol (Eq. 1) or reactions between
substrate and O$_2$ (Eqs. 8).

Cholesterol oxidation in aqueous sodium stearate dispersions is characterized by an induction period of about
10 hrs at 37°C but of about 15 m at 85°C [197]. Our own
work using rigorously purified cholesterol suggested an
induction period of at least 1 h at 85°C [2303], not altered by Cu (II) ions. Almost identical 1 h induction
periods at 85°C with very pure cholesterol have been reported
by others [1297]. A poorly defined but much shorter induction period of no more than 20 m was observed at 85°C with
an air-aged commercial cholesterol sample taken directly
from the bottle, cf. FIGURE 8. Here Cu (II) ion catalysis
was not marked, and one may conclude that sufficiently high
levels of performed peroxides are adequate to initiate
autoxidations without specific transition metal ion catalysis being necessary.

Another aspect of these cholesterol autoxidation in
the aqueous sodium stearate dispersion model system is that
of saturation. Reactions do not proceed to completion but

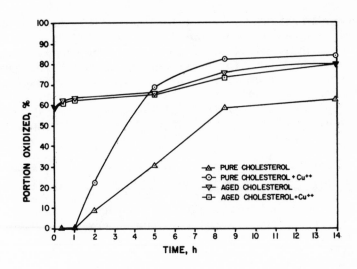

FIGURE 8.   Cholesterol autoxidation rates in aqueous
            sodium stearate dispersions at 85°C.   Re-
            printed with permission of Elsevier Scien-
            tific Publishing Co., Amsterdam, from
            J. Chromatog., 27,187(1967).

appear to level off, possibly dependent upon product inhibition of further autoxidation. As this phenomenon is
observed in this model system but not necessarily in others
or in air-aged crystalline cholesterol where total destruction of the sample occurs given enough time, a special condition must obtain in these dispersions. Interference with
oxygen uptake by the autoxidation products formed on the
surface of the sterol dispersed phase was proposed by Bergström [197], and as we now know that the common cholesterol
autoxidation products 14-16 have greatly different solubi-
lizing and dispersing properties than cholesterol, this may
serve to rationalize the saturation phenomenon observed.

      Systematic study of the catalysis effects of transition
metals or their salts on the autoxidation of crystalline
cholesterol in air or in other model or natural conditions
appear not to have been recorded. For example, although
heme Fe catalysis of cholesterol autoxidation in *post mortem*
tissues (barring microbial action) seems likely, controlled
observations on the matter have not been made. However,
were overt heme Fe (or other transition metal) catalysis of

autoxidation in such tissues prominent, one might expect
to observe the concommitant autoxidation of cholesterol
fatty acyl esters of blood and tissues. Although esterification protects cholesterol from autoxidation in the aqueous
stearate dispersions model system [197] and sample of cholesterol fatty acyl esters are not noticeably autoxidized by
shelf storage, cholesterol esters can be autoxidized by
heating [2455], by Fe-containing systems such as the
Fenton reagent and possibly by the microsomal NADPH-dependent lipid peroxidation system of liver.

However, very few observations of the recovery of
autoxidized cholesterol fatty acyl esters from tissues have
been reported. Indeed, as outlined in Chapter II, almost
all observations of autoxidized cholesterol derivatives
from tissues involve free sterols and not esters. In a few
cases, fatty acyl esters of the epimeric 3β,7-diols 14 and
15 have been detected under conditions which allow their
formation *in vivo* (i.e., during life) or possibly by catalyzed *post mortem* autoxidations in which ordinary air oxidations due to careless manipulations and ignorance are
excluded. Thus, the sterol fatty acyl ester fraction from
human aortal tissues [369,493,1404] and from other tissues
[338,343] have yielded the 3β,7-diols 14 and 15 upon saponification under conditions carefully designed to exclude
air. Moreover, in very recent unpublished work from my
laboratory, in an extensive reexamination of the sterols of
human plasma we have once more encountered evidence of the
presence of fatty acyl esters of the 3β,7-diols 14 and 15
from 7-ketone 16. This matter demands definitive examination not yet accomplished.

Furthermore, whereas traces of heme Fe might serve as
catalyst in tissue autoxidations, the autoxidation of pure
cholesterol free from tissue components is not so readily
rationalized. Nonetheless, there have been no studies reported on the trace metal composition of pure cholesterol
from different sources and processes. Indeed, no heavy
metals limit for USP cholesterol has been set up by the
United States Pharmacoepaeia [2535], and we have no information whatsoever regarding the levels of catalytic metals
which may be present in processed cholesterol.

Although Fe (II) and Cu (II) ions may be the most
likely transition metal ions putatively catalyzing cholesterol autoxidations other transition metals and metal

ions may be implicated in special cases. In that Zn is used in the classic Windaus debromination procedure for purification of cholesterol via the 5,6-dibromide [741], the presence of Zn in highly purified cholesterol samples is assured, even though the resultant levels may be quite low. However, the catalytic effects of Zn or of its salts on cholesterol autoxidation has not been studied.

Furthermore, although catalysis by trace levels of transition elements is a most obvious suggestion for initiation of autoxidations, other one-electron transfer agents could also be involved. Thus, elemental Br_2 used in the aforementioned Windaus bromination-debromination procedure for cholesterol purification is also an oxidant which may oxidize cholesterol to the corresponding 3-ketone cholest-4-en-3-one (8), 3β,7-diols 14 and 15, and other products.

As the oxidation potential of elemental I_2 (+0.4 V) is comparable to that of O_2, I_2 may also act as an oxidant or as an initiator of autoxidation of cholesterol. One-electron oxidations of cholesterol in which I_2 is implicated as an initiator have been described [2680]. These several modes of reaction with I_2 include formation of substrate-I_2 complexes which may be reversible, as the detection of cholesterol and other unsaturated sterols and lipids by I_2 vapors is an established method of visualization of such compounds on chromatograms [420,992,1726,1728]. Whereas this method is viewed as nondestructive and reversible [2363], insidious losses from radical cation formation and subsequent dimerizations, dehydrogenations, and reactions with O_2 as described here may also occur without recognition.

The oxidation of cholesterol by I_2 is viewed as involving formation of a radical cation as suggested in FIGURE 9 [2680], from which two degraded, partially dehydrogenated products suggested as substituted cyclopentanophenanthrene and picene derivatives 241 and 242 were formed. Other products included a partially degraded dimeric tetraene $C_{43}H_{68}$ and trimer $C_{69}H_{110}$, inter alia. The cyclopentanophenanthrene 241 was also obtained by Se dehydrogenation of cholesterol [2680].

The cholesterol radical cation implicated in cholesterol oxidations by I_2 was posited on mechanism requirements, but indirect experimental evidence suggests that

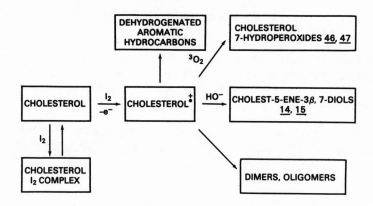

FIGURE 9.   Putative oxidation reactions of
cholesterol initiated by I_2.

241   242

such a radical cation may indeed form in other systems.
Photolysis of chloroform solutions of cholesterol and hematoporphyrin give electron spin resonance spectra dominated
by a single-line signal of the hematoporphyrin radical
anion formed by a one-electron transfer process involving
cholesterol as electron donor. The resultant cholesterol
radical cation which must have been formed was not observed
but escaped detection, possibly from very fast recombination
reactions [439].

Although no cholesterol autoxidation products were
sought in the report of I_2 catalyzed oxidation of cholesterol, the formulation of reaction between the cholesterol
radical cation of FIGURE 9 and O_2 was proposed [2680]. We
have examined the action of I_2 vapors on cholesterol adsorbed

on Silica Gel HF_{254} chromatoplates in air in hopes of detecting sterol hydroperoxides possibly formed thereby. However, the N,N-dimethyl-*p*-phenylenediamine reagent [2295] so useful for such purposes in other cases failed in this, for residual I_2 grossly interferred, causing the one-electron reduction of the reagent to Wurster's Red all over the chromatogram.

Other reactions of cholesterol effected by elemental I_2 are known, for instance condensation of cholesterol with itself to dicholesteryl ether (18) by I_2 is recorded [976].

Excited Oxygen Species

Yet another proposition may be examined as a means of deriving sterol peroxides or other free radical sources for initiation of cholesterol autoxidation. This process would be by the reaction of excited oxygen species nominally present in air or of other oxidants such as nitric oxide (NO) and nitrogen dioxide (NO_2), both being stable free radicals, peroxyacetyl nitrate ($CH_3COOONO_2$), etc. possibly present adventitiously in air but known to be present in some polluted air supplies. Any of these agents even though present in but very low levels in air might serve as an initiator of cholesterol autoxidation. Experimental evidence for such action is available in only a few cases, but in any event, the proposed initiation proceeds by free radical processes. The radical species implicated may include the activated oxidant itself if it be a radical or free radicals induced within the substrate molecule, or from homolysis of unstable peroxides formed.

The excited or activated oxygen species implicated in the initiation of cholesterol autoxidation, indeed of other autoxidations, may be monoatomic, diatomic, or triatomic. Experimental studies of higher, polymeric oxygen species related to cholesterol autoxidation are nonexistent.

Ozone. The most likely excited oxygen species of air which may initiate free radical reactions and cholesterol autoxidation is the triatomic species ozone (O_3), well known for its oxidizing power and for initiating autoxidations. Ozone is also present in air at relatively high levels, so that its reaction with cholesterol or other susceptible

substrate by projection could be involved in initiation of
generalized cholesterol autoxidation. This point has never
been proven however.

The reaction between O_3 and cholesterol in solution
has been studied since 1905 [1660], but early interests
were in the stoichiometry of the reaction. Depending on
conditions, more than one equivalent of O_3 reacts with
cholesterol [610,628-630,1426,1660,2602], but reaction in hexane solution occurs with but one equivalent of O_3 to yield
an isolable ozonide $C_{27}H_{46}O\cdot O_3$, demonstrated in the early
1908-1912 period [629,964] and more recently confirmed
[187,188,524]. The monoozonide is presumed to have the
structure 5,7ξ-epidioxy-5ξ-B-homo-6-oxacholestan-3β-ol
which may exist in both theoretical configurations 5,7α-
epidioxy-5α-B-homo-6-oxacholestan-3β-ol (243) and 5,7β-
epidioxy-5β-B-homo-6-oxacholestan-3β-ol (244) [187,1809].
Both isomers are postulated as possibly accounting for some
variations in physical properties of the ozonide, which is
also associated in solution [188].

If ozonization is conducted on cholesterol 3β-acetate
in polar solvents (alcohols), peroxyacetals such as 3β-
acetoxy-6ξ-alkoxy-5,6ξ-epidioxy-5ξ-5,6-secocholestan-5-ol
(245) are obtained. [1461,1462].

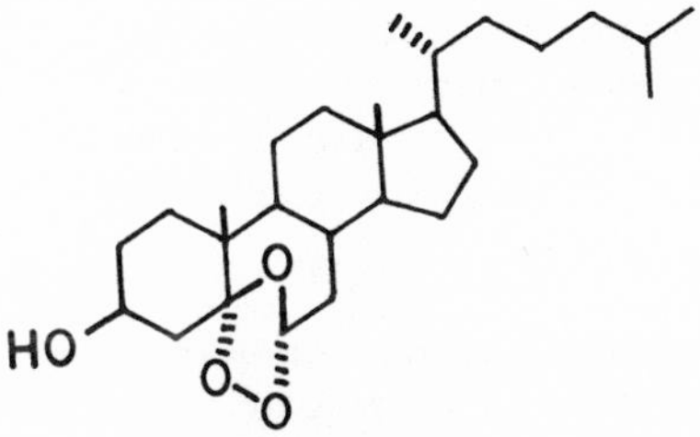

243 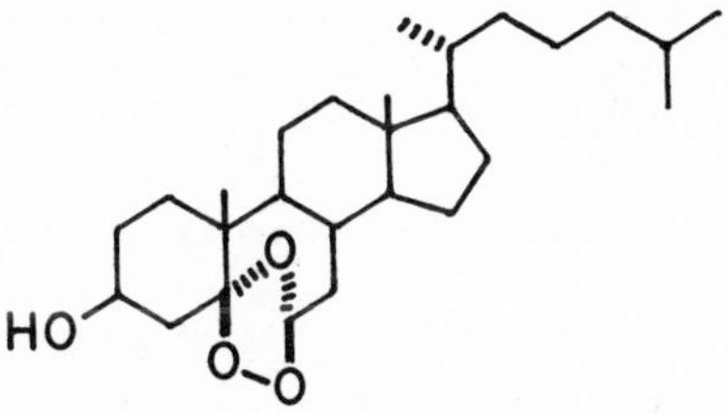   244

Reductions of the ozonides 243 and/or 244 with Zn and
acetic acid yields the secoaldehyde 3β-hydroxy-5-oxo-5,6-
secocholestan-6-al (246), whereas oxidative scission of
the ozonides with H_2O_2 yields the corresponding secoacid
3β-hydroxy-5-oxo-5,6-secocholestan-6-oic acid (103) [524].
Reduction of the peroxyacetals 245 with LiAlH₄ yields the
corresponding secoalcohol 5α-5,6-secocholestane-3β,5,6-triol
[1462].

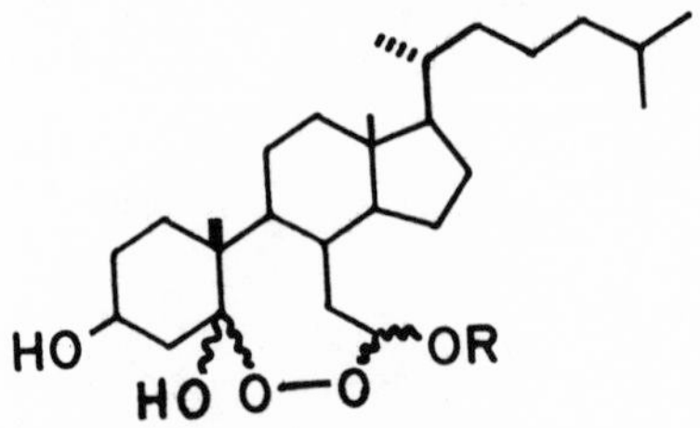

245 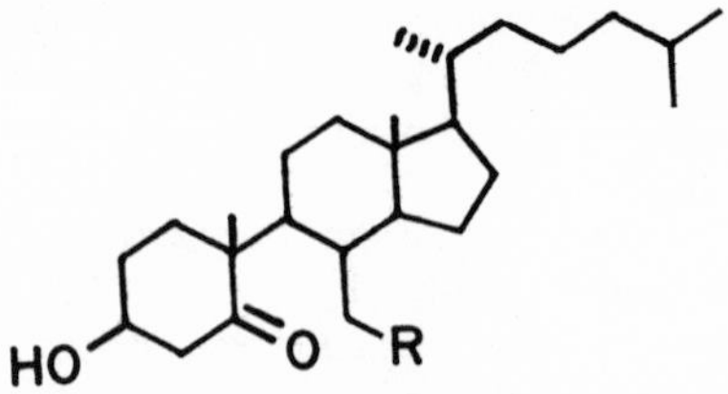   246 R = CHO

                                        103 R = COOH

Although no proper study of the action of O_3 on cholesterol in aqueous systems appears to have been reported, we
have recently demonstrated that cholesterol is rapidly con-
sumed in such systems with the formation of two major oxidized (unidentified) products. Nonetheless it has been
reported that cholesterol ozonides decompose in water, possibly liberating CO_2 and H_2O_2 [610,629].

We may project that O_3 will oxidize cholesterol in
aqueous systems and in tissues, possibly via a putative
peroxyhemiacetal analogous to the peroxyacetal 245 with
R = H.   Decomposition of such a putative peroxyhemiacetal
to the secoaldehyde 246 might then be expected.   In each
of these cases, the likelihood of decomposition of the
ozonides 243-245, etc. by peroxide bond homolysis or by
other free radical generating processes poses the chance
that general cholesterol autoxidation could be initiated.
This possibility does not seem very likely for pure cholesterol stored in closed bottles.

Molecular Oxygen. Whereas autoxidation is generally
considered to involve the diatomic molecular oxygen O_2 of
the air, other dioxygen species may be variously implicated
in the initiation of autoxidation. The nomenclature of
Vaska [2586] will be used in describing the several species
of dioxygen, whereby the term dioxygen be the generic de-
signation for diatomic oxygen (O_2) which comprises all
states and forms of O_2 in which there is an oxygen-oxygen
covalent bond, without regard to whether the dioxygen is
free, part of another compound, or carries an electronic
charge. Dioxygen species include the dioxygen cation (O_2^+),
ground state (triplet) molecular oxygen ($^3O_2$), electronically

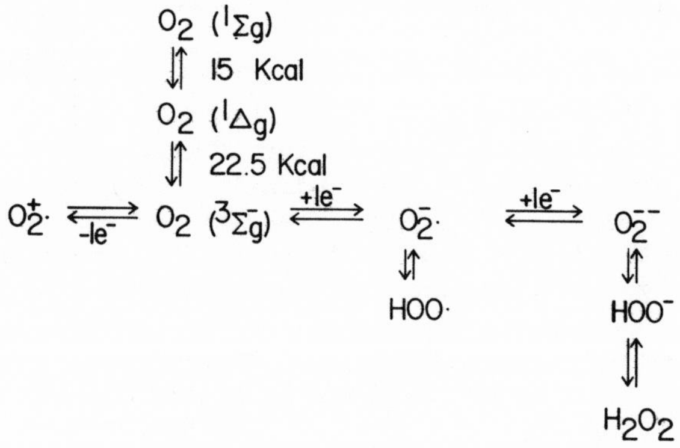

$$O_2 \ (^1\Sigma g)$$
$$\updownarrow \ 15 \ \text{Kcal}$$
$$O_2 \ (^1\Delta g)$$
$$\updownarrow \ 22.5 \ \text{Kcal}$$

$$O_2^{+}\cdot \ \underset{-1e^-}{\overset{}{\rightleftharpoons}} \ O_2 \ (^3\Sigma g) \ \overset{+1e^-}{\rightleftharpoons} \ O_2^{-}\cdot \ \overset{+1e^-}{\rightleftharpoons} \ O_2^{--}$$

$$\updownarrow \qquad\qquad \updownarrow$$

$$HOO\cdot \qquad\quad HOO^-$$

$$\updownarrow$$

$$H_2O_2$$

FIGURE 10.   Dioxygen species.

excited (singlet) molecular oxygen ($^1O_2$ in either $^1\Sigma g$ or $^1\Delta g$ states), one-electron reduced species such as the superoxide radical anion (O_2^-) or conjugate hydroperoxyl radical (HOO$\cdot$), or two-electron reduced dioxygen species such as the peroxide anion ($O_2^=$) or hydroperoxyl anion (HOO$^-$) or hydrogen peroxide (HOOH). The oxygen-oxygen covalent bond may be present in neutral or ionized species, in other covalent bonding as in peroxides and hydroperoxides, or as the free dioxygen species. The several dioxygen species are outlined in FIGURE 10.

Dioxygen Cation. The dioxygen cation O_2^+ is not likely to be implicated as an oxidant in biological systems where O_2 serves as an electron sink and not as electron donor. The high ionization potential (+12.1 eV) precludes formation of O_2^+ in all but the most energetic processes. Nonetheless, certain salts O_2PtF_6 of O_2^+ have been demonstrated [145], and O_2^+ may be formed under other conditions. The reaction of O_2^+ produced by electric discharge (together with mono-atomic oxygen species) with cholesterol has been recently demonstrated to yield the several common cholesterol autoxidation products, including the epimeric 3β,7-diols 14 and 15, the 7-ketone 16, and the isomeric 5,6-epoxides 35 and/or 36. No sterol hydroperoxides were detected, but the high ionization energy for O_2^+ exceeds that nominally expected to destroy peroxide bonds, and no direct evidence of the stability of the cholesterol hydroperoxides 46,47, and 51 in the system was provided [2071].

Calculations suggested that O_2^+ was a predominant species in these experiments but that the excited neutral dioxygen O_2 and monoatomic species O^+ and atomic oxygen were also formed [2071]. Accordingly, there is some uncertainty as to which oxygen species participated in these oxidations of cholesterol, and mechanisms involving hydroperoxides or not involving hydroperoxides may be involved.

By contrast small amounts of cholesterol hydroperoxides 46,47 or 51 were found together with the 3β,7-diols 14 and 15, the 7-ketone 16, and the 5,6-epoxides 35 and/or 36 when CO^+ (produced from CO) was used as the ionizing species instead of O_2^+. As a source of dioxygen is inferred in the formation of sterol hydroperoxides, these results suggest that ionized gases may initiate cholesterol autoxidation even with very low levels of dioxygen adsorbed onto surfaces of the equipment in this instance. The 3β-formate ester of cholesterol was also formed in experiments using CO^+ [2071,2072].

Ground State Dioxygen. The autoxidation of cholesterol by the ordinary molecular oxygen of the air is viewed as involving unexcited dioxygen in its ground state. It is necessary to give consideration to the electronic structure of molecular oxygen at this point in order to provide adequate understanding of the role of ground state $^3O_2$ and electronically excited $^1O_2$ as either may relate to cholesterol autoxidation. Only a simplified and superficial treatment will be attempted here [86,655,2067,2068].

As $^3O_2$ is paramagnetic, its electronic structure must account for this property. Dioxygen has sixteen extranuclear electrons of which four (two per atom) are of the $(1s)^2$ configuration of the K shells and may be disregarded in the present description. The remaining twelve electrons of the next L shell may participate in chemical bonding and electronic excitation and must give rise to the paramagnetism of ground state dioxygen $^3O_2$ by some means of unpairing of two electrons. The most simple representation ·O-O· of ground state dioxygen reflects its biradical nature, as do the simplified Lewis [1466] structures [:Ö::Ö], [Ö::Ö:], and [:Ö:Ö:] which account for all L shell electrons as single or double bonds and as electron pairs. Electronically excited, metastable dioxygen species which

are diamagnetic then may be depicted as [:Ö::Ö:], [:Ö̤:Ö:],
and [:Ö:Ö̤:] with all electrons paired.

Another means of accounting for the paramagnetism of
ground state dioxygen is in a three-electron bond structure
[:O⦂⦂⦂O:] proposed by Pauling [1827,1828]. This structure
involves one two-electron bond and two weaker three-electron bonds. This concept appears to be no longer useful in
the description of the electronic structure of dioxygen
expecially so in the face of molecular orbital descriptions
which may have become widely used.

The expression of Eq. 14 gives the molecular orbital
description of dioxygen as derived from two oxygen atoms of
the configuration $(1s^2 2s^2 2p^4)$. Antibonding character is
indicated by the asterisk. This description achieves minimum electrostatic repulsion between the oxygen nuclei and

$$2O(1s^2 2s^2 2p^4) \rightarrow O \; [KK(2s\sigma_g)^2 (2s\sigma_u^*)$$

$$(2p\sigma_g)^2 (2p\pi_u)^4 (2p\pi_g^*)^1 (2p\pi_g^*)^1] \quad \text{Eq. 14}$$

provides ready means of accounting for observed chemical
properties of dioxygen. Interaction of the two $2s\sigma$ orbital
electrons is cancelled by that of the antibonding $2s\sigma^*$ electrons, thus giving four $2s\sigma$ electrons participating in non-
bonding interaction. The two electrons of the $2p\sigma$ orbitals
are engaged in σ-bonding. Furthermore, the four electrons
in the $2p\pi$ bonding orbitals and two electrons in the $2p\pi^*$
antibonding orbitals balance to give two $2p\pi$ electrons
engaged in π-bonding. This molecular orbital description
of ground state dioxygen thus provides for one σ-bond,
one π-bond, and two outer orbital $2p\pi^*$ antibonding electrons
which have the same energies, i.e., are degenerate. It is
to these $2p\pi^*$ electrons that the paramagnetism and biradical
chemical nature of ground state dioxygen be attributed.
Moreover, the electronic states of dioxygen are determined
by the population pattern of this $2p\pi^*$ pair of degenerate
orbitals.

Of the L shell electrons the two outer $2p\pi^*$ electrons
are of highest energy, and these may occupy the same or
different orbitals, thus giving rise to the different
electronically excited states of dioxygen. With both $2p\pi^*$
electrons in the same orbital, the orbital angular momentum
of one must be in the same direction as the other, giving a

Δ state, and their spins must be in opposite directions, thus paired. With these electrons in separate orbitals, their angular momentums are opposed, giving a Σ state, and their spins may be opposite (antiparallel), thus paired (in a singlet state, expressed by the multiplicity term as superscript), or parallel, thus unpaired (triplet multiplicity). These three $2p\pi^*$ orbital electron configurations exhaust possibilities and give rise to the three dioxygen species of TABLE 7. The dioxygen species of lowest energy is that of highest multiplicity. Thus the $^3\Sigma_g^-$ state is the ground state, where the g subscript (for "gerade") denotes an unchanged wave function for the electronic state under an inversion operation (versus subscript u for "ungerade" denoting change of sign only upon wave function inversion). In this monograph the abbreviated term $^3O_2$ represents the $^3\Sigma_g^-$ ground state of dioxygen.

TABLE 7. Dioxygen Electronic States

Dioxygen Species	Spectral Term	Outer Orbital Occupancy	Energy Above Ground State (kcal/mole)
Ground State	$^3\Sigma_g^-$	↑ ↑	–
1st Excited state	$^1\Delta_g$	↑↓ ___	22.5 (0.97 eV)
2nd Excited state	$^1\Sigma_g^+$	↑ ↓	37.5 (1.62 eV)

The two next low energy dioxygen states are the first excited state $^1\Delta$ only 22.5 kcal/mole above ground state and the second excited state $^1\Sigma_g^+$ 15 kcal/mole above the first. Both of these electronically excited dioxygen species are designated as singlet molecular oxygen, abbreviated herein $^1O_2$, although only the $^1\Delta$ state is likely to be encountered in chemical reactions of cholesterol, for the lifetime of the $^1\Sigma_g^+$ state is very short (7.1 s versus 45 m for the $^1\Delta$ state) [1251,1268].

Yet other dioxygen and oxygen monoatomic oxygen states may be considered advantageously here. Beyond the first and second excited dioxygen states are the more highly excited $^3\Sigma_u^+$ and $^3\Sigma_u^-$ states of no particular relevance to the oxidation of cholesterol. Vibrational energy imparted to unexcited $^3O_2$ does not alter the electronic configuration of the antibonding electrons but yields excited $^3O_2^*$

which retains its paramagnetic properties, two unpaired electrons, and biradical chemistry. However, given sufficient vibrational energy, the internuclear distances of the several dioxygen species increase so as to lead to dissociation of the molecule into oxygen atoms.

Like the diatomic oxygen molecule, the oxygen atom has three low-lying energy states of some concern. The ^3P ground state is accompanied by two metastable, excited ^1D and ^1S states which are 1.967 eV and 4.188 eV above ground state respectively. All must be viewed as energetic species.

In connection with the issue of initiation of free radical autoxidation the biradical character of $^3O_2$ might be postulated as inducing bond homolysis in susceptible substrates as represented in Eq. 15 followed by the propagation reactions of Eqs. 2 and 3. However provocative this concept be, it remains undemonstrated.

$$R - H + \cdot O\text{-}O\cdot \rightarrow R\cdot + HOO\cdot \qquad \text{Eq. 15}$$

The actual features of cholesterol autoxidation by free radical processes involving $^3O_2$ will be covered in detail in ensuing sections of this chapter.

Singlet Molecular Oxygen. The electronic excitation of $^3O_2$ accomplishes a pairing of the outer antibonding $2p\pi^*$ electrons leading to formation of metastable species with special physical and chemical properties. As previously mentioned, these singlet dioxygen species are collectively termed $^1O_2$ in this monograph.

The electronic excitation of $^3O_2$ may occur under a variety of circumstances, including possibly production of $^1O_2$ in the air. Although the most investigated means of generation of $^1O_2$ under controlled conditions is via photosensitized excitation of $^3O_2$, other means include absorption of electric discharge, radio, or microwave energy, chemical methods (NaOCl oxidation of H_2O_2, decomposition of phosphite ester ozonides, etc.) [2090], and the more questionable cases of possible biological production of $^1O_2$ discussed later in Chapter VI.

By far the most used and readily operated methods of $^1O_2$ generation is by photosensitization. The detailed means by which this excitation is achieved is described in the

later section of this chapter. The controlled photosensitized oxygenation of unsaturated sterols began with the
work of Windaus with the 5,7-diene ergosterol, but studies
with monounsaturated sterols under like conditions generally
failed to give detectable oxidation products, both in
Windaus' day [2700] and in the earlier work by Schenck
[2097], who eventually succeeded in the first photosensitized oxygenation of a sterol cholesterol in 1957 [2098].
Photosensitized oxygenation of cholesterol in which $^1O_2$
is implicated yields the 5α-hydroperoxide 51 in good yield
[1751,2098,2101,2102,2104], with two minor hydroperoxide
products 3β-hydroxycholest-4-ene-6α-hydroperoxide (247) and
its epimer 3β-hydroxycholest-4-ene-6β-hydroperoxide (89)
also formed in much lower yields [1401]. The attack of $^1O_2$
on cholesterol thus yields sterol hydroperoxides which are
different from those found in free radical oxygenations of
cholesterol (the epimeric 7-hydroperoxides 46 and 47).

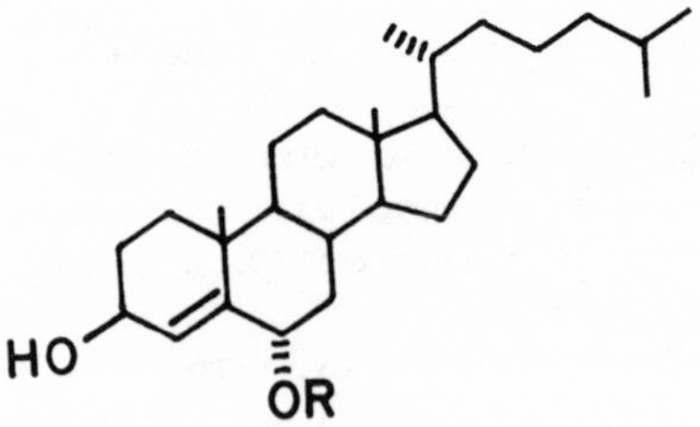

247 R = OH

248 R = H

The attack of $^1O_2$ on steroid olefins has attracted
much attention regarding the mechanism of the reaction. As
steroids offer great advantage as substrates for such mechanism studies, the substrate steroid molecules having well
established structure and stereochemistry, many have been
used in detailed studies of subtle influences, both electronic and stereochemical. Only results dealing with
naturally occurring sterols and closely related stenols
will be given close examination here.

The oxidation of cholesterol by $^1O_2$ proceeds by a cyclic ene reaction mode in which little ionic character is
indicated. A stereospecific abstraction of the quasiaxial
7α-hydrogen results [1751] with concommitant shift of the

Δ^5-double bond to the Δ^6-position, bonding of dioxygen at
the 5α-position, and hydrogen transfer to the developing
peroxyl oxygen atom, yielding the 5α-hydroperoxide 51.
So well accepted is this mechanism and stereospecific
abstraction of the 7α-hydrogen that $^1O_2$ oxidations of
cholesterol are used in the determination of ^3H distribution at the C-7 position in biosynthesis studies [35,
447,1900,2686].

A like mechanism is suggested for the formation of the
minor $^1O_2$ oxidation products of cholesterol, the epimeric
6-hydroperoxides 89 and 247. Approach of the $^1O_2$ molecule
to the β-face, with abstraction of the axial 4β-hydrogen
leading to formation of the axial 6β-hydroperoxide 89 is
consistent with present appreciation of the reaction mechanism. However, although it is generally considered that
the cyclic ene reaction mechanism favors allylic axial
hydrogen involvement, in the case of formation of the 6α-
hydroperoxide 247 from cholesterol some reaction at the
quasiequatorial 4α-hydrogen is inferred. The alternative
possibility of epimerization of the 6β-hydroperoxide 89
was not observed in control experiments [1401]. A final
possibility of conformational change in the A-ring of
cholesterol in solution prior to reaction remains untested.

Most reports of the formation of the 5α-hydroperoxide
51 from cholesterol have used protoporphyrin or hematoporphyrin as sensitizers in pyridine [439,584,626,709,1401,
1421,1751,1754,1900,2101,2102,2104,2399]. Methylene blue
in dimethylformamide [2286] and in aqueous buffer [785],
rose bengal in pyridine [2383] and in aqueous buffer [785],
and 9,10-dicyanoanthracene in acetonitrile [784] have been
used, the 5α-hydroperoxide 51 being formed in all cases.
However, a systematic study of sensitizer and solvent appears
not to have been conducted for cholesterol! Furthermore,
our experience with rose bengal in pyridine suggested that
a second transient product sterol hydroperoxide formed.
The unidentified product had slightly increased thin-layer
and high performance liquid chromatographic mobility than
the 5α-hydroperoxide 51, reverted to 51 on storage, and
could be formed from 51 reintroduced into the rose bengal-
pyridine system. The transient component appears to be a
complex between 51 actually produced in the oxygenation and
solvent, rose bengal dye, or an impurity in the 85% pure
commercially available rose bengal used in the experiment
[2383].

Other unsaturated sterols also react with $^1O_2$ in much the same manner as cholesterol. Cholest-4-en-3β-ol (79) reacts with $^1O_2$ to give a complex mixture of products from which the enone 8 and 4α,5-epoxy-5α-cholestan-3-one (249) have long been recognized as formed [665,1269,1270,1754]. However, our more recent reexamination of the action of $^1O_2$ on the 4-stenol 79 in photosensitized oxygenations has established that the isomeric 4β,5-epoxy-5β-cholestan-3-one (250) is also a product [2456].

Formation of the 4α,5α-epoxide 249 has been rationalized

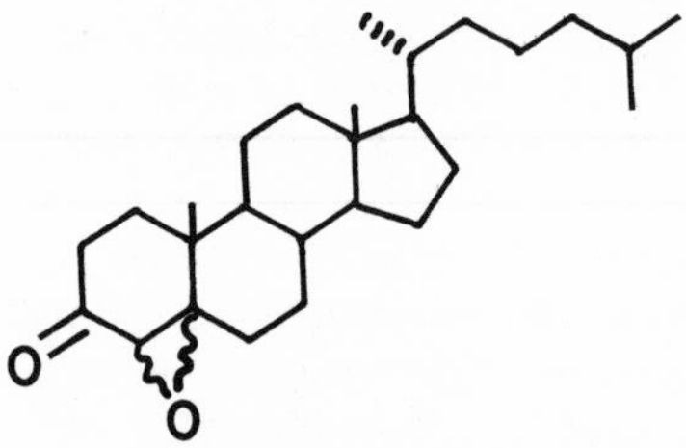

249 4α,5α

250 4β,5β

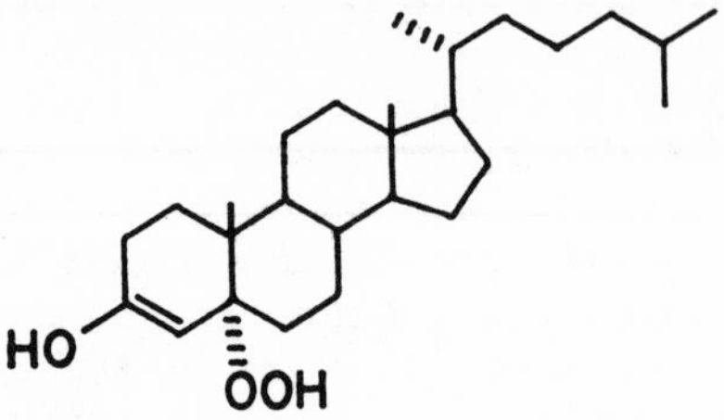

251

by attack of $^1O_2$ and abstraction of the 3α-hydrogen to give the formal intermediate 3-hydroxy-5α-cholest-3-ene-5-hydroperoxide (251) which leads to the stable epoxyketone 249 as product. This formulation cannot account for the formation of the 4β,5β-epoxide 250 as product, for there is no appropriate 3β-hydrogen for involvement in the usual cyclic ene mechanism. Moreover, we were unable to demonstrate an isomerization of the 4α,5α-epoxide 249 to the 4β,5β-epoxide 250. In that the 4β,5β-epoxide 250 is the chief product of the attack of alkaline H_2O_2 on the enone 8 which is also formed in the $^1O_2$ attack on the 4-stenol 79, it may be that the 4β,5β-epoxide 250 is derived from the enone 8 rather than from the 4-stenol 79 [2456].

The reaction of the 6-stenol 5α-cholest-6-en-3β-ol with $^1O_2$ yields by abstraction of the 5α-hydrogen the Δ^5-7α-hydroperoxide 46 also formed by allylic isomerization of the 5α-hydroperoxide 51 [1751] and by free radical autoxidation of cholesterol [2313].

Reaction between 5α-cholest-7-en-3β-ol (57) and $^1O_2$ involves the complication of the oxygenation of the initially

formed sterol hydroperoxide product, thus the consumption
of two equivalents of $^1O_2$ per 7-stenol [665]. Although
the three products formed from the 7-stenol 57 [2309,2383]
have not been individually isolated and identified, three
products derived from the 7-stenol 57 3β-acetate have been.
In view of the established consumption of two equivalents
of $^1O_2$ by the 7-stenol 57 and of the formation of only
three sterol hydroperoxide products, the three products
from the 7-stenol are assumed to have analogous structures
to those from the 7-stenol 57 3β-acetate.

The initial attack of $^1O_2$ on the 7-stenol 57 3β-ace-
tate yields 3β-hydroxy-5α-cholest-8(14)-ene-7α-hydroperoxide
(252) 3β-acetate which is then oxidized by the second equivalent of $^1O_2$ to the isomeric 7α,8-dihydroperoxides 3β-
hydroxy-5α,8α-cholest-14-ene-7α,8-dihydroperoxide (253)
3β-acetate and 3β-hydroxy-5α,8β-cholest-14-ene-7α,8-dihy-
droperoxide (253) 3β-acetate [2100]. By analogy, products
of $^1O_2$ oxidation of the 7-stenol 57 should be 252-253.

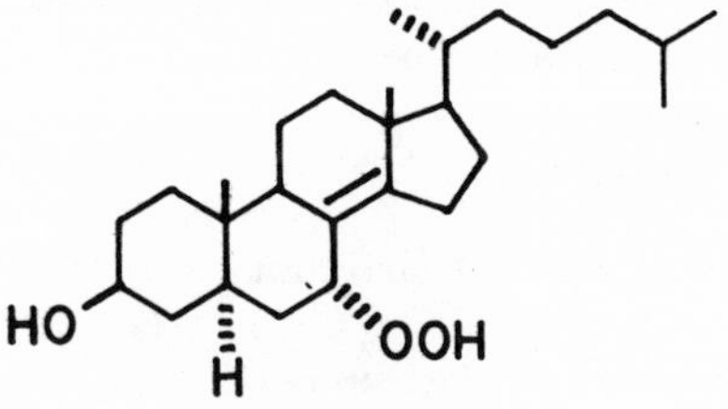

252 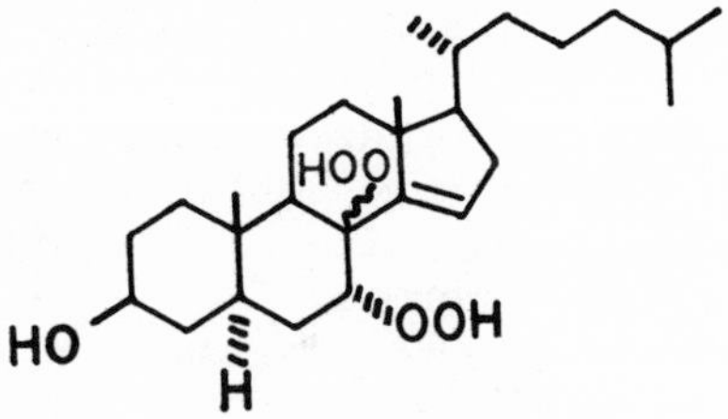   253

The attack of $^1O_2$ on the 8-stenol 68 has been examined
in our laboratory, but the chemical characterization and
identification of structure of the five sterol hydroperoxides formed has not been addressed [2309,2383]. It is clear
that none of these $^1O_2$ products corresponds with the three
established free radical autoxidation products 3β-hydroxy-
5α-lanost-8-ene-7β-hydroperoxide (254), 3β-hydroxy-5α-lanost-
8-ene-11β-hydroperoxide (255), and 3β-hydroxy-5α-lanost-8-
ene-7β,11β-dihydroperoxide (256), whose formation is given
more detailed treatment in Chapter VI.

The attack of $^1O_2$ on the 8-stenol 68 3β-acetate has
been described, but in this case only one sterol hydroperox-

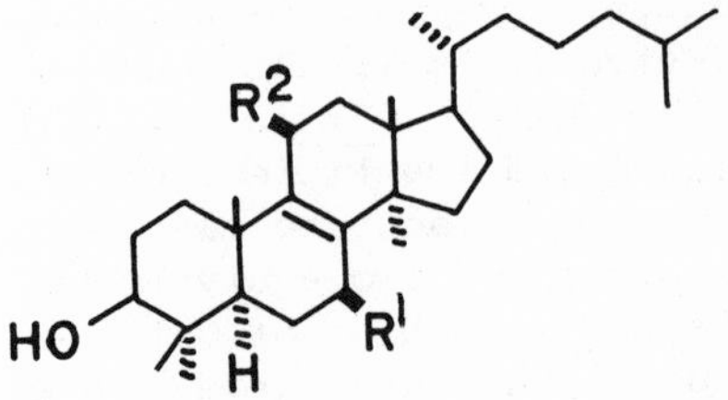

68 $R^1 = R^2 = H$ 257

254 $R^1 = OOH$, $R^2 = H$

255 $R^1 = H$, $R^2 = OOH$

256 $R^1 = R^2 = OOH$

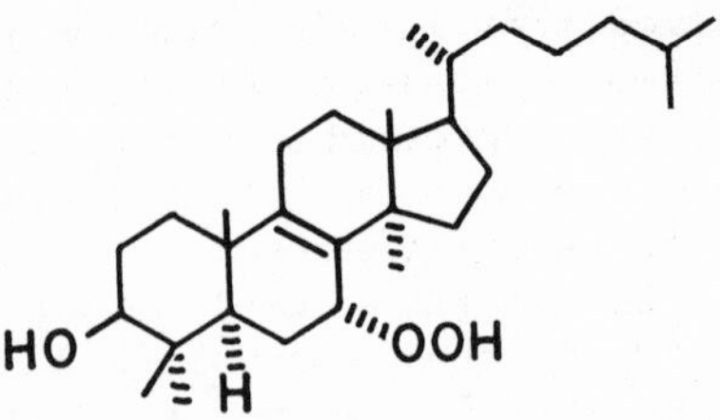

ide product 3β-hydroxy-5α-lanost-8-ene-7α-hydroperoxide
(257) 3β-acetate was identified *inter alia*. However, the
photosensitized oxygenation of substrate 68 3β-acetate was
conducted in pyridine solutions containing p-nitrobenzene-
sulfonyl chloride ostensibly to trap any sterol hydroperoxides formed by esterification [792,793]. This acid chloride
may well have altered the primary spate of $^1O_2$ products
putatively formed, for of the six isolated products, only
one was a hydroperoxide (the 7α-hydroperoxide 257). Moreover, the Δ^8-7α-hydroperoxide structural feature of the
product does not conform to any of the several recognized
modes of action of $^1O_2$ on cyclic olefins. Shift of the
Δ^8-double bond to the Δ^7- and/or $\Delta^{9(11)}$ - positions would
normally be expected. A thorough reexamination of the
reaction between the 8-stenol 68 and $^1O_2$ is in order.

Other naturally occurring sterols with which $^1O_2$ might
react include lanosterol (67) and the several other sterols
which are biosynthesis intermediates linking lanosterol and
cholesterol. Systematic examination of these reactions has
not been made, but as all are olefins or dienes reactions
with $^1O_2$ is to be expected. No studies of the action of
$^1O_2$ on lanosterol itself appear to have been recorded, but
the oxidation of 67 3β-acetate by $^1O_2$ previously discussed
in Chapter III involves preferential attack on the side-
chain Δ^{24}-double bond, with no oxidation about the nuclear
Δ^8-double bond [1704]. This pattern thus confirms that

observed for the 5,24-diene desmosterol (78) which also
yielded side-chain oxidation products and none from attack
on the nuclear Δ^5-double bond [1678,1704].

Studies on other endogenous dienes such as 5α-cholesta-
8,24-dien-3β-ol (zymosterol), 5α-cholesta-7,24-dien-3β-ol,
etc. and the various mono- and di-methylated sterol dienes
implicated in cholesterol biosynthesis appear not to have
been reported. The action of $^1O_2$ on the ultimate precursor
of cholesterol cholesta-5,7-dien-3β-ol (56) has also been
studied, but in this case the cyclic ene reaction mode does
not occur. Rather, a 1,4-cyclic addition to the 5,7-diene
feature occurs in the well recognized 2 + 2 cyclic addition
reaction mode of $^1O_2$ with conjugated dienes yielding the
5α,8α-peroxide 60 [2096]. Formation of the 5α,8α-pero-
xide 60 by reaction of substrate 56 with $^1O_2$ is not a reaction unique to $^1O_2$, for the 5α,8α-peroxide 60 is also
formed in reactions involving $^3O_2$ (cf. Chapter VI).

The common plant sterol sitosterol (20) may be oxidized
by $^1O_2$ in the same manner as cholesterol, for the photosensitized oxygenation of sitosterol or its 3β-acetate yields
the corresponding 5α-hydroperoxide homolog 3β-hydroxy(or
acetoxy)-5α-stigmast-6-ene-5-hydroperoxide [2098,2270].
Moreover, yet other unsaturated sterols may serve as substrates for $^1O_2$, the fungal sterol fungisterol (5α-ergost-
7-en-3β-ol) (258) yielding 3β-hydroxy-5α,8α-ergost-6-ene-
8-hydroperoxide (259) upon oxidation by $^1O_2$ [2098].

258   259

The preference for oxidation of the side-chain Δ^{24}-double
bond over the nuclear Δ^5- and Δ^8-double bonds does not
extend to the case of the 5,Z-22-diene stigmasterol (70),
where nuclear attack of $^1O_2$ on 70 3β-acetate occurs yielding 3β-acetoxy-5α-stigmasta-6,Z-22-diene-5-hydroperoxide
[2098].

The action of $^1O_2$ on numerous synthetic stenols, olefins, dienes, etc. has been examined as part of extensive
studies of the mechanism by which $^1O_2$ oxidizes unsaturated
substrates, with products uniformly being hydroperoxides
or cyclic peroxides [602,665,1751-1753,1755,1757]. As
these synthetic steroids do not occur in tissues, their
individual chemistries do not contribute further to our
understanding of cholesterol autoxidation initiation events.

The import of these oxidative attacks of $^1O_2$ on sterols
is that the reaction products from the naturally occurring
sterols cholesterol, the 7-stenol 57, and the 8-stenol 68
as well as of others are hydroperoxides whose peroxidic
oxygen-oxygen bond upon homolysis should provide corresponding oxyl radicals capable of initiation of generalized
autoxidation of cholesterol. Thus, homolysis of the 5α-
hydroperoxide 51 peroxide bond should generate the 3β-hy-
droxy-5α-cholest-6-en-5-oxyl radical (260). Homolysis of
the hydroperoxide oxygen-hydrogen bond though less likely
energetically might nonetheless occur, particularly in the
presence of one-electron oxidizing agents, yielding the
corresponding 3β-hydroxy-5α-cholest-6-en-5-peroxyl radical
(261). The 5α-peroxyl radical 261 may then enter the free
radical chain propagation reaction of Eq. 3 as well as the
reactions of Eqs. 5-7, and the 5α-oxyl radical 260 may,
through the hydrogen abstraction reaction of Eq. 13, continue the chain reactions. In these cases, ROO· of Eq. 3
would be the 5α-peroxyl radical 261, RO· of Eq. 13 would
be the 5α-oxyl radical 260 and the substrate RH in either
case would be cholesterol. The free radical R· formed from
cholesterol is suggested as the 3β-hydroxycholest-5-en-7-yl
radical (262) whose chemistry is discussed later in this
Chapter.

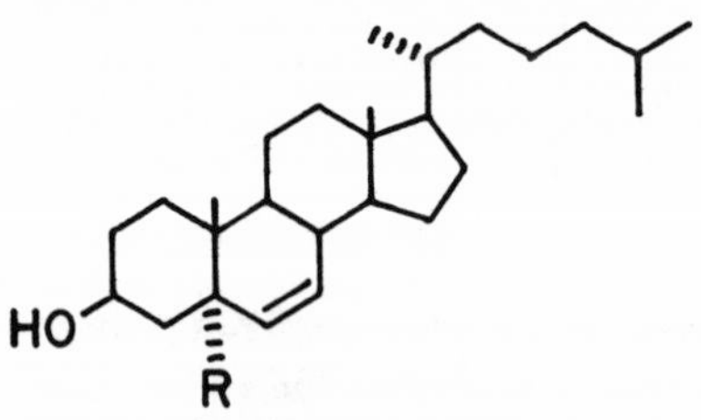

51 R = OOH   262

260 R = O·

261 R = OO·

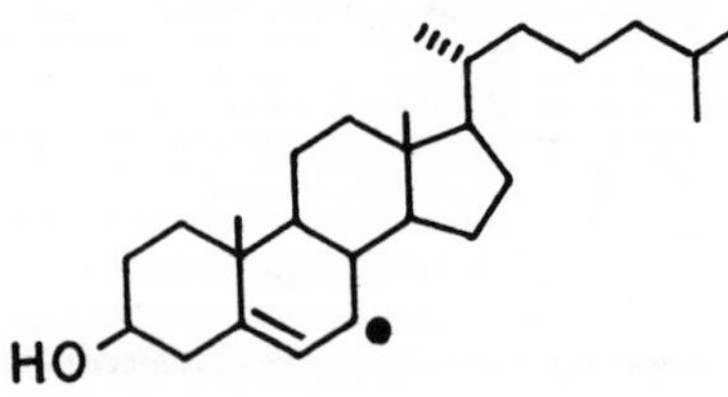

Although controlled experimental attempts to initiate cholesterol autoxidation using 5α-hydroperoxide 51 or either 5α-oxyl or 5α-peroxyl radicals therefrom have not been made, we must presume that these possibilities exist and that therefore the agency of $^1O_2$ attack on sterols poses a means of initiation of sterol autoxidation by free radical processes. A similar case of initiation of cholesterol autoxidation by homolysis of the peroxide bond of sterol hydroperoxides formed by the action of $^1O_2$ on trace level congeners 57,68,78, etc. as well as by homolysis of the cyclic peroxide bond of the 5α,8α-peroxide 60 formed from the 5,7-diene 56 may be posited.

Necessarily, homolysis of the hydroperoxide bond of other sterol hydroperoxides such as the epimeric 7-hydroperoxides 46 and 47, not products of $^1O_2$ action, also putatively yield 7-oxyl radicals such as the 7α- and 7α-oxycholest-5-en-3β-ols (263) that may serve to propagate the radical autoxidation chain.

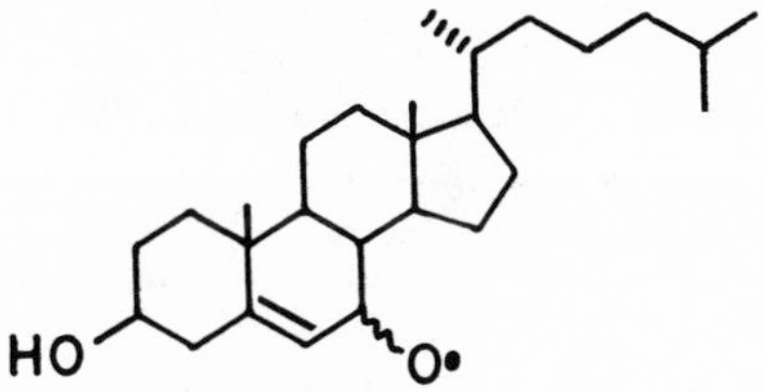

263

Superoxide Radical. The one-electron reduction product of $^3O_2$ is the superoxide radical anion $O_2^{\overline{\cdot}}$ formed by the process of Eq. 16. As $O_2^{\overline{\cdot}}$ may react with a proton to give the hydroperoxyl radical HOO· according to Eq. 17, characterized by pK_a 4.75-4.88 [136,216], present consideration of free radical autoxidation initiation by superoxide radical reactions must include the conjugate base or anion $O_2^{\overline{\cdot}}$ and the conjugate acid HOO·. For simplicity the term $O_2^{\overline{\cdot}}$ will be used hereafter with the understanding that both species be involved.

$$O_2 \quad + \quad 1e^- \quad \rightleftharpoons \quad O_2^{\overline{\cdot}} \qquad Eq.\ 16$$

$$O_2^{\overline{\cdot}} \quad + \quad H^+ \quad \rightleftharpoons \quad HOO· \qquad Eq.\ 17$$

The chemistry of $O_2^{\overline{\cdot}}$ has attracted much recent attention, both as a chemical reagent for chemical synthesis

[1446] and a previously unrecognized and neglected naturally
occuring one-electron reduction product of many enzymic
reactions dependent upon the oxygen of the air [323,803,804].
The $O_2^{\bar{\cdot}}$ may react in several modes, thus (i) as a base and
conjugate acid as indicated by Eq. 17, (ii) as a one-electron transfer agent, mainly as an electron donor indicated
by the reverse reaction of Eq. 16 but also as a very weak
oxidizing agent accepting a second electron according to
Eq. 18, (iii) as a weak nucleophile in substitution, elimination, and addition reactions, (iv) in disproportionations

$$O_2^{\bar{\cdot}} + 1e^- \; \overset{\rightarrow}{\underset{\leftarrow}{}} \; O_2^= \qquad \text{Eq. 18}$$

according to Eqs. 19 and 20 but not by Eq. 21 [216,1570],

$$O_2^- + HOO\cdot \rightarrow O_2 + HOO^- \qquad \text{Eq. 19}$$

$$2HOO\cdot \rightarrow O_2 + H_2O_2 \qquad \text{Eq. 20}$$

$$2O_2^{\bar{\cdot}} \rightarrow O_2 + O_2^= \qquad \text{Eq. 21}$$

(v) as a source of hydroxyl radical $HO\cdot$ according to the
process of Eq. 22 or of an analogoous one catalyzed by trans-

$$O_2^{\bar{\cdot}} + H_2O_2 \rightarrow HO\cdot + HO^- + O_2 \qquad \text{Eq. 22}$$

sition metal ions given by Eqs. 23, and (vi) in inducing
homolysis of susceptible covalent bonds as in Eq. 24.

$$Fe^{+++} + O_2^{\bar{\cdot}} \rightarrow Fe^{++} + O_2 \qquad \text{Eq. 23a}$$

$$Fe^{++} + H_2O_2 \rightarrow Fe^{+++} + HO\cdot + HO^- \qquad \text{Eq. 23b}$$

$$R-H + O_2^{\bar{\cdot}} \rightarrow R\cdot + HOO^- \qquad \text{Eq. 24}$$

Despite very great interest in $O_2^{\bar{\cdot}}$ chemistry, some
uncertainty in its simple chemistry continues. Although
disproportionation as the prevalent reaction of $O_2^{\bar{\cdot}}$ in
aqueous systems is well known [864], only very recently
have rate constants for the three possible modes of disproportionation of Eqs. 19-21 been recorded. The self-reaction
of two $O_2^{\bar{\cdot}}$ anions (Eq. 21) does not occur ($k < 0.3$ $M^{-1}s^{-1}$).
By contrast the reaction of Eq. 19 is characterized by the
rate constant $k = 8.86 \times 10^7$ $M^{-1}s^{-1}$, that of Eq. 20 by
$k = 7.61 \times 10^5$ $M^{-1}s^{-1}$ [216].

Moreover, some controversy exists regarding certain of the listed reaction modes. Whereas the properties of $O_2^{\bar{\cdot}}$ as a base, as an electron donor, and as a weak nucleophile are well documented, the behavior of $O_2^{\bar{\cdot}}$ as a free radical and as an oxidant needs careful examination. It is just these properties as radical and oxidant which are important to present consideration of $O_2^{\bar{\cdot}}$ as a potential initiator of cholesterol autoxidation.

Even though $O_2^{\bar{\cdot}}$ is an added reagent in certain reacting systems and products are attributed to the action of $O_2^{\bar{\cdot}}$, the possibilities that dismutation or other reactions of $O_2^{\bar{\cdot}}$ provide means of oxidation should not be disregarded. Thus, the nucleophilic addition of $O_2^{\bar{\cdot}}$ to cyclohex-2-en-1-one [612] may represent just such a formulation but might also involve dismutation of $O_2^{\bar{\cdot}}$ to $O_2^{=}$, to $^1O_2$, or transformation to other oxygen species as true active oxidants. Although $O_2^{\bar{\cdot}}$ is described as a "pitifully weak oxidizing agent" [2086], it may nonetheless be indirectly responsible for the generation of more highly active oxidants in certain instances.

One of these powerful oxidizing species is the hydroxyl radical HO· discussed in regard to its oxidation of cholesterol in a later section of this chapter. The proposal of Haber and Willstäter [936] later known as the Haber-Weiss reaction [935] represented in Eq. 22 suggests that HO· derive from $O_2^{\bar{\cdot}}$ in reaction with H_2O_2. As the uncatalyzed process of Eq. 22 has in fact not been experimentally demonstrated despite good efforts [265,1603,1955], catalysis by iron (III) or other transition metal ion as indicated in Eqs. 23, whose individual steps sum to the process of Eq. 22 [1369], may be necessary for HO· production. Moreover, an analogous reaction between $O_2^{\bar{\cdot}}$ and organic hydroperoxides (in lieu of H_2O_2) yields alkoxyl radicals according to Eq. 25 [1852]. The full extent of the interactions of $O_2^{\bar{\cdot}}$

$$O_2^{\bar{\cdot}} + ROOH \rightarrow O_2 + RO\cdot + HO^- \qquad \text{Eq. 25}$$

with other oxygen species to give HO· and related excited species and free radicals has yet to be realized.

The ability of $O_2^{\bar{\cdot}}$ to abstract hydrogen from susceptible substrates as indicated in Eq. 24 is supported in selected cases by product analysis and reaction kinetics data. Thus, it has been demonstrated that $O_2^{\bar{\cdot}}$ abstracts benzylic hydrogen from 9,10-dihydroanthracene, fluorene, diphenylmethane, and cumene. Subsequent reactions of the benzylic carbon free radicals from 9,10-dihydroanthracene gave anthraquinone (and some anthracene) as isolated product. Fluorene and diphenylmethane likewise yielded fluoren-9-one and benzophenone respectively as products. Product analysis and reaction kinetics supported a reaction sequence involving benzylic hydrogen abstraction, subsequent reaction of the benzylic carbon radical with $^3O_2$ to give the corresponding peroxyl radical, stabilized in turn by a second hydrogen abstraction from substrate to give a product benzylic hydroperoxide not isolated as such. Dehydration of the benzylic hydroperoxides to the corresponding ketones thus accounts for the products found. However, in the case of cumene, cumene hydroperoxide was found as the stable product. Furthermore, oxidation of cyclohexane by O_2^- was indicated, although products were not identified. These several oxidations may involve transition metal ion complexation of O_2^- and not the uncomplexed radical [2472].

In reaction with p-quinones O_2^- as electron donor reduces the quinone to semiquinone radical anion, whereas O_2^- as electron acceptor in reaction with secondary amines gives substituted nitric oxide radical products, according to Eq. 26. Although both semiquinone and nitric oxide products are radicals and might thereby serve to indicate radical processes, experimental evidence supporting such formulation is lacking.

$$R_2NH + O_2^- \rightarrow R_2N\text{-}O\cdot + HO^- \qquad Eq. 26$$

Finally, claims have been made that O_2^- initiate generalized free radical autoxidations of tissue unsaturated lipids, thus lipid peroxidations [418,1838,1839], but much evidence fails to support this contention [395,782,1250, 1306,1840,2532]. Indeed, data variously interpreted in favor of HO· [782,1418], $^1O_2$ [122,1276,1277], and perferryl ion FeO_2^{++} [1840] as the active initiators have been presented. Although it now appears that O_2^- does not act as an initiator of free radical autoxidations of endogenous lipids, the true nature of the complex phenomenon of lipid peroxidation in the several biological systems of common interest remains incompletely understood.

Faced with the question whether O_2^- oxidized cholesterol or supported the initiation of free radical autoxidation of cholesterol we attempted to provoke reaction between cholesterol and O_2^- generated by several means. However, O_2^- generated chemically from crown ether sequestered KO_2, photochemically, or enzymically with the xanthine-xanthine oxidase system, in a variety of anhydrous organic solvent or in aqueous system failed to react with cholesterol to give products which could be detected [2301].

In view of the comparative ease with which we have detected oxidation products of cholesterol formed in reactions involving $O_3, O_2^+, {}^3O_2, O_2^=$, HO·, and several dioxygenases our failure to demonstrate any transformation products with O_2^- may be regarded as definitive.

Hydrogen Peroxide. The long recognized oxidizing power of H_2O_2 and of its ionic forms peroxide anion $O_2^=$ and hydroperoxyl anion HOO^- linked to H_2O_2 by Eqs. 27 and 28 are of importance in considerations of initiation of autoxidations. Neither H_2O_2 nor its ionic species HOO^- and $O_2^=$

$$H_2O_2 \; \underset{\leftarrow}{\rightarrow} \; HOO^- \; + \; H^+ \qquad \text{Eq. 27}$$

$$HOO^- \; \underset{\leftarrow}{\rightarrow} \; O_2^= \; + \; H^+ \qquad \text{Eq. 28}$$

shows free radical properties, but their rapid reaction with a variety of oxidizable substrates leading to epoxides, etc., their disproportionation to ${}^3O_2$ and to ${}^1O_2$ [2298], and the omnipresent potential for peroxide bond homolysis to give HO· provides several modes of action suspected as serving initiation of autoxidation.

The action of H_2O_2 and its ionic forms, of organic hydroperoxides ROOH and peroxides ROOR, and of peracids $RCOO_2H$ on cholesterol must be examined together, as experimental evidence suggests they all act to epoxidize cholesterol. Considerable study of the epoxidation of cholesterol by these reagents has been made, mainly with a view of preparing the 5α,6α- or 5β,6β-epoxides 35 or 36 respectively but usually both 35 and 36 are formed in varying proportions. At the one extreme, peracid (perbenzoic, m-chlorobenzoic, peracetic) oxidation of cholesterol favors attack on the α-face, yielding the 5α,6α-epoxide 35, whereas at the other, air oxidation of cholesterol in aqueous media invariably favors β-face oxidation, yielding the 5,6-epoxides in 35:36 ratios ranging from 0:1 [479] to 1:11 [2297], thus the same as epoxidation of cholesterol in aqueous sodium

stearate dispersions containing H_2O_2 where a 35:36 ratio of
1:8 was found [2298,2299]. Air oxidation of dry [4-^{14}C]
cholesterol gave the 5,6-epoxides in the 35:36 ratio of
1:3.6 [93].

An early investigation of the attack of H_2O_2 on aqueous
sodium stearate dispersions of cholesterol gave the 3β,7β-
diol 15 isolated as the 3β,7β-dibenzoate from intractible
product material but failed to demonstrate formation of
5,6-epoxides [1721]. However, an unidentified product isolated as a benzoate ester, m.p. 177°C might possibly be the
5β,6β-epoxide 36 3β-benzoate, m.p. 172-173°C [1861].

Organic hydroperoxides such as cumene hydroperoxide
and the sterol hydroperoxides 46,47, and 51 likewise favor
β-face attack and 5β,6β-epoxide 36 formation from cholesterol in aqueous dispersions [2297], but in other cases
different ratios or products obtain where organic hydroperoxides or H_2O_2 are utilized with transition metal ions.
Thus, tris(acetylacetonato) iron (III) and H_2O_2 oxidize
cholesterol and its 3β-acetate to the corresponding 5α,6α-
and 5β,6β-epoxides in the ratio 1:4 [2489]; tert.-butyl
hydroperoxide with tris(acetylacetonato) iron (III) also
oxidizes cholesterol 3β-acetate to the epoxides 35
3β-acetate and 36 3β-acetate in 1:4 ratio [1301]. Cumeme
or tert.-amyl hydroperoxide with $MoCl_5$ oxidizes cholesterol
to 5-hydroxy-5α-cholestane-3,6-dione [652], cholesterol 3β-
acetate to equal proportions of the 5α,6α- and 5β,6β-epox-
ides [2492,2494]. More vigorous oxidation of cholesterol
3β-acetate by cumeme hydroperoxide and molecular oxygen
gave the 7-ketone 16 3β-acetate [2679]. The oxidation of
cholesterol 3β-acetate by tert.-butyl hydroperoxide and
tris(acetylacetonato) iron (III) to the epimeric 3β-ace-
toxycholest-5-en-7-yl tert.-butyl peroxides also occurs
[1301].

Epoxidation of cholesterol in conjunction with enzyme
action also yields both 5,6-epoxides, soybean lipoxygenase
giving 35:36 ratios of 1:3.7, the rat liver NADPH-dependent
microsomal lipid peroxidation system giving 1:3.3-1:3.9
ratios [93].

These results establish that epoxidation be the one
mode of oxidation of cholesterol affected by H_2O_2 and
related organic hydroperoxides. Nonetheless, other oxidation products of cholesterol have been encountered in

oxidations conducted with H_2O_2. One must project that other active oxidants are formed from H_2O_2 in these cases, a matter particularly obvious in the case of formation of the HO· by the Fenton reagent discussed in a later section of this chapter. However, even in the absence of specific additions of transition metal ions, such as in the model aqueous sodium stearate dispersions of cholesterol, other oxidation products derive from H_2O_2 action. The 3β,7β-diol 15 was isolated from such systems as the 3β,7β-dibenzoate in the earliest such study [1721], and we have isolated a spate of products in similar studies which evince the derivation of active oxidizing species from H_2O_2 in this system.

Chief among such products are the epimeric 7-hydroperoxides 46 and 47, together with the epimeric 3β,7-diols 14 and 15 and 7-ketone 16. Taken together these five products 14-16, 46, and 47 establish the pattern of free radical autoxidation of cholesterol by $^3O_2$, and the disproportionation of H_2O_2 (its ionic form HOO^- in these dispersions at pH 9.5) to $^3O_2$ according to Eqs. 29-31 is inferred thereby [2298,2299,2312]. The release of $^3O_2$ from other sorts of aqueous dispersions of cholesterol containing H_2O_2 has been demonstrated [1935-1937], and the overall impression that H_2O_2 serve to initiate cholesterol autoxidation in aqueous systems is inescapable.

$$H_2O_2 \ + \ HO^- \ \underset{\leftarrow}{\overset{\rightarrow}{}} \ HOO^- \ + \ H_2O \qquad \text{Eq. 29}$$

$$H_2O_2 \ + \ HOO^- \ \rightarrow \ HO^- \ + \ H_2O \ + \ ^3O_2 \qquad \text{Eq. 30}$$

$$2H_2O_2 \ \rightarrow \ 2H_2O \ + \ ^3O_2 \qquad\qquad \text{Eq. 31}$$

Moreover, among isolated products from these same experiments we found the dienone 12 and 3β,5α-diol 50, both of which are products formed by the thermal decomposition of the 5α-hydroperoxide 51 [2300,2454,2578]. No other means of derivation for the 3β,5α-diol 50 is known to me, but the dienone 12 may also be formed by the attack of $^1O_2$ on the enone 6 [1756] or by oxygen-dependent dehydrogenation of the 3β,7-diols 14 and 15, as previously mentioned in Chapter III. In any event, formation of 12 and 50 in these systems infers the transient presence of the 5α-hydroperoxide 51 therein also. Furthermore, a third product of this model system was isolated by us which also infers the presence of 5α-hydroperoxide 51 in the system. This product 7α-stearat-

oxycholest-5-en-3β-ol is formed in this system from cholesterol as substrate but also from the 5α-hydroperoxide 51
and/or from the 3β,5α-diol 50 as substrates. It follows
that the presence of the 7α-stearate ester among products
from cholesterol was formed from 5α-hydroperoxide 51 and/or
3β,5α-diol 50 that were also formed in the reaction. Thus,
three isolated products 12,50, and 7α-stearatoxycholest-5-
en-3β-ol infer the presence of the 5α-hydroperoxide 51 in
the system.

       Derivation of 7α-stearatoxycholest-5-en-3β-ol from
the 5α-hydroperoxide 51 and from the 3β,5α-diol 50 represents allylic rearrangement of the B-ring functional groups
affected by stearate anion. A related precedent is had in
the allylic rearrangement of the 3β,5α-diol 50 3β-acetate
in aqueous acetic acid yielding products 3β,7α-diol 14 3β-
acetate and 3β,7α-diol 14 3β,7α-diacetate [1668]. The
indicated allylic rearrangement of 50 and 51, the thermal
decomposition of 51 to 12 and 50, and other relevant subsequent transformations of these sterols are discussed more
fully in Chapter V.

       The presence of the 5α-hydroperoxide 51 in turn infers
the attack of $^1O_2$ on cholesterol, no other means of derivation of 51 being known. One formulates accordingly, the
formation of $^1O_2$ along with $^3O_2$ from the base-catalyzed disproportionation of H_2O_2 according to Eqs. 29-31 [2298].

       Other products isolated from these same experiments
with H_2O_2 and aqueous dispersions of cholesterol were the
3β,5α,6β-triol 13 recognized as a hydration product of the
5,6-epoxides 35 and 36 and 5α-cholestane-3β,6β-diol (264),
for which no specific chemical process has been defined
[2298]. The 3β,6β-diol is the major product of the LiAlH₄
reduction of the 5β,6β-epoxide 36, 5β-cholestane-3β,5-diol
(265) being the minor product [1869]. However, the 3β,6β-
diol was not formed in H_2O_2-containing aqueous sodium stearate dispersions of either 5,6-epoxide 35 or 36, indeed of
any product 12-16,35,36,46,47,50, or of the 5α-hydroperoxide
51. The 3β,6β-diol 264 is formally a hydration product of
cholesterol, but the photosensitized hydration of cholesterol has been shown to yield the 5β-cholestane-3β,5-diol
(265) together with the carbon-carbon bond cleavage product
3α,5α-cyclo-A-homo-4-oxacholestane (266) [1367,2641]. Moreover the 3β,6β-diol 264 is not among products of the attack

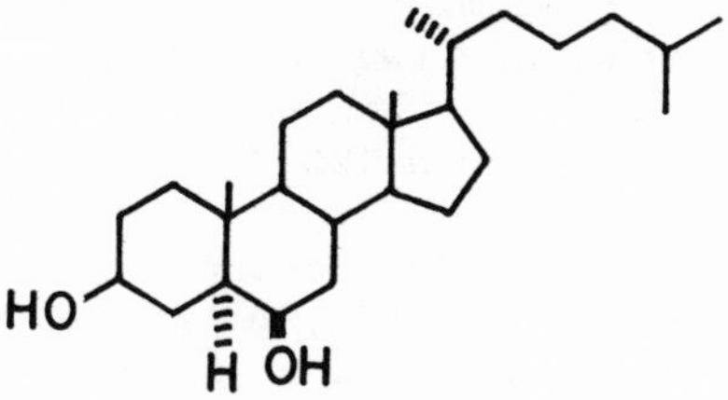

264

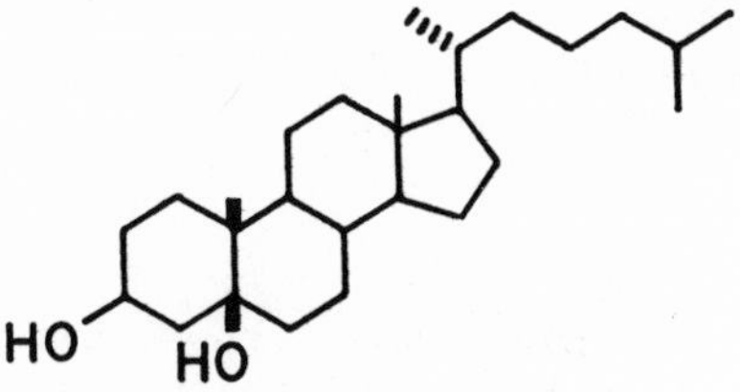

265

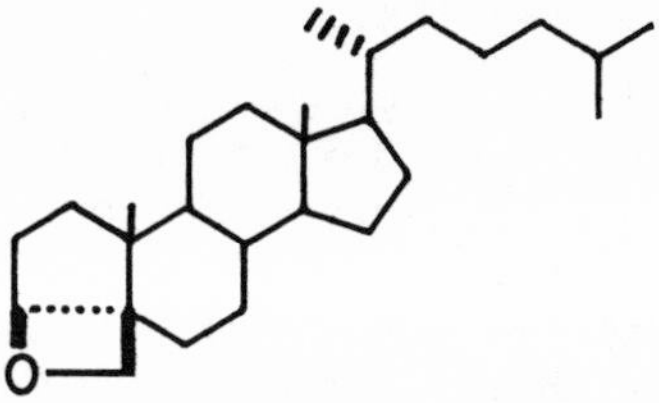

266

of HO· on cholesterol discussed in a later section of this chapter.

We may include the 3β,6β-diol 264 as an oxidation products of cholesterol derived by uncertain mechanism, in much the same manner as 3β-hydroxy-5α-cholestan-6-one (45) found in pig spleen [1888] appears to be autoxidation product but one for which no distinct chemical pathway has been suggested.

Monoatomic Oxygen Species. Some consideration of the possibilities that monoatomic oxygen species initiate cholesterol autoxidations is needed. These monoatomic species include the oxygen atom O· in its ground state (^{3}P), first excited state (^{1}D), and second excited state (^{1}S), the oxygen cation O$^+$, oxygen anion O$^-$, and the postulated oxene species ·O· formulated as the oxygen analog of carbene and nitrene species [948,949].

The previously mentioned studies of the action of the O_2^+ species on cholesterol also may have involved generation of excited monoatomic oxygen species. In the system involved the oxidation of cholesterol to the 3β,7-diols 14 and 15, the 7-ketone 16, the 5,6-epoxides 35 and/or 36, inter alia, could be attributed by calculations to the action of the cation O_2^+, to that of the oxygen atom, to that of the oxygen cation O^+, or to the actions of any two species or to all three, but the individual chemical processes leading to each observed product were not elucidated [2071].

Hydroxy Radical. The hydroxy radical HO· occupies a unique place among these excited oxygen species, being the conjugate acid of the monoatomic species $O^{\mathrel{\overline{\cdot}}}$. The radical is thought to be highly active oxidant capable of rapid and indiscriminant oxidation of a broad variety of organic compounds of biological interest [631].

Several reports of cholesterol oxidation by HO· have issued. One generally considers that HO· is produced by ionizing radiation, by the Fenton reagent involving $FeSO_4$-H_2O_2 in aqueous acetic acid, by related systems incorporating H_2O_2 and selected transition metal salts such as $Ti_2(SO_4)_3$, by the thermal homolysis or photolysis of H_2O_2, by radiolysis of water saturated with nitrous oxide N_2O, etc. Moreover, HO· is present in certain polluted air supplies [166,762,1866].

The radiolysis of water by energetic X-radiation or by ^{60}Co γ-radiation is considered to generate a variety of highly reactive species, as represented in Eq. 32 [217,631,2655]. In the absence of air oxidized cholesterol derivatives should have to be derived from attack by HO· (or H_2O_2, etc.

$$H_2O \rightsquigarrow \quad HO·, \ H·, \ e^-_{aq}, \ H_2, \ H_2O_2, \ H_3O^+, \ HO^-_{aq} \qquad Eq. \ 32$$

formed at low yields). An alternative generation of HO· by radiolysis of water utilizes N_2O as a scavenger of the hydrated electron e^-_{aq} formed, as given in Eqs. 33 [631].

$$N_2O + e^-_{aq} \rightarrow N_2 + O^- \qquad Eq. \ 33a$$

$$O^- + H_2O \rightarrow HO· + HO^- \qquad Eq. \ 33b$$

$$N_2O + e^-_{aq} + H_2O \rightarrow N_2 + HO· + HO^- \qquad Eq. \ 33c$$

By contrast the classic Fenton reagent of H_2O_2 and $FeSO_4$ in aqueous acetic acid generates HO· chemically by the process

$$H_2O_2 + Fe^{++} \rightarrow Fe^{+++} + HO^- + HO· \qquad Eq. \ 34$$

of Eq. 34 [2627]. The products have also been represented as HO· and the ion pair $(FeOH)^{++}$ [1304]. The companion system utilizing Ti (III) ions serves the same end, as shown in Eq. 35 [631]. These chemical systems are quite complex ones

$$H_2O_2 + Ti^{+++} \rightarrow Ti^{++++} + HO^- + HO· \quad \text{Eq. 35}$$

however, and several other reactions occur, those of Eqs. 36 being of importance [2627]. From the radicals R· and HO·

$$HO· + Fe^{++} \rightarrow Fe^{+++} + HO^- \quad \text{Eq. 36a}$$

$$RH + HO· \rightarrow R· + H_2O \quad \text{Eq. 36b}$$

generated there are formed various products, according to Eq. 37-39 [2627]. We are fortunate in having results of

$$R· + Fe^{+++} \rightarrow Fe^{++} + \text{products} \quad \text{Eq. 37}$$

$$2R· \rightarrow RR \text{ (dimer)} \quad \text{Eq. 38}$$

$$R· + Fe^{++} + H^+ \rightarrow Fe^{+++} + RH \quad \text{Eq. 39}$$

cholesterol oxidation from several laboratories using HO· generated by the several recognized means.

Oxidations of solutions of cholesterol in organic solvents by HO· generated X-radiolysis or by Fenton reagents variously implicated six common cholesterol autoxidation products 13-16,35, and 36, as summarized in TABLE 8. As most of the work with HO· generated by these means involved product isolation and not controlled monitoring of the reaction by chromatography, the discrepancies in product distributions are best reconciled on this basis and not from the differences in solvent or system used.

In the aqueous acetic acid systems some acetylation of substrate cholesterol and of oxidation products occurred [500,1275,1303]. Cholesterol 3β-esters are also oxidized by HO· [500].

In our own work we attempted to avoid solvent effects and other potential problems using H_2O_2 in Fenton systems by generating HO· by γ-radiolysis of simple water dispersions of cholesterol. By such means all six of the products 13-16,35, and 36 previously implicated were isolated. Radio-

TABLE 8. Oxidation of Cholesterol by Hydroxy Radicals

System	Oxidation Products	References
Radiolysis, 205 kV X-Rays:		
Methanol	13,15,16	[503]
Acetone or dioxane	15,35,36	[503]
Aq. acetic acid	13,16	[1275,2656]
Fenton Reagents with H_2O_2:		
FeSO$_4$, aq. acetic acid	13,13 3β-acetate,13 6β-acetate, 13 3β,6β-diacetate, 35,36	[500,1303]
FeSO$_4$, aq. CH$_3$CN	13,14,15,35,36	[1304]
Ti$_2$(SO$_4$)$_3$, aq. CH$_3$CN	13,14,15,35,36	[1304]
Unspecified, water	16,27	[169]
Radiolysis, ^{60}Co γ-Radiation:		
Water	14,15,16,35,36	[73,2291]
Water, N$_2$O	13,14,15,16,35,36	[73,2291]

lysis of aqueous cholesterol dispersions saturated with N_2O gave the same six products more rapidly. Major products were the 7-ketone 16, 3β,7β,-diol 15, and 5α,6α-epoxide 35; minor products were the 3β,7α-diol 14, 5β,6β-epoxide 36, and 3β,5α,6β-triol 13. As all six products were formed in experiments in which air was excluded and the 7-hydroperoxides 46 and 47 indicative of cholesterol autoxidation were not detected in experiments conducted in air, autoxidation cannot account for these results [73,2291]. We did not encounter the 3β,25-diol 27 found in one case of Fenton reagent oxidation of cholesterol in aqueous suspension [169].

The products found establish that HO· attacks cholesterol at or near the B-ring olefinic bond. Radical addition to the double bond yield the 5,6-epoxides 35 and 36 and the triol 13; radical abstraction reactions at the adjacent allylic C-7 site yield the products 14-16. The 5,6-epoxides 35 and 36 were formed in the ratio 3.5:1, thus quite different from the 1:8 to 1:11 ratio found in the oxidation of cholesterol in aqueous dispersions by H_2O_2, organic hydroperoxides, or air [2297-2299] or to the 1:3.6 ratio found in other cholesterol autoxidations [93]. The disparity between the 5,6-epoxide ratio obtained with H_2O_2 as specific oxidant and that in the radiolysis experiments in which HO· is implicated suggests that the 5,6-epoxides 35 and 36 be products of HO· oxidation of cholesterol and not of any H_2O_2 potentially generated in the radiolysis of H_2O. Cholesterol is also epoxidized by H_2O_2 in aqueous acetic acid solutions, but the addition of Fe(III) ions essential for the Fenton reagent greatly accelerate the reaction. An alternative concept of formation of peracetic acid as the active epoxidizing agent in such systems [1303] cannot be valid in accounting for the presence of the 5,6-epoxides in other systems devoid of acetic acid.

The 3β,5α,6β-triol 13 appears to be formed by the simple hydration of the 5,6-epoxides 35 and 36 initially formed by HO· attack on cholesterol [73,1303] and not by the addition of two HO· to the cholesterol double bond [500]. Other related systems incorporating H_2O_2 and transition metal ions likewise apparently yield HO· which oxidizes cholesterol to the 3β,5α,6β-triol 13 as chief product. For instance, the photolysis of H_2O_2 sensitized by UO_2SO_4 gave the triol 13 [2098].

The proportions of C-7 oxygenated products 14-16 are characteristic of those found in autoxidations. Thus, the ratio of 3β,7α-diol 14 to 3β,7β-diol 15 was 1:8, approximatley the same as found in many air oxidations of cholesterol. The more stable quasiequatorial 7β-hydroxyl group of 15 versus the less stable quasiaxial 7α-hydroxyl group of the 3β,7α-diol 14 appears to reconcile these data, but it is clear that neither 3β,7-diol derived by autoxidation but derived by HO· attack.

The autoxidation of cholesterol may also occur together with HO· oxidations or as consequence of HO· attack initiating autoxidation. In experiments in which aqueous cholesterol dispersions were sparged with O_2 during radiolysis the 7-hydroperoxides 46 and 47 were formed along with products 13-16,35, and 36 [73]. Furthermore, the greater yields of products 13 and 16 obtained in X-radiolysis of aqueous acetic acid products 13 and 16 obtained in X-radiolysis of aqueous acetic acid solutions of cholesterol conducted in air versus yields obtained in the absence of air [1275] may also be taken as suggesting that HO· initiate autoxidation of cholesterol.

Other Atmospheric Oxidants. Other oxidants present

variously in polluted air may include $NO, NO_2, HNO_3, CH_3COOONO_2$, etc., in addition to O_3, $^1O_2$, $HOO\cdot$, $HO\cdot$ $O(^3P)$, etc. already mentioned. Some of these species exist in unpolluted, natural air but it seems unlikely that these species serve to initiate cholesterol autoxidations except where tissues or sterol may be exposed to such oxidants deliberately.

The oxidation of cholesterol by NO or by $CH_3COOONO_2$ is not recorded, but the attack of NO_2 on monomolecular layers of cholesterol spread on water yields products of two sorts. Cholesterol esterification, giving cholesterol 3β-nitrate, is a prominent aspect of the reaction [1217,1529], but the common autoxidation products including the epimeric 3β,7-diols 14 and 15, the 7-ketone 16, and the 3β,5α,6β-triol 13 were also formed [1217]. However, as the exclusion of air from the experiments with NO_2 was not stipulated and the same monolayers of cholesterol yield these products 13-16 by autoxidation [1217,1218,2651], it is not certain that NO_2 is directly involved in these autoxidations.

On the basis of chromatographic evidence the dienone 12 and the 3β,5α-diol 50 were suggested as also being products of the experiments with NO_2 and of autoxidation as well [1217,1218]. In this matter, the chromatographic identifications cannot be considered as definitive, for neither 12 nor 50 have been associated with cholesterol autoxidation heretofore but rather with the decomposition of the 5α-hydroperoxide 51 formed from cholesterol by attack of $^1O_2$ [2300,2454,2578].

The oxidizing action of NO_2 on tissue cholesterol has been demonstrated in one *in vivo* study. Rats breathing 3-6.5 ppm NO_2 were found to have formed substantial levels of the 5,6-epoxides 35 and 36 in their lung tissue [2192, 2193]. Dietary antioxidant appeared to limit the extent of epoxidation [2191].

Although HNO_3 is known to be present in some polluted air, its action on cholesterol in such circumstances is not described, and only more harsh conditions involving HNO_3 have been investigated. Cholesterol 3β-acetate is converted by fuming HNO_3 to 6-nitrocholest-5-en-3β-ol 3β-acetate (267) [52,1596], whereas acetic acid solutions of sodium nitrite containing conc. H_2SO_4 or fuming HNO_3 yield the 6-nitro compound 267, and 3β-acetate of 3β,5-dihydroxy-5α-

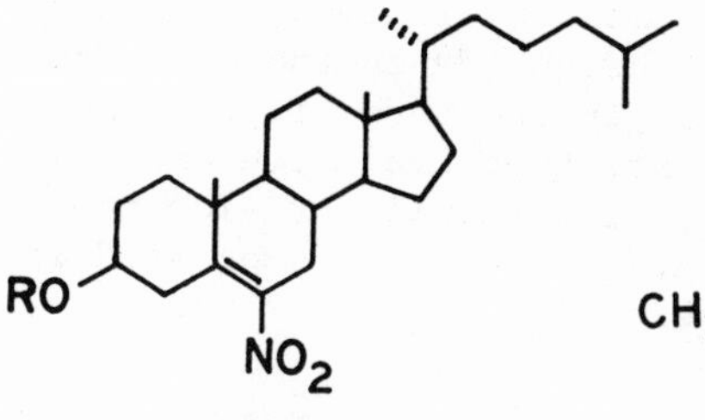

267 R = CH₃CO

268 R = NO₂

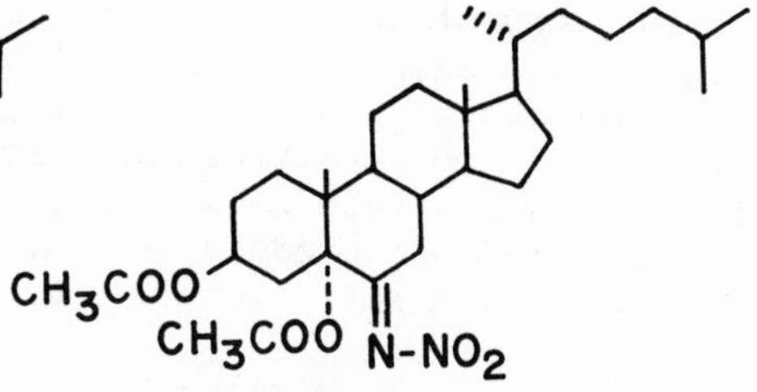

269

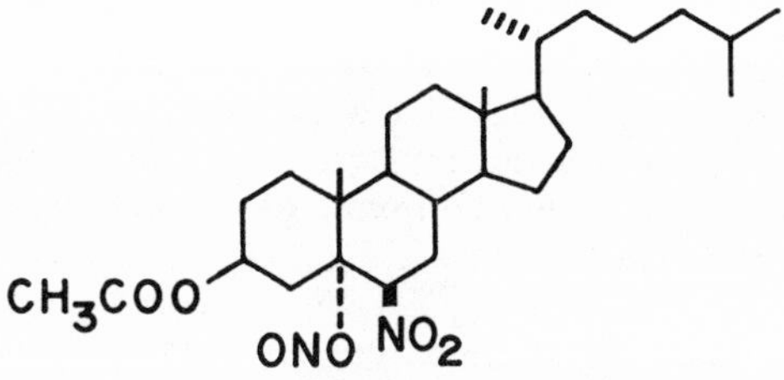

270

cholestane-6-one (44), and 6-nitrimino-5α-cholestane-3β,5-
diol 3β,5α-diacetate (269) [1717,1718,1798-1800]. Reaction
with a 2:1 mixture of N_2O_4 and O_2 (in diethyl ether) gave
3β-hydroxy-6β-nitro-5α-cholestane-3β,5-diol 3β-acetate 5α-
nitrate (270) [52]. Nitric acid acting upon cholesterol
yields the ester 6-nitrocholest-5-en-3β-ol 3β-nitrate (268)
[995,2687]. It thus appears that nitrogen oxides and HNO_3
may transform cholesterol into nitro derivatives but also
oxidize and esterify cholesterol as well.

Radiation-Induced Events

Considerable experimental evidence has been adduced in
support of the initiation of cholesterol autoxidation by
radiation, including infrared, visible, and ultraviolet
light and ionizing X- and γ-radiation. More direct demonstrations of cause and effect have been provided in these
cases than for the prior cases of organic peroxide homolysis,
transition element catalysis, and activated oxygen species.

<u>Radiation</u> <u>without</u> <u>Sensitizer</u>. Radiation-induced au-
toxidations of cholesterol in which $^3O_2$ is implicated may
proceed via several possible pathways, depending on whether
a sensitizer be present or not. In the case of pure cholesterol with no sensitizer absorption of radiation energy
quanta may increase rotational, vibrational, or electronic
energy levels to produce energized species as suggested in
Eq. 40. Given sufficiently energetic quanta for electronic

$$\text{Cholesterol} + h\nu \rightarrow \text{Cholesterol*} \qquad \text{Eq. 40}$$

excitation, the excited cholesterol species may then undergo bond homolysis or ionization, with subsequent reaction
with $^3O_2$ to give recognized cholesterol autoxidation
products.

Bond homolysis as initiation event with subsequent
radical propagation reactions involving $^3O_2$ and hydrogen
abstraction represented in Eqs. 1-3 accounting for formation
of the first stable hydroperoxide products $\underline{46}$ and $\underline{47}$ are
summarized in FIGURE 11.

Rather than bond homolysis, radiation-induced initiation events may involve ionizations. Ionization summarized
in Eq. 41 may be posited, these being effected by radiation
in the present discussion. However, ionizations to radical
carbanion or carbocation in strongly basic or acidic media
respectively or in transition element catalysis might also
ionize cholesterol under other circumstances.

$$\text{Cholesterol}^{+\cdot} \overset{-e^-}{\leftarrow} \text{Cholesterol*} \overset{+e^-}{\rightarrow} \text{Cholesterol}^{-\cdot} \qquad \text{Eq. 41}$$

$$\text{Cholesterol}^{-\cdot} + {}^3O_2 \rightarrow \text{Cholesterol-OO}^- \qquad \text{Eq. 42}$$

$$\text{Cholesterol-OO}^- + H^+ \rightarrow \text{Cholesterol-OOH} \qquad \text{Eq. 43}$$

Subsequent reaction of the radical carbanion with $^3O_2$
yielding the sterol hydroperoxide anion which upon protona-
tion yields the stable sterol hydroperoxide is suggested in
Eq. 42 and 43. Evidence supporting such radiation-induced
events is wanting, but as strongly alkaline solutions or
dispersions of cholesterol in aqueous alcohol (as from
alkaline tissue hydrolysates)are notoriously susceptible
to autoxidation, yielding the common products $\underline{14}$-$\underline{16}$, the
ionic processes of Eqs. 41-43 may be operative. Steroid
carbanions are regularly invoked in the autoxidation of other

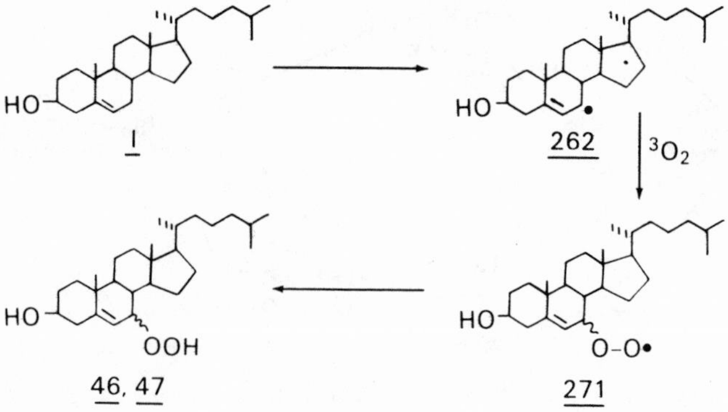

FIGURE 11.   Initial events of free radical
autoxidation of cholesterol.

steroids by air in strongly alkaline media (*cf.* Chapter VI).

More reasonably, the cholesterol carbocation may be
involved in radiation-induced initiation events. Subsequent
loss of a proton from the radical cation, as in Eq. 44, may
follow very shortly upon the radiation-induced ejection of
an electron, thereby yielding the same cholesterol 7-radical
intermediate 262 (FIGURE 11) implicated in cholesterol autoxidations. The two-step process of ejection of electron, then
of proton, may indeed be the initiating events caused by
radiation.

$$\text{Cholesterol}^{\ddagger} \rightarrow \text{Cholesterol}^{\cdot} + \text{H}+ \qquad \text{Eq. 44}$$

Radiation with Sensitizers. The formulations of Eq. 40
and Eqs. 1-3 or of Eqs. 40,41, and 44 and Eqs. 2 and 3 adequately account for the observed autoxidations of cholesterol induced by radiation, whether infrared, visible, or
ultraviolet light or X- and γ-radiation. However, in the
presence of an appropriate sensitizer or other species capable of energy transfer, other processes may operate. The
general processes of FIGURE 12 have been advanced as possible reactions occurring in photosensitized oxidations
involving $^{3}O_2$ [329,2120]. In the general scheme, although

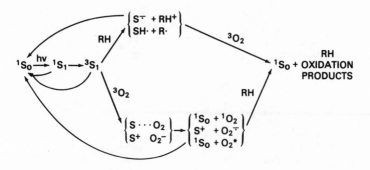

FIGURE 12. Generalized photosensitized oxidation scheme.
Substrate, RH; ground state sensitizer, $^1S_0$;
excited singlet sensitizer, $^1S_1$; triplet
sensitizer, $^3S_1$.

$^3O_2$ is the formal species involved, the ultimate reaction
may involve not only $^3O_2$ but also $^1O_2$, O_2^-, or vibrationally
excited dioxygen. The scheme encompasses both Type I
(Eqs. 45,46,2,3,47) and Type II (Eqs. 45,48,49) photosensitized oxidation processes [887] but includes charge transfer processes in which ionic as well as radical species
are formulated.

Type I: $^1S_0$ + $h\nu$ → $^1S_1$ Eq. 45

            $^1S_1$ + RH → R· + SH·   Eq. 46

            R· + $^3O_2$ → ROO·   Eq. 2

            ROO· + RH → ROOH + R·   Eq. 3

            ROO· + SH· → ROOH + $^1S_0$ Eq. 47

Type II: $^1S_1$ → $^3S_1$ Eq. 48

            $^3S_1$ + $^3O_2$ → $^1S_0$ + $^1O_2$ Eq. 49

        The Type I process involves reaction between the ex-
cited state sensitizer $^1S_1$ and substrate cholesterol in
the ground state to give radical species according to
Eq. 46. The product sensitizer radical is then returned to

ground state sensitizer $^1S_0$ by the termination reaction of
Eq. 47. The cholesterol radical, presumed to be the 7-radi-
cal 262,then reacts with $^3O_2$ to give cholesterol hydroperoxides according to Eq. 2 and 3 and FIGURE 11.

Type II photosensitized oxygenation of cholesterol
involves the key feature of Eq. 48 of intersystem crossing
in which the excited singlet state sensitizer $^1S_1$ with
paired electrons is transformed to the triplet sensitizer
$^3S_1$ with unpaired electrons. Energy transfer between $^3S_1$
and $^3O_2$ then occurs (Eq. 49), with formation of electronically excited $^1O_2$ which then reacts with substrate cholesterol as previously described.

These two well recognized photosensitized processes
acting on cholesterol give different hydroperoxide products,
the Type I process involving radical intermediates giving
the epimeric 7-hydroperoxides 46 and 47 [2313], the Type II
process in which $^1O_2$ is implicated giving the 5α-hydroperox-
ide 51 and the epimeric 6-hydroperoxides 247 and 89 [1401].

In the charge transfer process of FIGURE 12, reaction
between singlet sensitizer $^1S_1$ and substrate RH yields by
Eq. 50 the pair $(S^{\overline{\cdot}}, RH^{\overset{+}{\cdot}})$ which reacts with $^3O_2$ as in Eq. 51
to give the reduced species $O_2^{\overline{\cdot}}$ whose reaction with $RH^{\overset{+}{\cdot}}$
yields the product hydroperoxide as indicated in summary
Eq. 52. Superoxide radical anion is thus implicated and
not $^1O_2$.

$$^1S_1 \quad + \quad RH \quad \rightarrow \quad [S^{\overline{\cdot}}, RH^{\overset{+}{\cdot}}] \qquad\qquad \text{Eq. 50}$$

$$[S^{\overline{\cdot}}, RH^{\overset{+}{\cdot}}] \quad + \quad ^3O_2 \quad \rightarrow \quad ^1S_0 \quad + \quad O_2^{\overline{\cdot}} \quad + \quad RH^{\overset{+}{\cdot}} \qquad \text{Eq. 51}$$

$$O_2^{\overline{\cdot}} \quad + \quad RH^{\overset{+}{\cdot}} \quad \rightarrow \quad ROOH \qquad\qquad \text{Eq. 52}$$

Using 9,10-dicyanoanthracene as sensitizer oxidation
products of several substrates have been described which
mimic products formed in $^1O_2$ reactions, and these results
have been interpreted in terms of the charge transfer processes of Eqs. 50-52 [705,2091]. The photosensitized
oxygenation of cholesterol using 9,10 dicyanoanthracene as
sensitizer yields the 5α-hydroperoxide 51 as product [783,
784] with subsequent isomerization and epimerization leading to formation of both 7-hydroperoxides 46 and 47.
Whether the reaction involves $^1O_2$ or charge-transfer processes remains to be settled.

The Cholesterol 7-Radical. Many of these radiation-
induced processes are advanced for cholesterol autoxidation
by analogy to those extensively studied with other substrates for which much other supporting evidence is available. By product analysis it is reasonable to assume that
similar processes are at work in the specific case of cholesterol, but some ambiguity may attend the matter. Thus, if
charge-transfer processes indeed are implicated in the formation of the acknowledged $^1O_2$ product 5α-hydroperoxide 51,
then the presence of 51 would not infer per se the nature
of the oxidation process involved. Likewise, the epimeric
7-hydroperoxides 46 and 47 are products of free radical
autoxidation but might also be formed by putative ionic
processes. Moreover, the 7-hydroperoxides are established
products of the rearrangement of the 5α-hydroperoxide 51.

Thus mere product analysis, no matter how complete,
might not establish an oxidative process correctly. Other
supporting evidence is required, and the direct observation
of electron spin resonance of an unpaired electron of a
radical derived by irradiation of cholesterol would be
strong evidence supporting true free radical intermediates
in radiation-induced cholesterol autoxidation. Although
cholesterol autoxidation in air is increased by infrared,
visible, or ultraviolet light irradiations [1072,2303,2313],
electron spin resonance studies on such preparations have
not been described. However, a distinctive six-line electron spin resonance spectrum has been repeatedly recorded
on crystalline cholesterol irradiated in vacuum with ionizing radiation, by 50 kV X-rays [1004] or by ^{60}Co γ-rays
[661,899,1948].

The upper spectrum of FIGURE 13 is the first derivative
spectrum of pure cholesterol irradiated in vacuum with
^{60}Co γ-radiation [661], properly interpreted as deriving from
the 3β-hydroxycholest-5-en-7-ylradical (262). Prior interpretations of the spectrum in terms of a 3β-hydroxycholest-
6-en-5-yl radical [899,1948] do not properly account for the
observed hyperfine splittings, a matter which the 7-radical
262 formulation does achieve. The axial 4β- and 8β-hydrogens
of the 7-radical 262 are equivalent and give a hyperfine
splitting of 26 ± 1 Oe, whereas the 7ξ-hydrogen gives a
15 ± 1 Oe splitting to make a triplet of doublets pattern.
The equatorial 4α- and vinyl 6-hydrogens appear to have
hyperfine splitting constants of less than 6 Oe, and as the

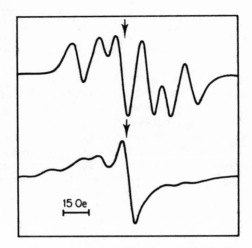

FIGURE 13. First derivative electron spin resonance spectra
of irradiated cholesterol in vacuum (upper) and
following introduction of O_2 (lower). Arrows
indicate the g-value of the free electron.

observed line widths are approximately 10 Oe, these effects
were not observed [1004].

The six-line signal of the cholesterol 7-radical 262
was stable for days in vacuum at room temperature [899],
but introduction of air into the system caused a rapid
degradation of the signal, (cf. FIGURE 13) which was lost
in a matter of minutes [661,1948]. The one-line signal was
recognized as being derived from the product 7-peroxyl radical 271 formed from the 7-radical 262 upon reaction with
$^3O_2$. These data clearly support the cholesterol autoxidation sequences depicted in FIGURE 11.

Similar spectra were obtained from the 5-stenols stigmasterol (70) and sitosterol (20) and from other sterols
[899,1948], but the spectrum obtained from the irradiated
5α-stanol 2 was of a different sort which was not analyzed
as to localization of the unpaired electron [1004].

Supporting electron spin resonance data have been

recorded for the radiation-induced bond homolysis of cholesterol in solution. Solutions of cholesterol in cyclohexane
containing CCl_4 and benzene as free radical initiators,
irradiated with 235-265 nm light in O_2, constitute a free
radical oxidation system according to Eqs. 53-55. The
electron spin resonance spectrum recorded for the system
(one-line, g = 2.0145 $\pm$ 0.0005) was identified as derived
from a peroxyl radical formed by reaction of the cholesterol radical with $^3O_2$ according to Eq. 2. As the epimeric 3β,7-diols 14 and 15 and 7-ketone 16 were isolated
from the reaction, it may be concluded that the 7-radical
262 and 7-peroxyl radical 271 were those formed in the
system [1005].

$$CCl_4 + h\nu \rightarrow CCl_4^* \qquad \text{Eq. 53}$$

$$CCl_4^* + C_6H_{12} \rightarrow HCl + Cl_3C\cdot + C_6H_{11}\cdot \quad \text{Eq. 54}$$

$$Cl_3C\cdot + \text{Cholesterol} \rightarrow \text{cholesterol}\cdot + CHCl_3 \quad \text{Eq. 55}$$

These spectral data for both crystalline cholesterol
irradiated in vacuum and solutions of cholesterol irradiated
in air [1004,1005] together with product isolations establish that the major free radical autoxidation process acting
upon cholesterol is via the 7-radical 262, 7-peroxyl radical
271, and epimeric 7-hydroperoxides 46 and 47. The 7-hydro-
peroxides are the first stable products of cholesterol
autoxidation and were postulated as such by Bergström in
1942 [197,198,203]. Kinetics studies in aqueous sodium
stearate dispersions of cholesterol support the formulation
in detail [1297].

Early Radiation Studies. Indeed, sterol hydroperoxides
were suggested as oxidation products of cholesterol associated with radiation-induced events from even earlier work
directed towards the cholecalciferol problem [2020].
Although sterol hydroperoxides were variously detected in
autoxidized cholesterol preparations using modern methods,
direct isolations of the epimeric 7-hydroperoxides 46 and
47 from air-aged cholesterol were not achieved until 1972,
when we isolated both the 7α-hydroperoxide 46 and 7β-hydro-
peroxide 47 from such sources and demonstrated their formation as the first stable products in ^{60}Co γ-radiation
experiments with cholesterol [2313,2455].

Several early reported oxidations of cholesterol with

or without sensitization bear mention at this point. Oxidation of cholesterol in benzene solution containing benzophenone as sensitizer gave an unidentified ketone [580],
but xylene solutions of cholesterol 3β-acetate containing
iron phthalocyanine gave the 7-ketone 16 3β-acetate [512]
in a reaction reminiscent of Type I photosensitized oxidation of cholesterol via the 7-hydroperoxides 46 and 47
which upon dehydration yield the 7-ketone 16. Irradiation
of thin layers of cholesterol on glass (without sensitizer)
gave the 3β,7β-diol 15, the 3β,6β-diol 40, and an unidentified stenediol [2701].

    More recent general studies of radiation-induced oxi-
dations of cholesterol have identified the common autoxidation products including the enone 10, 3β,5α,6β-triol 13,
3β,7-diols 14 and 15, 7-ketone 16, 3β,25-diol 27, and
5,6-epoxides 35 and/or 36 under a variety of conditions
[282,495,496,503,567,1072,1275,1522,2303,2313,2487,2656].
Moreover, the presence of these common autoxidation products as artifacts in animal tissues and plasma exposed to
light during manipulations has been demonstrated [281,459,
481,494,496,940-942]. Sterol hydroperoxides presumably
the epimeric 7-hydroperoxides 46 and 47 have been detected
chromatographically using KI-starch, $Fe(SCN)_2$, or N,N-
dimethyl-p-phenylenediamine reagents [1072,1789,2295,2303,
2313] in some of these cases and also in irradiated cholesterol-containing foodstuffs [12]. The presence of these
several autoxidation products in these radiation experiments is viewed as arising via the radiation-induced
autoxidation processes of FIGURE 12 in which the initially
formed 7-hydroperoxides 46 and 47 are thermally decomposed
to the products identified by transformations discussed in
the next chapter. However, products 13-16,35, and 36 also
arise by attack of HO· on cholesterol as well as by the
7-hydroperoxide pathway. Therefore, these products may
have dual chemical origins in systems where HO· forms.

                REACTIONS WITH GROUND STATE DIOXYGEN

    The features of the reaction between $^3O_2$ and the
B-ring of cholesterol have already been discussed in several
ways, but a more systemized discussion of this mode of
reaction together with other oxidation reactions is presented
here. Autoxidation in the B-ring allylic center may well
preceed, indeed cause the autoxidations at other centers,

but evidence in support of such a thesis has not been recorded.

A-Ring Dehydrogenation

The oxygen-dependent dehydrogenation of cholesterol results in formation of cholest-5-en-3-one (6) and presumably H_2O_2 in the sequence suggested by Eq. 56. This autoxidation step is thus quite different from the dehydrogenations obtained in the dry distillation of cholesterol in

$$RHOH + O_2 \rightarrow R=O + H_2O_2 \qquad Eq. 56$$

which hydrogen gas (and the enone 8) is product [611,768, 997]. The reaction may be viewed as the nonenzymic equivalent of the dehydrogenation of cholesterol by microbial cholesterol oxidases (cholesterol:oxygen oxidoreductase, EC 1.1.3.6) [2276-2278] of considerable interest for sterol analyses in plasma.

Dehydrogenation of cholesterol to the Δ^5-3-ketone 6 is inferred by isolation of the isomeric Δ^4-3-ketone 8 from cholesterol heated in air [1311] and by detection of 8 in various autoxidized cholesterol preparations [765,1523,1690, 2313]. Moreover, our isolation from air-aged cholesterol of 3-oxocholest-4-ene-6β-hydroperoxide (59) [2455] provides compelling indirect evidence that autoxidative A-ring dehydrogenation of cholesterol must occur. Nonetheless, recovery of pure enone 6 from oxidized cholesterol samples has not been achieved despite good try. The enone 6 has been detected by chromatography and mass spectrometry in naturally air-aged cholesterol samples and in pure cholesterol samples irradiated in air with [60]Co γ-radiation, but concommitant isomerization to the enone 8 and oxidation to the 3,6-diketone 108 precluded isolation of pure enone 6 [70].

The accumulated data are sufficient to establish A-ring dehydrogenation of cholesterol as a separate though quantitatively minor mode of autoxidation. As the transformation is oxygen dependent and evidence for formation of H_2O_2 from cholesterol samples irradiated under similar conditions has been recorded [1540,1600,2349], the formulation of Eq. 56 is advanced.

However, direct removal of hydrogen from cholesterol

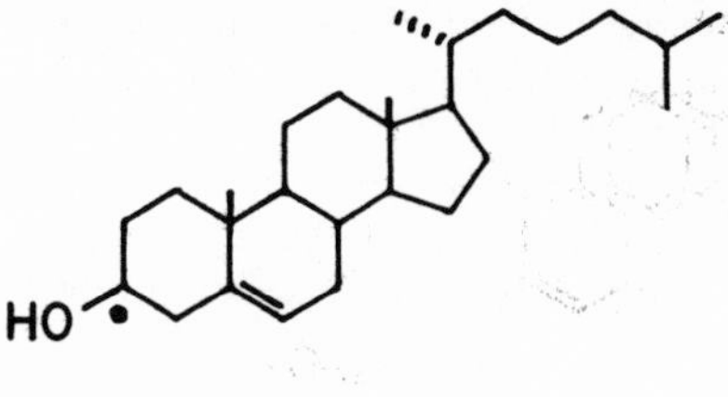

272 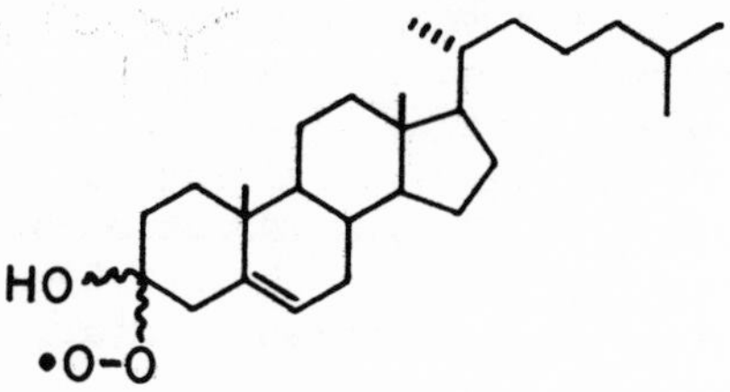   273

by $^3O_2$ may not occur. Rather, initial hydrogen abstraction
yielding the cholesterol 3-radical 272 followed by reaction
with $^3O_2$ to form the 3-peroxyl radical 273 which upon elimination of H_2O_2 gives the enone 6 may be the process implicated. Hydrogen abstraction from the C-3 carbon atom might
be a radiation induced event but also might be an event
induced by cholesterol 7-radical 262, 7-oxyl radical 263, or
7-peroxyl radical 271 variously implicated in free radical
chain propagation reactions of cholesterol autoxidations.

Other examples of oxygen-dependent alcohol dehydrogenations are provided in Chapter VI. However, there are other
means by which cholesterol can be oxidized at the 3β-hydroxyl
group that are not dependent on O_2. For instance, in acidified methanol the anodic oxidation of cholesterol yields
cholesterol cholest-5-en-3β-yloxymethyl ether [1017].

A-Ring Hydroperoxide Formation

We have never been able to identify a sterol hydroperoxide in air-aged cholesterol that would suggest that
A-ring hydroperoxides be formed in autoxidation. The
3-peroxyl intermediate 273 just suggested as a putative
intermediate in the oxygen-dependent dehydrogenation of
cholesterol to the enone 6 has been neither detected as
such nor isolated as a stabilized hydroperoxide. Moreover,
hydroperoxide formation at the allylic C-4 position of the
A-ring remains undemonstrated. Nonetheless, cholesterol
autoxidation at the C-4 position must occur, witness our
isolation of cholest-5-ene-3β,4β-diol (41) from air-aged
cholesterol (cf. TABLE 4). In exact analogy to the formulation of C-7 radical, 7-peroxyl radical, and stable 7-
hydroperoxides of FIGURE 12, one may posit formation of
a C-4 carbon-centered radical, 4-peroxyl radical, and

product 4-hydroperoxides 3β-hydroxycholest-5-ene-4α-hydro -
peroxide and 3β-hydroxycholest-5-ene-4β-hydroperoxide (274).
However, although one expects that both 4-hydroperoxides be
formed, we have evidence only for formation of the 4β-hydro-
peroxide 274 via isolation of the 3β,4β-diol 41, a search
for the epimeric cholest-5-ene-3β,4α-diol in air-aged cholesterol not being successful.

B-Ring Hydroperoxide Formation

Formation of the epimeric cholesterol 7-hydroperoxides
46 and 47 has already been described in an earlier section
of this chapter (cf. FIGURE 12), and these 7-hydroperoxides
may be regarded as the chief quantitatively major products
of cholesterol autoxidation. We have not encountered evidence of rearrangement of the first-formed 7-hydroperoxides,
of multiple attack of $^3O_2$ to form dihydroperoxides, or of
formation of other nuclear A-,B-, or C-ring hydroperoxides
other than that just mentioned for oxidation at the 4β- position.

The quasiequatorial 7β-hydroperoxide 47 predominates
over the epimeric quasiaxial 7α-hydroperoxide 46 in all our
studies [2295,2300,2303,2313,2454,2455], which relationship
may be understood in terms of the generally greater thermo-
dynamic stability of equatorial alcohols over their axial
epimers. We have consistently observed the 7β-hydroperoxide
47 in up to ten-fold increased levels over those of the
7α-hydroperoxide 46. Moreover, the facile epimerization of
the 7α-hydroperoxide 46 and 47 supports the greater stability of the 7β-hydroperoxide 47 [2454,2455]. The reverse
epimerization of 47 to 46 has not been observed.

The epimerization of the 3β,7α-diol 14 to the 3β,7β-
diol 15 under the same conditions is also demonstrated, with
insignificant epimerization of the 3β,7β-diol 15 to the quasiaxial epimer 14 [2454,2455]. However, in acidic solutions
the interconversion of the epimeric 3β,7-diols 14 and 15
occurs as does also the interconversion of epimeric 3β,7-diol
7-methyl and 7-ethyl ethers [1404] and 3,7-diacetate esters
[2051,2354].

Side-Chain Hydroperoxide Formation

In the absence of mitigating evidence this present

formulation appears to explain adequately the nature of the
major free radical autoxidation process in which the B-ring
of cholesterol is involved. By extension of the free radical chain reaction concept to include hydrogen abstraction
from side-chain carbon atoms instead of from the allylic
C-7 carbon atom, it is possible to formulate a related means
by which the side-chain cholesterol hydroperoxides are
formed. In this case, it is reasonable to expect that the
tertiary carbon atoms C-17, C-20, and C-25 would be more
susceptible to hydrogen abstraction then would be the secondary carbons C-22, C-23, and C-24, which would be more reactive than the primary carbon atoms C-21 and C-26. In general
this order was found in the yields of cholesterol hydroperoxides isolated from air-aged cholesterol wherein the 25-hydro-
peroxide 26 was of highest yield, followed by the isomeric
20-hydroperoxides 32 and 22 [2565,2576]. The postulated
tertiary 17-hydroperoxides 3β-hydroxycholest-5-ene-17α- and
17β-hydroperoxide 275 were not isolated from air-aged cholesterol, although their putative β-scission produce 3β-hy-
droxyandrost-5-en-17-one 86 was [2576]. Moreover, several

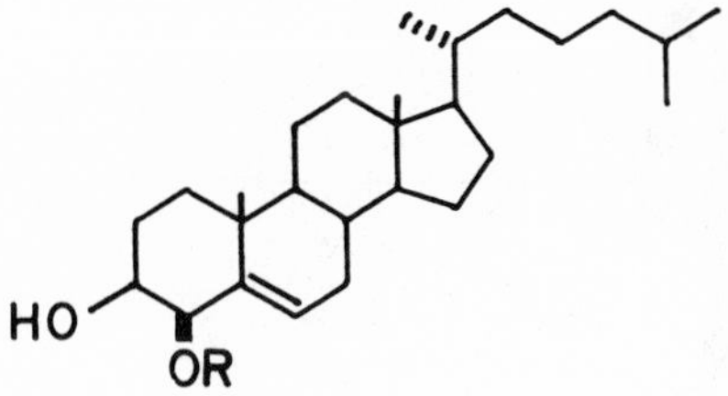

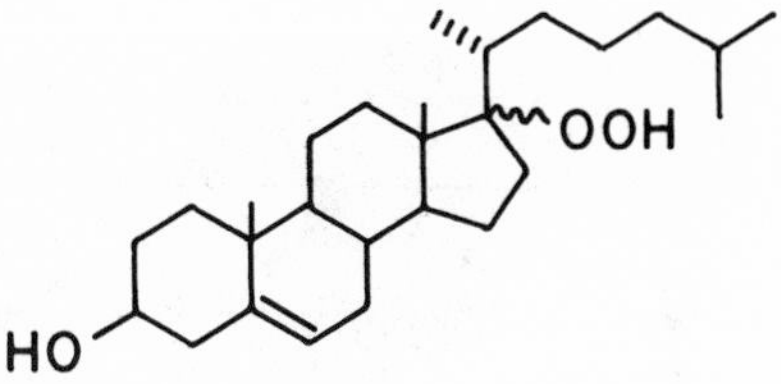

274 R = OH   275

41 R = H

unidentified sterol hydroperoxides were detected from air-
aged cholesterol, and the 17-hydroperoxides 275 may be among
these unidentified hydroperoxides. Finally, nuclear tertiary
8-,9-, and 14-hydroperoxides, for which no evidence exists,
conceivably might also be formed.

     The scheme of FIGURE 14 suggests a means by which the
initial B-ring oxidations leading to the 7-radical 262,
7-oxyl radical 263 and 7-peroxyl radical 271 may, instead of
continuing the chain reaction by abstraction of C-7 allylic
hydrogen, continue the chain reaction by abstraction of the

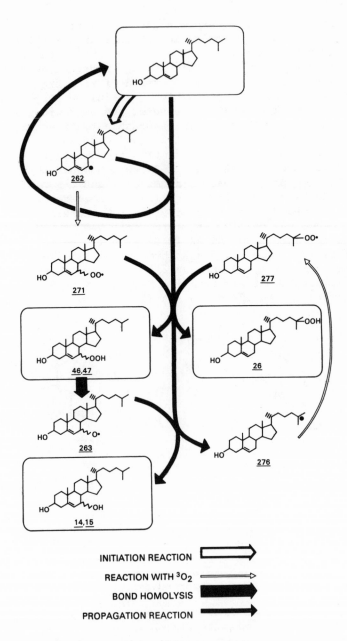

FIGURE 14. Postulated chain propagation reactions leading to cholesterol side-chain hydroperoxides.

C-25 hydrogen atom yielding putatively the 25-radical 276 which upon reaction with $^3O_2$ yields the 25-peroxy radical 277. Hydrogen atom abstraction from another cholesterol molecule, in the C-7, C-25, or other appropriate position, then leads to the stable 25-hydroperoxide product 26 and continuance of the chain propagation reactions involving other cholesterol radicals. Although this formulation seems reasonable, there is no direct experimental evidence in support.

By this mechanism one projects the formation of the several side-chain hydroperoxides of cholesterol, now recognized as numbering perhaps as many as thirteen. These are: the isomeric 17-hydroperoxides (275) whose formations are postulated on the basis of the isolation of the 17-ketone 86 from air-aged cholesterol [2576], the isomeric 20-hydroperoxides 22 and 32 isolated as such from air-aged cholesterol [2565,2576], the epimeric 22-hydroperoxides 3β-hydroxy-(22R)- and (22S)-cholest-5-ene-22-hydroperoxide (278) whose formation in air-aged cholesterol is implied by the presence of the putative C_{21} degradation products 110-113 isolated from air-aged cholesterol [2562,2563,2576],

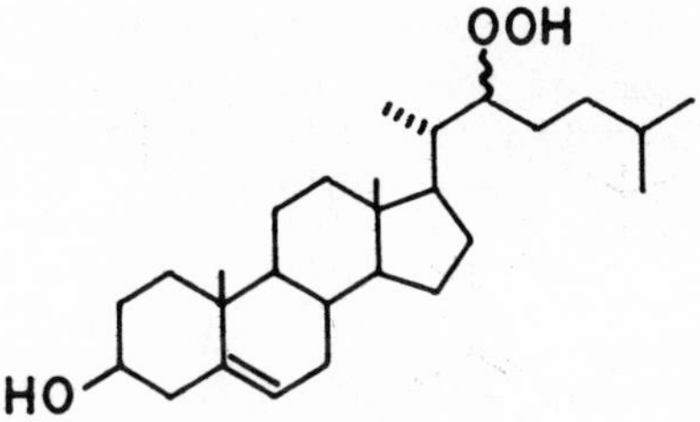

278 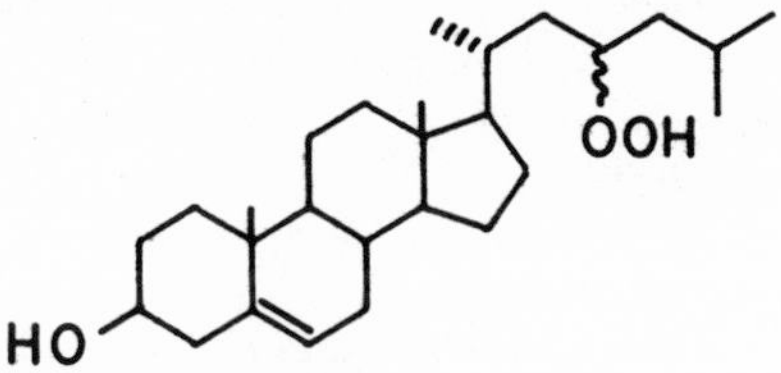   279

the epimeric 23-hydroperoxides 3β-hydroxy-(23R)- and (23S)-cholest-5-ene-23-hydroperoxide (279) not detected but implied by the presence of several degradation products discussed in Chapter V, the epimeric 24-hydroperoxides 3β-hydroxy-(24R)-cholest-5-ene-24-hydroperoxide (105) and 3β-hydroxy-(24S)-cholest-5-ene-24-hydroperoxide (106) isolated as an unresolved mixture from air-aged cholesterol [2579], and isomeric 26-hydroperoxides 3β-hydroxy-(25R)-cholest-5-ene-26-hydroperoxide (28) and 3β-hydroxy-(25S)-cholest-5-ene-26-hydroperoxide (30) also isolated as an unresolved mixture from autoxidized cholesterol [2559].

Whereas our initial view of the autoxidation of cholesterol at the C-20 position was one involving stereospefici-
ty, forming only (20S)-20-hydroperoxide 22 [2576], this was
not correct, and both C-20 isomeric hydroperoxides 22 and 32
are formed [2565]. Furthermore, both (24R)- and (24S)-24-
hydroperoxides and both (25R)- and (25S)-26-hydroperoxides
are formed, thus clearly establishing the generality of
autoxidative attack by $^3O_2$ at these centers. It is thus
reasonable to extend these findings to postulation that
both C-17, C-22, and other centers also oxidized in both
possible stereochemical sites.

It is thus possible to account for the formation of
all known hydroperoxide derivatives of cholesterol by the
initial formation of the 7-radical 262 and 7-peroxy radical
271 and the chain propagation reaction sequence of FIGURE 14.
However, there may be alternative means of derivation of
the requisite carbon radicals of the side-chain, for although hydrogen atom abstraction in the allylic C-7 position
is by far the predominant reaction of irradiated cholesterol
in vacuum as evinced by electron spin resonance spectra,
some direct C-25 (or other) hydrogen abstraction may also
occur at a much diminished level and thus be lost to detection by electron spin resonance spectroscopy. Formation of
an electron spin resonance spectrum of the 5α-stanol 2
[1004], while presently not interpreted, indicates that
other hydrogens of the stanol molecule are subject to
abstraction.

As no stanol hydroperoxides have been observed as being
formed from the parent stanol by ^{60}Co γ-irradiation, this
matter is unsettled. Very preliminary and incomplete experiments where stanol and stenol have been mixed in the solid
state with sterol hydroperoxides suggest that an extensive
oxidative reaction system obtains. However, we have not
identified any of the many oxidation products formed.

In TABLE 9 are listed the thirteen sterol derivatives
that we consider to be the first formed stable autoxidation
products of cholesterol. Only the 4β-hydroperoxide 274,
17-hydroperoxides 275, and 22-hydroperoxides 278 are inferred products. Other products listed have been identified
by isolation.

TABLE 9. Initial Autoxidation Products of Cholesterol

1. A-Ring Autoxidation
   Cholest-5-en-3-one (6)
   3β-Hydroxycholest-5-ene-4β-hydroperoxide (274)*   [70]

2. B-Ring Autoxidation
   3β-Hydroxycholest-5-ene-7α-hydroperoxide (46)   [2313,2576]
   3β-Hydroxycholest-5-ene-7β-hydroperoxide (47)   [2313,2455]

3. Side-Chain Autoxidation
   3β-Hydroxycholest-5-ene-17-hydroperoxides (275)*
   3β-Hydroxy-(20R)-cholest-5-ene-20-hydroperoxide (32)   [2565]
   3β-Hydroxy-(20S)-cholest-5-ene-20-hydroperoxide (22)   [2576]
   3β-Hydroxy(22RS)-cholest-5-ene-22-hydroperoxide (278)*
   3β-Hydroxy-(24R)-cholest-5-ene-24-hydroperoxide (105)   [2576,2579]
   3β-Hydroxy-(24S)-cholest-5-ene-24-hydroperoxide (106)   [2576,2579]
   3β-Hydroxycholest-5-ene-25-hydroperoxide (26)   [2576]
   3β-Hydroxy-(25R)-cholest-5-ene-26-hydroperoxide (28)   [2559,2562]
   3β-Hydroxy-(25S)-cholest-5-ene-26-hydroperoxide (30)   [2559,2562]

*Inferred by other products isolated.

CONDITIONS OF AUTOXIDATION

Discussion of cholesterol autoxidation to this point
has not emphasized the physical and chemical conditions
under which the oxidations occur. Rather, arguments suggesting a common reaction mechanism have been advanced.
Comparison of oxidation products found in a large variety
of controlled chemical systems as well as in animal and
plant tissues leave this conclusion undisputed. Nonetheless, brief examination of the actual conditions involved
in controlled study of cholesterol autoxidation is now in
order.

Controlled chemical conditions provoking cholesterol
autoxidation are many and will be treated according to the
physical state of the oxidizable substrate in the system.
Thus, imposed autoxidation conditions may involve cholesterol in the solid state, in the dispersed state, or in
solution. Experience teaches that cholesterol autoxidations occur most rapidly in dispersed systems, with autoxidations in the solid state and in solution occurring less
rapidly. Obvious factors of time, temperature, concentration, pH, presence or absence of radiation, photosensitizers,
catalysts, transition metal ions, etc., are also of importance, but so long as air be the source of $^3O_2$ for the oxidation, the products formed may be reconciled with initial
formation of the cholesterol 7-hydroperoxides 46 and 47.
Other states of cholesterol (liquid, vapor) exist, but
these extremes have not been examined for autoxidations,
which must surely occur if air be present.

Autoxidation of Solid Cholesterol

Cholesterol autoxidized as pure, crystalline material
has received much attention, for it is in this state that
the deterioration on storage is so obvious. Anyone opening
up a tightly closed bottle of once pure cholesterol which
has been stored on the shelf for several years is bound to
be aware of the ease with which cholesterol is oxidized in
the crystalline state. The matter of autoxidation on storage will be discussed more fully in Chapter IX.

The similarity between naturally air-aged solid cholesterol and cholesterol heated or irradiated in air is well

recognized, and analyses establish that common products are
formed in all cases [1072,2303]. An accelerated autoxidation of cholesterol which mimics that of natural air-oxida-
tion is provoked by mere heating in air. We have found that
heating crystalline cholesterol in an oven in the dark at
60°C for 48 d, at 65°C for 30 d, at 70°C for 28 d, or at
100°C for 42 h causes extensive autoxidation of cholesterol
[2295,2303,2313,2455]. Indeed, heating at 70°C for 28 d,
recrystallization of the material to remove unoxidized
crystalline cholesterol, with repeated heating of the recovered cholesterol at 70°C, etc., [2455,2562] constitutes a
suitable means of creating autoxidation products for isolation work. Heating at 105°C for a week leads to destruction
of approximately half of the cholesterol [1072]. There may
be substantial problems involved in the isolation of autoxidation products from such heated sterol samples!

   A companion model system involves thin layers or films
of cholesterol deposited from solutions upon glass plates
exposed to $^3O_2$ [1789], to visible light in air [169,2303,
2313], or to ultraviolet light in air [2295,2303,2313,2701].
The method has also been applied to foodstuffs enriched in
cholesterol [12]. The finely divided state of cholesterol
on dried paper chromatograms subjected to light and air is
also conducive to autoxidation [942,1632], and other steroids
on dried paper chromatograms are extensively decomposed in
light and air [2085]. These conditions have not been advanced as model systems for autoxidation studies. However,
the rapid autoxidation of monolayers of cholesterol adsorbed
on silica gel in the presence of linoleic acid (and light
and air) has been utilized for systematic studies of lipid
autoxidations [2725].

   A less complicated system yielding cholesterol autoxidation but without extensive secondary decompositions involves brief exposure of crystalline cholesterol to ^{60}Co
γ-radiation in air [1072,2295,2303]. Autoxidation induced
by ionizing radiation yields the 7β-hydroperoxide 47 as a
major product; indeed, irradiation of cholesterol with
approximately 2.2 Mrad may be a preferred means of access
of the 7β-hydroperoxide 47 with a minimum of purification
difficulties [2303]. Ionizing radiation also provokes air
oxidations of the 4-stanol 79, 7-stenol 57, and the 8-stenol
68 giving relatively readily resolved primary autoxidation
products [1402,2309].

There naturally arises the question whether these
accelerated autoxidation systems are proper models of natural air-aging of cholesterol. The same early products,
14-16,46, and 47 are implicated throughout. Moreover, oxidations in the side-chain and A-ring dehydrogenation are
common to natural and model system autoxidations. The only
apparent distinguishing characteristic is that of rate, the
natural autoxidation taking much longer time.

Nonetheless, issue might be taken that oxidizing species
other than $^3O_2$ be implicated, particularly in the case of
ionizing radiation. For instance ^{60}Co γ-radiation acting
on $^3O_2$ under certain conditions may form a steady-state
concentration of O_3 [2179]! However, O_3 formation does not
appear to be a recognized problem with ^{60}Co γ-radiation
sources, the odor of O_3 is never apparent, and we have not
detected recognizable O_3-oxidation products of cholesterol
in samples of cholesterol irradiated with ^{60}Co γ-radiation.
Furthermore, the formation of $^1O_2$ from $^3O_2$ by direct or
indirect energy transfer from ionizing radiation is not a
described process [1268], nor is O_2 reduction to $O_2^{\overline{\cdot}}$ or $O_2^=$
indicated. Finally, oxidations of cholesterol by HO· formed
by radiolysis of minute amounts of water possibly present
as cholesterol monohydrate or in ambient humid air is not
supported by product analysis. Although HO· oxidation products 14-16 are formed, products 13,35, and 36 are not, and
the early formation of 7-hydroperoxides 46 and 47 cannot be
explained by HO· oxidations.

As products indicative of generation of other active
oxygen species are not found in this model system utilizing
ionizing radiation, such other active species cannot be implicated directly in the formation of observed products.
There remains the possibility, untested experimentally,
that one or other of these active oxygen species be formed
and initiate free radical oxidation of cholesterol by $^3O_2$.

Autoxidations in Aqueous Dispersions

Although discovery of the instability of crystalline
cholesterol to storage preceeded recognition of autoxidations in the dispersed states as found in tissues, tissue
extract hydrolysates, and aqueous emulsions, biochemical
and physiological investigations were shortly faced with
the presence of cholesterol autoxidation products in aqueous

systems. Such studies have been regularly limited in many
ways by the very low solubility of cholesterol in water,
cholesterol solubilities ranging from low values 25-90 ng/mL
[669,1552,2054] through medium values 1.8-90 µg/mL [313,860,
937,1424,1537] to high value 2.6 mg/g [586] (surely not correct!) have been recorded. A solubility of 17 µg/mL in 0.9%
saline in indicated [1537].

     Cholesterol dissolved or dispersed in water undergoes
reversible self-association, forming rod-shaped micelles
with a critical micellar concentration of 20-40 nM [872,873,
937]. Moreover, cholesterol solubility is greatly influenced by many factors, including the purity of the sterol
and the presence of other agents. Water-miscible organic
solvents, protein, carbohydrate, and ionic and nonionic
detergents greatly affect cholesterol solubility. Salt and
pH effects are also important items.

     One of the early interests in stable aqueous chole-
sterol dispersions was that for intravenous injection as
antidote for hemolysis caused by saponins or snake venom
poisoning [586,1902,2135]. The stimulation of tumor growth
in rats by aqueous cholesterol dispersions was also noted
early [1966], and aqueous dispersions were used as media
for the culture of bacteria [2029]. Commercially available
aqueous cholesterol dispersions [25] and irradiated ergosterol dispersions [1567] appear to have been contemplated
for direct intravenous administrations. More modern interests in cholesterol dispersions is had in their use as
model systems for the study of cholesterol autoxidation but
also as reference preparations of cholesterol for clinical
assay of plasma cholesterol levels.

     Aqueous dispersions of cholesterol have been made by
the simple method of Porges and Neubauer since 1908 [1880]
by the addition of cholesterol solutions in acetone [572,
1693,1880,1929,2002] or alcohol [653,922,1271,1288,1666,
1693,1822,1929,1965,2359-2361] to water, with removal of
solvent by vacuum, by heating, or by dialysis. By further
evaporation, stable dispersions as concentrated as 10 mg/mL
may be obtained [572,922,2359]. The stability of the cholesterol emulsions is dependent upon pH and salt effects
[1272,1822,1880,2002,2360,2361]. Moreover, the ease of formation and the stability of such cholesterol dispersions is
influenced by the purity of the cholesterol used, pure sterol
being least readily emulsified, samples containing small

amounts of autoxidation products giving much improved emul-
sion [1288,1779,1780,1939,2135,2136]. Simple aqueous dispersions of cholesterol can be prepared without use of organic solvent by grinding cholesterol at liquid N_2 temperatures and shaking with water for 6 h [1940].

These simple aqueous cholesterol dispersions have been
used as vehicles for delivery of cholesterol to various
metabolizing systems, for study of cholesterol autoxidation,
but also for other matters. Among the items of interest
examined on aqueous cholesterol dispersions are dialyzabil-
ity [1931], dielectric constant [1939,1945], ζ-potentials
[1693,1930,1937,1944], oxidizing power towards diamines
and phenols [1933,1934,1937,1938,1941], reducing power
[1942] and hydrogen donor capacity [1943], and catalysis of
H_2O_2 disproportionation [1935-1938]. The electrophoretic
mobility of aqueous cholesterol sols in which movement
towards the anode occurs first suggested in 1911 [1822] has
attracted attention repeatedly [1693,1944] to the present
[2664].

Stable emulsions of cholesterol in water may be prepared using other dispersing agents. Lecithins recognized
as effective for the purpose since 1908 [1880] are important means of incorporating cholesterol into aqueous systems
[156,296,585,987,1271,2002], including synthetic liposomes
[2502]. Aqueous dispersions of cholesterol in protein
solutions (albumins, gelatin, etc.) [239,247,296,1288,1932,
2303], in dextrin or mannan solutions [23,296,1865,2478],
and in bile salt solutions [25,157,1298,1730] have also
been prepared for various biochemical applications. However, cholesterol autoxidation in these media is not generally observed. Indeed, these aqueous cholesterol sols
appear to be protected against autoxidation [197,247,296,
1298,2299,2303,2650]. Nonetheless, cholesterol autoxidations can be provoked in such dispersions, for Fe(III)
salts catalyze cholesterol autoxidation in phospholipid
dispersions [296]. Moreover, phospholipids or glycolipids
form good dispersions of cholesterol autoxidation products
[1602], and synthetic liposomes formed with phospholipids
have been used for administration of cholesterol autoxidation products to experimental animals [2661].

Because of protein, phospholipid, and antioxidant protective agents preventing autoxidations, biological fluids
have generally not been useful as model systems for autoxi-

dation studies. However, aerated dispersions of cholesterol
in blood are autoxidized [247,296]. Moreover, whereas human
urine is hardly a model system for the study of cholesterol
autoxidation, cholesterol is present in urine [422]. The
high levels of urinary cholesterol of 0.3-6.6 mg/L [318,1265,
1569] or 0.1-1.5 mg/24 h [8,9,2587,2588] (13.7 mg/L in steer
urine [1568]), possibly as a protein-bound complex [7] elevated in pregnancy and certain neoplasms [8,9,218], in fact
constitute a dispersed cholesterol system potentially subject to autoxidation. Indeed, the 7-ketone 16 has been
recovered from human urine [2153]. Cholesterol is also
present in other human body fluids at low levels:
5.9-33 ng/mL in parotid fluid [1169], 150-2030 μg/mL in
semen [672,881,1665,2178], 65-320 μg/mL in tears [2039,2555],
and 0.23-12.66 μg/mL in cerebrospinal fluid [778,1815,2657].
Nonetheless, other than our preliminary observation of cholesterol autoxidation products in a stored cerebrospinal fluid
sample, interests in these other biological fluids for such
studies have not materialized.

Neither have dispersions or suspensions of subcellular
organelles been of much interest. With the exception of
rat liver microsomes and their associated NADPH-dependent
lipid peroxidation system as discussed more fully in Chaperr
VII, no studies of such dispersions as model systems for
cholesterol autoxidation have been recorded. However, suspensions of human erythrocyte ghosts in aqueous buffer have
been used in model studies of photosensitized oxygenations
in which $^{1}O_{2}$ is implicated [584,1421] as have also synthetic
liposomes [1507,2399,2400]. Formation of the 5α-hydroper-
oxide 51 is indicated in such studies, thus the same product obtained from photosensitized oxygenations of cholesterol in pyridine [626,792,1401,1751,2101,2102,2104] or
dimethylformaide [2286] solutions.

Almost no systematic studies of cholesterol autoxidation under controlled conditions in the ultimate dispersed
system animal tissue have been made. However, the thrust of
the details discussed in Chapter III establish that such
autoxidation occur. Generally what cholesterol autoxidation
that is observed in tissues is not investigated systematically to demonstrate the ease, rate, inhibition, or mechanism
but rather is encountered as an unexpected artifact of
faulty manipulation, all too often not recognized as such.
We have examined human aortal tissue for the apparent increase in content of the 3β,25-diol 27 and have suggested

that the diol increase in amount during work with the tissues [2567] (a matter now needing reconsideration), and
extensive cholesterol autoxidation occurs in human tissues
stored frozen for years [103]. Moreover, autoxidation products 13-16,35, and 36 are formed in human and rat skin
irradiated with ultraviolet light in air [276,277,1523].
Otherwise, well planned systematic study of cholesterol
autoxidation in the dispersed phases of animal tissue have
not been conducted!

Tissue cholesterol is highly susceptible to autoxidation during alkaline saponification of its fatty acyl
esters in tissue samples. The general ease with which emulsions may be formed during hydrolysis and in solvent extraction procedures categorizes these conditions as involving dispersed phases. Despite careful exclusion of air in
many cases evidence of autoxidations can be adduced, although
it is uncertain whether the tissues examined contained traces
of fatty acyl esters of the 3β,7-diols 14 and 15, for example.
The facile autoxidation of cholesterol in such aqueous alkaline emulsions probably represents free radical oxidation
by $^3O_2$ and not the ionic process discussed in Chapter VI
promoted by strong alkali in which a putative C-7 carbanion
might be an intermediate.

By far, more studies of cholesterol autoxidation in the
dispersed phase have been conducted using aqueous soap solutions, the solubility of cholesterol in soap solutions having been discovered early [879]. Fatty acid salts from
sodium caprylate [667-669] and laurate [666] to stearate
[296,2477] and oleate [295,296,353,1966,2477] are effective
model systems, 1% aqueous sodium oleate dissolving
cholesterol to the extent of 2.25 mg/mL [353]. The
sodium stearate system of Blix and Löwenhielm [296] so
masterfully exploited by Bergström and Wintersteiner [197,198
201-204,2711] has become a standard system for most subsequent work. The conditions of Bergström and Wintersteiner
in which an aqueous solution of sodium stearate at pH 8.5
is diluted with an ethanol solution of cholesterol and
heated at 85°C for 5 h provide extensive cholesterol autoxidation and have been used by others [247,251,1298,1721,2650].
The modified conditions of Mosbach, et al. [1690] in which
an aqueous ethanol solution of stearic acid and Na_3PO_4 is
made, pH adjusted, and cholesterol added as an ethanol solution (final cholesterol concentration 0.5-2.0 mg/mL), with

heating at 70-85°C, has also been used widely [479,1690, 2295,2297-2299,2303,2456].

Aqueous sodium stearate dispersions of cholesterol have also been used as vehicle for administration of cholesterol to biological systems [489,571,624,1279,1280], but emulsions prepared with nonionic surfactants such as Tween 20 (polyoxyethylene (20) sorbitan monolaurate) or Tween 80 (polyoxyethylene (80) sorbitan monooleate) have become much more useful for such work. Nonetheless, these polymeric ethers also autoxidize in aqueous solutions [627]! Incorporation of cholesterol into serum lipoproteins [110] and into synthetic liposomes [2502] pose alternative means as well.

Various other detergents solubilize cholesterol in water [137,880], and autoxidation studies using several (sodium dodecyl sulfate [1777,1778], sodium cholesterol sulfate or hemisuccinate [1264,1298]) have been reported. Moreover, the autoxidation of aqueous dispersions of cholesterol is promoted by radiation, aqueous suspensions being autoxidized by ultraviolet light [282,1611] as are also cholesterol dispersions in water, pH 8.5 Tris buffer, 2N NaOH, or 5% trichloracetic acid exposed to sunlight [942]. Extensive autoxidation of cholesterol in colloidal dispersions can also be promoted by catalysts such as platinum black [1720].

The need for solubilized reference cholesterol preparations for clinical serum assays using enzyme methods has engendered use of Triton X-100 [520] and bile salt [4]. Alternatively, water-soluble 3β-hemisuccinate salts with amines [1328,2350], mixed adipate esters of cholesterol and Polyethylene Glycol 600 [1893], and lyophyllized serum containing cholesterol [2719] have been used.

Autoxidation in Solution

Cholesterol autoxidation in organic solvent solutions is not found as an unwanted artifact so much as it is deliberately provoked in model experiments. Perhaps the most simple system for cholesterol autoxidation is the mere heating of solutions without exclusion of air. Thus, heated aqueous alcohol solutions are autoxidized [296]. Refluxing benzene solutions of cholesterol are extensively autoxidized after 4 d, and the 5α-stanol 2 is dehydrogenated to the

corresponding ketone [754]. Refluxing toluene solutions
of cholesterol likewise result in cholesterol autoxidation
but also in toluene autoxidation to benzaldehyde [2295].

Other sterols in other solvents are likewise autoxidized. The 8-stenol 68 3β-acetate in ethyl acetate at 50°C
[2175], in cyclohexane at 40°C [2173], or in aerated benzene
solutions at 25°C irradiated with visible light [357] is
extensively autoxidized.

Irradiation of cholesterol solutions and the addition
of photosensitizers, free radical chain initiators, or transition metal ions, etc. further promote air oxidations.
Thus, cyclohexane solutions of cholesterol containing CCl_4
irradiated with ultraviolet light are autoxidized [1005],
as are also irradiated aqueous cholesterol suspensions [282].
Irradiation of steroid solutions in chlorinated solvents
may be a particularly effective system for initiating autoxidations, for the enone 8 resistant to autoxidation on storage and on ^{60}C γ-irradiation is readily autoxidized in
CH_2Cl_2, $CHCl_3$, or CCl_4 solutions. Related light-induced
oxidation of other Δ^4-3-ketones also occur, possibly giving
hydroperoxides and Δ^4-3,6-diketone products *inter alia*
[1961].

Inclusion of a photosensitizer in heated solutions of
cholesterol may promote autoxidation, cholesterol 3β-acetate
being transformed to the 7-ketone 16 3β-acetate in heated
xylene solutions containing iron phthalocyanine [512], whereas photosensitizers exciting $^3O_2$ to $^1O_2$ give $^1O_2$ reaction
products which may or may not be accompanied by autoxidation
products.

Inclusion of acknowledged free radical reaction initiators such as benzoyl peroxide harkens back to the early
work of Lifschütz previously described. Very complex mixtures of unidentified products are generally obtained in
such cases [2303], but the controlled use of benzoyl peroxide for the autoxidation of the enone 6 to the 6β-hydro-
peroxide 59 *inter alia* is recorded [528]. Addition of
other oxidizing agents to cholesterol solutions may lead to
better controlled specific oxidations simulating autoxidations, such as the epoxidation of cholesterol by H_2O_2 added
to acetonitrile solutions of cholesterol [2299]. Allylic
oxidations with *tert.*-butyl perbenzoate [2335], oxidations
to the Δ^4-3,6-diketone with *tert.*-butyl chromate [1618], and

oxidations with *tert.*-butyl hypochlorite [877] as well as of other similar oxidizing agents lie beyond the scope of further discussion.

Whereas these comments do not exhaust examples of sterol autoxidations in solutions, it is well established that cholesterol may be autoxidized under controlled conditions in solution as well as in the solid and dispersed states, with or without the presence of sensitizers or catalysts, but necessarily in the presence of O_2.

CHAPTER V.   SUBSEQUENT TRANSFORMATIONS

Whereas the initial events of cholesterol autoxidation
involve two distinct modes of reaction, alcohol dehydrogenation to the 3-ketone 6 and hydroperoxide formation yielding
7-,17-,20-,22-,24-,25, and 26-hydroperoxides, it is in the
subsequent transformations of these initial products that
the vastly complex mixtures of cholesterol autoxidation products obtain.   No set number of autoxidation products is
realistic, for only oxidation products generally less mobile
than cholesta-3,5-dien-7-one (10) but more mobile than
5α-cholestane-3β,5,6β-triol (13) have been systematically
examined.   A substantial amount of more mobile material and
more polar, acidic material from air-aged cholesterol has
not received attention [2576].   A formal count of cholesterol autoxidation products from a naturally air-aged
cholesterol sample resolved by two-dimensional thin-layer
chromatography numbered thirty-two [2303], and the number
of discrete subsequent alteration products derived from the
acknowleged initial products described in this chapter runs
to forty-six, thus accounting for approximately sixty-six
steroid oxidation products.   Not all of these have been
isolated from or detected in air-aged cholesterol, but from
the observed transformations of each initial product under
relatively mild conditions which simulate natural aging processes one may project that each recognized product might
be present among the myriad of yet unidentified products.

The subsequent transformation of the initial cholesterol autoxidation products involve several modes of chemical reaction, but all such reactions may be conveniently
classified as those involving carbon-carbon bond scission
and those which do not involve such scissions.   Bond scission reactions are confined to the several cholesterol
hydroperoxides oxidized in the side chain, and carbon-
carbon bond cleavage has not been discovered as a mode of
decomposition for any of the A- and B-ring oxidation products.

Subsequent product types derived from the initial
autoxidation products may be other hydroperoxides, alcohols,
ketones, aldehydes, epoxides, olefins, or unfunctionalized
alkyl chains or D-ring.   However, degradation to acidic
material is also indicated and carboxylic acids appear also
to be among the structural types derived ultimately by

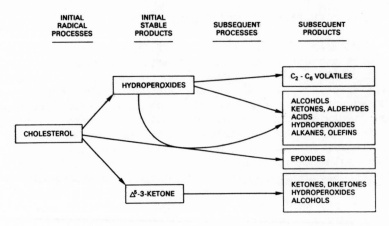

FIGURE 15. Schematic relationships among initial processes
and subsequent processes and products of cholesterol autoxidation.

autoxidations.

It is also from these subsequent transformations of
the side-chain substituted cholesterol hydroperoxides that
the volatile products conveying the acrid odor of aged cholesterol are derived. Identified volatile matter includes
carboxylic acids, ketones, alcohols, and olefins, thus
adding relatively little new as regards functionality to be
considered for these degraded fragments of autoxidized cholesterol.

The scheme outlined in FIGURE 15 affords a summary of
the presently recognized relationships among the initially
formed products and their several secondary and higher order
transformation products. Individual descriptions of subsequent events will be presented in the systematic numbering
order for cholesterol.

A-RING ALTERATIONS

Subsequent autoxidation reactions of the first formed
product enone $\underline{6}$ are of two types, one dependent upon $^{3}O_{2}$,

the other independent of oxygen. The oxygen-independent
transformation is simply the isomerization of the 5,6-double
bond into conjunction with the 3-carbonyl group to give
cholest-4-en-3-one (8). The lability of the 5-ene-3-ketone
system towards this isomerization, under acid or base catalysis or simply during manipulations, thin-layer chromatography, and storage is well documented, both for enone 6
isomerization to 8 and for isomerization of other Δ^5-3-ke-
tones to the corresponding Δ^4-3-ketones as well [70,396,
588,621,1957,2363].

The transformation of cholesterol to enone 6 and thence
to the conjugated enone 8 is the only pathway discovered for
the formation of 8 from cholesterol. An alternative process
involving isomerization of cholesterol to cholest-4-en-3β-ol
(79), which yields enone 8 upon irradiation in air [1402]
is unlikely, as no radiation-induced isomerization of cholesterol to the 4-stenol 79 is detected [70].

The oxygen-dependent transformation of the 5-ene-3-
ketone 6 is also previously described and well recognized
[1957]. In fact, the enone 6 is an unstable compound even
more so than is cholesterol, and commercially obtained samples of 6 are notoriously oxidized to the epimeric 6-hydro-
peroxides 6α-hydroperoxycholest-4-en-3-one (280) and the
6β-hydroperoxide 59 and to their decomposition products,
including the 3,6-dione 108, 6α-hydroxycholest-4-en-3-one
(281), and 6β-hydroxycholest-4-en-3-one (88) [2462]. The
oxidation is facilitated by transition metal ions [2599].

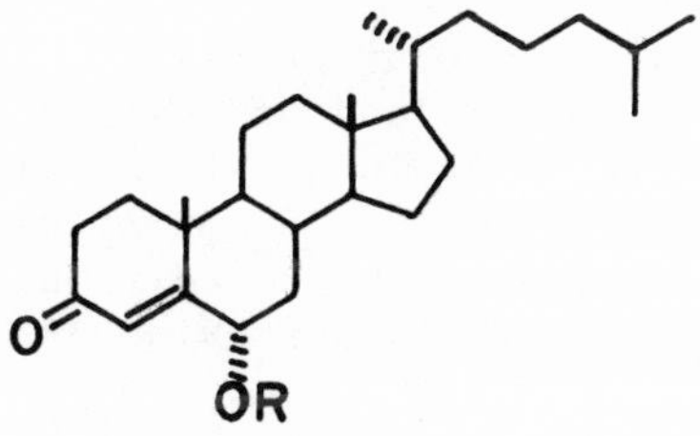

280   R = OH

281   R = H

The 6β-hydroperoxide 59 found as an autoxidation product of crystalline cholesterol heated in air [2455,2560]

represents initial dehydrogenation to the enone 6 followed
by reaction with $^3O_2$ to give the 6-hydroperoxides 59 and
280, in work where only the predominant 6β-hydroperoxide
was isolated. However, the transformation of enone 6 into
both epimeric 6-hydroperoxides 59 and 280 by free radical
processes and $^3O_2$ in aprotic and protic solvents, with and
without photosensitizer [528,588,752,1756,2101] and by controlled radiation-induced autoxidation of crystalline 6
[70,2462] amply demonstrates that the process may not be
stereospecific. Light-induced autoxidations of several
analogous C_{19}- and C_{21}-steroid 3,5-enol ethers also give
corresponding product 6α- and 6β-hydroperoxy-Δ^4-3-ketones
in which the axial 6β-substituent predominate [848].

The 6-hydroperoxide products 59 and 280 are derived
solely from the enone 6 and not from enone 8, as in our
experience no transformation products at all are found upon
extensive irriatiation in air of 8 with ^{60}C γ-radiation [70,
1402,2462]. Others have suggested that irradiation of the
enone 8 in air with ultraviolet light gave 5α-cholestane-
3,4-dione and unidentified products [2475,2721] and with
γ-radiation gave the 6β-hydroxyketone 88 [2487]. More
interestingly, the 6-hydroperoxides 59 and 280 are derived
from the enone 8 under other circumstances, as we have demonstrated their formation from 8 by the action of soybean
lipoxygenase in aerated buffered aqueous systems [2462].
Formation of 6-hydroperoxy-Δ^4-3-ketones from other parent'
Δ^5-3-ketones in aerated vegetative cell cultures of *Actinoplanes missouriensis* [1561] and *Flavobacterium dehydrogenans*
[899] are recognized instances of artificial nonenzymic
oxygenations.

The issue of the interconvertability of the epimeric
6-hydroperoxides 59 and 280 (and of their corresponding
6-alcohols 88 and 281) is of interest in this matter, as
the equatorial 6α-hydroperoxide 280 might be expected to be
the more stable [528,2455]. Indeed, although the analogous
3β-hydroxycholest-4-ene-6β-hydroperoxide (89) was not noted
to epimerize [1401], the ketone analog 59 is readily epimerized to the 6α-hydroperoxide 280 both in the solid state
and in solution. Moreover, the 6α-hydroperoxide also epimerizes but to a less extent [2462]. It is uncertain
whether the 6-hydroperoxides 59 and 280 are formed separate-
ly as initial products from the enone 6 or whether the 6α-
hydroperoxide 280 is derived from the 6β-hydroperoxide 59.
Chromatographic analysis of oxidized cholesterol preparations

containing the epimeric 6-hydroperoxides 59 and 280 involve bothersome variations in mobility of 59 and 280, which may be of greater, equal, or less mobility than cholesterol depending on minor changes in irrigation solvent [2459].

Thermal decomposition in solid state or in solution of the 6-hydroperoxides 59 and 280 proceeds by formal reduction and dehydration,by yielding the corresponding 6-alcohols 88 and 281 and the 3,6-diketone 108 [588,2462]. The 6α-hydroperoxide 280 was more prone to dehydration to the 3,6-diketone 108 than was the epimeric 6β-hydroperoxide 59. Pyrolysis of the 6β-hydroperoxide 59 gave as major products the 3,6-diketone 108 and 5α-cholestane-3,6-dione (42), the saturated diketone 42 being derived thermally from the 6β-alcohol 88 presumably formed first but unstable at pyrolysis temperature (250°) [2462]. The facile isomerization of epimeric 6-hydroxy-Δ^4-3-ketones by solvolysis to 5α-3,6-diketone and Δ^4-3,6-diketone derivatives is well recognized [1373].

Although the 5α-3,6-diketone 42 has not been detected in autoxidized cholesterol, the thermal process for its derivation from the enone 6 via the 6-hydroperoxides 59 and 280 implicates 42 as a cholesterol autoxidation product. The 5α-3,6-diketone 42 has only rarely been isolated from animal tissure [1274,1889].

The transformations of FIGURE 16 interrelate the several A-ring autoxidation products of cholesterol, including the enones 6 and 8, 6-hydroperoxides 59 and 280, 3,6-diketones 42 and 108, and 6-ketols 88 and 281. Of these eight products, specific evidence for the presence of 6,8,59, and 108 in air-aged cholesterol has been provided [70,2455], and we have since isolated the 6β-alcohol 88 from such material. The epimeric 6α-alcohol 281 has not been found in air-aged cholesterol but is inferred as a product from the demonstrated thermal decomposition of the parent 6α-hydroperoxide 280 to 281

For balance, FIGURE 16 also includes a summary of the $^1O_2$ attack on cholesterol yielding the minor product 6-hydroperoxides 89 and 247. These epimers are not interconverted as are the analogous 6-hydroperoxides 59 and 280. Thermal decomposition of the 6-hydroperoxides 89 and 247 follows the same pattern for the ketone analogs 59 and 280, thus

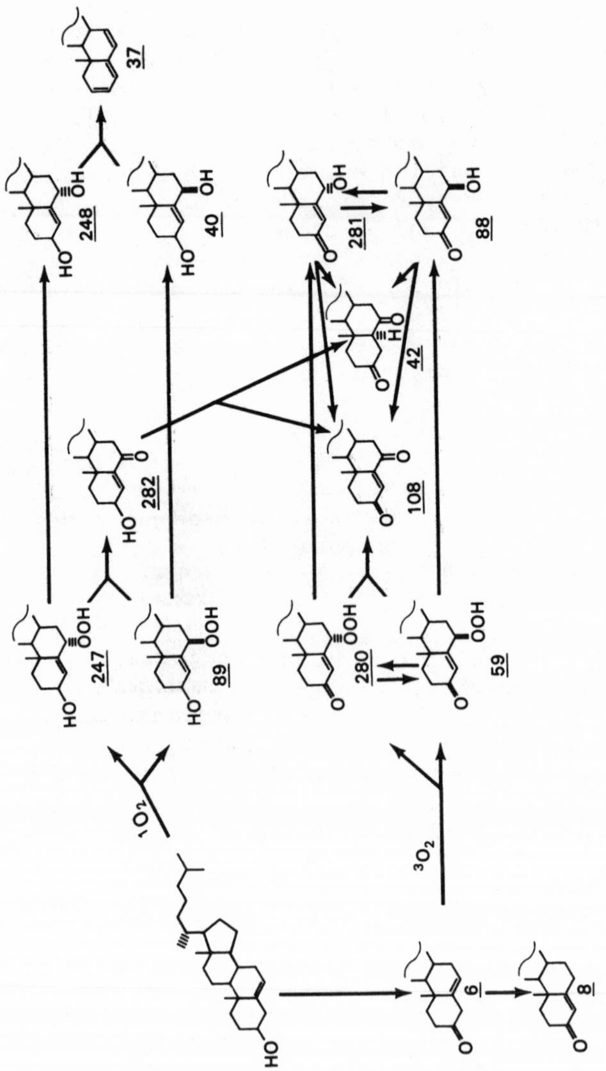

FIGURE 16. A-Ring ketone derivatives from cholesterol oxidations.

by formal reductions to epimeric 6-alcohols 40 and 248
respectively and by dehydration to 3β-hydroxycholest-4-en-
6-one (282). Thermal decomposition of the 6-ketone 282 in
turn yields the 3,6-diketones 42 and 108, thus the same
diketones obtained via thermal decompositions of the free
radical autoxidation products 59 and 280 derived from the
enone 6. Pyrolysis of the epimeric 3β,6-diols 40 and 248
gave a triene, most likely cholesta-2,4,7-triene (37) [1401].

Yet further interrelationships among these oxidation
products obtain, as the radiation-induced dehydrogenation
of the $^1O_2$ product 6β-hydroperoxide 89 to the radical
autoxidation product 6β-hydroperoxide 59 has been demonstrated. Moreover, the radiation-induced oxidation of
cholest-4-en-3β-ol (79) proceeds by dehydrogenation to
enone 8 and also to the epimeric 6-hydroperoxides 59 and
280, but 59 and 280 are not derived from 8. More likely,
59 and 280 formed from the 4-stenol 79 derive by initial
radiation-induced oxygenations at the allylic C-6 position,
followed by dehyrogenations to the 6-hydroperoxides 59 and
280 as represented in FIGURE 16. The transformation of the
6α-hydroperoxide 247 to the related 6α-hydroperoxide 280
is inferred by the observed analogous dehydrogenation of
280 to 59 [1401,1402].

The features of the A-ring oxidation chemistry of
cholesterol of FIGURE 16 thus account for fourteen oxidation
products, eight from autoxidation, six from $^1O_2$ attack. As
the 4-stenol 79 is not a congener of cholesterol and has
not been found in tissues as a natural product [2122], in
its absence and in the of $^1O_2$ action, products 40,89,247,
and 248 inter alia may not be found. However, the 3β,6β-
diol 40 has been isolated from pig spleen [1888] and rat
adrenal [1894] tissues.

Cholesta-4,6-dien-3-one (12) recognized as an autoxidation product of cholesterol (cf. TABLE 4) needs mention
as an A-ring oxidation product even though its autoxidation
origin appears to be from products initially oxidized in
the B-ring. There is no basis for postulating that dienone
12 derive from the conjugated enone 8, but derivation from
the isomeric Δ^5-3-ketone 6 deserves attention. Inconclusive
evidence suggests that dienone 12 derive from enone 6 by
the action of I_2 vapors [2363], and the facile oxidation of
6 to 12 by MnO_2 has been suggested [2324]. However, the
formation of 12 as chief product of the attack of $^1O_2$ on

the enone 6 is established. An unisolated intermediate
5-hydroperoxy-5α-cholest-6-en-3-one (283) is putatively
involved; the elimination of the elements of H_2O_2 from 283
then yields the dienone 12 [1756]. The origin of the dienone 12 from B-ring oxidized sterols is discussed more
fully in the next section of this chapter.

B-RING ALTERATIONS

The oxidation of the enone 6 yielding the epimeric
6-hydroperoxides 59 and 280 in fact involves B-ring alterations of the cholesterol molecule, but as these derivatives do not form without the prior A-ring dehydrogenation
they are treated under A-ring transformations. The B-ring
alterations discussed in this section include reactions of
the epimeric 7-hydroperoxides 46 and 47, epoxidations of
cholesterol, and formation of sterol 3β,5α,6β-triol and
3β,5α,6β,7α-tetraol derivatives.

Cholesterol 7-Hydroperoxides

The subsequent transformations of the cholesterol
7-hydroperoxides 46 and 47 follow upon the same chemical
pathways adumbrated in the previous section for the thermal
decomposition of the several 6-hydroperoxides 59,89,247,
and 280, namely formal reduction to the corresponding
alcohols and dehydration to the ketone. The 7α-hydroperoxide
46 yields the 3β,7α-diol 14 by sodium borohydride reduction
and also by thermal decomposition, the epimeric 7β-hydro-
peroxide likewise yields the 3β,7β-diol 15 by borohydride
reduction and by thermal decomposition. Both 7-hydroper-
oxides are dehydrated thermally to the 7-ketone 16 common
to both [2300,2454,2578].

The transformation of the 7-hydroperoxides 46/47 to
7-alcohols 14/15 and 7-ketone 16 may occur as separate
monomolecular processes. In the absence of reducing agents,
the formal reduction of the 7-hydroperoxides 46/47 to the
7-alcohols 14/15 may be regarded as involving thermal or
transition metal ion catalyzed homolysis of the peroxide
bond, yielding 7-oxyl radicals 263 that in turn abstract
hydrogen from other molecules to give the stable 3β,7-diols

14/15. The 7-oxyl radicals 263 may also be precursor of
the 7-ketone 16. However, the secondary products 14-16 may
be formed not from the isolable 7-hydroperoxides 46/47
but from the precursor 7-peroxyl radicals 271 via disproportionation. Kinetics data suggest that a bimolec-
ular decomposition of the 7-peroxyl radicals 271 yield
one equivalent of 3β,7-diols 14/15, one equivalent of
7-ketone 16, and one dioxygen molecule [1005]. The dismutation of 7-peroxyl radicals 271 may contribute to the
presence of the secondary products 14-16 in some cases
but cannot account for their presence in all cases, as
the 7-hydroperoxides 46/47 are also formed in cholesterol
autoxidations and the levels of 3β,7-diols 14/15 do not
usually equal the level of 7-ketone 16 as required by
a disproportionation mechanism.

These decompositions to alcohol and ketone dominate
the B-ring chemistry of these 7-hydroperoxides. However,
hydroperoxide isomerization is also possible, the quasiaxial 7α-hydroperoxide 46 being epimerized to the quasiequatorial 7β-hydroperoxide 47 in organic solvent solution, by aging in the crystalline state, and by pyrolysis
on gas chromatography [2454,2455]. The reverse epimerization has not been observed. In the same vein, the 3β,
7α-diol 14 is epimerized to the 3β,7β-diol 15 but the
reverse epimerization has not been observed except at
low level in pyrolysis [2454]. Under extreme conditions
it may be projected that the chief autoxidation products
46 and 47 should be transformed into the 7-ketone 16 and
the quasiequatorial 3β,7β-diol 15 as major final stable
products.

It is of interest to note here that for years the
presence of 3β,7β-diol 15 as a prominent product of cho-
esterol autoxidation could not be properly explained.
The present scheme of FIGURE 17 now completes a satisfying mechanism for these matters. Kinetics studies confirm most aspects of the scheme as also occurring in
heated aqueous sodium stearate dispersions of autoxidizing cholesterol [1297]. However, at least three other processes can account for the presence of the 3β,7-
diols 14 and 15 and the 7-ketone 16 in oxidized cholesterol preparations. The oxidation of cholesterol by
dioxygen cation O_2^+, oxygen atom, or oxygen cation O^+

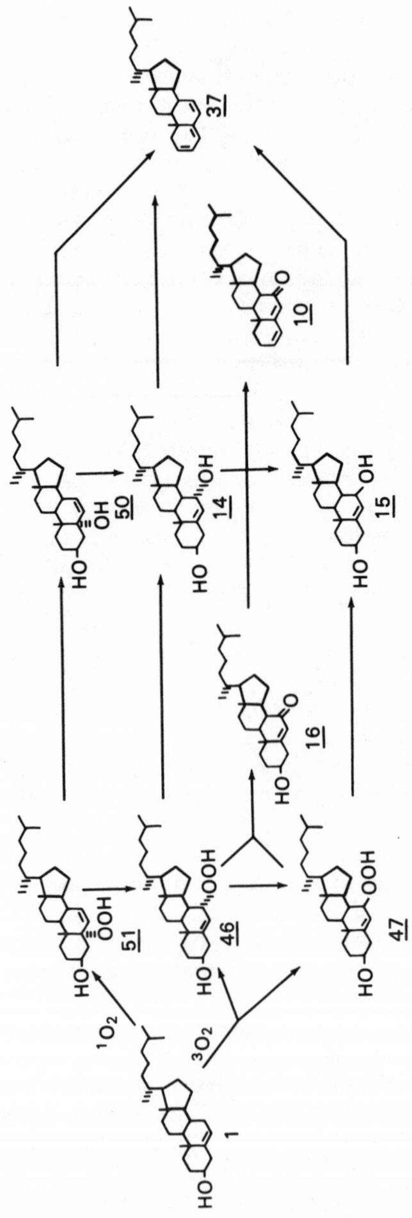

FIGURE 17. Transformations of cholesterol 7-hydroperoxides.

appears to yield these products [2071], as does also the oxidation of cholesterol by HO· [73,2291].

Yet another pathway not involving free radical autoxidations at all may account for the 7-oxygenated products 14-16,46, and 47. Attack of $^1O_2$ on cholesterol yields the 5α-hydroperoxide 51 as described already, but the 5α-hydroperoxide is readily isomerized to the 7α-hydroperoxide 46 in organic solvents [1544,2098,2103,2104,2576], in aqueous sodium stearate dispersions [2298,2299], and upon pyrolysis [2300,2454], in a reaction that appears to involve free radicals [807].

Furthermore, the 3β,5α-diol 50 is similarly isomerized in like systems to the 3β,7α-diol 14. The 7α-oxygenated derivatives 14 and 46 are further transformed to the corresponding 7β-epimers 15 and 47, and both 7-hydroperoxides 46 and 47 are dehydrated to the 7-ketone 16. The dynamic relationship between $^1O_2$ reaction products and autoxidation products is shown in FIGURE 17.

Thus, the rearrangement of 51 to 46 with subsequent epimerization of 46 to 47 also may account for the presence of 46 and 47 in a given system, together with the corresponding alcohols 14 and 15 and 7-ketone 16. In such dynamic systems levels of the quasiaxial epimers 14 and 46 may exceed those of the quasiequatorial 15 and 47, whereas free radical oxidation of cholesterol gives product mixtures in which the quasiequatorial 15 and 47 predominate. As the allylic rearrangement of 50 and 51 is not reversible, remnants of 50 or 51 detected in such systems would disclose the processes of FIGURE 17 as occurring in the system.

The allylic rearrangement of other C_{19}-, C_{21}-, and C_{29}-steroid Δ^6-5α-hydroperoxides to the corresponding Δ^5-7α-hydroperoxides has been recorded [1668,2102,2104], but not all such Δ^6-5α-hydroperoxides isomerize, as 3β,19-diacetoxy-5-hydroperoxy-5α-androst-6-en-17-one did not [1667,1669].

One of the technical limitations involved in the chromatographic analysis of oxidized cholesterol preparations for the initially formed hydroperoxides is the general failure to resolve the 7α-hydroperoxide 46 from the 5α-hydroperoxide 51. The quasiequatorial 7β-hydroperoxide 47 is resolved from the less mobile quasiaxial 7α-hydroperoxide

46, and the stenediols 14,15, and 50 are readily resolved
[2455,2576]. The problem case of the 7α- and 5α-hydroper-
oxides 46 and 51 has been resolved using ternary solvent
mixtures incorporating acetic acid, but hydroperoxide isomerization occurs in such systems, thereby thwarting their
application.

Gas chromatography of the sterol hydroperoxides 46,47,
and 51 in fact pyrolyze them, and although pyrolysis patterns of individual hydroperoxides are distinguished from
one another, mixtures of hydroperoxides would give indis-
tinguishable pyrolysis patterns from which the composition
of the mixture could not be determined [2454].

The advent of high performance liquid column chromatography using commercially available microparticulate silica
gel (μPorasil) adsorption columns irrigated with hexane-
isopropyl alcohol mixtures and of similar columns of organo-
silicon derivatives containing octadecanyl groups bonded to
silica   (μBondapak C_{18}) operated in the reverse phase mode
with aqueous methanol or aqueous acetonitrile resolves the
three hydroperoxides 46,47, and 51, thereby permitting analysis of their mixtures [72].

The 7-ketone 16 is perhaps the most frequently encountered cholesterol autoxidation product and derives by thermal decomposition of the 7-hydroperoxides 46 and 47. Although 7-ketone formation is a formal dehydration of the
7-hydroperoxide precursor, disproportionations yielding
7-ketone 16 and 3β,7-diol 14 or 15 also may account for
the transformations. Moreover, other free radical decomposition processes acting on the 7-hydroperoxides may give
the 7-ketone, as evinced by photolysis of $HgBr_2$ in the presence of the 7α-hydroperoxide 46 [806]. Furthermore, the
7-ketone 16 may be derived from the 5α-hydroperoxide 51
under conditions which cause rearrangement of 51 to the
7α-hydroperoxide 46 [2104].

The dienone cholesta-3,5-dien-7-one (10) derived from
the 7-ketone 16 by thermal processes [2569] and in alkali
[542,726,1069,1178,1632] is an artifact derived from an
autoxidation artifact!

The presence of the six 7-oxygenated steroids 10,14-
16,46, and 47 constitute a *prima facie* case for autoxidation of cholesterol. Only in a few instances where active

284

dioxygenases act upon cholesterol have these products been
found, and these examples obviously involve free radical
processes [2311,2458,2461].

Although the thermal and chemical dehydration of cholesterol yields the dienes cholesta-2,4-diene (284) and cho-
lesta-3,5-diene (11) [2123,2266,2267], as does pyrolysis of
the 4-stenol 78 [1796], these dienes do not represent cholesterol autoxidations. However, the analogous triene chole-
sta-2,4,6-triene (37) not heretofore recognized as an autoxidation product is a thermal dehydration product of the
7-hydroperoxides 46 and 47 and 3β,7-diols 14 and 15 and
thereby formally a cholesterol autoxidation product. Furthermore, thermal decomposition of the 3β,5α-diol 50 also
yields triene 37, whereas pyrolysis of the 5α-hydroper-
oxide 51 yields 37 and the isomeric 3,5,7-triene 38 [2300,
2454]. Neither triene 37 nor 38 has been found in air-aged
cholesterol or in animal tissues, but a cholestatriene
possibly 37 or 38 has been identified in recent marine sediments [834]. However, trienes detected under such circumstances more likely derive from dehydration of the 5,7-
dienol 56 rather than from cholesterol autoxidation products. The facile dehydration of C_{19}-Δ^5-3β,7-diols and
their 7-methyl ethers to the corresponding C_{19}-2,4,6-
triene is also recorded [24].

Finally, the dienone 12 previously mentioned as a
product of $^1O_2$ attack on enone 6 is both an autoxidation
product of cholesterol (cf. TABLE 4) and a secondary product of $^1O_2$ attack on cholesterol. The dienone 12 is a
minor product of pyrolysis of the epimeric 7-hydroperoxides
46 and 47 and 3β,7-diols 14 and 15 [2300,2454], and air-
aged samples of the 3β,7-diols 14 and 15 contain traces of
dienone 12. Moreover, exposure of the 3β,7α-diol 14 to
^{60}Co γ-radiation in air yielded the dienone 12 as a minor

product [72], thus formally demonstrating the pathway.
These results suggest that 12 derive autoxidatively via
3β-alcohol dehydrogenation of the 3β,7-diols 14 and 15.

In FIGURE 18 are summarized the three oxidation pathways by which the dienone 12 may be formed from cholesterol
via the agency of O_2. The longest process is that of autoxidation, passing through the 7-hydroperoxides 46 and 47 and
3β,7-diols 14 and 15 to the putative intermediates 7ξ-hy-
droxycholest-5-en-3-one (285) and 7ξ-hydroxycholest-4-en
3-one (286), the facile dehydration of which yield the
dienone 12 [918,1361]. Neither 7-hydroxyketone 285 nor
286 has been detected in air-aged cholesterol or in other
model systems, but the formulation is appealing.

The most direct pathway for formation of 12 is via $^1O_2$
attack on cholesterol yielding the 5α-hydroperoxide 51, the
thermal decomposition of which then yields 12 as a major
product [2300,2454,2578]. The third process involves initial autoxidation of cholesterol to the enone 6 followed
by $^1O_2$ oxidation, thus a process involving both ground-
state and excited species of dioxygen. As $^1O_2$ in our
experience is not involved in cholesterol autoxidations,
these two pathways are formal ones only with respect to the
presence of 12 in naturally air-aged cholesterol. However,
the facile oxidation of enone 6 by air to other products
[70] and by I_2 [2363] or MnO_2 [2324] to 12 raises the
possibility that heretofore undiscovered oxidations of
6 to 12 may also occur.

The dienone 12 is not a stable compound *per se* but appears to be quite sensitive to air oxidation. Commercially
available samples of 12 invariably contain several peroxidic
and several nonperoxidic components readily resolved by thin-
layer chromatography, some of which appear to be formed by
heating 12 at 50-100°C in air. Others have also noted the
sensitivity of dienone 12 to air [1232].

Cholesterol Epoxidation

The isomeric 5,6-epoxides 35 and 36 recognized as
cholesterol autoxidation products formed in air-aged cholesterol (*cf.* TABLE 4) do not form directly as thermal alteration products of any initially formed autoxidation product

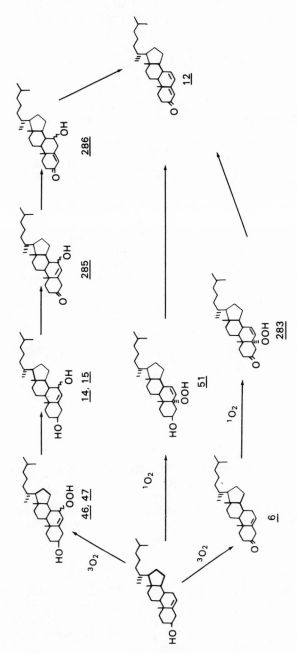

FIGURE 18.   Three oxygen-dependent pathways of formation of cholesta-
4,6-dien-3-one (12) from cholesterol.

listed in TABLE 8 but are derived by a less direct pathway.
Moreover, the 5,6-epoxides are not formed as first or early
autoxidation products from cholesterol but appear in controlled studies only after the initially formed 7-hydro-
peroxides 46 and 47 and their resultant thermal decomposition products 14-16. It follows that the 5,6-epoxides must
be formed by reaction of cholesterol with an oxidant other
than $^3O_2$, thus H_2O_2, sterol hydroperoxides, or possibly some
other unrecognized oxidizing agent formed during autoxidation.

The 7-hydroperoxides 46 and 47 cleanly epoxidize
cholesterol in aqueous sodium stearate dispersions (under
N_2) and in $CHCl_3$ solutions the isomeric 5,6-epoxides 35 and
36 being obtained in approximately 1:10 ratio. The epimeric 3β,7-diols 14 and 15 were also products. As the
levels of 7-hydroperoxides diminish as the amount of
5,6-epoxides increases and the reaction was conducted under
N_2, it follows that the sterol 7-hydroperoxides in fact
epoxidized cholesterol. The 5α-hydroperoxide 51 and cumene
hydroperoxides gave the 5,6-epoxides 35 and 36 in the same
ratio [2297].

Furthermore, the autoxidation of cholesterol in aqueous
sodium stearate dispersions by air without added organic
hydroperoxides also yields both 5,6-epoxides, the 5β,6β-
epoxide 36 predominating [479,2297]. As the 7-hydroperox-
ides 46 and 47 are the first observed cholesterol autoxidation products in such dispersions [2295], one deduces that
the autoxidation of cholesterol to 5,6-epoxides involves
initial autoxidation at the allylic C-7 position to form
the 7-hydroperoxides 46 and 47 which in turn oxidize cholesterol to the 5,6-epoxides 35 and 36. It is just this process which must account for the presence of the 5,6-epoxi-
des in air-aged cholesterol as well, although the ratio of
5α,6α-epoxide 35 to 5β,6β-epoxide 36 is 1:1 in air-aged
cholesterol [72,74], 1:3.6 in [4-14C] cholesterol that
had autoxidized [93].

Although this formulation is sound, it must be kept
in mind that cholesterol in aqueous dispersions is also
epoxidized by H_2O_2 [2297,2299] and by HO· [73,2291]. These
species if formed in autoxidation systems might contribute
accordingly to the 5,6-epoxide content. In alkaline media
cholesterol epoxidation probably proceeds by the accepted
ionic mechanisms involving hydroperoxide (or peroxide)

anion. Epoxidations in the solid state and in organic solvent solutions may not occur by an ionic but by radical or
other processes [1023].

Furthermore, other mechanisms are probable for cholesterol oxidation with organic hydroperoxides conducted with
MoO$_5$ or MoCl$_5$ catalysis, in which case overoxidation to
5-hydroxy-5α-cholestane-3,6-dione occurs [652,2492]. Related oxidations of cholesterol 3β-acetate with the same reagents yields the isomeric epoxides 35 and 36 in approximately equal proportions [2492,2494].

Epoxidations of other 5-stenols by organic hydroperoxides with MoO$_5$ or MoCl$_5$ catalysis also yields both isomeric
5,6-epoxides [2492,2494]. Epoxidations of other steroids
by organic hydroperoxides is recorded, *tert.*-butyl hydroperoxides transforming pregna-4,16-diene-3,20-dione in
alkaline medium to 16α,17α-epoxypregn-4-ene-3,20-dione
[2757]. More interestingly, self-epoxidations by steroid
hydroperoxides in basic media have been described. The
attack of 10β-hydroperoxy-17β-hydroxyestr-4-en-3-one (287)
upon itself yields 4β,5-epoxy-10β-hydroperoxy-17β-hydroxy-
5β-estran-3-one (288) and 10β,17β-dihydroxyestr-4-en-3-one
(289) [1590] in analogy to reaction between cholesterol and
its 7-hydroperoxides 46 and 47 to give the epoxides 35 and
36 and 3β,7-diols 14 and 15. A related example shows that
17α-ethinyl-10β-hydroperoxy-17β-hydroxyestr-4-en-3-one
yield 4β,5-epoxy-17α-ethinyl-10β,17β-dihydroxy-5β-estran-
3-one [1449].

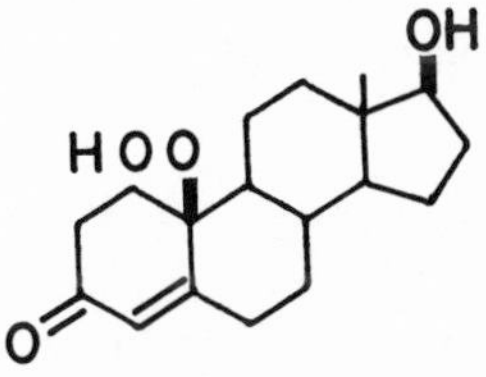

287 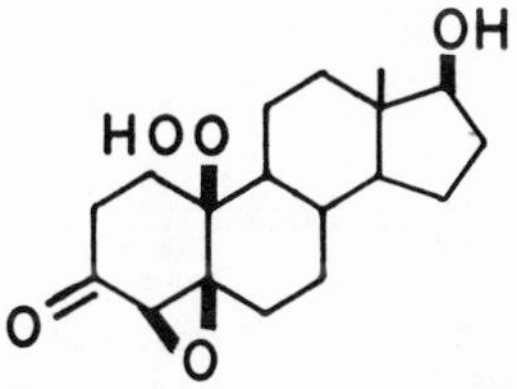   288

The epoxidation by air of dienones such as cholesta-3,
5-dien-7-one (10) in apparently free radical processes in
solution is described, giving 3α,4α-epoxycholest-5-en-7-one
(290) [965]. With only the one exception discussed later
in this chapter, the epoxidation of previously formed chole-

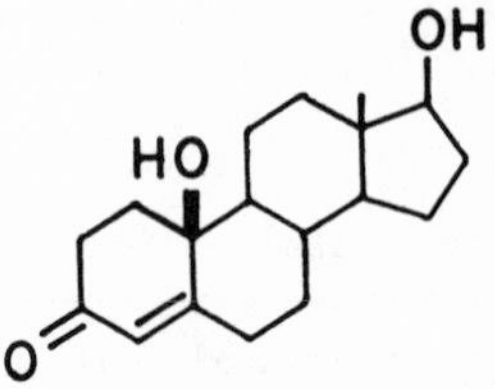

289 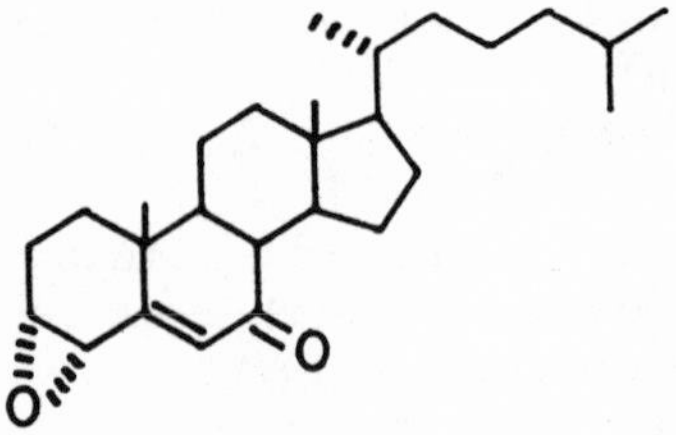   290

sterol autoxidation products has not been observed, either
in naturally air-aged cholesterol or in systematic studies
of known autoxidation products.

Epoxidations of cholesterol under different conditions
gives widely differing product ratios. The autoxidation of
cholesterol in aqueous dispersions by air, organic hydroperoxides, or H_2O_2 most favors the 5β,6β-epoxide 36 with a
35:36 ratio of 1:9 to 1:11 [2297-2299]. By contrast the
epoxidation of cholesterol by perbenzoic acid greatly favors
the 5α,6α-epoxide 35 [1871,2049,2660] as does also epoxidation during photolysis of 6-methylpyridazine 1-oxide most
favor the 5α,6α-epoxide 35, with a 35:36 ratio of 8:1
[2513]. Intermediate product ratios obtain in other circumstances: ferric acetylacetonate and H_2O_2 in acetonitrile,
1:4 [2489]; soybean lipoxygenase or rat liver enzyme incubations and air oxidation of solid cholesterol, 1:3.3 to
1:3.9 [93]; X-radiation of acetone solutions in which HO·
be implicated, 7:13 [503]; MoO_5-hexamethylphosphotriamide
complex, 3:2 [33]; and HO· in aqueous dispersions of cholesterol, 3.5:1 [73,2291].

Many studies of the 5,6-epoxides 35 and 36 have been
beset with technical difficulties, as analytical methods to
distinguish properly between 35 and 36 have only recently
been devised. Indeed, availability of authentic samples
further limits study, the 5α,6α-epoxide 35 being commercially available, the isomeric 5β,6β-epoxide 36 not so. Although the pure 5α,6α-epoxide 35 and impure 5β,6β-epoxide
36 were prepared early via perbenzoic acid oxidation of
cholesterol by Westphalen [2660], pure 5β,6β-epoxide 36 was
not described until 1940 [165,979]. A discussion of the
problem is more fully developed in Chapter IX.

Triol and Tetraol Derivatives

There are three more highly oxidized sterols 5α-chole-
stane-3β,5,6β-triol (13), 3β,5-dihydroxy-5α-cholestan-6-one
(44), and 5α-cholestane-3β,5,6β,7α-tetraol (291) that have
been isolated from cholesterol autoxidation systems and for
which a proper description of the relevant chemistry is
possible. Each of these three sterols is derived from
cholesterol by an extended multi-step process involving
initial hydroperoxide formation, epoxidation, and hydration
of the 5,6-epoxides to the polyols . The hydration of 35
3β-acetate and 36 3β-acetate to the same 3β,5α,6β-triol 13
3β-acetate occurs [1870], as does also the hydration of
unesterified 35 and 36 to the triol 13 [821,1655].

The 3β,5α,6β-triol 13 known from 1908 when it was prepared from cholesterol by H_2O_2 oxidation [1864] has been
frequently encountered in various oxidized cholesterol
preparations since. The triol has been found in air-aged
crystalline cholesterol [1072,2303,2576,2580], aqueous
sodium stearate dispersions of cholesterol oxidized by air
[1690] or by H_2O_2 [2298,2299], systems involving HO· (cf.
TABLE 7), animal tissues (cf. TABLE 1), and in other model
oxidation systems.

The companion 6-ketone 44 found in air-aged cholesterol
[2172,2303] is viewed as arising from the triol 13 by a relatively facile oxygen-dependent dehydrogenation, although
this specific transformation has not been examined systematically. Oxidation of the triol 13 by Br_2 during purification of cholesterol via the dibromide may account for the
6-ketone 44 [2172], but other processes may lead to 44
directly from the 5,6-epoxides 35 and 36, as suggested by
the transformation of the 5α,6α-epoxide 35 by periodic acid
to the 6-ketone 44 in a process not involving the triol 13
[2288] and by the acid hydrolysis of an analogous C_{19}-5α,6α-
epoxide yielding both 3,5,6-triol and 3β,5α-dihydroxy-6-
ketone derivatives [2653].

The 3β,5α,6β,7α-tetraol 291 is the latest more highly
oxidized cholesterol autoxidation product to be described.
Autoxidation of cholesterol in aqueous sodium cholesterol
3β-sulfate dispersions by air gave the tetraol 291 together
with the triol 13 and 3β,7-diols 14 and 15 [1264,1298].
The tetraol 291 may be viewed as derived by epoxidation of
the 3β,7α-diol 14 to a 5,6ξ-epoxy-5ξ-cholestane-3β,7α-diol

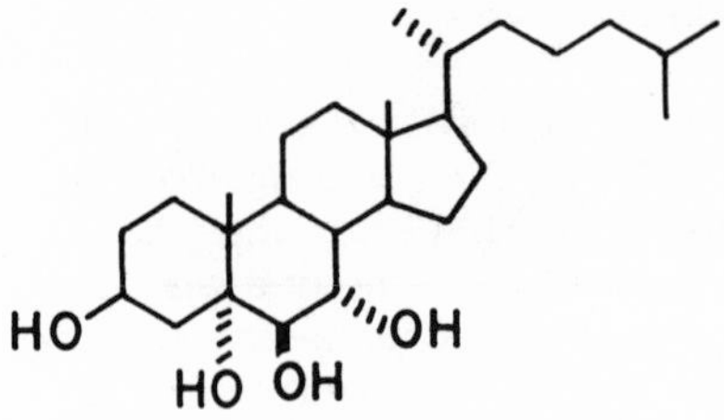

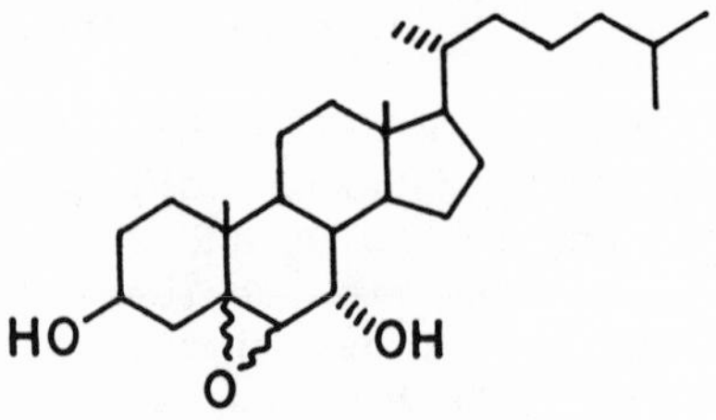

291 R = OH

13 R = H

292

(292), followed by hydration. The tetraol 291, the most
polar transformation product derived from cholesterol, has
not been identified yet among products of natural air-aging
of cholesterol. Neither the epimeric 5α-cholestane-3β,5,6β,
7β-tetraol potentially formed from the 3β,7β-diol 15 also
formed in the system nor the putative 5ξ,6ξ-epoxide 292
were detected in these experiments [1264,1298].

I have discussed thus far the A- and B-ring autoxidation chemistry of cholesterol whereby oxidative attack at
three sensitive sites (the allylic C-7 and C-4, C-3 alcohol,
and Δ^5-double bond) leads to twenty-one recognized products.
These include six 7-oxygenated derivatives (10,14-16,46,47),
one 4-oxygenated sterol (41), nine 3-ketones (6,8,12,42,59,
88,108,280,281) and five 5,6-dioxygenated sterols (13,35,36,
44,291). Additionally, following manipulations utilizing
acidified methanol or excessive heat, epimeric 7-methyl
ethers 52 and 53 and the 2,4,6-triene 37 respectively may
be encountered.

SIDE-CHAIN TRANSFORMATIONS

As the cholesterol side-chain is unfunctionalized, its
autoxidation involves only the formation and degradation of
hydroperoxides. As with the A- and B-ring hydroperoxides
with major decomposition processes or dehydration to ketone
and reduction to alcohol, so also are these processes major
ones for the decomposition of the side-chain hydroperoxides.
However, the additional major chemical features of β-scission of carbon-carbon bonds leading to degraded derivatives
bearing less carbon and volatile C_2- and C_6-compounds is a

prominent mode of decomposition of the side-chain hydroperoxides. Discussion of the chemistry of the side-chain
hydroperoxides involves directly observed reactions of the
20-hydroperoxides 22 and 32, the 24-hydroperoxides 105 and
106, the 25-hydroperoxide 26, and the 26-hydroperoxides 28
and 30 as well as reactions of 17-, 22-, and 23-hydroperox-
ides inferred by the presence of degradation products in air-
aged cholesterol.

17-Hydroperoxides

Though not isolated from air-aged cholesterol as
initially thought, the postulated cholesterol 17-hydroperox-
ides 3β-hydroxycholest-5-ene-17α- and 17β-hydroperoxide
(275) must have been formed in the course of cholesterol
autoxidation in order to account for the presence of the
C_{19}-17-ketone 86 among cholesterol autoxidation products of
air-aged cholesterol [2576]. Moreover, the 17-ketone 86
has not been observed as a thermal decomposition product of
the other recognized sterol hydroperoxides derived from
cholesterol. Although dehydrogenation of the corresponding
17β-alcohol 87 could account for the 17-ketone 86, this dehydrogenation is not observed as a pyrolysis reaction of
87.

Other hypothetical products of thermal decomposition
of putative cholesterol 17-hydroperoxides would include the
corresponding 17-alcohols cholest-5-ene-3β,17α-diol and
cholest-5-ene-3β,17β-diol and D-secosterols derived by
β-scission of C-13/C-17 or C-16/C-17 bonds. None of these
possible products have been encountered.

Support for some of the projections is found in the
obserbed chemistry of two accessible 17-hydroperoxides.
Thermal decomposition of 17α-hydroperoxypregn-4-ene-3,20-
dione gave androst-4-ene-3,17-dione and 17α-hydroxypregn-
4-ene-3,20-dione establishing β-scission of the C-17/C-20
bond and formal reduction of the hydroperoxide to alcohol.
17α-Hydroperoxy-3β-hydroxypregn-5-en-20-one in similar
manner gave the 17-ketone 86 and 3β,17α-dihydroxypregn-5-
en-20-one by the same processes [1091,2435].

However, other degradation processes were also apparent
for these 17α-hydroperoxides. Both were deoxygenated thermally, 17α-hydroperoxypregn-4-ene-3,20-dione yielding

progesterone (pregn-4-ene-3,20-dione), 17α-hydroperoxy-3β-
hydroxypregn-5-en-20-one yielding pregnenolone (23). Fur-·
thermore, from 17α-hydroperoxypregn-4-ene-3,20-dione apparently a D-homosteroid 17,17a-dihydroxy-D-homopregna-4,17
(17a)-dien-3-one and a 16,17-epoxyandrost-4-en-3-one were
derived, evincing further modes of alteration for these
17-hydroperoxides [1091,2435].

20-Hydroperoxides

Both the (20R)- and (20S)-20-hydroperoxides 32 and 22
have been isolated from air-aged cholesterol [2565,2576],
and both 20-hydroperoxides appear to undergo the same thermal decomposition reactions. The (20S)-20-hydroperoxide 22
is degraded via formal reduction to the alcohol (20S)-
cholest-5-ene-3β,20-diol (21) and by β-scission reactions
involving both C-20/C-22 and C-17/C-20 bonds to C_{21}- and
C_{19}-derivatives pregnenolone (23), androsta-5,16-dien-3β-ol
(191), androst-5-ene-3β,17β-diol, (87), and androst-5-en-
3β-ol (114) [2578].

Simple β-scission of the C-20/C-22 bond yielding the
20-ketone 23 is the predominant degradation mode of the
20-hydroperoxide 22. Alternatively, β-scission of the C-17/
C-20 bond leads to the three C_{19}-derivatives 87,114, and
191. Direct β-scission is formulated as giving the intermediate 3β-hydroxyandrost-5-en-17-yl radical (293) which is
stabilized by hydrogen atom abstraction to give the D-ring
olefin androsta-5,16-dien-3β-ol (191). Hydrogen atom ex-
pulsion from free radicals centered on carbon to give olefins also has been posed formally in our work with C_6- fragments derived from side chain hydroperoxides [2556,2576].

Formation of the 3β,17β-diol 87 from the 17-radical
293 is also postulated, with attack of HO· (derived from
peroxide oxygen-oxygen bond homolysis) on the 17-radical
yielding the 3β,17β-diol 87. These several transformations
are presented in FIGURE 19.

As no epimeric androst-5-ene-3β,17α-diol was found
this result suggest that retention of C-17 sterochemistry
occurs in these free radical reactions! However, an alternative possibility of preferential dehydration of the putative androst-5-ene-3β,17α-diol to the 5,16-diene 191 has
not been examined.

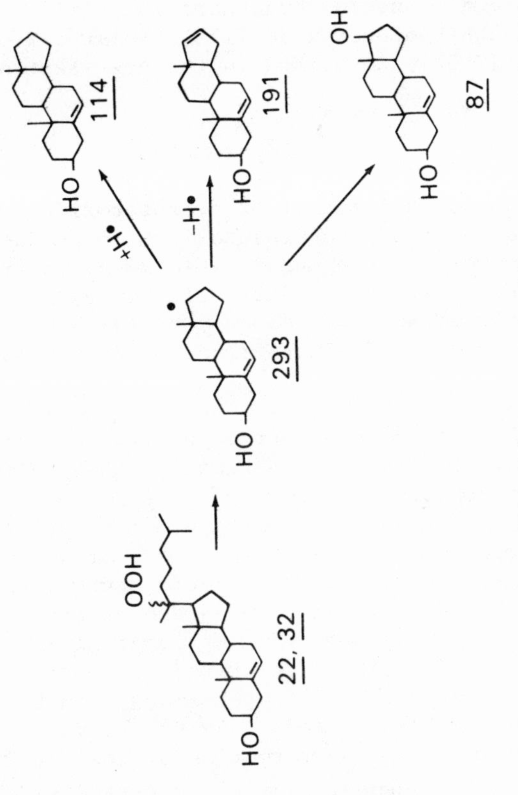

FIGURE 19. Postulated process leading to C$_{19}$-steroids.

The decomposition of the (20R)-20-hydroperoxide $\underline{32}$
has not been reported in detail, but its pyrolysis gives
the corresponding alcohol (20R)-cholest-5-ene-3β,20-diol
($\underline{33}$) as a prominent product [2565]. We may presume
that the (20R)-20-hydroperoxide also decomposes to the
20-ketone $\underline{23}$ and to the C_{19}-products $\underline{87},\underline{114}$, and $\underline{191}$.

The same C_{19}-products $\underline{87},\underline{114}$, and $\underline{191}$ are also formed
by the thermal decomposition of the C_{21}-20-hydroperoxides
3β-hydroxypregn-5-ene-20α-hydroperoxide ($\underline{112}$) and 3β-hydroxy-
pregn-5-ene-20β-hydroperoxide ($\underline{113}$) isolated from air-aged
cholesterol [2562], described in the next section.

22-Hydroperoxides

Although no cholesterol 22-hydroperoxides have been
identified among cholesterol autoxidation products, their
presence, as well as the undescribed 23-hydroperoxides,
appears likely. It is not difficult to overlook the presence of small amounts of isomeric sterol hydroperoxides in
the complex oxidized products mixtures from air-aged cholesterol.

One of the more interesting observations of hydroperoxide formation from cholesterol is that of the epimeric
C_{21}-20-hydroperoxides $\underline{112}$ and $\underline{113}$. As loss of the terminal
C_6-side-chain is involved, the C_{21}-20-hydroperoxides appear
to be secondary transformation products of an initially
formed C_{27}-autoxidation product rather than direct reaction
products of the attack of $^3O_2$ upon cholesterol. The β-scission of the C-20/C-22 bond without concommitant scission of
the peroxide bond is not a described process; therefore,
derivation of the C_{21}-20-hydroperoxides from the isomeric
C_{27}-20-hydroperoxides $\underline{22}$ and $\underline{32}$ is unlikely. However, no
direct observation of the formation of the C_{21}-20-hydroper-
oxides has been recorded, and the following formulations
are only speculations based on best present information.

On the basis that pregn-5-en-3β-ol ($\underline{110}$) was identified as a pyrolysis product of the 25-hydroperoxide $\underline{26}$
[2578] scission of the C-20/C-22 bond occurred, this cleavage being formulated as involving a six-membered cyclic
transition state and 1,5-hydrogen atom transfer, as suggested in FIGURE 20. Here the intermediate 3β,25-dihydroxy-
cholest-5-en-22-yl radical ($\underline{294}$) is stabilized by scission

of the C-20/C-22 bond yielding the C_{21}-steroid 110 and
2-methylpentan-2-ol. Both 2-methylpentan-2-ol and 2-methyl-
pent-4-en-2-ol have been recovered from air-aged cholesterol. Similar cyclic transition states have been suggested as means by which the C_{21}-20-hydroperoxides 112 and
113 form from the 25-hydroperoxide 26 [2558,2562] as well
as for the formation of the C_5- and C_6-fragments of the
side-chain isolated from air-aged cholesterol. However,
the 25-hydroperoxide 26 may not be the initial product
precursor of the C_{21}-20-hydroperoxides or of the related
pregn-5-ene-3β,20α-diol (111) isolated from air-aged cholesterol, for in controlled studies of pyrolysis of the 25-
hydroperoxide 26 the 3β,20α-diol 111 was not detected
[2578].

In that scission of the C-20/C-22 bond of a C_{27}-
sterol must occur for formation of the C_{21}-steroids 23,
110-113 it is reasonable to postulate hydroperoxide functionalization of the C-22 carbon atom as an alternative
means of C-20/C-22 bond cleavage by simple β-scission, as
outlined in FIGURE 21.

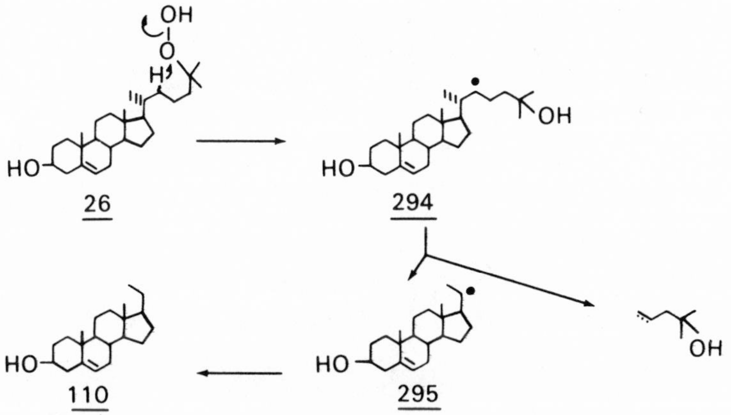

FIGURE 20. 1,5-Hydrogen atom abstractions
            of the cholesterol 25-hydro-
            peroxide 26.

In this formulation, β-scission of the 22-hydroper-
oxide 278 yields the intermediate 3β-hydroxypregn-5-en-20-yl

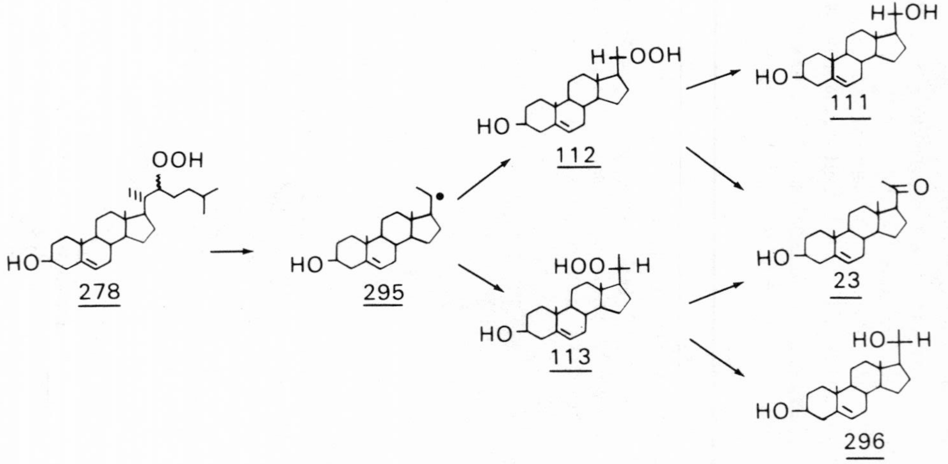

FIGURE 21. Derivation of C_{21}-steroids from putative 22-hydroperoxides.

radical (295) which is stabilized by reaction with a second
molecule of $^3O_2$ to give the C_{21}-20-hydroperoxides 112 and
113 whose thermal degradation gives the corresponding alcohols 3β,20α-diol 111 and pregn-5-ene-3β,20β-diol (296) and
20-ketone 23.

However, postulation of the 22-hydroperoxides 278 suggests that the corresponding 22-ketone 3β-hydroxycholest-
5-en-22-one (297) be a likely product also, but such a
product has not been found as a cholesterol autoxidation
product. Nonetheless, for the like case of the epimeric

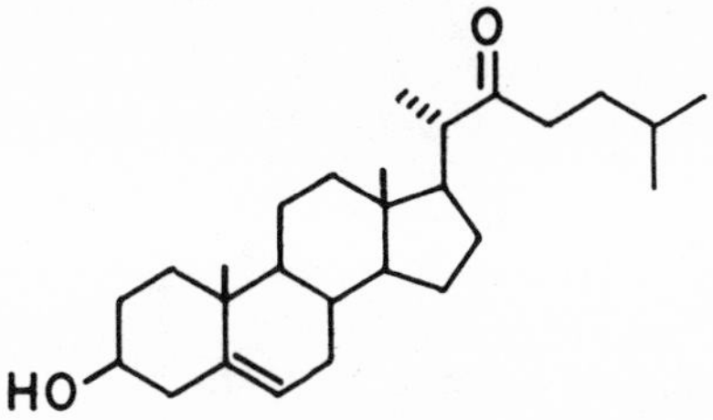

297

24-hydroperoxides 105 and 106 where 24-ketone 34 is a demonstrated pyrolysis product [2579], the 24-ketone 34 has not
yet been found as a cholesterol autoxidation product in
air-aged cholesterol.

These arguments suggest a possible dual origin for the
C_{21}-steroids 110-113 and 296 from the 25-hydroperoxide 26
or from the postulated epimeric 22-hydroperoxides 278. For
present purposes, the C_{21}-20-hydroperoxides 112 and 113 and
their corresponding 3β,20-diols 111 and 296 have been con-
signed as products from the 22-hydroperoxides. However, in
the absence of authentic 22-hydroperoxides, the matter cannot be concluded.

In that both (20R)- and (20S)-20-hydroperoxides 32 and
22 are formed autoxidatively from cholesterol, so also both
the C_{21}-(20R)- and (20S)-20-hydroperoxides 113 and 112 may
both be formed by whatever processes be implicated, that is,
whether from the 25-hydroperoxide 26 by a 1,5-hydrogen shift
and reoxidation by a second $^3O_2$ molecule or whether from the
postulated 22-hydroperoxides 278 by β-scission and reoxidation. However, derivation of the (20R)-20-hydroperoxide
113 may also occur by means of another perplexing feature
of sterol hydroperoxide chemistry, for the (20S)-20-hydro-

peroxide 112 epimerizes in solution to the (20R)-20-epimer
113 [2562]. In distinction to the epimerization of the
allylic quasiaxial 7α-hydroperoxide 46 to the quasiequatorial 7β-hydroperoxide 47 where the epimerization is aided by
the associated allylic double bond, the epimerization of
the (20S)-20-hydroperoxide 112 to the (20R)-20-hydroper-
oxide 113 is achieved without other associated functional
groups participation.

   The pyrolysis of the C_{21}-20-hydroperoxides leads as
anticipated to β-scission of the C-17/C-20 bond to yield
the same C_{19}-products 87,114, and 191 as derived by the
same β-scission process from the C_{27}-20-hydroperoxide 22
[2558,2562]. A dual origin for the C_{19}-steroids 87,114,
and 191 as well as of the 20-ketone 23 from C_{27}-20-hydro-
peroxides 22 and 32 and/or from C_{21}-20-hydroperoxides 112
and 113 is thus suggested.

   Our recent discovery of 3β-hydroxyandrost-5-ene-17β-
carboxylic acid (102) among cholesterol autoxidation products suggests that yet more extensive oxidations of initially formed cholesterol autoxidation product occurs.
The C_{20}-acid 102 might derive from the C_{27}-20-hydroper-
oxides 22 and 32, but derivation of 102 from the C_{21}-20-
hydroperoxides 112 and 113 seems more probable. It must be
emphasized that direct observation of formation of 102 from
either 20-hydroperoxide (or from any other precursor) has
not been achieved, so definitive asignment of an oxidative
origin for the acid 102 is not possible. It should be
noted that β-scission of the C-20/C-21 bond of the C_{21}-20-
hydroperoxides (not heretofore observed) [2562], should
provide a suitable intermediate C_{20}-20-aldehyde 3β-hydroxy-
androst-5-en-17β-aldehyde from which the C_{20}-acid could derive. On this speculative basis I have considered the
C_{20}-acid 102 to derive from the C_{21}-20 hydroperoxides 112
and 113 and thereby from the posited 22-hydroperoxides 278.

   Finally, the 22,23-bisnor acid 101 may be regarded as
a degradation product of postulated 22-hydroperoxides 278
via β-scission of the C-22/C-23 bond. A putative product
3β-hydroxy-22,23-bisnorchol-5-en-24-ol or -24-al would then
yield the 24-carboxylic acid 101 by further air oxidation.

   No cholesterol autoxidation product specifically oxygenated at the C-23 position has been recognized in any of
our studies. Thus, 3β-hydroxycholest-5-en-23-one or the

epimeric cholest-5-ene-3β,23-ols have not been found as
autoxidation products, although all are known from chemical
synthesis [2577]. Nonetheless, certain items imply that
autoxidation of cholesterol at the C-23 site occur. Three
volatile components 2-methylpropionic acid, 2-methylprop-
1-ene, and 2-methylpropan-1-ol recovered from air-aged
cholesterol discussed in a later section of this chapter
(cf. TABLE 9) suggest cleavage of the C-23/C-24 bond of a
cholesterol autoxidation products suitably functionalized
at either the C-23 or C-24 carbon atoms, yielding also
C_{23}-steroid fragments as well.

The thermal decomposition of the cholesterol 24-hydro-
peroxides 105 and 106 in fact yields the only C_{23}-steroid
24-norchol-5-en-3β-ol (196) heretofore recognized as a
cholesterol autoxidation product [2579]. The thermal decomposition of putative 23-hydroperoxides 279, in analogy
to that of the 24-hydroperoxides 105 and 106, would be
expected to yield by β-scission C_{23}-steroids functionalized
at the C-23 position and C_{22}-steroids. Such products have
not been discovered yet.

24-Hydroperoxides

Both (24R)- and (24S)-24-hydroperoxides 105 and 106
were isolated together unresolved from autoxidized cholesterol samples in the ratio 2:1, and both 24-hydroperoxides
gave their corresponding 24-alcohols on sodium borohydride
reduction but not on thermal decomposition [2579]. In this
respect the secondary 24-hydroperoxides differ in their
chemistry from the tertiary 20- and 25-hydroperoxides 22
and 26 where formal reduction to the corresponding 20- and
25-alcohols 21 and 27 respectively was a prominent aspect
of their thermal decomposition [2578].

Thermal decomposition of the 24-hydroperoxides did proceed by dehydration to the corresponding 24-ketone 3β-
hydroxycholest-5-en-24-one (34) and also by β-scission of
the C-24/C-25 bond to 3β-hydroxychol-5-en-24-al (298) and
of the C-23/C-24 bond to 24-norchol-5-en-3β-ol (196) [2579].
Neither the 24-ketone 34 nor either degraded product 298
or 196 have been found in air-aged cholesterol, but the related acid 3β-hydroxychol-5-enic acid (99) has been recently
found by us in air-aged cholesterol. The acid 99 is presumed to derive from the cholesterol 24-hydroperoxides as

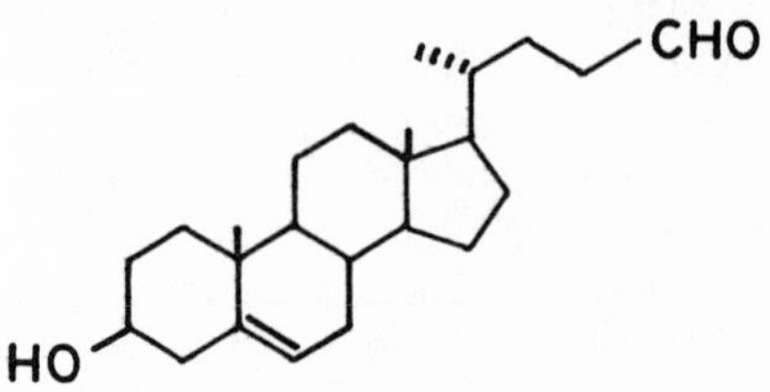

298 R = CHO   299

196 R = H

99 R = COOH

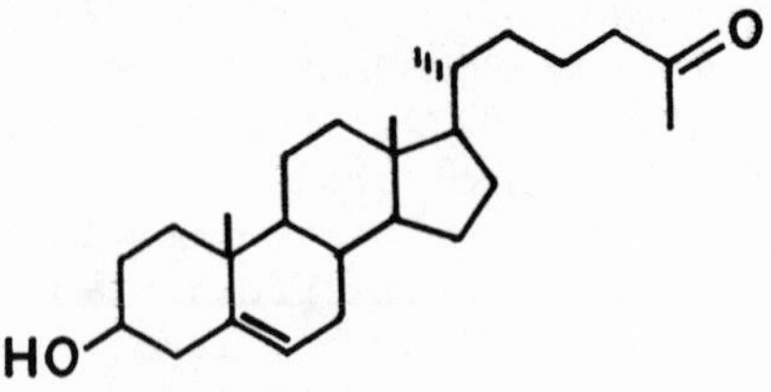

most likely precursors. However, as carboxylic acid formation suggests more extensive oxidations of other sterol derivatives may also occur, it may be that the C_{24}-acid 99 is
derived from the 25-hydroperoxide 26 by oxidative processes
not yet observed directly.

Nonetheless, as the 24-aldehyde 298 is a thermal degradation product of the 24-hydroperoxides 105 and 106
[2579], it is apparent that further air oxidation of the
aldehyde by recognized means may yield the carboxylic acid
99, and we postulate that the acid 99 is indeed derived by
these steps.

25-Hydroperoxide

The thermal decomposition of the 25-hydroperoxide 26
yields the corresponding 25-alcohol cholest-5-ene-3β,25-diol
(27) as major product, with products of β-scission and of
other reactions [2578]. Cleavage of the C-24/C-25 bond
gave the unfunctionalized alkyl side chain product chol-5-
en-3β-ol (109) as product. Scission of the other C-25/C-26
bond gave the norketone 3β-hydroxy-27-norcholest-5-en-25-
one (299). Thus, formation of these three products 27, 109,
and 299 conforms to the expected modes of reduction and
β-scission.

However, the minor product pregn-5-en-3β-ol (110) established that more extensive degradation of the 25-hydro-
peroxide 26 occurred. In this case, as previously suggested
in FIGURE 22 a six-membered cyclic transition state involving

a 1,5-hydrogen atom migration with concommitant homolysis
of the C-20/C-22 bond would provide the degraded C_{27}-radi-
cal 295 which upon stabilization by hydrogen abstraction
would yield the isolated product 110.

The matter of the apparent 1,5-hydrogen atom abstraction reactions of the 25-hydroperoxide 26 find precedent
in similar reactions of other hydroperoxides. Cyclic transition states involving 4-,5-,6-, and 7-membered rings have
been suggested where 1,3-,1,4-,1,5-, and 1,6-hydrogen atom
transfers have occurred respectively, the 1,5-hydrogen
transfer being the most favored [769].

26-Hydroperoxides

Although the cholesterol 26-hydroperoxides isolated
from air-aged cholesterol [2559,2562] have not been resolved into putative (25R)-and (25S)-26-hydroperoxide isomers
28 and 30, we may assume that both isomers are in fact
formed. Indeed, in exact analogy to the case for the presence of both (24R)- and (24S)-24-hydroperoxides 105 and
106 as established by studies on the 3β,24-dibenzoate esters
of the corresponding reduced alcohols [2575], so chromatographic analysis of 3β,26-diacetates of the 3β,26-diols derived from the cholesterol 26-hydroperoxides 28 and 30 suggests that approximately equal parts of (25R)- and (25S)-
isomers were present [1914].

The same suggestion may be made from melting point behavior though not so convincingly. The 3β,26-diol preparation derived from the mixed isomeric 26-hydroperoxides 28
and 30 had m.p. 169-170°C [2559], while that isolated directly from air-aged cholesterol from bovine spinal cord
and brain had m.p. 169-171°C [2580]. As pure (25R)-3β,26-
diol 29 has a higher melting point (m.p. 177-178°C [2095],
173-175°C [545], 175.5-176.5°C [2620]) as does also the
(25S)-3β,26-diol 31 (m.p. 171-173°C [2152], 171-172°C [2584]
173°C [1915]), and the melting point of (25RS)-3β-26-diol
preparations is in the range m.p. 168-173°C [575], 168.5°C
[2585], 171.5-172.5°C [2584], these 3β,26-diol preparations
probably represent autoxidations of cholesterol via the 26-
hydroperoxides 28 and 30. By contrast, our isolation of a
3β,26-diol of m.p. 161-166°C [2316] from fresh human brain
most likely represents inadequate purification via recrystallization, but the matter is not settled.

<u>300</u> R = CHO

<u>301</u> R = OH

<u>210</u> R = H

Thermal decomposition of the 26-hydroperoxides <u>28</u> and
<u>30</u> follows the same general reaction pathways already des-
cribed, with formal reduction to the corresponding 3β,26-
diols <u>29</u> and <u>31</u> being a major mode. Dehydration to the
corresponding 26-aldehyde 3β-hydroxy-(25RS)-cholest-5-en-
26-al (<u>300</u>) was also prominent [2559]. Two lesser products
of β-scission were also encountered, 27-norcholest-5-en-3β-
ol (<u>210</u>) and (25ξ)-27-norcholest-5-ene-3β,25-diol (<u>301</u>)
[2559]. Both of these C_{26}-products are considered to derive from the 26-hydroperoxide via β-scission to give an
intermediate 3β-hydroxy-27-norcholest-5-en-25-yl radical
stabilized by subsequent hydrogen abstraction to the 3β-
alcohol <u>210</u> or by addition of HO· to the 3β,25ξ-diol <u>301</u>

The origins of the oxygen atoms found in the 3β,25ξ-
diol <u>301</u> and in further degraded autoxidation products
androst-5-ene-3β,17β-diol (<u>87</u>) and pregn-5-ene-3β,20α-diol
(<u>111</u>) as well as of the dioxygen moieties found in the
C_{21}-20-hydroperoxides <u>112</u> and <u>113</u> are obscure. Although
the peroxidic features of the C_{21}-20-hydroperoxides are
best rationalized by reaction of a C_{21}-radical <u>295</u> with a
second $^3O_2$ molecule (*cf.* FIGURE 21), the alcohols <u>87</u>,<u>111</u>,
and <u>301</u> may not be formed in this manner. The diols <u>87</u> and
<u>111</u> have been isolated from air-aged cholesterol, so their
formation by reaction with $^3O_2$ is possible, the 3β,17β-diol
<u>87</u> forming possibly from putative intermediate 3β-hydroxy-
androst-5-ene-17-hydroperoxide formed in turn from reaction
between a C_{19}-radical <u>293</u> (derived by β-scission of the 20-
hydroperoxides <u>22</u> and/or <u>32</u>) and $^3O_2$, the 3β,20α-diol <u>111</u>
in like fashion from the corresponding C_{21}-intermediate
<u>295</u>. However, formation of the 3β,17β-diol <u>87</u> from pyrolysis

of the 20-hydroperoxides $\underline{22}$ and/or $\underline{32}$ must involve retention of the original peroxidic oxygen atoms of the parent
hydroperoxides and not a second $^3O_2$ molecule, as air was
excluded from these studies [2565,2578]. Likewise, the
$3\beta,25\xi$-diol $\underline{301}$ not yet isolated from air-aged cholesterol
but a thermal decomposition product of the 26-hydroperoxides
$\underline{28}$ and $\underline{30}$ must derive via processes which retain the original oxygen atoms of the 26-hydroperoxides. Formally, the
diols $\underline{87},\underline{111}$, and $\underline{301}$ may be viewed as formed by the addition of HO· to the appropriate 17-,20-, or 25-yl radical
formed upon β-scission of the parent hydroperoxide.

VOLATILE AUTOXIDATION PRODUCTS

Discussion thus far of the subsequent decomposition
reactions of the initially formed cholesterol autoxidation
products has emphasized the steroid moiety only. As scission of the side-chain hydroperoxides yields C_{19}-,C_{20}-,C_{21}-,
C_{22}-,C_{23}-,C_{24}-, and C_{26}-sterols, generation of C_1- to C_8-
fragments is to be expected as well. The well known rancid
smell of autoxidized cholesterol samples further establishes
the formation of odorous, volatile components, and we have
trapped volatile material from air-aged cholesterol in yield
of 2 mL/kg from which at least fourteen components ranging
from C_2- to C_6-composition have been identified (cf. TABLE
10). Other possible C_1-,C_7-, and C_8-components have not
been found [2556,2576].

Two volatile components acetone and acetic acid dominate the mixture. The predominant product acetone must derive from the 25-hydroperoxide $\underline{26}$ by β-scission of the
C-24/C-25 bond. However, derivation from the 24-hydroper-
oxides $\underline{105}$ and $\underline{106}$ may also occur, for pyrolysis of the
24-hydroperoxides yielded the 24-aldehyde [2579], and the
terminal isopropyl group might be a source of acetone
found. The dual origins of acetone are suggested in FIGURE
22.

Acetic acid as the next major product may obviously be
derived by scissions of the side chain in several sites,
but origins from the most abundant 25-hydroperoxide $\underline{26}$ or
from the 20-hydroperoxides $\underline{22}$ and $\underline{32}$ seems likely. As acetic acid must represent a higher order of oxidation of an
initially formed fragment, possibly of ethanol also detected as a minor product, the derivation of ethanol and acetic

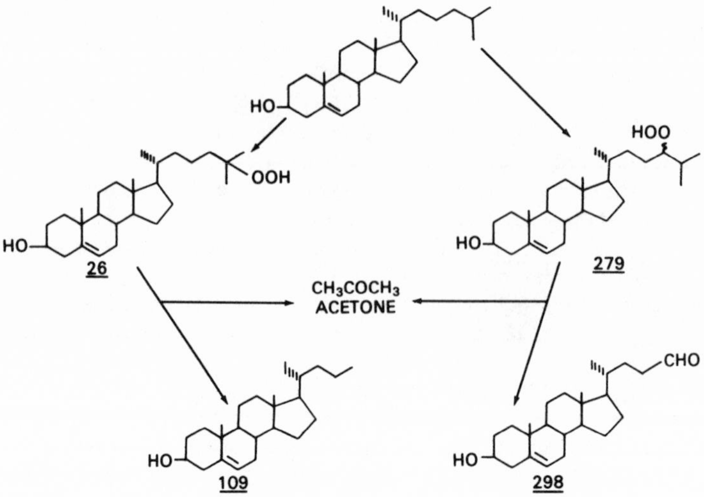

FIGURE 22.   Possible origins of acetone derived
              by cholesterol autoxidation.

acid may be linked together.

        Rationalization of the origins of the remaining eleven
C_4-C_6 products of TABLE 10 may emphasize obvious structural
features of the products or their total carbon content.
The total carbon content of each fragment indicates which
bond of the sterol side-chain was broken and may be suggestive of the structure of the putative precursor implicated.
Thus, the three C_6-compounds clearly derive via scission of
the C-20/C-22 bond of the side chain. By this consideration
only, 20- or 22-hydroperoxide precursors be suspect. Indeed,
β-scission of the (20S)-20-hydroperoxide 22 yields pregnenolone (23) and by difference a C_6-fragment. However,
such simple β-scission does not account for the additional
olefin or alcohol functionalization of the three C_6-frag-
ments. Rather, in this case the second consideration of
olefin or alcohol functionalization of the three C_6-frag-
ments more suitably suggests their origins from the 25-hydro-
peroxide 26 by 1,5-hydrogen transfer processes as outlined
in FIGURE 20.

TABLE 10.   Volatile Organic Products of
             Cholesterol Autoxidation

C_2-Compounds:
    Ethanol
    Acetic acid
C_3-Compound:
    Acetone
C_4-Compounds:
    Butan-2-one
    2-Methylprop-1-ene
    2-Methylpropan-1-ol
    2-Methylpropan-2-ol
    2-Methylpropionic acid
C_5-Compounds:
    Pentan-2-one
    2-Methylbutan-2-ol
    3-Methylbutan-2-one
C_6-Compounds:
    2-Methylpent-1-ene
    2-Methylpentan-2-ol
    2-Methylpent-4-en-2-ol

Similarly, two C_5-products 2-methylbutan-2-ol and
3-methylbutan-2-one may derive via C-22/C-23 bond scissions
from putative 22- and/or 23-hydroperoxide 278 and/or 279
precursors, but other processes must intervene also. Furthermore, the third C_5-product pentan-2-one must derive by
scission of two carbon-carbon bonds, thus of the C-20/C-22
and C-25/C-27 bonds.

The more numerous C_4-products also pose uncertainties
in the processes of their origins. Butan-2-one obviously
requires cleavage of two carbon-carbon bonds in analogy
with pentan-2-one, but the four branched chain C_4-products
all involve cleavage only of the C-23/C-24 bond. Clearly
more information must be gained to elucidate the chemistry
of these processes.

Other systematic study of volatiles from air-aged

cholesterol is not known to me. However, examination of
volatiles generated from cholesterol subjected to energetic
ionizing radiation has been recorded. Cholesterol
irradiated with Rn α-particles was decomposed by dehydrogenations, yielding H_2 and minor products H_2O, CO, CO_2,
and alkanes CH_4, C_2H_5, and C_3H_8 [349]. Irradiation of
cholesterol with ^{60}Co γ-radiation yielded alkanes from CH_4
to C_8H_{18} (C_3H_8 predominant) but also C_4- to C_8-isoalkanes
(2-methylbutane and 2-methylpentane predominating) [1620,
2487]. In neither case were the same volatile components
found which we identified in naturally air-aged cholesterol
[2556]. Sterol products were not examined in either of
these studies, and in these cases it is obvious that ionizing radiation decomposed cholesterol extensively beyond that
accomplished by natural aging or by our use of ^{60}Co γ-radiation (70 krad) to initiate air oxidations [2313].

     Yet another volatile component must derive cholesterol
subjected to radiation. Solid cholesterol, cholesterol in
solutions, and cholesterol-containing tissue preparations
irradiated in air with sunlight, ultraviolet light, X-radiation, or Ra β- and γ-radiation were reported over fifty
years ago to fog photographic plates [946,947,1100-1102,
1540,1600,1731,1975,1977,1979,1988,2378,2600], even to provide images on exposed photographic plates [1100]. These
effects were recognized as being oxidations dependent upon
oxygen of the air [947,1977,1979,2378] and upon other influences such as irradiation time and wavelength, temperature, humidity, etc. [1972,1973] and involving formation
of a volatile product suspected to be H_2O_2 from its positive peroxide reaction with KI-starch color test systems
[1540,1600,2600], thus to be a special case of the Russell
effect whereby H_2O_2 fogged photographic plates in the dark
[2408]. Later considerations of these effects also suggested formation of organic peroxides or ozonides of cholesterol
which liberated H_2O_2 with moisture [1600,2349], the H_2O_2
being viewed as the active agent in any event. Perfectly
dry cholesterol gave no photographic plate fogging following its irradiation [2349], nor did heating to 148° in the
dark or treatments in vacuum [1600].

     These several studies support the formation of volatile
H_2O_2 from cholesterol irradiated in air, but the sterol
products formed were but partially characterized [196,1630,
1983,1997,2000] and none identified. Our own recent demonstration of the oxidation of cholesterol irradiated in air

with ^{60}Co γ-radiation to cholest-5-en-3-one (6) [70] is
formally an oxygen-dependent dehydrogenation which might
yield H_2O_2 as volatile product, but we have not pursued
this possibility, and a satisfying correlation between
H_2O_2 (or other oxidizing volatile material) generated and
steroid 3-ketone (or other) product formed has not been
made.

RESUME OF PRODUCTS AND REACTIONS

The material discussed to this point now allows a complete treatment of the autoxidation of cholesterol as regards all recognized products found in air-aged cholesterol
or derived from recognized products by likely thermal processes. The earlier presentation of autoxidation products
isolated from naturally air-aged cholesterol (TABLE 4) may
now be expanded to include not only those products but also
those directly observed as subsequent alteration products.
In TABLE 11 are listed all recognized cholesterol autoxidation products in association with the initially formed
autoxidation products from which each may derive. There are
nine distinct sites of initial autoxidative attack, at the
C-3,C-4,C-7,C-17,C-20,C-22,C-24,C-25, and C-26 carbon atoms,
and a total of 66 autoxidation products are implicated.

In order to compose TABLE 11 it has been necessary to
make a few assumptions, all of which have been discussed in
the text. Although ten initially formed autoxidation products have been established by isolation, it has been necessary to assume that five others were formed but not discovered. These are the 4β-hydroperoxide 274, epimeric 17-hydro-
peroxide 275, and epimeric 22-hydroperoxides 278, all inferred by the presence of seconary autoxidation products
actually isolated.

The subsequent oxidation products listed in TABLE 11
number forty-six, all having been found in experimental
work that permits their arrangement under an appropriate
initially formed autoxidation product. However, in addition
to these sterols there are five others that have not been
demonstrated in any autoxidation system but that must be
formed in order to account for more highly oxidized autoxidation products actually found. For instance, the 7-hydroxy-
ketones 285 and 286 implicated in the transformation of epimeric 3β,7-diols 14 and 15 to dienone 12 have not been

TABLE 11. Established Cholesterol Autoxidation Products

Initial Autoxidation Products	Subsequent Products Formed
Cholest-5-en-3-one (6)	Cholest-4-en-3-one (8) 6α-Hydroperoxycholest-4-en-3-one (280) 6β-Hydroperoxycholest-4-en-3-one (59) 6α-Hydroxycholest-4-en-3-one (281) 6β-Hydroxycholest-4-en-3-one (88) Cholest-4-ene-3,6-dione (108) 5α-Cholestane-3,6-dione (42)
3β-Hydroxycholest-5-ene-4β-hydroperoxide-(274)*	Cholest-5-ene-3β,4β-diol (41)
3β-Hydroxycholest-5-ene-7-hydroperoxides (46, 47)	Cholest-5-ene-3β,7α-diol (14) Cholest-5-ene-3β,7β-diol (15) 3β-Hydroxycholest-5-ene-7-one (16) Cholesta-3,5-dien-7-one (10) Cholesta-2,4,6-triene (37) 5,6α-Epoxy-5α-cholestan-3β-ol (35) 5,6β-Epoxy-5β-cholestan-3β-ol (36) 5α-Cholestane-3β,5,6β-triol (13) 3β,5-Dihydroxy-5α-cholestan-6-one (44) 5α-Cholestane-3β,5,6β,7α-tetraol (291) Cholesta-4,6-dien-3-one (12)
3β-Hydroxycholest-5-ene-17-hydroperoxides (275)*	3β-Hydroxyandrost-5-en-17-one (86)
3β-Hydroxycholest-5-ene-20-hydroperoxides (22,32)	(20R)-Cholest-5-ene-3β,20-diol (33) (20S)-Cholest-5-ene-3β,20-diol (21) 3β-Hydroxypregn-5-en-20-one (23) Androsta-5,16-dien-3β-ol (191) Androst-5-ene-3β,17β-diol (87) Androst-5-en-3β-ol (114)
3β-Hydroxycholest-5-ene-22-hydroperoxides (278)*	3β-Hydroxypregn-5-ene-20α-hydroperoxide (112) 3β-Hydroxypregn-5-ene-20β-hydroperoxide (113)

(continued)

TABLE 11 (continued)

3β-Hydroxycholest-5-ene-24-hydroperoxides (105,106)	Pregn-5-ene-3β,20α-diol (111) Pregn-5-ene-3β,20β-diol (296) 3β-Hydroxypregn-5-en-20-one (23) Androsta-5,16-dien-3β-ol (191) Androst-5-ene-3β,17β-diol (87) Androst-5-en-3β-ol (114) 3β-Hydroxyandrost-5-ene-17β-carboxylic acid (102) 3β-Hydroxy-22,23-bisnorchol-5-enic acid (101) 3β-Hydroxycholest-5-en-24-one (34) 3β-Hydroxychol-5-en-24-al (298) 24-Norchol-5-en-3β-ol (196)
3β-Hydroxycholest-5-ene-25-hydroperoxide (26)	3β-Hydroxychol-5-enic acid (99) Cholest-5-ene-3β,25-diol (27) 3β-Hydroxy-27-norcholest-5-en-25-one (299) Chol-5-en-3β-ol (109) Pregn-5-en-3β-ol (110)
3β-Hydroxycholest-5-ene-26-hydroperoxides (28,30)	(25R)-Cholest-5-ene-3β,26-diol (29) (25S)-Cholest-5-ene-3β,26-diol (31) 3β-Hydroxy-(25RS)-cholest-5-en-26-al (300) (25ξ)-27-Norcholest-5-ene-3β,25-diol (301) 27-Norcholest-5-en-3β-ol (210)

*Sterol hydroperoxides not isolated but posited to explain the presence of certain subsequent autoxidation products listed.

detected but are implicated from the obvious chemistry. These and the 5,6-epoxy-3β,7α-diol 292 implicated in the derivation of the 3β,5α,6β,7α-tetraol 291 and postulated 3β-hydroxyandrost-5-ene-17β-aldehyde and 3β-hydroxy-22,23-bisnorchol-5-en-24-al intermediates suggested as being involved in the formation of the acids 102 and 101 respectively have not been listed in TABLE 11.

Several of the subsequent products have more than one origin. For instance, the 7-ketone 16 derives from both 7α- and 7β-hydroperoxides 46 and 47. Indeed, all ketones

derive from both epimeric parent hydroperoxides, thus the
20-ketone 23 from 22 and 32, the 24-ketone 34 from 105 and
106. The C_{19}-steroids 87,114, and 191 derive from the C_{27}-
20-hydroperoxides 22 and 32 but also from the C_{21}-20-hydro-
peroxides 112 and 113. Yet additional complexity of products not immediately obvious must obtain, for the 26-alde-
hyde 300 derived from the 26-hydroperoxides 28 and 30 must
surely be an unrecognized mixture of (25R)- and (25S)-iso-
mers. Moreover, the 27-nor-3β,25ξ-diol 301 derived from
28 and 30 may also be a mixture of 25-epimers. Given additional complexities, sixty-six steroids, fourteen volatile
organic compounds, and H_2O_2 make a total of eighty-one compounds thus far implicated in the natural air oxidation of
cholesterol.

Although TABLE 11 exhausts what is presently known of
cholesterol autoxidation products, the tabulation hardly
covers the obviously much more complex spate of products
which are in fact formed. The products listed in those
with chromatographic mobilities generally between those of
the dienone 10 as most mobile and the 3β,5α,6β,7α-tetraol
291 as most polar (excluding acids 99,101,102). Autoxidation products more mobile or less mobile than these limiting sterols have not received study, but it is obvious from
published thin-layer chromatograms that much nonpolar and
immobile material awaits investigation [2303,2576]. We have
preliminary data which suggest that unsaturated steranes
and possibly dicholesteryl ether (18) are amongst the most
mobile components, but whether these derivatives are autoxidation products or artifacts of cholesterol manufacturing
cannot be said. Likewise, the presence of much acidic
material in air-aged cholesterol suggests that more extensively oxidized cholesterol derivatives are formed in abundance.

In FIGURE 23 are summarized stylistically the several
pathways of cholesterol autoxidation demonstrated herein.
Nine primary sites of autoxidative attack and the special
case of 5,6-epoxidation are incorporated into the scheme,
thereby accounting for the generation of approximately
sixty-six distinct steroidal autoxidation products in up
to five consecutive reactions removed from substrate cholesterol at the center of the scheme.

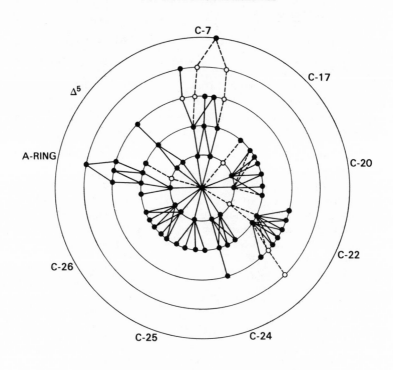

FIGURE 23.   Stylized relationships among cho-
             lesterol autoxidation products and
             processes.

The several chemical processes implicated in the autoxidation of cholesterol are summarized in TABLE 12.

TABLE 12.   Chemical Reactions Implicated in Cholesterol
            Autoxidation

Reaction	Typical Examples
A- and B-Rings:	
1. Alcohol dehydrogenation	1 to 6, 13 to 44
2. Double bond isomerization	6 to 8
3. Olefinic ketol isomerization	88 to 42
4. Hydroperoxide formation	
(1) With double bond isomerization	6 to 59/280
(2) Without double bond isomerization	1 to 46/47
5. Hydroperoxide epimerization	46 to 47
6. Hydroperoxide dehydration	47/47 to 16
7. Hydroperoxide reduction	46 to 14, 47 to 15
8. Olefin epoxidation	1 to 35/36
9. Epoxide hydration	35/36 to 13
10. Alcohol dehydration	16 to 10
Side-Chain:	
1. Hydroperoxide formation	1 to 26
2. Hydroperoxide dehydration	105/106 to 34
3. Hydroperoxide reduction	26 to 27
4. β-Scission	
(1) To unfunctionalized products	
(i) Without radical rearrangement	22 to 114, 26 to 109
(ii) With radical rearrangement	26 to 110
(2) To functionalized products	
(i) Olefins	22 to 191
(ii) Alcohols	28/30 to 301
(iii) Aldehyde	105/106 to 298
(iv) Ketone	26 to 299
(v) Carboxylic acid	99,101,102
(vi) Hydroperoxide	112,113
5. Hydroperoxide epimerization	112 to 113

CHAPTER VI. OTHER OXIDATIONS

Treatment of cholesterol autoxidation thus being complete, it is now suitable to examine related autoxidation
of other steroids in order to see what else may be learned
that could aid in further understanding of cholesterol autoxidation. Throughout the text to this point mention has
been made of various odd autoxidations of other steroids,
and the present chapter will examine the autoxidation of
the two important sterols ergosterol (65) and 5α-lanost-8-
en-3β-ol (68) for which considerable information is at hand
and also the oxidation of other steroids under a variety of
conditions related to simple air autoxidations.

OTHER STEROLS

Despite the importance of sterols other than cholesterol, ergosterol, and the 8-stenol 68 throughout Nature,
few details of the autoxidation chemistry of other sterols
has developed. For instance, the major plant sterols,
other than sitosterol (20), have not received attention.
Moreover, other than the 8-stenol 68, desmosterol (78) and
cholesta-5,7-dien-3β-ol (56) the several sterols implicated
in cholesterol biosynthesis have not been investigated in
this regard.

Other 5-Stenols

Even though systematic examination of the autoxidation
of other important 5-stenols has not been conducted, enough
information is at hand to suggest that other 5-stenols are
autoxidized in the same fashion as is cholesterol. Sitosterol heated in air yields both epimeric sitosterol 7-
hydroperoxides in the same manner as cholesterol, the sitosterol 7-hydroperoxides decomposing thermally in turn to the
epimeric 3β,7-diols 76 and 77 and 7-ketone 72. This trio
of secondary products 72,76, and 77 recovered from autoxidized sitosterol along with stigmasta-3,5-diene,stigmasta-
3,5-dien-7-one (143), stigmast-4-en-3-one (149), stigmast-
4-ene-3,6-dione (151), 6β-hydroxystigmast-4-en-3-one (150),
stigmast-4-ene-3β,6β-diol, 5-hydroxy-5α-stigmastane-3,6-
dione, and 5α-stigmastane-3β,5,6β-triol (145) characterize
sitosterol autoxidation as paralleling that of cholesterol
[615,1260,2758,2759].

The autoxidation of sitosterol in edible oil also occurs, as the 3β,7-diols 76 and 77 and 7-ketones 72 and 143 have been detected in processed edible oils (cf. TABLE 3). Stigmasterol (70) and brassicasterol (24-methyl-(24R)-cholesta-5,E-22-dien-3β-ol) in oils are likewise autoxidized to the corresponding 7-hydroxylated derivatives [1260]. Commercial processing of edible oils is characterized by formation of other artifacts from sterols too. Dehydration of cholesterol to dicholesteryl ether (18) and of phytosterols to the corresponding ethers occurs [1260], but elimination reactions of the sterols (or their esters) affected by bleaching earths used in clarifications yield the corresponding stera-3,5-dienes [1241,1259,1381,1760,1762, 1763,1765,1766]. Moreover, elimination reactions of the autoxidation product 3β,7-diols also yield steratriene products [1260,1764,1765].

Furthermore, autoxidation of sitosterol also occurs in the side-chain, as stigmast-5-ene-3β,25-diol (302) has been shown to form upon heating sitosterol in air [94]. The isolation of the steroid acids 3β-hydroxy-22,23-bisnorchol-5-enic acid (101) and 3β-hydroxy-23-norchol-5-enic acid (100) from samples of air-aged sitosterol [2404, 2405] also indicates a parallel autoxidation of the sitosterol side-chain.

The attack of $^1O_2$ on sitosterol yields 3β-hydroxy-5α-stigmast-6-ene-5-hydroperoxide as an exact 24-ethyl homolog of the 5α-hydroperoxide 51 derived thereby from cholesterol [2270].

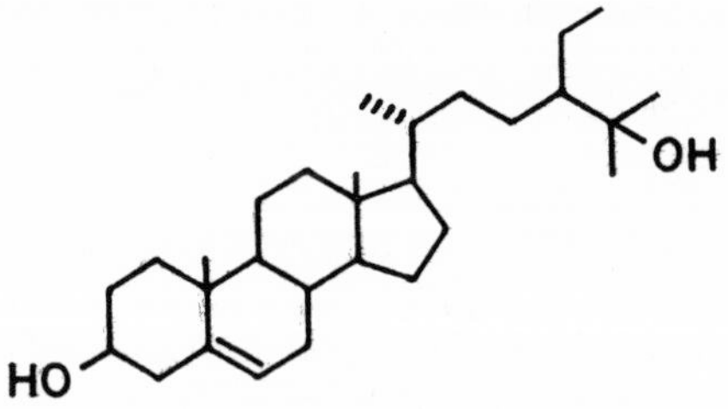

302

From these bits of evidence and from other matters previously outlined in Chapters II and III we may presume that other 5-stenols related to cholesterol have oxidation chemistry entirely analogous to that of cholesterol.

Although supporting free radical oxidation studies are not available, additional supporting evidence that various 5-stenols yield analogous Δ^6-5α-hydroperoxide in $^1O_2$ oxidations is recorded. Stigmasterol, diosgenin, and the 17-ketone 86 yield analogous 5α-hydroperoxides [2102], as also does 3β,19-diacetoxyandrost-5-en-17-one [1667,1668]. Photosensitized oxygenation of (25R)-cholest-5-ene-3β,26-diol (29) and subsequent allylic rearrangement yielding 3β,26-dihydroxy-(25R)-cholest-5-ene-7α-hydroperoxide [1107] further attests uniform action of $^1O_2$ on these 5-stenols.

These generalities notwithstanding, it is possible to derive the major cholesterol autoxidation product 7-hydroperoxides 46 and 47 in other ways. The isomeric 6-stenol 5α-cholest-6-en-3β-ol (not naturally occurring) has an oxidation chemistry which yields the same epimeric 7-hydroperoxides 46 and 47 as derived from cholesterol by autoxidation. Air oxidation induced by ^{60}Co γ-radiation gives the 7-hydroperoxides in the approximate ratio 46:47 of 1:2 [1402], whereas the 7α-hydroperoxide 46 is the sole product isolated from the attack of $^1O_2$ [1751]

The 5-stenols which are recognized cholesterol autoxidation products (cf. TABLE 10) have likewise received almost no attention regarding their further autoxidation, the case of 3β,5α,6β,7α-tetraol 291 from the 3β,7α-diol 14 via the 5,6-epoxide 292 already discussed being the only case of record. Further autoxidation of the Δ^5-3-ketone 6 has, of course, received additional study.

Oxidation of the Δ^5-3β,7β-diol 15 by $^1O_2$ gives 5,6-epoxy-3β-hydroxy-5α-cholestan-7-one and some 7-ketone 16 [1754] in analogy to the $^1O_2$ oxidation of the 4-stenol 79 discussed in a later section of this chapter.

Interestingly, the 11β-hydroxylated derivative of cholesterol (20R)-cholest-5-ene-3β,11β-diol was not found to be readily autoxidized. However, in distinction to other Δ^4-3-ketones such as 8, the related (20R)-cholest-4-ene-3, 11-dione deteriorated by autoxidation upon storage. Oddly, the isomeric (20S)-cholest-4-ene-3,11-dione was not autoxidized under the same conditions [2117].

Of the several important 5-stenols bearing additional olefinic unsaturation, including such sterols as the 5,22-diene stigmasterol (70), the 5,24-diene desmosterol (78),

and the 5,24(28)-diene fucosterol (75) little is known.
The sensitivity of desmosterol and fucosterol to air oxidations upon storage has been noted [2403,2480,2546], but full
studies have not been made on either. Cholesta-5,E-23-diene-
3β,25-diol (181) and (24ξ)-cholesta-5,25-diene-3β,24-diol
(179) have been detected in air-aged desmosterol samples;
similarly, 3β-hydroxy-24-methyl-(24ξ)-cholesta-5,28-diene-
24-hydroperoxide (164) and saringosterol (163) have been
recognized as fucosterol (75) autoxidation products [794].
The 24-ketone 34 is also recognized as a fucosterol autoxidation product. These items clearly indicate the greater
susceptibility to air oxidation of the side-chain double
bonds of desmosterol and fucosterol compared to the nuclear
Δ^5-double bonds.

Sterol Esters

    The oxidation of cholesterol 3β-fatty acyl esters,
indeed of all sterol esters, has not received systematic
attention, but it is generally held that sterol esters are
less sensitive to air oxidation than the corresponding free
sterol. This position derives in large part from the lore
of the sterol chemist that recognizes the greater stability
of sterol esters but also is supported by comparison studies
in several systems. Thus, the rate of autoxidation of
several cholesterol esters in aqueous sodium stearate dispersions was less than that of cholesterol [197,204], and
the slower autoxidation of cholesterol esters relative to
cholesterol has been noted in monomolecular layers of
sterols spread over water [1218,1413] and on silica gel
[2725].

    Samples of pure cholesterol 3β-myristate, 3β-palmitate,
and 3β-stearate stored over a decade at room temperature
have not yet developed odor or color indicative of autoxidative deterioration (during which time cholesterol surely
does!). Nonetheless, differential scanning calorimetry of
cholesterol 3β-myristate suggested deterioration over a
two year period of storage [2598]. Samples of unsaturated
fatty acyl esters are much more obviously sensitive to
deterioration, as samples of pure cholesterol 3β-linoleate
(234), 3β-linolenate, and 3β-arachidonate stored with
3β-palmitate and 3β-stearate samples previously mentioned
have deteriorated extensively by visual appearance and odor.

Exactly the same pattern obtains for sitosterol fatty acyl esters. Whereas sitosterol and sitosterol 3β-stearate stored at 20°C for 8 weeks were resistant to air oxidations, sitosterol 3β-linoleate was extensively autoxidized via peroxidic intermediates, yielding products oxidized in the linoleate moiety (hydroxyacids and ketoacids) and sterol nucleus (most probably at the C-7 position) [1804].

The autoxidation of cholesterol esters of polyunsaturated fatty acids appears to be the most rapid of all, cholesterol 3β-linoleate (234) oxidizing faster than cholesterol in aqueous sodium dodecyl sulfate dispersions and much more rapidly than cholesterol 3β-oleate, 3β-ricinoleate, and 3β-12'-hydroxystearate. Notably, the 3β-linoleate 234 is more rapidly autoxidized than its more highly unsaturated congeners cholesterol 3β-linolenate and cholesterol 3β-arachidonate in these aqueous dispersions [1777, 1778].

Initiation of autoxidation of cholesterol fatty acyl esters is effected in the same manner as for the free sterol, thus by radiation, free radicals, etc. The initiation of autoxidation of a series of lower fatty acyl esters of cholesterol by the frequently used radical initiator azobisisobutyronitrile has been described, but products were not identified [614,2583]. The rate of autoxidation of cholesterol 3β-benzoate appears to depend on the physical phase examined, rates increasing on going from solid to mesophase and from mesophase to isotropic liquid phase [1291].

Initiation of autoxidation of cholesterol fatty acyl esters by radiation, other than by heat [2455], has not attracted much attention. However, as certain cholesterol esters acting as cholesteric liquid crystals are now of economic importance and their optical properties are quite sensitive to radiation and the presence of impurities, interest in the matter is developing. Many reports deal with various mixtures of cholesterol fatty acyl esters in admixture with other cholesterol esters and derivatives, but a few deal with pure esters. The optical properties of a typical cholesteric liquid crystal cholesterol 3β-nonanoate are affected by various bulk impurities, the color transition of temperature decreasing with amount of impurity [618,1784]. Likewise, irradiation of solutions of cholesterol 3β-nonanoate or 3β-decanoate with ^{60}Co γ-

radiation causes a dose-dependent decrease in the color
transition temperature [40,730]. As the same effects were
had in the presence or absence of air, oxidation of the
sterol ester was not apparent, but hydrogen abstraction
from sterol allylic positions was indicated. Moreover,
natural aging of cholesterol 3β-nonanoate in air, particularly in combination with ultraviolet light irradiation,
greatly decreases intensity of color display, although the
color transition temperature is not affected [2087]. Although chemical reactions of sterol esters have not been
demonstrated in these several studies, adequate chromatographic methods were not applied.

The course of autoxidation of cholesterol esters is
probably the same as that for cholesterol, but the nature
of the fatty acyl moiety may direct the matter. Thus,
autoxidative attack in the polyunsaturated fatty acyl moiety
may precede that at the C-7 allylic nuclear position, a
matter suggested in several cases [1777, 1804]. For saturated fatty acid esters of cholesterol, autoxidation at the
C-7,C-20 and C-25 positions is established, as the 3β-ace-
tates of cholesterol 7β-, 20-, and 25-hydroperoxides have
been isolated from cholesterol 3β-acetate heated at 90-100°C,
along with the 3β-acetates of the 3β,7-diols 14 and 15 and
the 7-ketone 16 [2455]. Furthermore, autoxidation of monomolecular films of cholesterol 3β-acetate spread over water
apparently also yield the same 3β,7-diol 3β-acetates [1218].
More extensive study of the autoxidation of cholesterol
(or other sterol) fatty acyl esters has not been recorded.

Stanols

The stanols 5α-cholestan-3β-ol (2) and 5β-cholestan-
3β-ol (4) not having olefinic unsaturation are not noted
for their autoxidation reactions. Indeed, it is well recognized that these stanols are not as unstable as cholesterol
towards air and autoxidation, and samples of either stanol
2 or 4 stored for years do not contain demonstrable levels
of oxidized impurities.

Nonetheless, as much as cholesterol is autoxidized in
the side-chain to a variety of hydroperoxides, so might
also the stanols 2 and 4 given an appropriate initiation
process. To this end, irradiation of the 5α-stanol 2 with
[60]Co γ-radiation in vacuum causes a distinctive electron

spin resonance signal which is quite different from that
from irradiated cholesterol [1004]. The spectrum has not
been analyzed however. Moreover, we have irradiated the
5α-stanol 2 in air with ^{60}Co γ-radiation and obtained elu-
sive and weak positive peroxide color tests on the irradiated stanol with N,N-dimethyl-*p*-phenylenediamine. The
matter of induced stanol autoxidations has not been examined
systematically.

Cholest-4-en-3β-ol

As the 4-stenols are not recognized as naturally occuring, interest in their oxidation chemistry and stability
has not materialized. The air oxidation of cholest-4-en-
3β-ol (79) induced by ^{60}Co γ-radiation is dominated by
dehydrogenation to the enone 8, but allylic oxidations
yielding the epimeric 3β-hydroxycholest-4-ene-6α- and 6β-
peroxides 247 and 89 occur [1402]. These are the same
6-hydroperoxides found as minor $^1O_2$ oxidation products of
cholesterol [1401] whose further chemistry has been presented previously in FIGURE 17.

Photosensitized oxidation of the 4-stenol 79 is also
dominated by dehydrogenation to the enone 8, [665,1269,1270,
1754,2456], but as dehydrogenation of 79 to 8 by $^3O_2$ is
facile and alcohol dehydrogenation is not a recognized mode
of action of $^1O_2$, formation of 8 under these conditions does
not appear to involve $^1O_2$ [1402,2456]. Rather, as previously
discussed (*cf*. Chapter IV), the recognized product of $^1O_2$
attack on the 4-stenol 79 is the 4α,5α-epoxide 4α,5-epoxy-
5α-cholestan-3-one (249) [665,1269,1270,1754]. The isomeric
4β,5-epoxy-5β-cholestan-3-one (250) is a minor $^1O_2$ product.
The action of alkaline H_2O_2 on the enone 8 likewise yields
both 4,5-epoxides [2456].

5α-Cholest-7-en-3β-ol

The penultimate biosynthesis intermediate 5α-cholest-
7-en-3β-ol (57) which occurs in tissue cholesterol samples
to the extent of 0.3-3% [1712] could conceivably contribute
to the composition of sterol autoxidation products nominally
attributed to cholesterol. However, the air oxidation of
57 induced by ^{60}Co γ-radiation yields the anticipated allylic hydroperoxides 3β-hydroxy-5α-cholest-7-ene-6α-hydro-

peroxide (303) and 3β-hydroxy-5α-cholest-7-ene-6β-hydroper-
oxide (304). Thermal decomposition of these hydroperoxides
proceeds via the same dehydration and reduction processes
characteristic of other sterol hydroperoxides, yielding the
6-ketone 3β-hydroxy-5α-cholest-7-en-6-one (80) and the epimeric 3β,6-diols 5α-cholest-7-ene-3β,6α-diol (305) and 5α-
cholest-7-ene-3β,6β-diol (306) [1402].

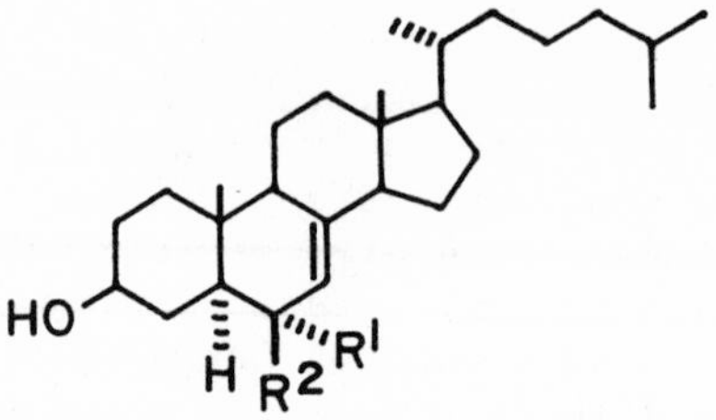

303 R^1 = OOH, R^2 = H

304 R^1 = H,R^2 = OOH

305 R^1 = OH,R^2 = H

306 R^1 = H, R^2 = OH

However, a new thermal decomposition reaction of sterol
hydroperoxides was encountered in our studies of the 6-hydro-
peroxides 303 and 304. Both epimeric 6-hydroperoxides were
subject to a ready thermal elimination of the hydroperoxide
moiety to yield cholesta-5,7-dien-3β-ol (56) as a major degradation product [1402]. A similar thermal elimination
has not been found in the cholesterol series. Although the
autoxidation products 303-306 and 80 derived from 57 have
not been found in air-aged cholesterol, the 5,7-dienol 56
known as a biosynthesis precursor of cholesterol commonly
present in most samples does have an autoxidation origin
from 57.

The oxidation of the 7-stenol by $^1O_2$ is also of interest, as products different from those derived by radiation-
induced oxidation are obtained. The attack of $^1O_2$ on the
7-stenol 57 involves uptake of two equivalents of $^1O_2$ per

equivalant of sterol as does also attack on 57 3β-acetate
[665]. Three prominent sterol hydroperoxides are formed
from 57 [2309] and from 57 3β-acetate [2100], the esterified
products having been identified as previously mentioned
(Chapter IV) as 3β-hydroxy-5α-cholest-8(14)-ene-7α-hydro-
peroxide (252) 3β-acetate and secondary products 3β-hydroxy-
5α-cholest-14-ene-7α,8α- and 7α,8β-dihydroperoxide (253)
3β-acetates. We presume that hydroperoxides 252, 253 form
by $^1O_2$ oxidation of the free sterol 57 [2309].

Sequential attack of two equivalents of $^1O_2$ on other
steroids is known: for instance, 11β-chloro-17α-ethinyl-
estr-4-en-17β-ol yields after hydroperoxide reduction 11β-
chloro-17α-ethinyl-5α-estr-3-ene-5,17β-diol and 11β-chloro-
17α-ethinylestr-5-ene-4β,17β-diol but also 11β-chloro-17α-
ethinyl-5α-estr-6-ene-4β,5,17β-triol [2127]

Other Stenols

Studies of the autoxidation of other unsaturaged sterols
such as 5α-cholest-8-en-3β-ol, 5α-cholest-8(14)-en-3β-ol
(84), 5α-cholest-14-en-3β-ol, etc. have not been made, but
evidence that the 8(14)-stenol 84 is autoxidized via hydroperoxides on storage has been recorded [2295]. Also the
instability of zymosterol on storage has been noted [2480].

However, interest in these and related stenols has developed in relation to the mechanism of action of $^1O_2$ on
cyclic olefins. Thus, attack of $^1O_2$ on 5α-cholest-8-en-3β-
ol 3β-acetate yields 3β-acetoxy-5α,9α-cholest-7-ene-9-hydro-
peroxide, 3β-acetoxy-5α,8α-cholest-9(11)-ene-8-hydroper-
oxide, and the dihydroperoxides 3β-acetoxy-5α,9α,14α-cholest-
7-ene-9,14-dihydroperoxide and 3β-acetoxy-5α,8ξ,9α-cholest-
14-ene-8,9-dihydroperoxide (derived from initially formed
3β-acetoxy-5α,9α-cholest-8(14)-ene-9-hydroperoxide). Oxidation of the 8(14)-stenol 84 yields 3β-hydroxy-5α,14α-cholest-
7-ene-14-hydroperoxide, 3β-hydroxy-5α,8α-cholest-14-ene-8-
hydroperoxide, and 3β-hydroxy-5α,14α-cholest-8-ene-14-hydro-
peroxide [665].

5α-Lanost-8-en-3β-ol

As a component of sheep wool fat 5α-lanost-8-en-3β-ol
(68) is of interest as a major naturally occurring sterol

known to be readily autoxidized. Sheep wool fat when fresh
contains very little or no autoxidation products [1069],
but wool on sheep constitutes an ideal medium in which the
8-stenol 68, cholesterol, and other congeners may autoxidize,
exposed as it is to air and sunlight over long periods of
time. Many autoxidation products of 68 have been found in
oxidized wool fat, including the ketones 3β-hydroxy-5α-
lanost-8-en-7-one (82), 3β-hydroxy-5α-lanost-8-ene-7,11-
dione (83), and 3β-hydroxy-5α-lanostane-7,11-dione (cf.
Chapter III). Nonetheless, systematic study of the autoxidation of 68 has not been recorded. Rather, studies of
autoxidation of the 8-stenol 68 3β-acetate in model systems
have been made. One may presume that the chemistry of 68
3β-acetate holds also for the free sterol as well.

    Free Radical Oxidations. Autoxidation of 68 3β-acetate
in solution by $^3O_2$ yields as first detected product a 7β-
hydroperoxide 3β-hydroxy-5α-lanost-8-ene-7β-hydroperoxide
(254) 3β-acetate. Autoxidative attack on the 8-stenol and
on cholesterol thus both give 7β-hydroperoxides (254 and
47 respectively) as initial product. However, in distinction
to the cholesterol example where the epimeric 7α-hydroper-
oxide 46 is also formed, a 7α-hydroperoxide 257 has not been
found as autoxidation product.

    A second monohydroperoxide 3β-hydroxy-5α-lanost-8-ene-
11β-hydroperoxide (255) 3β-acetate is formed more slowly
[357,2175], and the dihydroperoxide 3β-hydroxy-5α-lanost-8-
ene-7β,11β-dihydroperoxide (256) 3β-acetate has been identified among products [1071]. Several other sterol hydroperoxides formed remain unidentified [2175]. We have confirmed
that the autoxidation of the free sterol 68 induced by
^{60}Co γ-radiation also generates three sterol hydroperoxides
early, and we presume that these are the 7β- and 11β-mono-
hydroperoxides and the 7β,11β-dihydroperoxide derivatives
254-255 [2309,2383].

    The 7β-hydroperoxide 254 3β-acetate is dehydrated to
the corresponding 7-ketone 82 3β-acetate upon chromatography,
attempted acetylation, or treatment with Fe(II) salts [1071];
thus, 254 is obviously the parent of the 7-ketone 82 found
in autoxidized wool fat. Likewise, the 7β,11β-dihydroper-
oxide 256 3β-acetate is dehydrated readily to the 7,11-dike-
tone 83 3β-acetate [1071]. Moreover, the 11β-hydroperoxide
255 3β-acetate is also transformed readily to the corresponding 11-ketone 3β-acetoxy-5α-lanost-8-en-11-one, but this

11-ketone has not been identified as an autoxidation product of 68 3β-acetate.

As the 7,11-diketone 83 3β-acetate is a prominent product of 68 3β-acetate autoxidation in model systems as is
83 in autoxidized wool fat [1070,1071], studies of 83 3β-
acetate autoxidation yielding addition novel products is
of interest. Autoxidation of 83 3β-acetate gives 3β-acetoxy-
1β-hydroperoxy-5α-lanost-8-ene-7,11-dione (307), from which
is derived by dehydration 3β-acetoxy-5α-lanost-8-ene-1,7,11-
trione (308). An elimination reaction of the 1,7,11-trione
308 also gives 5α-lanosta-2,8-diene-1,7,11-trione [2174].

307   308

Yet other novel reaction sequences obtain in the autoxidation of 68 3β-acetate in model systems, as oxidative attack at the C-15 position and concommitant reduction of the
Δ^8-double bond occurs. Thus, products 3β-acetoxy-15α-
hydroxy-5α-lanostan-7-one, 3β-acetoxy-15β-hydroxy-5α-lanos-
tan-7-one, and 3β-acetoxy-5α-lanostane-7,15-dione suggest
such processes. Similarly, 3β-acetoxy-5α-lanostane-11,15-
dione also isolated evinces C-15 oxidation and concommitant
Δ^8-double bond reduction as well [2173].

Formation of the 1β-hydroperoxide 307 poses an example
of oxidation at a homoallylic site with an intervening
quarternary allylic carbon atom, thus a previously unrecognized mode of autoxidation. The quaternary carbon may exert
a slight activating effect on the adjacent methylene group
in the autoxidation of the 7,11-diketone 83 3β-acetate [2174].
Furthermore, one may presume that formation of the several
15-oxygenated products also involve initial formation of
Δ^8-15-hydroperoxides whose dehydration yield 15-ketones,
reduction yield epimeric 15-alcohols. Such oxidation at
the C-15 position would be another case of homoallylic attack of the kind suggested for formation of the 1β-hydro-

peroxide 307. However, these 15-oxygenated sterols bear no
Δ^8-double bond, and means by which this nuclear unsaturation
be removed remains an obscure matter.

A further reaction of the C_{30}-7β-hydroperoxide 254 3β-
acetate is that of hydroperoxide elimination to give 5α-
lanosta-7,9(11)-dien-3β-ol (81) 3β-acetate [357]. Sterol
hydroperoxide eliminations may thus occur with double bond
rearrangement but also without (cf. transformation 303 and
304 to the 5,7-dienol 56).

It is thus obvious that three additional modes of
autoxidative transformation need to be added to those of
TABLE 10, these being formation of homoallylic hydroperoxides (such as the 1β-hydroperoxide 307), hydroperoxide
elimination to the olefin (such as the 7β-hydroperoxide
254 to the 7,9(11)-diene 81), and Δ^8-double bond reductions.

Singlet Oxygen Reactions. The oxidation of 68 and of
68 3β-acetate by $^1O_2$ already discussed (cf. Chapter IV) has
been investigated, but a complete study of this system
remains to be made. In photosensitized oxygenations in
which $^1O_2$ is implicated 68 3β-acetate is oxidized extensively
to seven isolated products, only one of which is hydroperoxide 3β-hydroxy-5α-lanost-8-ene-7α-hydroperoxide (257)
3β-acetate. Other products include 7,9(11)-dienol 81 3β-
acetate and 7-ketone 83 3β-acetate, both of which may be
viewed as deriving from the 7α-hydroperoxide 257 3β-
acetate by elimination and dehydration reactions respectively.

It is obvious that formation of the 7α-hydroperoxide
257 3β-acetate from 68 3β-acetate does not conform with
the accepted modes of reaction between a cyclic olefin and
$^1O_2$. Given the usual restraints on the $^1O_2$ reaction of
preference for allylic axial hydrogen abstraction and the
cyclic ene reaction mode, 3β-hydroxy-5α-lanost-7-ene-9α-
hydroperoxide (309) 3β-acetate would be the anticipated
product. Indeed, in anticipation of obtaining this product but with no confidence that such a product could be
isolated, the experimental conditions used included p-nitro-
benzenesulfonyl chloride as a putative means of trapping
the 9α-hydroperoxide as a stable ester [792,793]. Instead,
it appears that the esterifying agent may have caused allylic rearrangement of the sought 9α-hydroperoxide product to
the 7α-hydroperoxide 257 3β-acetate actually recovered!

Other secondary oxidized products from this system
include 8a-oxa-5α-BC(8a)-homolanosta-7,9(11)-dien-3β-ol
3β-acetate, 7α,8-epoxy-5α, 8α-lanost-9(11)-en-3β-ol 3β-
acetate, 7α,8-epoxy-5α,8β-lanost-9(11)-en-3β-ol 3β-acetate,
and 9,11α-epoxy-5α,9α-lanost-7-en-3β-ol 3β-acetate [792,793].

In our examination of the attack of $^1O_2$ on 68 we have
found that five sterol hydroperoxides are formed, but the

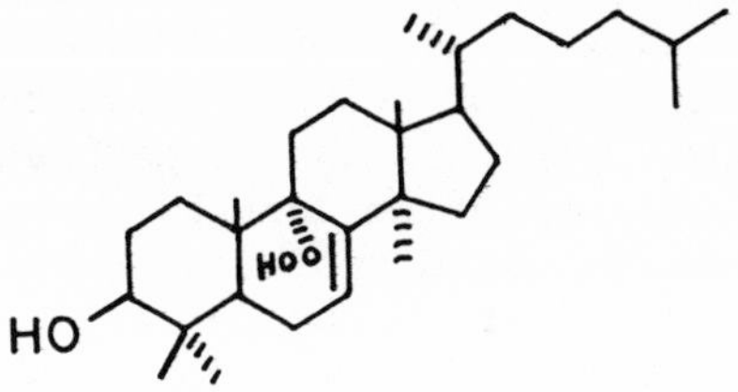

309

identities of none have been investigated [2309,2383].
Clearly, the postulated 9α-hydroperoxide 309 as well as 3β-
hydroxy-5α-lanost-9(11)-ene-8β-hydroperoxide as products of
the abstraction of quasiaxial 7α- and 11β-hydrogen in the
cyclic ene reaction mode may be among observed products, as
may also be other directly formed mono- and di-hydroperoxides
and allylically rearranged hydroperoxides.

CONJUGATED DIENES

The oxidation chemistry of steroid conjugated dienes
poses additional features of complexity not found in sterol
dienes in which the double bonds are widely separated as in
the nucleus and in the side-chain. By far the most developed
understanding of sterol diene oxidations is had for ergosterol (65), with its characteristic 5,7-diene feature.

Ergosterol

The notorious instability of ergosterol and the related
cholesta-5,7-dien-3β-ol (56) leading to formation of many
degradation products has been recognized since 1928 when
Bills addressed the matter of ergosterol purity [222]. The
established deleterious effects of light, heat, and air on
ergosterol [333,1439,1916] have probably influenced adversely

the systematic study of the oxidation chemistry of ergosterol so that to this day an incomplete understanding of
the problem obtains. Both 5,7-dienes are sensitive to decomposition during chromatography on silica gel [1329, 1330],
but few of the many oxidation products from either diene
have been identified. As the B-ring chemistry of ergosterol
is the same as that of 56, I shall career through these
matters with examples from either series as befits.

     The major oxidative reaction of ergosterol (and 56) is
that of 1,4-cycloaddition of O_2 to give the 5α,8α-peroxide
62 (or 60) [2700]. Other recognized oxidation products
appear to derive from the product 5α,8α-peroxides. However,
unidentified hydroperoxide products are also found together
with the 5α,8α-peroxides (62 or 60) [1167,2401]. Oxidation
of 56 by the NADPH-dependent microsomal lipid peroxidation
system of rat liver likewise yields the 5α,8α-peroxide 60
but also unidentified oxidation products [1225]. The cyclic
peroxides are relatively stable; indeed 5α,8α-peroxide formation stabilizes sterol 5,7-dienes so as to facilitate
their isolation from biosynthesis experiments [598,600,648,
649].

     By contrast, in the absence of O_2 ergosterol (and 56)
undergoes complex thermal and photo transformations, including dehydrogenations, dimerizations, isomerizations, and
degradations, yielding over thirty-five recognized products,
many of which are as highly sensitive to air oxidation as
the parent sterol 5,7-diene. In the presence of O_2, oxygenation of ergosterol to the 5α,8α-peroxide 62 should predominate, but in systems where parent sterol and thermal
and photo products occur together, very complex mixtures
of oxidation products may result.

     Photodehydrogenations. The photosensitized dehydro-
genation of ergosterol in the presence of O_2 yields ergosta-
5,7,9(11),E-22-tetraen-3β-ol (66) [566]. 5α,8α-Peroxides
may then be formed from both ergosterol and 66. In the
absence of air, photosensitized dehydrogenations of ergosterol or 56 yield dimeric pinacols [2096,2699,2726]. The
C_{28}-dimers have been variously viewed as mixtures of rotational isomers [773] or of C-7 stereoisomers [531], but
the C_{27}-dimers have been shown to be the three double bond
isomers 7α-(3β-hydroxycholesta-5,8-dien-7α-yl)cholesta-5,8-
dien-3β-ol (310), 7α-(3β-hydroxycholesta-5,8(14)-dien-

7α-yl)cholesta-5,8(14)-dien-3β-ol, and 7α-(3β-hydroxychole-
sta-5,8(14)-dien-7α-yl)cholesta-5,8-dien-3β-ol [319].

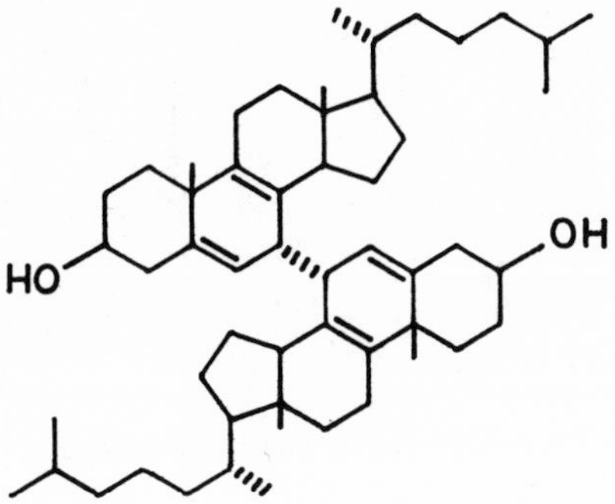

310

Ergosterol Isomerizations. Thermal decomposition of
the ergosterol dimers yields among other products (discussed
in the next section) 5α-ergosta-5,8(14),E-22-trien-3β-ol
[744]. The combined photodimerization and subsequent thermal reaction is thus a formal isomerization of ergosterol.
Moreover, although ergosterol is stable at below 350°C, at
400°C an isomerization to 5α-ergosta-7,9(11),E-22-trien-
3β-ol occurs along with degradation to the B-ring aromatic
sterol neoergosterol (19-norergosta-5,7,9,E-22-tetraen-
3β-ol)(311) [2722] discussed in the next section.

B-Ring Aromatic Sterols. Three types of B-ring aromatic sterols are derived from ergosterol, 56, and related
unsaturated steroids. These types include 19-norsteroids
such as neoergosterol derived by scission of the C-10/C-19
bond and 1,10-seco-B-ring aromatic sterols and anthrasteroids derived by scission of the C-1/C-10 bond (by different processes). Thermal degradation of the dimers of
ergosterol [1158,1691,2699] and of 56 [2096] yield B-ring
aromatic sterols. From the ergosterol dimers are obtained
neoergosterol (311) as major product, its dehydration product 19-norergosta-3,5,7,9,E-22-pentaene, 1,10-secoergosta-
5,7,9,E-22-tetraen-3β-ol (313) and its dehydration product
1,10-secoergosta-3,5,7,9,E-22-pentaene, and ergosta-5,8(14),
E-22-trien-3β-ol [774].

These B-ring aromatic sterols have not been implicated
in the air oxidation of ergosterol, but the action of

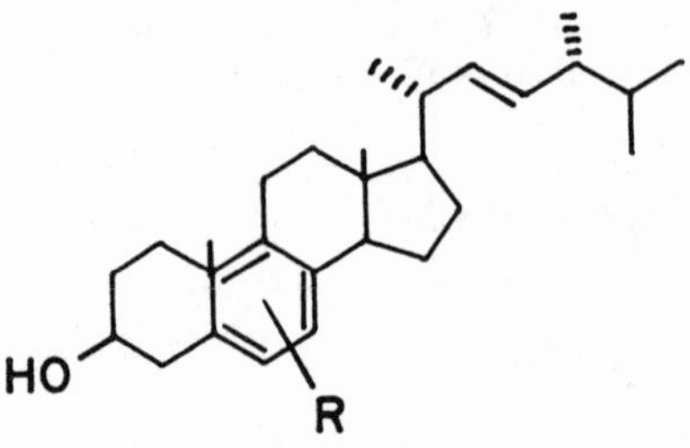

311   R = H   313

312   R = CH₃

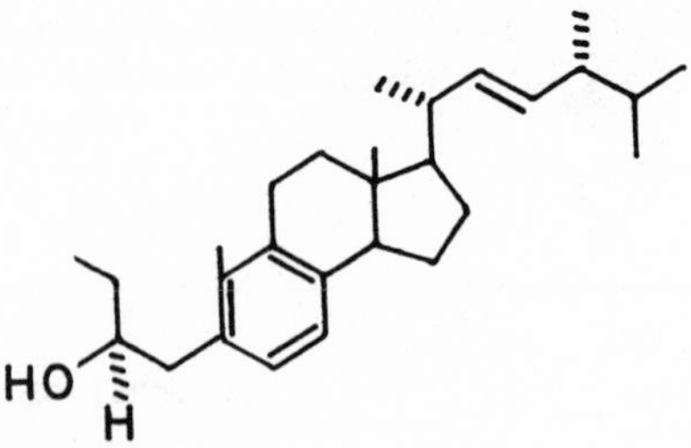

di-*tert.*-butyl peroxide on ergosterol yields neoergosterol
[2157], so this degradation might occur in some aerobic
systems. Furthermore, homogenates of the ameba *Acanth-
amoeba castellani* (but not intact cells) transform endogenous sterol by the action of enzymes released upon cell disruption in air into B-ring aromatic sterols, apparently
derived by migration of the 10β-methyl group to the C-6 or
C-7 position. Four such products 6(or 7)-methyl-19-nor-
ergosta-5,7,9,E-22-tetraen-3β-ol (312), 6(or 7)-methyl-19-
norergosta-5,7,9-trien-3β-ol, 24-ethyl-6(or 7)-methyl-(24R)-
19-norcholesta-5,7,9,E-22-tetraen-3β-ol, and 24-ethyl-
6(or 7)-methyl-(24S)-19-norcholesta-5,7,9-trien-3β-ol have
been characterized [1314].

Related enzymic B-ring aromatization but without retention of the 10β-methyl group is also demonstrated with
the C₁₉-17-ketone 86, the apparent sequence in horse liver
and placenta being initial hydroxylation of 86 by liver
microsomal 7α-hydroxylase, the product 3β,7α-dihydroxy-
androst-5-en-17-one then being dehydrated to the diene 3β-
hydroxyandrosta-5,7-dien-17-one aromatized by placental
microsomal enzymes to 3β-hydroxyestra-5,7,9-trien-17-one
[2338].

The anthrasteroids are another class of B-ring aromatic derivatives which may be viewed as 9(10 → 6)*abeo*-steroids
derived by ring closure following prior scission of the
C-1/C-10 bond. Strong acid acting on ergosta-5,7,9(11),E-
22-tetraen-3β-ol (66) 3β-acetate gives 1(10 → 6)*abeo*-
ergosta-5,7,9,14,E-22-pentaene (221), probably via initial
elimination and isomerization leading through ergosta-5,7,
9(11),14,E-22-pentaene [1735,1740]. An analogous anthra-

steroid 1(10 → 6)abeo-cholesta-5,7,9,14-tetraene is formed
from cholesta-5,7,9(11)-trien-3β-ol 3β-acetate [400].

The anthrasteroid 221 has been found among other oxidized ergosterol derivatives in yeast [824,827,829]. As
ergosterol undergoes photosensitized dehydrogenation to 66
[566] and 66 may undergo the anthrasteroid rearrangement,
this pathway could account for 221 under some conditions.
More pointedly, the anthrasteroid 221 is also a minor
decomposition product of the 5α,8α-peroxide 62, possibly
via the 3β,5α,8α-triol 222 discussed in a later section of
this chapter [825]. In either event, anthrasteroid 221
may be regarded as an ergosterol oxidation product in these
relationships.

A parallel set of transformations have been observed
in the C_{27}-series, the 5α,8α-peroxide 60 yielding 5α,8α-
cholest-6-ene-3β,5,8-triol in turn converted putatively
into a C_{27}-anthrasteroid [826]. However, C_{27}-anthra-
steroids have not been detected amongst cholesterol oxidation products.

Anthrasteroids retaining the 3β-hydroxyl function of
the parent sterol are formed under free radical oxidation
conditions. Photochemical dehydrogenation and rearrangement of 56 in the absence of air yields 1(10 → 6)abeo-
cholesta-5,7,9-trien-3β-ol; the analogous transformation
of ergosterol gives 1(10 → 6)abeo-ergosta-5,7,9,E-22-
tetraen-3β-ol (314) [2515]. Furthermore, oxidation of
ergosterol 3β-acetate by $^3O_2$ catalyzed by Lewis acid discussed in detail in a later section of this chapter also
yields the anthrasteroid 314 3β-acetate [148].

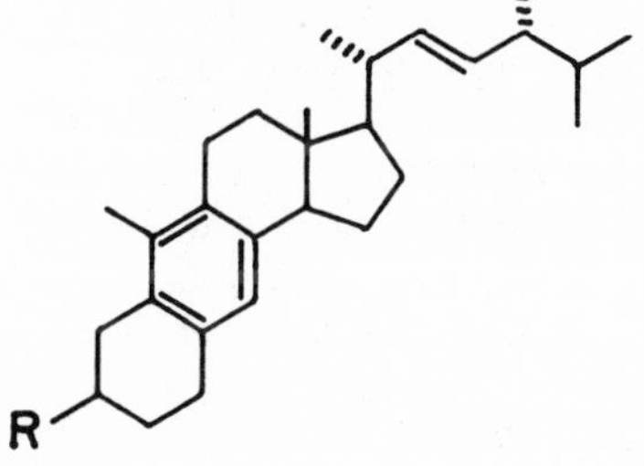

314   R = OH   315

221   R = H

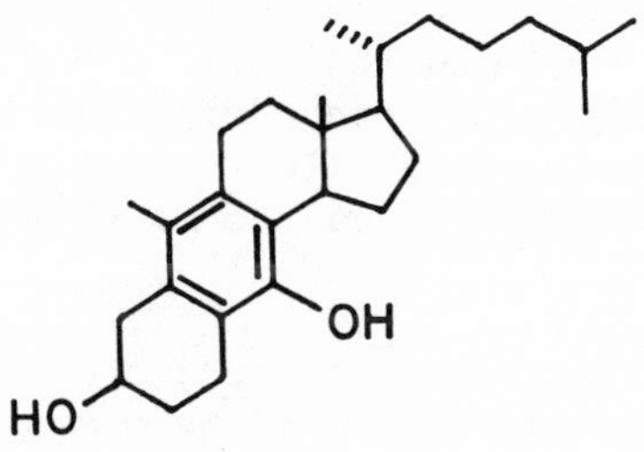

Yet higher oxidized anthrasteroids have been described.
For instance, rearrangement of 3β-acetoxycholesta-5,8-dien-
7-one yields 1(10 → 6)*abeo*-cholesta-5,7,9-triene-3β,7-diol
(315) [2514]. As 3β-hydroxyergosta-5,8,E-22-trien-7-one
(316) is a product of ergosterol oxidation via the 5α,8α-

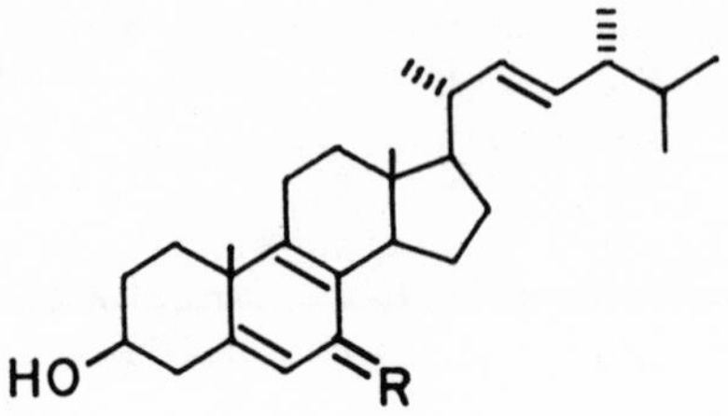

316   R = O

317   R = H$_2$

peroxide 62 as discussed in a later section, its putative
rearrangement could also provide an analogous anthrasteroid
3β,7-diol among oxidized ergosterol derivatives. However,
neither the alcohol 314 nor putative 3β,7-diol types have
been found in naturally oxidized sterol preparations.

The $\Delta^{5,8}$-sterol ergosta-5,8,E-trien-3β-ol (317) found
in the lichen *Xanthoria parietina* may be of related artificial origins [1454].

9,10-Secosterols. The photochemistry of ergosterol
in the absence of air is particularly complex, leading via
scission of the C-9/C-10 bond to a variety of 9,10-seco-
sterols isomeric with ergosterol. The first recognized
product preergocalciferol (precalciferol$_2$, previtamin D$_2$,
9,10-secoergosta-5(10),Z-6,8,E-22-tetraene-3β-ol) (318) in
equilibrium with ergosterol is also in equilibrium with
lumiergosterol (lumisterol$_2$, 9β,10α-ergosta-5,7,E,22-trien-
3β-ol) (319) and tachysterol$_2$ (9,10-secoergosta-5(10),E-6,
8,E-22-tetraen-3β-ol) (320). The four isomers ergosterol,
318,319, and 320 constitute an equilibrium mixture moderated by light, as suggested in FIGURE 24. Precisely the
same equilibrium transformations occur within the C$_{27}$-
series, 5,7-diene 56 yielding precholecalciferol (precalci-
ferol$_3$, previtamin D$_3$, 9,10-secocholesta-5(10),Z-6,8-trien-
3β-ol), lumicholesterol (lumisterol$_3$, 9β,10α-cholest-5,7-
dien-3β-ol), and tachysterol$_3$ (9,10-secocholesta-5(10),E-
6,8-trien-3 β-ol) [320,982].

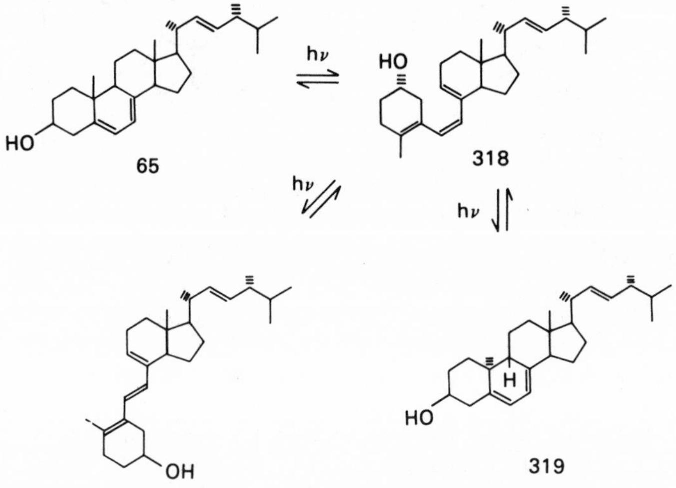

FIGURE 24.   Light-induced equilibria among ergo-
             sterol and related isomers.

Other 9,10-secosterols form from these equilibrium
mixtures; however, as the precalciferol links parent sterol,
lumisterol, and tachysterol derivatives, the formation of
subsequent 9,10-seco derivatives may be viewed as involving
the precalciferol.

The Calciferols. Subsequent thermal isomerization of
preergocalciferol 218 at temperatures between 20-80°C
yields ergocalciferol (calciferol$_2$, Vitamin D$_2$, 9,10-secoer-
gosta-Z-5,E-7,10(19),E-22-tetraen-3β-ol (321). An analogous
transformation in the C$_{27}$-series gives cholecalciferol
(Vitamin D$_3$, calciferol$_3$) (90), the endogenous vitamin D.
The active vitamins D 90 and 321 are in thermal equilibrium
with the corresponding previtamins D. The extent of thermal
interconversion of the calciferols and previtamins D depends on the physical state of the vitamin preparations,
solid glassy cholecalciferol resins isomerizing more slowly
than less viscous resins or solutions [953].

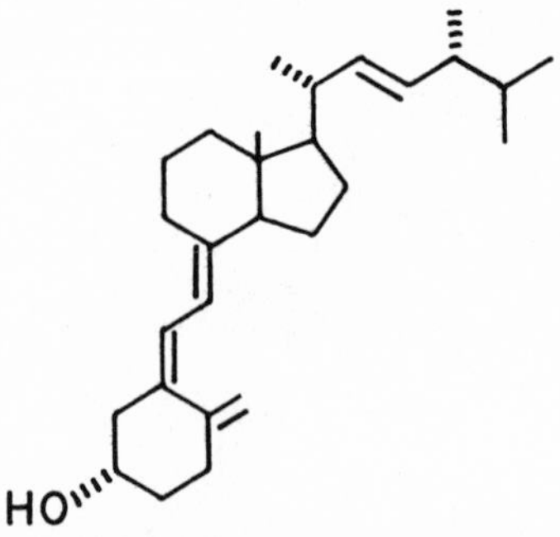

321 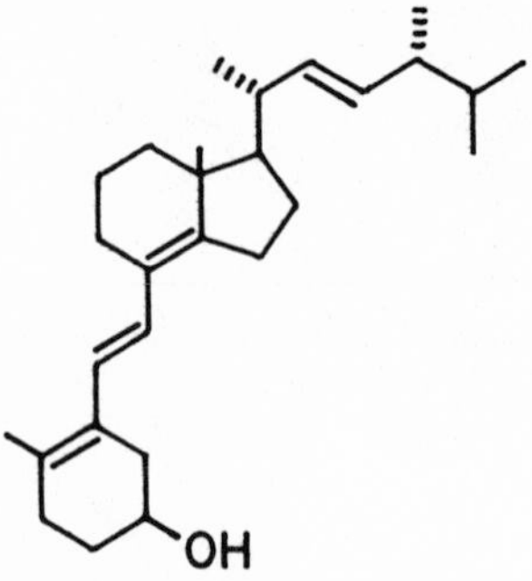   322

The vitamins D are subject to several further transformations yielding yet other types of isomers. Thus,
cholecalciferol (90) is isomerized to isotachylsterol$_3$
(9,10-secocholesta-5(10),E-6,8(14)-trien-3β-ol (322) and
to *trans*-Vitamin D$_3$ (9,10-secocholesta-E-5,E-7,10(19)-
trien-3β-ol (323). Commercial preparations of cholecalciferol contain besides the active vitamin 90 and parent

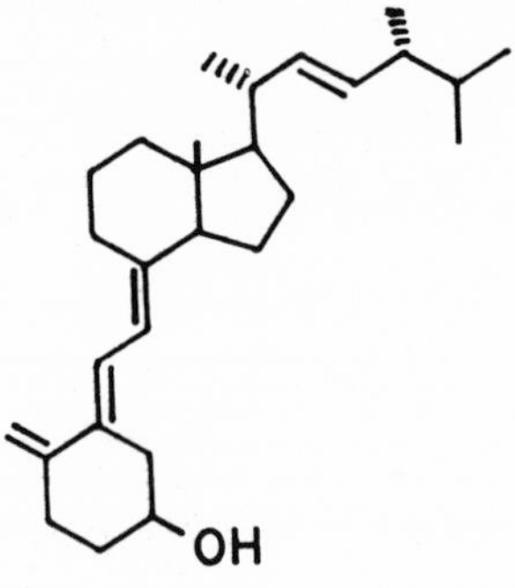

323

sterol 56 the lumisterol, tachysterol, isotachysterol, and
trans-Vitamin D$_3$ analogs all readily demonstrated chromatographically [2449].

Furthermore, ergocalciferol is unstable towards isomerization in the presence of acidic solid excipients
[2421,2424], ascorbic acid [2425], and strong acid [1349].
Products identified include the C$_{28}$-isotachysterol analog
9,10-secoergosta-5(10),E-6,8(14),E-22-tetraen-

3β-ol, 9,10-secoergosta-E-5,E-7,10 (1),E-22-tetraen-3β-ol, and 9,10-secoergosta-Z-5,E-7,10(1),E-22-tetraen-3β-ol [2423].

As these isomerizations occur in the presence of air to about the same extent, oxidations may not be prominent for all these isomers [2422]. However, the established sensitivity of the vitamins D to deterioration in air may not involve isomers derived by acid catalysis so much as other isomeric congeners.

Autoxidation is not emphasized in recent monographs devoted to the vitamins D [1782] any more than is cholesterol autoxidation in monographs on cholesterol [2055,2322]. Nonetheless, air oxidation of these interrelated 9,10-secosterols is an important process leading to deterioration of vitamins D preparations. Uncertainty exists as to reactions implicated, as vitamins D autoxidations with and without [51] peroxidic material being detected have been recorded. The vitamins D are also sensitive to oxidations by other agents, the hydroperoxide 2,6-di-*tert.*-butyl-4-hydroperoxy-4-methylcyclohexa-2,5-dien-1-one formed by air oxidation of antioxidant BHT (2,6-di-*tert.*-butyl-4-methylphenol) used to prevent vitamin D autoxidation during saponifications being an active oxidant of ergocalciferol [14]. Furthermore, reduced 9,10-secosterols are sensitive to autoxidations, witness the oxidation of a dihydrotachysterol derivative by peroxide-containing oils in oily preparations of the derivative [830].

It is most likely that prior studies of the vitamins D have been conducted on mixtures of vitamins. the several precursors, isomers, oxidation products, and subsequent decomposition products discussed further in this chapter. The spontaneous change in composition of irradiated ergosterol preparations (containing ergocalciferol) was recognized early [102,1159,1918,2697], and the sensitivity of the vitamins D to decomposition in air and light [810], in aqueous dispersions [474,1098,1814,2221], in organic solvent solution [1917,2060], or adsorbed on paper [1597] is well known. Although the vitamins D are sensitive to air oxidation *per se*, the many isomers possibly present in a given system may be even more sensitive [101,102,430] and thus act to initiate or promote autoxidation of the vitamins D. For instance, the sensitivity of ergocalciferol to air is increased in the presence of the tachysterol 320 [1159].

No systematic examination of these matters is known to
the author, the complexity of the case clearly discouraging
such study. Nonetheless, a few isolated observations have
been reported that suggest that free radical oxidations of
the calciferols by O_2 occur along pathways common to the
cholesterol case but also along pathways unique to the
secosterol case. Thus, the radiation-induced autoxidation
of crystalline ergocalciferol (321) is accompanied by chemi-
luminescence and a singlet electron spin resonance signal
is observed, clearly indicative of free radical processes
[2328,2329]. Aqueous dispersions of ergocalciferol undergo
autoxidation, yielding both peroxidic and ketonic products.
Oxidative scission of the C-5/C-6 and C-6/C-7 bonds appears
to occur [1377,2329,2331], but little else about the matter
has been discovered [2330].

A bit more chemical information is available for the
cholecalciferol (90) case. The presence of 50-150 ppm of
the 25-hydroxylated cholecalciferol derivative 91 in preparations of the parent sterol [271,272] suggests that
autoxidation occurs by attack at the C-25 position in the
same manner as for cholesterol. Moreover, several oxidized
9,10-secosterols have been described which may be air oxidation products of cholecalciferol or one of its isomers.
The 9,10-secoketone 3β,10ξ-dihydroxy-9,10-secocholest-5-en-
7-one (324) found in processed fish oils [1903,1904,1906]
and in aerial parts of several plants [1905] appears to be

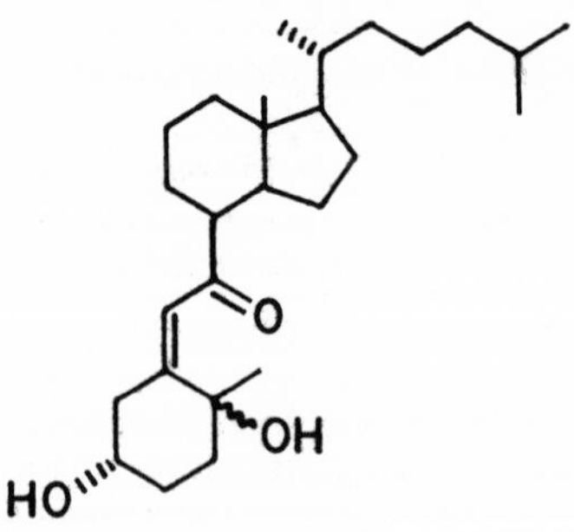

324 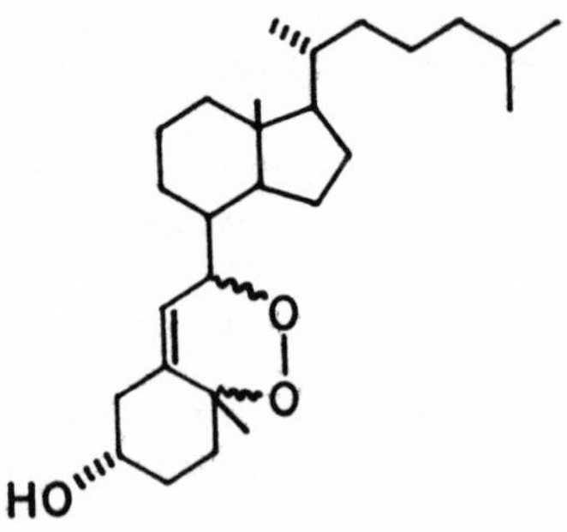   325

derived by air oxidation, possibly via the cyclic peroxide

7ξ,10ξ-epidioxy-9,10-secocholest-5-en-3β-ol (325),the decomposition of which yields the secoketone 324 [139]. A 9,10-
secocholesta-5,7-diene-3β,10ξ-diol may also form as a hydration product of cholecalciferol [140].

As with $^{3}0_2$ oxidation studies, studies of the oxidation of the calciferols by $^{1}0_2$ have been few. Methanol-
adducts of oxidation products were isolated from the reaction of cholecalciferol (90) with $NaOCl-H_2O_2$ in methanol
(in which $^{1}0_2$ is implicated) [292], but photosensitized
oxygenation in which $^{1}0_2$ is involved yielded peroxides
6,19-epidioxy-(6R)-9,10-secocholesta-5(10),E-7-dien-3β-ol
(326) and the corresponding C-6 epimer 6,19-epidioxy-(6S)-
9,10-secocholesta-5(10),E-7-diene-3β-ol. The (6R)-6,19-
peroxide 326 is as active as cholecalciferol in stimulating
Ca^{++} uptake in the duodenum in cholecalciferol-deficient
rats [1681]. Photosensitized oxygenation of ergocalciferol
(321) also yielded both C-6 epimeric cyclic peroxides 6,19-
epidioxy-9,10-secoergosta-5(10),E-7,E-22-trien-3β-ol [2738,
2739].

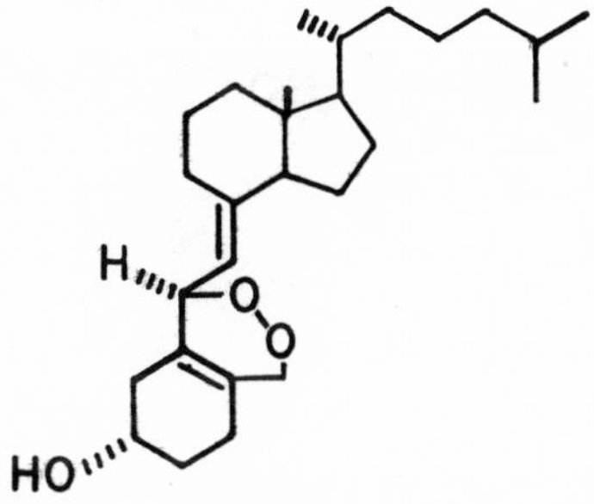

326

In addition to the obvious vitamin D activity of the
9,10-secosterols, other biological activities may be demonstrated for the 9,10-secosterols and their precursors 5,7-
dienol 56 and ergosterol respectively. Unidentified oxidation products of 56 (one designated as Rf 0.53 compound
[1682]) show hemolytic activity [1684,1685], effects on
Ca^{++} [1683,1686,2238], and other effects [1767] in experimental animals. Furthermore, an hypothesis has been constructed which attempts to rationalize vitamin D toxicity
in terms of initiation of generalized lipid peroxidations
by vitamin D or its oxidation products [2327,2330].

Whereas the thermal isomerization of previtamins D to calciferols occurs at low temperatures (20-80°C), at higher temperatures (100-180°C) ring closures to 9α,10α-ergosta-5,7,E-22-trien-3β-ol (pyrocalciferol₂) (327) and 9β,10β-ergosta-5,7,E-22-trien-3β-ol (isopyrocalciferol₂) (328) occur [421]. These 9,10-*syn*-isomers 327 and 328 are in turn subject to additional photoisomerizations, 327 to 5,8-cyclo-9α,10α-ergosta-6,E-22-dien-3β-ol (photopyrocalciferol₂) (329), 328 to 5,8-cyclo-9β,10β-ergosta-6,E-22-dien-3β-ol (photoisopyrocalciferol₂) (330) [566]. A similar thermal transformation of precholecalciferol (9,10-seco-cholesta-5(10),Z-6,8-trien-3β-ol to the C_{27}-9,10-*syn*-homologs 9α,10α-cholesta-5,7-dien-3β-ol (pyrocholecalciferol) and 9β,10β-cholesta-5,7-dien-3β-ol (isopyrocholecalciferol) occurs [1843].

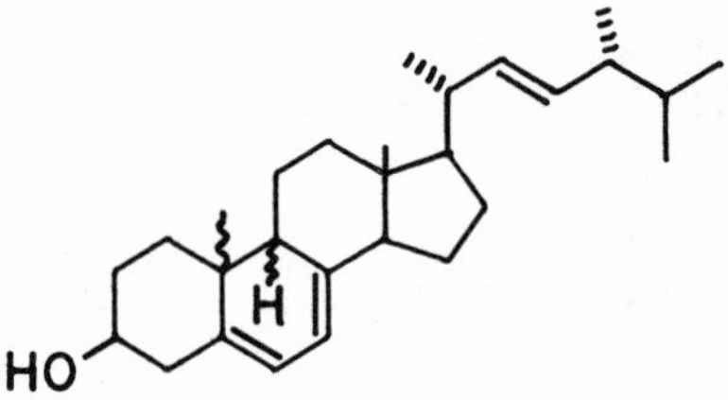

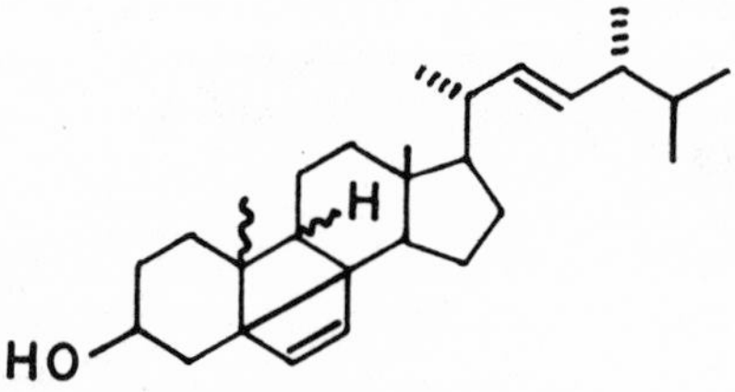

327 9α,10α	329 9α,10α
328 9β,10β	330 9β,10β

Over-irradiation Products. There are two sets of irreversibly formed over-irradiation products derived by extended irradiations of the previtamins D and of the calciferols. Over-irradiation of the previtamins D give the toxisterols, thirteen of which have been recognized in the

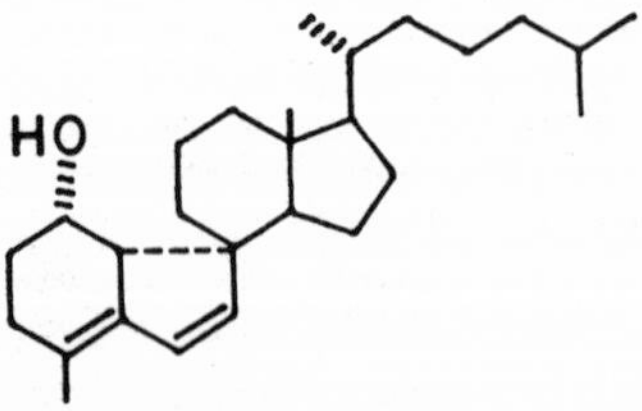

331 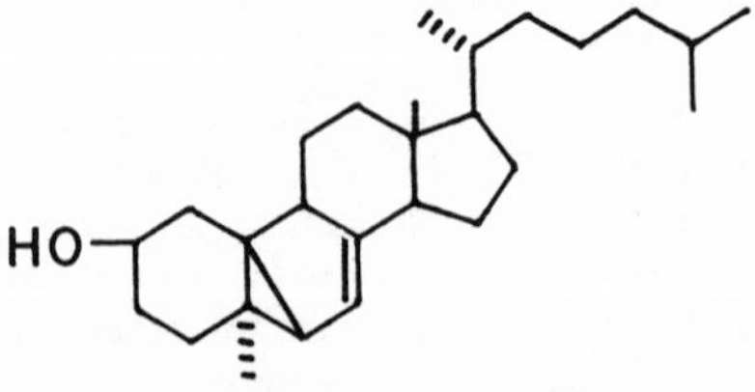   332

C_{27}-series. The group includes three C-8 spiro derivatives toxisterol A1 ($4\alpha,8\alpha$-cyclo-9,10-secocholesta-5(10),6-dien-3β-ol) (331), toxisterol A2 ($4\alpha,8\beta$-cyclo-9,10-secocholesta-5(10),6-dien-3β-ol), and toxisterol A3 ($4\beta,8\beta$-cyclo-9,10-secocholesta-5(10),6-dien-3β-ol); two bicyclo[3.1.0]hexeno derivatives toxisterol C1 (5-methyl-5α-19-nor-6β,10β-cyclo-cholest-7-en-2α-ol) (332) and C2 (5-methyl-5β,10α-19-nor-6α,10α-cyclocholest-7-en-2α-ol); three 9,10-secotrienes toxisterol D1 (9,10-secocholesta-Z-5,8,10(19)-trien-3β-ol (333), toxisterol D2 (9,10-secocholesta-E-5,E-7,10(1)-trien-3β-ol), and toxisterol D3 (9,10-secocholesta-5(10),E-6,8(14)-trien-3β-ol); one D3 cyclobuteno derivative toxisterol E1 ($6\beta,9\beta$-cyclo-9,10-secocholesta-5(10),7-dien-3β-ol) (334); and one reduced 9,10-secosterol toxisterol R1 (9,10-secocholesta-5(10),E-7-dien-3β-ol). As these transformations are commonly conducted in methanol (or ethanol) three alcohol addition products toxisterol B1 (10β-methyoxy-

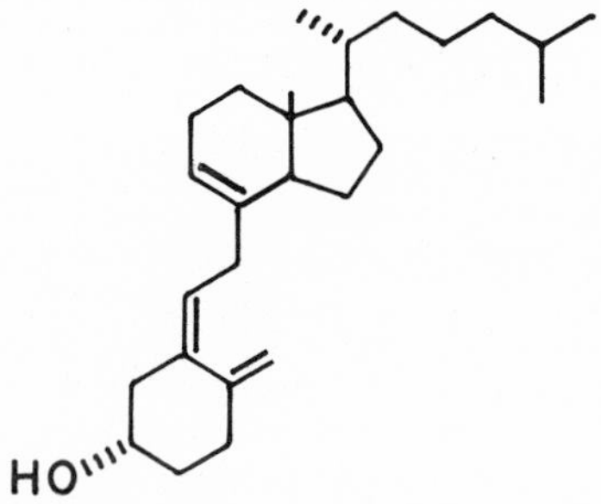

333 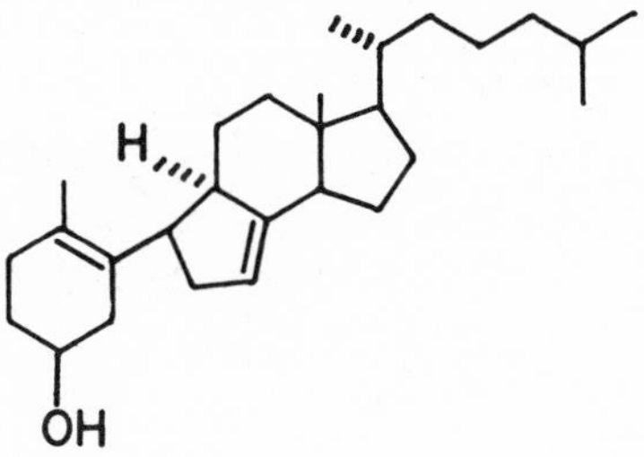   334

10α-9,10-secocholesta-E-5,E-7-dien-3β-ol), B2 (10α-methoxy-9,10-secocholesta-E-5,E-7-dien-3β-ol) and B3 (10β-methoxy-10α-9,10-secocholesta-Z-5,E-7-dien-3β-ol) complete the thirteen products [320,1168]! A similar complex spate of toxisterols is also found in the ergosterol series, of which three toxisterols A have been identified. [142].

Extended irradiation of cholecalciferol yields three sets of isomeric products, the suprasterols 9,10-seco-6α,8α; 7,19-biscyclocholest-5(10)-en-3β-ol (335) and 9,10-seco-6β,8β;7,19-biscyclocholest-5(10)-en-3β-ol, the isomeric cyclobutenes 9,10-seco-6α,19- and 6β,19-cyclocholesta-5(10), E-7-dien-3β-ol (336), and the isomeric allene 9,10-seco-cholesta-5(10),6,7-trien-3β-ols (337) [125,982].

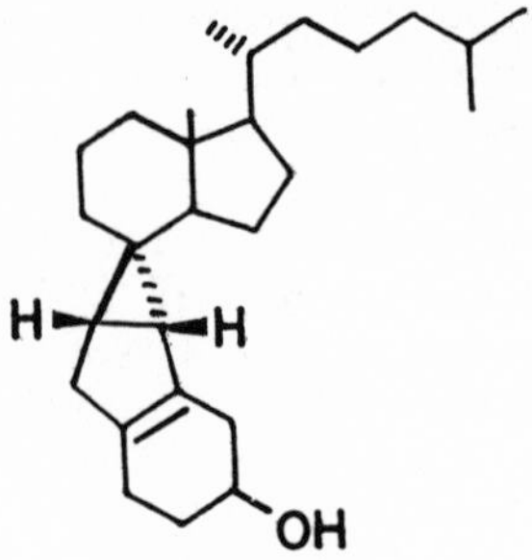

335

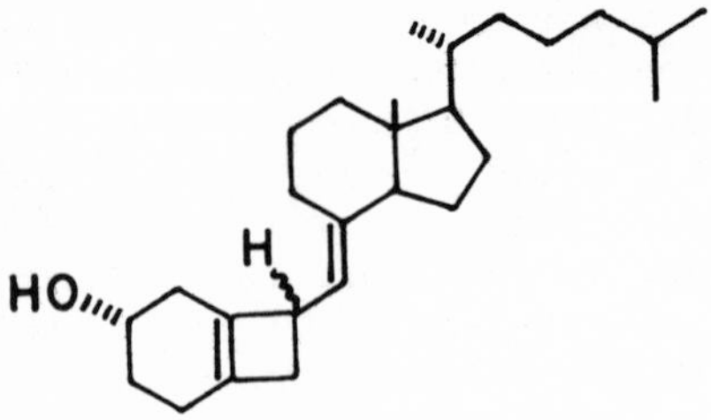

336

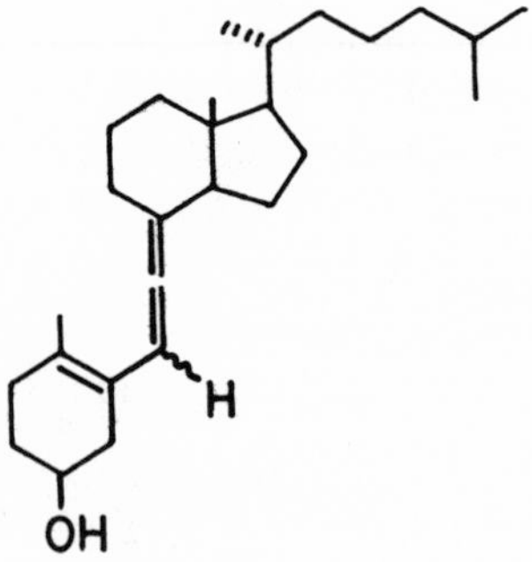

337

Photofragmentations also occur in extended irradiations of ergosterol, scission of the C-6/C-7 and C-7/C-8 bonds yielding C_{19}-C_{20} perhydroindene derivatives [143].

Ergosterol 5α,8α-Peroxide. The major oxidation product of ergosterol is the 5α,8α-peroxide 62, whose distribution in nature has already been discussed (cf. Chapter III). Three other oxidation products occur with 62 in a variety of systems, these being 3β-hydroxyergosta-5,8,E-22-trien-7 one (316), 5α,8α-ergosta-6,E-22-diene-3β,5,8-triol (222), and cerevisterol (154). The 5α,8α-peroxide 62 is formed from ergosterol in photosensitized oxygenations [1167,2700] and H_2O_2-NaOCl oxidations [1607] in which $^1O_2$ is implicated but also in systems involving solutions of ergosterol in xylene containing biacetyl in which $^3O_2$ is implicated

[1393-1395] and in systems involving Lewis acids catalyzing 1,4-cycloaddition of $^3O_2$. Corresponding $5\alpha,8\alpha$-peroxides are also major products of oxidation of sterol 5,7-dienes 56 [2096], 64 [2702], 66 [1697,2704] and others, as well as from the C_{19}-dienes androsta-5,7-diene-3β,17β-diol [426] and 3β-hydroxyandrosta-5,7-dien-17-one [1193].

Because of the ready synthesis of these $5\alpha,8\alpha$-peroxides by photosensitized oxygenations in which $^1O_2$ is implicated, it has been assumed that the $5\alpha,8\alpha$-peroxides whereever encountered are $^1O_2$ products. However, the established oxidation of ergosterol 3β-acetate by $^3O_2$ to the $5\alpha,8\alpha$-peroxide 62 3β-acetate now dispels such concept. Many agents catalyzing the oxidation of ergosterol 3β-acetate to 62 3β-acetate have been described by Barton and colleagues, the trityl cation (as BF_4^- salt) catalyzing the oxidation with light radiation, amine cation radicals such as tris(p-bromophenyl) ammoniumyl (BF_4^- or $SbCl_6^-$ salts) acting as catalysts by thermal processes in the dark. The oxidation is also catalyzed by a variety of Lewis acids, some acting without light ($VOCl_3$, $FeCl_3$, $MoCl_5$, WCl_6), some requiring light for catalysis (BF_3, $SnCl_4$, $SnCl_5$, $SbCl_5$, WF_6, $TiCl_4$) [148,153]. Liquid SO_2 also serves as a Lewis acid catalyzing the oxidation in light [989].

These catalyzed oxidations involve $^3O_2$ directly, and the excitation of $^3O_2$ to $^1O_2$ does not occur! Sterols which serve as substrates in $^1O_2$ reactions, such as 66 3β-acetate, are not oxygenated, and lumisterol (319) 3β-acetate is oxidized at a rate of 1/6000th that of ergosterol 3β-acetate. Cholesterol 3β-acetate is not oxidized [148,153].

Several mechanisms involving radical species as intermediates have been proposed as means by which these catalysts overcome the spin barrier inherent in reactions between $^3O_2$ and singlet state sterol substrates. The formation of complexes involving sterol, $^3O_2$, and catalyst have been invoked, as for instance in the photochemical process involving the trityl cation as catalyst. In this formulation, initial photochemical generating of the triplet trityl cation $^3(T^+)$, reaction with $^3O_2$, addition of the trityl peroxyl radical to the unsaturated sterol (R), and hydrolysis of the sterol-O_2-catalyst complex as shown in Eqs. 57-60, yield products 62 3β-acetate and triphenylmethanol [153]. However, other mechanisms also appear to operate, with oxygen insertion occurring both before and after spin

$$^1(T^+) \xrightarrow{\text{h}\nu} {}^3(T^+) \qquad\qquad \text{Eq. 57}$$

$$^3(T^+) \;+\; {}^3O_0 \longrightarrow T\text{-}O\text{-}O\cdot \qquad\qquad \text{Eq. 58}$$

$$T\text{-}O\text{-}O\cdot \;+\; R \longrightarrow T\text{-}O\text{-}O\text{-}R^{\underline{+}} \qquad\qquad \text{Eq. 59}$$

$$T\text{-}O\text{-}O\text{-}R^{\underline{+}} \;+\; H_2O \longrightarrow T\text{-}OH \;+\; RO_2 \qquad \text{Eq. 60}$$

inversion [148]. Yet another mechanism advanced for Lewis acid (LA) catalysis involves complex formation and associated charge transfer processes as suggested in Eq. 61-62. Reactions between the sterol cation ($R^{\underline{+}}$) and $^3O_2$ followed by reaction with the Lewis acid radical anion ($LA^{\overline{\cdot}}$) as in Eqs. 63-64, and/or the alternative reaction between Lewis acid radical anion and $^3O_2$ to give $O_2^{\overline{\cdot}}$ which then reacts with the sterol radical cation as in Eqs. 65-66 then account for formation of the sterol peroxide [988]

$$R \;+\; LA \;+\; {}^3O_2 \longrightarrow [R\text{-}LA\text{-}{}^3O_2] \qquad \text{Eq. 61}$$

$$[R\text{-}LA\text{-}{}^3O_2] \longrightarrow R^{\underline{+}} \;+\; LA^{\overline{\cdot}} \;+\; {}^3O_2 \qquad \text{Eq. 62}$$

$$R^{\underline{+}} \;+\; {}^3O_2 \longrightarrow [R\text{-}O\text{-}O\cdot]^+ \qquad \text{Eq. 63}$$

$$[R\text{-}O\text{-}O\cdot]^+ \;+\; LA^{\overline{\cdot}} \longrightarrow RO_2 \;+\; LA \qquad \text{Eq. 64}$$

$$LA^{\overline{\cdot}} \;+\; {}^3O_2 \longrightarrow LA \;+\; O_2^{\overline{\cdot}} \qquad \text{Eq. 65}$$

$$R^{\underline{+}} \;+\; O_2^{\overline{\cdot}} \longrightarrow RO_2 \qquad \text{Eq. 66}$$

However, complex formation between sterol, O_2, and catalyst may not occur. Rather, the oxidation now appears to be one in which free radical chain processes participate. The catalyst tris (*p*-bromophenyl) ammoniumyl radical cation ($X^{\underline{+}}BF_4^-$) functions in this case to initiate the radical reactions, as in Eq. 67. Electrol spin and ^1H and ^{13}C nuclear magnetic resonance spectra indicate that a sterol radical cation is formed, one involving the four carbon atoms C-5, C-6, C-7, and C-8. Subsequent reaction of the sterol radical cation with $^3O_2$ and chain propagating reactions as in Eqs. 68-69 then yield the product sterol peroxide and continue the radical chain [2444].

5α,8α-Peroxide Transformations. The $\Delta^5,{}^8$-7-ketone <u>316</u> and 3β,5α,8α-triol <u>222</u> found as minor oxidation products

$$R + X \overset{+}{\cdot} BF_4^- \longrightarrow R \overset{+}{\cdot} BF_4^- + X: \qquad Eq. 67$$

$$R \overset{+}{\cdot} BF_4^- + {}^3O_2 \longrightarrow RO_2 \overset{+}{\cdot} BF_4^- \qquad Eq. 68$$

$$RO_2 \overset{+}{\cdot} BF_4^- + R \longrightarrow RO_2 + R \overset{+}{\cdot} BF_4^- \qquad Eq. 69$$

of ergosterol appear to derive from the 5α,8α-peroxide 62, although the matter is far from settled. The 7-ketone 316 is found with 62 in photosensitized [1167,2517] and biacetyl-catalyzed [1394] oxygenations of ergosterol. The analog 3β-acetoxycholesta-5,8-dien-7-one is formed in similar photosensitized oxygenations from 56 3β-acetate [2401].

The 7-ketone 316 may be viewed on the one hand as a product derived by the attack of $^1O_2$ on ergosterol in the cyclic ene mode or, on the other, as a subsequent transformation product of the major product 5α,8α-peroxide 62. In the first formulation, abstraction of the axial 9α-hydrogen and bond migration would give a $\Delta^{5,8}$-7-hydroperoxide whose subsequent dehydration would yield the requisite $\Delta^{5,8}$-7-ketone 316. Alternatively, 316 may derive from the 5α,8α-peroxide 62, as the thermal transformation of the C_{19}-analog 5,8-epidioxy-3β-hydroxy-5α,8α-androst-6-en-17-one to 3β-hydroxyandrosta-5,8-diene-7,17-dione has been observed [1193]. The conversion of the 5α,8α-peroxide 62 to 7-ketone 316 has not been observed [300,905], but the product could have escaped recognition.

The 3β,5α,8α-triol 222 identified as an ergosterol oxidation product formed in systems containing biacetyl [1394] has not otherwise been found in ergosterol oxidizing systems. The triol 222 has not been found as a thermal decomposition product of the 5α,8α-peroxide 62 but is a reduction product of 62,being formed by Zn/alkali [641,2698, 2704], $CrCl_2$ [2442], and $LiAlH_4$ in pyridine [556] but not in ether [541,1437,2265].

The minor yeast sterol 5α-ergosta-6,14,E-22-triene-3β,5-diol (218) appears to be derived by acid dehydration of the 3β,5α,8α-triol 222, via 5α-ergosta-6,8(14),E-22-triene-3β,5-diol (338) as an intermediate [823]. This 3β,5α-diol 338 has also been found among $LiAlH_4$ reduction products of the 5α,8α-peroxide 62 [2265].

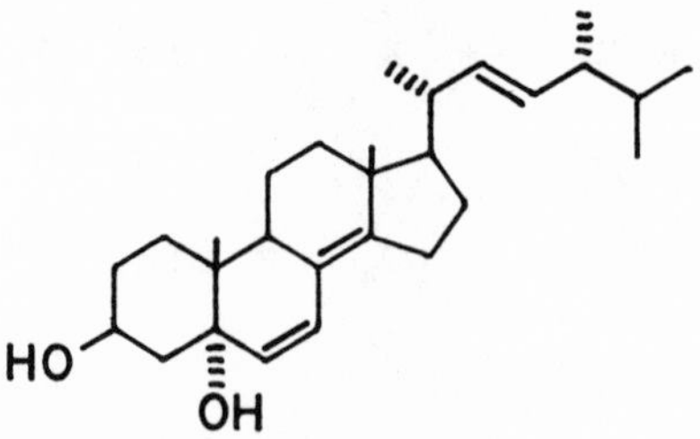

338 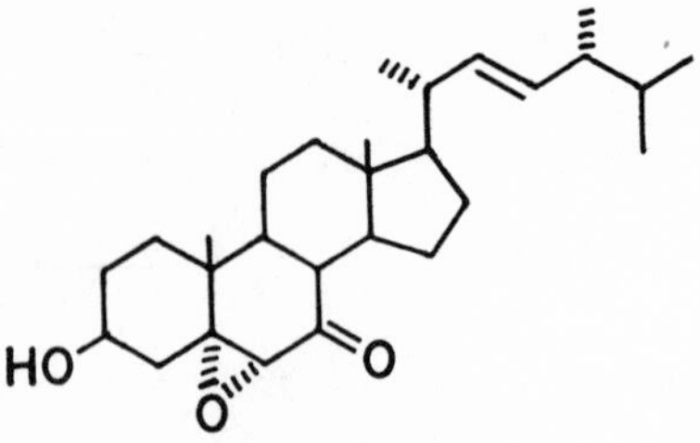   339

The 3β,5α,8α-triol 222 is implicated in the formation
of cerevisterol discussed in the next section and may serve
as an alternative precursor of the $\Delta^{9(11)}$-analog 66 of ergosterol, as thermal processes or acid dehydrate the·triol
222 to 66 [641,2043].

The established thermal degradation products of the
5α,8α-peroxide 62 are the 5,6-epoxides 5,6α-epoxy-3β-hydroxy-
5α-ergost-E-22-en-7-one (399), 5,6α-epoxy-5α-ergosta-8,E-22-
diene-3β,7α-diol (219), and 5,6α-epoxy-5α-ergosta-8(14),E-
22-diene-3β,7α-diol (220) [195,300,905,2698]. Thus the
exoxydiol 219 found with the 5α,8α-peroxide 62 in dried
yeast must be derived from the peroxide [824,827,829].
Neither the isomeric epoxydiol 220 nor the epoxyketone 339
has been detected among yeast sterols.

Cerevisterol. The most frequently encountered ergosterol companion sterol that is also an artifact of áir oxidations on ergosterol is cerevisterol (154) previously discussed in this relation (cf. Chapter III). Two possible
pathways linking ergosterol and cerevisterol may be proposed, one proceeding from the 3β,5α,8α-triol 222 just described. In this formulation [827] the 3β,5α,8α-triol must
undergo allylic rearrangement to 5α-ergosta-7,E-22-diene-
3β,5,6α-triol (340) whose subsequent epimerization yield
cerevisterol. Whereas the requisite rearrangement of
3β,5α,8α-triol 222 to 3β,5α,6α-triol 340 has been demonstrated in acidified solutions [11,641,823,2042] and some
cerevisterol is also formed in such systems [823], direct
observation of the epimerization of 340 to cerevisterol has
not been addressed.

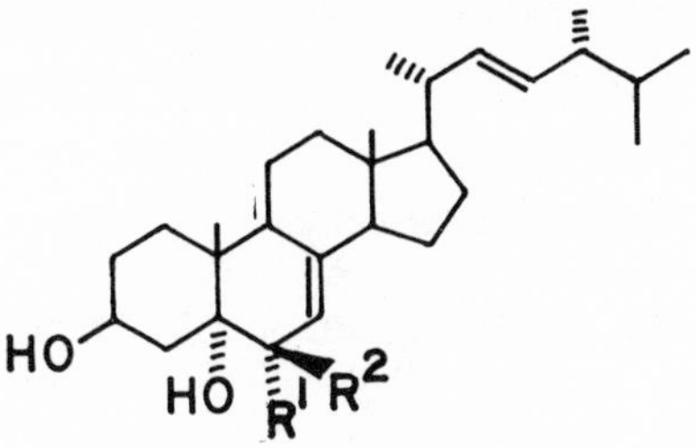

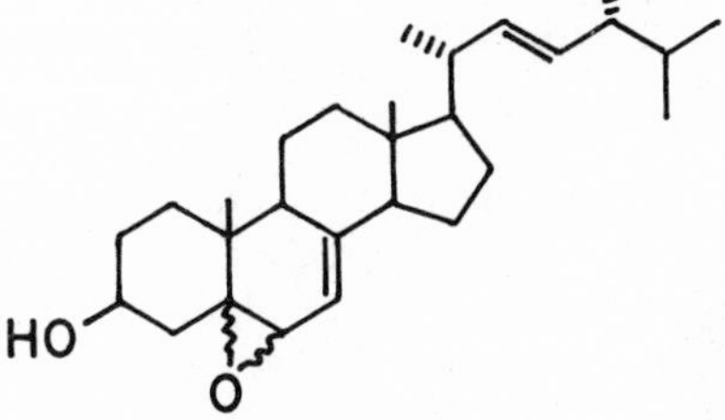

$\underline{340}$ R^1 = OH, R^2 = H $\underline{341}$

$\underline{154}$ R^1 = H, R^2 = OH

    3β-Acetates of both epimeric 3,5,6-triols $\underline{340}$ and
cerevisterol are also formed from ergosterol 3β-acetate by
perphthalic acid [43]. Only the cis-diol $\underline{340}$ 3β-acetate
is obtained by OsO₄ [536] or perbenzoic acid [641,2707]
oxidations, the cis-diol $\underline{340}$ 3β,6α-diacetate by lead
tetracetate oxidations [2709].

    The alternative oxidative process for cerevisterol
formation from ergosterol is speculative but offers a means
for derivation for both cerevisterol and the 3β,5α,8α-triol
$\underline{222}$. The demonstrated selectivity of peracid oxidation of
the Δ^5-double bond of ergosterol 3β-acetate, yielding epimeric 3β,5α,6-triols $\underline{340}$ and $\underline{154}$ 3β-acetates probably
involves unrecognized isomeric 5,6ξ-epoxy-5ξ-ergosta-7,E-22-
dien-3β-ol ($\underline{341}$) 3β-acetate derivatives as intermediates,
the subsequent dehydration of which then yield the 3,5,6-
triols. One postulates similar selective epoxidation of
the Δ^5-double bond of ergosterol by other non-peracid epoxidation reagents such as H₂O₂ or sterol hydroperoxides also
yield isomeric 5ξ,6ξ-epoxides $\underline{341}$ yielding cerevisterol upon
hydration. This formulation is patterned after our demonstrated epoxidation of cholesterol by cholesterol 7-hydro-
peroxides $\underline{46}$ and $\underline{47}$ formed initially in autoxidations [2297].
The requisite epoxidizing agent for cerevisterol derivation
may well be the 5α,8α-peroxide $\underline{62}$. In this formulation,
epoxidation and subsequent hydrations might provide both
cerevisterol (from ergosterol) and the 3β,5α,8α-triol $\underline{222}$
(from $\underline{62}$). Specific demonstration of the oxidation of
ergosterol by $\underline{62}$ has not been sought, but support of the
concept may be found in the attack of 5α,8α-peroxide $\underline{62}$
3β-acetate on ergosterol 3β-acetate, catalyzed by Lewis acid

$TiCl_4$, yielding two equivalents of 5,6α-epoxy-5α-ergosta-7,E-22-dien-3β-ol 3β-acetate as a putative intermediate [147]

This description of the oxidation chemistry of ergosterol, now completed, accounts for the derivation of the major product 5α,8α-peroxide 62, the $\Delta^{5,8}$-7-ketone 316 and 3β,5α,8α-triol 222 closely associated with 62, and cerevisterol. Moreover, it is by now obvious that the autoxidation of ergosterol and, by extension, of the C_{27}-analog 56 proceeds via reactions unique to 5,7-dienes to products totally different from those found in the autoxidation of cholesterol.

Other 5,7-Dienes

The oxidation chemistry of cholesta-5,7-dien-3β-ol (56) has been discussed with that of ergosterol, and little additional material is to be gleaned from studies with the 5,7-dienes 64, 66, etc. that also yield 5α,8α-peroxide products. Moreover, the enol ester ergosta-3,5,7,E-22-tetraen-3-ol 3-acetate yields the corresponding 5α,8α-peroxide 5,8-epidioxy-5α,8α-ergosta-3,6,E-22-trien-3-ol 3-acetate but also ergosta-3,5,7,9(11),E-22-pentaen-3-ol and 3-acetoxy-5,14-dihydroxy-5α,14α-ergosta-3,7,E-22-trien-6-one [291].

All of the ergosterol isomers stereoisomeric at the C-9 and C-10 centers form 5,8-peroxides in the same manner as does ergosterol [566]. Isomeric 5β,8β-peroxides are formed from the lumisterol 319 3β-acetate in systems which generate $^1O_2$. Both photosensitized oxygenations [284] and the H_2O_2-NaOCl system [1607] yield 5,8-epidioxy-5β,8β,9β,10α-ergosta-6,E-22-dien-3β-ol (342) 3β-acetate and 3β-acetoxy-10α-ergosta-5,8,E-22-trien-7-one. The photosensitized reaction also gives dehydrogenation product 10α-ergosta-5,7,9(11),E-22-tetraen-3β-ol 3β-acetate and the 5β,8β-peroxide 5,8-epidioxy-5β,8β,10α-ergosta-6,9(11),E-22-trien-3β-ol 3β-acetate therefrom [284].

As previously mentioned the C_{19}-steroid 3β-hydroxyandrosta-5,7-dien-17-one also forms a 5α,8α-peroxide 5,8-epidioxy-3β-hydroxy-5α,8α-androst-6-en-17-one, the further phototransformation of which gives 3β-hydroxyandrosta-5,8-diene-7,17-dione, 5,6α-epoxy-3β,7α-dihydroxy-5α-androst-8-en-17-one, and 5,6α-epoxy-3β,7α-dihydroxy-5α-androst-

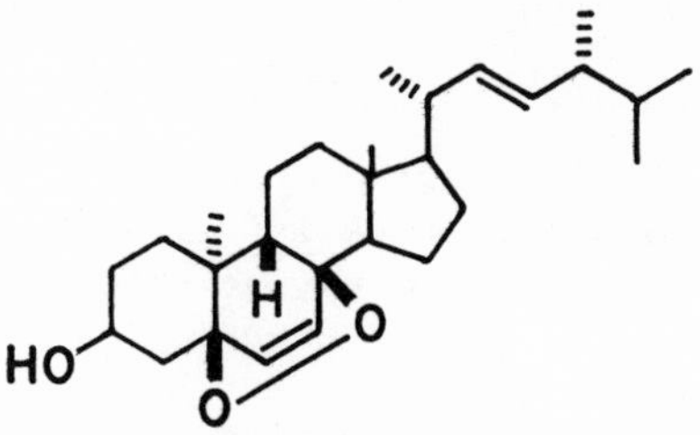

342 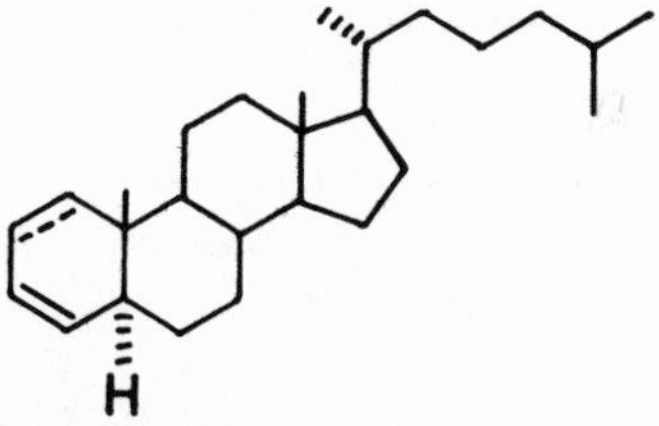   343 Δ^1

344

8(14)-en-17-one [1193].   These transformations are thus
analogous to those established for the 5α,8α-peroxide 62.

Other Conjugated Dienes

Several other conjugated steroid diene systems have
been subjected to photooxygenation conditions in which $^1O_2$
is implicated. 5α-Cholesta-1,3-diene (343) gives cholesta-
1,4-dien-3-one (115) as product, with no cyclic peroxide
being found [1099]. It is uncertain whether this $^1O_2$ reaction proceeds via the cyclic ene mode of attack on the
Δ^3-double bone 343 or via cycloaddition to a putative
Δ^2-1,4-peroxide which is further transformed to the isolated
product in analogy with the C_{19}-example just described. By
contrast, cholesta-1,3,5-trien-7-one yields, among other
photo products, 1α,4α-epidioxycholesta-2,5-dien-7-one [1010].

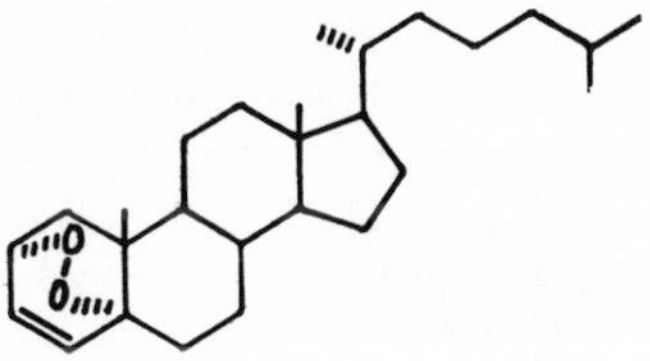

345

The isomeric cholesta-2,4-diene (284) reacts with $^1O_2$ generated by the H_2O_2-NaOCl system [1607] or photochemically [425,1099,2266,2267] to give the anticipated cyclic peroxide 2α,5-epidioxy-5α-cholest-3-ene (345). Further photo and thermal transformations of the 2α,5α-peroxide 245 provide the 4α,5-epoxy-5α-cholestan-2-one and 5-hydroxy-5α-cholest-3-en-2-one [193,506].

Steroid heteroannular cisoid dienes also give cyclic peroxides in photosensitized oxygenations. The 7,14-diene 5α-ergosta-7,14,E-22-trien-3β-ol 3β-acetate gives an unidentified peroxide or hydroperoxide *inter alia* [152], and 5α-ergosta-7,14-dien-3β-ol 3β-acetate gives 7ξ,15ξ-epidioxy-5α-ergost-8(14)-en-3β-ol 3β-acetate (346) which in turn

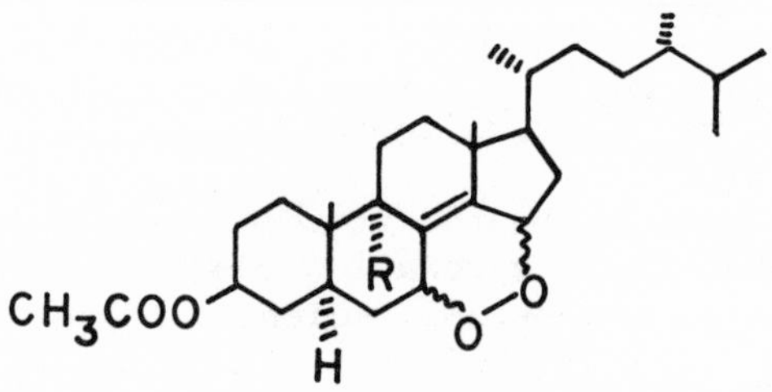

346   R = H

347   R = OOH

reacts with a second equivalent of $^1O_2$ [665]. The cisoid 8(14),9(11)-diene system of 5α-ergosta-6,8(14),9(11),E-22-tetraen-3β-ol 3β-acetate yields a cyclic peroxide 11α,14-epidioxy-14α-ergosta-5,8,E-22-trien-3β-ol 3β-acetate [1431, 1432], the rearrangement of which gives 8,14;9,11α-bisepoxy-5α,8α,9α,14α-ergosta-6,E-22-dien-3β-ol 3β-acetate [29].

Moreover, cyclic peroxide formation from heteroannular transoid dienes is also observed in cases where two equivalents of $^1O_2$ attack the substrate. Oxidation of 5α-ergosta-8,14-dien-3β-ol 3β-acetate gives 3β-acetoxy-5α,9α-ergosta-7,14-diene-9-hydroperoxide (348), further oxidized to 3β-acetoxy-7ξ,15ξ-epidioxy-5α,9α-ergost-8(14)-en-9-hydroperoxide (347) [665]. Furthermore, oxidation of 3,3-ethylene-

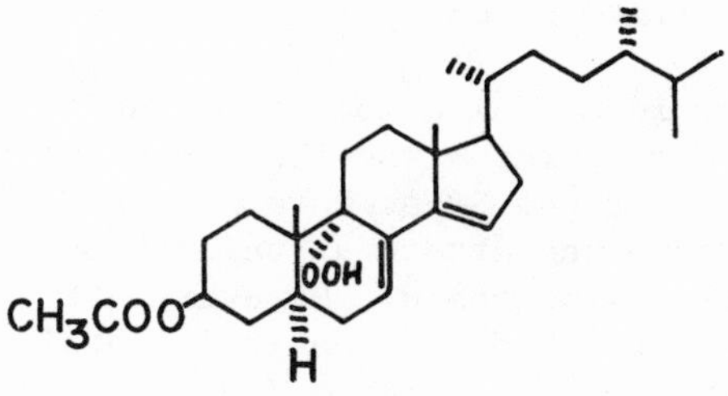

348 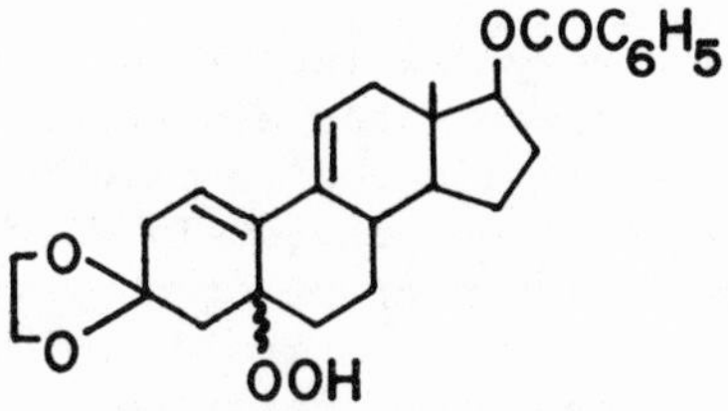   349

dioxyestra-5(10),9(11)-dien-17β-ol 17β-benzoate yields isomeric 5-hydroperoxides 17β-benzoyloxy-3,3-ethylenedioxy-
5α- and 5β-estra-1(10),9(11)-diene-5-hydroperoxides (349)
which add a second equivalent to $^1O_2$ to give all four
possible cyclic 1,11-peroxides 17β-benzoyloxy-1ξ,11ξ-epi-
dioxy-3,3-ethylenedioxy-5α- and 5β-estr-9-ene-5-hydroper-
oxides (350,351) [1591].

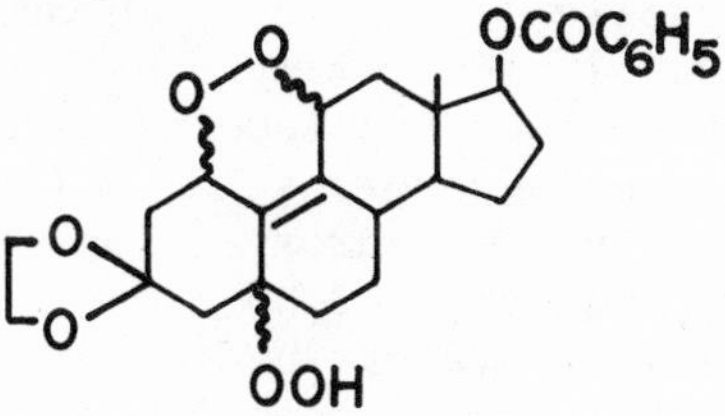

350, 5α

351, 5β

OTHER STEROIDS

Although this monograph is devoted to cholesterol autoxidation, there is much to be learned from examination of
the autoxidation of other types of steroids, even though
most other classes of steroids are more highly and differ-
ently substituted than is cholesterol. However, the unidentified more highly oxidized autoxidation products of cholesterol that we know are formed may well involve some of
the additional modes of autoxidation evinced for other
classes of the more highly oxidized steroids.

Unfunctionalized Sites

Very few reports of steroid autoxidation at unfunctionalized sites are known. Among these are the homoallylic
sites implicated in oxidations of the 3β-acetate of the
8-stenol 68 previously discussed and attack at unfunctionalized carbon atoms of the sterol side-chain. However, autoxidations of steroids by $^3O_2$ promoted by transition metal
ions have been described. In aqueous buffered (pH 6.8)
solutions of Fe(II) salts bile acids are oxidized by $^3O_2$
selectively in the 15α-position. Thus, deoxycholic (3α,
12α-dihydroxy-5β-cholanic) acid is transformed to 3α,12α,
15α-trihydroxy-5β-cholanic acid; taurodeoxycholic acid
(3α,12α-dihydroxy-5β-cholan-24-oyltaurine), taurocholanic
acid (5β-cholan-24-oyltaurine), and 3α,12α-dihydroxy-5β-23-
norcholan-24-oic acid to their corresponding 15α-hydroxy-
lated derivatives [1296,1299,1300,1302]. In a related
system utilizing Fe(II) salts, $^3O_2$, ascorbic acid, and
ethylenediaminetetracetic acid transformation of deoxycholic acid to cholic (3α,7α,12α-trihydroxy-5β-cholanic)
acid is reported. In this example H_2O_2 could be used
instead of $^3O_2$ to effect the same transformation [1584].
The system utilizing ascorbic acid, Fe(II) salts, and $^3O_2$
is also effective in the hydroxylation of other steroids,
17α,21-dihydroxypregn-4-ene-3,20-dione being oxidized to
hydrocortisone (11β,17α,21-trihydroxypregn-4-ene-3,20-
dione), with further dehydrogenation of the 11β-hydroxyl
group to the 11-ketone also occurring [487,1289,1772,1947].

Functionalized Sites

By far, examples of autoxidations of other steroids
involve oxidations of functionalized sites of the steroid
molecules. The most prominent reactions are those of alcohol dehydrogenation and allylic oxidations of olefins.

Alcohol Dehydrogenation. The O_2-dependent dehydro-
fenation of cholesterol to the 3-ketone 6 has been discussed
in Chapter IV. Dehydrogenations of other 3-hydroxysteroids
by air are also known, dehydrogenation of the 5α-stenol 2
to 5α-cholestan-3-one by air being a key example [754]. A
3-hydroxyandrost-5-en-17-one is transformed to androst-4-
ene-3,17-dione by heating in air [616], and both epimeric

3-hydroxy-5α-androstan-17-one derivatives are dehydrogenated
to 5α-androstane-3,17-dione during chromatography on alumina
[317].

    Dehydrogenations of other steroid alcohols are known.
Dehydrogenation of 3β,7α,11α-trihydroxypregn-5-en-20-
one to 3β,11α-dihydroxypregn-5-ene-7,20-dione by heating
in air at 100°C or by mere storage in air for several weeks
is recorded [1360]. Spontaneous dehydrogenation of the 11β-
hydroxyl group of crystalline 21-acetoxy-11β-hydroxy-17α,
20ξ-isopropylidenedioxypregn-4-en-3-one isomers [1465] and
of crystalline 21-esters of hydrocortisone [350] occur on
storage of samples in air. A sample of the 21-*tert*.butyl-
acetate ester of hydrocortisone contained up to 40% of the
corresponding 11-ketone after 1-2 years, 80% after 15 years.
Samples of the analogous 21-*tert*.-butylacetate ester of
9α-fluoro-11β,17α,21-trihydroxypregn-4-ene-3,20-dione by
contrast were not dehydrogenated even after 15-17 years!
Solvation effects appear to moderate the autoxidation,
crystal polymorphs not readily desolvated and nonsolvated
forms being more stable to autoxidation then polymorphs
which are readily desolvated [350].

    The 21-hydroxyl group of hydrocortisone is also spon-
taneously dehydrogenated to the 21-aldehyde 11β,17α-dihy-
droxy-3,20-dioxopregn-4-en-21-al by air in aqueous media.
Similarly, cortisone (17α,21-dihydroxypregn-4-ene-3,11,20-
trione) and prednisolone (11β,17α,21-trihydroxypregna-
1,4-diene-3,20-dione) are dehydrogenated to corresponding
21-aldehydes. Analogs corticosterone (11β,21-dihydroxy-
pregn-4-ene-3,20-dione) and deoxycorticosterone (21-hydroxy-
pregn-4-ene-3,20-dione) were much less sensitive [1661].

    Olefins and Enols. Oxidations of steroid olefins,
enols, enol esters and ethers, and hydrazones by dioxygen
species come within this class. Oddly, very few examples
of autoxidation of simple steroid olefins by $^3O_2$ are available, but many examples of $^1O_2$ oxidations are recorded.

    In photosensitized oxygenations of 5α-cholest-3-ene
(344) both autoxidation and $^1O_2$ reactions occur. The major
product (after hydroperoxide reduction) is cholest-4-en-3α-ol
(352), viewed as a $^1O_2$ product of the cyclic ene mode of
attack, whereas minor product cholest-4-en-3β-ol (79) may
be viewed as an autoxidation product. Similar oxidation of
the isomeric 5β-cholest-3-ene gives (after reduction) the

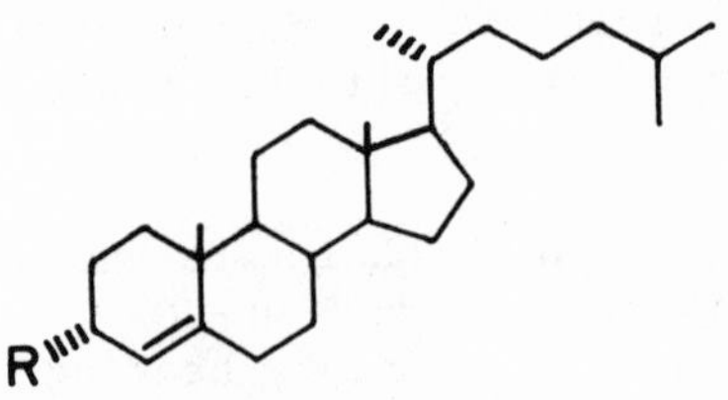

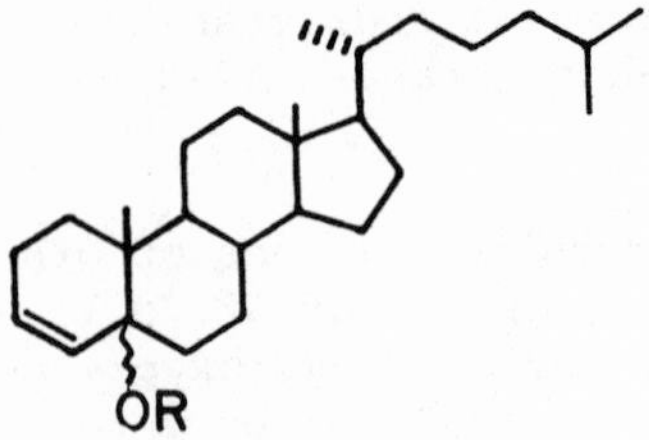

352 R = OH	353	R = H,5α
9 R = H	354	R = H,5β
	355	R = OH,5α
	356	R = OH,5β

same products 79 and 352. Small amounts of isomeric cholest-2-en-4ξ-ols also formed from both Δ^3-olefins [1757]. Oxidation of isomeric cholest-4-ene (9) by $^1O_2$ yields after reduction of initial products 5α-cholest-3-ene-5-hydroperoxide (355) and 5β-cholest-3-ene-5-hydroperoxide (356) the corresponding alcohols 5α-cholest-en-5-ol (353) and 5β-cholest-3-en-5-ol (354) but also cholest-5-en-4β-ol and the epimeric 79 and 351 [1755].

Other A-ring olefins react similarly, with preferential abstraction of allylic quasiaxial hydrogen and associated double bond shift. 2-Methyl-5α-cholest-2-ene yields 2-methyl-5α-cholest-1-en-3α-ol, 2-methylene-5α-cholestan-3α-ol, and 2-methylene-5α-cholestan-3β-ol; isomeric 3-methyl-5α-cholest-2-ene yields 3β-methyl-5α-cholest-1-en-3α-ol and 3-methylene-5α-cholestan-2α-ol [1752,1753].

Attack on $^1O_2$ at other isolated double bonds gives similar allylic hydroperoxides. 3β,17α-Diacetoxy-6-methylpregn-5-en-20-one gives 3β,17α-diacetoxy-5-hydroperoxy-6-methylene-5α-pregnan-20-one [1453]. Oxidation of 17α-ethinylestr-5-ene-3β,17β-diol yields 17α-ethinyl-3β,17β-dihydroxy-5α-estr-6-ene-5-hydroperoxide and 17α-ethinyl-3β,17β-dihydroxyestr-4-ene-6β-hydroperoxide [1851]. 3β-Acetoxy-5α-cholest-8(14)-en-6-one gives after reduction 3β-acetoxy-14-hydroxy-5α,14α-cholest-7-en-6-one [831]. 9β,10α-Pregna-4,16-dien-3-one yields 20ξ-hydroperoxy-9β,10α-pregna-4,16-dien-3-one [1392].

Photooxidation of a 20(22)-double bond is exemplified in that of the enamine 22-(1'-morpholinyl)-23,24-bisnorchola-4,20(22)-dien-3-one yielding progesterone, possibly via 1,2-cycloaddition of $^1O_2$ to the 20(22)-double bond, with subsequent decomposition of the putative intermediate dioxetane to the ketone product [1097]. Photooxygenations of the Δ^{24}-double bond of desmosterol (78) and of lanosterol (67) have been described previously.

Additional examples of both autoxidations and $^1O_2$ reactions of enols, enol ethers, and enol esters are known. Air oxidation of the stable acyclic enol 3α,20-dihydroxy-4α,4β,14-trimethyl-10α,14α-19-norpregna-5,17(20)-dien-11-one (357) yields the hydroperoxyketone 17α-hydroperoxy-3α-hydroxy-4α, 4β,14-trimethyl-10α,14α-19-norpregn-5-ene-11,20-dione (358), from which the 17-ketone 3α-hydroxy-4α,4β,14-trimethyl-10α,14α-19-norandrost-5-ene-11,17-dione is derived by thermal degradation [686]. Likewise, air oxidation of 3β-acetoxy-5α-pregna-9(11),17(20)-dien-20-ol yields 3β-acetoxy-17α-hydroperoxy-5α-pregn-9(11)-en-20-one [107].

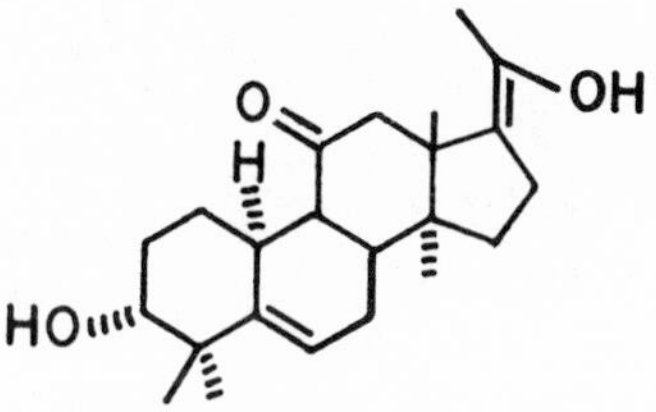

357 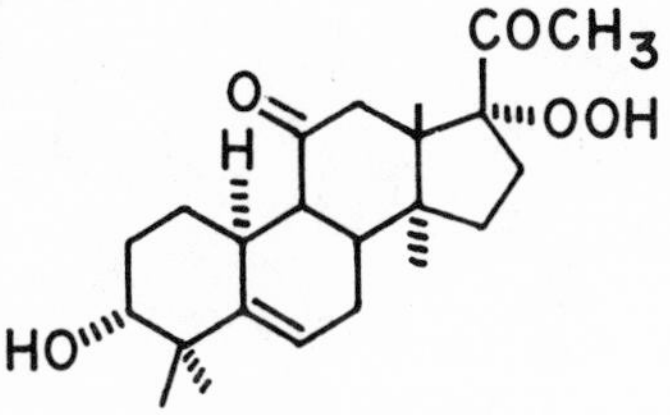   358

The cyclic enol ether pseudosolasodine 3β,N-diacetate (26-acetylamino-(25R)-furosta-5,20(22)-dien-3β-ol 3β-acetate) (359) is also oxidized by air to the exocyclic hydroperoxide 3β-acetoxy-26-acetylamino-(20β,25R)-furosta-5,22-diene-20-hydroperoxide (360) [1559]. Analogous autoxidation occurs with (25R)-furosta-5,20(22)-diene-3β,26-diol (pseudodiosgenin) 3β,26-diacetate [1558].

Light-induced autoxidation of the enol ether 3-ethoxy-cholesta-3,5-diene (361) gave 6β-hydroxycholest-4-en-3-one (88), probably via the intermediate 6β-hydroperoxide 59

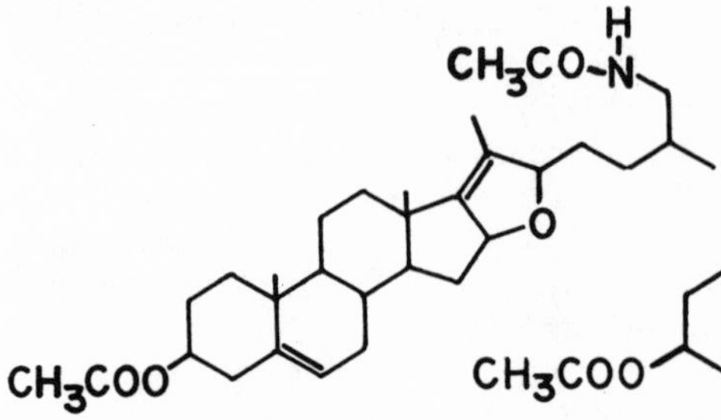

359 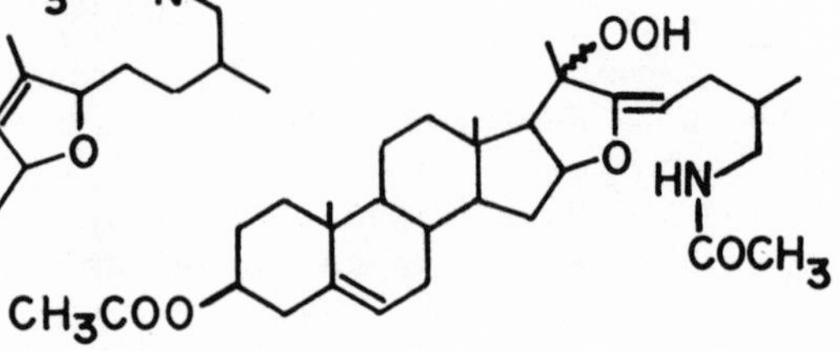   360

[848]. Similar air oxidations of a variety of $C_{19}-C_{21}$
$\Delta^{3,5}$-enol ethers likewise gave the corresponding 6β-hydroxy-
Δ^{4}-3-ketones accompanied by small amounts of epimeric 6α-
hydroxy-Δ^{4}-3-ketones [460,848,2520].

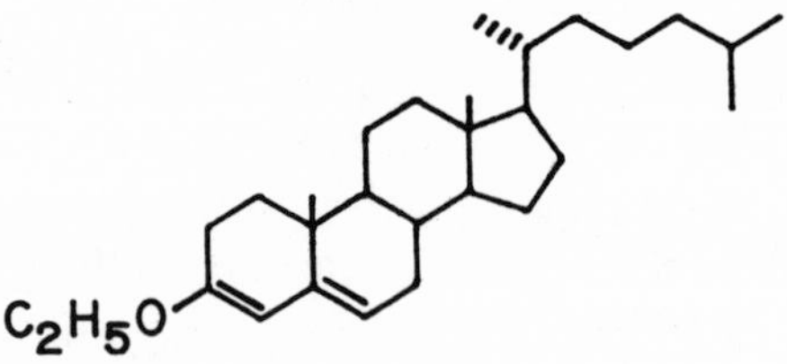

361

Photosensitized oxygenation of the enol ether 3β-eth-
oxy-20-methoxypregna-5,17(20)-dien-21-ol 21-acetate (362)
yields 21-acetoxy-3β-ethoxy-20-methoxypregna-5,16-dien-20-
hydroperoxide (363), the reduction of which yields the
20-ketone 21-acetoxy-3β-ethoxypregna-5,16-dien-20-one [1723].

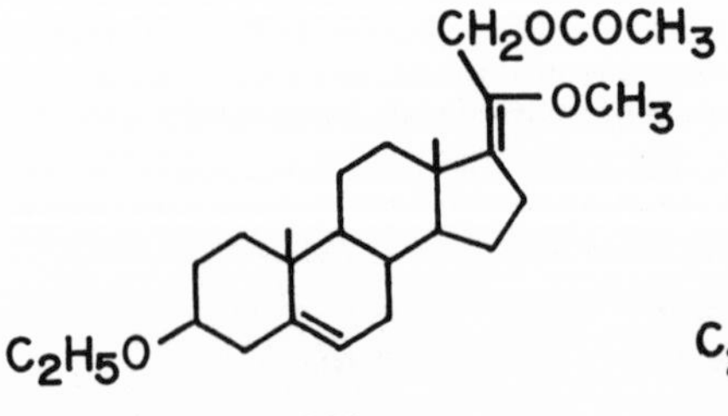

362 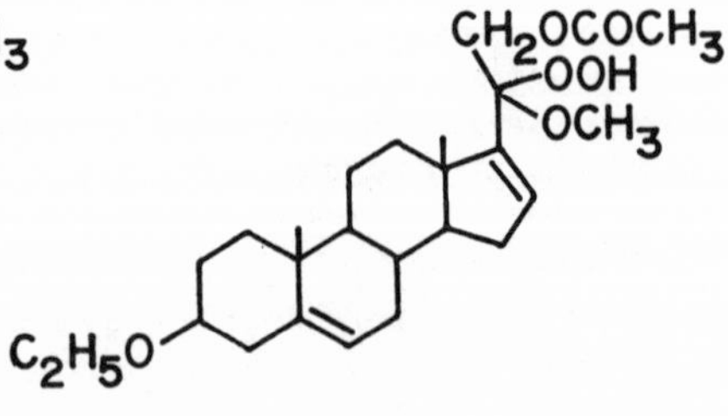   363

Photosensitized oxygenation of the enol ester 3-acetoxy-5α-cholest-2-ene yields the enone 8 and the isomeric 5α-cholest-1-en-3-one. Formation of the latter Δ^1-3-ketone is rationalized by quasiaxial 1ξ-hydrogen abstraction by $^1O_2$ in the cyclic ene mode and formation of putative intermediate 3ξ-acetoxy-5α-cholest-1-ene-3ξ-hydroperoxide, the decomposition of which yields product. Alternative abstraction of quasiaxial 4ξ-hydrogen yielding putative intermediate 3-acetoxy-5α-cholest-3-ene-2ξ-hydroperoxide, with subsequent allylic rearrangement to 3-acetoxy-5α-cholest-2-ene-4ξ-hydroperoxide and decomposition to the enone 8 completes the matter [1896].

Photosensitized oxygenation of the dienol 3-acetoxy-5α-cholesta-1,3-diene yields cholesta-1,4-dien-3-one (115) as major product, thus the same product as derived from cholesta-1,3-diene (343) [1099] previously discussed. Minor products found included 3ξ-acetoxy-1α,2α-epoxy-5α-cholestan-4-one, 3-acetoxy-1α-hydroxy-5α-cholest-2-en-4-one, and 5-hydroxy-5ξ-4-oxacholest-1-en-3-one [1896]. The transoid dienol ether 3-ethoxyandrosta-3,5-dien-17-one yields by radical processes the epimeric 6α- and 6β-hydroperoxyandrost-4-ene-3,17-diones [848].

The shift of double bond concommitant with so many free radical and photosensitized oxygenations of olefins, enols, etc. is also obtained in the autoxidation of ketone phenylhydrazones. For instance, 5α-cholestan-3-one phenylhydrazone 364 is readily autoxidized to both possible epimers of 3ξ-phenylazo-5α-cholestane-3ξ-hydroperoxide (366). Similar autoxidation of 3β-acetoxy-5α-cholestan-7-one phenylhydrazone yielded 3β-acetoxy-7ξ-phenylazo-5α-cholestane-7ξ-hydroperoxide, whereas the phenylhydrazone of the 17-ketone 86 gave 3β-hydroxy-17β-phenylazoandrost-5-ene-17α-hydroperoxide [392].

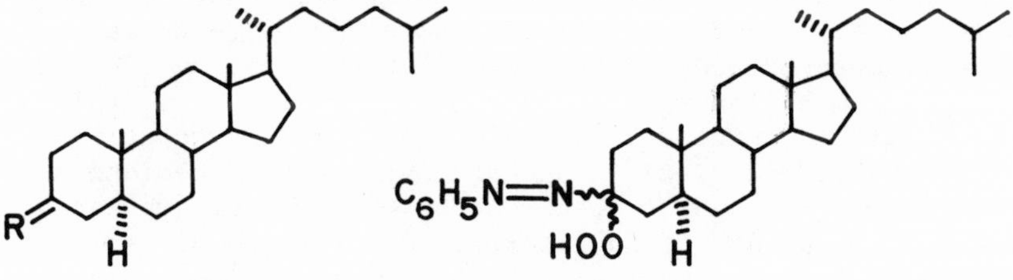

364   R = NNHC$_6$H$_5$ 366
365   R = O

Enones. Conjugated enones such as cholest-4-en-3-one
(8) are not sensitive to autoxidation nor to oxidation induced by [60]Co γ-radiation [70,1402]. However, 6-methyl-Δ^4-
3-ketone homologs may be. Dimethisterone (17β-hydroxy-6α-
methyl-17α-(1-propynyl)androst-4-en-3-one monohydrate)
(367) is autoxidized upon short-term storage in air to the
epimeric 6α- and 6β-hydroperoxy-17β-hydroxy-6-methyl-17α-
(1-propynyl)androst-4-en-3-one derivatives together with
the corresponding epimeric 6-alcohols [1556]. Formation of
the 6α- and 6β-hydroperoxide 280 and 59 from enone 8 in
soybean lipoxygenase incubations [2462] also suggests that
Δ^4-3-ketones can be oxygenated under special curcumstances.

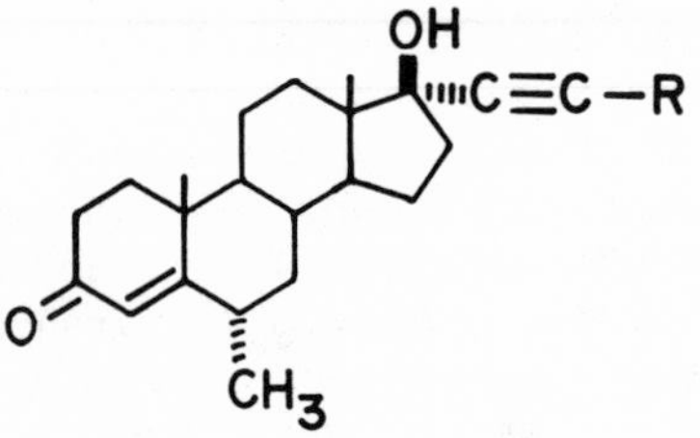

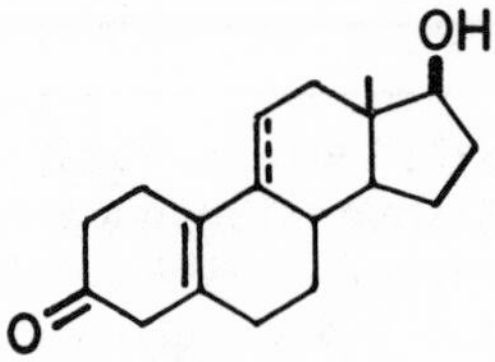

367   R = CH$_3$ 369

368   R = COOH   370 $\Delta^{9(11)}$

In the case of the autoxidation of 367 oxidation of the
side-chain methyl group to carboxyl also occurs, yielding
17β-hydroxy-6α-methyl-3-oxoandrost-4-en-17α-ylpropiolic acid
(368) [1556], this being an autoxidative transformation not
herefore encountered with other steroids.

Autoxidations of β,γ-unsaturated ketones are much more
facile, the extreme sensitivity of cholest-5-en-3-one (6)
to autoxidation having been previously discussed (cf. Chapter IV). 17β-Hydroxy-17α-methylandrost-5-en-3-one is
readily autoxidized to the epimeric 6-hydroperoxy-17β-hy-
droxy-17α-methylandrost-4-en-3-one derivatives, as anticipated [2197]. Dibenzoyl peroxide catalyzed oxygenations
of pregn-5-ene-3,20-dione and androst-5-ene-3,17-dione gave
the corresponding 6β-hydroperoxides 6β-hydroperoxypregn-4-
ene-3,20-dione and 6β-hydroperoxyandrost-4-ene-3,17-dione
[2765]. 17β-Hydroxyestr-5(10)-en-3-one (369) is transformed by air [375] and by photosensitized oxygenation [831,

1593,2197] to 10β-hydroperoxy-17β-hydroxyestr-4-en-one (287), the latter condition also according the isomeric 10α-hydroperoxide 10-hydroperoxy-17β-hydroxy-10α-estr-4-en-3-one [1593]. Photooxygenation of 17β-acetoxy-17α-ethinylestr-5(10)-en-3-one under free radical conditions yields 17β-acetoxy-17α-ethinyl-10β-hydroperoxyestr-4-en-3-one [1450,2196].

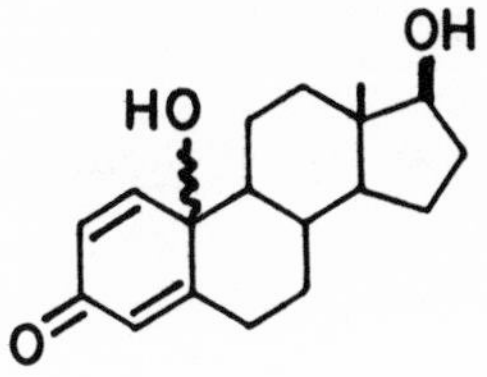

371 10α   373 R = CH₃COO

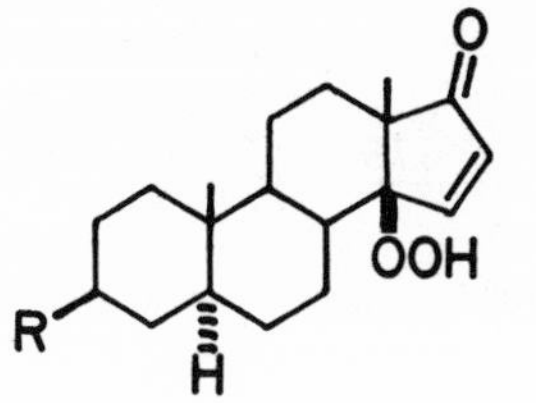

372 10β   374 R = H

The thermal decomposition of the 10β-hydroperoxide 287 and the isomeric 10α-hydroperoxide accord us a new mode of transformation of steroid hydroperoxides not heretofore discovered in sterol hydroperoxide chemistry, namely that of concommitant dehydrogenation and reduction. Thus, the 10β-hydroperoxide 287 yielded 10,17β-dihydroxy-10β-estra-1,4-dien-3-one (372), and the epimeric 10-hydroperoxy-17β-hydroxy-10α-estr-4-en-3-one yielded the epimeric 10,17β-dihydroxy-10α-estra-1,4-dien-3-one (371) [1593].

Unsaturated 17-ketones are also sensitive to air oxidations, the β,γ-enone 3β-acetoxy-5α-androst-14-en-17-one giving 3β-acetoxy-14-hydroperoxy-5α,14β-androst-15-en-17-one (373) in air [27]. Similarly, air oxidation of the α,β-enone 5α,14α-androst-15-en-17-one gives 14-hydroperoxy-5α,14β-androst-15-en-17-one (374) [42,437].

Dienones. Related dienones are also readily autoxidized by air to hydroperoxydienones, 17β-hydroxyestra-5(10), 9(11)-dien-3-one (370) yielding 11β-hydroperoxy-17β-hydroxyestra-4,9-dien-3-one (375) [375,577], estr-5(10),9(11)-diene-3,17-dione yielding 11β-hydroperoxyestra-4,9-diene-3,17-dione [1201,2236]. The conjugated dienone 17β-hydroxyestra-4,9-dien-3-one (376) is oxidized under photosensitized conditions to epimeric 3,17β-dihydroxyestra-1,3,5(10)-triene-9-hydroperoxides [1592].

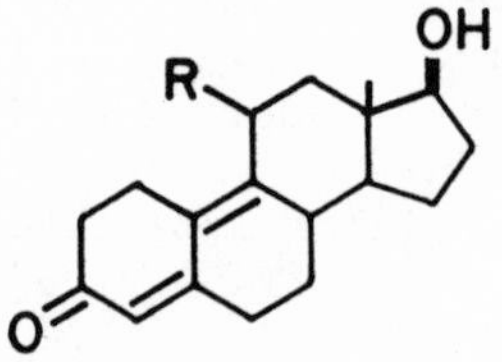

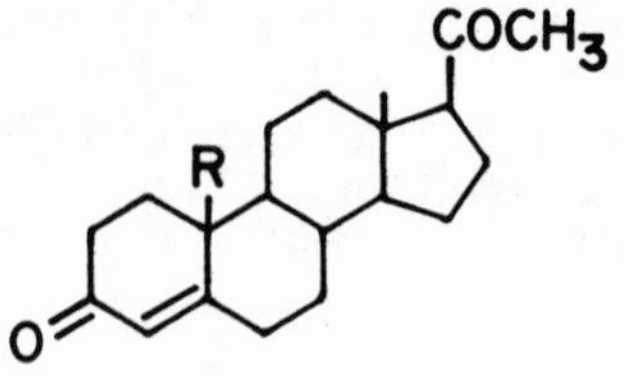

375   R = OOH   377   R = COOH

376   R = H   378   R = OOH

19-Oxosteroids. A variety of 19-oxosteroids, including 10β-aldehydes and 10β-carboxylic acids, are readily
oxidized by air in free radical type reactions to corresponding 19-nor-10β-hydroperoxides and/or 10β-alcohols.
Thus, 3,17-dioxoandrost-4-en-19-al yields 10β-hydroperoxy-
estr-4-ene-3,17-dione [921,2249]; 3,20-dioxopregn-4-en-19-
oic acid (377) yields 10β-hydroperoxy-19-norpregn-4-ene-
3,20-dione (378) [167].

The class of 19-oxocardenolides including strophan-
thidin (3β,5,14-trihydroxy-19-oxo-5β,14β-card-20(22)-
enolide)(379), its glycosides cymarin and convallotoxin
calotropagenin (3β,12β,14-trihydroxy-19-oxo-5α,14β-card-
20(22)-enolide), and its glycoside calotropin are susceptible to air oxidations by free radical processes in aqueous solutions. Dehydrogenation to the corresponding 10β-
carboxylic acid followed by scission of the C/10-C-19 bond,
probable hydroperoxide formation, and ultimate formation of
19-nor-10β-alcohol products, such as 3β,5,10,14-tetra-
hydroxy-5β,10β,14β-card-20(22)-enolide (380) from stroph-
anthidin, occurs [227,1020,1946,2603]. The bond cleavage
reaction for these 19-oxocardenolides is thus like that of
the C_{19}- and C_{21}-aldehydes and acids previously described.

                    AUTOXIDATIONS IN ALKALI

Autoxidations or organic compounds by $^3O_2$ in systems
containing strong alkali, with or without transition metal
ions capable of undergoing electron transfer reactions,
have been observed. The initiating event is formulated
as ionization of susceptible substrate (RH) by base (B⁻),

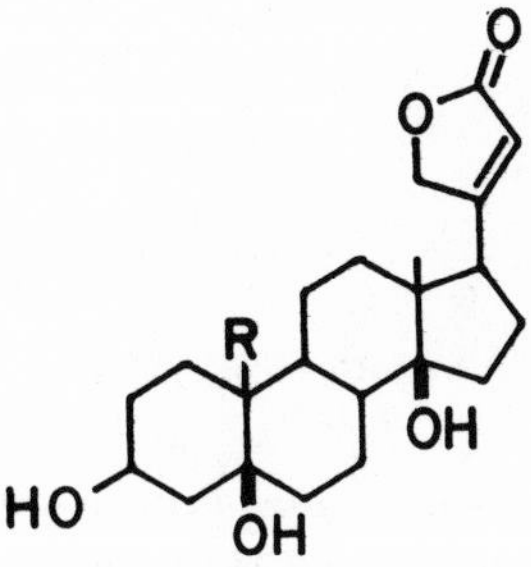

<u>379</u>   R = CHO

<u>380</u>   R = OH

with subsequent reaction of the anion with $^3O_2$ as suggested
in Eqs. 70-72. Example of this process acting on unfunctionalized sites of a steroid molecule has escaped my
notice. Moreover, the ready oxidation by air of cholesterol and the calciferols in alkaline media more probably
represents free radical autoxidations rather then the ionic
process of Eq. 70-72.

$$RH + B^- \longrightarrow R^- + BH \qquad \text{Eq. 70}$$

$$R^- + {}^3O_2 \longrightarrow R\text{-}O\text{-}O^- \qquad \text{Eq. 71}$$

$$R\text{-}O\text{-}O^- + BH \longrightarrow R\text{-}O\text{-}O\text{-}H + B^- \qquad \text{Eq. 72}$$

Most examples of the autoxidation of steroids in alkali involve steroid enolizable ketones, yielding as isolable products either an α-hydroperoxyketone or α-diketone.
Such alkali autoxidations have been commonly conducted
with potassium tert.-butoxide in the parent alcohol tert.-
butanol but more effectively in a binary system (tert.-
butanol and tetrahydrofuran) [2582]. Other bases KOH [564,
1417,1425], NaH [850], and triphenylmethyl sodium [120]
have also been used as well as other solvents such as benzene and dimethylsulfoxide.

The oxidation may proceed via a mechanism involving
carbanion, carbon-centered radical, and $O_2^{\overline{\cdot}}$ as suggested
by Eqs. 73-76 [120]. The suggested mechanism thus involves
free radical chain reactions, which may be moderated by the
alcohol solvent used. The generation of $O_2^{\overline{\cdot}}$ has not been
demonstrated.

$$-CO-CH= \quad + B^- \quad \rightarrow \quad -CO-(C^-)= \quad + BH \qquad \text{Eq. 73}$$

$$-CO-(C^-)= \quad + \ ^3O_2 \quad \rightarrow \quad -CO-(C\cdot)= \quad + O_2^{-} \qquad \text{Eq. 74}$$

$$-CO-(C\cdot)= \quad + \ ^3O_2 \quad \rightarrow \quad -CO-C(OO\cdot)= \qquad\qquad \text{Eq. 75}$$

$$-CO-C(OO\cdot)= \ + \ -CO-CH= \quad \rightarrow \quad -CO-C(OOH)= + \ -CO-(C\cdot)= \qquad \text{Eq. 76}$$

A more appealing mechanism not involving O_2^{-} posits that $^3O_2$ add to the ionized enolic form of the steroid ketone, yielding α-hydroperoxyketone as product, as suggested in Eqs. 77-79 [2431,2432]. Subsequent decomposition of the α-hydroperoxyketone in the basic media then yields an α-diketone product, as in Eq. 80.

$$-CO-CH= \ + B^- \quad \rightarrow \quad -C(O^-)=C= \quad + BH \qquad \text{Eq. 77}$$

$$-C(O^-)=C= \ + \ ^3O_2 \quad \rightarrow \quad -CO-C(OO^-)= \qquad\qquad \text{Eq. 78}$$

$$-CO-C(OO^-)= \ + BH \quad \rightarrow \quad -CO-CH(OOH)- \ + B^- \qquad \text{Eq. 79}$$

$$-CO-CH(OOH)- \quad\qquad \rightarrow \quad -CO-CO- \ + \ H_2O \qquad \text{Eq. 80}$$

Although both α-hydroperoxyketone and α-diketone products have been isolated from air oxidations of steroid monoketones, other reactive species are also formed in these alkaline systems, and more extensive alteration may occur. Electron spin resonance data support formation from steroid monoketones of semidione derivatives $R^1-C(O\cdot)=C(O^-)-R^2$ and of other unidentified radical anions [2044-2047,2431,2432]. The semidione species may derive from product α-diketones in a radical generating reaction suggested in Eq. 81, where the electron donor R^- may be the enolate anion of Eq. 77 [2431,2432].

$$-CO-CO- \ + \ R^- \quad \rightarrow \quad -CO(O\cdot)=C(O^-)- \ + R\cdot \qquad \text{Eq. 81}$$

α-Hydroperoxyketones

Autoxidations in alkali of enolizable steroid monoketones in which the α-carbon atom involved in enolization is tertiary result in the isolation of α-hydroperoxyketone products. Oxidation of 3β-hydroxy-5α-cholestan-6-one (45) yields 5-hydroperoxy-3β-hydroxy-5α-cholestan-6-one [120].

Oxidation of progesterone gives 17α-hydroperoxypregn-4-ene-
3,20-dione [120,121].17α-Hydroperoxides have also been isolated from oxidations of pregnenolone (23) [2247,2435], 3α,
12α-dihydroxy-5β-pregnan-20-one [2202], 3β-acetoxy-16β-
methylpregn-5-en-20-one[850],3β-acetoxy-6,16β-dimethylpregn-
5-en-20-one [851],and 3-methoxy- and 3-ethoxypregna-3,5-
dien-20-one [120,712,2247,2434,2582]. Alkaline autoxidation
of 3β-hydroxycholest-5-en-22-one yields a 20-hydroperoxy-3β-
hydroxy-(20ξ)-cholest-5-en-22-one, that of the 24-ketone 34
25-hydroperoxy-3β-hydroxycholest-5-en-24-one [108,2564].

Relatively few transformations of steroid α-hydro-
peroxyketones have been recorded. Thermal decomposition
of 17α-hydroperoxy-20-ketones yields as expected the corresponding 17-ketones [2247,2439], but in the case of pyrolysis of 17α-hydroperoxy-3β-hydroxygpregn-5-en-20-one elimination of the 17α-hydroperoxyl group, giving the parent
20-ketone 23, is suggested [2439]. Decomposition of 20-
hydroperoxy-3β-hydroxy-(20ξ)-cholest-5-en-22-one yields the
20-ketone 23 but also 3β,20-dihydroxy-(20S)-cholest-5-en-22-
one and C₁₉-products [108,2564]. Attempted acetylation of
3β-acetoxy-17α-hydroperoxy-16β-methyl-5α-pregnan-20-one results in the rearranged product 3β-hydroxy-16β-methyl-5α-D-
homo-17a-oxapregn-16-en-20-one [850]. Nitrosyl chloride in
pyridine acting on 17α-hydroperoxypregn-4-ene-3,20-dione
yields the 17α-nitrate ester of 17α-hydroxypregn-4-ene-3,
20-dione [151].

α-Diketones

With steroid monoketones for which product α-hydroper-
oxyketones are not isolated, α-diketones may result. The
products are viewed as forming via putative α-hydroperoxy-
ketones which are dehydrated directly to α-diketones isolated in enolic forms. Although products of autoxidation
from steroid 1-ketones have not been described, electron
spin resonance data for 5α-androstan-1-one and 5α-cholestan-
1-one evince slow formation of corresponding 1,2-semidione
radical anions [2432], these products thereby suggesting
that enolic 1,2-diones form.

Autoxidation of 5α-cholestan-2-one in alkali yields
5α-cholestane-2,3-dione isolated as a mixture of enols
3-hydroxy-5α-cholest-4-en-2-one (381) and 2-hydroxy-5α-
cholest-1-en-3-one (382). The same 2,3-dione and mixture

of enols is obtained by autoxidation of 5α-cholestan-3-one
(365) [120,1708,1710,2073] and of 2α-hydroxy-5α-cholestan-
3-one (or its 2α-acetate) [1417]. An analogous mixture of
enolic 2,3-diones is also recovered from autoxidations of
5α-(25R)-spirostan-2-one and 5α-(25R)-spirostan-3-one in
alkali [1708,1710]. Moreover, electron spin resonance data

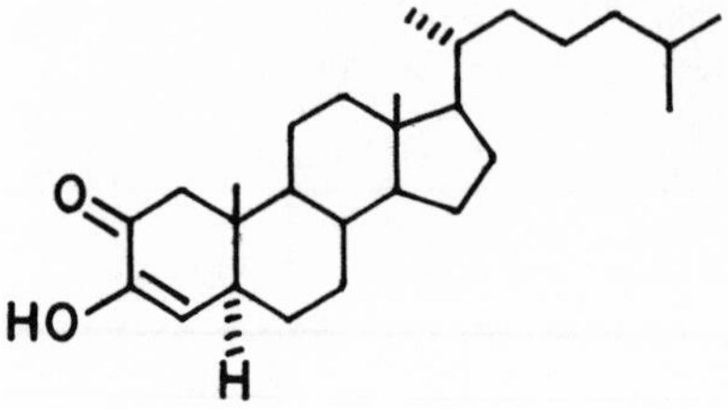

381 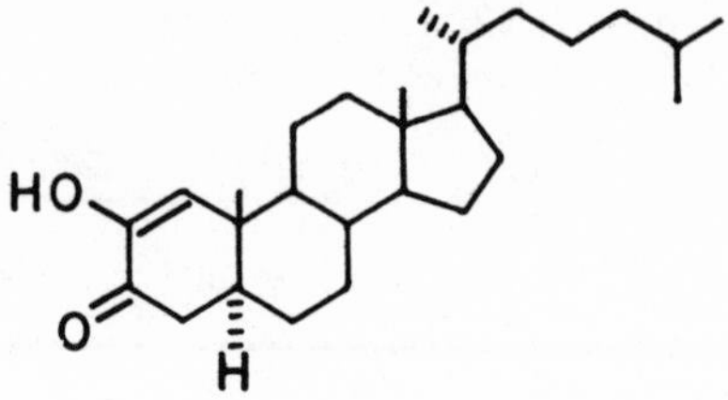   382

support   formation of 5α-2,3-semidione radical anions from
a variety of steroid 5α-2- and 5α-3-ketones [1708,1709,
2045,2046,2432], the 5α-cholestane-2,3-semidione radical
anion (383) giving a characteristic 14-line pattern [2045,
2432].

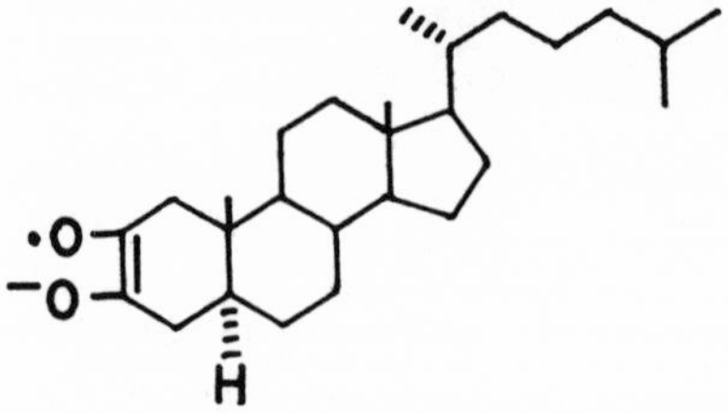

383 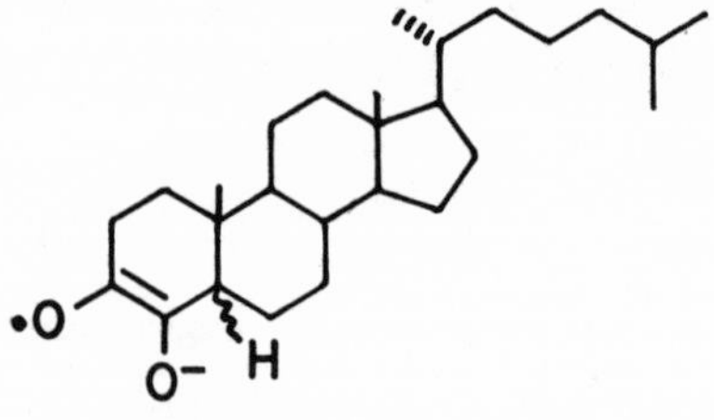   384 5α

                       385 5β

    Steroid 5α-2-ketones appear not to give 1,2-semidione
but only the 2,3-semidione radical anions, whereas electron
spin resonance spectra of 5β-2-ketones such as a 5β-spiro-
stan-2-one exhibit equal parts of 14-line and 4-line patterns interpreted as deriving from isomeric 5β-2,3-semi-
dione and 5β-1,2-semidione radical anions respectively
[1976,2046].

Autoxidation of 5α-3-ketones yields enolic 2,3-dike-
tones as isolable products in several instances [120,1709,
2045], but electron spin resonance data suggest that both
2,3-semidione and 3,4-semidione radical anions form. Thus,
5α-cholestan-3-one (365) gives the 14-line spectrum of the
5α-2,3-semidione 383 predominantly but also a small proportion of an 8-line pattern associated with the 5α-cholestane-
3,4-semidione (384). Autoxidation of the isomeric 5β-chole-
stan-3-one yields 4-hydroxycholest-4-en-one *inter alia*
[215]; autoxidation of 17β-hydroxy-5β-androstan-3-one likewise gives 4,17β-dihydroxyandrost-4-en-3-one [434,435].
Electron spin resonance spectra from these reactions display predominantly an 8-line spectrum of the 5α-3,4-semi-
dione 384 evincing substantial C-5 isomerization, with
minor contribution from the 5β-3,4-dione radical anion
(385) [2432].

Autoxidation of 4,4-dimethyl-3-ketones yields but the
one possible enol, 4α,4β-dimethylcholest-5-en-3-one yielding 2-hydroxy-4α,4β-dimethylcholesta-1,5-dien-3-one [954],
17β-acetoxy-4α,4β-dimethylestr-5-en-3-one yielding 2,17β-
dihydroxy-4α,4β-dimethylestra-1,5-dien-3-one [955]. 5α-
Lanost-8-en-3-one likewise gives the enolic product 2-hy-
droxy-5α-lanosta-1,8-dien-3-one [120]. Electron spin resonance spectra in this case reveal a simplified 4-line
pattern of the 2,3-semidione radical anion [2045]. Related
examples of pentacyclic triterpene 3-ketone autoxidation in
alkali also are known [687].

Autoxidation of the isomeric 5α- and 5β-cholestan-4-
one derivatives has not been investigated as regards recovery of products, but electron spin resonance spectra
characteristic of 6:1 mixtures of the 5α-3,4-semidione
384 and 5β-3,4-semidione 385 are recorded [2432].

Autoxidation of steroid ketones in alkali at other
sites are also known. The isomeric B-ring ketones 3α,5α-
cyclocholestan-6-one and 3α,5α-cyclocholestan-7-one yield
the same ketol 7-hydroxy-3α,5α-cyclocholest-7-en-6-one but
also 6α-hydroxy-3α,5α-cyclocholestan-7-one *inter alia* [452].
Paramagnetic species have not been observed with other B-
ring ketones, but a 6,7-semidione radical anion is observed
from 6α-bromo-5α-cholestan-7-one [2432]. Autoxidation of
hecogenin (3β-hydroxy-5α-(25R)-spirostan-12-one) yields an
enol 3β,11-dihydroxy-5α-(25R)-spirost-9(11)-en-12-one [2432].

Oxidations of D-ring ketones give analogous products but with an additional reaction occurring with 17-ketones. Autoxidations of 5α,14β-androstan-15-one and of 5α,14α-androstan-16-one yield electron spin resonance spectra of a common 5α,14β-androstane-15,16-semidione [2047]. Although the autoxidation of the 17-ketone 3β-hydroxy-13α-androst-5-en-17-one yields an isolated product enol 3β,16-dihydroxy-13α-androsta-5,15-dien-17-one [482], electron spin resonance data from autoxidation of 5α-androstan-17-one suggest that a dimeric radical anion form, perhaps 17-(5α'-androstan-17'-yl)-5α-androstane-17,17'-semidione radical anion (386). The dimeric radical anion 386 is converted to 5α-androstane-16,17-semidione which can be reconverted by O$_2$ back to the dimeric 386 [2044,2047].

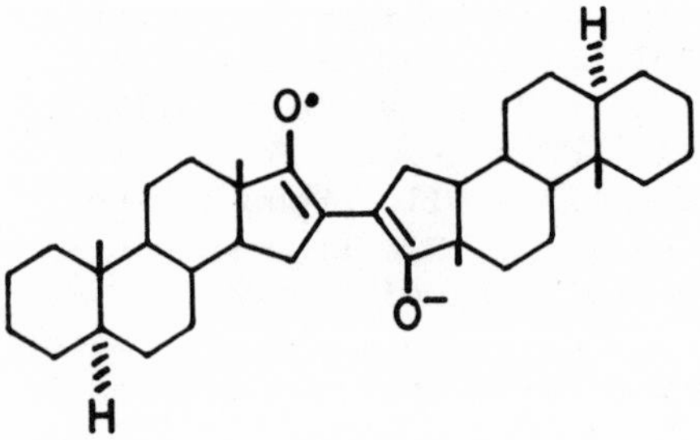

386 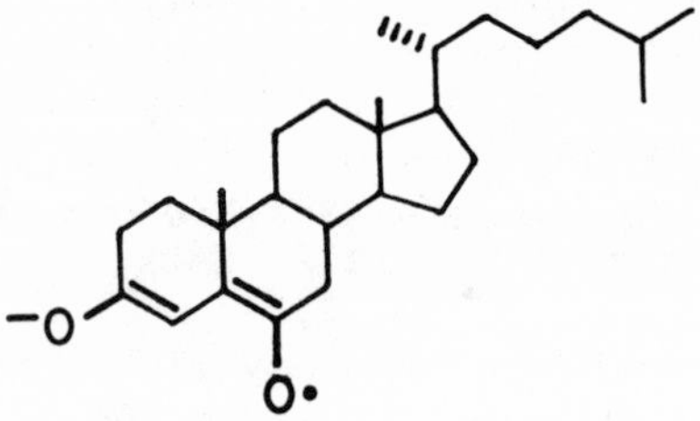   387

Autoxidations in alkali are not limited to simple ketones as α,β-unsaturated ketones and 1,4-diketones also undergo autoxidation. Cholest-4-en-3-one (8) not heretofore found sensitive to air oxidations is autoxidized in alkali to 4-hydroxycholesta-4,6-dien-3-one and the Δ^4-3,6-diketone 108 as isolated products [434,436]. 23,24-Bisnorchol-4-en-3-one is readily autoxidized in alkali to the analogous product 4-hydroxy-23,24-bisnorchola-4,6-dien-3-one [1203], but progesterone in alkali yields pregn-4-ene-3,6,20-trione [1425].

1,4-Diketones are dehydrogenated, the 5α-3,6-diketone 42 yielding the Δ^4-3,6-diketone 108, 5α-A-norcholestane-2,6-dione yielding A-norcholest-3-ene-2,6-dione [564].

Electron spin resonance spectra of the related 3,6-dike-
tones 42 and 108 establish that a common heteroannular radical anion 387 is formed. Moreover, the enone 8 forms the
same radical anion 387 but also a second radical anion,
either cholest-6-ene-3,4-semidione or cholest-4-ene-2,3-
semidione [2047].

The autoxidation of the enone 8 in alkali may now be
rationalized as involving initial enolization to cholesta-
3,5-dien-3-ol, with 1,2-addition of $^3O_2$ yielding via intermediate cholest-6-ene-3,4-dione isolated enol product 4-
hydroxycholesta-4,6-dien-3-one. The second product 108
results from alternative 1,4-addition of $^3O_2$, yielding the
5α-3,6-diketone 42 as intermediate which is directly dehydrogenated to the isolated product 108.

Further Transformations

It is obvious from the C-5 and C-14 isomerizations and
time-course of change of electron spin resonance spectra of
ketones in alkali that a complex, dynamic state exists under
these autoxidation conditions, leading not only to simple
α-hydroperoxyketone and α-diketone products but also to
more extensively altered material [1709,2432]. Isolation
studies establish that carbon-carbon bond scissions, further
oxidations, and ring closures occur. Simple scission of the
C-2/C-3 bond of 5α-cholestane-2,3-dione (or the enolic
forms 381 and 382) in methanolic KOH yields both monomethyl
esters of the secoacid 5α-2,3-secocholestane-2,3-dioic acid
[1417]. Prolonged treatment gives as benzilic rearrangement
product the isomeric 2ξ-hydroxy-5α-A-norcholestane-2ξ-
carboxylic acids (388) [1702]. Moreover, the same noracids
388 are formed from 5α-cholestan-3-one (365) and from the
5α-stanol 2 under more harsh autoxidizing conditions using
triphenylmethyl sodium and potassium tert.-butoxide in benzene respectively [120]. Further autoxidation of 4-hydroxy-
cholest-4-en-3-one formed from 5β-cholestan-3-one yields an
isomeric A-noracid 3β-hydroxy-5α-A-norcholestane-3α-carboxy-
lic acid [215,2073].

Corticosteroids with the dihydroxyacetone side-chain
undergo bond scissions in base-catalyzed air oxidations.
From prednisolone (11β,17α-dihydroxypregna-1,4-diene-3,20-
dione) is formed 11β,17α-dihydroxy-3-oxopregna-1,4-diene-
17β-carboxylic acid [484,933].

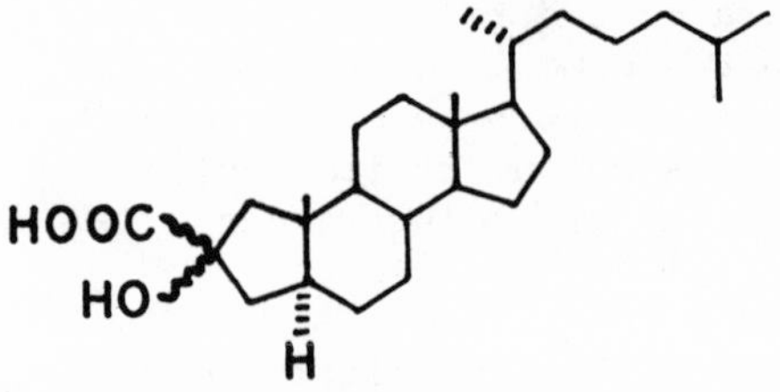

388 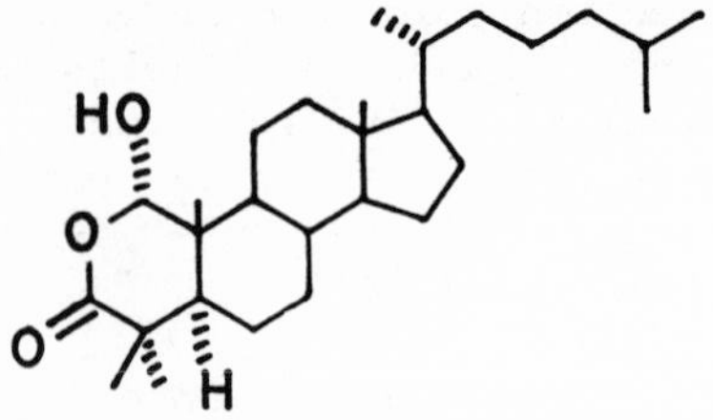   389

Autoxidation of the enol 2-hydroxy-4α,4β-dimethyl-5α-
cholest-1-en-3-one yields 1α-hydroxy-4α,4β-dimethyl-5α-2-
oxacholestan-3-one (389) [955]. The A-norketone 4α,4β-
dimethyl-A-norcholest-5-en-one yields the related 1α-hydro-
xy-4α,4β-dimethyl-2-oxacholest-5-en-3-one [954].

SOME COMPARISONS

Review of the oxidations of 5,7-dienes and other steroids and of subsequent transformations of initially formed
products suggests that no specific products form that could
be confused with products of cholesterol autoxidations.
Furthermore, although a number of chemical transformations
are common to cholesterol oxidation and to oxidations of
other steroids, oxidations of these other steroids generally follow other pathways and involve other processes. Thus,
alcohol dehydrogenations and hydroperoxide dehydrations to
ketones occur in all series. Bond scissions occur, in the
side chain for cholesterol, in the A- and B- rings for 5,7-
dienes.

The three endogenous sterols cholesterol, 5α-cholest-
7-en-3β-ol (57), and 5α-lanost-8-en-3β-ol (68) have unique
features of oxidation chemistry such that products from 57
and 68 are clearly distinct from those of cholesterol. However, the 7-stenol 57 may be oxidatively transformed to the
5,7-diene 56, thereby leading to autoxidation products of
56 under selected circumstances. Otherwise, the chemistry
of cholesterol and of 57 and 68 remain distinct from that
of 56 and ergosterol, where 1,4-cycloaddition of O_2 predominates.

Only the 3,6-diketones 42 and 108 appear to have dual autoxidation origins from cholesterol. The air oxidation of cholesterol via the enone 6 to the epimeric 6-hydroperoxides 59 and 280 and epimeric 6-alcohols 88 and 281 affords a pathway to 42 and 108, whereas isomerization of the enone 6 to enone 8 and subsequent autoxidation in alkali also yields the 3,6-diketones 42 and 108 by the ionic process.

CHAPTER VII.   METABOLIC TRANSFORMATIONS

Although interest in the metabolic fate of cholesterol, exemplified in its bioconversion to the 5β-stanol 4 [316], and sterol autoxidation date from the same time (1896), interest in the metabolism of cholesterol autoxidation products is a much more recent matter. Interim period studies in the metabolic fate of oxycholesterol preparations intravenously injected into dogs [678] notwithstanding, relatively little interest in the class developed. In the present chapter I examine the processes by which recognized cholesterol autoxidation products might be formed enzymically and also the means by which these products are transformed by enzymes and metabolized by living cells. Whereas interests in the metabolism of cholesterol autoxidation products are now expanding, with few exceptions the enzymic formation of these autoxidation products by natural *in vivo* processes remains undemonstrated.

BIOSYNTHESIS OF AUTOXIDATION PRODUCTS

Many claims to genuine metabolite or natural product status for the several cholesterol autoxidation products of TABLE 1 and TABLE 2 have been made. However, the genuine natural occurrence of these products infers their enzymic derivation. This inference left unstated throughout the literature needs careful attention, and it is disturbing that so little record of biosynthesis in tissues or in enzyme incubations has been made [2563]. This paucity of evidence may represent *bona fide* absence of enzyme systems for biosynthesis of these products, but few investigators have chosen to address these issues. Moreover, the low levels of radioactivity from specific precursors incorporated into traces of minor products in biosynthesis experiments rarely attract attention.

All genuine sterol metabolites should have specific enzyme systems associated with their biosynthesis. However, the question of possible enzyme origins for individual cholesterol autoxidation products cannot now be answered to satisfaction for want of experimental evidence. Of the recognized cholesterol autoxidation products (TABLE 10) only six (the 3β,7α-diol 14, the (20S)-3β,20-diol 21, the 3β,25-diol 27, the (25R)-3β,26-diol 29, the 5α,6α-epoxide 35, and the C_{24}-acid 99) have demonstrated enzymic origins! This

is not to say that other autoxidation products may not also
be formed enzymically; present evidence is simply unpersua-
sive. In the face of the obvious need for experimental evidence, it is simply not adequate to posit the existence of
an undemonstrated (but possible) cholesterol 7β-hydroxylase
giving rise to the 3β,7β-diol 15, for example.

Where specific cholesterol hydroxylases have been suspected, such as hepatic 7α- and (25R)-26-hydroxylases, adrenal cortex 20α$_F$- and 22β$_F$-hydroxylases, or brain 24β$_F$-hydro-
xylase, the usual enzymology approaches suffice to provide
adequate evidence of such enzymes. Moreover, recently a hepatic mitochondrial cholesterol 25-hydroxylase activity (not
resolved from the associated (25R)-26-hydroxylase) has been
demonstrated [1837]. In the absence of supporting evidence
one might be conservative in promulgating claims of metabolic status for recognized autoxidation products.

In order to emphasize the essential enzyme component of
naturally occurring metabolic events, I have arranged discussion of the biosynthesis of cholesterol autoxidation products to follow along lines of the enzymes necessarily implicated in the bioconversion.

Dioxygenases

As hydroperoxides are the first products of cholesterol
autoxidation, so attention is first focused onto the possible derivation of sterol hydroperoxides enzymically. In
order to form sterol hydroperoxides or peroxides enzymically
it is necessary to invoke the action of dioxygenases on the
sterol substrate, there being no known enzymic process in
which the oxygen-oxygen peroxide bond is formed from two
oxygen atoms in separate chemical bondings. By definition
dioxygenases utilize $^3O_2$ as substrate and introduce both
atoms of the oxygen together into the second (sterol) substrate. Dioxygenase products would then be sterol hydroperoxides such as the epimeric 7-hydroperoxides 46 and 47
formed enzymically according to Eq. 8c or cyclic peroxides
such as ergosterol 5α,8α-peroxide 62 formed by the 1,4-
cycloaddition of dioxygen to the 5,7-diene system.

The enzymic hydroxylation of steroids has been speculatively viewed as possibly being a two-step process, involving an initial hydroperoxidation followed by reduction

of the hydroperoxide to the corresponding alcohol [457,913, 2563]. Although the hydroperoxide reduction phase of this sequence has been repeatedly demonstrated in a variety of enzyme systems, the requisite dioxygenase phase has not been supported, and the hydroxylation of steroids by monooxygenases in which cytochrome P-450 is implicated has now displaced this alternative. Dioxygenases unrelated to steroid hydroxylation have been demonstrated using cholesterol, the isomeric 4-stenol 79 and 7-stenol 57, the 8-stenol 68, the 5,7-dienol 56, and ergosterol.

The oxidation of cholesterol by dioxygenases to yield the characteristic 7α- and 7β-hydroperoxides 46 and 47 has been demonstrated with three separate enzyme systems (soybean lipoxygenase, horseradish peroxidase, and the microsomal NADPH-dependent lipid peroxidation system of rat liver). Moreover, inconclusive evidence suggests that the (20S)-20-hydroperoxide 22 be formed enzymically in incubations of adrenal cortex mitochondria [2310,2572], but here the contribution of autoxidation was not excluded.

Soybean Lipoxygenase. The oxidation of cholesterol by soybean lipoxygenase (linoleate:oxidoreductase, EC 1.13.11. 12, formerly EC 1.13.1.13 and EC 1.99.2.1.) in incubations including ethyl linoleate as prime substrate, buffered at pH 6.6 or at 9.0, gives the epimeric 7-hydroperoxides 46 and 47 in low yields in the proportion 1:3 to 2:3 [2458,2461]. The 7-hydroperoxides were the first formed oxidation produces of cholesterol, with the epimeric 3β,7-diols 14 and 15 and the 7-ketone 16 previously found by others in the reaction [93,251,1192] being secondary products in the same fashion as observed in autoxidations of cholesterol. Cosubstrate ethyl linoleate essential for the oxidation of cholesterol was also peroxidized in the transformations [2458, 2461]. The isomeric 5,6-epoxides 35 and 36 and the 3β,5α,6β-triol 13 have also been found in soybean lipoxygenase incubations of cholesterol [93,1192]. The 5,6-epoxides may represent secondary reactions of the initially formed 7-hydroperoxides on substrate cholesterol, as previously described; the 3β,5α,6β-triol 13 may be viewed as the hydration product of both 5,6-epoxides.

Yet an additional complication occurs in this oxidation. The 5α-hydroperoxide 51 was detected in our studies at very low levels [2457,2461]. As formation of the 5α-hydroperoxide 51 from cholesterol suggests participation of $^1O_2$,

the generation of this excited species in these lipoxygenase
incubations might be concluded. However, the sterol hydroperoxides 46, 47, and 51 added individually as substrates
in place of cholesterol were all interconverted by soybean
lipoxygenase and by boiled enzyme to a mixture of 46, 47,
and 51 as well as of 14,15,16, and 50 [2461]. This interconversion of 46,47, and 51 in aqueous buffered protein dispersions contrasted markedly with the behavior of these
sterol hydroperoxides in organic solvent solutions where
the 5α-hydroperoxide 51 isomerizes to the 7α-hydroperoxide
46 (which in turn epimerizes to the 7β-hydroperoxide 47) but
in which no reverse isomerization have been observed.

     It thus appears that the presence of the 5α-hydroper-
oxide 51 in soybean lipoxygenase incubations of cholesterol
result from a nonspecific ion catalyzed interconversion of
sterol hydroperoxides and not from attack of $^1O_2$ on cholesterol. These same conclusions are supported by results of
soybean lipoxygenase action on cholest-4-en-3β-ol (79), the
7-stenol 57, and the 8-stenol 68 discussed shortly.

     Another interesting feature of the oxidation of chole-
sterol by soybean lipoxygenase to the 3β,7α-diol 14 (presumably via the 7α-hydroperoxide 46) is the presence of an
isotope effect which suggest the scission of the 7α-H bond
is rate-limiting [251]. As an isotope effect is also observed in the autoxidation of cholesterol in aqueous sodium
stearate dispersions but not for liver microsomal enzymic
7α-hydroxylation of cholesterol, the oxidation of cholesterol by the dioxygenase is clearly by different means than
that by the monooxygenase. This fundamental difference between the actions dioxygenase and monooxygenase on cholesterol is matched by similar results in the oxidation of
cyclohexane to cyclohexanol, where an isotope effect was
measured for peracid oxidation [2537] but not for rat liver
microsomal cytochrome P-450 oxidation [2536].

     Soybean lipoxygenase also oxidizes other sterols. Sito-
sterol (20) is oxidized, but the requisite sitosterol hydroperoxides disclosing dioxygenase action were not sought.
The common secondary oxidation products including 3β-hydroxy-
stigmast-5-en-7-one (72), the epimeric 3β,7-diols 76 and 77,
the isomeric 5,6-epoxides 5,6α-epoxy-5α-stigmastan-3β-ol and
5,6β-epoxy-5β-stigmastan-3β-ol, and 5α-stigmastane-3β,5,6β-
triol were detected. [93].

However, dioxygenase action of soybean lipoxygenase on sterols may be demonstrated where appropriate attention to detection of the hydroperoxide products is had. Cholest-4-en-3β-ol (79) is oxidized by soybean lipoxygenase to the enone 8 which is then in turn transformed to a mixture of the epimeric hydroperoxides 6α-hydroperoxycholest-4-en-3-one (280) and 6β-hydroperoxycholest-4-en-3-one (59). These epimeric 6-hydroperoxides were readily interconverted and thermally decomposed to the corresponding 6-alcohols 6α-hydroxycholest-4-en-3-one (281) and 6β-hydroxycholest-4-en-3-one (88) and to the 3,6-dione cholest-4-ene-3,6-dione (108) [2462]. These conversions were not observed with boiled lipoxygenase.

The 7-stenol 57 and 8-stenol 68 are also oxidized by soybean lipoxygenase, with sterol hydroperoxide products indicative of free radical oxidation formed. Incubations of the 7-stenol 57 gave two products 3β-hydroxy-5α-cholest-7-ene-6α-hydroperoxide (303) and 3β-hydroxy-5α-cholest-7-ene-6β-hydroperoxide (304), whereas the 8-stenol yielded three hydroperoxides tentatively suggested as being 3β-hydroxy-5α-lanost-8-ene-7β-hydroperoxide (254), 3β-hydroxy-5α-lanost-8-ene-11β-hydroperoxide (255), and 3β-hydroxy-5α-lanost-8-ene-7β,11β-dihydroperoxide (256) [2309,2383].

The precise mechanism by which these sterol hydroperoxides are formed by soybean lipoxygenase is uncertain, but since cooxidation of ethyl linoleate is required, it may be that radicals formed in this enzymic process in fact initiate the free radical oxidation of cholesterol, the 7-stenol 57, and the 8-stenol 68 by $^3O_2$ to give the recognized free radical oxidation products. Likewise, the 4-stenol 79 may suffer the same oxidation processes, with dehydrogenation to the enone 8 preceeding hydroperoxide formation. The formation of enzyme-substrate complexes between lipoxygenase and sterol has not demonstrated.

A few observations of the oxidative action of soybean lipoxygenase on other steroids have been recorded. Thus, soybean lipoxygenase catalyzes the oxidation of pregnenolone (23) and of the 17-ketone 86, in the presence of cosubstrate linoleate, to the corresponding 7α- and 7β-alcohol and 7-ketone derivatives. Suspected 7-hydroperoxide intermediates were not sought in these experiments [1192].

Horseradish Peroxidase. Horseradish peroxidase (donor:
hydrogen peroxide oxidoreductase, EC 1.11.1.7) also acts as
a formal dioxygenase on cholesterol. In this case hydrogen
peroxide serves as nominal substrate for the enzyme, and
the same epimeric 7-hydroperoxides 46 and 47 (with 47 predominating) form as initial products. Subsequent thermal
decomposition of 46 and 47 accord the epimeric 3β,7-diols
14 and 15 and the 7-ketone 16 [2458,2461]. As in the case
of soybean lipoxygenase, active enzyme was required, as was
the nominal substrate H_2O_2. Non-specific ion catalyzed
interconversion of the sterol hydroperoxidases 46,47, and
51 also occurred [2461].

A similar set of results has been obtained with the
7-stenol 57 and 8-stenol 68 in incubations with horseradish
peroxidase. Incubations at pH 5.5 with the 7-stenol 59
give the epimeric 6-hydroperoxides 303 and 304 indicative
of free radical oxidation, with the 8-stenol 68 the 7β- and
11β-hydroperoxides 254 and 255 and the 7β,11β-dihydroper-
oxide 256, thus the same products formed from these stenols
in soybean lipoxygenase incubations [2309,2383].

Horseradish peroxidase acting on H_2O_2 also oxidizes
other steroids. Ergosterol is transformed into the 5α,8α-
peroxide 62; 17α-ethinyl-17β-hydroxyestr-5(10)-en-3-one is
oxidized to 17α-ethinyl-10β-hydroperoxy-17β-hydroxyestr-4-en-
3-one and 17α-ethinyl-10β,17β-dihydroxyestr-4-en-3-one
[1315]. Furthermore, pregn-5-ene-3,20-dione is oxidized
to 6β-hydroperoxypregn-4-ene-3,20-dione [2765].

Liver Dioxygenases. Attempts have been made to demonstrate the action of the dioxygenases of mammalian tissue
on cholesterol. Weak evidence suggests the formation of
the (20S)-20-hydroperoxide 22 from $[1,2-^3H]$ cholesterol in
aerobic incubations of bovine adrenal cortex mitochondrial
acetone powers [2572], but attempts to oxidize cholesterol
(and poluynsaturated fatty acid cosubstrates) with sheep
vesicular gland protaglandin synthetase preparations were
not successful [1571]. Only one mammalian enzyme system,
the hepatic microsomal NADPH-dependent lipid peroxidation
system of rat liver, has demonstrated dioxygenase activity
towards cholesterol, as evinced by our detection of the
epimeric cholesterol 7-hydroperoxides 46 and 47 as initially
formed oxidation products [2311,2461].

The generalized lipid peroxidation system of liver microsomes has been extensively studied with polyunsaturated fatty acids as substrates, but the system also oxidizes cholesterol, yielding besides the 7-hydroperoxides 46 and 47 the well known trio of 7-oxygenated sterols 14-16 [92, 93,126,251,258,265,341,867,1192,1363,1640,1653-1655,2126, 2311,2461], the isomeric 5,6-epoxides 35 and 36 [92,93,1640, 1655], the 3β,5α-6β-triol 13 [92,93,1655], and the 6-ketone 44 [93]. With the exception of the 6-ketone 44, these same products have been variously detected in oxidations of cholesterol by soybean lipoxygenase, and it is just this pattern of oxidized sterols 13-16,35,36,46, and 47, which characterizes free radical autoxidation of cholesterol by air.

Inasmuch as formation of the trio of products 14-16 characterizes rat liver microsomal lipid peroxidation of cholesterol their formation in incubations of human [259] and guinea pig [257] liver microsomes and the formation of the 5α,8α-peroxide 60 from the 5,7-dienol 56 in mouse liver microsomes incubations [1225] suggests a broad distribution of the microsomal lipid peroxidation system in mammalian liver. However, liver microsomal lipid peroxidations have been demonstrated only in in vitro incubations, and these transformations have not been detected in living tissues. Moreover, microsomal enzyme systems are quite impure and contain other enzymes which potentially compromise interpretation of results. Thus, liver microsomes contain the extablished cholesterol 7α-hydroxylase implicated in bile acid biosynthesis and also peroxidases acting to destroy sterol hydroperoxide products of dioxygenase actions.

Cholesterol peroxidation may also occur in incubations of other liver subcellular fractions. Thus, rat liver mitochondria variously yield products 14-16,35,36,46, and 47 [94,268,1655,2461] though to an extent less than that of microsomes. Incubations of bovine adrenal cortex mitochondria also have yielded 14-16,35 and 36 [327,2255].

The peroxidation of cholesterol in incubations of liver microsomes may be linked to the enzymic peroxidation of endogenous polyunsaturated fatty acids and their derivatives. Although an absolute requirement of cholesterol peroxidation for polyunsaturated fatty acid has not been demonstrated, microsomal peroxidation of cholesterol is stimulated by

added ethyl linoleate, and the addition of ethyl linoleate
hydroperoxides greatly increases the peroxidation [2461].
Cumeme hydroperoxide also appears to stimulate the lipid
peroxidation of cholesterol by rat liver microsomes [552].
One may speculate that the enzymic peroxidation of polyunsaturated fatty acid, yielding fatty acyl hydroperoxides,
is the initiating event for the subsequent peroxidation of
cholesterol, in much the same manner as may be formulated
for the action of soybean lipoxygenase.

        Given means of deriving the requisite cholesterol 7-
radical 262 formation of the products 13-16, 35, 36, 46,
and 47 may be rationalized by initial oxidation of cholesterol to the 7-hydroperoxides 46 and 47, their subsequent
oxidation of cholesterol to give the 5,6-epoxides 35 and
36 whose hydration yields the triol 13, their subsequent
reduction to the corresponding 3β,7-diols 14 and 15, and
their dehydration to the 7-ketone 16 in the exact manner
of autoxidation. In this formulation, the 3β,5α,6β-triol
13 serves as precursor of the 6-ketone 44 found in one
case. In support of this formulation is the demonstrated
presence of the 7-hydroperoxides 46 and 47 [2311,2461] and
the demonstrated transformation of the 7α-hydroperoxide 46
to the 3β,7α-diol 14 and 7-ketone 16 by rat liver microsomes [1655]. The formulation accounts for the observed
results, but other free radical processes may also be involved. Detection of the requisite 7-hydroperoxides 46 and
47 inferring dioxygenase action was not achieved with all
ages or strains of rats examined [2461], and the secondary
products 13-16, 35, and 36 are also formed from cholesterol
by attack of HO· [73,2291] and rat liver microsomal preparations utilizing oxidants such as NaIO₄ which contain but
one atom of oxygen [552]. Nonetheless, as most investigations do not attempt to detect sterol hydroperoxide products,
detection only of the secondary products 13-16, 35, and 36
may merely reflect this lapse or very rapid transformations
of the hydroperoxides.

        However, the lipid peroxidation system is in fact one
catalyzed by enzymes, and although autoxidation may contribute to the oxidation of cholesterol in such systems, microsomal lipid peroxidation is a separate process. Lipid peroxidation is dependent upon an electron transport system
moderated by NADPH and a cytochrome reductase. Heat-inactivated microsomes preparations and reduced NADPH levels decrease the extent of cholesterol oxidation [1192,1640,2461].

Lipid peroxidation is dependent on Fe(II), and liver microsomes prepared in the presence of chelating agents (EDTA)
are inactive with respect to cholesterol peroxidation [93,
126,1640,1655,2212]. Lipid peroxidation is also inhibited
by xenobiotic thiols such as 2-mercaptoethanol, 2-mercapto-
ethyl amine, cysteamine, and dithiothreitol [93,341,342,
376,867,868,1640,1655,2125,2126,2552], by ascorbic acid
[867], by antioxidants such as BHA [253], and by thermo-
stable components of liver cytosol [341,920,1655].

Lipid peroxidation is also distinct from the hepatic
cholesterol 7α-hydroxylase which also requires NADPH and
$^3O_2$. Whereas the monooxygenase utilizes a cytochrome P-450
enzyme which is sensitive to CO, lipid peroxidation is insensitive to CO [1192,1640], and other distinctions obtain.
Cholesterol autoxidation and enzymic peroxidation in liver
microsomal incubations are now routinely suppressed by
thiols so that studies of monoxygenase 7α-hydroxylation may
be conducted without undue difficulties.

There are thus at least three recognized processes
contributing to the oxidation of cholesterol in liver microsomal incubations: the cholesterol 7α-hydroxylase yielding
the 3β,7α-diol 14 as sole specific product, autoxidation
yielding the 7-hydroperoxides 46 and 47 and their subsequent
transformation products, and generalized lipid peroxidation
yielding the same products but involving other catalysis.

There may also be other hepatic enzymes which oxidize
cholesterol in the B-ring, possibly a system forming the
3β,7β-diol 15 and the 7-ketone 16 by yet unrecognized means.
The presence of the 7β-hydroxyl group in ursodeoxycholic
acid (3α,7β-dihydroxy-5β-cholanoic acid) suggests that a
sterol 7β-hydroxylase might exist, but the one examination
of a putative cholesterol 7β-hydroxylase system [1616]
surely dealt with the lipid peroxidation system and not a
monooxygenase yielding the 3β,7β-diol 15 uniquely. Moreover, a microsomal oxidoreductase utilizing NADP$^+$ reduces
the 7-ketone 16 to the 3β,7β-diol 15 (but not to the epimeric 3β,7α-diol 14) [265,1653,1655,2548,2552].

Although the requisite sterol 7-hydroperoxides inferring dioxygenase action have not been demonstrated, the peroxidation by rat liver microsomal enzymes of other Δ^5-sterols
is indicated. From sitosterol are formed the epimeric 3β,7-

diols 76 and 77, 7-ketone 72, the isomeric 5,6-epoxides
5,6α-epoxy-5α-stigmastan-3β-ol and 5,6β-epoxy-5β-stigmastan-
3β-ol, and 5α-stigmastane-3β,5,6β-triol in the same fashion
as does cholesterol yield homologous products [92,93,340].
Moreover, campesterol (70) is oxidized to epimeric 24-methyl-
(24R)-cholest-5-ene-3β,7-diols and 3β-hydroxy-24-methyl-(24R)-
cholest-5-en-7-one (74) derivatives, and cholest-5-en-3α-ol
to epimeric cholest-5-ene-3α,7-diols and 3α-hydroxycholest-
5-en-7-one, all oxidations being stimulated by Fe(II) [91].
Notably, rat liver mitochondria likewise oxidize sitosterol
and campesterol to the corresponding epimeric 3β,7-diol,
7-ketone, and 5β,6β-epoxide derivatives [94,2385].

The oxidation of pregn-5-en-3β-ol (110) to unidentified products by rat liver microsomal enzymes appears to be
by peroxidation [376] but oxidation of the C_{21}- and C_{19}-Δ^5-
3β-alcohols 23 and 86 [1192] and of several other steroids
[91] in the C-7 positions may involve specific monooxygenases and not the lipid peroxidation system.

Other types of steroids are also subject to lipid peroxidations by murine liver microsomal enzymes. The 5,7-
dienol 56 is transformed into the 5α,8α-peroxide 60 by
mouse liver microsomes in a reaction dependent upon NADPH
and Fe(II) ions but ergosterol is not peroxidized in the
same system [1225]! Moreover, 17α-ethinyl-17β-hydroxyestr-
5(10)-ene-3-one is peroxidized by rat liver microsomal enzymes to 17α-ethinyl-10β-hydroperoxy-17β-hydroxyestr-4-en-
3-one [462,1445], the same product formed by horseradish
peroxidase [1315] and by photosensitized oxygenations of
the substrate [1450,2197]. The 10β-hydroperoxide product
has progestational and contraceptive activity [2642] and
may be implicated in metabolic processes associated with
the presence of 10β-hydroxylated metabolites of 17α-ethinyl-
17β-hydroxyestr-5(10)-en-3-one in human urine [81,1438].

Some evidence suggesting dioxygenase action on C_{21}-
steroids in adrenal subcellular fractions has been recorded.
Thus, some [^{14}C] radioactivity was trapped by 17α-hydroper-
oxypregn-4-ene-3,20-dione in aerated incubations of [^{14}C]
progesterone with bovine adrenal preparations [1051], but
autoxidation processes were not ruled out. Further, bovine
adrenal microsomes appear to peroxidize progesterone to 6β-
hydroperoxypregn-4-ene-3,20-dione, a product also formed by
bovine adrenal mitochondria acting on the same substrate or
on pregn-5-ene-3,20-dione [2765].

The rat liver peroxidation system may operate as a dioxygenase on a variety of xenobiotic aromatic substrates also. Formation of the benzylic hydroperoxides tetralin hydroperoxide and fluorene hydroperoxide from tetralin and fluorene respectively is indicated [463,464,1508,2040], and formation of the allylic hydroperoxide 2,5-di-*tert*.-butyl-4-hydroperoxy-4-methylcyclohexa-2,5-dien-1-one from the antioxidant BHT (2,5-di-*tert*.-butyl-4-methylphenol) [465, 2203] also occurs. Yet added dimension to dioxygenase action on xenobiotic aromatic substrates obtains in the oxidation by rat liver microsomal enzymes of 9,10-dimethyl-1,2-benzanthracane (390) to the endo-peroxide 9,10-dimethyl-9,10-dihydro-1,2-benzanthracene-9,10-epidioxide (391) and corresponding dihydrodiol 9,10-dimethyl-9,10-dihydro-1,2-benzanthracene-9,10-diol [466].

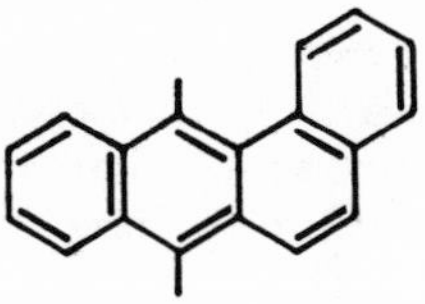

390 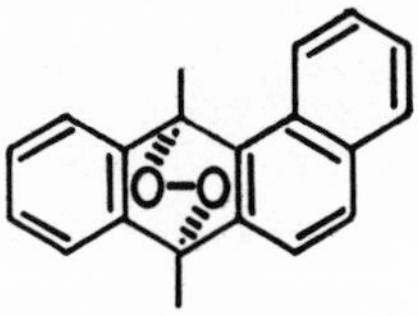   391

Microbial Dioxygenases. Some evidence of the presence of sterol dioxygenases in flourishing vegetative cell growth of microorganisms exists. Cultures of *Penicillium rubrum* and *Gibberella fujikuroi* oxidize ergosterol to its 5α,8α-peroxide 62 [160]. Moreover, formation of ergosta-4,6,8(14),E-22-tetraen-3-one (223) from ergosterol and from the 5α,8α-peroxide 62 in *P. rubrum* cultures also suggests that an ergosterol dioxygenase is present in this mold [2666].

The complex balance between enzymic and nonenzymic oxidation of ergosterol is well illustrated in cultures of *P. rubrum* or *G. fujikuroi*. A substantial amount of ergosterol was oxidized to the 5α,8α-peroxide 62 by the viable culture grown in the dark, but peroxidation of ergosterol continued in such cultures which had been autoclaved! Autoclaved cultures maintained in the dark were inactive, but killed cultures held in light gave appreciable amounts

of 5α,8α-peroxide 62.   Both organisms are pigmented, mito-
rubrin from *P. rubrum* and bikaverin from *G. fujikuroi* being
good photosensitizers apparently capable of promoting the
photooxygenation of ergosterol without the action of the
dioxygenase [160].   The involvement of fungal pigments in
promoting the formation of the 5α,8α-peroxide 62 as a nonenzymic product is established in other cases [16].   As are
the cases with cholesterol, none of these indicated enzymic
oxidations of sterol 5,7-dienes to 5α,8α-peroxides have been
studied adequately.   For instance, demonstration of an enzyme-substrate complex has not been attempted.

Nonenzymic oxidations may occur in other aerated microbial fermentations, yielding products that mimic those from
dioxygenase action.   The transformation of 17α-hydroxy-6,16β-
dimethylpregn-5-ene-3,20-dione to 6β-hydroperoxy-17α-hydroxy-
6α,16β-dimethylpregn-4-ene-3,20-dione by *Flavobacterium
dehydrogenans* [851] is an example.   Furthermore, in certain
other cases an interplay between enzymic and nonenzymic
events may occur in microbial cultures.   In order to account
for the formation of the epimeric products 6α- and 6β-
hydroxyandrosta-1,4-diene-3,17-dione (396) from the Δ^3-3β-
alcohol 86 in aerated incubations with *Actinoplanes* sp. No.
431 (*A. missourienses*) which 6-alcohols 396 were not formed
from androst-4-ene-3,17-dione (392) or androsta-1,4-diene-

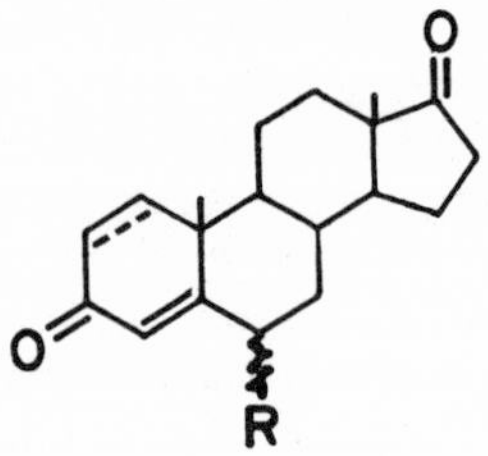

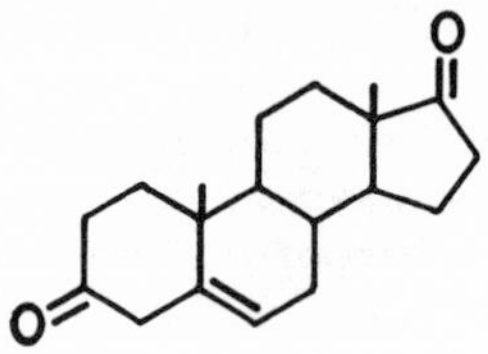

392   R = H   397

393   R = OH

394   R = OOH

395   R = H,Δ^1

396   R = OH,Δ^1

3,17-dione (395) as substrates [1561], the enzymic dehydrogenation of substrate to androst-5-ene-3,17-dione (397) was
postulated, with subsequent nonenzymic oxygenations, yielding epimeric 6α- and 6β-hydroperoxyandrost-4-ene-3,17-diones
(394), each in turn then subject to dehydrogenation and reduction to the corresponding $\Delta^{1,4}$-3-ketone-6-alcohols 396
[34]. Moreover, in autoclaved cultures of Rhizopus arrhizus
ATCC 11145 the Δ^5-3-ketone 397 is isomerized to 392 and
oxidized to the epimeric 6-alcohols 393 and to androst-4-
ene-3,6,17-trione [1060], these results probably deriving
via the intermediate 6-hydroperoxides 394.

Monooxygenases

     The biosynthesis of cholesterol autoxidation products
bearing but one added oxygen atom conceivably might proceed
via the action of monooxygenases. In general such enzymes
introduce the hydroxyl group stereospecifically into the
steroid molecule and are also implicated in steroid epoxidations. Accordingly, such autoxidation products as the
monohydroxylated derivatives 14,15,21,27, etc. and the 5,6-
epoxides 35 and 36 need examination in this light. In general the biosynthesis of these products is closely associated with the biosynthesis of bile acids [255,263].

     Cholesterol 7α-Hydroxylase. By far the most thoroughly
investigated case of monooxygenase action on cholesterol is
that of the formation of the 3β,7α-diol 14 in mammalian
liver, where demonstration of a cholesterol 7α-hydroxylase
(cholesterol, reduced NADP:oxygen oxidoreductase (7α-hy-
droxylating), EC 1.14.13.17) associated with hepatic biosynthesis of bile acids has been achieved. The matter of
the enzymic 7α-hydroxylation of cholesterol to yield the
3β,7α-diol 14 as an initial and rate-limiting step in the
bioconversion of cholesterol to bile acids presents a clas-
sical paradigm case history [1701], with early surmise that
7α-hydroxylation must be involved [1518], that 7α-hydroxy-
lation was stereospecific and occurred with retention of
configuration [199], but that convincing demonstration of
enzymic 7α-hydroxylation of cholesterol could not be had
because of the ubiquitous interference of autoxidation
yielding the same 3β,7α-diol 14 product [543,548]. Selective inhibition of interferring autoxidation and lipid
peroxidation reactions of cholesterol by additions of

sequestering agent EDTA or thiols allowed convincing demonstration of the enzymic 7α-hydroxylation of cholesterol in
liver.

The cholesterol 7α-hydroxylase was further characterized as a post-mitochondrial [548] or more precisely microsomal [1192,1640,1655] enzyme involving a microsomal electron transport system utilizing NADPH and O_2 [342,1640,2623]
in which cytochrome P-450 is implicated as terminal oxidase.
Additionally, inhibition of the enzyme by carbon monoxide
[342,1192,1640,2126,2552,2623] with release from inhibition
by 450 nm light [342], inhibition by cytochrome c [342] and
by antibody against NADPH-cytochrome c reductase [2623],
inhibition by heavy metal ions [2551] and by p-chloromer-
curibenzoate anion [342,2551] but not by cyanide anion
[342] and stimulation by phenobarbital pretreatment of some
strains of rats [346,2213,2623] and by liver cytosol factors
[532,920] characterize the hepatic cholesterol 7α-hydroxy-
lase. A reconstituted system consisting of partially purified rat liver cytochrome P-450 and cytochrome P-450 reductase, phosphatidylcholine, and NADPH has been found active
[262,957].

Still further characterization of the 7α-hydroxylase
includes absence of an isotope effect [251] and examination
of kinetics, where K_m for oxygen of ca. 20 μM [342], a K_m
for cholesterol of 100 μM [868,2552], and an activation
energy of 22 kcal/mole [342] have been reported. Moreover,
numerous sophisticated means of assay of liver microsomal
cholesterol 7α-hydroxylase activity have recently been
innovated [254,870,1195,1560,1598,1759,2215,2551,2552].
Assay of microsomal cholesterol 7α-hydroxylase activity is
a complicated matter subject to differential hydroxylation
rates for endogenous and exogenous substrate [127,256] and
displaying a circadian rhythm in rats [547,869-871,1641,
2547,2550] apparently under regulation by the hypophysis-
adrenal cortex axis [1599,2137,2549] among other factors.

Although these many studies clearly establish the presence of a cholesterol 7α-hydroxylase in liver no such evidence has ever been adduced in support of a related 7β-
hydroxylase. A cholesterol 7β-hydroxylase in rat liver
homogenates has been posited [1616], but the 3β,7β-diol 15
formed more probably by lipid peroxidation. Indeed, enzymic 7β-hydroxylation of cholesterol has yet to be demonstrated as a specific event. The possible biosynthesis

origins of the 3β,7β-diol 15 and the 7-ketone 16 may be considered together to advantage, as both appear not to be
genuine endogenous metabolites of cholesterol but to be
nonphysiological products. At least evidence in support
in vivo biosynthesis is not adequate.

The 7-hydroperoxides 46 and 47 are obvious potential
precursors of the 3β,7β-diol 15 and the 7-ketone 16. Thermal decomposition of the 7β-hydroperoxide 47 yields both
15 and 16, and the enzymic reduction of 47 by rat liver
microsomal cytochrome P-450 has been demonstrated [1088].
The 7-ketone 16 (and the 3β,7α-diol 14) also derive from
the 7α-hydroperoxide 46 in incubations with rat liver microsomes [1655]. Whether the indicated dehydration is enzyme
catalyzed or not cannot be answered, but hydroperoxide dehy-
dratases (transforming organic hydroperoxides to ketones)
have not been recognized as a class of enzymes.

The 3β,7β-diol 15 might also arise by epimerization of
the 3β,7α-diol 14. The epimerization may be enzyme catalyzed or not, as the 3β,7α-diol 14 is epimerized nonenzymically
in both aqueous and organic solvent solutions [1404,2051,
2354,2458,2461], and the enzymic interconversion of the
related epimeric pair 3β,7α-dihydroxyandrost-5-en-17-one
and 3β,7β-dihydroxyandrost-5-en-17-one by rat liver homogenates has been observed [952,2341].

The 3β,7β-diol 15 and 7-ketone 16 are linked by other
means also. It has been repeatedly shown that liver microsomal enzymes do not dehydrogenate the 3β,7α-diol 14 to 16
nor do they reduce 16 to 14. Rather, liver microsomal dehydrogenases transform the 3β,7β-diol 15 to the 7-ketone 16
using $NADP^+$ as hydrogen acceptor and reduce the 7-ketone 16
to the 3β,7β-diol 15 in the reverse action [265,1653,1655,
2548,2552]. The combination of 3β,7α-diol 14 epimerization
to 3β,7β-diol 15 and 3β,7β-diol dehydrogenation to 7-ketone
16 may well be an alternative biosynthesis pathway for
three sterols

Rat liver microsomal monooxygenases 7α- and 7β-hydroxy-
late other sterols too. Cholesterol fatty acyl esters are
not substrates for 7α-hydroxylation [128,1255,1598], but
5α-cholestan-3β-ol (2) is 7α-hydroxylated [91,266,2212].
Moreover, a reconstituted system involving cytochrome P-450
and a NADPH-cytochrome P-450 reductase 7α-hydroxylates
cholesterol and the 5α-stanol 2 [957].

However, the 5α-stanol 2 is also 7β-hydroxylated by
rat liver microsomal preparations [91]. As the 5α-stanol
has no allylic positions sensitive to autoxidation and
should thereby not be a substrate for lipid peroxidation,
the demonstrated 7β-hydroxylation of 2 supports the
presence of a sterol 7β-hydroxylase in rat liver microsomes,
one which may also oxidize cholesterol. Other C_{27}-sterols
and stanols are also 7α- and/or 7β-hydroxylated by microsomal monooxygenases are also several $24\alpha_F$-methyl and $24\alpha_F$-
ethyl homologs [91]. Short side-chains sterols pregn-5-en-
3β-ol (110) and 23,24-bisnorchol-5-en-3β-ol (194) are likewise oxidized by liver microsomal monooxygenases [98,340].

Other 7-Hydroxylases. Enzymic 7-hydroxylations of
cholesterol and of other Δ^5-3β-alcohols by monooxygenases
from a variety of sources, including systems of intact
cells and tissue subcellular fractions, have been demonstrated. Aerated incubations of cholesterol with microbial
vegetative cell cultures pose again the difficult problem
of whether enzymic 7-hydroxylation or autoxidation processes
account for products, as inadequate control studies characterize these reports. The early observation that Proactinomyces roseus oxidize cholesterol to the 4-ene-3-ke-
tone 8 and to a 7ξ-hydroxy-derivative 14 and/or 15 over 14
days (!) of aeration at 34° [1384] is highly suspicious.
However, autoxidation has not generally been encountered
in microbial transformations of cholesterol [1154,1577].

The enone 8 has been recovered as a metabolite in
incubations of cholesterol with fecal mycobacteria [1527]
and is an established early intermediate in the degradation
of cholesterol by soil microorganisms [393,569,1153,1155,
1467,1854,2526,2641,2767]. However, 7α- and 7β-hydroxyla-
tion of cholesterol is not a recognized microbial degradation process of sterols [2289], and the 3β,7ξ-diol 14
and/or 15 formed by P. roseus may be considered an autoxidation product despite reported control (no organism) studies [1385]. The autoxidation of cholesterol in other
studies using Fusarium diversisporum Sherb [2380] and a
soil Penicillium sp. has been reported, and autoxidation
during heat sterilization of media containing cholesterol
is clearly likely [1855].

A related example obtains for tissue culture transformations of cholesterol conducted in the dark with callus
tissue of the plant *Evonymus europaea*. Products identified
included the epimeric 3β,7-diols 14 and 15 and the 7-ketone
16. In this case control experiments utilized heat inacti-
vated cells, and no products, 14,15, or 16 were formed.
Callus tissue from *Digitalis purpurea* may have yielded some
7-oxygenated products but at ca. 0.2% yield only [677].

More complicated cases of cholesterol oxidation products have been reported, and fungus *Coriolus hirstus* oxidizing cholesterol to the epimeric 3β,7-diols 14 and 15
and 7-ketone 16 but also to the epimeric pair 6α-hydroxy-
cholest-4-en-3-one (281) and 6β-hydroxycholest-4-en-3-one
(88). Moreover, fucosterol (75) was oxidized by the same
fungus to the homologous product 24-ethylcholesta-5,E-
24(28)-diene-3β,7α-diol, 24-ethylcholesta-5,E-24(28)-diene-
3β,7β-diol, 3β-hydroxy-24-ethylcholesta-5,E-24(28)-dien-7-
one, 24-ethyl-6α-hydroxycholesta-4,E-24(28)-dien-3-one and
24-ethyl-6β-hydroxycholesta-4,E-24(28)-dien-3-one [2624].

Other odd soil microorganism transformations of cholesterol resembling autoxidations include formation of
cholest-4-ene-3,6-dione (108) by *Mycobacterium* sp. [2334]
and 4-hydroxycholest-4-en-3-one (398) by *Streptomyces*
14PH8 [383], the enone 8 being also formed in both instances. Both 8 and 108 have demonstrated autoxidation origins,
but 4-hydroxycholest-4-en-3-one has not been implicated in
cholesterol autoxidation but is a product of the autoxidation of 5β-cholestan-3-one in alkaline media [215].

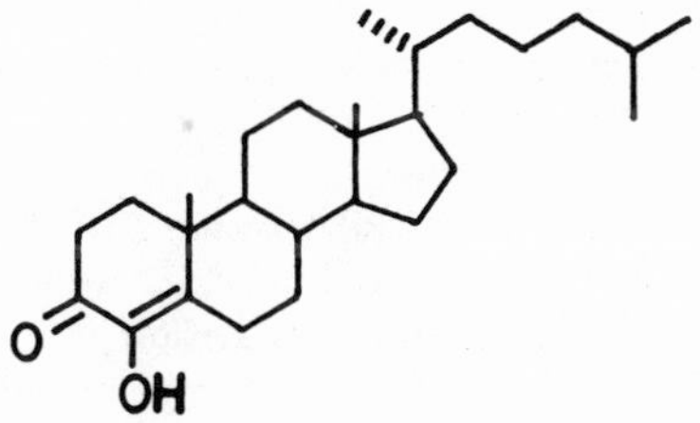

398

Formation of 7-oxygenated derivatives of other Δ^5-3β-
hydroxysteroids by microbial action has also been observed,
such oxidations having the appearance of autoxidations.
However, in most cases either control observations, accompanying 7-oxidation of other non-Δ^5-3β-hydroxysteroids, or
additional metabolic transformations suggest enzymic processes.   Thus, the 7β-hydroxylation of diosgenin by *Cunninghamella blakesleeana* is accompanied by additional 11α- and
12β-hydroxyations to provide products (25R)-spirost-5-ene-
3β,7β-diol, (25R)-spirost-5-ene-3β,7β,11α-triol, and (25R)-
spirost-5-ene-3β,7β,12β-triol [1240] whereas *Helicostylium
piriforme* ATCC 8992 yields (25R)-spirost-5-ene-3β,7β,11α-
triol and 3β,11α-dihydroxy-(25R)-spirost-5-en-7-one [984,
985,2082].   Moreover, *H. piriforme* acting on the nitrogenous analog (22R,25R)-tomat-5-enin-3β-ol (solasodine) yields
inter alia a 7β-hydroxylated product (22R,22R)-tomat-5-
enine-3β,7β-diol [2080-2082].

The 7-hydroxylation of a variety of C_{19}-C_{23}-Δ^5-3β-
alcohols by microbial agents has also been reported.   Hydroxylation of 3β-hydroxycarda-5,20(22)-dienolide by *Mucor
griseo-cyanus* ATCC 1207 yielded the 7α-alcohol 3β,7α-dihy-
droxycarda-5,20(22)-dienolide [440], whereas oxidation of
connessine (3β-dimethylaminocon-5-enine) by *Aspergillus
ochraceus* [1407], *Cunninghamella echinulata* [1824], and
Gloeosporium fructigenum [1580] afforded both epimeric
7-alcohols 3β-dimethylaminocon-5-enin-7α-ol and 3β-dimethyl-
aminocon-5-enin-3β-ol.

Microbial oxidations of pregnenolone (23) also yield
7-hydroxylated derivatives, *Rhizopus nigricans* yielding
both 3β,7α,11α-trihydroxypregn-5-en-20-one and 3β,11α-
dihydroxypregn-5-ene-7,20-dione.   In this case transformation of the product 7α-alcohol to product 7-ketone on storage in air or on drying at 100°C was noted [1360].   Oxidation of 23 by *Circinella muscae* gave 3β,7α,9α-trihydroxy-
pregn-5-en-20-one [901].

The C_{19}-substrates, 3β-hydroxyandrost-5-en-17-one (86)
or its 3β-acetate have also been oxidized in the C-7 position to yield various products bearing 7α- and 7β-hydroxyl
and 7-ketone groups.   Such oxidations by *Absidia orchidis*
[2167], *C. blakesleena* QM631 [2118], *Cunninghamella elegans*
[530] *Fusidium* sp M61-1 [625], *Gibberella saubinetti* [1795],
Rhizopus arrhizus ATCC 11145 and *R. nigricans* ATCC 6277b
[2118], and *Rhizopus* sp M2045 [625] have been reported, as

have also oxidations by fruit body slices or cultivated
mycelium of several *Basidiomycetes* [1892]. Additionally,
the 17a-azasteroid 3β-acetoxy-17a-aza-D-homoandrost-5-en-
17-one (399) was oxidized by *C. elegans* to 3β,7β-dihydroxy-
17a-aza-D-homoandrost-5-en-17-one (400), and 3β-hydroxy-17a-
aza-D-homoandrost-5-ene-7,17-dione (401) [530].

   In the oxidations of the C_{19}-substrates by *C. elegans*
not only were both epimeric 7-alcohols and 7-ketones formed
but 5β,6β-epoxides were also isolated. Thus, the 3β-acetate
of 86 gave 3β,7α-dihydroxyandrost-5-en-17-one, 3β,7β-dihy-
droxyandrost-5-en-17-one, 3β-hydroxyandrost-5-ene-7,17-
dione, *inter alia* but also 5,6β-epoxy-3β,12α-dihydroxy-5β-
androstan-17-one, and the 17a-azasteroid 399 gave 5,6β-
epoxy-3β-hydroxy-5β-17a-aza-D-homoandrostan-17-one (402)
[530]. Although these transformations are particularly
suspicious as regards possible autoxidation, control experiments indicated dependence upon viable enzymes.

   Turning to mammalian metabolism, a flourishing enzymic
7-hydroxylation of the C_{19}- and C_{21}-Δ^5-3β-alcohols 86 and
23 is evinced by discoveries of 7-oxygenated metabolites
in urine and plasma and by *in vitro* studies as well. Human
urine contains sulfate esters of the 7-oxygenated derivatives 3β,7α-dihydroxyandrost-5-en-17-one and 3β,7β-dihydroxy-
androst-5-en-17-one [1181,2118,2344], and of 3β,7α,16α-tri-
hydroxyandrost-5-en-17-one [1794] as well as of the 7-ke-
tones 3β-hydroxyandrost-5-ene-7,17-dione [814,837,1471] and
3β,16α-dihydroxyandrost-5-ene-7,17-dione [1794]. Whereas
artifact status for the 7,17-diketones was clearly considered [814], metabolite status seems assured [838]. However,
such is not the case for urinary androsta-3,5-diene-7,17-
dione, which is formed by acid hydrolysis of the 3β-hydroxy-
androst-5-ene-7,17-dione 3β-sulfate [2155]. Moreover, sulfate esters of 3β,7α-dihydroxyandrost-5-en-17-one and of
3β-hydroxyandrost-5-ene-7,17-dione have been found in human
plasma [161,162,1541,2268,2340] and in an adrenal viriliz-
ing carcinoma [2056]. 3β-Hydroxyandrost-5-ene-7,17-dione
has been found in venous blood of rhesus monkeys infected
with a hemorrhagic fever [889].

   Incubations of the C_{19}-steroid 86 *in vitro* with a
variety of mammalian tissue preparations variously yield
the epimeric 7-alcohol and 7-ketone products. Tissues
examined include human liver microsomes [264,2337,2391],
testis and epididymis [2395], adrenal [1745,2337,2391],

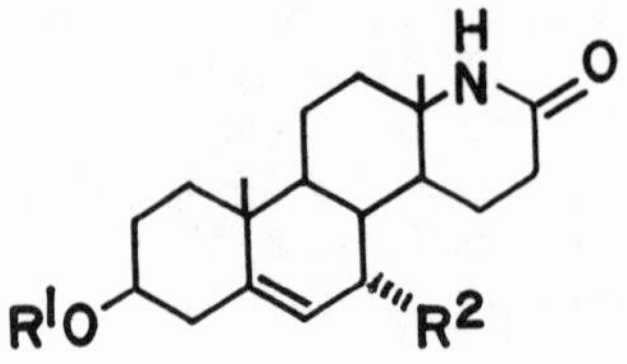

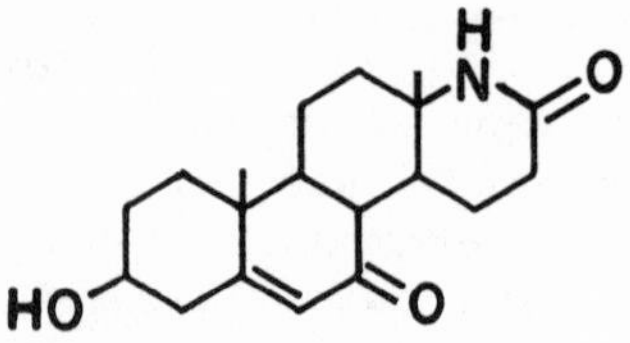

399 R^1 = CH_3CO, R^2 = H   401

400 R^1 = H, R^2 = OH

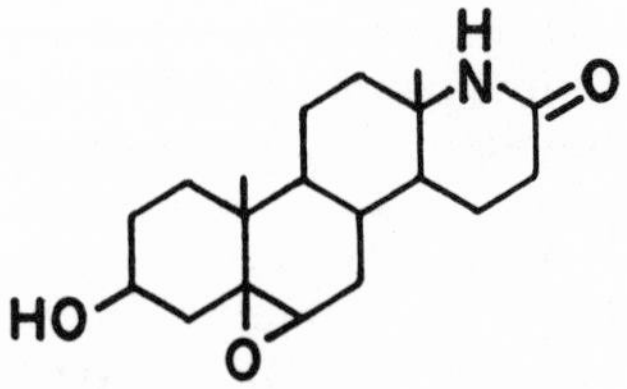

402

skin [716,1209], amnion epithelium [2392], chorion [2391], lymphocytes [2731], placenta [1636], and mammary tissue and tumors [525,1468], rat liver homogenate [2342] and micro-somes [1192,2339,2393,2394,2412,2413], testis microsomes [1156], rabbit liver microsomes [1001], calf lens [2593], horse liver microsomes [2338], and pig liver homogenates [760]. In the case of rat and rabbit liver studies, addi-tional enzymic 16α-hydroxylation also occurs [1001,1192,2413].

A related case obtains with the C_{21}-steroid 23. Both 3β,7α-dihydroxypregn-5-en-20-one and 3β,7β-dihydroxypregn-5-en-20-one are products of rat liver microsomal oxidation of 23 [551,1192,2343], but this case differs from that of the C_{19}-analog 86 in that epimeric 2-hydroxylated products 2α,3β-dihydroxypregn-5-en-20-one and 2β,3β-dihydroxypregn-5-en-20-one are also formed [551].

Although a formal similarity obtains in the 7α-hydroxy-lation of cholesterol in mammalian liver microsomal systems and the 7α-hydroxylation of the C_{19}- and C_{21}- substrates 86 and 23 it is evident that different processes are in fact involved. The rat liver microsomal hydroxylations of substrates 23,86, and androst-5-en-3β-ol (114) exhibit Michaelis-Menten kinetics from which the K_m may be derived [376]. Moreover, these hydroxylations are inhibited by CO but not by 2-mercaptoethyl amine [376,1192], thus clearly differentiating the two processes.

Cholesterol 20-Hydroxylase. As for the 3β,7α-diol 14 both a genuine cholesterol metabolite implicated in import-ant metabolic processes and an autoxidation product, so also (20S)-cholest-5-ene-3β,20-diol (21) implicated in steroid hormone biosynthesis has dual origins of autoxida-tion and of monooxygenase action on cholesterol. Also in exact analogy to the case of C-7 oxidations, the chole-sterol 20-hydroxylase implicated transforms cholesterol stereospecifically to the (20S)-3β,20-diol 21, where autoxidative attack is not specific but yields products of both C-20 configurations, thus both (20R)- and (20S)-20-hydroperoxides 32 and 22 respectively. From the (20S)-20-hydroperoxide 22 is then derived the (20S)-3β,20-diol 21. Furthermore, as for 3β,7α-diol 14 formation in in vitro experiments, the relative contributions of autoxidation versus enzymic 20S-hydroxylation has not been properly assessed, although C-20 autoxidations are much less in evi-dence than are C-7 autoxidations and probably do not com-promise interpretations of results.

It has long been posited that the (20S)-3β,20-diol 21 be the initial oxidation product of cholesterol in the scission of the isohexyl moiety of the side-chain in the biosynthesis of the C_{21}-20-ketone 23 by adrenal cortex mitochondria. This formulation is supported by isolation of the (20S)-3β,20-diol 21 from adrenal tissue [1963], trap-ping of radioactivity from labelled cholesterol in the diol [1112,1113,2323] and direct observation of formation of the (20S)-3β,20-diol 21 from cholesterol in incubations utilizing partially purified adrenal cortex mitochondrial cytochrome P-450 [1680]. Nevertheless, it is now clear that the predominant initial oxidative transformation of cholesterol by adrenal cortex mitochondria is hydroxylation at the C-22 position, yielding (22R)-cholest-5-ene-3β,22-diol (24), which is not a known autoxidation product of

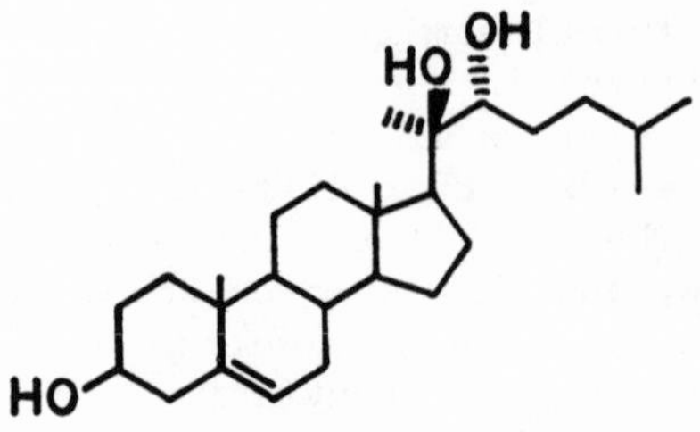

<u>403</u>

cholesterol. Both the (20S)-3β,20-diol <u>21</u> and the (22R)-
3β,22-diol <u>24</u> are then subject to subsequent hydroxylations
to yield the common product (20R,22R)-cholest-5-ene-3β,20,
22-triol (<u>403</u>) from which the 20-ketone <u>23</u> is then derived.
Kinetics [407-411,413,416] and isotope ($^{18}O_2$) incorporation
[414,415] data establish these features and suggest dual
routes to the 3β,20,22-triol <u>403</u> from cholesterol in which
both diols <u>21</u> and <u>24</u> participate [642,643].

Although sequential hydroxylation to the 3β,20,22-
triol <u>403</u> is well supported experimentally, several alternative processes have variously been proposed. Chief among
these are proposals for formal rearrangement of the (20S)-
20-hydroperoxide <u>22</u> by adrenal cortex mitochondrial cytochrome P-450 preparations to the 3β,20,22-triol <u>403</u> [2565,
2566,2573,2574], for the epoxidation of a cholesta-5,20(22)-
dien-3β-ol to cholesterol 20,22-epoxides whose hydration
yield the 3β,20,22-triol <u>403</u> [1378,1380,1381], and for unisolated sterol-enzyme complexes [1054,1056,1470,1543].
However, as discussed later in this chapter, the issue of
side-chain scission via the 20-hydroperoxide <u>22</u> as an
in vivo process is discounted, and experimental work testing the 20,22-epoxide possibility fails to support the
formulation [405,1675,2542,2632]. Moreover, as the monohydroxylated cholesterol derivatives <u>21</u> and <u>24</u> do indeed
form as intermediates, speculations of unisolable enzyme
complexes are wholly unsupported.

All three phases of side-chain cleavage (initial hydroxylase, subsequent hydroxylase, and C-20/C-22 lyase
activities) are dependent upon O_2 and an electron transport

system involving NADPH and utilize cytochrome P-450 sensitive to CO as terminal oxidase, in exact analogy with liver microsomal cholesterol 7α-hydroxylase just described. The mixed function oxidase character of the cholesterol side-chain cleavage enzyme system located in the inner membrane [2736] of adrenal cortex mitochondria was recognized early [2255,22561], with NADPH-dependent adrenodoxin reductase flavoprotein, iron-sulfur protein (adrenodoxin), and cytochrome P-450 components [391,1246,1324,2257]. The side-chain cleavage activity resolved from associated steroid 11β-hydroxylase activity [341,1161,1184,1186,1901,2224, 2225,2390,2426,2638] appears to be an aggregate of 46,000-53,000 dalton subunits [2225,2427,2485,2628] and reconstituted systems active in side-chain cleavage have been described [2430,2484,2628]. Although most progress towards purified side-chain cleavage systems has been made with bovine adrenal cortex mitochondria, human placenta [2259] and rat and bovine corpora lutea [339,445] micochondria also yield active purified enzymes.

Although all of these purified systems cleave the cholesterol side-chain, none has been resolved into individual monooxygenase components. However, a purified 46,000 dalton cytochrome P-450 transforming cholesterol to the (20R,22R)-3β,20,22-triol 403 has been reported [32].

Even though the (20S)-3β,20-diol 21 is clearly an enzymic product and autoxidation of cholesterol in these investigations has generally not been a problem, autoxidation may nonetheless occur in *in vitro* studies of the side-chain scission, as the characteristic products 14-16,35 and 36 have been observed in amounts equalling those of the scission product 23 [327].

Cholesterol 24-Hydroxylase. Whereas both (20S)-3β,20-diol 21 and (22R)-3β,22-diol 24 have enzymic origins, much less work has been done on other cholesterol hydroxylases which might monohydroxylate cholesterol in other sites in the side-chain. The occurrence of several bile acid derivatives and the detection of a cholest-5-ene-3β,23-diol sulfate in human meconium [680] suggest that a cholesterol 23-hydroxylase may exist, but no studies of the matter have been attempted. Like circumstantial evidence for a cholesterol 24-hydroxylase yielding the (24S)-3β,24-diol 25 need not now be adduced, as direct evidence of a cholesterol (24S)-24-hydroxylase in murine and bovine brain microsomes

has been recorded. The enzyme yields the (24S)-3β,24-diol
<u>25</u> as sole product, requires O_2, and is stimulated by NADPH,
although a dependence of NADPH was not demonstrated [607,
1511]. Furthermore, a rat liver mitochondrial hydroxylation of cholesterol yielding a (24ξ)-cholest-5-ene-3β,24-
diol (either or both epimers) has been described [94].

In any event, the (24S)-3β,24-diol <u>25</u> is not a recognized cholesterol autoxidation product but is a true companion sterol with an enzymic origin. As the 3β,24-diol <u>25</u> is
recovered along with autoxidation products from air-aged
cholesterol, it is important to retain the distinction.

<u>Cholesterol 25-Hydroxylase</u>. The matter of the bio-
synthesis of the 3β,25-diol <u>27</u> from cholesterol via the
action of cholesterol 25-hydroxylase is complicated by low
or uncertain yields and the everpresent question of the
contribution of autoxidation to the conversion. The formation of the 3β,25-diol <u>27</u> together with a 3β,26-diol <u>29</u>
and/or <u>31</u> in mouse [797,798] and rat [94,268,927,1247,1248,
1390,1639,1837] liver mitochondria incubations has been
repeatedly demonstrated, but failure to detect the 3β,25-
diol <u>27</u> in one case [545] and results with rat liver microsomes in which autoxidation may have occurred [1615] left
the issue unsettled. More recently, there have been made
assertions that rat liver mitochondria do have a cholesterol 25-hydroxylase [94,268,271,927,928], the evidence
for which being that boiled mitochondria controls did not
give the transformations, and in experiments where cholesterol autoxidation increased considerably, levels of the
3β,25-diol <u>27</u> did not [94,268].

Moreover, a convincing demonstration of rat liver mitochondrial cholesterol 25-hydroxylase activity has also been
reported. A reconstituted system incorporating a partially
purified solubilized mitochondrial cytochrome P-450, ferredoxin, NADPH-ferredoxin reductase, and NADPH transforms
cholesterol to the 3β,25-diol <u>27</u> and to a 3β,26-diol <u>29</u>
and/or <u>31</u> in product ratio of 1:9, with neither product
formed in the absence of ferredoxin. It is uncertain whether separate cholesterol 25- or 26-hydroxylase systems or
but one hydroxylase of diminished substrate specificity are
implicated [1835,1837]. It is important to concede that a
liver mitochondrial cholesterol 25-hydroxylase does appear
to function, at least in *in vitro* incubations, and
this monooxygenase action along with autoxidation may

account for the presence of the 3β,25-diol 27 in mammalian tissue.

   Cholesterol 26-Hydroxylases. The biosynthesis of a 3β,26 diol 29 and/or 31 from cholesterol by mouse [207,545, 797,798], rat [94,260,267,268,927,928,1247,1248,1390,1639, 1834,1836,1837], and human [269] liver mitochondria is well established. Moreover, it is now apparent that the 3β,26-diol formed is the result of a stereospecific hydroxylation of the (25-pro-S)-methyl group of cholesterol yielding the (25R)-3β,26-diol 29. Furthermore, the 3β,26-diol accumulated in the human aorta is now recognized as being predominantly the (25R)-3β,26-diol 29 as well [1914].

   As adequate means of distinguishing between the isomeric 3β,26-diols 29 and 31 were not available for most of these studies, assignment of the C-25 stereochemistry had to be made indirectly. Direct chromatographic analysis of mixtures of the isomeric 3β,26-diols 29 and 31 may now be made (using their 3β,26-diacetates) [1915].

   The means by which enzyme stereospecificity in hydroxylations of the terminal methyl groups is demonstrated is based on the recognition that the terminal groups of the sterol side-chain are magnetically and enzymatically non-equivalent and that each is derived in biosynthesis from a different carbon atom of precursor mevalonate. Thus, the (25-pro-R)-methyl group is derived from the C-2 carbon atom of mevalonate, the (25-pro-S)-methyl from the C-3' carbon [846,1876]. Hydroxylation of the (25-pro-S)-methyl group yields the (25R)-3β,26-diol 29; hydroxylation of the (25-pro-R)-methyl group yields the isomeric (25S)-3β,26-diol 31. It has been suggested that the (25-pro-R)-methyl group of cholesterol (derived from the mevalonate C-2 carbon) be henceforth termed the C-26 carbon atom, thereby making the (25S)-3β,26-diol 31 the one 3β,26-diol cholest-5-ene-3β,26-diol (31). The (25-pro-S)-methyl group (derived from the mevalonate C-3' carbon atom) would then be the C-27 carbon of cholesterol and the (25R)-3β,26-diol 29 cholest-5-ene-3β,27-diol (29) [930,1206,1876]. This suggestion is not used in this monograph.

   This method of selective isotope labeling of the (25-pro-R)-carbon atom via biosyntheses using substrate [2-^{14}C] mevalonate has been used to establish that sapogenin biosyntheses involved stereospecific hydroxylations. As the

C-25 stereochemistry of the product sapogenins is fully
known, degradations to locate the isotope establish directly
which terminal methyl group has been oxidized in biosynthesis. Thus, the mevalonate C-2 carbon atom is retained as
the (25-*pro*-R) 27-methyl group of (22R,25R)-tomat-5-enin-
3β-ol (solasodine) derived from *Solanum aviculare* [926] and
5α-(25R)-spirostan-3β-ol (tigogenin) from *Digitalis lanata*
EHRH [1200], as the C-27 methylene group of 5β-spirost-25(27)-
ene-1β,3β-diol (convallamarogenin) from *Convallaria majalis*
[2003], and as the (25-*pro*-R) 26-methyl group of 5β-(25S)-
spirostan-3β-ol (sarsapogenin) from *Agave attenuata* Solm.
[891], and 5α-(22S,25S)-tomatanin-3β-ol (tomatidine) and
5α-(25S)-spirostan-3β-ol (neotigogenin) from *Lycopersicon
pimpinellifolium* [2004,2005].

    The same approach has been used for cholesterol, where
the stereochemistry of oxidized products is not so easily
determined. The (25-*pro*-R)-methyl group of labeled cholesterol derived from [2-^{14}C]mevalonate is not the one hydroxylated by mouse liver mitochondria [207] nor is the
(25R-*pro*-R)-methyl group oxidized by rat liver mitochondria
to propionate [1643]. By inference, murine liver mitochondria oxidize the (25-*pro*-S)-methyl group to form the (25R)-
3β,26-diol 29.

    On the other hand, oxidation of labeled cholesterol by
Mycobacterium smegmatis SG346 is at the (25-*pro*-R)-methyl
group, yielding (25S)-26-hydroxycholest-4-en-3-one [446,
846,2768]. Given these results and the many known (25R)-
and (25S)-sapogenins, it is obvious that oxidative attack
at either (25-*pro*-R)- or (25-*pro*-S)-methyl group may occur.
Moreover, evidence for the presence of both (25R)- and (25S)-
3β,26-diols 29 and 31 respectively in biological samples
has been recorded. Thus, 3% of (25S)-3β,26-diol 31 has
been found with the major component (25R)-3β,26-diol 29
recovered from human aortal tissues, 10% of (25S)-3β,26-diol
29 derived from 3β,26-dihydroxy-(25R)-cholest-5-ene-16,22-
dione (kryptogenin) [1914]. Given correct component identities (and no C-20 isomerizations, for instance, during
transformation of kryptogenin to the (25R)-3β,26-diol 29),
the (25S)-3β,26-diol 31 in material derived from kryptogenin could have been derived from isomeric 3β,26-dihydroxy-
(25S)-cholest-5-ene-16,22-dione (barogenin) recently discovered in *Solanum tuberosum* [1241] but possibly also present but unrecognized as such in *Dioscorea* species from
which kryptogenin is recovered.

The presence of 3% of (25S)-3β,26-diol 31 with 97%
(25R)-3β,26-diol 29 isolated from human aortal tissue suggests a dual origin for the 3β,26-diols 29 and 31. Although weak evidence for the biosynthesis of a 3β,26-diol
29 and/or 31 in human aortal segments has been reported
[1563], our own unpublished investigation of such matters
does not support aortal biosynthesis. In fact, although
material corresponding to the 3β,26-diol 29 isolated from
incubations of human aorta homogenates with [1,2-³H]chole-
sterol was selectivley radioactive, further purification
reduced isotope levels to those of controls. The matter
of human aortal biosynthesis of 3β,26-diols 29 and 31 has
not been settled, but it may be that the sterols are accumulated from plasma (where a 3β,26-diol has been found as
a sulfate ester [2396], with ultimate liver biosynthesis
implicated. However, although human liver mitochondria
transform cholesterol to the (25R)-3β,26-diol 29 [269],
no enzymic origin for the minor component (25S)-3β,26-diol
31 has been demonstrated. Thus, liver mitochondria apparently do not give the (25S)-3β,26-diol 31, and liver microsomes do not transform cholesterol to either isomeric 3β,26-
diol 29 or 31 [94,267]. However, liver microsomal sterol
26-hydroxylases have been described. Both C-25 isomers
5β-(25R)-cholestane-3α,7α,26-triol and 5β-(25S)-cholestane-
3α,7α,26-triol are formed from 5β-cholestane-3α,7α-diol by
microsomal enzymes of human, rat, guinea pig, and rabbit
liver [2207,2208], and rat liver acting on 5β-cholestane-
3α,7α,12α-triol gives 5β-(25S)-cholestane-3α,7α,12α,26-
tetraol [267,930]. It is tempting to speculate that small,
undetected amounts of (25S)-3β,26-diol 31 be formed by liver
microsomal sterol hydroxylases and that these be accumulated in human aortal tissues along with the (25R)-3β,26-
diol 29.

It may be that the liver mitochondrial ω-hydroxylase
system which nominally yields the (25R)-3β,26-diol 29 from
cholesterol also yield the 3β,25-diol 27 and a (24ξ)-cholest-
5-ene-3β,24-diol as consequence of diminished product specificity. Furthermore, the same mitochondrial ω-hydroxylase oxidizes campesterol to a 24-methyl-(24R,25ξ)-cholest-
5-ene-3β,26-diol, sitosterol to a (25ξ)-stigmast-5-ene-
3β,26-diol and to stigmast-5-ene-3β,29-diol *inter alia*
[94,2385].

Besides the biosynthesis of the (25R)-3β,26-diol 29 from cholesterol in mammalian liver, the 26-hydroxylation of cholesterol by enzymes of etiolated potato sprouts has also been demonstrated, but the C-25 configuration of the product has not been investigated [994].

In summary, evidence supports the enzymic hydroxylation of cholesterol on the 20-,22-,24-,25-, and 26-carbon atoms to give the (20S)-3β,20-diol 21, the (22R)-3β, 22-diol 24, the (24S)-3β,24-diol 25, the 3β,25-diol 27, the (25R)-3β,26-diol 29, and the (25S)-3β,26-diol 31. Furthermore, sterols of unassigned stereochemistry, including a (22ξ)-cholest-5-ene-3β,22-diol, (23ξ)-cholest-5-ene-3β,23-diol, and (24ξ)-cholest-5-ene-3β,24-diol, have been found in various biological systems such that yet other specific hydroxylases may exist.

Dehydrogenases and Oxidases

The biosynthesis of the enones 6,8,16,59, dienones 10, 12, and 6-ketones 42,44,45,108 imply the actions of hydroxysteroid dehydrogenases or oxidases. Hydroxysteroid dehydrogenases are implicated in a variety of other metabolic processes involving steroids, including the biosynthesis of stanols and bile acids from cholesterol, of cholesterol from lanosterol, and of steroid hormones from the C_{21}-3β-alcohol 23, and the biosynthesis in man of the enone 8 from cholesterol, presumably via enone 6, is indicated by indirect evidence [2011]. Moreover, the oxidation of cholesterol to enone 8 has been directly demonstrated using rat liver microsomal enzymes supplemented with NAD$^+$ [273,2751], thus establishing dehydrogenase (3β-hydroxy-Δ^5-steroid:NAD$^+$ oxidoreductase, EC 1.1.1.45) activity as a genuine means of derivation of 8 from cholesterol.

In distinction to this action of mammalian hydroxysteroid dehydrogenase on cholesterol, the action of microbial cholesterol oxidases (cholesterol:oxygen oxidoreductase, EC 1.1.3.6) have been actively studied, as these enzymes have found use in routine clinical serum cholesterol analyses. In an oxygen-dependent removal of hydrogen from cholesterol the enone 6 is formed and isomerized directly to enone 8. Hydrogen peroxide is evolved, and the 4β-hydrogen of cholesterol appears as the 6β-hydrogen of enone 8 [2276-2278]. Purified cholesterol oxidases have

been prepared from *Nocardia* sp. NCIB 10554 (*N. rhodocrous*)
[394,1951], *Nocardia erythropolis* [777,2273,2274,2369],
Streptomyces violascens [813,1214,2495], *Streptomyces
griseocarneus* [1287], *Brevibacterium sterolicum* ATCC 21387
[2541,2542], and *Scizophyllum commune* [815,1286,1773]. The
oxidation of cholesterol to enone 8 by other microbial systems has also been repeatedly observed. Cultures of higher
plants also oxidize cholesterol to the enone 8 [2367].

A biosynthesis process involving hydroxysteroid dehydrogenase action has also been demonstrated for the
7-ketone 16. As previously noted the 3β,7β-diol 15 is
transformed to the 7-ketone 16 in incubations of rat liver
microsomal preparations supplemented with NAD$^+$ and the reverse transformation occurs with supplementation by NADH$^+$.
Neither appropriate alcohol substrate nor dehydrogenase or
oxidase enzymes have been proposed as biosynthesis means
for the dienone 10, but the isomeric dienone 12 is product
of the action of mouse liver microsomal hydroxylated dehydrogenase action on cholest-4,6-dien-3β-ol [1226]. However,
as cholesta-4,6-dien-3β-ol cannot be regarded as a naturally
occurring sterol, this metabolic process does not establish
12 as a natural product. Likewise, the several 6-ketones
42,44,45, and 108 do not have acceptable enzyme and substrate systems for their biosynthesis. The presence of the
6-ketone 44 in rat liver microsomes incubations in company
with 13-16,35, and 36 [93] is best rationalized in terms of
autoxidative dehydrogenation of the 3β,5α,6β-triol 13, a
transformation known to be facile.

Several other ketonic derivatives of cholesterol, including the 24-ketone 34 found in marine sources and 3β-
hydroxycholest-5-en-22-one (297) possibly detected in human
urine [654] (if they be genuine metabolites) are reasonably
formulated as products of alcohol dehydrogenations of precursor alcohols, but no examination of such possibilities
has been made.

Sterol 5,6-Epoxidases and Epoxide Hydratases

One of the more preplexing aspects of biosynthesis of
common cholesterol autoxidation products is whether the
isomeric 5,6-epoxides 35 and 36 have enzymic origins and
whether the 3β,5α,6β-triol 13, the common hydration
product of 35 and 36 , be a metabolite. The distribution

of 5,6-epoxides 35 and 36 in biological material (TABLE 2)
and the ready epoxidation of cholesterol by sterol hydroperoxides, organic hydroperoxides, H_2O_2, and HO· requires
that sound evidence be advanced before positing enzymic
epoxidation, and careful evaluation of possible enzymic
epoxidations of steroids of several kinds in a variety of
biological systems [1210] leaves the matter unsettled.

The 5,6-epoxides 35 and 36 have been found as products in a variety of *in vitro* incubations, including soybean lipoxygenase [93, 1192], rat liver microsomal preparations [92-94,1640,1655], and bovine adrenal cortex mitochondrial preparations [327]. In view of the presence of
accompanying oxidation products 13-16,46,47, in these experiments no inference of enzymic epoxidation may be made.
Indeed, in systems where the 7-hydroperoxides 46 and 47
and possibly other epoxidizing agents are formed, the formation of both 5,6-epoxides 35 and 36 from cholesterol may
be viewed as a nonenzymic or artificial process. Moreover,
the formation of a cholesterol epoxide 35 and/or 36 in human
and rat skin irradiated with ultraviolet light [276,277,
281,1523] is also not an enzymic process, although this
radiation-induced oxidation of skin cholesterol may be a
means by which the 5,6-epoxides 35 and 36 derive.

Inspiration by rats of air containing 3-6.5 ppm NO_2
leads to a dramatic generation of the isomeric 5,6-epoxides
35 and 36 in lung tissue, the 5α,6α-epoxide 35 predominating [2191-2193]. Whereas the matter cannot be viewed as
enzymic, low levels of 5,6-epoxides in controls suggest
that these sterols be normal components of lung, whether
formed enzymically or not.

There are a few cases of cholesterol metabolism in
which the 5α,6α-epoxide 35 alone is implicated which do
support the possibilities of enzymic 5α,6α-epoxidation.
As the 5β,6β-epoxide 36 predominates over the 5α,6α-epoxide
35 in cholesterol autoxidations [2297-2299], cases involving predominance of the 5α,6α-epoxide 35 invite speculations of enzymic formation, particularly if differential
loss of the 5β,6β-epoxide 36 in analysis or selective
metabolism of 36 be ruled out from consideration. However,
HO· yields the 5α,6α-epoxide 35 in greater abundance than
the 5β,6β-epoxide 36, so the participation of HO· in derivation of the 5α,6α-epoxide 35 must be evaluated before
positing enzymic epoxidation.

These provisos aside, demonstration of the formation
of the 5α,6α-epoxide 35 (the 5β,6β-epoxide 36 though suspected was not demonstrated) in incubations of [4-^{14}C]
cholesterol 3β-palmitate with a 12,000 x g supernate of rat
brain homogenate in which cholesterol autoxidation and
lipid peroxidation was suppressed with antioxidants implies enzymic epoxidation. Sterol ester hydrolysis appeared
to preceed epoxidation, and 3β,5α,6β-triol 13 formation
occurred at yet longer times [1578]. As added sequestering
agent EDTA for Fe(II) ions suppressed epoxidation, Fe(II)-
dependent processes may be implicated.

A more persuasive example lies in the discovery of the
5α,6α-epoxidation of cholesterol by bovine adrenal cortex
microsomal preparations under conditions where the isomeric
5β,6β-epoxide 36 is not formed. The epoxidation was dependent on cytochrome P-450, O$_2$, and NADPH and was not inhibited by EDTA, the well known inhibitor of generalized
lipid peroxidations. Only after prolonged incubations was
the 5β,6β-epoxide 36 also formed, in this case suppressed
by EDTA [2632,2633]. The 5α,6α-epoxide 35 is clearly a
cholesterol metabolite.

Bovine adrenal cortex microsomal enzymes also catalyze
the hydration of the 5α,6α-epoxide 35 to the 3β,5α,6β-triol
13 [2632,2633]. Furthermore, 5α,6α-epoxide hydratases have
been demonstrated in rat brain microsomes [1578], liver
microsomes and mitochondria [92,93,1655], and in hairless
mouse skin and liver [455,1525]. Thus, the 3β,5α,6β-triol
13 may also be viewed as a cholesterol metabolite, given
the 5α,6α-epoxide 35 substrate.

In contrast to the demonstration of enzymic 5α,6α-
epoxidation of cholesterol by bovine adrenal cortex microsomes, the 5,6-epoxidation of cholesterol by bovine liver
microsomal preparations is clearly result of lipid peroxidation. Both 5,6-epoxides 35 and 36 were formed in the
ratio 1:4 along with their common hydration product 3β,5α,6β-
triol 13. In the same system pregnenolone (23) gave both
5,6-epoxides 5,6α-epoxy-3β-hydroxy-5α-pregnan-20-one and
5,6β-epoxy-3β-hydroxy-5β-pregnan-20-one [2631]. In mouse
liver mitochondria incubations desmosterol may have been
transformed to the corresponding triol 5α-cholest-24-ene-
3β,5,6β-triol [550].

Stereospecific epoxidation of other steroid olefins by other mammalian tissues and enzymes and by microbial vegetative cell cultures is beyond present interest, but it is clear that the same kind of problems may arise in selected cases. Thus, the Δ^5-3β-alcohol 3β-hydroxy-5α-B-norandrost-5-en-17-one is transformed by *R. nigricans* to 5,6α-epoxy-3β-hydroxy-5α-B-norandrostan-17-one and 3β,5,6α-trihydroxy-5β-B-norandrostan-17-one *inter alia* [1207].

Lyases

Scission of carbon-carbon bonds of the cholesterol side-chain requires the action of lyases or desmolases, generally in oxidative metabolism. Other than for the cleavage of the isohexyl moiety of the side-chain already discussed, for which considerable attention has been given to the enzymes involved, the lyases are uncharacterized. Their existence is postulated solely on the derivation of products of diminished carbon content in studies of cholesterol metabolism. The degraded sterols of interest include 3β-hydroxychol-5-enic acid (99) and the several short side-chain sterols such as 109,110,114,191-197.

The biosynthesis of the C_{24}-acid 99 from cholesterol, probably via intermediate formation of a 3β,26-diol 29 and/or 31, in rat liver mitochondria has been demonstrated [1639,1645], although the enzymes involved in scission of the C-24/C-25 bond have not been studied.

Enzymic derivation of sterols with short unfunctionalized or olefinic side-chains is another matter, one for which no evidence at all exists. Only for the derived C_{19}-sterol 191 is there evidence of enzymic origins, in this case from steroids other than cholesterol. The biosynthesis of the 5,16-dienol 191 from the C_{21}-steroids 23,111, and 296 in incubations with boar testis microsomes supplemented with NADPH is well established [518,909,1256,1526, 1583,2233], and the transformation of 23 to 191 has been demonstrated in man [371,910]. Boar testis and adrenal tissues transform 23 to androsta-4,16-dien-3-one [30,907], which may then be transformed to 5,16-dienol 191 [372,373].

By imposing the restraint that biosynthesis processes
for the common cholesterol autoxidation products be demonstrated under conditions where autoxidation, lipid peroxidation, or other possible artificial processes are not evident, it may now be concluded that the 3β,7α-diol 14, (20S)-
3β,20-diol 21, 3β,25-diol 27, (25R)-3β,26-diol 29, 5α,6-
epoxide 35, and C_{21}-acid 99 have genuine enzymic origins
and that the cases of the 7-hydroperoxides 46 and 47,
3β,7β-diol 15, 7-ketone 16, 5β,6β-epoxide 36, triol 13, and
6-ketone 44 remain uncertain, requiring further study.

METABOLISM OF AUTOXIDATION PRODUCTS

Studies of the metabolism of cholesterol autoxidation
products fall easily into two categories, the one involving
those products 8,14,21,27,29,35, and 99 also identified as
endogenous metabolites of cholesterol, the other including
all the other autoxidation products for which no genuine
endogenous metabolite role has yet been identified. Until
such role be recognized, these other cholesterol autoxidation products must be viewed as xenobiotic substances not
nominally found in living cells.

Indeed, the metabolism of the first group of cholesterol autoxidation products leads to the biosynthesis of
other classes of steroids which serve special functions in
the animal. Thus, metabolism of the 3β,7α-diol 14, the
(25R)-3β,26-diol 29, and the C_{24}-acid 99 in liver leads to
bile acids, whereas metabolism in endocrine tissues of the
(20S)-3β,20-diol 21 leads to steroid hormones.

The metabolism of the xenobiotic cholesterol autoxidation products, like their biosynthesis, is but poorly developed at present. The one departure from this condition
is, perversely, the study of metabolism of steroid by hydroperoxides, which has received considerable attention.

Hydroperoxides

In sharp distinction to the study of biosynthesis of
steroid hydroperoxides, numerous studies of their metabolism have been recorded, The data of TABLE 13 summarize
results obtained with various steroid hydroperoxides and
partially purified enzyme systems. Although there are four

different modes of transformation of steroid hydroperoxides
recorded (hydroperoxide reduction, hydroperoxide dehydration,
carbon-carbon bond scission, and hydroperoxide rearrangement),
the data establish that sterol hydroperoxide reduction to the
corresponding alcohol is the predominant mode of metabolism.
Dehydration, bond scission, and rearrangement reactions are
much less frequently encountered.

Hydroperoxide reduction by subcellular fractions appears to be associated with cytochrome P-450 species acting
as peroxidases [1087,1088], and the interaction between some
steroid hydroperoxides and cytochrome P-450 preparations is
indicated [1088,2440,2574,2765]. Moreover, when incubations
of enzyme systems capable of other steroid transformations
are conducted aerobically, other oxidative transformations
also occur. For instance, 17α-hydroperoxypregn-4-ene-3,20-
dione yielded 17α-hydroxypregn-4-ene-3,20-dione in anaerobic
incubations of adrenal cortex microsomes but 17α,21-dihy-
droxypregn-4-ene-3,20-dione under O_2 [1091].

Steroid hydroperoxide reduction by pig erythrocyte
glutathione peroxidase (glutathione:H_2O_2 oxidoreductase,
EC 1.11.1.9) was effective in reducing several hydroperoxides. In the case of 17α-hydroperoxypregn-4-ene-3,20-dione
saturation kinetics and an apparent Km 40μM were found. However, the enzyme was denatured by 5-hydroperoxy-3β-hydroxy-
5α-cholestan-6-one and did not reduce this substrate. Moreover, the 25-hydroperoxide 26 (which did not denature the
enzyme) was not a substrate for the enzyme [1521]! A glu-
thathione peroxidase from human or pig aortas reduced
cholesterol 3β-hydroperoxyoctadecadienoate esters to the
corresponding 3β-hydroxyoctadacadoeniates [2282].

The rapid reduction of steroid hydroperoxides by vega-
tative cell cultures of microorganisms also noted for their
capacity for steroid hydroxylations is now well established.
Thus, 6β-hydroperoxypregn-4-ene-3,20-dione is reduced to the
corresponding 6β-alcohol and 11α-hydroxylated, yielding 6β,
11β-dihydroxypregn-4-ene-3,20-dione [2765]. Likewise, 10β-
hydroperoxy-17β-hydroxyestr-4-en-3-one is reduced by *Curvu-
laria lunata* NRRL 2380 to 10β,17β-dihydroxyestr-4-en-3-one,
which is further transformed to 10β-hydroxyestr-4-ene-3,17-
dione and 10β,11β,17β-trihydroxyestr-4-en-3-one [1403].
17α-Hydroperoxypregn-4-ene-3,20-dione and 17α-hydroperoxy-
3β-hydroxypregn-5-en-20-one are also reduced and 11α-hydroxy-
lated by *Aspergillus ochraceus* NRRL 405 [711,2434,2435].

TABLE 13. Metabolism of Steroid Hydroperoxides

Steroid Hydroperoxide	Reaction	Enzyme System	Reference
Cholesterol Hydroperoxides:			
7α-Hydroperoxide (46)	Reduction	Rat liver microsomes	[265,1088,1655]
		Bovine adrenal cortex microsomes	[1088]
7β-Hydroperoxide (47)	Reduction	Rat liver microsomes	[1087-1090]
		Bovine adrenal cortex microsomes	[1087,1088,1090,2557]
		Bovine adrenal cortex mitochondria	[2557]
		Pig erythrocyte gluta-thione peroxidase	[1521]
(20R)-20-Hydroperoxide (32)	Rearrangementa	Bovine adrenal cortex mitochondria	[2565,2566]
(20S)-20-Hydroperoxide (22)	Reduction	Rat liver microsomes	[1088]
		Bovine adrenal cortex microsomes	[1088]
	Rearrangementb	Bovine adrenal cortex mitochondria	[2563,2573,2574]
		Murine adrenal cortex homogenate	[2573]
	Scissionc	Bovine adrenal cortex mitochondria	[2563]

(Continued)

TABLE 13. (continued)

25-Hydroperoxide (26)	Reduction	Rat liver microsomes	[1087,1088]
		Rat liver, kidney	[2465]
		Bovine adrenal cortex microsomes and mitochondria	[1087,1088,2557,2573]
		Calf liver	[2465]
26-Hydroperoxides (28,30)	Reduction	Rat liver microsomes	[1088]
		Bovine adrenal cortex microsomes and mitochondria	[1088,2557]
6β-Hydroperoxycholest-4-en-3-one (59)	Reduction	Rat liver microsomes	[1088]
		Bovine adrenal cortex microsomes	[1088]
3β-Hydroxy-5α-cholest-6-ene-5-hydroperoxide (51)	Reduction	Rat liver microsomes	[1655]
20-Hydroperoxy-3β-hydroxy-(20ξ)-cholest-5-en-22-one	Scission c	Bovine adrenal cortex mitochondria	[108,2564]
Other Steroid Hydroperoxides:			
20α-Hydroperoxypregn-5-en-3β-ol (114)	Reduction	Bovine adrenal cortex microsomes and mitochondria	[2561,2562]
	Dehydration d	Bovine adrenal cortex microsomes and mitochondria	[2561,2662]
20β-Hydroperoxypregn-5-en-3β-ol (115)	Reduction	Bovine adrenal cortex microsomes and mitochondria	[2561,2562]

(Continued)

TABLE 13. (continued)

20β-Hydroperoxypregn-5-en-3β-ol (115)	Dehydration[d]	Bovine adrenal cortex microsomes and mitochondria	[2561, 2562]
17α-Hydroperoxy-3β-hydroxypregn-5-en-20-one	Reduction	Rat liver microsomes	[1088-1090]
		Bovine adrenal cortex homogenates	[2438]
		Bovine adrenal cortex microsomes and mitochondria	[1088-1091, 2440, 2557]
		Pig erythrocyte glutathione peroxidase	[1521]
	Scission[e]	Bovine adrenal cortex microsomes	[1091]
		Bovine adrenal cortex homogenates	[2438]
17α-Hydroperoxypregn-4-ene-3,20-dione	Reduction	Rat liver microsomes	[1087-1090]
		Rat testis microsomes	[2437]
		Bovine adrenal cortex microsomes and mitochondria	[1087-1091, 2438, 2440, 2557]
		Pig erythrocyte glutathione peroxidase	[1521]
		Guinea pig liver cytosol	[2244]
	Scission[f]	Rat testis microsome	[2437]
		Bovine adrenal cortex microsomes and mitochondria	[1091, 2557]

(Continued)

TABLE 13. (continued)

17α-Hydroperoxy-3β-hydroxy-5α-pregnan-20-one	Reduction	Rat liver microsomes	[1087-1090]
		Bovine adrenal cortex microsomes	[1087,1088]
		Pig erythrocyte gluta-thione peroxidase	[1521]

a Product (20S)-cholest-5-ene-3β,20,21-triol (404).
b Product (20R,22R)-cholest-5-ene-3β,20,22-triol (403).
c Products suggested: pregnenolone (23) and 3β,20-dihydroxy-(20S)-22,23-bisnorcholenic acid.
d Product pregnenolone (23) in aerobic incubations.
e Product 3β-hydroxyandrost-5-en-17-one (86).
f Product androst-4-ene-3,17-dione (392)

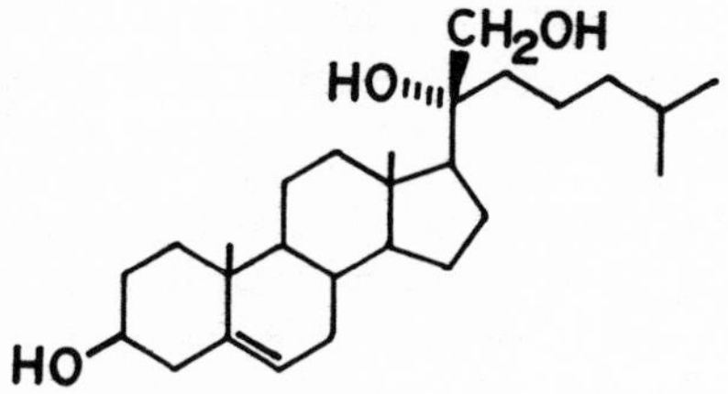

404

Bond scission reactions occur with steroid 17α- and 20-hydroperoxides, the 17-ketones androst-4-ene-3,17-dione (392) and 86 being formed from the 17α-hydroperoxides, the 20-ketone 23 from several 20-hydroperoxides. Although these side-chain scissions appear to be enzymic, nonenzymic thermal decomposition reactions lead to the same degraded products in all cases and may contribute to the transformations recorded.

By far the most interesting of these metabolic transformations is the formal rearrangement of the isomeric 20-hydroperoxides 22 and 32 to vicinal diols. The 20-hydroperoxide 22 of natural C-20 configuration yields in incubations of bovine adrenal cortex mitochondria in the absence of oxygen the sole product (20R,22R)-cholest-5-ene-3β,20,22-triol (403), an acknowledged intermediate in the biosynthesis of the 20-ketone 23 from cholesterol in endocrine tissues. The (20R)-20-hydroperoxide 32 likewise is rearranged to the sole product (20S)-cholest-5-ene-3β,20,21-triol (404) in separate incubations. It was initially thought that the 3β,20,22-triol 403 and a cholest-5-ene-3β, 20,21-triol be products of rearrangement of the (20S)-20-hydroperoxide 22, but inadvertant use of a mixture of 20-hydroperoxides 22 and 32 as substrate (unrecognized at the time) in fact accounted for the presence of two triols. The matter has now been corrected, and it is certain that the (20S)-20-hydroperoxide 22 yield the (20R,22R)-3β,20,22-triol 403, the (20R)-20-hydroperoxide 32 the (20S)-3β,20,21-triol 404 [2563,2565,2566,2573,2574].

The rearrangement of the 20-hydroperoxides 22 and 32
to the vicinal diols 403 and 404 respectively does not require NADPH or oxygen, and the oxygen of water is not incorporated into the triol products. As the rearrangements
are catalyzed by cytochrome P-450 enriched fractions from
mitochondria and binding of substrate 20-hydroperoxides
with cytochrome P-450 components is suggested [2565,2574],
the rearrangement may be effected by cytochrome P-450 associated with the side-chain cleavage enzyme activity of these
mitochondria.

The rearrangement may occur by transfer of the distal
oxygen atom of the hydroperoxide function to the pro-22R
position in the side-chain with an intermediate (20R,22R)-
20,22-epidioxide or dioxetane posited, the reduction of
which would yield the 3β,20,22-triol 403 [1379]. A similar
argument may be proposed for formation of the 3β,20,21-triol
404 from the (20R)-20-hydroperoxide 32.

However, a more likely mechanism in which an intermolecular transfer of oxygen occur may be the case. The
hydroxylation of steroids and other organic compounds by
monooxygenases utilizing cytochrome P-450 requires $^3O_2$ and
two equivalents of NADPH as electron donor. However, other
dioxygen species may spare the $^3O_2$ requirement as well as
that for one equivalent of NADPH. Thus, $O_2\dot{\overline{}}$ spares need
of $^3O_2$ in the liver microsomal cytochrome P-450 hydroxy-
lation of several organic compounds [2379], as does also
cumene hydroperoxide [1211]. Moreover, cumene hydroperoxide and other organic hydroperoxides are similarly effective in the specific hydroxylation of steroids. Such rat
liver microsomal cytochrome P-450 catalyzed hydroxylations
of androst-4-ene-3,17-dione (392) in the 6β-,7α-,15-, and
16α-positions, 17β-hydroxyandrost-4-en-3-one in the 6β-,
7α-, and 16α-positions, progesterone in the 2α-,6β-,7α-,
15α-,15β-, and 16α-positions, and estra-1,3,5(10)-triene-
3,17β-diol in the 6α-,6β-, and 16α-positions are effected
by cumene hydroperoxide [1084-1086]. Bovine adrenal cortex
cytochrome P-450 catalyzed hydroxylations are likewise effected. Progesterone is hydroxylated by microsomal enzymes
in the 6β-,7β-, and 21-positions, by mitochondrial enzymes
in the 1β-,6β-, and 15β-positions, whereas the 3,17-dione
392 is hydroxylated by mitochondrial enzymes in the 6β-,
11β-,16β-, and 19-positions [929].

Moreover, steroid hydroperoxides have been found to serve as oxygen donors as well. 17α-Hydroperoxy-3β-hydroxy-pregn-5-en-20-one acts as oxygen donor in rat liver microsomal [1086] and in bovine adrenal cortex microsomal [929] incubations of progesterone as substrate.

These items allow a reinterpretation of the metabolic transformation of the isomeric 20-hydroperoxides 22 and 32 to the product vicinal diols 403 and 404 respectively. It may be that the adrenal cortex mitochondrial cytochrome P-450 utilizes exogenous 20-hydroperoxide 22 or 32 both as substrate for specific hydroxylation and as a dioxygen species donor. Whether the transformation of (20S)-20-hydroperoxide 22 to product 3β,20,22-triol 403 be intra-molecular, that is, with the same 20-hydroperoxide molecule serving as both oxidizable substrate and dioxygen donor species, or intermolecular, where two 20-hydroperoxide molecules reciprocally hydroxylate one another, the resulting product 3β,20,21-triol 403 be formed. A like argument for the (20R)-20-hydroperoxide 32 leads to the separate product 3β,20,21-triol 404. Indeed, the metabolic transformation of these 20-hydroperoxides to vicinal diols may be regarded as a prior, though unrecognized at the time, discovery of the sparing effects of reduced dioxygen species on the $^3O_2$ requirement of cytochrome P-450 hydroxylating enzymes.

A second possible example of the utilization of steroid hydroperoxides as both oxidizable substrate and donor of oxygen in cytochrome P-450 systems may be the transformation of [17α-$^{18}O_2$]17α-hydroperoxypregn-4-ene-3,20-dione in aerobic (!) incubations with bovine adrenal cortex microsomes, yielding product 17α,21-dihydroxypregn-4-ene-3,20-dione apparently retaining two atoms of ^{18}O [2436].

Sterol 5α,8α-Peroxides

In the case of the sterol 5α,8α-peroxides 60 and 62 the balance between biosynthesis and metabolism is the inverse of that for steroid hydroperoxides, there being more information about the biosynthesis of 60 and 62 than about their metabolism. Indeed, suitable studies of the metabolism of 60 or 62 in mammalian systems appear not to have been conducted! Metabolism studies of the C_{27}-5α,8α-peroxide 60 have been made only to test whether 60 serve as an

intermediate in the biosynthesis of cholesta-5,7-dien-3β-ol
(56) in rat liver. However, aerobic incubations of rat
liver homogenates did not reduce 60 to the dienol 56 [1813].

    Studies of the metabolism of ergosterol peroxide 62
have been confined to microbial systems, where both reductive and oxidative transformations have been observed.
Anaerobic cultures of yeast reduce 62 to ergosterol (also
formed in such cultures from 5α-ergosta-7,E-22-diene-3β,5-
diol) [2498]. Aerated vegetative cell cultures of *Mycobacterium crystallophagum, Proactinomyces restrictus, Bacillus
lentus,* and other molds transform 62 into 5,6α-epoxy-5α-
ergosta-8,E-22-diene-3β,7α-diol (219) and 5,6α-epoxy-5α-
ergosta-8(14),E-22-diene-3β,7α-diol (220) [1858]. However,
control experiments were not reported, and as these epoxides
are thermal decomposition products of 62 (cf. Chapter V)
there remains some uncertainty as to whether these are
indeed enzymic transformations.

    The oxidation of 62 *P. rubrum* yields ergosta-4,6,8(14),
E-22-tetraen-3-one (223) as a genuine metabolic product
[2666,2667].

                    Common Autoxidation Products

    Despite broad distribution of the common cholesterol
autoxidation products in biological materials and their
interference in many studies of cholesterol biosynthesis
and metabolism, systematic investigations of the metabolism
of the autoxidation products have yet to be made. Nonetheless, all the common autoxidation products appear to be
subject to metabolic transformations, but questions of
absorption, transport, tissue distribution, mode of excretion, etc. remain obscure.

    7-Oxygenated Sterols. The sterol trio 14-16 is con-
veniently treated together with the dienone 10. Evidence
for the general metabolism of sterols 14-16 in intact rabbits [1280,2077], rats [1212], and mice [2661] and in cultured human fibroblasts [884] is relatively little, and
most interest is focused to their liver metabolism in relation to bile acid biosynthesis. Administrations of the 3β,
7α-diol 14 to bile fistula rats, rabbits, and hens, and to
man clearly establish metabolism to bile acids characteristic of each animal [59,114,1518,2596,2746,2748,2749].

Similarly administered 3β,7α-diol 14 to carp showed metabolism to bile acid and bile alcohols which were found in gall-bladder bile [1073].

In vitro metabolism of the 3β,7α-diol 14 may proceed along three pathways, all ultimately implicated in bile acid biosynthesis. Incubations with mammalian liver enzymes supplemented with NAD+ uniformly yield 7α-hydroxycholest-4-en-3-one (286) as a product of 3β-alcohol dehydrogenase action. Human [259], mouse [544], rat [209,250, 1107,1617,1654,2750,2751], and guinea pig [257] liver microsomes all conduct this transformation, the rate-limiting step appearing to be removal of the 3β-hydrogen from 14, yielding 7α-hydroxycholest-5-en-3-one (285) as most probable intermediate. The 4β-hydrogen of 14 appears as the 6β-hydrogen product of the 7α-hydroxyketone 286. Whether individual 3β-alcohol dehydrogenase and $\Delta^5 \rightarrow \overline{\Delta^4}$-isomerase activities exist has not been settled [250].

Incubations of rat liver microsomes with 3β,7α-diol 14 but without added NAD+ also yield a cholest-5-ene-3β, 7α,26-triol [1107], whereas rat and human liver microsomal preparations with added NADPH transform 3β,7α-diol 14 into cholest-5-ene-3β,7α,12α-triol [208,259,549,663]. Rat liver mitochondria supplemented with NAD+ also yield 7α-hydroxycholest-4-en-3-one (286) [1107], 7α-hydroxychol-5-enic acid, and other bile acids [113,2747]. Oxidation of the terminal methyl groups of the 3β,7α-diol 14 to CO_2 is indicated [576].

The microbial sterol oxidases of Nocardia sp [2227] and B. sterolicum [1129] that transform cholesterol to the enone 8 also transform the 3β,7α-diol 14 to the corresponding 7α-hydroxyenone 286. Further, 3β,7β-diol 15 and 7-ketone 16 are oxidized by the B. sterolicum oxidase as well [1129].

The metabolism of the 3β,7β-diol 15 by rat liver microsomal systems has already been mentioned as leading by dehydrogenation to the 7-ketone 16. Dehydrogenation to 7β-hydroxycholest-4-en-3-one has not been observed. However, 3β,7β-diol 15 metabolism in bile-fistula rats nonetheless leads to bile acids, including all four stereoisomers of 3,7-dihydroxychol-5-enic acid and the epimeric 3-hydroxy-7-oxochol-5-enic acids. The anticipated product ursodeoxycholic (3α,7β-dihydroxy-5β-cholanic) acid was but a minor product [265,1781,2747].

Metabolism of the 7-ketone 16 is also not extensively
studied. Besides the indicated reduction of 16 to the
3β,7β-diol by rat liver microsomal preparation in vitro,
studies with bile fistula rats and guinea pigs suggest that
bile acids be formed [265]. Thus, all three sterols 14-16
may be degraded to steroid acids by pathways adumbrated
only for the 3β,7α-diol 14 which is subject to A-ring dehydrogenation, 12α-hydroxylation, and 26-hydroxylation.

The 7-ketone 16 in isologous plasma administered
intravenously to rabbits or to perfused pig arteries
in vitro appears to be rapidly removed by absorption into
erythrocytes (in vivo) or into lipoproteins (in vitro)
[225].

Yet another mode of metabolism is indicated for trio
14-16, that of esterification with fatty acids. Esterification of the 3β,7α-diol 14 by rat liver cytosol and microsomal fatty acyl transferases [128,1107,2059] has been
demonstrated in vitro. Moreover, as the oxidation of cholesterol fatty acyl esters by liver does not yield 3β,7α-diol
14 3β-esters [128,1363,1598,1792], the presence of such
esters in tissues implies esterifications of 14. Fatty
acyl esters of the epimeric 3β,7-diols 14 and 15 have been
found in human serum and plasma [338,343,494-496,2314],
aorta [369,493,1404,2314], and liver from Wolman's disease
[103] and from rat serum, skin, and liver [343]. Also,
fatty acyl esters of the 7-ketone 16 have been found in
human liver [103] and plasma [2314], and sulfate esters of
3β,7α-diol 14 have been detected in human meconium [1435],
all inferring metabolic esterification of the free sterols
14-16 in as yet unrecognized tissue sites.

Although cholesterol acyl esters are not oxidized to
the 3β,7α-diol 14, fatty acyl esters of 14 may be further
oxidized, witness the transformation of 3β-stearatoxy-
cholest-5-en-7α-ol in bile-fistula rats to bile acids
[1793].

The metabolism of cholesta-3,5-dien-7-one (10), the
fourth 7-oxygenated cholesterol autoxidation product, has
also been examined in rats and cockerells. The dienone 10
is poorly absorbed but is found in liver and intestines.
No metabolites have been recognized [1243-1245].

Ketone Derivatives. Metabolism studies of the several cholesterol autoxidation products (excluding the 7-ketones 10 and 16 previously discussed) are generally confined to enzyme-catalyzed isomerization of enone 6 to enone 8 and enzymic reductions of 8 to the stanols 2 and 4. Indirect evidence suggests that enone 8 be precursor of the 5α-stanol 2 in man and experimental animals [2010,2011,2356,2496], and *in vitro* demonstrations of reduction of 8, via 5α-cholestan-3-one (365) to 2 in rat and rabbit liver microsomes [963, 2210,2211,2214] establish the matter.

A long suspected analogous reduction in mammals of the enone 8 to the 5β-stanol 4 moderated by enteric microorganisms [55,504,2022] is supported by direct evidence in man [2009,2010,2124]. Moreover, *in vitro* demonstrations of the reduction of 8 via 5β-cholestan-3-one to the 5β-stanol 4 by rat caecal microorganisms [270,281] and by pure cultures of *Eubacterium* sp. ATCC 21408 isolated from rat intestinal microflora [1819] have been recorded. Moreover, reduction of 8 to cholesterol by human feces suspensions also has been observed [2009].

However, oxidative metabolism of the enone 8 via side-chain cleavages has been demonstrated. Rat testis mitochondria transform 8 into the 17-ketone 392; a rat adrenal 8500 x g sediment degrades 8 to progesterone and 17α-hydroxy-pregn-4-ene-3,20-dione [95]. Metabolism of the enone 8 in other systems is also indicated, including probable reduction to cholesterol by *Labyrinthula vitellina* [2595], transformation in the rat to unidentified acidic fecal products [963], and metabolism to unidentified products in mouse L cell cultures [2027]. There are also extensive investigations of the degradation of cholesterol via the enone 8 by a variety of microorganisms [2289].

Metabolism studies of other ketonic cholesterol autoxidation products such as 42,44,45,59,108, etc. have not been made. Although the 24-ketone 34 has been shown to be oxidized by the sterol oxidase of *N. erythropolis* [368], its metabolism in mammalian systems has not attracted interest. However, the 22-ketone analog 297 (not a recognized cholesterol autoxidation product) is metabolized to isovaleric acid and (tentatively) the bisnor acid 101 in intact guinea pigs [1261], but metabolism in bovine adrenal cortex mitochondrial preparations is not indicated [458].

5,6-Epoxides. The metabolism of the 5α,6α-epoxide 35
was investigated but superficially since early conjecture
of its possible role in bile acid biosynthesis [678]. More
recently concern over the potential toxicity and carcinogenicity of the 5α,6α-epoxide 35 has revived interests in
its metabolism. Some absorption and metabolism of the
5α,6α-epoxide 35 occurs in rats fed the sterol [763], and
a slow metabolism from subcutaneous deposits of epoxide in
hairless mice may occur [237]. However, other mice administered the 5α,6α-epoxide 35 per os or upon whose skin the
sterol was painted excreted the sterol rapidly via the
feces, but the sterol was also widely distributed in low
amounts throughout body tissues as well. Following the
oral dose relatively high levels of sterol were found in
intestinal contents, liver, and blood; from skin painting
higher levels were found in intestinal contents, skin, and
liver [335]. These results suggest that the 5α,6α-epoxide
35 whether administered orally or cutaneously is rapidly
eliminated via the gastrointestinal tract, as such or
metabolized. Transformation of the 5α,6α-epoxide 35 to the
3β,5α,6β-triol 13 has been demonstrated in rat gastrointestinal tract [764], and cholesterol 5α,6α-epoxide hydratase [EC 4.2.1.63] activity yielding the 3β,5α,6β-triol 13
as product has been demonstrated in incubations of human
gastrointestinal microflora [1109] as well as a variety of
mammalian tissues, including rat liver [92,93,1655], brain
[1578], and lung [2192,2193], mouse liver [274,280] and
skin [455,1525], and bovine adrenal cortex [2632,2633].

Besides the metabolic addition of the elements of
water to the 5α,6α-epoxide 35 addition of glutathione to
35 moderated by a rat liver cytosol S-glutathione transferase yielding 3β,5-dihydroxy-5α-cholestan-6β-yl-S-gluta-
thione (405) has been described [2634].

Metabolism investigations of the isomeric 5β,6β-epox-
ide 36 are confined to observations that the 3β,5α,6β-triol
13 be formed in vitro. However, both 5,6-epoxides 35 and
36 may be subject to esterification, as fatty acyl esters
of both have been found in aged liver from Wolman's disease victims [103].

5α-Cholestane-3β,5,6β-triol. The metabolism of the
3β,5α,6β-triol 13 has been of interest as a possible source
of bile acids in bile fistula dogs [678] but more recently
in relation to past interest in use of the triol as a

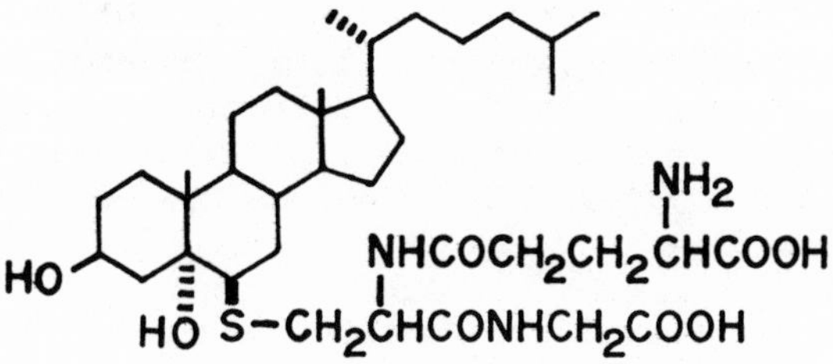

405

hypocholesterolemic agent. In the intact rat, esterification yielding fatty acyl esters of 13, dehydrogenation yielding the 3β,5α-dihydroxy-6-ketone 44, and side-chain degradation yielding bile acids, including 3β,5,6β-trihydroxy-5α-cholan-24-oic acid has been demonstrated, these metabolites being excreted via biliary and fecal routes [1294,2006,2007]. A fourth mode of metabolism to conjugated enone derivatives in rat liver homogenates has been suggested, but little support of this formulation is available [599,2717].

Side-Chain Alcohols. Metabolism of the cholesterol autoxidation products substituted by hydroxyl in the side-chain ((20S)-3β,20-diol 21, 3β,25-diol 27 and the 3β,26-diols 29 and 31) in mammalian systems is indicated, as also is metabolism of the related (22R)-3β,22-diol 24, (22S)-cholest-5-ene-3β,22-diol, the epimeric cholest-5-ene-3β,23-diols, the (24S)-3β,24-diol 25, and (24R)-cholest-5-ene-3β,24-diol not presently recognized as cholesterol autoxidation products. Metabolism of the (20S)-3β,20-diol 21 (both an endogenous metabolite of cholesterol and an autooxidation product) to C_{21}-steroids in a variety of *in vitro* mammalian [99,417,511,582,943,1582,2201,2228-2232,2428,2453, 2685] and plant [2368] systems is well known, the transformations in adrenal cortex mitochondrial systems proceeding via 22R-monohydroxylation to the 3β,20,22-triol 403, as previously described. Metabolism of 21 by bovine adrenal cortex mitochondria inhibited by aminoglutethimide may be via initial product (20S)-cholest-5-ene-3β,20,25-triol to 3β,20-dihydroxy-(20S)-chol-5-en-24-al [582].

Adrenal cortex mitochondria also metabolize synthetic analogs of the (20S)-3β,20-diol 21 to pregnenolone (23). From (20R)-20-phenylpregn-5-ene-3β,20-diol was obtained 23 and 17-methyl-18-norandrosta-5,17(13)-dien-3β-ol; from (20S)-20-cyclopropylmethylpregn-5-ene-3β,20-diol, pregnenolone [1055,1057,1076].

By contrast, studies of metabolism of the 3β,25-diol 27 are few. For instance, only indirect evidence is recorded testing whether the 3β,25-diol 27 be implicated in bile acid biosynthesis in rats [2465]. Unidentified metabolites of 3β,25-diol 27 have been detected in squirrel monkeys [2474] and in perfused rat liver [703]. More directly, adrenal cortex mitochondrial preparations oxidize the 3β,25-diol 27 to the C_{21}-20-ketone 23 [341,409,582, 1186,1582]. Malondialdehyde and acetone have been suggested as other products, and in adrenal cortex mitochondrial oxidations of 27 inhibited with aminoglutethimide 3β-hydroxychol-5-en-24-al (298) and acetone were formed [582]. Isolated rat adrenal cell cultures oxidize 27 completely to 11β,21-dihydroxypregn-4-ene-3,20-dione (corticosterone) [714,715,1103]. Despite these bioconversions of 27 *in vitro* there is no evidence to suggest that the 3β,25-diol 27 be a natural substrate for C_{21}-steroid biosynthesis.

Metabolism of the 3β,26-diols 29 and/or 31, both having putative autoxidation and biosynthesis origins, is also established in close association with liver bile acid biosynthesis. *In vitro* metabolism of the (25RS)-3β,26-diols 29 and 31 by rat liver mitochondria yields a 3β-hydroxycholest-5-en-26-oic acid and CO_2 [575,576], that of (25R)-3β,26-diol 29 the C_{24}-acid 102 [1639]. Moreover, the (25R)-3β,26-diol 29 is transformed in bile fistula rats and in humans with external biliary drainage to cholic acid (3α,7α,12α-trihydroxy-5β-cholanic acid) *inter alia* [59]. Adrenal cortex mitochondria oxidize the (25R)-3β,26-diol 29 to pregnenolone (23) [99].

In plants the 3β,26-diols are thought to be precursors of sapogenins, and the metabolism of the (25R)-3β,26-diol 29 to (25R)-spirost-5-en-3β-ol (diosgenin) has been demonstrated in *Dioscorea floribunda* [184].

The oxidized cholesterol derivatives which are not known to be autoxidation products are metabolized also. The (22R)-3β,22-diol 24 is hydroxylated by adrenal cortex mitochondria to the 3β,20,22-triol 403 further oxidized to the C_{21}-20-ketone 23 as previously mentioned. Moreover, the epimeric (22S)-cholest-5-ene-3β,22-diol is also oxidized by adrenal cortex mitochondria to 23, presumably via (20R,22S)-cholest-5-ene-3β,20,22-triol [409,2428,2453].

Metabolism of the cholest-5-ene-3β,23-diols, other than that suggested in TABLE 14, is an unknown matter. The (24S)-3β,24-diol 25 and the epimeric (24R)-3β,24-diol 107 appear to have a special *in vivo* metabolism in rat brain [1513]. The 3β,24-diols also are oxidized to the 20-ketone 23 *in vitro* by adrenal mitochondria [99,1582].

Besides these special adrenal, liver, and brain metabolic processes for the stenediols 21,24,25,27, and 29, there are several other metabolic processes inferred by product distributions or by other evidence, these being listed in TABLE 14. The presence in tissues of fatty acyl or sulfate esters of these stenediols is taken here to support enzymic metabolism of the parent sterols. However, an autoxidative pathway to these oxidized sterol esters, particularly those of the 3β,25-diol 27, cannot be totally disregarded [495,496]. Furthermore, the stereochemistry of the side-chain hydroxyl group in most cases is uncertain; only in the one case of the sulfate esters of the (22R)-3β,22-diol 24 is stereochemistry assigned.

Metabolism of the esters of these stenediols may also occur, as lysosomal hydrolysis of 3β-oleatoxycholest-5-en-25-ol to 27 is indicated in cultures of human skin fibroblasts [1387].

Yet other metabolic processes involving these stenediols may await discovery. The mixed function oxidase oxidation of these sterols is of present interest as a means of delving into the details of cytochrome P-450 function and oxygen utilization. Many of these stenediols interact characteristically with adrenal cortex mitochondrial cytochrome P-450 to give Type I (λ_{max}385 nm,λ_{min}420 nm) or inverted Type I (λ_{max}420 nm,λ_{min}385 nm) (also designated as Type II) induced difference spectra suggesting low-to-high and high-to-low spin interconversions respectively of the hemoprotein. Oxidized cholesterol derivatives

TABLE 14. Generalized Metabolism of Cholest-5-ene-3β,X-diols

X	Metabolic Process	Tissue	Reference
20S-OH	Esterification (fatty acyl)	Rat adrenal	[1963]
	Sterol oxidase	*N. erythropolis*	[368]
22-OH	Esterification (sulfate)	Human meconium,feces	[680,931,1435]
		Human plasma	[654]
23-OH	Esterification (sulfate)	Human meconium	[680]
24-OH	Esterification (fatty acyl)	Human aorta	[369,2460]
	Esterification (sulfate)	Human meconium,feces	[680,931]
		Human plasma	[2396]
	Sterol oxidase	*N. erythropolis*	[368]
25-OH	Esterification (fatty acyl)	Human aorta,plasma	[495,496,2314,2463]
	Esterification (sulfate)	Human plasma	[654]
	Sterol oxidase	*N. erythropolis*	[368]
26-OH	Esterification (fatty acyl)	Human aorta	[362,369,874,2460,2463]
	Esterification (sulfate)	Human meconium,feces	[680,931,1435]
		Human plasma,urine	[2396]
	Sterol oxidase	*N. erythropolis*	[368,2273,2274]

giving Type I spectra include the (20R)-3β,20-diol 33 [406, 412], (24R)-3β,24-diol 107 [2684], the 3β,25-diol 27 [406, 412,1183,1185-1187,1292,2684], the 3β,20,22-triol 403 [406, 412,1182,1185,2684], and 3β,20-hydroxy-(20R)-cholest-5-en-22-one [406,412]. Inverted Type I spectra are given by the (20S)-3β,20-diol 21 [178,356,406,412,485,1161,1182-1185, 1187,1292,1364,1605,1637,1638,1801,2668-2670], the (22R)-3β,22-diol 24 [406,412,1183,1801,2684], (22S)-cholest-5-ene-3β,22-diol [406,412,2684], the 22-ketone 297 [406,412,2684], (20S)-cholest-5-ene-3β,20,21-triol (404) [406,412], and (20R,22S)-cholest-5-ene-3β,20,22-triol [412]. The spin state changes of the cytochrome P-450 heme iron suggested by optical spectra are confirmed by electron spin resonance spectra as well for the (20S)-3β,20-diol 21 and (22R)-3β,22-diol 24 [1637,1801]. However, binding of oxidized sterols to cytochrome P-450 preparations may occur without induced spectral changes or with variable changes dependent on the preparation [178,406,412,2574] and on other factors [1182,1183,1185,1801]. Binding per se may not lead to metabolism, and binding to other adrenal cytosol protein also occurs for the (20S)-3β,20-diol 21 and 3β,25-diol 27 [1447,2381,2382].

Degraded Sterols. Cholesterol autoxidation products and related derivatives with short (or no) side-chains are also subject to metabolism in several systems. In TABLE 15 are listed various metabolic processes directly examined using the several substrates 109,110,114,191,193,194,210, and 301 as well as a few inferred by the recovery of metabolites from biological specimens. For most substrates side-chain scission in adrenal or testis systems is observed, and the expected esterification, 3β-hydroxysteroid oxidation, and Δ^5-reduction reactions are represented.

Not listed in TABLE 15 is the C_{24}-acid 99 which is both an endogenous metabolite of cholesterol and also a cholesterol autoxidation product. Metabolism of 99 and of 99 3β-acetate in bile fistula rats and in rat liver mitochondria in vitro incubations has been repeatedly shown to proceed to bile acids [1130,1642,1644,1646,1707]. Less extensive metabolism of 99 to 3β,7α-hydroxychol-5-enic acid found in human, rat, and hen bile and clearly an intermediate in bile acid biosynthesis [114,958,1131,2749] was not observed. The sterol oxidase of B. sterolicum oxidizes 99 [1129].

TABLE 15.   Metabolism of Degraded Cholesterol Autoxidation
            Products

Autoxidation Product	Metabolism
27-Norcholest-5-en-3β-ol (210)	Rat adrenal mitochondria, side-chain scission [1582]
(25 ξ)-27-Norcholest-5-ene-3β,25-diol (301)	Microbial sterol oxidase, 3-ketone formation [2273,2274]
Chol-5-en-3β-ol (109)	Rat adrenal mitochondria, side-chain scission [1582]
23,24-Bisnorchol-5-en-3β-ol (194)	Bovine adrenal slices, 20- and 21-hydroxylations [2415,2416]
	Dog adrenal homogenate, steroid hormone biosynthesis [2417]
	N. rippertii, Δ^5-reduction [2533]
	Ps. porosa, esterification [444,1877]
Pregn-5-en-3β-ol (110)	Rat testis microsomes, testosterone formation [1053]
	Rat liver microsomes, unidentified metabolites [376]
Pregna-5,20-dien-3β-ol (193)	Rat testis microsomes, not metabolized [1053]
Androst-5-en-3β-ol (114)	Rat liver microsomes, unidentified metabolite [376]
	N. rippertii, Δ^5-reduction [2533]
Androsta-5,16-dien-3β-ol (191)	Boar testis homogenates (with NAD)$^{+}$ 3-ketone formation [373,1256]
	Boar testis homogenates (with NADPH), Δ^5-reduction [1256]
	In vivo human, esterification (sulfate) [2041]
	In vivo human, glucuronide formation [908]

Consideration of the metabolism of other degraded cholesterol autoxidation products, such as the C_{19}-steroids 86 and 87 and C_{21}-steroids 23,111, and 296, which are certainly involved in steroid hormone biosynthesis and metabolism, is beyond present interest.

IMPLICATIONS

The question whether cholesterol autoxidation products have a genuine role in metabolism may now be addressed. It is quite clear that such products administered as xenobiotic substances to living systems are metabolized. However, the issue of their biosynthesis from cholesterol as endogenous metabolites must be qualified, as some products are metabolites with recognized cellular function but others are surrounded by uncertainties.

The preponderance of evidence from *in vitro* studies in a variety of biological systems establishes the enzymic biosynthesis from cholesterol of at least seven cholesterol autoxidation products. Among these are six primary oxidation products, including the enone 6, 3β,7α-diol 14, (20S)-3β,20-diol 21, 3β,25-diol 27, and (25R)-3β,20-diol 29, and 5α,6α-epoxide 35 and two products of further transformations, enone 8 and bile acid 99. Moreover, evidence for the biosynthesis from cholesterol of the (25S)-3β,26-diol 31 suggest that the sterol also be viewed as metabolite of cholesterol under special settings. Furthermore, the 7-hydroperoxide 46 and 47 are products of enzyme-catalyzed lipid peroxidation processes. Barring direct evidence to the contrary, other cholesterol autoxidation products from *in vitro* studies must be regarded as artifacts of nonenzymic oxidations or of manipulations, as nonenzymic alteration products of metabolites or artifacts, as metabolites or artifacts, or possibly as products derived from other processes. This same proviso holds for such sterols encountered in biological material not under prior direct control.

Problem Complexity

Arguments advanced throughout this monograph point out the serious problem of cholesterol autoxidation and intrusion of artifacts into studies of cholesterol metabolism.

The attendant uncertainties in results obtained have been
so great as to preclude timely demonstration of hepatic
microsomal cholesterol 7α-hydroxylase activity in early
studies of bile acid biosynthesis [543,548]. These experimental difficulties have now been overcome for *in vitro*
studies, and it is obvious from other studies of cholesterol metabolism not cited in detail that artificial oxidations need not compromise results where awareness of the
problem and proper care are taken.

The complexity of autoxidized cholesterol derivatives
possibly encountered in uncontrolled metabolizing systems
includes artifacts of artifacts (10 from 16, 13 from 36),
and metabolites of artifacts (16 from 15, 23 from 27) and
of metabolites (23 from 21) *inter alia,* all obviously possible. Moreover, extended transformation sequences involving metabolism and artificial processes also exist, the
oxidation of cholesterol via the (20S)-3β,20-diol 21,
(20R,22R)-3β,20,22-triol 403, and 20-ketone 23 to the epimeric 3β,20-diols 111 and 296 being an example of multiple
enzymic processes in operation, the same 111 and 296 being
also derived enzymically or nonenzymically from the corresponding 20-hydroperoxides 112 and 113 formed autoxidatively
from cholesterol.

In FIGURE 25 are presented several selected pathways
which have been demonstrated for the several key cholesterol autoxidation products and metabolites for which
demonstrated processes exist. For simplicity, the metabolites 29,31, and 99 have not been included, but side-chain
oxidations at the C-20 and C-25 positions are represented.
Also, only selected secondary transformation products are
included. Cholesterol oxidations by HO· yielding products
13-16,35, and 36 are not included in FIGURE 25.

The correspondence between enzyme processes accounting
for products 6,8,14,21,27 and 35 is perfect, all acknowledged metabolites having major autoxidation pathways also.
Moreover, the epimeric 7-hydroperoxide 46 and 47 are also
indicated as enzyme products, although some qualifications
for these compounds may be in order. Other demonstrated
enzymic processes variously linking the metabolites (as 21
to 23 to 111 and 296, 21 to 23 to 191, 21 to 23 to 86 and
87, 35 to 13, etc.) all have equally established pathways
to match. Several demonstrated enzyme processes removed
from direct metabolites are also shown (14 to 15 to 16, etc.).

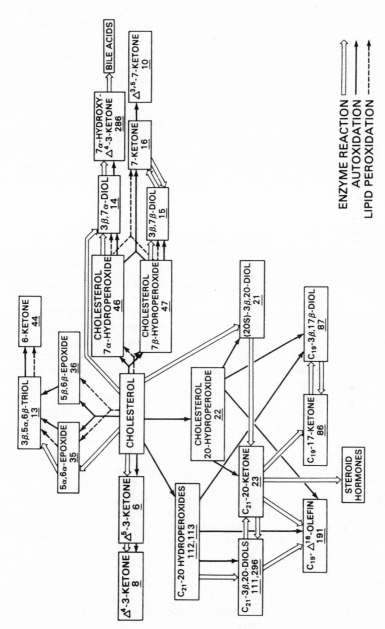

FIGURE 25.   Demonstrated enzymic and nonenzymic oxidations

From FIGURE 25 it is obvious that there exist many
combinations of metabolic and nonenzymic transformations
which may be posed to account for formation of most of the
cholesterol autoxidation products of TABLE 10 in some
fashion. It is also certain that the mere presence of a
given oxidized cholesterol derivative of a biological sample may not infer the oxidative processes for its formation from cholesterol. Given proven processes of biosynthesis as well as of autoxidation, one questions whether
resolution of origins for the many cases of isolation of
cholesterol autoxidation products from biological material
(cf. TABLE 1 and TABLE 2) be possible. As experimental
evidence is generally limiting and a *posteriori* arguments
may not reconcile opposing possibilities, proper conclusions generally cannot be drawn.

Questions and Answers

Several aids in evaluating many prior accounts of
cholesterol metabolism and autoxidation are at hand. For
studies in which cholesterol autoxidation products were
encountered, one may ask: (i) are the oxidized sterols not
artifacts but genuine metabolites, (ii) is the history of
the specimen sufficiently well known to warrant investigation, (iii) do the investigations have flawed design or
conduct which generate oxidized products artifically, and
(iv) are analysis methods too sensitive and thereby detect
low levels of oxidation products present ubiquitously in
all samples whether   recognized or no. For studies in
which cholesterol autoxidation products are not encountered,
one asks: (i) are such studies devoid of artifacts because
of experimental care, or (ii) are artifacts not detected
because of lack of awareness of or disinterest in the matter or because of inadequate analysis methodology. These
same issues apply to chemical as well as to enzyme systems.
As exclusion of air is not always the case in the use of
chemical oxidants, it is equally important to consider the
contribution of O_2 from air in chemical oxidations, a matter rarely addressed.

Metabolite Status. Answers to the questions posed
must tend upon yet other considerations. Answer to the
eternal question "Natural product (or metabolite) or artifacts?" necessitates careful distinction among the terms.
Whereas one has a good general concept about their meaning,

there are instances where uncertain or inexact meaning and
facts leave the issue unsettled. Metabolite status infers
derivation from cholesterol via enzyme-substrate complex
formation characterized by kinetics and other properties,
from which specific products derive. Metabolites are thereby natural products, but natural products derive by processes of Nature, thereby via enzymic and via other processes.

The presence of cholesterol autoxidation products in
freshly collected biological material thus evinces their
natural product status, but ultimate origins may be from
diet, ingestion of environmental components by other means,
or photochemical processes unrelated to enzyme actions of
the living specimen. Actions of microorganisms and symbionts, aberrant metabolism associated with infection,
disease, or dysfunction from loss or regulatory control,
and deranged metabolism following trauma and death may also
contribute as possible factors. Almost nothing is known
about these factors, but it is clear that the specific past
history of a given biological sample may very much influence the composition of oxidized sterols found on analysis.
Moreover, as the several factors mentioned are natural
ones, all products of their effects on cholesterol oxidation would be natural products. As the autoxidation of
cholesterol in tissues exposed to air is also very much a
natural event, cholesterol autoxidation products by this
train of thought are thus natural products! Such a broad
definition becomes meaningless for present interest.

The mere formation of oxidized sterol derivatives
from cholesterol in aerated *in vitro* incubations with enzyme preparations, cells, or tissues doe not establish
origins, as indirect reactions may in fact contribute to
product formation, along with or instead of enzyme actions.
There appear to be at least four distinct oxidative processes, *cf*. FIGURE 26, which may give rise to sterol oxidation products in biological material. Other than the
apparently uncatalyzed free radical autoxidations discussed
in depth in this monograph and the obvious enzyme-catalyzed
processes yielding specific products, there are two more
obscure processes which may contribute.

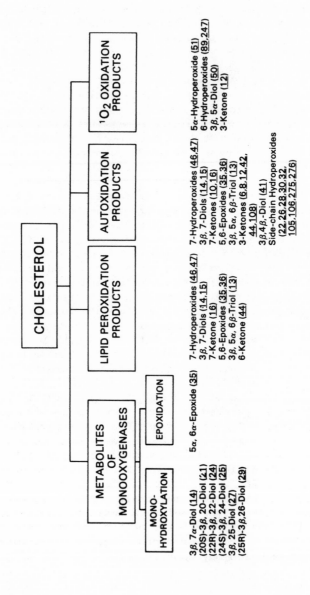

FIGURE 26. Recognized cellular oxidation processes

The oxidation of sterols by photosensitized oxygenations using cellular pigments as sensitizers and in which $^1O_2$ may participate is implicated in the transformation of ergosterol to its $5\alpha,8\alpha$-peroxide 62, as previously discussed, such photosensitized oxidation being viewed as a biological process [2217] though a nonenzymic one. The specific participation of $^1O_2$ in these reactions remains to be demonstrated, as does also any participation of $^1O_2$ in the oxidation of cholesterol in biological systems.

The oxidation of cholesterol by other dioxygen species, HO·, or other powerful oxidizing species formed from $^3O_2$ in biological systems, whether photochemically, enzymically, or by other means, would be biological processes in the same sense as that involving $^1O_2$. Careful distinction among these terms and processes must be made.

The fourth oxidation process of interest is that of generalized lipid peroxidation, wherein enzyme-catalyzed reduction of Fe(III) provides catalytically active Fe(II) and/or HO· or other active oxidant. The terms "biological autoxidation" [1365] and "in vivo autoxidation" [444] appear to encompass the same biochemistry, and our suggestions that dioxygenases peroxidize cholesterol [2311,2458,2461] may also be interpreted in terms of a generalized lipid peroxidation as well. Unsaturated lipids are oxidized by $^3O_2$ to hydroperoxides; cholesterol is transformed to the 7-hydroperoxides 46 and 47. Whereas autoxidation (apparently uncatalyzed) is easily distinguished from enzyme-catalyzed lipid peroxidation, for cholesterol products are the same. Moreover, enzyme-catalyzed lipid peroxidation is readily demonstrated in vitro, but the extent of in vivo lipid peroxidation remains uncertain. Finally, peroxidation may also occur at diminished levels in lipid peroxidation systems where the requisite enzyme action has been destroyed by heat.

Given these aspects of complexity, it is no longer necessary to argue natural product versus artifact, enzyme versus nonenzyme action, etc. for cholesterol autoxidation products encountered. Rather, by the doctrine res ipsa loquitur one may conclude that most cases are at best indeterminable for want of evidence and are more likely examples of unrecognized autoxidation and artifact formation.

Sample History. It is disturbing that so little attention has been paid to the possibilities of artifact
formation in biological specimens before isolation procedures are applied. Autoxidation during isolation has regularly been considered but not the past history of the sample! It is important to consider the effects of the trauma
of specimen collection, death, conditions of preservation
and storage, actions of lysosomal enzymes released *post
mortem*, and possible microbial contaminations.

In the process of derivation of oxidized sterol derivatives there is a time when the tissue of origin is alive,
within its natural endogenous host, performing its biological function, but most certainly respiring. Upon collection of the specimen from the natural habitat or of exci-
sion of tissue at surgery the system is thereafter in the
hands of the investigator, and all alterations of sensitive
unsaturated lipids should be under control in order to
avoid the many problems of origins of oxidized lipid products found.

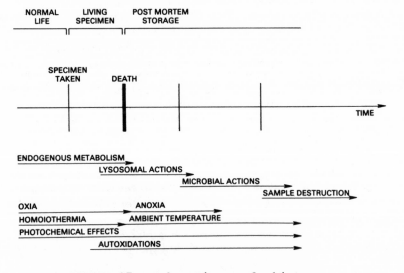

FIGURE 27. Schematic sample history

The scheme of FIGURE 27 represents the progression in time from natural life through specimen collection and preservation prior to analysis, and points out where much uncertainty has been generated in prior studies which have not paid heed to each phase of tissue component alteration. During normal life, respiration and endogenous metabolism proceeds, cholesterol is oxidized enzymically according to the capacity and need or the organism, and specimens taken from the natural habitat and processed directly whether for analysis of sterol composition or for preparation of enzymes for *in vitro* studies should yield, with care, sound experimental results from which valid conclusions may be drawn.

It is at the second phase and thereafter that uncertainties are introduced. Specimens taken alive from the natural habitat may be subjected to various contrived storage conditions which may simulate natural conditions but which are surely quite different ones, and uncertain metabolic changes may occur. For example, mere removal of the sponge *Psammaplysilla pupurea* from sea water into air causes a color change [115]. Attempt to transport marine creatures alive in cold aerated sea water [444] represent the extreme which has been taken to overcome uncertainty in such matters, but most cases pass over this phase directly to the death of the organism or excised tissue, with preservation in alcohol or formalin, freezing or cooling, or drying in sunlight (!) being frequently used processes. As none of these means involves exclusion of air, none is recommended for careful work! Where there is no need to keep the organism or tissue alive, immediate freeze-drying under vacuum and subsequent storage frozen under vacuum pending isolation and analysis, all conducted with due care to limit exposure of sample to air and light or heat, is by far the preferred means of avoiding adventitious air oxidations.

At death respiration and regulatory processes cease and a change from aerobic to anaerobic state begins. Enzymes protecting against oxidative attack fail and lysosomal enzymes controlling degradation processes are released. Material from homoiothermic animals drops in temperature to ambient. In this changing environment it is uncertain just what transformations cholesterol undergo, but the cholesterol ester hydroperoxides in human autopsy aortas may derive as consequence of such changes [961].

As tissues stored in air, no matter the temperature,
may be subject to insidious autoxidations, autoxidation products as well as those of microbial action, lysosomal enzymes, and of other processes may be found. Analysis of human
tissues collected (aortas [960]) or stored frozen (Wolman
disease tissues [103]) over several years and of other biological materials stored for years under less safe conditions
(nonfat dried milk at ambient temperature, anhydrous milk
fat at 5°C [776]) cannot be accepted without major reservations about sample history.

Although the degradation of cholesterol by a variety
of microorganisms occurs and several individual pathways to
C_{27}-,C_{24}-,C_{19}-, and other derivatives have been elucidated
[2289], no study is known which addresses alterations of
cholesterol or its oxidation products during *post mortem*
decomposition and putrefaction. The closest comparable
case lies in the metabolism of cholesterol by enteric microflora to the 5β-stanol 4. Although frank microbial degradation of cholesterol in biological material is probably not
likely in most cases, the recovery of the C_{27}-3,4-secoacid
43 from whale ambergris [1440] may well be such a case.

Given enough time beyond death a given specimen may be
totally degraded, but under favorable conditions cholesterol
is preserved for millenia, witness its presence in 2000 y
old human coproliths [1509] and 1500-6000 y old Egyptian
mummy tissues [109,1399,1555]. However, over the ages cholesterol may also be altered by esterification and autoxidation
as is evinced by analyses of mummy brain of great age [1305,
1428,1555]. Thus it is crucially important to have an adequate sample history and to know at what stage of natural
degradation a given biological specimen may be at the time
of analysis. Lacking such information proper conclusions
cannot be made.

All biological material unprotected from light may also
be subject to photochemical effects and in air to photooxidations. Photosensitized oxidations of fucosterol (75) to
the 24-ketone 34 in *A. nodosum* [1338], of ergosterol to the
5α,8α-peroxide 62 in fungi containing anthraquinone pigments
[16], and of sterol Δ^{23}-25- and Δ^{25}-24-alcohols from
Δ^{24}-sterols of Spanish moss (*cf.* Chapter III) are cases in
point. The cholesterol of human serum exposed to light and
air is oxidized to the 3β,7-diols 14 and 15 and 3β,25-diol
27 as are also possibly their fatty acyl esters [495,496].

Exposure of human and murine skin to radiation in air yields
a variety of cholesterol autoxidation products.

   Experimental Design. In all natural product work most
attention is paid to isolation of components from the sample
by solvent extraction, chromatography, crystallization, and
derivitization means, with quality control analyses by
chromatographic or spectral methods at appropriate stages.
These methods are so well established that no concern for
their adequacy need be raised. Given suitable biological
material for isolations and analyses, the present methods
are adequate, even for nanogram and picogram amounts of
sterols.

   It is important to realize that poorly designed or
conducted work not using adequate or sensitive analysis
methods easily fails to encounter cholesterol autoxidation
products which are nonetheless present, thereby appearing
to be free from artifacts, whereas sound work utilizing
sensitive methods encounters the autoxidation products as
a matter of course, thereby receiving unwarranted stigmata
associated with artifact formation.

                        Control Measures

      There arises the natural question of how to tell arti-
facts from metabolite for the general case of cholesterol
metabolism as well as for the specific case. *In vitro*
studies under direct control are much more easily conducted
to settle this issue than are studies where full control of
the work is not possible, but care in devising and conducting controlled experiments must be had in all studies lest
uncertainty about the biological significance of analysis
results be the case. The usual criteria of classic enzymology apply throughout; thus, products formed only in the presence of active enzyme may be considered to be enzyme products, and levels of enzymic metabolites from experiments
with active enzyme must be higher than levels of products
in proper control experiments. Indeed, the problem may not
be in experimental means but in awareness of complexity of
the case and of need to devise suitable controls so as to
establish the nature of the active oxidizing species implicated.

The disposition to see cholesterol autoxidation products as metabolites only has been emphasized, and a plethora of sophisticated arguments discounting air oxidation as explanation for observed oxidation products in biological material has accumulated over the years. In general these arguments fall into three classes: (i) those erected on disputation alone, (ii) those attended by analogy or some experimental evidence, no matter how weak, and (iii) those based upon adequate experimental control measures. A fourth class in which no argument or evidence is given nor conclusions drawn also deserves mention.

Most of the arguments based on assertion   or disputation alone fail to persuade and should be dismissed out of hand. Nonetheless, their influences must be recognized. Whereas instances of unsupported claims generally occurred long before present realizations and are easy to spot, such strong statements that the 3β,5α,6β-triol 13 isolated from formalinized human brain "can hardly be an artifact" [746] need citation here to emphasize an extreme position. Arguments to the same end based on lack of precedence [1894], inadequate time [1632], or other factors are of the same sort.

Arguments presenting supporting experimental data are also abundant, but care must be exercised in interpreting individual cases, as arguments may be specious. Arguments based on failure to detect oxidation products in certain cases are suspect. Thus, discovery of the C_{29}-7α-hydroperoxide 58 but not of C-7 oxidized sitosterol derivatives in *A. hyppocastanum* leaves [767] poses an argument against autoxidation of uncertain validity. Arguments based on more detailed product distributions still must be carefully evaluated. Simple differences in autoxidation products distribution in human erythrocytes from those from other tissues [1160] can no longer be regarded as appropriate for conclusions. Nor can the argument that the oxidation products 14-16,35, and 36 be enzymic metabolites of bovine adrenal cortex mitochondrial incubations of cholesterol and not artifacts because only 14-16,35, and 36 were found and the (20S)-3β,20-diol 21 and 3β,25-diol 27 also recognized cholesterol autoxidation products were not found [327] be accepted as a valid basis. Indeed, the products found support exactly the opposite conclusion, that unrecognized autoxidations or lipid peroxidations account for the products formed. Moreover, formation of oxidation products in one medium but not in another, as suggested by the formation of the C_{28}-5α,8α-peroxide 216 in aerated vegetative cell incubations of soy fungus but not in solid soy meal on which the same fungus grows [2115], poses no valid basis for making conclusions.

Most of these diffuse and unconvincing arguments and others like them have been applied to static analyses of biological material not under prior direct control and for which past history is uncertain. Moreover, all too often control measures in such cases have proceeded with the misguided concept that air oxidations occur only during isolation and analysis and not before. Controls subjecting pure cholesterol fatty acyl esters to saponification and recovery procedures [2012] or pure cholesterol to a full analysis protocol [776] attest only the utility or lack thereof of the processes operated, as also is the case of exclusion of O_2 by working under an atmosphere of Ar [2533].

Adequate control measures for *in vitro* studies of cholesterol metabolism include experiments with heat-inactivated or no enzyme, and such obvious precautions have been broadly used with mammalian [2343], microorganism [1385,2624], and plant callus tissue [677] systems, where transformation products were not found in controls. Other control systems employing dried defatted blood [2136], acetone-denatured enzyme [525], or albumin instead of enzyme [2392] have been variously recorded. For the early issue of enzymic cholesterol 7α-hydroxylation, levels of product 3β,7α-diol 14 exceeded those in the artifacts 3β,7β-diol 15 and 7-ketone 16 [543,548]. A refinement of data analysis involved a ratio of 3β,7α-diol 14 to all autoxidation products including 3β,7β-diol 15, 7-ketone 16, 5,6-epoxides 35 and 36, and 3β,5α,6β-triol 13 [1655].

Demonstration of cholesterol 25-hydroxylation by hepatic mitochondrial enzymes was similarly supported by increased levels of the 3β,25-diol 27 over levels of interferring cholesterol autoxidation products 13-16,35, and 36, both in experiments where autoxidation product levels were low and in experiments where a several-fold increase was allowed [94]. These approaches utilizing differential levels of metabolites and artifacts infer complete analysis of all oxidized sterols in an experiment. A total absence of cholesterol autoxidation products from control experiments is not necessarily the case, as more sensitive analysis methods using radioactivity readily detect very low levels of products. However, control levels of autoxidation products may be quite high relative to the experimental value [1759]. Moreover, as the difference in experimental and control levels measures the extent of enzyme-catalyzed cholesterol oxidation, so also the difference between control levels and zero measures the extent of uncontrolled or unavoidable cholesterol autoxidation in the experiment!

Control measures which involve comparison of several subcellular fractions from the same tissue in different combinations are effective in demonstrating enzyme actions. Comparison of cytosol, microsome, and mitochondria preparations for cholesterol 7α-hydroxylase activity and for associated lipid peroxidation and autoxidation effects is a case in point [1655]. Deletion of a required component for enzyme action from experimental incubations involving purified enzyme preparations is a highly effective means of demonstration of enzyme action. For monooxygenases requiring NADPH, omission of NADPH affords a strong control measure. Other required factors may also be omitted. An elegant and convincing example lies in the deletion of the iron-sulfur protein ferredoxin from incubations of solubilized rat liver mitochondrial cytochrome P-450 dependent upon ferredoxin and NADPH-ferredoxin reductase components for 25- and 26-hydroxylation of cholesterol [1837]. This simple control fully justifies acceptance of the 3β,25-diol 27 (and (25R)-3β,26-diol 29 also formed) as metabolites of cholesterol.

Many additional control measures may be imposed on *in vitro* studies of cholesterol metabolism, including means of selective stimulation or inhibition of enzymes, selective suppression of lipid peroxidation and autoxidation via use of antioxidants, use of tissues from donors subjected to various pretreatment protocols (nutrition, phenobarbital, hormones, circadian rhythms, etc.).

It is obviously not possible to apply such control measures to the case where oxidized sterols are found in biological material not under adequate direct control. However, with awareness of these matters, it is also entirely possible to use restraint in interpretation of results and to avoid unfounded claims not supported by the data.

Biochemical Implications

Other than the pervasive issue of biosynthesis versus autoxidation origins for the oxidized sterols just discussed, there are several other topics which are linked tenuously to cholesterol autoxidation or to selected cholesterol autoxidation products. The previously mentioned concept that enzyme-catalyzed steroid hydroxylations be a two-step process involving initial hydroperoxide formation with subsequent reduction to the alcohol, possibly appropriate at the time, cannot now be advanced as an alternative to well established monooxygenase hydroxylations in which cytochrome P-450 be terminal oxidase.

Although this biochemical mechanism matter is now resolved for monooxygenase hydroxylations where hydroperoxides are not implicated and for *in vitro* lipid peroxidations where hydroperoxides are formed, there remains the question whether lipid peroxidation of cholesterol be an endogenous metabolic process occurring *in vivo*. Such lipid peroxidation might occur at low levels as a normal matter under metabolic control or as aberrant, adventitious oxidation not so limited and posing deleterious effects to the organism.

Demonstration of *in vivo* lipid peroxidation of cholesterol would be technically demanding, as protective peroxidative enzyme systems destroying sterol hydroperoxides may be presumed. Barring direct demonstration of formation of the 7-hydroperoxides 46 and 47 (or of the secondary products 13-16,35, and 36 by HO· attack) in appropriate living systems, the matter remains unsettled. One may posit that *in vivo* lipid peroxidation of cholesterol be implied by the presence of the secondary products 14-16 possibly in company of the 5,6-epoxides 35 and 36 and their hydration product triol 13, but reliable analyses on fresh tissue or plasma are lacking.

Detection of the spate of cholesterol autoxidation products 13-16,35, and 36 with or without the 7-hydroperoxides 46 and 47 in properly collected samples from living donors would be sound evidence for *in vivo* lipid peroxidation of cholesterol, possibly moderated by nutrition, health state, age, etc. of the donor. Suspicion that such be the case is raised by detection of elevated levels of the 7-ketone 16 in plasma from a rhesus monkey afflicted with a haemorrhagic

fever [888] and of the 5α,6α-epoxide 35, enone 8, and
dienone 12 from serum from human patients afflicted with
various hypercholesterolemias [914].   In these few cases,
the oxidized sterols found did not match the required pattern of 13-16,35, and 36, but selective subsequent metabolism of the set of cholesterol autoxidation products may
account for the altered pattern found.   For instance,
esterification of the 3β,7α-diol 14 by fatty acids has been
demonstrated in vitro [128,1107], and this esterification
might be selective, as surely is the dehydrogenation of the
3β,7β-diol 15 to the 7-ketone 16.   These formulations are
speculative, and other excursions into fancy may also be
constructed, given the paucity of sound evidence.

Adequate analysis of mammalian tissues for evidence of
in vivo lipid peroxidation of cholesterol is lacking, as
data of TABLE 1 and TABLE 2 cannot be taken as proper evidence.   The presence of esterified cholesterol oxidation
products in tissues summarized in TABLE 16 provides a better
basis but one in need of further careful evaluation.   It
appears that fatty acyl esters of the common cholesterol
products 14-16,35, and 36 be endogenous metabolites, formed
by esterification of oxidized sterols, by oxidation of cholesterol esters, or by both processes.   Evidence against
in vitro liver 7α-hydroxylation of cholesterol esters and
much chemical evidence speaks against oxidation of sterol
esters, but photochemical oxidations of serum cholesterol
esters may occur [495,496].   In vitro studies of lipid peroxidation of cholesterol esters have not been conducted.

The alternative esterification of cholesterol oxidation
products appears more likely the case, perhaps for transport
or further endogenous metabolism for utilization and for
excretion.   As the cholesterol autoxidation products 14-16
and others have cytotoxic properties (cf. Chapter VIII),
esterification may be a means of protective metabolism, a
matter yet to be resolved.   The place of sulfate esters of
the oxidized sterols (cf. TABLE 15) in endogenous sterol
metabolism also has yet to be understood.

The implied in vivo lipid peroxidation of cholesterol
and subsequent protective metabolism by esterification may
be a serious matter.   From the high levels of fatty acyl
esters of 14-16,35, and 36 found in liver of Wolman disease
patients, ranging 1-8 mg/g liver for each component (total
of all such esters 17-18 mg/g) [103], a high level of sterol

oxidation is suggested. However, the tissues analyzed were
stored frozen for years prior to analysis, and although
such esters were not found in control liver of unspecified
past history, sound biochemical implications cannot be made.

A matter related to the issue of *in vivo* lipid peroxidation of cholesterol as a normal endogenous metabolic
process is the question of endogenous metabolism in general
of sterol oxidation products. Those oxidation products
obviously directed into established metabolic pathways to
required agents, such as bile acids or hormones, are not
of interest here, but rather xenobiotic (dietary) sterols
whose disposition requires metabolism for transport, utilization, and excretion are of concern.

The presence of cholesterol autoxidation products in
human foodstuffs is just now under scrutiny, particularly
in view of the cytotoxic properties of several derivatives
and impuned carcinogenicity of the $5\alpha,6\alpha$-epoxide 35 discussed in Chapter VIII. Cholesterol autoxidation induced by
irradiation in air has been demonstrated for various dried
egg products, the 7-hydroperoxides 46 and/or 47, $3\beta,7$-diols
14 and 15, 7-ketone 16, isomeric 5,6-epoxides 35 and 36.
and $3\beta,5\alpha,6\beta$-triol 13 being variously found [12,481,2507,
2508]. Levels of the $5\alpha,6\alpha$-epoxide 35 of 1-33 µg/g have
been measured in dried egg products [2508]! It is not
known whether such dietary oxidized sterols would be absorbed whether they pose a toxicity burden if absorbed, or
whether protective metabolism (possibly esterification)
detoxifies these agents, but interest is mounting [2450].

Divers other examples of derivitized sterol oxidation
products might also be mentioned. The ergosterol $5\alpha,8\alpha$-
peroxide 3β-divaricatinate ester 215 from lichen [387],
fatty acyl esters of the C_{30}-$3\beta,25$-diol 169 from a higher
plant [1875], and fatty acyl esters of 23,24-bisnorchol-5-
en-3β-ol (194) from *Ps. porosa* [444,1877] raise the question whether esterification preceed or follow oxidation.
The short side-chain 194 is here considered as an oxidatively degraded sterol even though 194 is not an established
cholesterol autoxidation product. 5ξ-Androstan-3β-ol and
5ξ-23,24-bisnorcholan-3β-ol, as 5ξ-dihydro derivatives of
114 and 194 respectively, found with 114 and 194 *inter alia*
in certain termites [2533] pose the same enigma for reduced
B-ring olefinic sterols.

TABLE 16. Esterified Cholesterol Autoxidation Products in Tissues

Sterol	Fatty Acyl Esters	Sulfate Esters
Cholest-5-ene-3β,7α-diol (14)	Rat serum, skin, liver [338,343] Human serum, plasma [494-496,2314] Human aorta [369,493,1404,2314] Human liver (Wolman disease) [103]	Human meconium [1435]
Cholest-5-ene-3β,7β-diol (15)	Human serum, plasma [495,496,2314] Human aorta [369,493,1404,2314] Human liver (Wolman disease) [103]	—
3β-Hydroxycholest-5-en-7-one (16)	Human plasma [2314] Human liver (Wolman disease) [103]	—
5,6α-Epoxy-5α-cholestan-3β-ol (35)	Human liver (Wolman disease) [103]	—
5,6β-Epoxy-5β-cholestan-3β-ol (36)	Human liver (Wolman disease) [103]	—

There is yet another interest in the presence of the
B-ring oxidized sterols 14-16 and their fatty acyl esters
and for the side-chain oxidized sterols 21,25,27,29 and
their esters in human tissues. The sterols appear to suppress the biosynthesis of cholesterol by inhibition of the
regulatory 3-hydroxy-3-methylglutaryl coenzyme A reductase
step (cf. Chapter VIII), and on this basis it has been
postulated by Kandutsch that perhaps these sterols be implicated in as yet unrecognized regulatory processes involving cholesterol biosynthesis [472,1229,1232-1234].
Proper test of the hypothesis has yet to be devised, but
the presence of inhibitory sterols selectively in aorta and
brain where only low levels of cholesterol biosynthesis
occur appear to support the formulation. Whether other
tissue and plasma sterols and their esters are implicated
or whether these derivatives are transport and excretory
forms of products of unfavorable *in vivo* lipid peroxidation remain to be examined.

The participation of cholesterol autoxidation products
in endogenous steroid hormone biosynthesis has been suggested by certain metabolism studies. The metabolic transformation of the (20S)-20-hydroperoxide 22 by adrenal cortex mitochondrial cytochrome P-450 to the 3β,20,22-triol
403 suggested at one point that cleavage of the isohexyl
moiety of the cholesterol side-chain via dioxygenase action
involve the 20-hydroperoxide 22 [2572-2574]. However, as
already mentioned, this formulation is no longer tenable.

A second equally innovative but unsupported suggestion
proposes that the sesterterpenoid 23,24-bisnorchol-5-en-
3β-ol (194) be precursor of corticosteroid hormones, this
formulation being based on the metabolic transformation of
194 into corticosteroids by bovine and canine adrenal preparations [2415-2417]. The C_{22}-substrate 194 is apparently
a natural product found in several invertebrates (cf. TABLE
5) and not an established cholesterol autoxidation product,
but the matter is of interest for its biological implications at this point. Instead of derivation of 194 from
cholesterol in turn derived from the triterpenoid squalene,
it was noted that the C_{22}-sterol be potentially an end product of cyclization of a putative sesterterpene lower homolog of squalene with subsequent elaboration of nuclear
features. In such event, a wholly separate biosynthesis
for corticosteroids in which neither squalene, lanosterol,
cholesterol, nor 20-ketone 23 be implicated would exist!

Both proposals that the 20-hydroperoxide 22 and that the sesterterpenoid derivative 194 be intermediates in steroid hormone biosynthesis were built on substrate metabolism to C_{21}-products, with carrier trapping results in support. However, in neither case is the biosynthesis of the putative sterol 22 or 194 demonstrated, nor does all evidence in the cases meet stringent criteria advanced for establishing intermediacy of a putative sterol precursor in the biosynthesis of other steroids [2138]. The sesterterpenoid biosynthesis pathway is discounted on other bases as well [213].

A final biological problem now unresolved is raised by the presence of several short-chain sterols in divers invertebrates (cf. TABLE 5). Of the ten sterols, only one (191) has a demonstrated biosynthesis; five (109,110,114, 191,196) are established cholesterol autoxidation products. Autoxidation processes for the others (193-196,197) have been suggested though not demonstrated [444]. The question of interest is whether these short side-chain sterols serve as natural components to moderate the structural integrity and function of biological membranes as does cholesterol. In such event, an argument based on function might be advanced for the biosynthesis origin of these sterols over an autoxidative origin. Other discussion of membrane effects of cholesterol autoxidation products is deferred to Chapter VIII.

CHAPTER VIII. BIOLOGICAL EFFECTS

The high concentration and ubiquitous distribution of cholesterol in human tissues generate concern for the sterol's function. Searches for function reviewed by Flint in 1862 [779] continue to this day. The role of cholesterol as precursor of bile acids and steroid hormones is clearly established as is also its role in membrane integrity and function, and the association of cholesterol or its esters with several major health afflictions is well demonstrated, such matters as gallstones, hyperlipidemia, atherosclerosis, etc., being prominent. However, a polemic continues to exist regarding whether cholesterol be cause or effect in human disease.

Other implausible claims of important biological activity attributed to cholesterol such as that of brain hormone activity is the silkworm *Bombyx mori* [1316] also exist. However, recently evidence of specific biological activities for several oxidized cholesterol derivatives has been recorded, this evidence implicating circumstantially the common cholesterol autoxidation products as well as or instead of cholesterol in a variety of effects ranging through all forms of life and including influences on endogenous metabolism, cell growth, and human disease! The various biological effects of cholesterol and its esters will not be discussed except where additional dimensions are provided treatment of the effects of cholesterol autoxidation products.

As all early studies involving cholesterol and many recent studies as well were conducted with cholesterol preparations of uncertain purity despite description of means of purification, citation of melting point, USP quality, etc., it may be that activities attributed to cholesterol be more correctly those of low levels of unrecognized cholesterol congeners and autoxidation products almost certainly present in the cholesterol samples actually tested. This issue was recognized by Duff in 1935 [636,637] and received experimental evaluation by Schwenk in 1959 [2170], unfortunately without resolution. The matter has been variously reemphasized [50,2659].

Interest in possible physiological activities of cholesterol expressed in the early twentieth century included studies of the action of cholesterol on isolated frog heart

preparations dating from 1907 [553,811], of cholesterol on
phagocytosis [2384], and of dietary cholesterol on the
induction of aortal atherosclerosis in rabbits [65,67,2622].
Attention to possible cytotoxicity actions of cholesterol
oxidation products was also expressed in this early period,
although interest was based on a suspected relationship between the toad poison bufotalin as an oxidation product of
cholesterol [720-722,811].

Moreover, some individual secoacid cholesterol oxidation products were evaluated in 1911 for their cytotoxic
effects on isolated frog heart and gastrocnemius muscle
preparations and for their hemolytic action on defibrinated
blood [780]. By 1928 Seel had examined the physiological
effects of oxycholesterol and oxysitosterol (the sitosterol
analog of oxycholesterol) and of several chemically defined
cholesterol autoxidation products, including the 3β,5α,6β-
triol 13, the 5α,6α-epoxide 35, and the 3,6-diketones 42
and 108 inter alia in similar isolated frog heart preparations. Oxycholesterol samples were particularly active in
lengthening the duration of heart beat and increasing
pulse frequency, in detoxifying poisoned frog heart preparations, and in increasing contractions of guinea pig and
swine uterus preparations [2180,2181]. Oxycholesterol
preparations were also shown to prolong the time necessary
for coagulation of blood [1548].

During this same period intense activities directed
toward clarification of the antirachitic activity induced
in cholesterol by irradiation developed. As these irradiations were conducted in air, radiation-induced oxidations
of cholesterol surely occurred. Indeed, positive Lifschütz
color tests for oxycholesterol noted in some irradiated
preparations [211,2205] clearly established the occurrence
of oxidative processes. Radiation-induced antirachitic
activities in sterol samples [786,1018,1019,1382] later
demonstrated to derive from cholesta-5,7-dien-3β-ol (56)
and not from cholesterol [2021] were of such intensity that
the biological response could be detected in sterol samples
too small for chemical studies [786]!

Question whether impurities, congeners, or transformation products of cholesterol were in fact the antirachitic
agent beset investigations of the period, reports that
highly purified cholesterol be activated [223,1351] or not
be activated [1366] confusing matters. However, heating

and irradiation in air of highly purified cholesterol
thought devoid of the 5,6-diene 56 also yielded antirachitic preparations [978,1350]. The process by which pure
cholesterol can be transformed into antirachitic derivatives remains unresolved, but air oxidation of cholesterol
to the epimeric 7-hydroperoxides 46 and 47 whose thermal
decomposition yields the corresponding epimeric 3β,7-diols
14 and 15 and the 7-ketone 16 offers an explanation. Both
the 3β,7β-diol 15 and the 7-ketone 16 display antirachitic
properties when injected subcutaneously or administered
per os to rachitic rats on a depletion diet [1430]. Whether 15 and 16 are rachitic *per se* or are transformed metabolically to the diene 56 from which cholecalciferol derives has not been settled.

In 1944 Haslewood called for an evaluation of the physiological activities of the common cholesterol autoxidation
products, particularly those directly associated with growth
processes [975]! The matter has yet to be addressed systematically. The studies by Fieser [742,743] and by Bischoff
[231,232] emphasized possible carcinogenic properties of
oxidized cholesterol derivatives, discussed more fully later
in this chapter. A few investigations failed to promote
interest in the matter, and there developed deleterious effects in experimental animals fed oxidized steroids rather
than beneficial effects for which commercial exploitation
seemed likely.

Thus, adrenal hypertrophy in rats [2355,2356] and other
toxic effects in other animals [2496] fed cholest-4-en-3-one
(8), liver toxicity in rabbits [47-49] and loss of appetite
and of body and liver weight in rats [701] fed the 7-ketone
16 as well as cytotoxic effects of these and other oxidized
sterols in tissue and cell systems discussed in detail later
attest such matters.

Suggestions that specific cholesterol autoxidation
products such as the enone 8, the triol 13, 7-ketone 16,
and 3β,25-diol 27 be evaluated clinically because of their
hypocholesterolemic actions in experimental animals have
recurred periodically. Moreover, interest in the cytotoxic
effects of oxidized sterols for control of tumor growth is
developing [476,1456,1704,2766]. Suggestion that oxidized
cholesterol derivatives constitute a natural regulatory
system for sterol biosynthesis have been made, and the regulatory effect of 3β,7α-diol 14 on bile acids biosynthesis

from cholesterol is well recognized. New interests in the
atherogenicity of oxidized cholesterol derivatives is developing. Although the carcinogenicity of oxidized cholesterol derivates remains uncertain, very recent data demonstrate the mutagenicity of cholesterol autoxidation products
towards bacterial test systems. Finally, interest in oxidized sterols as moderators of biological membrane function
and integrity and of certain membrane-bound enzymes associated with sterol biosynthesis and metabolism is continuing.

   Barring synergism and related interactions the biological activity of naturally autoxidized cholesterol should
reflect the individual activities of the several autoxidation products. Thus, air-aged cholesterol might possess
weak androgenicity by dint of the weak androgens 86 and 87
present therein.

   A number of unrelated miscellaneous biological activities of cholesterol autoxidation products, not particularly
linked to the more fully recognized biological actions, have
been recorded. The 7-ketone 16 has "anti-cortisone" (17α,21-
dihydroxypregn-4-ene-3,11,20-trione) properties in that 16
apparently blocks the cortisone-induced fulmination of
Coxsackie virus infections in mice [1572,1573], but allusion
to antiinflammatory and immunosuppresant activities for 16
has also been made [1325].

   Several 6-ketosteroids related to the ecdysterols have
been tested for their effects on growth and development of
the ovaries of the fly *Musca domestica*. The ketonic cholesterol autoxidation products cholest-4-ene-3,6-dione (108),
5α-cholestane-3,6-dione (42), and 3β,5-dihydroxy-5α-chole-
stan-6-one (44) allowed egg development and egg laying but
reduced the number of viable flies hatched. Other cholesterol autoxidation products similarly active included
cholesta-3,5-dien-7-one (10), the 7-ketone 16 3β-acetate,
and the 5β,6β-epoxide 36. Notably, the 3β,5α,6β-triol 13
gave a normal hatch rate [1949].

   Both 3β-hydroxycholest-5-en-24-one (34) and a cholest-
5-ene-3β,24-diol fed in the first instar at 0.1% levels in
the diet to *B. mori* were lethal [1677].

Although the mechanisms by which these miscellaneous effects are expressed is unknown, it may be that some are related to the cytotoxic actions of oxidized sterols discussed in the next section or to the specific inhibition of intracellular enzymes implicated in sterol biosynthesis, as discussed later in this chapter.

CYTOTOXICITY

Cytotoxic effects of oxidized cholesterol derivatives have dominated studies of biological activities from the very early period, with observations made in 1911 of the necrotizing effects on frog heart preparations of the chemically derived secoacids 3,4-secocholest-5-ene-3,4-dioic acid (406), 7-oxo-3,4-secocholest-5-ene-3,4-dioic acid (407), and 5,7-dioxo-3,4;5,6-bissecocholestane-3,4,6-trioic acid (408) [780] introducing these interests.

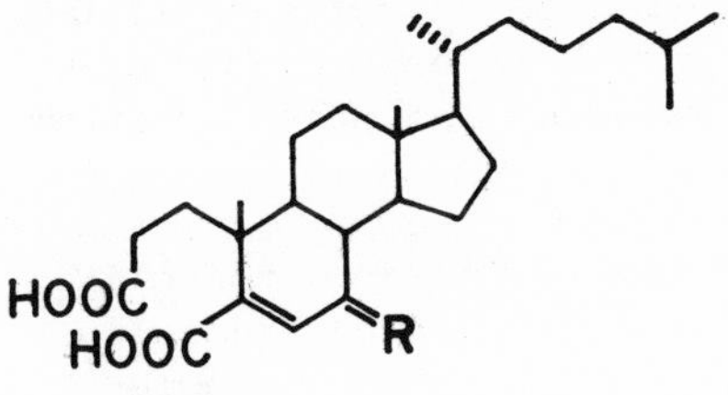

406   R = H$_2$

407   R = O

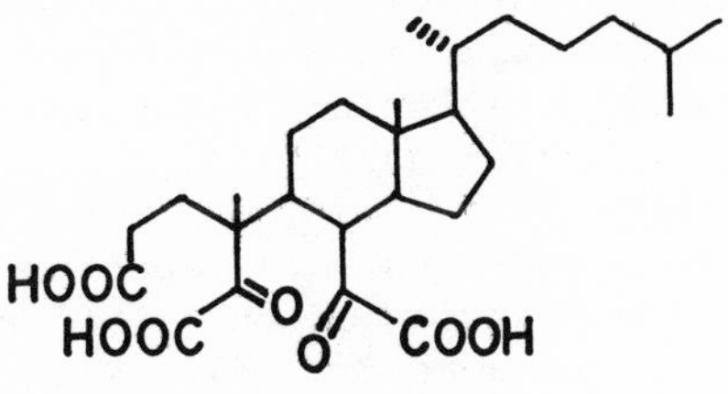

408

In addition to general cytotoxicity matters, such special cases as atherogenicity, mutagenicity, and carcinogenicity will be discussed in separate sections of this chapter. Generalized cytotoxic effects in intact animals and in cultured cells occupy present interests.

A few studies have shown that individual cholesterol autoxididation products fed experimental animals provoke generalized toxic responses. Rats fed 1% levels of cholest-4-en-3-one (8) became lethargic, lost appetite, and did not grow well. Adrenal hypertrophy also occurred [2355,2356]. Dogs and chickens fed the enone 8 likewise were affected [2496]. Moreover, rats fed 0.5% levels of the 7-ketone 16 exhibited loss of appetite and decreased liver and body

weight [701], and signs of hepatic toxicity and cirrhosis
in rabbits and liver necrosis in guinea pigs fed 16 have
been noted [47-49,225]. Single instances of renal cortex
malignancy [49] and lung granulomatous angiitis [2075] were
observed in rabbits fed the 7-ketone 16. Cockerells fed
the enone 10 had elevated serum cholesterol levels [1245].
Extensive acute, subacute, and chronic toxicity studies in
mice, rats, dogs, and monkeys of the 3β,5α,6β-triol 13 administered orally further suggested growth retardation and
organ atrophy [1266].

Some of the deleterious effects on skeletal muscle and
on liver and spleen of animals fed cholesterol, particularly
cholesterol preparations heated in air [45,46], may be attributed to cholesterol autoxidation products formed in the
prepared diets, not recognized at the time. More recent
electron microscopic examination of ultrastructural changes
in liver from New Zealand white rabbits fed or injected
intravenously with the 3β,5α,6β-triol 13, the ketone 16,
the 3β,25-diol 27, or the 5α,6α-epoxide 35 suggest diffuse
vesiculation of hepatocytes, hyperplasia of smooth endoplasmic reticulum, cytoplasmic vacuoles and other cytotoxic
manifestations [1148].

Interests have repeatedly been advanced in the clinical evaluation of these and related cholesterol autoxidation products for their possible utility as hypocholesterolemic agents capable of inhibiting sterol biosynthesis
from acetate and lowering serum cholesterol levels. However, early recognition of cytotoxic actions of the enone
8 [2355] and the 3β,5α,6β-triol 13 [1266] as well as more
recent data derived using cultured cells and inhibitory
actions on specific cellular enzymes discussed in subsequent sections of this chapter preclude serious interest
in these steroids. Moreover, more recent suggestions of
the same sort, that the 7-ketone 16 and the 3β,25-diol 27
be evaluated clinically for treatment of familial hypercholesterolemias in human patients [351,882], are equally
naive in the face of unacceptable toxic side-effects in
experimental animals fed 16 [47-49,225,701,702,903,1235]
and 27 [1235] and in a variety of tissue and cell culture
systems discussed shortly. Nonetheless, programs directed
to chemical synthesis and testing of analogs of the
3β,5α,6β-triol 13 [2717,2718] and of the 7-ketone 16 and
3β,25-diol 27 [651,885] for such purposes have been mounted.

Cytotoxicity has also been observed for phytosterol derivatives in their abortifacient actions in mice. Extracts of unripe fruit and leaves of the common pineapple *Ananas cosmosus* constitute an Indian folk remedy for termination of pregnancy, and sitosterol (20) and stigmast-5-ene-3β,7α-diol (76) found in the air-dried leaves [10,1811, 1812] and other oxidized sitosterol derivatives (including the 3β,7β-diol 77 3β,7β-dibenzoate, 7-ketone 72 3β-benzoate, and 3β,5α,6β-triol 145 3β-benzoate were effective abortifacients before implantation, the 3β,7α-diol 76,77 3β,7β-dibenzoate, and 145 3β-benzoate as well after implantation. The 5α,8α-peroxide 62 is also abortifacient in mice [1812] and toxic in rats, causing death at 2 μmole/kg [36].

In another direction the recognized cytotoxicities of oxidized sterols have been examined for possible utility in the control of growth of tumor cells. Evaluation of growth inhibitory properties has been conducted using cultured cells in media incorporating the test steroid. Cytotoxicity towards a number of normal (non-neoplastic) as well as towards several neoplastic cell lines has been demonstrated.

Thus, the derivatives 5β-5,6-secocholestane-3β,5α,6-triol (409) and 5β-5,6-secocholestane-3β,5α-diol (410) not implicated in cholesterol autoxidation but prepared via ozonization of cholesterol inhibit growth of chicken fibroblasts in culture [1456,1458,1463] and the 3β,5α,6-triol 409 exhibits weak activity against HeLa cells [1457]. Likewise, the 5α,8α-peroxide 62 of ergosterol exhibited cytotoxicities against normal mouse 3T3 fibroblasts in culture but also against neoplasm cells as well [476].

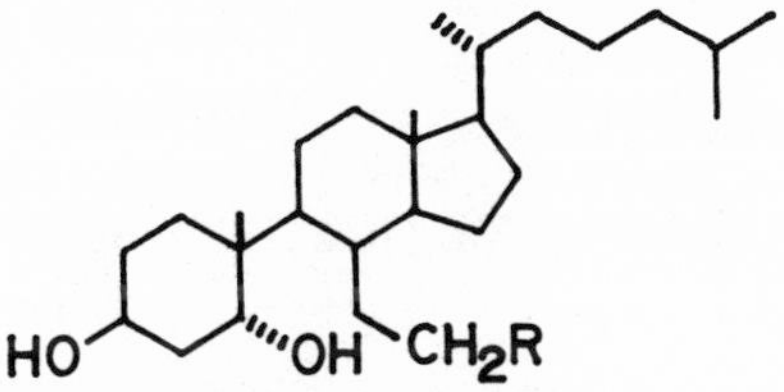

409   R = OH

410   R = H

Whereas the seco-3β,5α,6-triol 409 was quite toxic, neither the sitosterol analog 5β-5,6-secostigmastane-3β, 5α,6-triol nor 5α-cholestane-3β,5,6β-triol (13) was cytotoxic in these assays. However, synthetic derivatives such as 6-dimethylamino-5β-5,6-secocholestane-3β,5α-diol and 25-azacholest-5-en-3β-ol were cytotoxic [1456,1458, 1463].

Furthermore, cytotoxic effects on cultured mouse L cells expressed in inhibition of growth and cell lysis have been described for the enone 8, being very toxic at 5 µg/mL but causing lysis at 20 µg/mL concentration [2027], and for the common cholesterol autoxidation products 14,15, 16,21, and 27. The 3β,25-diol 27 at 1 µg/mL was the most inhibitory, with (in decreasing order of potency) the (20S)- 3β,20-diol 21, the 7-ketone 16, the 3β,7β-diol 15 and the 3β,7α-diol 14 being inhibitory [473].

Additional systematic studies in the cytotoxicity of oxidized sterols for the inhibition of tumor cell growth utilize Morris rat hepatoma HTC strain 7288c and Zajdela rat ascites hepatoma cells in culture [476,1704,2766]. Moreover, these interests have roots in folk medicine. The Russian anti-cancer drug "tchaga" derived from extracts of the mushroom *Inonotus obliquus* Pil. appears to include oxidized lanosterol derivatives, possibly 5α-(22R)-lanosta-8, 24-diene-3β,22-diol (inotodiol) and 5α-(22R)-lanost-8-ene- 3β,22-diol the latter being cytotoxic towards HTC cells *in vitro* [603,1704,1883,2237]. Still more interesting is the analysis of the Chinese drug *Bombyx cum Botryte* shown to contain several cytotoxic sterol autoxidation products [476].

Of the several oxidized sterols examined for their cytotoxicities against HTC cells, the following listing in TABLE 17 gives potencies for 33 µg/mL amounts of sterol, ranging from very potent to inactive (i.e., same as control, 200-800% growth cells over a three-day period). Similar cytotoxicities were found using the ZHC cell line as well [1704].

Only a few of these oxidized sterols are acknowledged cholesterol autoxidation products, the 3β,7β-diol 15 and 6-ketone 282 being among the most potent, the 3β,25-diol 27 and 3β,7α-diol 14 being weakly active cholesterol autoxidation products. However, other potent oxidized sterols are

TABLE 17. Cytotoxicities of Oxidized Sterols against HTC
           Cells in Culture*

Very Potent Sterols (0-50% growth):
    Cholest-5-ene-3β,7β-diol (15)
    (22R,24RS)-Cholesta-5,25-diene-3β,22,24-triol
    3β-Hydroxycholest-4-en-6-one (282)
    (22R)-Cholesta-5,24-diene-3β,22-diol
    3β-Hydroxycholesta-5,24-dien-7-one
    3β-Hydroxycholesta-5,25-dien-7-one
    Cholesta-5,24-diene-3β,7β-diol
    Cholesta-5,25-diene-3β,7α-diol
    Cholesta-5,25-diene-3β,7β-diol
    5α-(22R)-Lanost-8-ene-3β,22-diol
    Stigmasta-5,E-22-diene-3β,7α-diol
    Stigmasta-5,E-22-diene-3β,7β-diol
    24-Methyl-(24R)-cholest-5-ene-3β,7β-diol
    24-Methyl-(22S,24R)-cholest-5-ene-3β,22-diol

Active Sterols (50-100% growth):
    (22R)-Cholest-5-ene-3β,22-diol (24)
    (22S)-Cholest-5-ene-3β,22-diol
    3β-Hydroxycholest-5-en-7-one (16)
    3β,25-Dihydroxycholest-5-en-7-one
    5α-Cholest-7-ene-3β,6α-diol (305)
    3β-Hydroxycholesta-5,24-dien-22-one
    (23S)-Cholesta-5,24-diene-3β,23-diol
    Cholesta-5,Z-20(22)-diene-3β,25-diol
    5,8-Epidioxy-5α,8α-ergosta-6,E-22-dien-3β-ol (62)

Weakley Active Sterols (100-200% growth):
    Cholest-5-ene-3β,25-diol (27)
    Cholest-5-ene-3β,7α-diol (14)
    Cholest-5-ene-3β,7β,25-triol
    5α-Cholest-7-ene-3β,6β-diol (306)
    5α-Cholestane-3β,6α-diol
    5α-Cholestane-3β,6β-diol (264)
    Cholesta-5,24-diene-3β,7α-diol
    (20S)-Cholesta-5,24-diene-3β,20-diol
    (23R)-Cholesta-5,24-diene-3β,23-diol
    Cholest-5,23-diene-3β,25-diol (181)

(continued)

TABLE 17.   (continued)

Inactive Sterols (200-800% growth):
        Cholesterol (1)
        Cholest-5-ene-1α,3β-diol
        Cholest-5-ene-3β,7α,25-triol
        (22R)-Cholest-5-ene-3β,22,25-triol
        Cholesta-5,24-dien-3β-ol (desmosterol)(78)
        (22S)-Cholesta-5,24-diene-3β,22-diol
        Cholesta-5,E-20(22)-diene-3β,25-diol
        5α-(22R)-Lanosta-8,24-diene-3β,22-diol
        5α-(22S)-Lanosta-8,24-diene-3β,22-diol

---

*Data from [475,476,1704]

formal autoxidation products of congeners of cholesterol,
such as the desmosterol (78) derivatives 3β-hydroxycholesta-
5,24-dien-7-one and the epimeric cholesta-5,24-diene-3β,7-
diols, etc.

The selective cytotoxicity of the 3β,7β-diol 15 towards HTC and ZHC neoplasm cell lines but not towards the
normal mouse 3T3 fibroblast line is in distinction to that
of ergosterol 5α,8α-peroxide (62) which was toxic to all
three cell lines [476]. Moreover, other peculiar specifi-
cities were found to be the case with HTC and ZHC cells in
the same manner previously found with chicken fibroblasts,
the 3β,7β-diol 15 being quite toxic, the campesterol analog
24-methyl-(24R)-cholest-5-ene-3β,7β-diol also active [475]
but the sitosterol analog stigmast-5-ene-3β,7β-diol (77)
being inactive [476].

These cytotoxic effects appear to be related to the
balance of cholesterol (or other sterol) levels in affected
HTC and ZHC cells, as approximately equal amounts (25 μg/mL)
of exogenous cholesterol added to the test systems blocked
the toxic effects of the active sterols such as the 3β,7β-
diol 15 and 3β,25-diol 27. However, here a difference in
effects obtained; the 3β,7β-diol 15 retained cytotoxicity
whereas the 3β,25-diol 27 exhibited a mitostatic effect in
the presence of added exogenous cholesterol [2766].

Cytotoxic effects of oxidized sterols also appear to parallel their inhibitory effects on 3-hydroxy-3-methyl-glutaryl coenzyme A (HMG-CoA) reductase, the key regulatory enzyme implicated in cholesterol biosynthesis from acetate. Desmosterol (78) likewise offers some protection to HTC and ZHC cells, but sitosterol (20) gave no sparing effect [2766]. However, impure sitosterol is cytotoxic towards the Walker 256 intramuscular carcinosarcoma [969], the cytotoxicity likely arising from the presence of unrecognized sitosterol autoxidation products.

The protective effect of added exogenous cholesterol or of mevalonate to other cultured cells under inhibition by oxidized sterols further supports the conclusion that HMG CoA reductase inhibition be the site of inhibition. This matter is treated more fully in a latter section of this chapter.

These findings suggest that the oxidized cholesterol derivatives may interfere with the normal processes by which cholesterol is formed, processed, and metabolized in these cell lines. It is postulated that endogenous sterol, whether cholesterol, desmosterol, or other sterol, be essential to the integrity and functioning of cellular membranes, and in the imbalance in sterol biosynthesis thought to follow upon selective inhibition of HMG CoA reductase the composition of vulnerable membranes be altered, leading to dysfunction and cell death. Specificity of effects then obtains with the more rapidly proliferating neoplastic cells in contrast with normal cells.

An added factor in this rationalization is the stimulation of sterol esterification within the cell affected by the same oxidized cholesterol derivatives 16 and 27 [378, 723,885]. Esterification helps to deplete cellular cholesterol levels. The sparing action of exogenous cholesterol might be explained in terms of pinocytosis, the sterol thereby entering the cell being available for membrane maintenance. Moreover, oxidized sterols that block the action of HMG CoA reductase but do not display cytotoxicity to the cultured cells may be incorporated into membranes themselves, thereby sparing the need for cholesterol as such.

TABLE 18. Cytotoxicities of Oxidized Cholesterol Derivatives and Congeners in Tissue Cultures

Sterol	Chick Heart Explants*		Rabbit Aorta Explants**
	At 63 µg	At Other Levels	
5α-Cholestane-3β,5,6β-triol (13)	Very toxic	No growth at 13 µg	Highly toxic
Cholest-5-ene-3β,7α-diol (14)	Toxic	Slightly toxic at 13 µg	Nontoxic
Cholest-5-ene-3β,7β-diol (15)	Very toxic	Slightly toxic at 6 µg	Moderately toxic
(24S)-Cholest-5-ene-3β,24-diol (25)	Toxic	Slight growth at 31 µg	—
Cholest-5-ene-3β,25-diol (27)	Toxic	Toxic at 6 µg	Moderately toxic
(25R)-Cholest-5-ene-3β,26-diol (29)	Toxic	Toxic at 31 µg	Highly toxic
Cholest-5-en-3-one (6)	Toxic	Toxic at 13 µg	Moderately toxic
Cholesterol (1)	Nontoxic	—	Moderately toxic
5α-Cholest-7-en-3β-ol (57)	Toxic	Toxic at 13 µg	Highly toxic
5α-Cholestan-3β-ol (2)	Toxic	Nontoxic at 13 µg	Highly toxic
Fucosterol (75)	Very toxic	Slight growth at 13 µg	—
Ergosterol (65)	Very toxic	Nontoxic at 31 µg	—
6α-Hydroperoxycholest-4-en-3-one (59)	Very toxic	Slight growth at 31 µg	—

* Data from [249]
** Data from [1546,1547]

Other cytotoxicity data for several cholesterol autoxidation products and cholesterol congeners in tests involving tissue cultures are summarized in TABLE 18. Here toxicities towards chick heart explants of a standard 63 µg of sterol and of other levels are listed [249], together with more limited toxicity data obtained using New Zealand white rabbit aorta explants [1546,1547].

Several sterols not listed in TABLE 18 were also very toxic, such derivatives as 5α-cholesta-8,24-dien-3β-ol (zymosterol) 3β-acetate, cholesterol 3β-propionate, lumisterol (10α-ergosterol, 10α-ergosta-5,7,E-22-trien-3β-ol), 4α-methyl-5α-cholest-8-en-3β-ol, and 4α-methylcholesta-5,7-dien-3β-ol being noted [249].

Several other sterols were also weakly toxic towards the chick heart explants, stigmasterol (70) being an example of nonoxidized sterols. However, many sterols were nontoxic, including cholesterol, 5β-cholestan-3β-ol (4), cholest-4-en-3β-ol (79), campesterol (71), sitosterol (20), desmosterol (78), α-spinasterol, dihydrolanosterol (68), dicholesteryl ether (18), the 7-ketone 16 3β-acetate, the 3,5-diene-7-ketone 10, and the 4,6-diene-3-ketone 12, inter alia [249].

The toxicity effects elicited in the rabbit intimal explants involved the absence of nucleii from medial smooth muscle cells, the presence of abnormal basophilic granules therein, cellular death, and autolysis [1546,1547].

ATHEROGENICITY

One of the most important ramifications of cytotoxicity of oxidized cholesterol derivatives is the issue of atherogenicity: whether these sterols induce or promote atherosclerosis. The most widely recognized biological activity of cholesterol is that of a nutritional agent capable of inducing atheromas of the aorta and other arteries and of hypercholesterolemias. Induction of aortal athersclerosis as specific consequence of cholesterol feeding to rabbits first reported by Anitschkow in 1912 was confirmed directly thereafter [65-67,2622], and the method has survived to the present day as a routine means of provoking experimental atherosclerosis in rabbits which resembles natural human atherosclerosis.

However, the atherogenicity ascribed to cholesterol may well be due to other sterol congeners, cholesterol autoxidation products, etc. Thus, rabbits fed egg yolk rich in cholesterol congeners and precursors develop atheromas more extensively than rabbits fed pure cholesterol [1391]. That cholesterol used in feeding experiments to provoke artery injury and hypercholesterolemia is not a single chemical substance was recognized by 1935 [636,637], and over the years awareness that cholesterol *per se* may not be the active agent evolved. Altschul clearly set the problem down in 1950 [48].

Cholesterol preparations most commonly used for the dietary induction of atherosclerosis in experimental animals are of USP grade, thus containing the usual amounts of congeners 5α-cholestan-3β-ol (2), 5α-cholest-7-en-3β-ol (57), cholesta-5,7-dien-3β-ol (56), (24S)-cholest-5-ene-3β,24-diol (25), desmosterol (78), etc., as well as autoxidation products of each. Generally only the autoxidation products of cholesterol have received attention, but products derived from all congeners could be implicated. Finally, as diets of experimental animals may also contain grain or other vegetable material, congeners and autoxidation products of sitosterol, stigmasterol, campesterol, etc., also need consideration.

Even were highly purified cholesterol used for preparation of diets, the means by which the food be prepared and stored may severely compromise conclusions regarding cause and effect. Thus, standard rabbit chow is variously coated with solid USP cholesterol or with oily solutions thereof, or wetted with diethyl ether solutions (!) of cholesterol in order to attain stipulated levels of cholesterol in the diet. Treated food has not been stored away from air. Some of these methods virtually guarantee that cholesterol autoxidation products will be present in the prepared diet. The simple experiment where highly purified cholesterol, free from congeners and autoxidation products and maintained free from autoxidation over the course of the feeding period, is fed experimental animals, with a view of determining whether cholesterol *per se* be atherogenic appears not to have been conducted!

Congeners of Cholesterol

Evaluation of recognized congeners of cholesterol and of a few cholesterol autoxidation products as atherogenic agents in feeding experiments has been quite restricted, again probably because of the relative inaccessibility of other sterol derivatives. Thus, the 5α-stanol 2 found ubiquitously as a congener and metabolite of cholesterol is atherogenic when fed to rabbits [516,1748] and to chicks [1747] but prevents hypercholesterolemia and atherosclerosis in cholesterol-fed chickens [2264] and cholesterol-type fatty liver in mice [174]. The 5α-stanol 2 also suppresses the biosynthesis of cholesterol from acetate [2429].

The cholesterol biosynthesis precursors cholesta-5,7-dien-3β-ol (56) and 5α-cholest-7-en-3β-ol (57) were also atherogenic when fed rabbits. In these studies both sterols were absorbed from the intestines, but the bioconversion of 57 and 56 to cholesterol may have occurred, leaving the issue of their inherent atherogenicity unsettled [516].

Stigmasterol fed rabbits up to 116 days failed to induce artery damage or other pathological effects. However, the poor intestinal absorption of the sterol is noted as an obvious influence on the matter [48].

Cholesterol Metabolites

Concern for the potential atherogenicity of other cholesterol biosynthesis precursors and/or metabolites as direct causes of arterial lesions prompted the evaluation of over one hundred steroids in chick heart explants [249] already discussed in the previous section of this chapter. Indeed, the concept that an endogenous sterol present in the blood or formed in the aorta be capable of initiating or promoting human aortal plaque formation is supported by the discovery of such trace level sterol in human aortal tissues. Other than cholesterol and its fatty acid and sulfate esters, human aortal tissue is known to contain the 5α-stanol 2 [816,875,1601,1689,2133] and the 5β-stanol 4 [493] as obvious metabolites of cholesterol but also (25R)-cholest-5-ene-3β, 26-diol (29) (and its fatty acyl esters) and fatty acyl esters of a cholest-5-ene-3β,24-diol (25 and/or 107) and of the 3β,25-diol 27 are present in human aortal tissues [362,365,369,493,816,874,961,962,1404,2315,2351,2460,2567],

where these derivatives accumulate preferentially with increasing age and severity of atherosclerosis [816,2315]. As
the several side-chain diols 25, 27, and 29 were among
the oxidized cholesterol derivatives found to be toxic toward chick heart explants [249] and the (25R)-3β,26-diol 29
was highly toxic towards rabbit aorta explants [1546,1547],
the concept of the possible induction of aortal lesions by
these endogenous sterols derived. However, the 5α-stanol 2
also present in human aortal tissues does not appear to
accumulate preferentially there with increasing age or severity of atherosclerosis [816] but is toxic to both chick
heart and rabbit aorta explants [249,1546,1547].

     Moreover, cholesterol autoxidation products have uni-
formly been detected in human aortal tissues and plaques.
Among the most prominent autoxidation products are the 3β,7-
diols 14 and 15 [960,1012,2567], the 7-ketone 16 [364,2567],
the 3,5-diene-7-ketone 10 [493,960,2567], the 3β,5α,6β-triol
13 [960,1012,2567], and the 3β,25-diol 27 [2567], but fatty
acyl esters of the 3β,7-diols 14 and 15 [369,493,1404] and
of the 3β,25-diol 27 [2463] have also been detected in human
aortal tissue! The uncertainty associated with observations
of the free sterols 13-16 leaves the question open whether
artifact or endogenous metabolite status applies, and a similar status obtains for these same sterols in blood. However,
the fatty acyl ester derivatives of 13-16 and 27 suggest that
presently unrecognized metabolic processes involving these
sterols occurs in the human aorta.

     An even more preplexing issue is raised by the discovery
of highly oxidized fatty acyl derivatives of cholesterol
among human aortal plaque sterols. These derivatives include 9'- and 13'-hydroperoxy derivatives of cholesterol
3β-linoleate (234) as well as corresponding 9'- and 13'-alco-
hol and ketone derivatives [961,962] and may be formed by
peroxidative (or autoxidative) processes occurring immediately *post mortem* but before autopsy and tissue collection.
The loss of enzymic protection processes against peroxidations following death has been suggested as likely means of
origin for these derivatives, but whether they exist in living tissue or not cannot be addressed.

     Nonetheless, enough circumstantial evidence is available
to support the concept that some natural or uncontrolled
peroxidation or oxidation of cholesterol or its fatty acyl
esters may occur *in vivo* and that these oxidized derivatives

may accumulate in the aorta where their putative toxic effects may be expressed in aortal lesions. It is therefore
important to examine the issue of the atherogenicity of established endogenous cholesterol precursors and metabolites
and of the recognized autoxidation products as well.

Unfortunately, very little experimental work has been
conducted along these lines. A feeding experiment involving the (25R)-3β,26-diol 29, as the most prominent candi-
date, being both present in human aortal tissue [364,816,
2315,2351,2567] and of demonstrated cytotoxicity towards rabbit aortal tissue [1546,1547], has not been considered.
Neither has an experiment using the 3β,25-diol 27 also of
toxicity [249,1546,1547] and also present in human aortal
tissues [2567]   been recorded.

Cholesterol Oxidation Products

The question whether cholesterol autoxidation products
be atherogenic has recurred repeatedly without resolution
over the past three decades. Whereas intravenous injection
of aqueous sodium stearate dispersions of cholesterol and
its common autoxidation products into rabbits caused lipid
droplets within intimal cells within 24 h and intimal cell
proliferation within 72 h [1279], rabbits fed cholesterol
subjected to hydrogen peroxide oxidations (probably containing many unidentified oxidation products) caused relatively
little artery damage in comparison with that induced by
cholesterol which had been heated in air [48].

Other similar studies suggested that cholesta-3,5-diene
(11), cholest-4-ene-3-one (8), and the 7-ketone 16 not be
atherogenic when fed to rabbits, although liver toxicity was
noted for the ketones 8 and 16 [48]. The enone 8 is known
from other studies to inhibit the biosynthesis of cholesterol
from acetate [1805,2355,2497] and to reduce serum cholesterol
levels in experimental animals, but adrenal hypertrophy in
rats fed 1% levels of 8 [2355,2356] and other effects in dogs
and chickens [2496] clearly ruled further clinical interest
"out of the question" [2355].

The apparent greater severity of aortal lesions in rabbits fed cholesterol preparations which had been heated in
the air (thereby probably autoxidized more) has been noted
[45,48].

    Schwenk addressed the general question whether other
impurities in dietary cholesterol preparations were the
atherogenic agents rather than cholesterol. Thus, cholesterol was purified every day (!) by forming the oxalic
acid complex followed by the bromination-debromination
procedure of Fieser to give purified cholesterol [2170].
However, the Fieser procedure is accompanied by autoxidation and oxidation by bromine, and unless proper care be
exercised in subsequent recrystallizations to remove these
products, some oxidation products might escape notice.
After 12 weeks of feeding the purified cholesterol was
as active as USP cholesterol in inducing aortal lesions
and hypercholesterolemia, but the impurities fed were inactive [2170]. In the face of more recent information,
these findings cannot be accepted as presently meaningful,
both as regards the atherogenicity of pure cholesterol and
the inactivity of cholesterol autoxidation products.

    Only in the case of the readily available 3β,5α,6β-triol
13 has the Anitschkow type of feeding experiments have been
conducted. By feeding rabbits the triol 13 at 0.1% levels
in a basal diet (30 mg/kg/d up to 350 d) aortal lesions resembling those found naturally in man were produced [517].
Whereas intestinal absorption of exogenous sterols administered *per os* is always of concern for such experiments, the
triol 13 was absorbed by rabbits in sufficient amounts to
be found in other tissues [517,2007]. Nonetheless, in other
studies levels of the triol 13 up to 0.5% in diets also containing cholesterol did not produce aortal lesions in rabbits [82].

    However, hypocholesterolemic effects in experimental
animals were also elicited by feeding the triol 13. Lowered
serum cholesterol levels in rabbits [82,517,599,2716], chickens [82], Rhesus monkeys [96], White Carneau pigeons [2630],
etc., are recorded, but in rats liver cholesterol levels
were diminished rather than serum levels [82,1152]. The
triol 13 appears to achieve these effects by interferring
with intestinal absorption and thereby increasing fecal
excretion of sterols [1152,1164,2630].

    Intragastric administration of the triol 13 to lymph
duct-cannulated rats resulted in reduced cholesterol absorption into lymph but in little effect on fatty acid absorption. A reversible enlargement of mucosal cell mitochondria
and marked increase in length of microvilli accompanied triol
feeding [1164].

The 3β,5α,6β-triol 13 appears also to interfere at
several points with the biosynthesis of cholesterol from
acetate. Whether acetate or mevalonate be used as substrate
in rat liver homogenates, the triol 13 greatly inhibited
sterol biosynthesis, and the elaboration of the B-ring of
cholesterol is also affected. Both partially purified
Δ^5-dehydrogenase implicated in the bioconversion of the
7-stenol 57 to the diene 56 and Δ^7-reductase catalyzing
transformation of 56 to cholesterol were inhibited by the
triol 13 [599,2717,2718]. The triol 13 also may interfere
with the transformation of 4-methylsterols to 5α-cholest-8-
en-3β-ol in the biosynthesis of cholesterol [2088].

Considerations for clinical use of the triol 13 for
control of cholesterol biosynthesis appear to have been
contemplated [2717], as several analogs [82,2717,2718]
were made and variously tested, triol 13 metabolism studied
[1294,2006,2007,2717], and acute and chronic toxicities
evaluated in several experimental animals [1266].

Recently a rebirth of interest in the inherent atherogenicity of cholesterol has surfaced, with tests in a variety of systems using USP cholesterol, highly purified
cholesterol, and concentrates of cholesterol autoxidation
products prepared therefrom. Mere recrystallization of USP
cholesterol from methanol in the time-honored manner affords
methanol mother liquors in which the cholesterol autoxidation
products are concentrated and from which much of our own
chemical work has proceeded. These mother liquor preparations of cholesterol autoxidation products contain all the
recognized derivatives 13-16,27,35,36,46, and 47 as well as
all the many other recognized autoxidation products.

Such mother liquor concentrates have become standard
for provoking artery damage in experimental animals, both
oral administration and intravenous injection being effective. Within 24 h of administration per os of 250 mg/kg
cholesterol mother liquor autoxidation products (or within
6 h following 100 mg/kg intravenous injection) to New Zealand
white rabbits focal edema of the aorta and random smooth
muscle cell degeneration throughout the media were observed.
Three 100 mg/kg doses by gavage over 5 weeks caused slightly
elevated focal intimal lesions [1146,1443]. These aortal
lesions were distinct from those commonly induced by cholesterol feeding, the concentrated autoxidation products preparations causing lesions characterized by diffuse intimal

proliferation of smooth muscle cells and fibrous stroma
[1151]. Additionally, the aortas of New Zealand white
rabbits fed air-aged cholesterol at 1 g/kg or mother liquor
concentrate therefrom at 250 mg/kg exhibited intimal and
medial cell pyknosis and cell death with 24 h. Moreover,
accompanying a cytoplasmic diffuseness in affected cells
were dense mineral deposits that contained Ca, P, and Cl
[1147]. Many of these same effects were also obtained with
New Zealand white rabbits fed (1 g/kg) with cholesterol
that had been heated at 98°C for 6 d to effect extensive
autoxidation [1143]. There is also an indication that oxidized cholesterol obtained by heating in air may be more
angiotoxic in Japanese quail than in pure cholesterol
[2287].

These preparations of oxidation products are also angiotoxic on intravenous administration in saline. Injections of 5 mg/kg caused damaging effects in rabbits, including fibromuscular thickening and cell death in the
aorta and segmental thickening in both major and minor
branches of the pulmonary artery. Of the identifiable
components of the autoxidation products mixture used, the
3β,25-diol 27 and 3β,5α,6β-triol 13 were as potent as the
concentrated mixture in inducing these angiotoxic effects
in rabbits, the 5α,6α-epoxide 35 and 7-ketone 16 less so
[1144,1145,1149,1150]. Intravenous injections of the
3β,25-diol 27 and 3β,5α,6β-triol 13 in ethanol likewise
caused aorta damage in rabbits [2451].

Cholesterol purified via the dibromide did not cause
the damaging effects on rabbit arteries [1146,1150,1443]
that air-aged cholesterol, concentrated autoxidation products or specific oxidized sterols such as the 3β,25-diol
27 or 3β,5α,6β-triol 13 induced in vivo. Interestingly,
cholecalciferol (90) fed rabbits at 1 mg/kg also caused
cell injury and death [1142].

These toxic effects may also be demonstrated in vitro
using tissue and cell cultures. The toxicities of the 3β,
5α,6β-triol 13, 3β,7β-diol 15, and 3β,25-diol 27 have already been mentioned with respect to chick heart and rabbit
aorta explants (cf. TABLE 18). Moreover, monolayer cultures
of aorta smooth muscle cells from New Zealand white rabbits
are sensitive to autoxidized cholesterol [1846,1847] and to
several defined cholesterol autoxidation products, as shown
in TABLE 19 [1848]. The marked cytotoxicity of the

TABLE 19. Cytotoxicity in Aortal Smooth Muscle Cell Cultures*

Sterol	Cytotoxicity (dying or dead cells), %				Inhibition of HMG CoA Reductase, %
	10 µg/mL	20 µg/mL	50 µg/mL	100 µg/mL	3 µg/mL
Very Toxic:					
Cholest-5-ene-3β,25-diol (27)	5-25%	25-50%	75-100%	75-100%	83.2
(20R)-Cholest-5-ene-3β,20-diol (21)		5-25	25-50	75-100	83.0
5α-Cholestane-3β,5,6β-triol (13)	5-25	25-50	75-100	75-100	63.4
3β,5-Dihydroxy-5α-cholestan-6-one (44)		5-25	25-50	50-75	47.2
3β-Hydroxy-5α-cholestan-6-one (45)		5-25	25-50	75-100	78.3
Moderately Toxic:					
Cholest-4-en-3-one (8)			5-25	25-50	22.8
Cholest-5-ene-3β,7α-diol (14)			5-25	25-50	59.4
Cholest-5-ene-3β,7β-diol (15)			5-25	50-75	50.5
3β-Hydroxycholest-5-en-7-one (16)			5-25	50-75	78.1
Nontoxic:					
Cholesta-3,5-dien-7-one (10)					15.1**
5,6α-Epoxy-5α-cholestan-3β-ol (35)					55.0
5α-Cholestane-3,6-dione (42)					71.8
Pure cholesterol					12.0

* Data from [1848]

** Stimulation observed

3β,25-diol 27 and 3β,5α,6β-triol 13 in the monolayer cell
cultures correlates directly with prior results from rabbit aortal tissue explants and from intact animal studies.

    A suggestion as to how these toxic sterols exert their
effects may be had from other data of TABLE 19 demonstrating
inhibitory actions on HMG CoA reductase previously mentioned
as correlating with the cytotoxicities of some oxidized
sterols in other cultured cells, a matter discussed in more
detail in a later section of this chapter. The very toxic
sterols (20R)-3β,20-diol 21, 3β,25-diol 27, 3β,5α,6β-triol
13, and 6-ketone 44 are potent inhibitors of HMG CoA reductase and accordingly of *de novo* biosynthesis, this inhibition
then putatively affecting cell functions, growth, and reproduction adversely [1848,2473]. However, the potent enzyme
inhibitory actions of the 7-ketone 16 and 3,6-diketone 42
are not reflected in great cytotoxicity, and several sterols
with about the same enzyme inhibitory activity range from
very toxic (the 6-ketone 44) to moderately toxic (the
3β,7-diols 14 and 15) to nontoxic (the 5α,6α-epoxide 35),
thus mitigating the simple correlation.

    Furthermore, the dynamics of onset of toxicity may be
different for the several toxic sterols, suggesting that
perhaps more than one mechanism be implicated. These items
contribute to the hypothesis that some oxidized sterols
exert some of their effects at the cellular membrane level,
a matter dealt with more fully at the end of this chapter.
Other features of the biological actions of the 7-ketone 16
appear to be related to atherosclerosis. The 7-ketone 16
inhibits the uptake of cholesterol from perfusate in perfused human and pig coronary arteries and in perfused rabbit
aorta [224,2348]. The effect was not demonstrated in intact
rabbits fed the 7-ketone 16 and but uncertainly upon intravenous administration of 16 in different vehicles [2076,2078].
However, infusion of lipoprotein-bound 16 in isologous blood
did inhibit the uptake of cholesterol in carotid and femoral
arteries of intact rabbits [2348].

EFFECTS ON SPECIFIC ENZYMES

    There is evidence that cholesterol autoxidation pro-
ducts exert inhibitory effects on several important enzymes
implicated in the biosynthesis and metabolism of cholesterol
in mammalian cells. At least three regulatory enzymes

3-hydroxy-3-methylglutaryl coenzyme A (HMG CoA) reductase, hepatic cholesterol 7α-hydroxylase, and the adrenal cortex mitochondrial enzyme system implicated in the scission of the cholesterol side-chain in the biosynthesis of pregnenolone (23), of obvious importance in sterol, bile acid, and steroid hormone biosynthesis respectively, are inhibited. In addition several other key enzyme steps implicated in the bioconversion of lanosterol (67) to cholesterol are also affected.

These inhibitory effects have generally not been observed in cell-free enzyme systems but appear to be dependent upon intact functioning mammalian cells for their expression. Moreover, the requirement for intact cells may be interpreted in terms of cellular membrane phenomena, with putative receptor sites on the plasma membrane of affected cells. Indeed, hypotheses involving the possibility that oxidized sterol derivatives function as endogenous regulators of *in vivo* sterol biosynthesis and metabolism [471,472,1229,1232-1234] have been advanced.

HMG CoA Reductase

The key regulatory step implicated in the biosynthesis of sterols from acetate is the reduction of 3-hydroxy-3-methylglutaryl coenzyme A to mevalonate by 3-hydroxy-3-methylglutaryl coenzyme A (HMG CoA) reductase (mevalonate: NADP$^+$ oxidoreductase (CoA acylating), EC 1.1.1.34). The enzyme is membrane-bound within the endoplasmic reticulum and is under product (cholesterol) feedback control. Generally, inhibitors of sterol biosynthesis from acetate in different biological systems are more specifically associated with inhibitions of HMG CoA reductase. The inhibition is not accompanied by inhibitions of fatty acids biosynthesis, CO_2 formation, ribonucleic acid or protein biosynthesis, etc. [1228,1229] or of cyclic adenosine monophosphate levels [1230]. A satisfying correlation between inhibition of sterol biosynthesis from acetate and specific inhibition of HMG CoA reductase in an established mouse liver cell culture has been provided [1228].

Dietary Experiments. The suppression of sterol biosynthesis from acetate in experimental animals administered cholesterol autoxidation products stems from early observations of inhibitions by cholest-4-en-3-one (8) in incubations

of liver slices from rats fed 8 [2355,2497] and from mice
fed 8 or injected intraperitoneally with 8 [1236,1805] and
also in incubations of mouse fibroblast L cells [2027].

The inhibition of HMG CoA reductase by cholesterol, the
enone 8, cholesta-4,6-dien-3-one (12), and other steroids
fed mice at levels as low as 0.25% was complete within one
day following feeding, thus suppressing the biosynthesis of
cholesterol from acetate but not from mevalonate. Mice injected with the steroids were similarly affected. However,
various odd responses were observed in different genetic
strains of mice, and high (4%) levels of the enone 8 in the
diet stimulated rather than inhibited cholesterol biosynthesis from acetate [1236,1805].

Other studies have extended this means of suppression
of cholesterol biosynthesis to different modes of administration. Feeding mice chow containing the 7-ketone 16, the
3β,25-diol 27, the (20S)-3β,20-diol 21, or 3β-hydroxy-5α-
cholestan-6-one (45) depressed the incorporation of acetate
into sterols in liver slices from fed animals [470]. However, in the intact C57BL/6J mouse fed 16 or 27 intestinal
cholesterol biosynthesis was suppressed but not liver biosynthesis [1235].

Dietary experiments wherein the 7-ketone 16 was fed
rats at 0.1-0.5% levels for 18 h likewise resulted in substantial inhibition of liver HMG CO CoA reductase, as did
also feeding the 3β,25-diol 27 at 0.1% levels. In either
case feeding extended to 66 h did not result in increased
inhibitions, but a tolerance to both sterols 16 and 27
appeared the case [701,703]. An indication that the ratio
of HDL to LDL in plasma of guinea pigs administered an oxidized sterol analog orally may be increased has been reported [1191].

Inhibition of hepatic HMG CoA reductase was also had
in perfused rat liver where 7-ketone 16 or 3β,25-diol 27 was
added to the perfusate [701,703].

With respect to such feeding experiments, the same
reservations which hold for feeding of cholesterol in chronic tests apply to feeding experiments in which the cholesterol autoxidation products are employed. As with cholesterol, it is usual to mix oxidized sterols as diethyl ether
or acetone solutions with animal food to remove the solvent

by slow drying or stirring in air [703,1235,1236]. Although
the autoxidation of the 3β,25-diol 27 has not been demonstrated, there is little doubt that this sterol may autoxidize
in a fashion similar to that of cholesterol to give putatively such 7-oxygenated analogs as epimeric cholest-5-ene-3β,
7,25-triols and 3β,25-dihydroxycholest-5-en-7-one, the latter being also a likely product of the autoxidation of the
7-ketone 16.

    Cultured Cells. Yet other test systems for demonstra-
tion of the inhibitory effects of cholesterol autoxidation
products on HMG CoA reductase and sterol biosynthesis from
acetate have been devised. Thus, inhibitions of these processes by the 7-ketone 16 and by the 3β,25-diol 27 (as well
as by pure cholesterol) has been demonstrated in organ cultures involving dog ileal mucosa explants [858].

    However, it is in the realm of monolayer culture of
individual cell lines that recent progress in the demonstration of enzyme inhibitory effects of cholesterol autoxidation products has been made. The description of a human
skin fibroblast culture in which cholesterol biosynthesis
from acetate was suppressed by additions of exogenous
sterols, including cholesterol, sitosterol (20), the 7-stenol
57, demosterol (78), lanosterol (67), dihydrolanosterol (68),
stigmasterol, etc. [112] ushered in the period of present
sustained interest in evaluation of various sterols as inhibitors of cholesterol biosynthesis in cell cultures and as
specific inhibitors of HMG CoA reductase, matters which have
since received great attention.

    The suppression of cholesterol biosynthesis from ace-
tate in cultured L cells containing as little as 1 µg/mL of
the enone 8 [2027] evinced the potency of cholesterol autoxidation products for this purpose. More recently cholesterol autoxidation products listed in TABLE 20 have been
demonstrated to be inhibitory towards microsomal HMG CoA
reductase and/or sterol biosynthesis from acetate in a wide
variety of mammalian cells in culture.

    In addition to the data of TABLE 20, a large number of
natural and synthetic sterols have been tested for their
capacity to inhibit de novo sterol biosynthesis and HMG CoA
reductase in more limited test systems. Human skin fibroblasts, mouse L and primary liver cells, and Chinese hamster
lung cells have been most employed. In some cells pure

TABLE 20. Inhibitions in Cultured Cells of Sterol Biosynthesis from Acetate and/or of HMG CoA Reductase

Sterol	Cell Line	Range of Inhibitory Concentrations, µg/mL	References
Cholest-5-en-3-one (6)	Human fibroblasts	5	[381]
	Rat hepatoma HTC	3	[179]
Cholest-4-en-3-one (8)	Human fibroblasts	5	[381]
	Mouse L cells	1	[2027]
	Primary mouse liver	–	[1228]
	Rat hepatoma HTC	0.70	[179]
	Rabbit aorta smooth muscle	3	[1848]
Cholesta-3,5-dien-7-one (10)	Human fibroblasts	–	[884]
	Rabbit aorta smooth muscle	3	[1848]
	Mouse L cells	15	[1232]
Cholesta-4,6-dien-3-one (12)	Rat hepatoma HTC	2	[179]
	Mouse L cells	15	[1232]
5α-Cholestane-3β,5,6β-triol (13)	Human lymphocytes	(3 µM)	[2735]
	Hamster lung cell	–	[448]
	Rat hepatocytes	–	[980,2643]
	Rabbit aorta smooth muscle	3	[1848]
Cholest-5-ene-3β,7α-diol (14)	Human fibroblasts	3	[884]
	Human lymphocytes	(0.1 µM)	[2735]
	Mouse L cells	1	[1228]
	Primary mouse liver	5	[1228]
	Rat hepatoma HTC	0.60	[179]

(continued)

TABLE 20. (continued)

Cholest-5-ene-3β,7β-diol (15)	Rabbit aorta smooth muscle	3	[1848,2473]
	Hamster lung cells	(11.5 μM)	[448]
	Human fibroblasts	3	[884]
	Human lymphocytes	(3 μM)	[2735]
	Mouse L cells	1	[1228]
	Primary mouse liver	5	[1228]
	Mouse fibroblast 3T3	5	[2766]
	Rat hepatoma HTC	0.60-5	[179,2766]
	Rat hepatoma ZHC	5	[2766]
	Rabbit aorta smooth muscle	3	[1848,2473]
	Hamster lung cells	(3.1 μM)	[448]
3β-Hydroxycholest-5-en-7-one (16)	Human fibroblasts	0.1-5	[351,377-379,381,884,1922]
	Human lymphocytes	(1 μM)	[2735]
	HeLa S3G cells	1-10	[449]
	Mouse L cells	2-5	[471]
	Mouse FL83 liver	5-10	[471]
	Mouse SWR/J embryo	5-10	[471]
	Mouse AKR/J leukemia	2-10	[471]
	Mouse LM fibroblasts	5	[2645]
	Rat hepatoma SFHMT	2-10	[471]
	Rat brain G-6 glial astro-cytoma	1-5	[2601]
	Rabbit aorta smooth muscle	1-5	[1848,1909]
	Bovine adrenal cortex	2	[1375]
	Calf media smooth muscle	10	[1635]
	Hamster lung cells	(2 μM)	[448]

(continued)

TABLE 20. (continued)

(20S)-Cholest-5-ene-3β, 20-diol (21)	Human lymphocytes	(0.1 μM)	[2735]
	Mouse L cells	(1.5 μM)	[1229,1231]
	Rabbit aorta smooth muscle	3	[1848,2473]
	Calf media smooth muscle	10	[1635]
	Hamster lung cell	(0.5 μM)	[448]
	Human fibroblasts	0.1-5	[351,378,380,381, 1922]
Cholest-5-ene-3β,25-diol (27)	Human lymphocytes	(0.3 μM)	[2735]
	HeLa S3G cells	1-10	[449]
	Mouse L cells	1-2	[471,1229,1231,1239]
	Mouse PHA-stimulated lympho- cytes	1-5	[468,471]
	Primary mouse liver	1-5	[471]
	Mouse FL83 liver	2-5	[471]
	Mouse SWR/J embryo	2-5	[471]
	Mouse LM fibroblasts	1	[2645]
	Rat AKR/J leukemia	1-2	[471]
	Rat ovary granulosa cells	3	[2156]
	Rat hepatoma HTC	0.45-5	[179,2766]
	Rat hepatoma ZHC	5	[2766]
	Rat hepatocytes	5-10	[703,980,2643]
	Rabbit aorta smooth muscle	3	[1848,2473]
	Guinea pig lymphocytes	(80 nm)	[1860-1862]
	Chinese hamster ovary CHO-Kl	0.5	[2260]
	Hamster lung cells	(0.6 μM)	[448]
	Calf media smooth muscle	10	[21,1635]

(continued)

TABLE 20. (continued)

3β-Hydroxycholest-5-en-24-one (34)	Hamster lung cells	(1.5 µM)	[448]
5α,6-Epoxy-5α-cholestan-3β-ol (35)	Mouse L cells	9.3	[1232]
	Primary mouse liver	17.3	[1232]
	Rabbit aorta smooth muscle	3	[1848,2473]
Cholest-5-ene-3β,4β-diol (41)	Human lymphocytes	(4.5 µM)	[2735]
5α-Cholestane-3,6-dione (42)	Rabbit aorta smooth muscle	3	[1848]
3β,5-Dihydroxy-5α-cholestan-6-one (44)	Human lymphocytes	(3 µM)	[2735]
	Hamster lung cells	(10 µM)	[448]
	Rabbit aorta smooth muscle	3	[1848]
3β-Hydroxy-5α-cholestan-6-one (45)	Human fibroblasts	1-5	[378,381]
	Human lymphocytes	(1 µM)	[2735]
	Hamster lung cells	(2 µM)	[448]
	Rabbit aorta smooth muscle	3	[1848]
3β-Hydroxycholest-5-en-22-one (297)	Human lymphocytes	(0.35 µM)	[2735]
	Hamster lung cells	(2.3 µM)	[448]
3β-Hydroxy-27-norcholest-5-en-25-one (229)	Human fibroblasts	5	[381]

cholesterol is not inhibitory, but in others not only is
cholesterol (as low density lipoprotein) inhibitory at
5 μg/mL but desmosterol (78) and lanosterol (67) are also
[381]. Kryptogenin (3β,26-dihydroxy-(25R)-cholest-5-ene-
16,22-dione) is inhibitory in L cells, but diosgenin ((25R)-
spirost-5-en-3β-ol) and tigogenin (5α-(25R)-spirostan-3β-ol)
were not. Moreover, ecdysterone (2β,3β,14α,20,22,25-hexa-
hydroxy-5β-(20R,22R)-cholest-7-en-6-one) was not inhibitory
[1229]. Yet other diverse structural types are also inhibitory. Although cholecalciferol (90) was not inhibitory in
guinea pig lymphocyte cultures, its metabolites 25-hydroxy-
cholecalciferol (9,10-secocholesta-Z-5,E-7-diene-3β,25-diol)
(91) and 1,25-dihydroxycholecalciferol (9,10-secocholesta-Z-
5,E-7-diene-1ξ,3β,25-triol) (92) were inhibitors. Notably,
cholesta-5,7-diene-3β,25-diol was an even more powerful inhibitor of HMG CoA reductase in these lymphocytes [1861].

        In these tests the matter of sterol purity is always at
issue. Thus, although purified cholesterol was not inhibitory in mouse L cells or in mouse liver cell cultures, impure USP cholesterol was very active, as were also mixtures
of sterols from mouse preputial gland tumors. Moreover, in
extended incubations the possibility that some sterol autoxidation might occur, thereby giving rise to inhibitory oxidized sterols, has been recognized [1228]. Incorporation of
an appropriate antioxidant such as α-tocopherol in prolonged
incubations may serve to reduce such adventitious autoxidations [1231]. However, antioxidant incorporation, also suggested for isolation and chemical work with sterols [14,
2612], is not innocuous, witness the oxidation of ergocalciferol by oxidation products of antioxidant added to protect
the vitamin [14].

        In TABLE 21 the inhibitory potencies of a series of
oxidized cholesterol derivatives, putative biosynthesis
precursors, and synthetic sterols are compared. These data
clearly establish such inhibitions as broad phenomena not
restricted to cholesterol autoxidation products. Of particular interest are the inhibitory actions of the C_{28}-C_{30}
sterols 5α-30,31-bisnorlanost-7-ene-3β,32-diol, 5α-lanost-7-
ene-3β,32-diol, 3β-hydroxy-5α-lanost-7-en-32-al, 5α-lanost-
8-ene-3β,32-diol, and 3β-hydroxy-5α-lanost-8-en-32-al, all
variously metabolized to cholesterol by selected enzyme
systems [866,2148]. That these sterols (which may be natural biosynthesis intermediates linking lanosterol and cholesterol) act to suppress *de novo* cholesterol biosynthesis

TABLE 21. Inhibition of Sterol Biosynthesis by Sterols

| Sterol | Concentration Giving 50% Inhibition, μM | | | | | | References |
| | Sterol Synthesis* | | | HMG CoA Reductase* | | | |
	I	II	III	I	II	III	
Oxidized Cholesterol Derivatives							
(20S)-Cholest-5-ene-3β,20-diol (21)	1.2	5.7	0.2	1.5	3.2	0.4	[448,471,1229]
(22R)-Cholest-5-ene-3β,22-diol	1.0	6.0	–	3.5	7.5	–	[471,1229]
(22S)-Cholest-5-ene-3β,22-diol	3.7	6.0	–	3.5	5.8	–	[471,1229]
(24RS)-Cholest-5-ene-3β,24-diol	0.5	9.0	–	0.3	6	–	[471]
(24R)-Cholest-5-ene-3β,24-diol (107)	0.9	–	–	–	–	–	
(24S)-Cholest-5-ene-3β,24-diol (25)	1.8	–	–	–	–	–	
Cholest-5-ene-3β,25-diol (27)	0.07	1.0	0.25	0.05	3.0	0.26	[448,471,1229]
Cholesta-3,5-dien-7-one (10)	15.0	57	–	–	60	–	[1232]
Cholesta-4,6-dien-3-one (12)	15.0	52	–	>25	>75	–	[1232]
Cholesta-4,7-dien-3-one	2.5	11.0	–	9.5	>75	–	[1232]
3β-Hydroxycholest-5-en-7-one (16)	–	–	0.8	4.5	1.3	0.9	[448,1229]
3β-Hydroxycholest-5-en-22-one (297)	1.7	37	0.9	3.2	62	2	[448,471,1229]
3β-Hydroxycholest-5-en-24-one (34)	0.7	2.5	0.6	1.3	16	0.6	[448,471]

(continued)

TABLE 21. (continued)

(20S)-Pregn-5-ene-3,20-diol (111)	>30.0	–	–	–	–	–	[1229]
23,24-Bisnorchol-5-ene-3β, 20-diol	>30.0	–	–	–	–	–	[1229]
(20S)-24-Norchol-5-ene-3β,20-diol	>30.0	>75	–	>30.0	–	–	[1229]
(20S)-Chol-5-ene-3β,20-diol	15.0	>75	>15	>30.0	–	>15	[448,1229]
(20S)-26,27-Bisnorcholest-5-ene-3β,20-diol	2.7	16.0	–	6.4	75	–	[1229]
(20S)-27-Norcholest-5-ene-3β,20-diol	1.2	3.3	–	5.4	10	–	[1229]
3β-Hydroxy-27-norcholest-5-en-25-one (299)	1.3	28	–	1.6	>75	–	[1229]
27-Norcholest-5-ene-3β,25ξ-diol (301)	0.6	12	–	1.0	26	–	[1229]
Putative Biosynthesis Intermediates							
5α-Lanost-8-ene-3β,32-diol	3.4	7.1	2.6	2.5	6.7	2.8	[448,866]
3β-Hydroxy-5α-lanost-8-en-en-32-al	1.9	3.7	0.7	2.8	4.9	1.4	[866]
5α-Lanost-7-ene-3β,32-diol	2.7	3.6	1.2	1.7	3.4	2.4	[866]
3β-Hydroxy-5α-lanost-7-en-32-al	–	2.0	1.2	1.8	5.2	2.8	[866]
5α-30,31-Bisnorlanost-7-ene-3β,32-diol	2	10	–	3	8	–	[2148]

(continued)

TABLE 21. (continued)

Synthetic Sterols

Compound							Ref.
5α-Cholest-8(14)-ene-3β, 15α-diol	3.7	31	—	8.8	—	—	[2140,2141,2144]
5α-Cholest-8(14)-ene-3β, 15β-diol	1.8	10.3	—	2.5	16.1	—	[2140,2141,2144]
3β-Hydroxy-5α-cholest-8(14)-en-15-one (412)	0.1	4.0	—	0.3	4.0	—	[2140,2141,2144]
3α-Hydroxy-5α-cholest-8(14)-en-15-one	0.5	—	—	3.0	—	—	[2144]
5α-Cholest-8(14)-ene-3,15-dione	0.4	—	—	0.4	—	—	[2144]
5α-Cholest-8(14)-ene-3β, 7α,15α-triol	5.0	4.8	—	1.9	18.0	—	[2141,2518]
5α-Cholest-8(14)-ene-3,7,15-trione	0.4	—	—	1.0	—	—	[2144]
15α-Hydroxy-5α-cholest-8(14)-en-3-one	5.8	—	—	6.0	—	—	[2144]
7α,15β-Dichloro-5α-cholest-8(14)-en-3β-ol	2	—	—	0.6	—	—	[2142]
9α-Fluoro-3β-hydroxy-5α-cholest-8(14)-en-15-one (413)	0.2	—	—	—	—	—	[2145]
9α-Fluoro-5α-cholest-8(14)-ene-3,15-dione	0.2	—	—	—	—	—	[2145]
3β-Hydroxy-5α-cholesta-6,8(14)-dien-15-one	1.0	—	—	0.3	—	—	[2144]
5α-Cholesta-6,8(14)-diene-3,15-dione	0.8	—	—	0.3	—	—	[2144]

(continued)

TABLE 21. (continued)

5α-Cholest-8-ene-3β,15α-diol	0.3	-	0.3	-	[2144]
15α-Hydroxy-5α-cholest-8-en-3-one	0.6	-	0.3	-	[2144]
5α-Cholest-7-ene-3β,15α-diol	0.3	-	0.5	-	[2144]
15α-Hydroxy-5α-cholest-7-en-3-one	0.3	-	0.4	-	[2144]
5α,14β-Cholest-7-ene-3β,15α-diol	3.2	7.5	6.7	12.5	[2141,2144]
5α,14β-Cholest-7-ene-3β,15β-diol	1.0	2.5	4.5	6.0	[2141,2144]
15α-Hydroxy-5α,14β-cholest-7-en-3-one	2.0	-	0.33	-	[2144,2147]
15β-Hydroxy-5α,14β-cholest-7-en-3-one	0.25	-	0.25	-	[2144,2147]
5α-Cholestane-3β,15α-diol	0.2	-	0.5	-	[2144]
15α-Hydroxy-5α-cholestan-3-one	0.2	-	0.8	-	[2144]
5α,14β-Cholestane-3β,15β-diol	0.6	-	4.0	-	[2144]
14-Methyl-5α,14α-cholest-7-ene-3β,15α-diol	0.3	1.8	0.3	2.0	[2140,2141,2146]
14-Methyl-5α,14α-cholest-7-ene-3β,15β-diol	0.5	-	1.2	4.3	[2140,2141,2146]
3β-Hydroxy-14-methyl-5α,14α-cholest-7-en-15-one	0.3	4.5	0.3	2.8	[2140,2141,2146]
3β-Acetoxy-14-methyl-5α,14α-cholest-7-en-15α-ol	0.03	-	0.13	-	[2146]
5α-30,31-Bisnorlanost-6-ene-3β,32-diol	0.2	3	0.5	2	[2148]

(continued)

TABLE 21. (continued)

14-Ethyl-5α,14α-cholest-7-ene-3β,15α-diol	0.05	0.06	—	0.2	2.3	—	[1816,2143,2146]
14-Ethyl-5α,14α-cholest-7-ene-3β,15β-diol	0.4	0.8	—	3.5	7.9	—	[2143,2146]
14-Ethyl-15α-hydroxy-5α,14α-cholest-7-en-3-one (411)	0.006	—	—	0.05	—	—	[2146,2149]
14-Ethyl-3β-hydroxy-5α,14α-cholest-7-en-15-one	0.3	2.1	—	1.9	2.1	—	[2146]
14-Ethyl-3α-hydroxy-5α,14α-cholest-7-en-15-one	0.2	—	—	0.6	—	—	[2146]

*Cultured cell test systems used: I, mouse L cells; II, primary liver cells; III, Chinese hamster lung cells.

suggests the possibility that a natural control system may
exist which utilizes 3β,32-dioxygenated sterols as regulatory agents.

The hypothesis that oxidized cholesterol metabolites
serve in such regulatory action, the 3β,7α-diol 14 in liver,
the (20S)-3β,20-diol 21 in adrenal cortex, the (24S)-3β,24-
diol 25 in brain, etc. previously suggested [471,472,1229,
1232-1234] must then be expanded to include not only oxidized cholesterol metabolites but also oxidized lanosterol
derivatives that are putatively cholesterol biosynthesis
intermediates. Furthermore, in that escape from regulatory
control via cellular mutation is a recognized process occurring in animals and related to metabolic disorders, the
finding that some cultured cell lines give rise to defined
mutant strains that are insensitive to the inhibitory
actions of oxidized sterols may be interpreted in terms
supportive of the hypothesis [2262]. Several different
mutant cell lines have been isolated from Chinese hamster
ovary [2260-2262] and lung [448,467,866] cells and from
rat liver cells [1958] that are resistant to the inhibitory
effects of the 3β,25-diol 27 and 7-ketone 16 on *de novo*
sterol biosynthesis.

Whereas the inhibition of the specific enzyme HMG CoA
reductase implicated in *de novo* sterol biosynthesis is well
established, there may be other sites of inhibition beyond
the mevalonate state. With but the single site of inhibition, one would expect the inhibitory potency of an active
sterol be about the same for both the specific enzyme inhibition and for the inhibition of the full biosynthesis process. However, data for several oxidized sterols of
TABLE 21 clearly demonstrate that such a circumstance is
not the case, that a higher level of inhibitory sterol is
required for 50% inhibition of HMG CoA reductase than for
de novo sterol biosynthesis. Such data suggest additional
enzyme steps that are sensitive to the inhibitory effects
of the active agents involved [2144].

Support for this interpretation is also had from studies with mutant Chinese hamster lung cell lines, the wild
strain of which being sensitive to oxidized sterols with
respect to HMG CoA reductase and lanosterol 14α-demethylase
actions, a mutant strain thereof being insensitive with respect to inhibition of HMG CoA reductase but remaining sensitive with respect to lanosterol 14α-demethylation [866].

Furthermore, in Chinese hamster ovary CHO-Kl cells the inhibition of *de novo* sterol biosynthesis by 14-ethyl-5α,14α-
cholest-7-ene-3β,15α-diol is accompanied by an accumulation
of lanosterol and dihydrolanosterol (68), again pointing to
inhibitions in sterol demethylations [1634].

Several synthetic 15-oxygenated sterols related to possible biosynthesis intermediates linking lanosterol and
cholesterol listed in TABLE 21 have also been shown to be
inhibitors of HMG CoA reductase and of sterol biosynthesis
from acetate in cultures of L cells and primary liver
cells. The most potent of these, 14-ethyl-15α-hydroxy-
5α,14α-cholest-7-en-3-one (411), inhibited sterol biosynthesis at 6 nM concentration! Moreover, 3β-hydroxy-5α-
cholest-8(14)-en-15-one (412) inhibiting 50% of sterol
biosynthesis from acetate at 0.1 μM in L cells was also
active in intact animals. Thus, subcutaneous administration of the 15-ketone 412 to rats resulted in depressed
sterol biosynthesis from acetate (but not from mevalonate)
in liver homogenates and in diminished serum sterol levels
[1907]. Incorporation of the 15-ketone 412 at 0.1-0.2%
levels in the diet also gave a marked hypocholesterolemic

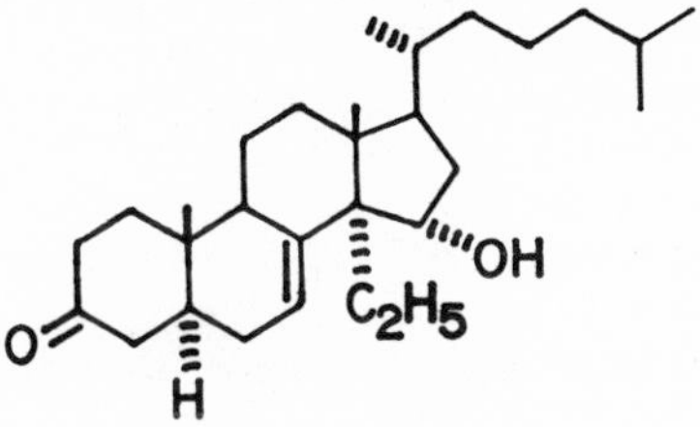

411 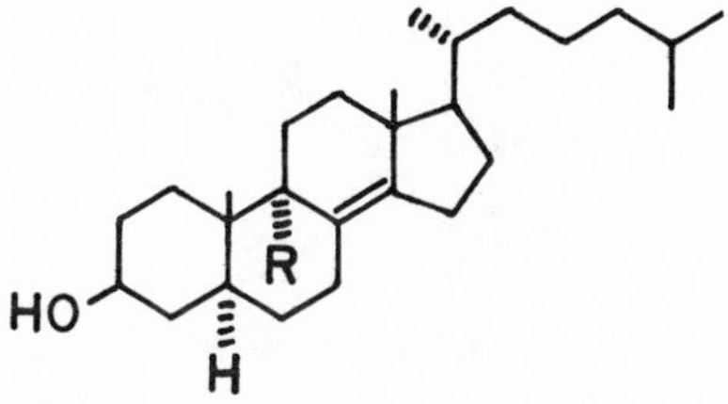   412 R = H

                       413 R = F

response in rats and mice [2139], and subcutaneous injection of the 3β-palmitate ester of the same 15-ketone 412
was significantly hypocholesterolemic in rats   [1317].
Finally, the fluorosterol analog 9α-fluoro-3β-hydroxy-5α-
cholest-8(14)-en-15-one (413) fed rats at 0.15% in the diet
was also hypocholesterolemic, but fed animals had decreased
food consumption and body weight gain [2150].

Whereas most highly potent inhibitors of sterol biosynthesis and of HMG CoA reductase are dioxygenated sterols,
the monoalcohol 7α,15β-dichloro-5α-cholest-8(14)-en-3β-ol is
an active inhibitor. Other synthetic analogs are also inhibitory in other cell systems, 25-azido-27-norcholest-5-en-
3β-ol (414) inhibiting HMG CoA reductase by 50% at 2-4 μM
concentration in baby hamster kidney BHK 21 cells. However,
the analogous cholest-5-en-3α-yl azide was not inhibitory
[1003,2366]. Several other synthetic analogs are also potent
inhibitors, 3β-hydroxy-17β-isohexyloxyandrost-5-en-7-one
(415) inhibiting HMG CoA reductase by 50% in cultured human

414   415

fibroblasts at approximately 1.5 μM concentration. The
analogs 25-methyl-(22ξ)-cholest-5-ene-3β,22-diol and 23-
methyl-(23ξ)-21-norcholest-5-ene-3β,23,25-triol are also
active [651,885].

Nonetheless, the use of these oxidized sterols as standard means of inhibition of HMG CoA reductase and related cell
processes is of increasing interest. So accepted has the
use of the 7-ketone 16 or 3β,25-diol 27 become for the specific inhibition of cellular HMG CoA reductase that these
sterols have been incorporated into protocols of studies of
the effects of drugs and other agents of interest. Interactions among added agents are of some interest from the
viewpoint of enzyme regulation and also from that of drug
action. Thus, the stimulatory effects of dexamethasone (9α-
fluoro-11β,17α,21-trihydroxy-16α-methylpregna-1,4-diene-
3,20-dione on HMG CoA reductase in HeLa and rat liver
cells is abolished by 16 and by 27 [449,1510], but the induction of rat liver microsomal HMG CoA reductase by 20,25-
bisazacholest-5-en-3β-ol is not appreciably affected by subcutaneous administration of the 7-ketone 16 in intact animals, possibly because of too rapid catabolism [1423].

The mechanisms by which the cholesterol autoxidation products exert their effects on intracellular sterol biosynthesis are as yet unclear, but intact cells are required for inhibitions and specific diminution of microsomal HMG CoA reductase activity without general impairment of other cellular metabolism appears the case. Only a few weaker inhibitory steroids (dienone 12, cholesta-4,7-dien-3-one) appear to suppress fatty acid biosynthesis and CO_2 production as well as sterol biosynthesis [1232]. Cell-free microsomal HMG CoA reductase activities from mouse liver and L cells [1228,1229], human fibroblasts [381], and rat liver [701,703] are not inhibited by the 7-ketone 16 or 3β,25-diol 27 as is the case for the corresponding intact cell enzyme. Indeed, enhancement of enzyme activity has been demonstrated with partially purified HMG CoA reductase in dipalmitoyl glycerophosphocholine dispersions treated with the 7-ketone 16 or 3β,25-diol 27 [186].

The sterols appear to be taken up into the affected cells, the 7-ketone 16 with an apparent Km 1.1 μM, the 3β, 25-diol 27 with Km 3.0 μM in L cells [1230], more rapidly than cholesterol in cultured human fibroblasts [884]. There may follow a rapid metabolism of the inhibitory sterols, but the inhibitory effects of a given sterol have not been related to metabolism or to any putative metabolites in a given cell structure. However, a rapid metabolism of the 7-ketone 16 in human fibroblasts [884] and of the 3β,25-diol 27 in rat liver hepatocytes [703] is indicated. Moreover, it appears that it is the free sterol and not fatty acyl esters which are effective inhibitors of sterol biosynthesis in cultured cells. The 3β-oleate of the 3β,25-diol 27 inhibits sterol biosynthesis in the same manner as the 3β,25-diol 27 in cells capable of enzymic hydrolysis of the ester, but in cells whose lysosomal sterol fatty acyl ester hydrolases be inhibited by chloroquine suppression of HMG CoA reductase did not occur [1387].

It may be posited that the inhibition of cellular HMG CoA reductase by the potent oxidized sterols 16, 27, etc. involve a common step [467], possibly that of binding of the sterol to receptor sites in the plasma membrane or intracellular structures [1230,1232,1233]. A sequence of unknown events may then follow, the result being suppression of HMG CoA reductase activity. Cyclic nucleotides do not appear to be implicated [1230]. However, it has been possible to demonstrate inhibitory effects of oxidized sterols on cell-

free microsomal HMG CoA reductase of rat liver in the presence of a cytosolic protein which stimulates the enzyme.
Impure cholesterol, the 6-ketone 45, 5α,6α-epoxide 35, and
a cholest-5-ene-1ξ,3β,25-triol *inter alia* were active in
this respect [2325]. A similar inhibition of cell-free
microsomal 4-methylsterol oxidase activity of rat liver has
also been demonstrated [857].

The endogenous regulation of microsomal HMG CoA reductase is a complex matter *per se*, one not yet satisfactorily
elucidated. Factors such as enzyme synthesis and degradation, allosteric effects, and reversible inactivation by
phosphorylation, etc. may be involved. The inhibition of
HMG CoA reductase by endogenous cholesterol (whole serum)
appears to be related to a decreased rate of biosynthesis
of the enzyme [176,179], but the observed rates of inhibition by cholesterol autoxidation products appear to be too
rapid for protein biosynthesis to be a major factor [172].
Evidence more clearly suggests that a different inhibitory
process be the case, and that the suppression of HMG CoA
reductase be consequence of a decline in the amount of enzyme or of an increased rate of inactivation or degradation
of enzyme [176,179,378,703,1228,1239]. As the 3β,25-diol
27 suppresses both the active and inactive (phosphorylated?)
forms of HMG CoA reductase of L cells, a reversible inactivation does not figure in the process [2083].

The matter may be much more complicated, as there are
suggestions that the potent sterol biosynthesis inhibitors
act at sites other than HMG CoA. Both 7-ketone 16 and 3β,
25-diol 27 appear to diminish the activity of HMG CoA synthetase (3-hydroxy-3-methylglutaryl coenzyme A acetoacetyl
coenzyme A lyase (coenzyme A-acetylating), EC 4.1.3.5) in
cultured HeLa cells, as does also cholesterol, serum, and
LDL [1899]. Therefore, both the synthesis and metabolic
disposition of HMG CoA may be under regulation by oxidized
sterols. Moreover, biosynthesis of other isoprenoids is affected by cholesterol autoxidation products. The biosynthesis of cholesterol from mevalonate in cultured human
fibroblasts is suppressed by the 3β,25-diol 27 at the same
time that the biosynthesis of ubiquinone Q_{10} from mevalonate is stimulated, thus evincing the possibility of other
control points in the metabolism of mevalonate [724]. Furthermore, studies with clones of rat liver cell line GAI
suggest that the 7-ketone 16 affect not only HMG CoA reductase but also a later step in sterol biosynthesis [1958].

Similar suggestions have been made with respect to dolichol and glycoprotein biosynthesis. As the dolichols
mediate the assembly of glycoproteins via intermediate doli-
chyl pyrophosphoryl oligosaccharides, suppression of dolichol
biosynthesis also depresses glycoprotein formation. The
biosynthesis of dolichols (and of sterols) from acetate in
cultured calf aorta smooth muscle cells is inhibited by the
3β,25-diol 27, (20S)-3β,20-diol 21, and 7-ketone 16 [21,
1635] and in cultured mouse L cells by the 3β,25-diol 27
[1172]. Although both dolichol and sterol biosynthesis in
L cells are under regulation of the key enzyme HMG CoA reductase common to both biosynthesis pathways, the two processes appear to vary independently, thus suggesting additional control points in dolichol biosynthesis [1172].
Evidence suggesting other additional regulatory points in
sterol biosynthesis is discussed in a later section of this
chapter.

Finally, the biosynthesis of tetrahymanol by the cili-
ated protozoan *Tetrahymena pyriformis* W. is not inhibited by
the 3β,25-diol 27, 7-ketone 16, epimeric 3β,7-diols 14 and
15, or isomeric 3β,20-diols 21 and 33 *inter alia* , but the
(20S)-3β,20-diol 21 and 7-ketone 16 do cause surface irregularities in the cell [507]. Clearly these observations add
complexity in need of address to the matter of interpretation
of mechanisms of action of the potent cholesterol autoxidation products.

Furthermore, other biochemical effects not specifically
involving isoprenoids or HMG CoA reductase are exerted by
cholesterol autoxidation products. Cellular fatty acyl coenzyme A:cholesterol acyl transferase discussed in a later
section of this chapter is stimulated, but the high affinity
binding of plasma low density lipoprotein (LDL) at the cell
surface is suppressed, an effect also exerted by exogenous
cholesterol. Binding of high density lipoprotein (HDL) is
not affected [159,382,883,1374,1386]. Mixtures of cholesterol and the 3β,25-diol 27 are now used routinely to suppress the activity of high affinity receptor sites for LDL in
fibroblasts and lymphocytes [158,454,724,725,886,1052,2119].

A specific cytosol protein from mouse L cells has been
found to bind the 3β,25-diol 27 and possibly the 7-ketone
16 and other oxidized sterols which are inhibitory towards
microsomal HMG CoA reductase. A similar binding protein may
also be present in fetal mouse liver cell cytosol [1234].

The relationship between cytosolic proteins which bind inhibitory sterols and regulation of HMG CoA reductase or other
cellular process has not been addressed. However, a new
cytosol protein from rat liver which binds cholesterol and
which speculatively act as a cholesterol receptor or transport protein in the hepatocyte has been discovered. This
binding protein differs from previously described sterol
proteins, and bound cholesterol could be displaced by the
epimeric 3β,7-diols 14 and 15, the 7-ketone 16, or desmosterol (78) [704]. In yet another case involving a similar
cholesterol binding protein from sheep adrenal tissue cytosol, the 3β,25-diol 27 and the (20S)-3β,20-diol 21 displaced
bound cholesterol [1447].

    Subsequent Effects. A number of derivative effects are
attendant upon administration of inhibitory cholesterol autoxidation products, some of which appear to be consequence
of the specific inhibition of HMG CoA reductase and of
sterol biosynthesis. As already mentioned, depletion of cellular sterol occurs on suppression of sterol biosynthesis,
and it follows that processes dependent on continuing supplies of de novo biosynthesized sterols may be affected.
A major, direct consequence of depletion of sterols is the
cessation of cell growth and proliferation [468,473,804,
1227,1233].

    The growth of cultured cells under inhibition by the
3β,25-diol 27 or 7-ketone 16 is retarded. Among such sensitive cell lines are human fibroblasts from embryonic lung
[523] or skin [379], rat myogenic cell line L_6 [523], mouse
L cells [473], and Chinese hamster ovary CHO cells [1734].
However, suppression of growth is overcome by additions of
mevalonate or of sterol. In mouse L cells for which desmosterol is the predominant sterol, added desmosterol counters
of the inhibitory effects on growth of the oxidized sterols
14-16,21, and 27 [473], and in cultured human fibroblasts
added exogenous cholesterol or LDL (but not HDL) likewise
overcomes the inhibition of growth and of the 7-ketone 16
[379].

    The growth of cultured cells may be precisely control-
led by manipulation of sterol and lipid supplies. The inhibition of sterol biosynthesis by the 3β,25-diol 27 in con-
cert with suppression of de novo biosynthesis of fatty acids
and of phospholipids and sphingolipids by depletion of bio-
tin and choline respectively leads to a prereplicative Gl

cell cycle arrest in human fibroblasts and in rat myogenic
cells which is reversed upon restoration of the required
lipids or of their biosynthesis pathways. A causal relationship between the supply of these lipids and passage of
the cells through the G1 stage is thereby suggested [523].

Yet other derivative effects have variously been noted
in cultured mammalian cells under suppression by cholesterol
autoxidation products. Repression of endocytosis in mouse
L cells by the (20S)-3β,20-diol 21, 3β,25-diol 27, 7-ketone
16, and 6-ketone 45 relieved by exogenous mevalonate appears
directly related to suppression of sterol biosynthesis
[1000]. Another intriguing response to the 3β,25-diol 27
is the diminution of cytolytic activity of cytolytic T
lymphocytes from mice. In this case, the depressed cytolytic activity could be restored by additions of mevalonate,
thereby implicating suppressed sterol biosynthesis in the
effect [999].

The effects on cell growth of cholesterol autoxidation
products are not limited to mammalian cells in monolayer
culture, as these oxidized sterols also exert nutritional
and toxic effects on certain microorganisms. Species of
the fungal genera *Pythium* and *Phytophthora* may produce vegetative growth in the absence of sterols, but with sterols
such as fucosterol, ergosterol, or cholesterol added, growth
and sexual reproduction occurs, with oogonia, antheridia,
and oospore formation. Interestingly, growth of *Phytophthora cactorum* is inhibited by the (20S)-3β,20-diol 21, sar-
gingosterol 163 (a putative oxidation product of fucosterol),
and 5α-cholestan-3-one (365) [1741], whereas the short side-
chain sterol chol-5-en-3β-ol (109), dienone 12, and enones
6 and 8 promote growth and formation of oogonia and antheridia but not of oospores [673]. By contrast, the enones 6
and 8 inhibit growth of *Mycoplasma mycoides,* where growth
is stimulated by cholesterol [1971]. Furthermore, growth of
Staphylococcus aureus is inhibited by the 3β,5α,6β-triol 13
[1409]. It is not now known whether these effects are con-
sequences of inhibitions of HMG CoA reductase or of isopre-
noid or sterol biosynthesis.

The inhibitory effects of cholesterol autoxidation products on sterol biosynthesis and cell growth also involve
a diminution of deoxyribonucleic acid (DNA) biosynthesis.
In mouse L cells treated with the 3β,27-diol 27 DNA biosynthesis declined progressively and ultimately ceased. Protein

biosynthesis decreased, as did growth, apparently as consequence of diminished DNA biosynthesis and not of direct inhibition of protein biosynthesis or of other vital cellular
metabolic processes. These effects being reversed by added
mevalonate or cholesterol, they may be attributed to suppressed sterol biosynthesis [1231]. Suppression of DNA biosynthesis and associated lymphoblastic transformations are
also observed in mitogen-stimulated mouse lymphocytes
treated with the 3β,25-diol 27 [468] and in stimulated
human lymphocytes treated with the (20S)-3β,20-diol 21, 3β,
25-diol 27, and other oxidized sterols [104,1885,2373,2732-
2734]. These effects are partially reversed by exogenous
cholesterol or by mevalonate [2733].

The 3β,25-diol 27 is highly active, provoking 50% inhibition of DNA synthesis of mitogen-stimulated human lymphocytes at 3.9 μM concentration, versus 50% inhibition at
7.4 μM for the (20S)-3β,20-diol 21 and 9.2 μM for the 3β,25-
diol 27 3β-acetate. Other cholesterol autoxidation products
were less effective, the 3β,7-diols 14 and 15, 7-ketone 16,
and 3β,5α,6β-triol 13 exhibiting 50% inhibitory concentrations of about 25 μM [2733].

The suppression of DNA and cholesterol biosynthesis in
cultures of mitogen-stimulated human lymphocytes was more
pronounced in media depleted of sterol and lipids, 50% inhibitory concentrations of 0.3 μM for the 3β,25-diol 27 being
the case for both sterol and DNA synthesis [2733]. Human
bone marrow granulocytic progenitor cells responded similarly
to the 3β,25-diol 27 [1058].

The crucial importance of timely biosynthesis of sterols
for these subsequent events in cell growth and proliferation
is amply supported by these several observations.

Still other subsidary biochemical effects of the inhibition of sterol biosynthesis by cholesterol autoxidation products are indicated, among which are vital transport processes. The uptake of cholesterol by cultures of rabbit aorta
smooth muscle cells is inhibited by 100 μg/mL of the 7-oxy-
genated sterols 14-16, the 3β,25-diol 27, and the 3β,5α,6β-
triol 13 [1849], but mouse L cells treated with the 7-ketone
16 or the 3β,25-diol 27, thereby depleted of sterols, exhibited increased ouabain-sensitive uptake of Rb$^+$ (thereby of
K$^+$) and also ouabain-insensitive efflux of Rb$^+$ [469].

The inhibitory affects of these cholesterol autoxidation products on cell growth *in vitro* are also expressed in
a few cases *in vivo* in rats and mice fed the sterols. Suppression of growth and loss of body weight in mice fed either
sterol has been noted, the loss in body weight being counter-
acted for the 7-ketone 16 by feeding cholesterol. Intestinal
sterol biosynthesis was rapidly inhibited but hepatic sterol
biosynthesis not so. Appetite was diminished, probably accounting for body weight loss [1235]. In rats fed either
sterol 16 or 27 hepatic HMG CoA reductase was suppressed
early, but loss in weight and an apparent tolerance for the
oxidized sterols later occurred [701,703]. Serum cholesterol
levels in rats fed the 7-ketone 16 were not depressed [1423],
but cholesterol biosynthesis from acetate was inhibited in
rats administered the 7-ketone intravenously [1212]. It is
apparent that the *in vivo* inhibitory effects of 16 and 27
are much less than their *in vitro* effects, therefore that
these sterols are not useful for suppression of sterol biosynthesis in experimental animals or man, toxic effects
aside [701,703,903,1235,1423].

Cholesterol 7α-Hydroxylase

The initial biosynthesis step in the transformation of
cholesterol into bile acids by the liver is that of 7α-hy-
droxylation by hepatic microsomal cholesterol 7α-hydroxylase
(cholesterol, reduced NADP:oxygen oxidoreductase (7α-hydroxy-
lating), EC 1.14.13.17), forming cholest-5-ene-3β,7α-diol
(14) in what appears to be a rate-limiting step. The 7α-
hydroxylase enzyme is typical of sterol monooxygenases
(mixed function oxidases) in which cytochrome P-450 is implicated [1701].

Rat liver cholesterol 7α-hydroxylase is experimentally
difficult of assay, witness a great deal of activity directed towards such assay. Among other problems is that of ac-
cessibility of exogenous labeled substrate cholesterol versus that of cholesterol endogenous to the endoplasmic reticulum found in the microsomal enzyme preparations to be as-
sayed. Thus, relative rate data reflecting these different
compartment availabilities may be of but limited value in
comparing different experiments. It does appear however,
that the rate by which cholesterol 7α-hydroxylase preparations 7α-hydroxylate cholesterol *in vitro* are subject to
product feedback inhibition and to inhibition by other

cholesterol autoxidation products as well. Thus the 3β,7α-
diol 14 suppresses the 7α-hydroxylation of cholesterol with
a 50% inhibition of enzyme achieved with 20 μM concentrations [342] and thereby bile acid biosynthesis in bile fistula rats [1254]. The 3β,7α-diol 14 also may inhibit the
sterol 12α-hydroxylase of cultured hepatocytes. These inhibitory effects notwithstanding, in cultured hepatocytes
the 3β,7α-diol 14 in 68 μM concentration appears to stimulate the biosynthesis of conjugated bile acids [328].

The epimeric 3β,7β-diol 15 also inhibits the 7α-hydroxy-
lase in like manner [342,1254], as does also the 7-ketone 16
[2548,2551]. The inhibition by the 7-ketone 16 appears to
have K_i 7 μM, which in comparison with the K_m 100 μM for the
7α-hydroxylation of cholesterol evinces a strong inhibition
[2548]. The nitrogenous analog 22-amino-(22R)-cholest-5-en-
3β-ol fed to rats strongly inhibited the liver microsomal
cholesterol 7α-hydroxylase [932]. Liver microsomal cholesterol 7α-hydroxylase is also inhibited by other steroids,
such diverse steroids as the 5,6-epoxides 35 and 36, cholesta-
5,7-dien-3β-ol (56), 3β-hydroxyandrost-5-en-17-one (86),
pregnenolone (23), and androst-5-en-3β-ol (114) being inhibitory, but pregn-5-en-3β-ol (110) and the triol 13 were not
[93,376,2551]. Bile acid sodium salts also inhibit 7α-hy-
droxylase, possibly via detergent action disrupting the membrane-bound enzyme [342]. Peroxidized dietary lipids do not
appear to affect hepatic cholesterol 7α-hydroxylase [253].
A rat liver microsomal cholestanol 7α-hydroxylase, leading
to 5α-cholestane-3β,7α-diol (48) as product, is also inhibited by the 3β,7β-diol 15 and by the 7-ketone 16 [2212].

In distinction to these several inhibitions of liver
cholesterol 7α-hydroxylase, the synthetic analog 23-methyl-
(23ξ)-21-norcholest-5-ene-3β,23,25-triol stimulates the
enzyme [1749]!

Cholesterol Side-Chain Cleavage

Scission of the terminal isohexyl moiety of the cholesterol side-chain occurs in the biosynthesis of pregnenolone
(23) in adrenal cortex mitochondria. This specific side-
chain cleavage enzyme has been recognized as involving a
specific cytochrome P-450 system, inhibited by the product
pregnenolone [1111, 1371,1372,1898].

Moreover, the several side-chain monohydroxylated cholesterol derivatives have been shown to be inhibitors of the system. The (20S)-3β,20-diol 21 once thought to be the initially formed oxidation product of cholesterol in the biosynthesis sequence has been demonstrated to be a noncompetitive inhibitor of the side-chain cleavage enzyme in acetone-dried bovine adrenal cortex mitochondria [944] and in native bovine adrenal cortex mitochondria (K_i 17 μM versus K_i 130 μM for pregnenolone) [1898] but a competitive inhibitor in incubations of a 100,000 x g supernatant from bovine adrenal cortex mitochondria sonicate [2558]. The (20S),3β,20-diol 21 affects a 50% inhibition of side-chair cleavage activity for partially purified bovine adrenal cortex mitochondrial cytochrome P-450 preparations at 7.0 μM concentration, and the (22R)-3β,22-diol 24, the (20R,22R)-3β,20,22-triol 403, all four isomeric 20,22-epoxycholest-5-en-3β-ol derivatives, cholesta-5,E-20(22)-dien-3β-ol, cholesta-5,Z-20(22)-dien-3β-ol, and (20S)-cholest-5-en-3β-ol affect 50% inhibitions over concentrations of 5-20 μM [2452].

Yet other oxidized sterols which inhibit the bovine adrenal cortex mitochondria side-chain scission reaction include (24S)-3β,24-diol 25, 3β,25-diol 27, (25R)-3β,26-diol 29, 3β-hydroxy-27-norcholest-5-en-25-one (299), (25ξ)-27-norcholest-5-ene-3β,25-diol (301), and 27-norcholest-4-ene-3,25-dione [1186,1898,2258]. However, a stimulation of side-chain cleavage activity has been noted for the acids 3β-hydroxychol-5-enic acid (99) and 3β-hydroxy-22,23-bisnorchol-5-enic acid (101) [1898], and the stimulation of isolated rat adrenal cells in the absence of adrenocorticotrophic hormone (ACTH) to produce corticosterone (11β,21-dihydroxy-pregn-4-ene-3,20-dione) by the (20S)-3β,20-diol 21, the (22R)-3β,22-diol 24, and the 3β,25-diol 27 has been recorded [714]. Furthermore, the biosynthesis of pregnenolone (23) in rat testis mitochondria is stimulated by the (20S)-3β,20-diol 21 and 3β,25-diol 27 [124] and in rat luteal mitochondria by 21,27, and 3β-hydroxy-27-norcholest-5-en-25-one (299) [2486]. It is uncertain whether oxidized sterols 21,24,27, and 299 serve as substrates in these systems, thereby increasing levels of C_{21}-steroid products or whether other mechanisms are implicated.

The cholesterol side-chain enzyme of hog adrenal mitochondria is also inhibited by oxidized steroids. Equimolar (17 μM) amounts of substrate cholesterol or 7-ketone 16 or enones 6 and 8 or 5α-ketone 365 inhibit side-chain

scission competitively [1348]. However, the effect is not
one which may be attributed solely to oxidized steroids,
for 5α-cholestan-3β-ol (2) and desmosterol (78) were also
inhibitory. Moreover, both the stanol 2 and desmosterol
were inhibitory towards the cleavage of the side-chain of
substrate (20S)-3β,20-diol 21 in hog adrenal mitochondria,
whereas the ketosteroids 6,8,16, and the 5α-3-ketone 365
were not [1348].

    A variety of nonsterol nitrogenous agents also inhibit
the cleavage of the sterol side-chain, but several azasterol
analogs including 20-azacholest-5-en-3β-ol, 22-azacholest-5-
en-3β-ol, 25-azacholest-5-en-3β-ol, inter alia are potent
inhibitors of adrenal cortex mitochondrial side-chain
cleavage [526,675,1539].

    Although the specific site of inhibition is not evident,
the (22S)-3β,22-diol 24 inhibits the ACTH-stimulated production of corticosterone by isolated rat adrenal cells [1103].
The 3β,25-diol 27 inhibits the formation of corticosterone
from sterols, including 27 as substrate, in isolated rat
adrenal cells stimulated by ACTH and at the same time inhibited by aminoglutethimide. Chol-5-ene-3β,24-diol also inhibits corticosterone production in cells stimulated by
ACTH [715].

    Yet another mode of inhibition of utilization of chole-
sterol for steroid hormone biosynthesis in endocrine tissues
may exist. A cytosol protein specifically binding cholesterol has been found in sheep adrenal, testis, and ovary
tissues, and although no evidence for participation of such
protein in the utilization of cholesterol for hormone biosynthesis is available, the distribution and specificity of
the protein is suggestive speculatively of a possible role.
In this matter, both the (20S)-3β,20-diol 21 and the 3β,25-
diol 27 inhibited cholesterol binding to the protein [1447,
1448].

                    Enzymes of Sterol Biosynthesis

    Cholesterol oxidation products have been demonstrated
to be inhibitory towards several enzymic steps implicated
in the biosynthesis of cholesterol from lanosterol. The
greater amount of 14-ethyl-5α,14α-cholest-7-ene-3β,15α-diol
required for inhibition of HMG CoA reductase in cultured

fetal mouse liver or L cells than for inhibition of sterol biosynthesis from acetate [1816] as well as other evidence suggests that the inhibitory action of oxidized sterols on biosynthesis is not limited to suppression of HMG CoA reductase alone. Inhibition by the $3\beta,5\alpha,6\beta$-triol 13 of the stenol Δ^5-dehydrogenase and sterol 5,7-diene Δ^5-reductase of rat liver implicated in the transformation of the 7-stenol 57 to the 5,7-diene 56 and of 56 to cholesterol has previously been mentioned [599,2717,2718]. Furthermore, inhibition of the sterol Δ^{24}-reductase of cultured rat hepatocytes by the $3\beta,5\alpha,6\beta$-triol 13 and $3\beta,25$-diol 27 has been demonstrated [980].

An intriguing case of sterol reductase inhibition has also been described for the sterol Δ^{24}-reductase implicated in the transformation of desmosterol (78) to cholesterol in the tobacco hornworm *Manduca sexta* and in the rat. The two steroid acids 3β-hydroxy-23-norchol-5-enic acid (100), and 3β-hydroxy-22,23-bisnorchol-5-enic acid (101) found to be present as impurities in commercial sitosterol samples fed hornworm larvae were active inhibitors of the Δ^{24}-reductase in hornworm and in rats [2404,2405]! 3β-Hydroxychol-5-enic acid (99) was inactive. These particular findings vindicate in exemplary fashion the reservations expressed throughout this chapter with respect to the purity of sterols used in nutritional and pharmacological evaluations in experimental animals!

More remote from cholesterol and nearer the ultimate sterol precursor lanosterol 67, the oxidative removal of the 4-methyl groups of several putative biosynthesis intermediates linking lanosterol and cholesterol are affected by hepatic microsomal 4-methylsterol oxidase activity. In two cases the inhibitory action of several cholesterol autoxidation products has been recorded. Incubations of the 4-methyl-8-stenols $4\alpha,4\beta$-dimethyl-5α-cholesta-8,24-dien-3β-ol, $4\alpha,4\beta$-dimethyl-5α-cholest-8-en-3β-ol, 4α-methyl-5α-cholesta-8,24-dien-3β-ol, and 4α-methyl-5β-cholest-8-en-3β-ol with rat liver 20,000 x g supernatant oxidase preparations were inhibited by the $3\beta,5\alpha,6\beta$-triol 13 [2088].

A similar inhibition by the triol 13 of the microsomal methyl sterol oxidase of rat liver acting on 7-stenol substrates $4\alpha,4\beta$-dimethyl-5α-cholest-7-en-3β-ol or 4α-methyl-5α-cholest-7-en-3β-ol occurs. In this case the 7-stenol 57,

the 3β, 25-diol 27, cholesterol 3β-hemisuccinate, and crude
commercial cholesterol also were highly effective inhibitors of the oxidase. Pure cholesterol derived from the active batch of crude cholesterol was a weak competitive inhibitor of the oxidase as were also several cholesterol fatty acid esters. The 3β,7α-diol 14 and 7-ketone 16 were completely inactive. It should be noted that these inhibitions were effected on partially purified microsomal oxidase
preparations free from intact cells, and that the extent of
inhibition by oxygenated sterols was greater in the presence of a soluble protein from liver cytosol which stimulated the oxidase activity [857].

However, inhibition of the oxidative removal of the 4-
and 14-methyl groups of lanosterol and other methylsterols
has been demonstrated in cultured rat hepatocytes by the
3β,25-diol 27 [980] and in cultured Morris hepatoma cells
by the epimeric 3β,7-diols 14 and 15 [1802].

Enzymes implicated in the biosynthesis of the 5α-stan-
ol 2 from cholesterol are also affected by cholesterol oxidation products. Reduction of the enone 8 by a rat liver
microsomal reductase (3-oxo-5α-steroid:$NADP^+$ Δ^4-oxidoreduc-
tase, EC 1.3.1.22) yielding 5α-cholestan-3-one (365) is inhibited by the product 5α-3-ketone 365 but appears to be
stimulated by the dienones 10 and 12 [2210]. Furthermore,
the subsequent reduction of the 5α-3-ketone 365 by a rat
liver microsomal dehydrogenase (3β-hydroxysteroid: $NADP^+_5$
oxidoreductase) to the 5α-stanol 2 is inhibited by the Δ -
3-ketone 6 and also by 3β-hydroxy-5α-cholestan-7-one and
5α-cholestane-3β,7α-diol (48). The enone 8 and 5β-chole-
stan-3-one were weakly inhibitory [2211]. It thus appears
that the two distinct microsomal oxidoreductases implicated
in biosynthesis of the 5α-stanol 2 are both sensitive to
inhibition by oxidized sterol derivatives, as are also rat
liver microsomal reductions of several other steroid 3-ke-
tones [261].

Yet other inhibitory effects on the biosynthesis of
other sterols are recorded. For instance, the enone 8 inhibits the reduction of 7α-hydroxycholest-4-en-3-one (286)
to 5β-cholestane-3β,7α-diol by rat liver supernate reduc-
tases [1108]. The C_{19}-sterol androsta-5,16-dien-3β-ol (191)
inhibits its own biosynthesis in boar testis homogenates
from pregnenolone (23) [1526].

Sterol Acylases

Another important biological activity of oxygenated sterols is that as moderator of the esterification of cholesterol by fatty acids catalyzed by both extracellular and intracellular enzymes. As example of the former case, the esterification of cholesterol by rat pancreatic juice cholesterol esterase *in vitro* was substantially inhibited by graded additions of the 3β,5α,6β-triol 13 to incubations [1164]. Moreover, progesterone inhibits fatty acyl Coenzyme A:cholesterol O-acyltransferase of human fibroblast homogenates, but the 3β,25-diol 27 did not [885].

A soluble acid cholesterol ester hydrolase probably of lysosomal origin from pig aorta appears to be inhibited by the enone 8 but stimulated by cholesterol, the (25R)-3β,26-diol 29, the 3β,5α,6β-triol 13, and other sterols [2279].

Of more interest is the intracellular fatty acyl-Coenzyme A:cholesterol O-acyltransferase (EC 2.3.1.26) of cultured human fibroblasts which is stimulated by several common cholesterol autoxidation products. Monolayer cultures of fibroblasts exhibited markedly increased rates of cholesterol ester formation in the presence of exogenous 7-ketone 16, 3β,25-diol 27, or 3β-hydroxy-5α-cholestan-6-one (45) [378,885]. The ketone 16 likewise stimulates the esterification of cholesterol in mouse adrenal cell cultures [723]. These stimulations were achieved with 2-5 μg/mL sterols 16 or 27, thus at the same concentrations effective in inhibiting cellular HMG CoA reductase in the same cells. An elevation of liver cholesterol ester levels in rats fed 0.5% levels of the 7-ketone 16 has also been observed [701].

The fatty acyl-Coenzyme A:cholesterol O-acyltransferase also appears to be stimulated by the 3β,25-diol 27 in cell-free incubations of rat liver microsomes [1469]. The stimulation by the 3β,25-diol 27 of cholesterol esterification in fibroblasts is in turn inhibited (K_i 20 μM) by 3β-poly-oxyethylated derivatives of cholesterol [818,319].

Although the 7-ketone 16 and 3β,25-diol 27 exhibit HMG CoA reductase and stimulate acyl Coenzyme A:cholesterol acyltransferase, several synthetic analogs of these sterols inhibit both HMG CoA reductase and acyl Coenzyme A: cholesterol acyltransferase in fibroblasts. Thus, the 20-oxa-21-

TABLE 22. Enzyme Effects Patterns of Synthetic Analogs

Sterol	Enzyme Effects*			References
	I	II	III	
3β-Hydroxy-17β-isohexyloxyandrost-5-en-7-one (415)	(-)	(-)		[651,885]
25-Methyl-(22ξ)-cholest-5-ene-3β,22-diol	(-)	(-)		[651,885]
23-Methyl-(22ξ)-21-norcholest-5-ene-3β,23,25-triol	(-)	(-)	(+)	[651,885,1749]
25-Azido-26-norcholest-5-en-3β-ol (414)	(-)	(+)		[1003]
Cholesterol Polyethoxy-ethyl ether	(-)	(-)		[819]

*Enzymes are: I, HMG CoA reductase (EC 1.1.1.34); II, acyl-Coenzyme A:cholesterol acyl transferase (EC 2.3.1.26); III, cholesterol 7α-hydroxylase (EC 1.4.13.17). Inhibition (-); stimulation (+).

nor analog 415 of the 7-ketone 16, a 25-methyl-(22ξ)-cholest-
5-ene-3β,22-diol, and a 23-methyl-(23ξ)-21-norcholest-5-
ene-3β,23,25-triol inhibit both HMG CoA reductase and acyltransferase in cultured human fibroblasts reversibly, and
the 20-oxa-21-nor analog 415 also inhibits the acyltransferase in cell free homogenates of fibroblasts [651,885].

An obviously complex matter is at hand. Moreover, although HMG CoA reductase of both cultured rat hepatocytes
and rat hepatoma (HTC) cells is sensitive to suppression by
oxidized cholesterol derivatives, only the sterol acylases
of rat hepatocytes are stimulated by the 3β,25-diol 27
[980]. Neither 3β,25-diol 27, 3β,7α-diol 14, nor 7-ketone
16 stimulate sterol acylases of the rat hepatoma cell line
[179]. Moreover, the 20-oxa-21-nor analog 415 was not effective in intact rats whether administered orally or via
intravenous injection [651], but the 25-azido analog 414
administered rats intravenously inhibited liver HMG CoA reductase and stimulated cholesterol ester formation [1003].
These items summarized in TABLE 22 support hope that synthetic analogs may be found which selectively regulate the
different aspects of cellular cholesterol biosynthesis and
metabolism, a goal not now attained.

Cytochrome P-450

Interactions between adrenal cortex mitochondrial cytochrome P-450 involved in the scission of the cholesterol
side-chain and inhibitions of the reaction by a variety of
oxidized cholesterol derivatives have already been mentioned. Additionally, a few isolated observations suggest that
cholesterol autoxidation products moderate the activities
of other cytochrome P-450 enzymes which oxidize other substrates. Thus, naturally air-aged cholesterol fed weanling
albino mice at 1-2% of their diet stimulated the demethyl-
ation of 3-methyl-4-methylaminoazobenzene by a mixed function oxidase system of mouse liver. In this study, purified cholesterol did not stimulate this system nor did
freshly manufactured cholesterol. The stimulatory activity
was concentrated into the mother liquor from recrystallizations of the crude cholesterol. Finally, oxidation of
cholesterol by H_2O_2 in acetic acid gave an oxidized sterol
preparation which was also active in this respect [384,1925].

Moreover, cholesterol and sitosterol oxidized by H_2O_2
in acetic acid fed rats at 0.1-0.5% levels in their diet

provoked an increase in the amount of cytochrome P-450 induced by phenobarbital treatment of the animals. Pure sitosterol at 5.0% and cholesterol at 1% levels did not affect increases [1575]. Furthermore, the (20S)-3β,20-diol
21 reduced the rate of complex formation between reduced
cytochrome P-450 from bovine corpus luteum and CO [1605].

The amount of antimycin necessary for 50% inhibition
of the succinate oxidase system (succinate dehydrogenase,
EC 1.3.99.1; cytochrome c oxidase, EC 1.9.3.1; other intermediate carriers) of mouse liver is also increased in
mice fed naturally air-aged cholesterol. The effect might
be from stimulated increase of one or more components of
the succinate oxidase system or, as cytochrome P-450 enzymes also appear to be stimulated as well, from increased
metabolism of antimycin. Notably, the 5α-stanol 2 oxidized
with H_2O_2 also affected these results, but ergosterol 5α,
8α-peroxide (62) did not [1925].

The chronic feeding to rats of cholesterol also results in a stimulation of hepatic metabolism of xenobiotic
organic compounds by cytochrome P-450 systems. Although
these effects have been attributed to modifications in the
membranes in which the enzymes reside [1038,1039,1419,1422],
the insidious effects of cholesterol autoxidation products
likely to be present in the prepared diet should also be
considered.

Certain other oxidizing enzymes not involving cytochrome P-450 systems may also be affected by oxidized sterols. The deamination by rat liver mitochondrial monoamine
oxidase (amine:oxygen oxidoreductase(deaminating) (flavin-
containing), EC 1.4.3.4.) of several amino acids is diminished in liver of rats administered ergosterol 5α,8α-perox-
ide (62) or ergocalciferol (321)[36].

MUTAGENICITY

The issue of mutagenicity of cholesterol oxidation is
linked to the two major human disorders of cancer and atherosclerosis. In the first case, the concept that mutagenic
actions lead eventually to carcinogenic effects, though un-
proven, serves to make this speculative connection. In the
case of atherosclerosis the suggested monoclonal character
of human atherosclerotic plaques implies that an endogenous

mutagen acting within the intima and/or media of the aorta
may have a possible role in the etiology of atherosclerosis
[180,181]

In both of these speculative formulations attention has
been focused upon one cholesterol autoxidation product 5,6α-
epoxy-5α-cholestan-3β-ol (35). Animal data discussed in the
next section of this chapter and evidence of cell transformation by the 5α,6α-epoxide 35 [1278] support a relationship
of the sterol with cancer, and detection of the 5α,6α-epox-
ide 35 in human hyperlipemic blood [914] and in foods [2507]
provide additional circumstantial evidence. Nonetheless,
the 5α,6α-epoxide 35 has not been demonstrated to be mutagenic. In our hands neither the 5α,6α-epoxide 35 nor the
isomeric 5β,6β-epoxide 36 are mutagenic towards several test
strains of *Salmonella typhimurium* [69,2307], and several
others have likewise not found the 5α,6α-epoxide 35 mutagenic
in this test (with or without liver enzymes [1210,1278,1912].
In one case cytotoxicity precluded evaluation of mutagenicity for 35 [283]. Mutagenicity of 35 towards eukaryotic
cells or in intact animals has not been evaluated. Nonetheless, strong physical complexes between the 5α,6α-epoxide 35
and DNA form, possibly involving covalent bonding [283].

However, autoxidized samples of cholesterol have now
been shown to be active as frameshift mutagens towards *S.
triphimurium*. Samples of once pure cholesterol stored for
indeterminate periods of years on the shelf exhibit dose-
response mutagenicity and in some cases cytotoxicity towards
S. typhimurium. The dose-response mutagenicities towards
three test strains of the bacterium are shown in FIGURE 28
for six naturally autoxidized samples of once pure cholesterol. The considerable variations in response may reflect
the different ages and autoxidative decomposition of the
samples. Mutagenicity is lacking in the oldent, most com-
posed sample (No. 6) [69,2307].

Furthermore, pure cholesterol, which is nonmutagenic,
subjected to autoxidation by heating (70°C) in air for several weeks or by irradiation for several days with ^{60}Co γ-
radiation became mutagenic, the unidentified mutagenic components being concentrated in all cases along with recognized cholesterol autoxidation products in mother liquor
from methanol washes. High performance liquid chromatography of the mutagenic material showed that the activity
resided in very polar fractions, thus in regions much more

polar than those occupied by the presently recognized autoxidation products of cholesterol that are chromatographically more mobile than the 3β,5α-6β-triol 13. Several mutagens appear to be present among the 35 or so components
resolved in preliminary studies [75].

None of the well known cholesterol autoxidation products described thus far in the monograph appears to be mutagenic towards *S. typhimurium*. Specific tests on the 7-oxy-
genated derivatives 10,14-16,46, and 47, the 3-ketones 6,
8,12,90, and 108, the 6-ketones 44 and 45, the 5,6-epoxides
35 and 36, triol 13, the diols 21,25,27, and 41, the acids
99,101, and 102, the 27-nor-25-ketone 299, chol-5-en-3β-ol
(109), and pregnenolone (23) were negative. Other common
cholesterol oxidation products that were nonmutagenic include the 3β,5α-diol 50, the 5α-hydroperoxide 51, the dienone 115, the steroid olefins 7,11, and 284, and chol-5-ene-
3β,24-diol.

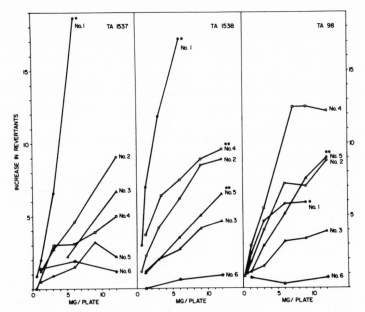

FIGURE 28.   Mutagenicity of six air-aged samples of
              cholesterol towards *Salmonella typhimurium*
              strains TA 1537, TA 1538, and TA 98.   Re-
              printed with permission of Elsevier/North
              Holland Biomedical Press, Amsterdam, from
              Mutation Res., 68, 23 (1979).

As the *S. typhimurium* mutagenicity tests are able to detect active mutagens to the part per million it is very important that the mutagenicity of known sterols be evaluated only with pure samples. A paradigm illustrating pitfalls in the use of impure samples follows. A commercially available sample of 3β,5-dihydroxy-5α-cholestan-6-one (44) tested as received exhibited dose-response mutagenicity (1.2-4.5 mg/plate) against *S. typhimurium* test strains TA 1538 and TA 98. Purification of the mutagenic sample by high performance liquid column chromatography gave pure 6-ketone 44 devoid of mutagenicity, the nonmutagenic 3β,5α,6β-triol 13 precursor of 44, and traces of a highly mutagenic material apparently not steroid in nature. Commercially available 6-ketone 44 samples from other sources tested as received were nonmutagenic as were also samples of 44 prepared in our laboratory from cholesterol using N-bromosuccinimide [757,1920]. Exactly the same kind of results were obtained on testing a commercially available sample of the 3,6-diketone 108. Tested as received the material was mutagenic; upon purification there was no mutagenicity response. Pure 3,6-diketone 108 prepared from the sample or synthesized and purified via chromatography likewise was nonmutagenic.

In regard to their possible implication in the etiology or progression of bowel cancer, several bile salts and bile acids, including the cholesterol autoxidation product 3β-hydroxychol-5-enic acid (99), have also been evaluated as mutagens against *S. typhimurium*, but these oxidized derivatives are not mutagenic [1262,1545,2253].

Although teratogenicity may or may nor be linked to mutagenicity, mention of the reported teratogenicity of cholesterol in rats receiving subcutaneous injections of cholesterol in oil must be made [398,399]. Teratogenicity attributed to cholesterol might be another example of unrecognized biological activity of cholesterol autoxidation products.

CARCINOGENICITY

The issue of carcinogenicity of cholesterol and its derivatives has been controversial for over fifty years, an issue yet to be accorded definitive treatment. Attribution of carcinogenicity to cholesterol *per se* has figured in such

matters. With assignment of the correct structure to cholesterol the close resemblence of the sterol to polycyclic
aromatic hydrocarbons just being recognized as carcinogenic
was noted, with speculations following which suggested that
cholesterol be transformed into such polycyclic aromatic
hydrocarbon carcinogens.

Early interests centered upon experimental gastric
cancer putatively caused by dietary ingestion of carcinogenic transformation products of cholesterol subjected to
heat or oxidation, upon skin cancer speculatively caused by
radiation-induced air oxidations of skin cholesterol to
carcinogens, and upon local sarcomas associated with cutaneous or subcutaneous deposits of cholesterol preparations.
More recently, concern has developed over the possibilities
that cancer of the lower bowel be linked to endogenous metabolism of cholesterol and bile acids by intestinal microflora associated with putrefraction. Although the case has
never been established for these earlier speculative no-
tions, it is of interest to note that four presently held
views of cancer initiation (dietary, radiation, solid-state,
endogenous metabolism) were embodied in these various early
formulations.

The proper testing of cholesterol and its derivatives
as carcinogens, as low-grade carcinogens, or as cocarcino-
gens may not have been conducted yet, as many factors compromise results with these slow acting agents. Obvious issues of compound identity, purity, and stability during
testing have not been treated definitively. Issues of test
animal, strain, sex, age, vehicle, mode and site of administration, dose, treatment, protocol, etc., of how long the
experiment must run, survival patterns, question of latent
period and cocarcinogenicity, and of the experimental plan
and statistical management of data are factors presently
compromising definitive conclusions. Moreover, question
whether tumors are malignant, are transplantable, etc. have
generally not been properly addressed.

Many different experimental approaches have been taken
in prior investigations of the cholesterol-cancer problem,
but these can be examined to advantage under three categories: (i) the presence in cancerous tissues of carcinogens that can be isolated therefrom and identified, (ii) the
demonstration of carcinogenicity of cholesterol prepara-

tions subjected to various chemical and physical treatments, and (iii) the demonstration of carcinogenicity of
specific cholesterol derivatives.

Carcinogens in Tissues

The notion that carcinogens be present in tumors led
to attempts to isolate putative carcinogens from cancer
tissues. Cholesterol had been isolated from human carcinoma fat in 1913 [2621], and carcinogenicity testing by
Hieger over the period 1940-1949 of cancer tissue extracts
enriched in cholesterol in which positive results were had
[1026-1031] served to focus attention on cholesterol or
traces of related materials as carcinogens. Yea, although
Hieger initially posed the question whether trace impurities in cholesterol were carcinogenic [1030,1037], his later conclusions were that cholesterol *per se* be carcinogenic
[1032-1037]! Moreover, positive results sustaining
Hieger's conclusions have been reported as late as 1973
[44,1859].

Cholesterol as Carcinogen. Attribution of carcinogenicity to cholesterol was more directly suggested by discoveries of Kennaway between 1925-1930 that tars from pyro-
lyzed human skin and from cholesterol were carcinogenic in
mice [1025,1281-1283,1285]. Also contributing to the suspicion that cholesterol be carcinogenic were numerous reports of A.H. Roffo, who noted by 1930 the correspondence
between increased incidence of skin cancer and elevated
levels of skin cholesterol and who suggested that irradiation of skin cholesterol by sunlight be a causative factor
[706,1976-1982,1999,2414].

The previously mentioned recognition of structure similarities between carcinogenic polycyclic aromatic hydrocarbons and cholesterol dated from 1932 [144,513,1284].
Roffo was aware of the matter and considered that cholesterol was transformed by irradiation or by heat into polycyclic aromatic hydrocarbons that acted as ultimate carcinogens [1976,1983,1985,1987]. Indeed, Roffo claimed to have
formed carcinogenic polycyclic aromatic hydrocarbons from
cholesterol by distillation and by irradiation [1984,1988,
1995,1996] and to have detected such components as fluorescent material in hyperkeratotic skin lesions, attributed to oxidized or aromatized cholesterol derivatives

[734,1990,2000]. Interests in the matter have continued,
with demonstrations of pyrolytic and chemical transformations of cholesterol to polycyclic aromatic hydrocarbons
[554,555,635].

   The prospects for transformation of bile acids into
carcinogenic polycyclic aromatic hydrocarbons were supported by synthesis of "dehydronorcholene" (416), a pentacyclic
steroid analog formed from C_{24}-steroid acids chemically
[513,2676], the dehydrogenation of which by Se gave 20-
methylcholanthrene (417), an acknowledged carcinogen [513,
2675]. Other chemical syntheses of 20-methylcholanthrene
from cholic acid [756] and from the 5α-3-one 365 [2026]
have also been described.

     416   417

   By contrast, Kennaway wondered whether abnormal metabolism of cholesterol *in vivo* might transform the sterol
into such carcinogens [1284,1285]. Expressions of similar
thoughts implicate ergosterol via its 5α,8α-peroxide 62 and
derivative anthrasteroids as precursors of potentially carcinogenic polycyclic aromatic hydrocarbons [1737,2265].
The metabolic transformation of sterols and bile acids by
intestinal microflora into polycyclic aromatic hydrocarbon
is a current concept as well [1045-1047,2729].

   The structure resemblences are all that can be adduced
in support of the general notion that steroids can be
transformed to such materials, and what little experimental
work has been attempted to demonstrate biological conversions has been unsuccessful [423]. It must be stated here
that evidence supporting the transformation of cholesterol
in vivo into polyclic aromatic hydrocarbons is nil and that
evidence suggesting transformation of tissue cholesterol
into carcinogens via metabolism, irradiation, or by some
other process, is so diffuse that a satisfactory case cannot now be made. In any event, recovery and identification

of a sterol carcinogen from cancer tissues has not been
demonstrated, nor would success in such ventures be likely,
given present concepts regarding covalent bonding of ultimate carcinogens to nuclear DNA.

Nonetheless, observations of tumors (generally at the
site of injection) have been made in mice and rats administered cholesterol in several different protocols, some of
which are summarized in TABLE 23. The ultimate causes of
these tumors remain obscure, but concerns about numbers of
test and control animals, statistical treatment, test protocols, tumor pathology, and identity of cholesterol as a
causative agent has engendered a great reluctance to accept
a conclusion that cholesterol be carcinogenic.

Some reviews of data continue to support the thesis of
carcinogenicity for cholesterol or its alteration products
[231,232,235,238,981], but other critical evaluations dis-
count such claims [87,100,163,189,1327,2720]. Formal evaluation of all data by the Internal Agency for Research on
Cancer concludes that experimental evidence for assessing
the carcinogenicity of exogenous cholesterol does not exist
[1110]! Furthermore, in many studies in which cholesterol
has been administered to experimental animals for other
purposes, ranging from the studies of Anitschkow of 1913
to the present, no evidence of carcinogenicity has been
recorded [483,968,1441,2245,2246].

However, there are aspects of the matter which bear
attention. If the tumors observed after cholesterol administration were caused by congeners or cholesterol autoxidation products whose presence was not noticed, then these
congeners or autoxidation products might serve to explain
the reported tumor development. As the presence of autoxidation products could be highly variable depending on the
nature of prior treatment of the sterol in air, some of
the conflicting data about cholesterol might be thereby
rationalized.

Yet another mitigating circumstance obtains in the precise physical state of the cholesterol preparation administered. Solid-state carcinogenesis is a recognized mode of
initiation of cancer for which relatively little is understood as to mechanisms, a mode of carcinogenesis not particularly in present fashion. As the injection of crystalline

TABLE 23.   Cholesterol Carcinogenicity Test Data

Administration Mode	Test Animals	Results	References
Subcutaneous*	Stock, BALB/C, CBA, & C57 mice	Sarcomas	[1029,1031-1037]
	Swiss mice	Fibrosarcomas	[44]
	NMRI mice	Fibromas	[1859]
	Marsh-Buffalo mice	No tumors	[230,240]
Intravenous**	Strain A mice	Lung nodules	[1530]
Intraperitoneal	Rats	Sarcoma	[1867]
Diet***	TM mice	Mammary cancer	[2410]
		Lung adenoma	[2410]
Intrarectal	Fisher rats	No colon tumors	[1912]

　　*In oily vehicle
　　**In Aerosol OT aqueous dispersion
　　***In lard

cholesterol in any liquid vehicle is likely to result in
solid cholesterol deposits at the site of injection, the
issue of solid-state effects needs attention [231,232,235].

　　Subcutaneous implantation of pure or autoxidized cholesterol or of autoxidation products 13,27, or 35 leads to
granulomas [15,234]; such implantation of cholesterol pel-
lets is highly sclerogenic [2] but not carcinogenic [2204,
2206].   However, pellet implantation in the urinary bladder
results in bladder carcinoma [41,389,390,497].

　　Body fluids transform anhydrous cholesterol into the
monohydrate [235], and anhydrous and hydrous forms are interconverted by body fluids so mixtures of both coexist
under some conditions [1786,2348]. The possibility that different physical states of cholesterol provoke differential
responses *in vivo* progressing ultimately to carcinogenesis

provide another basis for rationalization of some of the
conflicting test data.

Yet another aspect of carcinogenicity of cholesterol
needs mention, namely that of cocarcinogenicity or promoter
activity. Previously described positive carcinogenicity
test results for cholesterol preparations could have been
obtained in tests inadvertantly incorporating unrecognized
carcinogens, the cholesterol then acting as cocarcinogen
to promote the action of the unrecognized carcinogen. This
formulation is merely another facet of the concept that impurities in cholesterol be the carcinogen and not cholesterol.

The cocarcinogenic activity of dietary cholesterol in
augmenting the induction of colon tumors has been demonstrated in rats treated with 1,2-dimethylhydrazine [539].
Moreover, high fat diets and elevated fecal bile acids and
sterol levels are correlated with human colon cancer and
certain other disorders associated with colon cancer [1045-
1047,2729]. Among neutral sterols found is the 3β,5α,6β-
triol 13, elevated somewhat in high fat diets [1910]. From
these items it has been speculated that cholesterol metabolism by intestinal microflora via the 5,6-epoxides to the
3β,5α,5β-triol 13 be implicated in colon carcinogenesis
[1911,1913,2046]. However, intrarectal instillations of
either 5α,6α-epoxide 35 or triol 13 did not lead to colon
tumors in rats nor did the 5α,6α-epoxide 35 act as cocarcinogen in promoting colon tumors caused by N-methyl-N'-nitro-
N-nitrosoguanidine [1912].

Moreover, cholesterol appears to act in other circumstances as a cocarcinogen in increasing the incidence of
tumors in mice injected with benzo[a] pyrene [608] but also
as an inhibitor of tumorigenesis induced by the same agent
benzo [a] pyrene in mouse skin [2635]. These matters of co-
carcinogenesis are as unclear as those of carcinogenesis of
cholesterol.

In this regard, the necessity for cholesterol for cellular growth and replication already described clearly implicates cholesterol in the etiology of malignant tumors [472],
though not as a carcinogen.

Sterol Esters. Sterol fatty acyl esters of tissues

have also been implicated as carcinogens, but supporting
evidence is slim. Fieser considered that cancer in amimals
administered heated cholesterol (discussed in the next section) might have been caused by transesterification giving
abnormal (putatively carcinogenic) cholesterol fatty acyl
esters [743,758]. Although a search for such esters in
tissues was not attempted, synthesis and test of cholesterol 3β-isoheptylate as example was done. Tumors were obtained in Marsh-Buffalo mice administered the esters subcutaneously in sesame oil [230,2246].

The case of an anteiso fatty acyl ester of cholesterol
is of more but uncertain interest. Cholesterol 3β-(+)-14'-
methylhexadecanoate (carcinolipin)[1080] isolated by Hradec
from divers animal tissues is reported to stimulate protein
biosynthesis [1077-1079,1082,1083] but also to be carcinogenic in rats and mice [1081,2195]! Furthermore, several
common phytosterols and their fatty acyl esters have been
recovered from human breast cancer tissues and an osteoly-
tic activity associated with such sterols [573,895,896].
However, as phytosterols also occur in normal human breast
tissues as well as in other tissues, it is unlikely that a
meaningful relationship between human breast cancer and
phytosterols exists [939,1614].

Carcinogenic Cholesterol Preparations

In distinction to the notion that carcinogenic cholesterol derivatives be present in cancer tissues, the question whether cholesterol preparations be transformed into
carcinogenic material by chemical or physical manipulations
was also examined. These studies extending from the early
1930s have been responsible for much of the concern and the
confusion associated with the issue. Weaknesses in experimental protocols included generally unimpressive numbers of
test animals or controls and test preparations that were
operationally defined (and poorly at that) and for which no
analysis or composition could possibly be given.

Roffo expanded his early studies on cholesterol carcinogenesis previously mentioned to include the treatment of
various cholesterol containing foodstuffs with radiation or
heat, the feeding of which to experimental animals caused
gastric and other malignant tumors [1983-1985,1987,1989,

1992,1993,1998]. Roffo also boldy suggested dietary regi-
mens for reduction of endogenous cholesterol levels in patients at risk from cancer [1991,1994] and attempted to implicate fatty acid oxidation as well in cancer [1986].
Criticism of the work of Roffo began at once, with questions raised whether the tumors observed be malignant and
whether polycyclic aromatic hydrocarbons be formed from
cholesterol [189,304,713,966,967,1327]. These issues have
never been resolved [84,87,100,554,743,937,1294,2720].

The matter was compounded by results of Waterman, who
fed heated cholesterol 3β-oleate to mice and reported gastric carcinomas [2636,2637,2639,2640]. Heated fats in general were also implicated in gastric cancer [1831,1832].

Waterman thought that heated cholesterol esters were
transformed by elimination to cholesta-3,5-diene (11)
which was the ultimate carcinogen [2638-2640]. However, it
must be noted that Waterman may have had a very complex
mixture of sterol derivatives, including oxidation products, for his synthesis of cholesterol 3β-oleate involved
heating cholesterol and oleic acid at 200°C under vacuum in
a stream of air [2636,2637]! Feeding experiments with the
3,5-diene 11 were also reported to yield stomach tumors in
mice [2589].

However, Kirby was unable to confirm tumor formation
in rats fed cholesterol or cholesterol 3β-palmitate, 3β-
stearate, or 3β-oleate heated at 300°C or fed the 3,5-diene
11 [1311-1314]. Results of other heating and testing protocols summarized in TABLE 24 do not support on balance the
certain induction of carcinogenicity in heated sterols.

The transformation of sterols to carcinogens has also
been recorded using other means. Irradiated ergosterol administered to mice is reported to yield adenomatous tumors
[1872]; intraperitoneal administration of irradiated ergosterol to rats gave a squamous cell carcinoma [1867]. Solutions of cholesterol in lard irradiated with X-rays are reported to yield spindle cell sarcomas in mice administered
the preparations subcutaneously [404], whereas rats administered irradiated cholesterol intraperitoneally develop fibrosarcoma [1867]. However, cholesterol irradiated by ultraviolet light as a solid and painted on the skin of mice
[196] or by X-rays as a benzene solution and administered

TABLE 24. Carcinogenic Testing of Heated Sterols

Temperature	Test System	Results*	References
Cholesterol:			
700–920°C	Mice, painted	(+)	[1282]
810°C	Rats	(+)	[1314]
430°C	Rats, fed	(−)	[1313]
	Mice, injected	Papillomas	[168]
	Mice, painted	(−)	[168]
	Mice	Sarcomas	[1831]
	Mice, painted	Skin epithelioma	[1285]
	Rabbits, painted	Papilloma	[1285]
300°C	Rats, fed	(−)	[1312]
	Mice, injected	(−)	[2358]
	Mice	Sarcomas	[1831]
270–300°C	Rats, fed	(−)	[1311]
	Mice, injected	Sarcomas	[168]
	Mice, painted	(−)	[168]
	Mice	Sarcomas	[1831]
275°C	Mice, painted	Sarcomas	[2673]
200°C	Mice, injected	(−)	[2358]
Phytosterols:			
800°C	Mice, painted	(+)	[2730]

(continued)

TABLE 24. (Continued)

Cholesterol esters**:

300°C		
Rats, fed	(−)	[1313]
Mice, injected	(−)	[168]
Mice, painted	(−)	[1314]

*Results are expressed (+) where carcinogenicity was claimed or where tumors were found in excess of control levels, as (−) where such was not the case.
**Esters included cholesterol 3β-palmitate, 3β-stearate, 3β-oleate, 3β-linoleate.

in olive oil to mice [2503,2504] was not carcinogenic.

As is the case for the question whether cholesterol be carcinogenic *per se* so also the issue whether irradiation or heat treatments of cholesterol lead to carcinogenic preparations must remain uncertain. The uncertainties of these test data are reminiscent of the uncertainty of the extent of autoxidation that a given sample of cholesterol may have undergone. Putatively carcinogenic pyrolysis and/ or autoxidation products in some heated sterol preparations may have survived subsequent thermal decomposition so as to exert their tumorigenic actions in the tests conducted at one time but may have been degraded thermally to inactive products in other cases. At best one may posit that some manipulations of cholesterol result in generation of car- cinogenic components which may be revealed under some test protocols in certain experimental animals.

Cholesterol Autoxidation Products

A bit more satisfactory state of affairs has developed for the evaluations of cholesterol autoxidation products than is the case for cholesterol or heated cholesterol. As once pure cholesterol may have autoxidized before or during carcinogenicity testing and heated cholesterol may have been transformed by heating and by exposure to air, the prior positive results suggesting carcinogenicity of cho- lesterol may have derived instead from the presence of highly variable trace levels of carcinogenic autoxidation products. Thus, this same concern as previously expressed for all the other demonstrated biological activities of cholesterol bears on carcinogenicity as well.

The concept of Roffo that carcinogenic oxidized cho- lesterol derivatives formed in skin [1977–1979,1999], and food [1987,1989] subjected to irradiation in air cause skin and gastric cancer respectively, taken with the demonstra- tion of carcinogenicity in Marsh–Buffalo mice of a subcuta- neously injected preparation of crude progesterone derived from cholesterol by oxidation [242] focused attention to cholesterol autoxidation products as potential carcinogens. Subsequent evaluation of specific cholesterol autoxidation products as carcinogens led Fieser to propose that choles- terol autoxidation via cholest-5-en-3-one (6) yield carcinogenic hydroperoxides such as 6β-hydroperoxycholest-4-en-3-one (59) [742,743,745]. At the time when Fieser was concerned with cholesterol autoxidation as putative source of steroid carcinogens, others thought to implicate lipid peroxidation in general [981] and photooxidations of cholesterol [2097] in the cancer process. Indeed, an attempt to correlate fibrosarcoma formation provoked by skin painting of mice with human liver extracts and the amount of cholesterol hydroperoxides present has been made [1040]. Schenck demonstrated that cholesterol could be photooxidized to the 5α-hydroperoxide 51 and 6β-hydroperoxide 89 in benzene solutions using benzo [a] pyrene or 20-methylcholanthrene as sensitizers [2097].

Evaluation of individual cholesterol autoxidation products for carcinogenicity has not been systematically conducted to satisfaction, and conflicting and uncertain data have been recorded. The data of TABLE 25 summarize information available about carcinogenicity testing for eight other sterols, five C_{27}-hydrocarbons, dicholesteryl ether, and eighteen sterol oxidation products, including three steroid hydroperoxide and peroxide derivatives, two extensively oxidized steroids, and thirteen common cholesterol autoxidation products.

The data of TABLE 25 gleaned from the literature are presented for direction to the prior tests, but the interpretations of results offered be qualified ones! Except where noted to the contrary these tests were conducted with subcutaneous injections of steroid in an oily vehicle, but amounts of steroid, precise location of injection, number are timing of injections, other treatment protocol, etc. are not always expressed, and no attempt to correlate these important factors has been made. Likewise, no assessment of mortality, duration of experiment, pathology, or statistical treatment of data is made here, these matters being beyond the pale.

Moreover, initial interpretations have been variously revised and reinterpreted, and in general the degree of uncertainty on the data is sufficient to suggest on balance that no case is established for carcinogenicity in any of these steroids, Bischoff to the contrary notwithstanding [231,232,238]. As in the case for pure cholesterol, so also tests with some common sterols (5α-stanol 2, 7-stenol 57,

TABLE 25. Carcinogenicity Testing of Individual Sterols

Sterol	Test Protocol*	Results**	References
Cholest-5-ene (7)	Stock mice (skin)	No tumors	[144]
5α-Cholest-6-ene	Albino & C57 mice	No tumors	[587]
Cholesta-2,4-diene (284)	Mice (skin)	No tumors	[425]
	CBA mice	No tumors	[2267]
Cholesta-3,5-diene (11)	Mice (per os)	Stomach papilloma	[2589]
	AB & C57 mice	No tumors	[232,808,2151]
	C3H mice (pellet)	No tumors	[1429]
	Wistar rats (per os)	Stomach papilloma	[1429]
		No tumors	[1312]
	Wistar rats (pellet)	No tumors	[1429]
	Albino & C57 mice	No tumors	[587]
5α-Cholesta-1,3,6-triene (419)			
2α,5-Epidioxy-5α-cholest-3-ene	Mice (skin)	No tumors	[425]
5α-Cholestan-3β-ol (2)	NMRI mice	No tumors	[1859]
5β-Cholestan-3β-ol (4)	Marsh & AB mice	No tumors	[244,934]
5α-Cholest-7-en-3β-ol (57)	March-Buffalo mice	No tumors	[230,240]
	C57 mice	No tumors	[1035]

(Continued)

TABLE 25. (Continued)

Compound	System	Result	Ref.
Cholesta-5,7-dien-3β-ol (56)	NMRI mice	Fibrosarcoma	[1859]
	XVII nc/Z mice	Fibrosarcoma	[1416]
	Mice	No tumors	[1031]
Desmosterol (78)	XVII nc/Z mice	No tumors	[1416]
Cholecalciferol (90)	Hamster (cheek pouch)	No tumors	[1441]
Ergocalciferol (321)	Hamster (cheek pouch)	No tumors	[1441]
Ergosterol (65)	Mice	No tumors	[1031,2246]
Dicholesteryl ether (18)	Albino rats (per os)	No tumors	[1260,1311]
	C3H mice (pellet)	No tumors	[1429]
Cholest-5-en-3-one (6)	Marsh-Buffalo mice (aq)	No tumors	[230]
	C57 & stock mice	No tumors	[1034,1035]
	Evans rats (aq)	No tumors	[232]
Cholest-4-en-3-one (8)	Marsh mice	Oleomas	[242]
	Marsh-Buffalo mice	No tumors	[230]
	Albino rats (per os)	No tumors	[1311]
	Marsh-Buffalo mice	No tumors	[934]
Cholesta-3,5-dien-7-one (10)	Marsh-Buffalo mice	Fibrosarcoma	[230]
3β-Hydroxycholest-5-en-7-one (16)	AB & C57 mice	No tumors	[2772]
	Mice	No tumors	[1031]
	Rabbit (fed)	Renal cortex malignancy	[49]
6β-Hydroxycholest-4-en-3-one (88)	Marsh-Buffalo mice	Fibrosarcoma	[230,241]
	Marsh mice (aq)	Fibrosarcoma	[236,240]
	C57 & stock mice	No tumors	[1034,1035]

(Continued)

TABLE 25. (Continued)

Compound	Organism	Result	References
5,6α-Epoxy-5α-cholestan-3β-ol (35)	Swiss mice	Carcinomas, leukemia, lung adenoma	[231,232]
	Evans rats (aq)	No tumors	[232]
	Marsh-Buffalo mice	Fibrosarcoma	[229-232,241]
	Marsh-Buffalo mice (aq)	Fibrosarcoma, pancreas adenoma, lung carcinoma	[231,240,246] [2182]
	C57 mice (per os)	No tumors	[231,232]
	Swiss mice	Tumors	
	Mice (intratesticular, aq)	Fibrosarcoma	[231]
	Evans rats	Tumors	[231-233]
	Fisher rats (intra-rectal)	No colon tumors	[1912]
	Wister & Sprague-Dawley rats (per os)	No tumors	[763,764,861,2770]
5α-Cholestane-3β,5,6β-triol (13)	Marsh-Buffalo mice (aq)	No tumors	[230,240]
	Fisher rats (intra-rectal)	No colon tumors	[1912]
3β,5-Dihydroxy-5α-cholestan-6-one (44)	Marsh-Buffalo (aq)	No tumors	[240]
Cholest-4-ene-3,6-dione (108)	Marsh-Buffalo mice	Fibrosarcoma	[230,241]

(continued)

TABLE 25. (Continued)

Compound	Test animal	Tumor result	References
	C57 & AB mice	Extraordinary tumors	[1397]
	C57 & stock mice	No tumors	[1034,1035]
	AB mice	No tumors	[388,934]
	Marsh–Buffalo mice	No tumors	[388]
	Swiss mice	No tumors	[231,232]
3β-Hydroxy-5α-cholest-6-ene-5-hydroperoxide (51)	Mice (IP)	No Tumors	[1534,1357]
	Mice	No tumors	[231,232,554,1356]
	Irradiated rats (IP)	No tumors	[1355]
3β-Hydroxycholest-5-ene-7α-hydroperoxide (46) 3β-Hydroxycholest-5-ene-7α- and 7β-hydroperoxides (46,47)	Mice (IP)	No tumors	[1354,1357]
6α-Hydroperoxycholest-4-en-3-one (280)	Swiss mice	No tumors	[640]
	Marsh mice (aq)	Local sarcoma	[232,238]
6β-Hydroperoxycholest-4-en-3-one (59)	Marsh–Buffalo mice	Local sarcoma, fibrosarcoma	[229–231,236,243,752]
	Marsh mice (aq)	No tumors	[230,752]
	C57 mice	Sarcoma	[1034–1036]
	Swiss mice	No tumors	[640,1041]
	Evans rat	Local sarcoma	[231]
	Evans rat (aq)	No tumors	[232]
	Rats & mice (injection into stomach wall)	No tumors	[231]

(continued)

TABLE 25. (Continued)

Compound	Animal	Result	Ref.
6β-Hydroperoxystigmast-4-en-3-one	Marsh mice	Fibrosarcoma	[231]
Lanosterol hydroperoxides	Swiss mice	No tumors	[640]
3α,5;3β,4-Bisepoxy-5α-3,4-secocholestan-6-one	Marsh-Buffalo mice	Fibrosarcoma	[231]
	Marsh-Buffalo mice (aq)	No tumors	[231]
3β-Acetoxy-22,23-bisnor-chol-5-enic acid	XVII nc/Z mice	Fibrosarcoma	[1416]

*Mode of administration is via subcutaneous injection in oily vehicle except as noted otherwise by parenthetic notation: (aq), aqueous vehicle; (skin), skin painting with oily vehicle; (pellet), subcutaneous implanted pellet; (per os), fed in diet; (IP), intraperitoneal injection.

**Tumor incidence not greater than that of controls is marked as "No tumors".

ergosterol, the calciferols) did not lead to tumor development. However, conflicting results have been obtained on
other sterols, the 5β-stanol 4 and 5,7-dienol 56 being reported to give fibrosarcomas in some but not in other
tests. Moreover, the 3,5-diene 11 for which the early concern about steroid carcinogenicity was expressed likewise
gave conflicting results.

In that neither pure cholesterol nor congeners appear
to be carcinogenic, the observed tumor production using
chemically oxidized cholesterol or air-aged pure cholesterol [230] could well be laid to cholesterol autoxidation
products. However, the notion that a specific structural
feature such as hydroperoxide, peroxide, ketone, or expox-
ide group confer carcinogenicity [232] is not upheld by
test data, nor is a more recent suggestion comparing the Δ^6-
olefins   5α-cholest-6-ene (418) or 5α-cholesta-1,3,6-tri-
ene (419) with the electron-rich K-region of carcinogenic
polycyclic aromatic hydrocarbons [587].

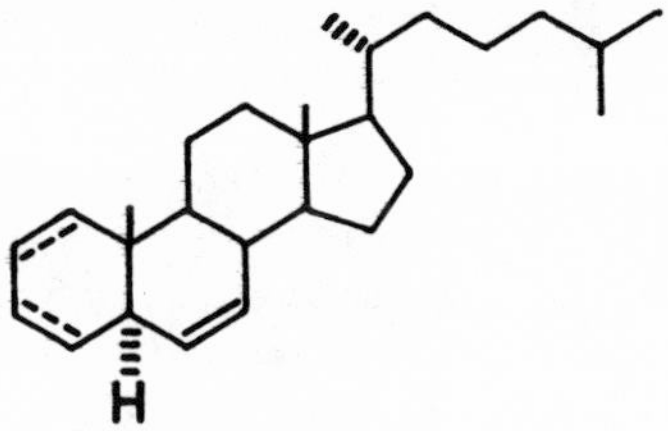

418

419 $\Delta^{1,3}$

Five cholesterol autoxidation products 6β-hydroxycho-
lest-4-en-3-one (88) and the corresponding 6β-hydroperoxide
59, the 5α,6α-epoxide 35, the Δ^4-3,6-diketone 108 and the
5α-hydroperoxide 51 have been advanced as carcinogens.
Various later qualifications, reinterpretations, and assessment of other data leave these cases unsettled. The 6β-al-
cohol 88 has been viewed by Bischoff sequentially as carcinogenic [230,236], of uncertain carcinogenicity [232], and
as carcinogenic [238]. The corresponding 6β-hydroperoxide
59 is regarded as highly carcinogenic [230-232,238].

The carcinogenicity of the 3,6-diketone 108 in three
strains of mice [230,241,1397] is not supported upon critical examination [388], and Bischoff has revised his appre-

ciation of this case [232].

The 5α-hydroperoxide 51 shown to provoke tumors in fe-
male mice at incidences considerably greater than in male
mice or in controls has variously been viewed as not carcinogenic (and as radioprotective agent) in irradiated rats
[1353,1355], as cocarcinogenic but not a true carcinogen
[1354,1356,1357], and as carcinogenic [232]! Thermal decomposition products of 51 (12 and 50) have not been tested
for carcinogenicity, nor apparently has the instability of
51 (or of 59) been of concern in these matters.

Clearly conflicting results obtain for these four putative carcinogens 51, 59, 88, and 108. Different experimental animals, strains, sexes, etc. as well as different
means of administration including oily versus aqueous vehicle, and experimental protocols different in other ways,
to say nothing of the general failure to demonstrate that
any tumors produced be transplantable all tend to mitigate
against acceptance of claims of carcinogenicity for these
sterols.

The fifth cholesterol autoxidation product for which a
stronger case has been made is the 5α,6α-epoxide 35. Bischoff has asserted the carcinogenicity of the 5α,6α-epoxide
35 based on his data in mice and rats and subcutaneous administrations with oily or aqueous vehicles [230-232,238].
These claims have been repeated by others as well, so that
a litany on the matter now obtains.

However, carcinogenicity of the 5α,6α-epoxide 35 administered in the diet [763,764,2182,2770] or intrarectally
[1912] was not demonstrated, and in rats fed the 5α,6α-ep-
oxide 35 at 1% levels in their diets for two years tumor
incidence was not greater than in controls. Decreased
growth and other signs of toxicity (including LD_{50} 1.82 g/kg
in acute toxicity tests) were manifested, but rats fed 35
actually had lower death rates than did control or cholesterol-fed animals [861].

Circumstantial evidence does implicate the 5α,6α-ep-
oxide 35 as a putative carcinogen, as 35 and/or isomeric 5β,
6β-epoxide 36 appear to be formed in human [281,1523] and
hairless mouse [275-277] skin subjected to ultraviolet
light irradiation, a treatment that leads in hairless mice

to squamous cell carcinoma. Moreover, antioxidants retard
these events [278,456,1524]. Additionally, the 5α,6α-ep-
oxide 35 has been detected in human serum [914], and either
35 and/or the 5β,6β-epoxide 36 has been detected in the
liver of hairless mice [279]. Both isomeric 5,6-epoxides
35 and 36 are formed from cholesterol in rat lung by oxidations initiated by inspired NO_2 [2191-2193]. Both liver
and skin of hairless mice and human intestinal microflora
have an epoxide hydratase transforming 35 into the 3β,5α,
6β-triol 13, possibly as a protective measure [280,455,
1109,1525].

Moreover, subcutaneous deposits of the 5α,6α-epoxide
35 appear to remain *in situ* for over a year. From the
amount of 35 remaining at the site of injection in Marsh
mice after 420 d, 7-22 μg per mouse per day was calculated
to be released, thus the same amount found in irradiated
hairless mice [237,238,246,277]. However, subcutaneous deposits of xenobiotic sterols may persist in any event, as
both the 6β-alcohol 88 and 5β-stanol 4 have been found at
the site of injection up to 18 months later [238].

Yet other information bears on the matter. Although
the 5α,6α-epoxide 35 is not mutagenic towards *S. typhimurium* nor is the 5β,6β-epoxide 36 [69,2307], the 5α,6α-epox-
ide 35 appears to be active as a transforming agent against
cultured ELa/ENG hamster embryo cells, but the capacity of
the transformed cells to provoke tumors upon transplanta-
tion was not examined [1278]. Furthermore, the 5α,6α-epox-
ide 35 induces chromosome aberrations, chromatid breaks or
deletions, and initiation of DNA repair synthesis in cultured human fibroblasts, these damaging effects being like
those caused by ultraviolet light irradiations [1820]. The
apparent covalent binding of the 5α,6α-epoxide 35 with DNA
[283] may be of further interest in this issue.

There is always the question whether impurities in
sterol preparations cause any tumor responses observed. Although this issue is foremost for cholesterol *per se*, the
same concern exists for the cholesterol autoxidation products themselves. Bischoff's concern on this point was carried even to question whether boric of phthalic acids possibly formed from the parent peracid used in the chemical
synthesis of the 5α,6α-epoxide 35 from cholesterol be present in his carcinogenic 5α,6α-epoxide preparations [232].

Furthermore, the question whether the isomeric 5β,6β-epoxide
36 (likely to be present with the 5α,6α-epoxide 35 whatever the chemical synthesis used) be carcinogenic has not
been addressed. As none of the studies heretofore reported
provided evidence of a pure 5α,6α-epoxide 35 for testing,
it is unlikely that pure 35 free of isomeric 36 has in fact
been evaluated for carcinogenicity.

     Besides test data with mice and rats, as previously
mentioned, data with other animals assembled in experiments
directed to other interests have also been examined for
possible carcinogenesis as well [2246]. Uniformly no support for tumor induction is evident. However, several
cholesterol autoxidation products have been tested as putative carcinogens in the newt *Triturus cristatus*. The significance of testing in this animal is uncertain, but cholesterol, the enone 6, isomeric 5,6-epoxides 35 and 36, epimeric 6-hydroperoxides 281 and 59, the 6β-alcohol 88 were
not active. By contrast, positive test results were obtained with 5-hydroxy-5α-cholestane-3,6-dione and 3,6-dike-
tone 108[1743], and cholesterol heated in air at 350°C or
photooxidized gave positive responses [88,89].

                         MEMBRANE EFFECTS

     It is now recognized that biological membranes of
eukaryotic cells require sterols for their function and
integrity [597,1736]. The long standing question of biological function of cholesterol [2241,2243] may surely be settled in this light. Indeed, the bulk of sterol in nature
is associated with cellular membranes, the resultant stability of which suggests that some membrane sterols present
in living organisms may be as old (10 Gy) as the sterol
biosynthesis process itself [1736]!

     A recent synthesis of the properties of sterols in
biological membranes of eukaryotes provides a functional
definition of a sterol. A sterol must: (i) reduce the effective molecular area of membrane phospholipids (the sterol
condensing effect), (ii) mobilize phospholipid fatty acyl
chains for phospholipids in the ordered gel state but reduce mobility of such chains for phospholipids in the fluid
liquid crystalline state, (III) reduce nonionic permeability

for phospholipids in the liquid crystalline state but increase nonionic permeability for phospholipids in the ordered gel state. The net effect of these items is the increase of membrane fluidity and stability and the modification of membrane permeability in a way essential to life.

A functional definition of a sterol requires these
three effects, whereas a chemical definition of a membrane
sterol requires the intact tetracyclic ring system not in
the 5β-configuration, a 3β-hydroxyl group, and a C-17 alkyl
side-chain [444,597,1736]. It is recognized that some non-
sterols may functionally be sterols and that some chemically defined sterols may not function as sterols in this con-
text. Functional sterols serving the stipulated purposes
possess a side-chain of at least five but preferably eight
to nine carbon atoms [533,2364,2387]. In some systems even
the C_{19}-sterol androst-5-en-3β-ol (114) shows some ordering
influence on membrane lipid components [427,2386,2387].

A similar measure of sterol side-chain effectiveness
is suggested by interactions between a specific human erythrocyte protein and sterols of divers side-chains, the C_{19}-
sterol 114 being least effective but with the C_{21}-sterol
110, C_{24}-sterol 109, etc. binding to the protein effectively [1326].

The effect of sterol side-chain structure on but one
measurable parameter, that of the condensing effect on egg
phospholipid liposome membranes, suggests that the cholesterol C_8-side-chain be the optimum. Synthetic sterols with
longer side-chains (26,26-dimethyl-27-norcholest-5-en-3β-ol,
27-n-propyl-26-norcholest-5-en-3β-ol) are less effective
than cholesterol [533], as are also sterols with short side-
chains, including the C_{26}-sterol 26-norcholest-5-en-3β-ol,
C_{25}-sterol 197, C_{24}-sterol 109, and C_{21}-sterol 110 [533,
2387]. Notably, the derivative with no side-chain androst-
5-en-3β-ol (114) is effective as a condensing agent but much
less so [427,2387]. A similar effect of side-chain structure on the level of incorporation of fatty acyl esters of
testosterone (17β-hydroxyandrost-4-en-3-one) into liposomes
is observed, the C_8-ester (17β-octanoate) being maximally
incorporated into liposomes [2364]. However, no experimental studies have been conducted with short side-chain sterols
and phospholipids from the various invertebrates in which
the short side-chain sterols have been found, and very

little else is known about invertebrate membranes.   No answer to this problem can now be posed.

     Further interests will be focused on possible biolog-
ical effects of cholesterol autoxidation products at the
membrane level, and no other discussion of the membrane effects of cholesterol will be attempted here.

     Very few biological responses to oxidized sterols in
intact animals have been associated with membrane effects.
One such is an obscure note that 24-hydroxy-(24S)-cholest-
4-en-3-one administered intraperitoneally (18 mg/kg) to
rats tended to synchronize cortical potentials within 10 m,
with return to normal by 30 m, an effect not exhibited by
cerebrosterol (25) or the 24-ketone 34 [696].

                        Model Systems

     Effects of cholesterol autoxidation products in model
systems have been examined with different simple means.
Perhaps the simplest study is the demonstration that cyclohexane solutions of cholesterol subjected to air oxidation
exhibit diminished interfacial tension between the cyclohexane solution and aqueous KH_2PO_4 [1853].   In other more
sophisticated studies the incorporation of cholesterol autoxidation products into synthetic liposomes has been
achieved with the enones 8 and 16 and dienone 12.   Little
effect of these oxidized sterols on ordering of phospholipids or of disruption of the liposome is evident [386,917,
1092].   The effects of cholesterol and of oxidized cholesterol preparations and derivatives upon surface pressure
area relationships of monomolecular layers of lipids spread
on aqueous media has been of interest for half a century
[18,804].   Indeed, such surface pressure-surface area measurements on cholesterol oxidation products suggested by
1930 that the then held structures for sterols might be in-
correct [19].   It was also recognized early that the oxidized cholesterol derivatives tended to be more extensively
anchored into the aqueous phase [17].

     It must be recalled that monomolecular layers of cho-
lesterol are quite sensitive to air oxidations [1217,1218,
2651], the autoxidized cholesterol products then exerting
an expansion effect on the surface area-surface pressure

curves, thus an increased area/molecule ratio proportional
to the extent of cholesterol autoxidation [865,1217,1218].
Similar expansion effects occur with autoxidized monomolecular films of cholesterol 3β-linoleate (234), 3β-linolenate,
and 3β-arachidonate [1413]. Moreover, individual cholesterol autoxidation products give monomolecular layers that
are more expanded (greater area/molecule) than cholesterol.
Such expanded films have been observed for the 3-ketones
6,8,10, and 12, for the 7-oxygenated derivatives of 14-16,
and for the 3β,5α,6β-triol 13 as well as for several other
cholesterol oxidation products [17,19,595,865,1216]. Moreover, those cholesterol autoxidation products that are
further sensitive to autoxidations, such as the Δ^5-3-ketone
6, are observed to yield yet further expanded films upon
exposure to air [1217].

Furthermore, these expansion effects are also observed
in mixed sterol-egg lecithin films, where the expansion effects of the several cholesterol autoxidation products 6,
8,10,12-16 as well as of cholesterol per se increase as the
mole fraction of lecithin increases in the mixtures [1216].
Other chemical alteration of cholesterol such as iodination
[2411] also gives expansion effects on monomolecular films.

In that autoxidation tends to increase measurably the
surface area/molecule ratio from 38.2 sq. Å for pure cholesterol it has been suggested that surface area determina-
tions be used as means of assessing purity [401]! However,
it must be recalled that condensation effects on surface
pressure-surface area relationships are also possible,
witness the action of NO_2 on monomolecular cholesterol
films [1215,1217,1529].

The surface pressure-surface area relationships of
model monomolecular films are not greatly altered by cholesterol autoxidation products, thus suggesting that the
autoxidation of cell membrane cholesterol in vivo might
not alter properties of the cell membrane significantly
[1216]. However, other measures of membrane properties
are much more obviously altered by inclusion of cholesterol autoxidation products into model membranes or generation of autoxidation products in situ. For example, incorporation of 10 mole % of the 3β,25-diol 27 into synthetic liposomes increased their permeability to Ca^{++}, efflux
from the liposomes being increased six-fold [1405]! Other

premeability effects in synthetic liposomes may be variously influenced by the length of the sterol side-chain [1711].

The model "black" lipid membrane of Tien has become a
very popular system for many membrane property studies.
These "black" lipid membranes form readily with naturally
oxidized cholesterol (but not with pure cholesterol) as
sole lipid component; thus, phospholipid is not required
for membrane stability [2483]. Cholesterol autoxidized in
organic solvents [1968,2483] and in aqueous dispersions in
which the 7-oxygenated products 14-16 are formed to the extent of 5-10% are also effective [1531]. Cholesta-5,7-dien-
3β-ol (56) very sensitive to autoxidation also forms such
stable "black" membranes. The model membranes are approximately 7 nm thick, are regarded as the closest model to
biological membranes yet, and have received much attention
with respect to ion transport phenomena [2482]. Special
batches of oxidized cholesterol are presently sold for
making these membranes [68].

Specific cholesterol autoxidation products may exert
yet other effects in model membrane systems. Thus, the 3β,
25-diol 27 appears to create phase separations in dilauroyl
and dimyristoyl lecithins, possibly because of inadequate
solubility in the lecithin phase [2592].

                          Membrane Enzymes

Demonstration of the potential for influencing membrane effects in biological membranes is found in the
in situ enzymic oxidation of membrane cholesterol, with
demonstration of attendant alterations of specific membrane
enzymes. A clear example is that of dehydrogenation of
cholesterol of human erythrocyte ghosts with a cholesterol
oxidase (EC 1.1.3.6) from N. erythropolis, yielding the
enone 8 produced within the membrane. Under these conditions membrane-bound Na^+, K^+-dependent ATPase is inhibited
roughly in proportion to the extent of sterol dehydrogenation [2251]. Oxidase action may occur only at the exposed
inner membrane of erythrocyte ghosts [902] and may be influenced by membrane phospholipid [1826].

Exchange of erythrocyte cholesterol for desmosterol
(78) has the opposite effect of stimulating membrane-
bound ATPase [736].

Effects on Cells

There are several effects that may be related to the
specific suppression of HMG CoA reductase activity in cultured cells. For many cells cultured in the absence of
exogenous cholesterol the several cholesterol products
exert marked influences on many vital cellular functions,
some of which have been already described. Thus, the regulatory enzyme HMG CoA reductase in suppreseed, thereby
limiting *de novo* sterol biosynthesis. The biosynthesis of
DNA and mitosis cease, and a reversible arrest of cell
replication at the G1 state of the mitotic cycle is effected. Cation transport in such cells is severely compromised. Endocytosis is impaired, and mouse lymphocytes
no longer inflict lymphocyte-mediated cytotoxicity on
certain target cells. The plasma membranes of such cultured cells exhibit a muchly diminished sterol/phospho-
lipid ration.

A second set of responses affected by oxidized sterols
has been found in a variety of mammalian cell systems.
Among the actions observed are suppression of an immune
response in mouse spleen cells, inhibition of E-rosette
formation between lymphocytes and alien erythrocytes,
inhibition of chemotaxis and cell motility of polymorphonuclear leukocytes, and transformation of erythrocytes
into echinocytes with attendant changes in osmotic fragility, all to be discussed in detail.

As previously outlined, the effects of the first
class of biological actions appear to be manifested following the specific inhibition of *de novo* sterol biosynthesis by suppression of the regulatory enzyme HMG CoA
reductase. These effects are reversed or mitigated substantially by additions of sterol or of mevalonate. Activities of the second group cannot be so readily explained
in these terms, as the effects occur too rapidly for
de novo biosynthesis to be implicated. Moreover, exogenous
mevalonate or sterol in some cases fail to reverse effects
as in the case for the first group of actions. Finally,
structure-activity relationships among the second set of
responses are quite different from those associated with
HMG CoA reductase inhibitions.

The several biological actions to be described all
appear to follow upon action of the oxidized sterol with or
at the plasma membrane of the cells. Indeed, uptake and
retention of oxidized sterol within the plasma membrane has
been demonstrated in some cases, and this incorporation of
sterol·into membrane may differentiate this class of biological actions from those involving suppression of HMG CoA
reductase.

Immunosuppression. The capacity of some cholesterol
autoxidation products to act as immunosuppressants is suggested for the epimeric 3β,7-diols 14 and 15 and the 7-ke-
tone 16 [1325] and fully demonstrated for the 25-hydroper-
oxide 26 and 3β,25-diol 27 [1104]. Incubations of mouse
spleen cells with liposomes containing air-aged cholesterol
gave almost total suppression of the plaque-forming cell
response to a synthetic challange acting as an *in vitro* T
cell-independent antigen. Pure cholesterol had no such im-
munosuppressant effect. The active component of such air-
aged cholesterol was identified as the 25-hydroperoxide 26,
and the 3β,25-diol 27 was equally active. As the immuno-
suppressant effect was not mitigated by added mevalonate,
the effect appears not to involve *de novo* sterol biosynthesis [1104].

E-Rosette Formation. Another action of cholesterol
autoxidation products on the immune system is the inhibition of formation of E-rosettes between human T-lymphocytes
and sheep erythrocytes. Human lymphocytes cultured in lipoprotein-depleted media are affected by 25 μM concentrations
of the 3β,7-diols 14 and 15, (20S)-3β,20-diol 21, or 3β,20-
diol 21, or 3β,5α-dihydroxy-6-ketone 44 such that subsequent incubations with sheep erythrocytes do not elicit full
E-rosette formation. The 3β,25-diol 27 so potent in suppression HMG CoA reductase is but weakly inhibitory of E-rosette formation. Several other cholesterol autoxidation
products show little or no inhibitory effect (cf. TABLE 26).

The inhibition of E-rosette formation is apparent after 15 m for the 6-ketone 44! Exogenous cholesterol, lipoprotein, or serum abolish the inhibition, but mevalonate
does not. As the inhibition thus appears to be independent
of *de novo* sterol biosynthesis, the effects are attributed
to incorporation of the autoxidation products into the
lymphocyte membrane [2373,2374,2735].

Chemotaxis. The migration of human polymorphonuclear
leukocytes toward a chemotaxin is inhibited within 5 m by
prior incubation of the leukocytes in lipoprotein-depleted
media with several cholesterol autoxidation products at
concentrations as low as 6.25 µM. Random cell motility
is unimpaired; cell viability is not significantly affected.
However, the capacity to migrate towards a chemical stimulus
is lost.

The 3β,5α-dihydroxy-6-ketone 44 is the most potent
inhibitor of chemotaxis in this system (cf. TABLE 26).
Incubations in lipoprotein-containing media partially protected the granulocytes against the inhibition but not
completely. As mature human polymorphonuclear leukocytes
do not synthesize sterols *de novo*, the inhibition of chemotaxis cannot be regarded as involving such biosynthesis,
and a plasma membrane effect is suspected. The rapidity
of onset of inhibition of chemotaxis and the lack of correspondence between the effect and the inhibition of HMG
CoA reductase activity further support the thesis [897,989,
2735].

Erythrocyte Effects. The effects of cholesterol autoxidation products on erythrocytes are several, all of which
appear to be moderated by plasma membrane alterations.
Chief among effects is that of stability of the erythrocyte
towards hemolysis. The enones 6 and 8 exchanged for human
erythrocyte cholesterol increase osmotic fragility [385,
1581],but the dienone 12 does not; the 7-ketone 16 may
diminish hemolysis [917]. Incubations of human erythrocytes
in 25 µM 3β,25-diol 27 or 3β,5α-dihydroxy-6-ketone 44
increase osmotic fragility, but 25 µM 3β,7α-diol 14 is
ineffective. Moreover, incubations of the (20S)-3β,20-diol
21 prevent development of osmotic fragility and cause the
appearance of a population of erythrocytes that are resistant to osmotic shock [2735].

Furthermore, incorporation of the 5α-hydroperoxide 51
into human erythrocytes also leads to increased osmotic
fragility, but an initial induction period appears to be
involved, during which time osmotic fragility is decreased.
Subsequent increase in fragility and attendant hemolysis
follows [1421].

Cholesterol and certain of its autoxidation products
exert a protective action against the lysis of human

erythrocytes by thiol-activated cytolytic toxins. The 6-
ketone 45 is weakly inhibitory of the lytic action of
cereolysin from *Bacillus cereus* [527] and of metridiolysin
from the sea anemone *Metridium senile* [206]; cholesterol
is protective in both cases as well. The C_{24}-acid 99 is
weakly inhibitory of the hemolytic activity of streptolysin
O from *Streptococcus pyogenes,* but other cholesterol autoxidation products 8,23,27,29,86,110, and 114 are not [2644].
By contrast, cultured mouse L cell fibroblasts become increasingly resistant to the lytic effects of streptolysin
O following incubations with the (20S)-3β,20-diol 21 or
3β,25-diol 27 [639]. The 3β-sulfate esters of cholesterol,
the 7-ketone 16, and 22-ketone 297 protect erythrocytes
against hemolysis also [293,294].

Membrane permeability is also affected by cholesterol
oxidation products. Human and pig erythrocytes become more
permeable to glycerol when treated with enone 8, dienone
12, or 5α-3-ketone 365 [385], more permeable to glucose
when treated with enone 8, 6-ketone 45, or 5α-3-ketone 365
[1581]. These ketosteroids also increase the permeability
of egg lecithin liposomes to glycerol and glucose [596].

A striking effect of cholesterol autoxidation products
is the transformation of normal biconcave disk-shaped human
erythrocytes into echinocytes, cells having lost their cen-
tral concavity and gained numerous spiked protrusions.
Human erythrocytes incubated with 25 μM concentrations of
several common cholesterol autoxidation products in lipoprotein-depleted serum become echinocytic within 2 m! The
effect attains fastigium at 5-150 m, subsequent nadir at
2-8 hrs, with a second phase of echinocyte formation over
the next 40 hr. Both cholesterol autoxidation products
eliciting the response and those that are inactive appear
in the erythrocyte membrane within 1 hr. Incorporation
of oxidized sterol into erythrocyte membrane is impeded
by added lipoprotein but not by added cholesterol. Echinocyte formation is reversible, as removal of the oxidized
sterols by washing and reincubation with serum returns
the cells to their usual biconcave disk shape.

Cholesterol autoxidation products may be grouped into
three separate classes depending upon their potency in provoking echinocyte formation. The most potent oxidized

sterols (cf. TABLE 26) are the common autoxidation products 3β,7β-diol 15, 3β,5α,6β-triol 13, and 3β,5α-dihydroxy-6-ketone 44 and the 6-ketone 45, the 5α-3β,6β-diol 264, and 3β-hydroxy-5α-cholestan-7-one. The 3β,7α-diol 14 is much less effective than its 3β,7β-epimer 15. Least effective in transforming erythrocytes to echinocytes are the potent HMG CoA reductase inhibitors (20S)-3β,20-diol 21 and 3β,25-diol 27. It thus appears that the B-ring oxidized sterols are effective echinocytic agents and side-chain oxidized sterols are not [1093,2735].

The extent of echinocyte formation parallels the amount of oxidized sterol incorporated into the erythrocyte membrane for the potent agents 3β,7β-diol 15 and 3β,5α,6β-triol 13. However, incorporation of the oxidized sterol into the erythrocyte membrane is not a sufficient condition for echinocyte formation, as the (20S)-3β,20-diol 21 is incorporated into erythrocyte membranes to a greater extent than are several other oxidized sterols which are more potent echinocytogenic agents. As these effects are so rapid and the mature erythrocyte does not synthesize sterols *de novo*, an involvement of HMG CoA reductase suppression and of *de novo* sterol biosynthesis in the effect is not indicated [1093].

Echinocyte formation characterizes certain human disorders, such as severe hepatocellular disease, and it has been speculated that perhaps elevated plasma levels of cholesterol autoxidation products, particularly the 3β,7β-diol 15, might be implicated in such diseases [1093,2735].

Sterol Insertion Hypothesis. The several biological effects discussed in this section all have aspects which suggest involvement at the membrane level. A summary of data of Yachnin [2735] for a dozen cholesterol oxidation products is given in TABLE 26, together with summarized data for effects attributed to intracellular effects associated with suppression of *de novo* sterol biosynthesis.

It is seen that the biological activities of this dozen oxidized sterols vary widely and that no consistent structure-activity relationship obtains for all the data. Those oxidized sterols most active in suppressing *de novo* sterol biosynthesis in stimulated lymphocytes are the 3β,7α-diol 14, (20S)-3β,20-diol 21, and 3β,25-diol 27, and the

TABLE 26. Divers Cellular Effects of Oxidized Cholesterol Derivatives

Sterol	Membrane Effects					Intracellular Effects	
	I	II	III	IV	V	VI	VII
Cholesterol (1)	0	0	0	0	2.6	1.6	>25
Cholest-5-ene-3β,7α-diol (14)	14.3	85.9	91.4	0	34.7	48.1	0.11
Cholest-5-ene-3β,7β-diol (15)	14.9	51.5	68.4	31.3	98.5	-	3.0
3β-Hydroxycholest-5-ene-7-one (16)	5	0	0	21.1	61.7	25.5	1
5α-Cholestane-3β,5,6β-triol (13)	5.4	0	0	56.4	100	34.1	3.0
Cholest-5-ene-3β,25-diol (27)	8.2	0	7.8	0	1.4	93.9	0.29
(20S)-Cholest-5-ene-3β,20-diol (21)	4.4	0	50.1	54.6	9.1	90.1	0.11
Cholest-5-ene-3β,4β-diol (41)	2.3	0	0	0	78.8	18.8	4.5
3β,5-Dihydroxy-5α-cholestan-6-one (44)	34.2	78.8	82.6	89.4	99	38.6	2.9
3β-Hydroxy-5α-cholestan-6-one (45)	1.3	0	0	81.1	98.5	48.8	1.2
3β-Hydroxycholest-5-ene-22-one (297)	7.2	0	14.5	52.5	17.5	29.8	0.35
5α-Cholestane-3β,6β-diol (264)	3.5	0	0	38.8	98	30.1	3.1
3β-Hydroxy-5α-cholestan-7-one	3.0	0	0	19.4	89.1	42.6	0.94

I   Lymphocyte cytotoxicity, %
II   Lymphocytolysis, %
III   Inhibition of E-rosette formation, %
IV   Inhibition of granulocyte chemotaxis, %
V   Echinocytic transformation of erythrocytes, %
VI   Inhibition of DNA biosynthesis in lymphocytes, %
VII   50% Inhibitory concentration of sterol biosynthesis, μM

diols 21 and 27 are the most inhibitory towards DNA synthesis. However, neither 21 nor 27 lyse lymphocytes and
both are but mildly cytotoxic, whereas the 3β,5α-dihydroxy-
6-ketone 44 is cytotoxic, and 44 and the 3β,7α-diol 14
cause marked lysis.

Granulocyte chemotaxis is inhibited maximally by the
6-ketones 44 and 45, and the 6-ketone 44 and the 3β,7α-diol
14 exert maximal inhibition on E-rosette formation by
lymphocytes. Echinocyte forming activity is expressed
maximally by the 3β,7β-diol 15, 3β,5α,6β-triol 13, 6-ketones
44 and 45, 5α-cholestane-3β,6β-diol (264), and 3β-hydroxy-
5α-cholestan-7-one.

Those oxidized sterols exhibiting high activities of
this second type appear to be oxidized in the B-ring and
not in the side chain. Thus, the 6-ketone 45 is the most
active in affecting morphology, viability, and function of
human lymphocytes, granulocytes, and erythrocytes. The
(20S)-3β,20-diol 21 and 3β,25-diol 27 highly inhibitory towards *de novo* sterol biosynthesis and DNA synthesis are
much less effective in these other matters.

It is thus obvious that other factors operate in the
exertion of some of these actions. In that the plasma membrane of erythrocytes has been demonstrated to incorporate
oxidized sterol [1093], it may be posited that action at
the membrane rather than at intracellular sites be the
means by which these actions are initiated. The insertion
of oxidized cholesterol derivatives into plasma membranes
of affected cells has been proposed as a means of exerting
the several actions described. As B-ring oxidized sterols
are the most effective in these matters, it appears that
steric effects of the added oxygen-containing functional
groups may limit the closeness with which the sterol molecules may approach one another, thus effecting an expansion
of the membrane in much the same manner as proposed for
such actions in model systems involving spread monomolecular films. The possibility that hydrogen-bonded aggregates
of the highly active 3β,5α-dihydroxy-6-ketone 44 form within
a membrane is advanced as an additional means of accounting
for the potency of this sterol [2735].

The capacity of a sterol nitroxide electron spin
resonance probe to inhibit *de novo* sterol biosynthesis in

guinea pig lymphocytes and to effect echinocytic changes in
human erythrocytes and diminish erythrocyte osmotic fragility has likewise been related to incorporation of the
sterol nitroxide into plasma membranes [1860].

As membrane morphology and function are vital aspects
of cellular viability, it may well be that as more becomes
known of the mechanisms of cytotoxicity, atherogenicity
and angiotoxicity, etc., that membrane level effects of active cholesterol autoxidation products will prove to be the
case.

CHAPTER IX. CHOLESTEROL PURITY AND STABILITY

At this point I have established that cholesterol is subject to insidious autoxidation under relatively mild conditions. It is essential to continued progress in the study of cholesterol that this fact not be ignored, as cholesterol autoxidation products may confound analyses and obscure details of sterol metabolism. The matter is all the more important as it is now certain that cholesterol autoxidation products exert divers deleterious biological actions in select test systems and may play a role in regulation of endogeneous metabolism and etiology of disease.

It is thus appropriate to close this monograph with a chapter devoted to the issue of cholesterol purity and stability. One seeks for definition of what constitutes pure cholesterol, for reliable methods for gaining such pure material, for analytical procedures of certification of such achievement, and for suitable means to maintain purity over periods of experimental work and storage.

CHOLESTEROL PURITY

Before directly addressing issues of cholesterol purity it is useful to consider the physical forms of cholesterol that may be encountered and aspects of the phase behavior of cholesterol. Besides dissolved cholesterol in true solutions, cholesterol may exist in at least three solid states: the amorphous form, an anhydrous crystal, and a crystalline monohydrate. Under special conditions cholesterol also may exist as a liquid crystalline modification. Liquid cholesterol may exist between 155-430°C [219], and cholesterol in the vapor state is obviously implicated in routine gas chromatography of the sterol.

The two crystalline forms are easily distinguished under the microscope, the anhydrous form occurring as needles, the monohydrate as plates [235,285,1205,1359,1528]. As the monohydrate loses its water of hydration at 86°C [1528], the melting points of the two (initial) forms should be the same, and reports of different melting point and ultraviolet absorption spectra [2111] must reflect other factors. Crystalline anhydrous cholesterol and cholesterol monohydrate are further distinguished by X-ray diffraction spectra [205,312,1331,1528,1856], anhydrous cholesterol being triclinic with eight molecules per unit cell [2223], the monohydrate also being triclinic with eight cholesterol and eight water molecules per unit cell [535]. Besides X-ray diffraction methods anhydrous cholesterol may be distinguished from the monohydrate by Raman spectra [441, 710], by differential scanning calorimetry [1528], and by the simple expedient of measurement of water content.

Differential scanning calorimetry shows that anhydrous cholesterol exhibits two exothermic transitions when heated over the range 0-200°C. An enantiotropic modification involving 0.66-0.91 Kcal/mole occurs in the range 35-40°C [219,781,852,1415,1528,1566,1694,2222,2326,2418,2581], and the anhydrous needles melt at 151-152°C, with heat of fusion 6.59-7.1 Kcal/mole. The low endotherm appears to vary over the range 35-40°C depending on particle size [2222,2418], crystallizing solvent [852], and extent of autoxidation that may have occurred [2418]. A cooling curve shows two exothermic transitions, one at 116°C, one at 13°C [1528, 1566]. Over an extended range 0-900°C, two additional exothermic transitions occur, one at 430°C and a later one at 610°C. These high temperature transitions are accompanied by weight losses and must represent chemical decomposition [219].

By contrast, cholesterol monohydrate undergoes different transitions upon heating, losing water at 86°C. Further heating in the presence of water causes the dehydrated crystal to undergo a crystalline to liquid crystalline transition at 123°C, the liquid crystal in contact with water then melting at 157°C [1528].

It is cholesterol monohydrate that is recovered from tissues [311,312,1258,2272], as the equilibrium state of cholesterol in contact with water (below 86°C) is the monohydrate [1528]. Anhydrous cholesterol in the presence of water undergoes a transition to the monohydrate, and cholesterol monohydrate loses its water in air, yielding anhydrous crystalline and/or amorphous cholesterol [311,312,1694]. A given sample of cholesterol may in fact be a variable mixture of monohydrate, anhydrous crystal, and amorphous material, depending on the precise history of the sample.

Moreover, the interrelationships between anhydrous needle and hydrated plate crystalline forms of cholesterol in biological fluids may not be so direct. The hydration

of anhydrous cholesterol needles added to aqueous solutions
or to plasma or serum appears to depend on the presence of
triacylglycerol [231,235,2347,2348], and both needle and
plate forms of cholesterol have been observed in human peri-
cardial and synovial fluids, the amount of needles increasing on storage of the fluids [1786]. However, these cholesterol crystals were not pure and may not represent anhydrous and hydrated crystals of cholesterol. Other reports of other crystalline modifications of cholesterol
from tissure exist [2397,2398], but questions of purity
and of data interpretations obtained in these cases.

Other remarks about cholesterol solubility and solution properties are given in a later section of this chapter. Little is known of cholesterol in its liquid and
vapor phases.

Pure Cholesterol

As with all matters, the purification of cholesterol
to the absolute level of purity where all molecules in
the sample are the same, with the absence of all other
molecular species, cannot now be attained. Furthermore,
it is very unlikely that ultrapure cholesterol (99.95%
purity by weight) has ever been prepared, as the analysis
methods suitable for measurement of the 500 ppm impurities
allowed by this definition have yet to be perfected to this
goal. Nonetheless, with relatively simple methods it is
now routinely possible to prepare cholesterol at purities
in the range 99.0-99.9%.

It must be understood that definition of cholesterol
purity is made operationally and that our description of
pure cholesterol is only so good as the methods of preparation and analysis applied. Specific attention must be given to each of the several likely impurities in cholesterol,
and no one process or analysis procedure should be relied
on for studies requiring pure cholesterol.

It has long been recognized that cholesterol recovered
from animal tissues is impure and that further purification
is necessary [61,1501]. Even after gross impurities are
removed, other lipids and sterol esters separated, and
crystalline cholesterol obtained there remain substantial
amounts of other companion sterols the presence of which

may influence experimental work with the sample. Thus
besides the bulk of cholesterol there are present several
biosynthesis precursors (the 7-stenol 57,5,7-dienol 56,
desmosterol (78), etc.), reduced metabolites (stanols 2
and 4), oxidized metabolites (the (24S)-3β,24-diol 25,
(25R)-3β,26-diol 29, etc), common autoxidation products
(10,13-16, etc.), and manufacturing process artifacts
(sterenes, steradienes, dicholesteryl ether, etc.) in
most commercial cholesterol samples. Depending on the
case, sterol homologs such as sitosterol (20), campesterol
(71), etc. as well as biosynthesis precursors such as lanosterol (67) and related methylsterols may be present.

     However, nonsterol impurities potentially present in
commercial cholesterol may need attention in some situa-   .
tions. Such items as heavy metals, salts, ash, etc., environmental contaminants such as plasticizers, and solvent
residues may ultimately attract concern. Furthermore, in
some uses the purity of cholesterol from microorganisms
and their toxins, viruses, and spurious radioactivity may
require examination.

     There are at least three purity levels that deserve
attention here, that of high purity for scientific interest,
that of officially pure cholesterol, and that of pure cholesterol for reference purposes in clinical analysis. Criteria for one use are not necessarily the same for other
uses.

     Physical Properties. Using classical melting point and
optical rotation measurements sterol chemists have repeatedly prepared highly pure cholesterol by simple methods.
Only recently have modern sophisticated analytical methods
been applied that demonstrate the power of the simple
methods in providing pure material. A physical description of pure cholesterol of fifty years ago is little
different from one given today, but today we can be more
certain.

     The most highly purified samples of cholesterol may
be about 99.9% pure. Present methodology is entirely suitable for measurement of impurities in cholesterol at this
level, the combination of thin-layer and high performance
liquid chromatography and gas chromatography coupled with
mass spectrometry being capable of routine detection of impurities at the 0.1% level [94]. No attempt has really been

made to prepare cholesterol in ultrapurity, as needs for
and interests in such material have not emerged.

Anent melting point as measure of purity, a brief
career through such data is given in TABLE 27. The slow
rise in melting point over the years reflects both higher
purity of cholesterol and improvement in technique for
melting point determination. Measurements in open capillaries (exposed to air) are potentially depressed by several degrees from autoxidation occurring during the determination, and capillaries sealed under vacuum are necessary to
prevent this problem [501,740,744]. The Kofler technique
where the sample is protected from exposure to air by a
coverglass is our preferred method, the Kofler block being
stationed under a microscope for use with a minimum of sample and for close study of melting behavior. More recently,
melting behavior has been determined from differential scanning calorimetry.

In order to have very high purity the melting point
of cholesterol should exceed 150°C, a point attained over
fifty years ago [61]. Our own best sample melted at 152.0-
153.5°C [2569], and a sample attributed to Fieser melted at
152.5-154°C [1520], the highest reported save for cholesterol differential calorimetric data suggesting approximately 155°C for cholesterol [219] and 157°C for cholesterol
in equilibrium with water [1528].

The other classic criterion of sterol purity, specific
rotation $[\alpha]_D$, has been advanced as a preferred measure of
purity for some sterols [222], and rotations have characterized highly pure cholesterol preparations over the years.
Typical $[\alpha]_D$ values in chloroform are -39.71° [61], -39.9°
[740], -40.0° [501]; a best $[\alpha]_D$-39.5 ± 0.5° has been set
[1170]. However, specific rotation is solvent, temperature,
and concentration dependent and is measured at the D line
in the spectrum of Na vapor (589 nm), thus at a wavelength
where optical activity is generally diminished. Moreover,
the measurement is insensitive to the presence of optically
inactive impurities, and fortuitously good values could be
had where impurities of opposite rotations were present and
cancelled one another.

TABLE 27. Melting Points of Pure Cholesterol

Year	Investigators	Melting Point, $C°$	Reference
1815	Chevreul	137	[477]
1846	Gobley	145	[879]
1862	Flint	145(293°F)	[779]
1872-73	Schulze	144.5-145	[2161,2162]
1885	Liebermann	145-146	[1472]
1908	Windaus	148.5	[2690]
1926	Rosenheim & Webster	149	[2020]
1927	Anderson	150-151	[61]
1928	Bills *et al.*	149	[223]
1937	Engel	148-149	[684]
1938	Robberecht	150.1	[1962]
1953	Fieser	149.5-150.0*	[740]
1953	Smith & States	149.0-150.0**	[2308]
1965	Spier & van Senden	150.8	[2326]
1967	Smith *et al.*	149.5-150.5**	[2303]
1968	van Lier & Smith	152.0-153.5**	[2569]
1968	Cohen	148.8 ± 0.1*	[501]
1969	Gent	149.2(422.4°K)***	[838]
1975	Marfey & Schultz	152***	[1566]
1979	Loomis *et al.*	151***	[1528]

  * Sealed capillary
 ** Kofler block
*** Differential scanning calorimetry

     Optical rotatory dispersion spectra of cholesterol are
relatively simple, with rotation increasing with decreasing
wavelength over the range 250-750 nm as expressed by a two-
term Drude equation [345,1516,1520,2511,2512].

     The use of optical rotation for assessment of purity
of cholesterol has disappeared from recent work. Other
than for differentiating natural cholesterol from racemic
cholesterol prepared by total synthesis [1197,1267], specific rotations serve only to define USP cholesterol as described in the next section of this chapter.

Furthermore, the ultraviolet light absorption spectrum of cholesterol is indistinctive [996]. Indeed, as with provitamin D activity, early investigators had difficulty deciding whether absorption above 220 nm be attributed to cholesterol or to impurities, but it is now clear that such absorption measures absorbing impurities and not cholesterol [1059,1606,2591]. However, cholesterol does possess discrete absorption bands at 189.5 nm (ϵ9,900) in cyclohexane [1626], 190 nm (ϵ7,900) in hexane [2529], 191 nm (ϵ9,400) in ethanol [1626] (or 203 nm [290] or at 206 nm [2663]), but these bands cannot be used for purity assessment. Neither are circular dichroism measurements for cholesterol (λ_{max} 200 nm,$\Delta\epsilon$2.0) [1451] of use. The circular dichroism of cholesterol may be substantially altered in the presence of certain compounds such as tetracyanoethylene [2176].

Infrared absorption spectra also characterize cholesterol, spectra being recorded on solutions of cholesterol in CS_2 [123,1414], CCl_4 [303], and $CHCl_3$ [1376], on solids as films [1196,2024] or dispersed in Nujol [1882] or KBr [1742,2089]. Details of spectra depend on the physical state of the sample, but generally there is agreement that there are bands associated with C-H stretching at 2860 and 2940 cm^{-1}, C-H bending at 1450-1700 cm^{-1}, C-CH_3 deformation at 1370-1390 cm^{-1}, C-O stretching at 1020-1060 cm^{-1}, and C-H out-of-plane bending at 800-840 cm^{-1} [228,287,1050, 1063,1196,1204,2024,2089,2371,2652].

Although fine distinctions exist among infrared spectra of different sterols, their great similarity has restricted application of the method for assessment of purity. Thus, spectra of cholesterol and sitosterol are but little different [175,812]. Other than special applications for measurement of cholesterol in serum [799,800] and of isomeric cholestanols [1376], little use of infrared quantitative analyses of sterols has developed.

Cholesterol has been fully characterized by nuclear magnetic resonance methods. Proton spectra of cholesterol first examined by Shoolery at 30 MHz in 1953 [212] now include data to 251 MHz [64,1123,2089,2271,2479] for the sterol. Proton spectra of cholesterol 3β-acetate are also recorded over the range 40-220 MHz [2035,2177,2239]. Protons of the angular C-18 and C-19 methyl groups are sharp 3-proton singlets, of the C-21 and terminal C-26 and C-27 methyl groups are doublets derived from signal splitting from the adjacent tertiary carbon atom hydrogen. The vinyl (C-6) hydrogen and hydroxyl hydrogen are also resolved, but most of the other hydrogens appear unresolved.

By contrast, ^{13}C nuclear magnetic resonance spectra of cholesterol disclose 26 lines, thus resolving signals from most of the 27 carbon atoms of the sterol. Only the C-7 and C-8 carbon atoms have the same chemical shift. Moreover, the terminal C-26 and C-27 methyl carbon atoms are resolved [64,78,308,1061,1206,1362,1562,1876,1921,2177,2318, 2519,2528,2723]. Spin relaxation times have also been obtained from ^{13}C spectra [79].

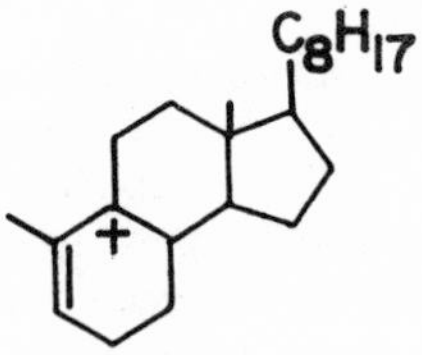

420 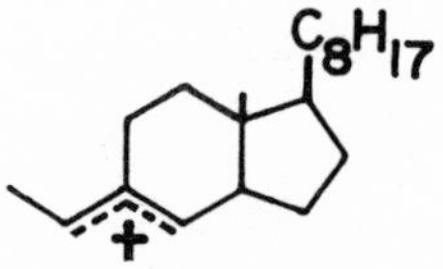   421 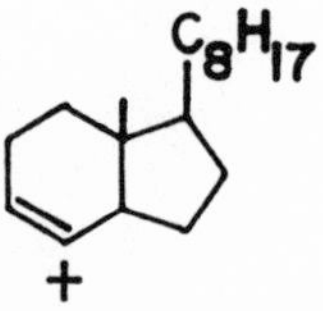   422

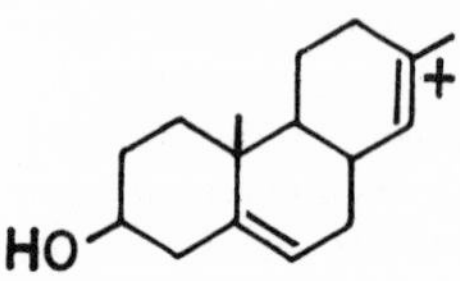

423 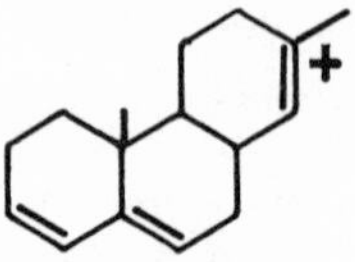   424 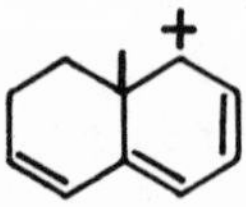   425

Cholesterol has also been fully characterized by mass spectrometry, including electron impact [761,771,805,1334, 1337,2053,2727] and chemical ionization [761,1514,1515,2033] modes. The electron impact mass spectrum of cholesterol includes the molecular ion $(M)^+$ m/z 386 and several high mass ions of simple elimination, thus m/z 371 $(M-CH_3)^+$, 368 $(M-H_2O)^+$, and 273 (M-side chain)$^+$. Combinations of these processes give ions m/z 353 $(M-CH_3-H_2O)^+$ and 255 $(M-H_2O-$ side chain)$^+$. Fragmentations of nuclear carbon-carbon bonds also occur, with loss of the A-ring giving ion m/z 301

(420) , of the A- and B-rings giving ions m/z 275 (421) and
247 (422),of the D-ring giving ions m/z 231 (423) and 213
(424), and of C- and D-rings giving ion m/z 145 (425) [1334,
2727,2769]. More extensive fragmentations also occur.

Cholesterol fatty acyl esters have also been characterized by electron impact mass spectrometry. Generally
no molecular ion is found; otherwise, spectra of the esters exhibit the same ions as the free sterol [843].

Additional mass spectral characterization of cholesterol is provided by spectra of the O-ethers, including
the 3-O-methyl ether [1117,1716] but more readily the 3β-
O-trimethylsilyl ether. Spectral details similar to those
of cholesterol mass spectra are obtained, including molecular ion m/z 458, elimination ions m/z 443 $(M-CH_3)^+$ and 368
$(M-C_3H_9SiOH)^+$,and fragmentation ions m/z 301,275,247, and
213 (420,421,422, and 424). However, the spectrum of the
trimethylsilyl ether is dominated by ions unique to this
derivative of Δ^5-sterols, namely m/z 329 $(M-129)^+$ and m/z
129 $(C_3H_9SiOCHCHCH_2)^+$, formulated as loss of the trimethyl-
silyloxy moiety along with A-ring carbon atoms C-1, C-2,
and C-3 [609,681,1337].

By contrast, chemical ionization mass spectral of cholesterol are greatly simplified, and several high mass ions
occur. With methane or isobutane the protonated molecular
ion m/z 387 $(M+H)^+$, molecular ion m/z 386 (M)$^+$, hydrogen
abstraction ion m/z 385 $(M-H)^+$, and dehydration ions m/z
369 $(M+H-H_2O)^+$ and 367 $(M-H-H_2O)^+$ are generated [336,761,
1514,1627,1698]. Additionally, an adduct ion m/z 415
$(M+C_2H_5)^+$ and a fragment ion m/z 353 $(M-CH_3-H_2O)^+$ have been
recorded with methane [761,1627], and adduct ions m/z 425
$(M+C_3H_3)^+$, 401 $(M+CH_3)^+$, and 399 $(M+CH)^+$ have been found
at very low abundances with isobutane as reagent gas [336].
Using ammonia as reagent gas cholesterol forms an ammonium
adduct ion m/z 404 $(M+NH_4)^+$ and ions m/z 386 and 369. However, the m/z 386 ion is not the molecular ion but is a
substitution ion $(M-OH+NH_3)^+$. The m/z 369 is formulated
as $(M-OH)^+$ and not as $(M+H-H_2O)^+$ [1514,1515].

Cholesterol has also been characterized by a negative
ion chemical ionization mass spectrum containing hydride
abstraction ion m/z 385 $(M-H)^-$ and ions m/z 367 $(M-H-H_2O)^-$
and 365 $(M-H-H_2-H_2O)^-$ and adduct ion m/z 429 $(M-H+N_2O)^-$
from the reagent gas [2033].

These several sophisticated spectral characterizations of cholesterol serve to define this sterol molecule uniquely and distinguish cholesterol from other known compounds.

Official Definition. The use of cholesterol in pharmaceuticals and cosmetics has required an official definition in the United States Pharmacopeia since 1947. In the United States, cholesterol is officially "white or faintly yellow, almost odorless, pearly leaflets or granules" [1803,2534,2535], but is a "white waxy powder or leaflets" in the United Kingdom [360]. Identification is warranted by positive Salkowski and Liebermann-Burchard color tests! Purity is defined by the criteria of TABLE 28. Considering the intended use of officially defined cholesterol is as an emulsifying agent for cosmetics and in hydrophilic petrolatum [1535], these standards may suffice. Such cholesterol is also suitable for manufacturing purposes, including synthesis of liquid crystals [1536] and for such odd potential applications as ice nucleation [990]. However, these criteria are inadequate for pure cholesterol to be used in bioassay work and as a reference standard in clinical analyses.

All too many nutritional studies have been conducted with USP cholesterol thusly defined. With the melting point and specific rotation values allowed it should be possible to have several percent of other sterols present that could not be detected.

TABLE 28.   Officially Defined Cholesterol

Property	USP XIX [2535]	British Pharmacopoeia [360]
M.p.	147–150°	145–149°
$[\alpha]_D$	−34 to −38°*	−38.5° to −40.5°**
Loss on drying	0.3%	0.5%
Ash	0.1%	0.1%
Acidity	0.03 meq/g	−

* In dioxane.   USP XIII gives −28° to −31° in ethanol
** In chloroform

Standard Reference Samples. Precisely for this reason USP cholesterol is inadequate as a reference standard for serum cholesterol analyses by clinical laboratories. The importance of making comparisons of data from different clinical research laboratories has been emphasized [1519], and the need for a highly pure reference standard cholesterol preparation is obvious. To this end the National Bureau of Standards has issued a highly pure cholesterol sample as Standard Reference Material No. 911 since 1967, the purity of which being about 99.9% [501].

The standard reference sample of cholesterol prepared via the dibromide as described in the next section of this chapter and recrystallized from methanol was fully characterized by all modern criteria save high performance liquid chromatography and ^{13}C nuclear magnetic spectra. The reference sample had m.p. 148.8 ± 0.1°C and $[\alpha]_D$ -40.0° (chloroform), was correctly analyzed for carbon and hydrogen by combusion methods, and was characterized suitably by ultraviolet, infrared, and proton spectra, no spectral evidences of impurity being evident. By electron impact mass spectrometry three high mass ions $(M+14)^+$, $(M+28)^+$, and $(M+44)^+$ suggesting impurities were found! Conventional gas chromatography revealed four impurities totalling approximately 0.6%, assuming equal detector responses for cholesterol and impurities. Thin-layer chromatography using 160 μg of sample revealed the possibility of traces of polar impurity only under 366 nm light following sulfuric acid spraying. The SeO_2 test for the 7-stenol 57 was negative. Under specified conditions of the Liebermann-Burchard test, a molar absorbance of 590 ± 1 L mole^{-1} cm^{-1} at 535 nm was obtained. The most revealing purity tests using differential scanning calorimetry and phase solubility analysis suggested that the reference sample was 99.84% and 99.96% pure respectively [501]!

The obviously high purity of this reference sample points out the lack of adequate analysis techniques for such work, the impurities suggested by thin-layer and gas chromatography and mass spectrometry not being quantitatively estimated with any accuracy and no hint of any extraneous impurities by the less sensitive methods. Moreover, the calorimetric and solubility methods that are inherently very sensitive to impurity have not been described fully and may not be properly calibrated for sterols. The Standard Reference Material No. 911 choles-

terol is probably the most pure material routinely available outside of special preparations made in individual laboratories, for which most of the analytical data provided
by the National Bureau of Standards is not ordinarily accumulated.

Purification Methods

Whereas the custom has been for so long to use cholesterol as received for many studies, it is now recognized
that cholesterol samples must be purified before use, the
purity properly assessed, and conditions of storage for the
purified sterol imposed that secure its stability over the
period of experimental work. A recent case reported involved triple recrystallization of cholesterol from ethanol,
storage at 4°C und N_2, and thin-layer chromatography,
melting point,and differential scanning calorimetry being
used to assess purity at greater than 99.5% and Raman spectra to establish that the sample was anhydrous cholesterol
[441]. Moreover, it is worth note that these precautions
were not excessive given the use of the sample in measurements of cholesterol solubility in model systems simulating
bile!

Evolutionary pressures have narrowed down interest in
purifying cholesterol to two simple physical methods, recrystallization and chromatography, and one chemical one,
bromination-debromination. Purification by forming complexes with a variety of other agents and by other chemical derivitization are either ineffective for resolution
or are more expensive of material, time, and effort without
particular advantage.

Recrystallization Methods. In order to examine simple
recrystallization methods of purification of cholesterol it
is well to review some points about the solubility behavior
of cholesterol. The water solubility of cholesterol is
quite low (*cf.* Chapter IV) a matter that limits use of water in all purification systems whatever their nature. Recrystallizations from aqueous systems where cholesterol is
dissolved with the aid of a miscible organic solvent such
as acetone appear to return fairly pure cholesterol with
no impurities detected by chromatography but with a lowered
melting point (m.p. 145.5-146.0 C) [252]. The use of solu-

TABLE 29. Cholesterol Solubilities (Selected)

Solvent	T,°C	Solubility,mg/g*	Reference
Acetone	20	40	[1532]
	25	24**	[2112]
	30	50	[1532]
	40	90	[1532]
Benzene	20	55-142	[1532,2648]
	30	70	[1532]
	40	135	[1532]
Chloroform	25	>40**	[1537,2112]
Cholesterol			
3β-linoleate	37	(48)	[1783]
		80	[2272]
		36**	[1173]
Cyclohexane	38	210	[2347]
		(70)	[2347]
Dioxane	20	90-113	[1532,2648]
	30	170	[1532]
	40	230	[1532]
Ethanol	20	15.5	[2648]
	78	100	[1532]
Ethanol (95%)	27	(23.4-23.7)**	[781]
	37	(31.1-32.5)**	[781]
Ethyl acetate	20	45	[1532]
	30	60	[1532]
	37	(76.4)*	[781]
	40	95	[1532]
Formamide	37	(1.41)**	[781]
Hexane	20	19.2	[2648]
	25	6.5**	[1537,2112]
Isopropyl			
myristate	21	52.6	[970]
	37	(67.4)**	[781]
Methanol	20	6.5-7.5	[1532,2648]
	25	5.8**	[1537,2112]
	27	(5.2-6.1)**	[781]
	30	10	[1532]
	37	(6.2-7.9)**	[781]
	40	15	[1532]
Propan-2-ol	37	(48.9)**	[781]
Pyridine	20-25	681	[586]
Toluene	38	262	[2347]
		(90)	[2347]

(Continued)

TABLE 29.   (Continued)

Triacetin	37	3.3	[1389]
		(7.17)**	[781]
Tricaprylin	37	43.8	[1389]
		(52.0)**	[781]
Triolein	21	28	[1173]
		(21)	[1173]
	37	43	[1173]
		(37.7)	[1173]

*Values in parenthesis are for cholesterol monohydrate
**Solubility expressed as mg/mL

bilizing agents for recrystallization from water is not indicated though.   Indeed, deposition of cholesterol from 10%
aqueous solutions made using lecithin as solubilizer is as
a gel [2084], obviously unsuited for purifications.

In contrast to water, many common organic solvents dis-
solve cholesterol.   The data of TABLE 29 contain inconsis-
tencies but give a guide to the solubility of cholesterol
in neat solvents.   In binary mixtures of solvents the solu-
bulity of cholesterol may pass through a maximum, diminish-
ing for either neat solvent component [2648].   Cholesterol
tends to associate via hydrogen bonding with solvents [728,
2371] and also to self-associate, forming dimeric species
[728,1817,1967].   Moreover, the solubility of cholesterol
in solvents is very much dependent on the form of the sterol,
cholesterol monohydrate being much less soluble than the anhydrous crystal.   Furthermore, the presence of water as a
third component also affects cholesterol solubility [1173,
1783,2223,2347].

Any of the listed solvents and others probably could be
used to recrystallize cholesterol, but the one solvent most
used is methanol, which has a relatively undistinguished
dissolving power.   Since 1904 [562] the time-honored method
of the sterol chemist in improving the quality of cholesterol has been recrystallization from methanol.   Ethanol, acetone, acetonitrile, and a few other solvents have also been
used, but no advantage in product quality obtains and recov-
eries are best from methanol.   Recrystallization from an-

hydrous solvent yields anhydrous cholesterol, from 95% ethanol or other moist solvent cholesterol monohydrate.

As a crystalline cholesterol-methanol complex forms, extensive vacuum drying of cholesterol recrystallized from methanol to constant weight is indicated. By these means beautifully crystalline cholesterol needles are obtained, free from color and the common autoxidation products. Although repeated crystallizations from alcohols yield material of constant melting point and specific rotation, it was recognized as early as 1927 that recrystallized cholesterol was not necessarily pure and that recrystallization did not remove congeners such as the dienol 56 or 5α-stanol 2 [61, 2131].

We now recognize that recrystallization of cholesterol from most solvents cannot remove traces of congeners 2, 56, 57, and 78 or of homologs such as sitosterol. Recrystallization from methanol or ethanol must accordingly be reserved for use where the presence of these analogs is unimportant or in conjunction with purification via the bromination-debromination procedure for cholesterol of very high purity.

It is well to dwell on the possibility that autoxidation occur during recrystallization from alcohol [862,2016]. Even though the common autoxidation products are not nominally detected in cholesterol recrystallized from methanol, the methanolic mother liquors from even highly purified samples thought to be free from autoxidation contain traces of the common autoxidation products. As a systematic study has not been made, it is not now possible to decide whether such traces can be formed during recrystallization or whether such impurities were present in the cholesterol before processing. In order to forestall the possibility of autoxidation during final recrystallizations, operations conducted as rapidly as possible using the cleanest apparatus and deoxygenated solvent, and under an inert atmosphere are in order. Incorporation of antioxidants at this point [2198] may reduce autoxidation but may introduce impurities from the antioxidant and its oxidized products.

These items nonwithstanding, repeated recrystallizations from alcohol yield highly pure cholesterol preparations, at least by criteria of melting point and colorimetric responses [1897,2326].

Recrystallization of cholesterol from glacial acetic
acid is also a procedure that has attracted attention. The
solubility of cholesterol in cold acetic acid is not great,
and in heated solutions some cholesterol is lost to acetylation, yielding cholesterol 3β-acetate that may ultimately
contaminate the recrystallized free sterol product. A crystalline equimolar complex of cholesterol and acetic acid
forms, and the purified sterol is recovered from the complex by dissociation with water or by vacuum removal of solvent. The cholesterol thereby recovered may then be saponi-
fied to remove vestages of the acetate ester and the product
recrystallized from alcohol [740].

This process is not suited to larger batches because
of the problems of handling hot acetic acid and the concomitant acetylation that occurs, but small batches of cholesterol may be purified simply by this method. However, a
modern assessment of the purity of such material using chromatographic methods has not been done, and we do not know
whether companion sterols such as 2, 56, 57, and 78 are removed. The cholesterol purified by the acetic acid process
is of high purity by melting point and colorimetric assay
standards [1897].

Via Cholesterol Dibromide. Cholesterol readily forms
an insoluble dibromide derivative, 5,6β-dibromo-5α-cholestan-
3β-ol (426) that can be recovered, recrystallized, and debrominated to yield cholesterol freed from many of its
natural congeners. It has become accepted practice over

HO   Br Br   HO   Br
                        426   Br
                                                               427

the years [61,223,1351,2705] to purify cholesterol by variations of this procedure originated by Windaus in 1906 [2688,
2689], and samples of high purity are now routinely prepared
by this means. Only one bromination-debromination sequence
appears needed for removal of common impurities [61,223,
1897].

There are numerous chemical methods described for debromination of the 5,6-dibromide 426, but Fieser's modification of the Windaus procedure using Zn and acetic acid
[741,744,749] and Schoenheimer's modification using NaI
and ethanol [2121,2131] have received the most attention.
Schwenk has devised a modification of Fieser's process for
the routine purification of cholesterol in biosynthesis procedures [2171]. Other debromination processes such as that
using FeCl$_3$ [352] are not as promising and are no longer
used.

Fieser stated that the common companion sterols such
as 5α-stanol 2, biosynthesis precursors 56, 57, and 78, oxidized metabolite cerebrosterol (25), and the autoxidation
product 3β,25-diol 27 were removed by the procedures [741,
749], and subsequent analyses of ours and of others on such
preparations support this conclusion. Cholesterol purified
via the dibromide and properly recrystallized has not been
shown to be contaminated by any known specific companion
sterol, but demanding analyses for sterols 2, 25, 27, 56,
78, etc. or for the 3β,26-diols 29/31 or homologs sitosterol, campesterol, etc. at the part-per-million levels have
not been conducted. Nor do we know whether traces of bro-
mosterols remain.

Moreover, the procedure is not without potential limitations. The initially formed 5α,6β-dibromide 426 in solution undergoes a spontaneous double epimerization, yielding
5,6α-dibromo-5β-cholestan-3β-ol (427) as the more stable
product. Furthermore, debromination of the 5β,6α-dibromide
427 3β-benzoate is much slower than is that of the 5α,6β-di-
bromide 426 3β-benzoate [155]. Accordingly, any inadvertent
isomerization of the dibromide 426 coupled with retarded debromination of the 5β,6α-dibromide 427 might result in retention of brominated sterol in recovered cholesterol.

There is also a question whether some oxidation of the
3β-hydroxyl group of cholesterol occurs by the action of elemental bromine, but it is clear that bromination of cholesterol generates more than one product [2303,2682]. Quick
recrystallization of the 5α,6β-dibromide 426 at this point
may be called for [684]. Furthermore, an extensive autoxidation of cholesterol during debromination also occurs,
whether the Zn-acetic acid or the NaI-ethanol procedure
be utilized [2288]. Although mere washing with methanol
appears to yield a highly pure cholesterol [2681], it is

critically important that repeated recrystallization from
alcohol of cholesterol recovered from debromination procedures be had in order that these autoxidation products be
properly removed!

    Purification of cholesterol via the dibromide is in-
tegrated into the protocol for preparation of the National
Bureau of Standards reference cholesterol sample of 99.9%
purity [501], indeed for all high purity preparations of
cholesterol. The importance of conducting all procedures
rapidly, at lower temperatures, in the absence of light
and air (or under inert atmosphere or vacuum), possibly
in the presence of added autoxidants, and of other precautions has been emphasized [2198].

    The differential stabilities of highly pure choles-
terol prepared by the two different debromination methods,
Zn-acetic acid versus NaI-ethanol, is another matter of
prime interest discussed in a later section.

    Comparison of Samples. As highly pure reference stan-
dards are of great importance in the measurement of serum
cholesterol in clinical laboratories, comparison of cholesterol purified by crystallizations from ethanol, from acetic acid, and via the dibromide is in order. Data of TABLE
30 summarize such studies in which melting point, colorimetric responses to the Liebermann-Burchard and acid-$FeCl_3$
reagents, and spectrophotometric measurement of contami-
nating autoxidation products 3β-hydroxycholest-5-en-7-one
(16), and gas chromatographic responses were determined
[1897,2625]. Under the colorimetry conditions imposed,
the molar absorbance with the Liebermann-Burchard reagent
at 620 nm was 1,750 L mole^{-1} cm^{-1}, to an acid-$FeCl_3$ reagent
at 535 nm was 11,500 L mole^{-1}cm^{-1}[1897]. Under other
conditions of the Liebermann-Burchard assay not adjusted
for maximum color development but for reproducibility the
National Bureau of Standards reference sample of cholesterol
gives a molar absorbance at 535 nm of 590 L mole^{-1} cm^{-1}
[501].

    It is clear from data of TABLE 30 that original sam-
ples of cholesterol are greatly improved by all three purification processes, but that the bromination-debromination
procedure is superior. Moreover, purification of cholesterol to maximum colorimetric response to the Liebermann-
Burchard or other reagent cannot be viewed as necessarily

TABLE 30. Comparisons of Purified Cholesterol Samples

Analysis Method	Purity, %				Reference
	Original Sample	From Ethanol	From Acetic Acid	Via Dibromide	
Liebermann-Burchard	92.0-98.3	94.9-99.4	95.4-99.4	96.5-100	[1897]
Acid-Iron Reagent	85.2-98.3	95.6-99.1	97.4-100	98.2-100	[1897]
Spectrophotometry	94.2-99.5	98.7-99.9	99.7-99.9	99.7-99.9	[1897]
Gas Chromatography	94.5-100.3	-	-	98.2-103.5	[2681]
Melting Point, °C**	148.7-150.5	149.7-151.8	150.3-150.9	149.9-150.9	[1897]
	149.0-149.7	-	-	150.0-150.6	[2681]

* Purity expressed versus best or select sample

** Best (highest, sharpest) melting point

providing highly pure cholesterol free from its congeners,
for a sample containing unreactive stanols such as 2, fast-
reacting stenols such as 56, 57, and epimeric 3β,7-diols 14
and 15, and esterified forms fortuitously could have an im-
pressively high colorimetric response that could be very
misleading with respect to sterol contaminants [2319].

   Complexation Methods. The notorious capacity of cholesterol to form stable complexes with a variety of substances can be turned to advantage in some cases for the
recovery and purification of cholesterol. Cholesterol
forms complexes with both inorganic and organic compounds,
example of the former being cholesterol monohydrate already
discussed. Moreover, the simple act of recrystallization of
cholesterol from glacial acetic acid as previously described
results in formation of the stable cholesterol-acetic acid
(1:1) complex [1516,1517].

   Cholesterol also forms addition compounds with other
carboxylic acids, including propionic and butyric acids,
inter alia [601] (not used in purification methods) and
with dicarboxylic acids discussed later. Cholesterol also
forms crystalline complexes with alcohols, the methanol
solvate being implicated in recrystallization of cholesterol from methanol in the common purification process.
Stable alcohol solvates have also been prepared with
propan-2-ol, 2-methylpropan-1-ol, and pentan-1-ol [13,
2648]. The complex with propan-2-ol is also described
in a gel state [13]. Moreover, liquid crystal forms of
cholesterol with higher aliphatic alcohols have been described [1436]. However, no application of these alcohol-
cholesterol complexes for cholesterol purification has
been made.

   Cholesterol also forms complexes with several dicarboxylic acids, 1:1 complexes with pimelic and methylmalonic
acids [2724] and a 2:1 complex with oxalic acid [1596,1628]
that has been used for purification of cholesterol on a relatively large scale [740,1863,2170]. The crystalline choles-
terol-oxalic acid complex may be recovered and decomposed into its components by water, giving a purified cholesterol.
However, the quality of the cholesterol thereby produced has
not been subjected to modern analyses; thus, it is uncertain
whether removal of congeners and autoxidation products is a-
chieved. The method has not been recently advocated.

Formation of stable complexes between cholesterol and other steroids also occurs. Complexes between cholesterol and other sterols such as sitosterol are notoriously troublesome, but complexes between cholesterol and steroid glycosides such as digitonin have been of great use in isolation of the sterol via the insoluble digitonide complex [2691]. However, precipitation of cholesterol as the digitonide, tomatide, etc. does not provide pure cholesterol, a matter long understood [61], nor resolve cholesterol from its congeners 2, 56, 77, etc. or homologs such as sitosterol. Moreover, some cholesterol autoxidation products may remain with precipitated cholesterol, and artifacts may be formed in decomposition of the complex. Nonetheless, digitonin-precipitation has been used as means of isolation of pilot plant amounts of cholesterol from wool fat [2764].

Cholesterol forms complexes with complex carbohydrates that are very strongly associated and render the sterol water soluble! The polysaccharides binding cholesterol appear to be cell-wall mannans and have been found in yeast [22, 23,1865,2478], protozoan *Euglena gracilis* [347], and fungi *Rhizopus arrhizus* and *Penicillium roquefortii* [1865]. The water solubility of cholesterol is also increased by pectin and acacia gum [1552]. These water soluble complexes complicate the recovery of cholesterol from systems containing them and offer no practical use for purification of the sterol.

Cholesterol forms a series of complexes with certain metal salts. A 1:1 complex is formed with LiCl [2773], and a series of 2:1 complexes, variously hydrated or solvated, with $CaCl_2$, $MnCl_2$, and $MgCl_2$ [644,645,938,1565,1612] and with chlorides of Al, Cd, Co, Cr, Fe, Sn, and Zn [359]. Although the $CaCl_2$ and $MnCl_2$ salts appear to have utility in the recovery of cholesterol from its natural sources, metal complexes have not been developed for further purification of cholesterol.

Other Chemical Methods. Purification of cholesterol by the dibromide derivative has so much captured interest that other chemical derivatives have fallen from use. In fact, use of most such other chemical derivatives was made before adequate awareness of the problem of companion sterols and autoxidation products was had.

The most obvious chemical derivitization of cholesterol

is that of acylation, particularly acetylation or benzoyla-
tion. The product acetate or benzoate esters are crystalline products the properties of which aid materially in
characterization and identification (by chemical means) of
natural sterols. Moreover, acetylation and subsequent alkaline saponification have been incorporated into purification schemes for cholesterol [61,1037,1352], and this manipulation is inherent in the purification of cholesterol
with glacial acetic acid, where some acetylation invariably
occurs [740]. However, the saponification of steryl esters
in strong alkali must be conducted in the absence of air
lest autoxidation intrude. The esterification and saponification procedures for purification of cholesterol have
received no recent attention, not have analyses using modern demanding methods been applied to the purified products.

    Chromatography. The purification of cholesterol by
chromatography is obviously possible, as chromatographic
procedures have been devised that resolve cholesterol from
each of its several companion sterols. All that is needed
in such case is the scale up of chromatographic methods
that are effective in analysis of the same sterols. However, the problems of purification of gram amounts of cholesterol from companion sterols by chromatography are such
that the approach cannot be recommended, and not one chromatographic process can presently be advanced as a method of
choice to compete with bromination-debromination and subsequent recrystallization procedures now accepted as standard purification means. Indeed, the costs of chromatographic media, solvents, and apparatus for preparative operations as well as of time and effort make the chromatographic purification of cholesterol a relatively unattractive
prospect, one limited to work within the milligram-centigram
range only.

    Whereas cholesterol has been variously purified by
chromatographic means in many studies directed towards
other ends, a systematic study of chromatography for the
expressed purpose of obtaining highly pure cholesterol in
good amount has not been reported. All manner of chromatography techniques have been used, thus simple adsorption
and partition column, paper partition, thin-layer adsorption and partition, gas-liquid partition, and high performance liquid column (adsorption and partition) chromatographic techniques. Also other chromatographic supports and

processes have been used, such as argentation chromatography and chromatography on Sephadex LH-20.

Simple adsorption column chromatography using alumina or silica gel as adsorbants is clearly capable of resolving cholesterol from its more polar autoxidation products, but resolution from precursors, stanols, and C-24 homologs is another matter. Although the removal of radioactive sterol impurities from [^{14}C] cholesterol from *de novo* biosynthesis (possibly including 56, 57, 78, etc.) by chromatography on silica gel may be about as complete as is bromination-debromination [2317], such operations cannot be recommended as secure. Likewise, paper partition chromatography of such [^{14}C] cholesterol, though yielding similarly purified material [2308] is obviously not of use.

Partition chromatography, particularly reversed phase operations, appears to be the method of choice for resolution of cholesterol from the 5α-stanol 2, precursors 56, 57, and 78, and C-24 homologs, but the low loading capacity of such system limits their use for preparative work. Preparative gas chromatography of cholesterol is also possible, but some thermal deterioration of the sample may occur and resolution from companion sterols 2, 56, 57, 78, etc. is unlikely. Indeed, although 1 mg amounts of cholesterol containing 10% 5β-stanol 4 (readily resolved from cholesterol on analytical gas chromatography) can be purified by gas chromatography, single passes of material do not achieve desired purity [2569].

Although, sublimation in vacuum is not a chromatographic process, the elevated temperatures of sublimation and gas chromatography subject cholesterol to similar thermal stresses, thereby inviting comparison. Sublimination under vacuum has been used to prepare pure cholesterol [501,1359,1579], but ultraviolet light absorption measurements on sublimed material clearly indicate the presence of impurities most likely formed during sublimation [501]. Neither sublimation nor gas chromatography serve the goal of preparation of highly pure cholesterol.

Preparative column and thin-layer chromatography using adsorbants that have been impregnated with AgNO$_3$ thereby to form complexes between Ag$^+$ and olefinic sterols offers a much better means of purification of cholesterol from its companion sterols. The sterol-Ag$^+$ complexes are viewed as

involving a π-bond formed by the overlap of a filled π-
orbital of the olefin with a free s-orbital of Ag^+ and a
π-bond obtained by overlap of a vacant antibonding π-or-
bital of the olefin with filled d-orbitals of Ag^+ [1687].
Stanols are not complexed, monoolefinic stenols are retarded
in their mobility by complexation, and dienols are retarded
even more so, thus according resolution of cholesterol from
stanol, stenol, and dienol congeners, as the free sterols
and as fatty acyl derivatives.

Use of argentation chromatography for the preparation
of multimilligram amounts of sterol has not been exploited,
and the present increasing cost of silver diminishes substantially ardor for the process for preparative work despite the merits.

Remarkable progress in sterol resolutions has been
recently made using high performance liquid column chromatography. With this advent one returns once more to adsorption and partition column operations but with a much improved prognosis for preparative separations of pure cholesterol. Although applications of high performance liquid
column techniques hold great promise, relatively little has
been reported using preparative apparatus and operations,
and this in conjunction with other modes of chromatography
[1104]. Other evidence of the promise of the method lies
in the preparative recovery of microgram amounts of cholesterol from milligram amounts of congener sitosterol [1106].

Certain other column supports and processes have merit
for purification of cholesterol in small lots. Chromatography on Sephadex LH-20 separates cholesterol from its
autoxidation products [2570], but resolution from stanols
2 and 4 and from sitosterol is not achieved [682]. Other
more detailed remarks about the application of chromatography to the analysis of cholesterol preparations will be
made in the next section of this chapter.

                        Analytical Methods

Mention has been made of several advanced analysis
methods that are of importance to special issues of identity and purity of cholesterol. Among these are sophisticated spectral techniques such as nuclear magnetic resonance, mass spectra, chromatographic methods (including

thin-layer, gas, and high performance liquid chromatography),
but also differential scanning calorimetry, X-ray diffraction, and colorimetry. Selection of analysis methods should
be made in consideration of the components or processes
sought, thus whether cholesterol is present and at what
absolute level, whether autoxidation of the sample has occurred, whether the sample has been purified from congeners, whether extraneous impurities of other sorts be present, etc. Necessarily one or even a few individual analysis methods may not suffice to satisfaction, depending upon
the extent of the demands made.

Analysis of pure cholesterol may be narrowed down to
its examination for seven distinct items: cholesterol precursors, reduced cholesterol metabolites, oxidized cholesterol metabolites, esterified sterols, C-24 alkylated homologs, cholesterol autoxidation products, and other processing artifacts. At present, detection and measurement of
these companion sterols must be achieved by chromatography
as no other analysis means for such achievement has ever
been addressed, let alone perfected. The additional matter of water of hydration, solvent residues, heavy metals,
ash, etc. are obviously outside of this consideration and
must be managed by other customary methods.

Simple Methods. Despite the need for sophisticated and
extensive instrumentation for some analysis procedures that
may be applied to cholesterol purity, it is fortunate that
the quality of cholesterol (once identified as such) may be
assessed by four relatively simple measures. Of the seven
classes of congeners listed above, the autoxidation products
are the most likely ones to be present in all cholesterol
samples, and it is the whole issue of autoxidation that is
clearly of interest in these analyses. To this end, examination of cholesterol samples for color and odor are perhaps
the most critical that can be made. These obvious means
should not be underestimated. Highly pure cholesterol is
colorless and does not display an odor, pharmacoepial standards to the contrary notwithstanding! Odor is probably
the first analytical measure of autoxidation that can be
had, one that develops before other indications.

The presence of any color or any odor in pure cholesterol samples is an immediate and absolute indication of
the presence of impurities in the sample, one that should
not be ignored. However, odor in samples of pure choles-

terol may not be from autoxidation but from other sources,
(possibly solvent residues from manufacturing), such odor
being readily removed by vacuum drying [2682].

Development of coloration in cholesterol samples requires more time; samples prepared by Engel in 1937 and by
myself in 1953 remain colorless in 1980. Nonetheless, the
development of yellow coloration in cholesterol well recognized since 1901 [1960,2163,2164] has been repeatedly noted
over the years, and the weak coloration permitted by pharma-
copial standards reflects tolerance for manufacturing and
aging problems. However, the color of USP cholesterol must
derive from other ingredients of manufacture or catalysis
of cholesterol autoxidation induced by extraneous ingredients, for highly pure cholesterol does not become colored
if previously freed from impurities associated with its
manufacture, as noted above.

Other than these two costless organoleptic analyses it
is fortunate that two other inexpensive analysis methods are
also of great utility. The melting point of cholesterol is
still a very sensitive measure of its purity, and analyses
by thin-layer chromatography are very effective in detection,
identification, and estimation of autoxidation products.
Melting point data have already been discussed (*cf*. TABLE
27), and pure cholesterol should not melt below 150°C.

Chromatographic Methods. Thin-layer chromatography is
the most versatile simple analysis method that can be applied to excellent effect. Indeed, thin-layer chromatography is that one technique that enabled resolution of the
cholesterol autoxidation problem. Although the companion
stanols, biosynthesis precursors, and C-24 homologs may or
may not be resolved from cholesterol, depending on the system, resolution of the polar autoxidation products is an assured matter, and simple systems can be devised that resolve
the other companion sterols as well. Specific comments a-
bout the resolution of each class of companion sterols will
be made in a later section of this chapter.

Although resolution of specific companion sterols from
cholesterol is required for analyses of purity, it is the
means of detection that ultimately limits the estimation
of purity by direct inspection of chromatograms. Sterols
have generally been detected on thin-layer chromatograms
by spraying with aqueous H_2SO_4, warming to full color dis-

play under visible light, and optionally charring by more extensive heating. As the several companion sterols give different color responses to aqueous H_2SO_4, color is of great utility in securing the identity of components re- solved. Thus, the stanols 2 and 4 and ketones 6, 8, 10, 12, and 16 give poor colorations with acid and must be detected by charring, whereas the 7-hydroperoxides 46 and 47 and 3β,7-diols 14 and 15 are uniquely detected by the immediate intense blue color developed rapidly without heating. Sterols retaining the Δ^5-bond (but not substitu- ted at C-7) display red, magneta, violet, blue-grey, blue- green, or other pretty colors; sterols devoid of nuclear olefinic substitution such as the 3β,5α,6β-triol 13, 6- ketone 44, and 5,6-epoxides 35 and 36 tend to give yellow or brown colors.

Besides this general means of detection of all sterols present, the detection of steroid contaminants is aided fur- ther by examination under 254 nm light of ultraviolet light absorbing components such as enones 8 and 16, and dienones 10 and 12 prior to spraying with acid. These ketosteroids may also be detected with the usual, 2,4-dinitrophenylhydra- zine reagent, but more sensitively by prior $NaBH_4$ reduction to the corresponding chromogenic alcohols. Thus, the 7-ke- tone 16 is reduced to the epimeric 3β,7-diols 14 and 15 that are easily and sensitively detected by H_2SO_4 [2305]. Reduc- tion of the dienone 10 similarly yields products that are readily detected by their intense blue colors with H_2SO_4, these products presumably being the uncharacterized choles- ta-3,5-dien-7-ols [1404].

Detection of many unsaturated sterol components may also be conducted using I_2 vapors before spraying with H_2SO_4. Mere spraying with H_2O detects sterol components that are not wetted by their transluscence.

Peroxidic components diagnostic of autoxidation may be detected most reliably with N,N-dimethyl-p-phenylenedia- mine or N,N,N',N'-tetramethyl-p-phenylenediamine, less re- liably with the usual KI-starch reagents or with NH_4SCN- $FeSO_4$ reagent [12,357,1072,1744,1789,2295,2303]. These peroxide detection methods can be applied prior to spraying with H_2SO_4, which is necessarily the last chemical treatment that can be applied.

There arises the issue of sensitivity of detection for

all probable sterol contaminants, as the simplest assay of
cholesterol purity lies in the failure to detect given components at the established level of sensitivity of the
visualizing process.

Data of TABLE 31 are assembled to suggest lower limits
of detection by the several reagents that have found general
utility. The phosphomolybdic acid reagent is a very useful
and sensitive reagent for detecting sterols but suffers
from the lack of color display for individual sterols. All
sterols respond alike. The $SbCl_3$ and $SbCl_5$ reagents so
often used in the past are sensitive and may give different
colors with different sterols, but their use is so troublesome and advantage so limited that these reagents are not
recommended.

It appears possible to detect under favorable circumstances as little as 50 ng of sterol impurity in cholesterol
using H_2SO_4 or phosphomolybdic acid reagents for all sterols, N,N,N',N'-tetramethyl-p-phenylenediamine for sterol
hydroperoxides, and absorption under 254 nm light for enone
and dienone derivatives. The amount of cholesterol sample
to be analyzed via thin-layer chromatography then sets the
overall sensitivity of the assay. Amounts of 160-500 μg
cholesterol sample per spot have been variously suggested
[179,441,501,765,1529,1783], these values equating to detection of 0.01-0.03% or 100-300 ppm of impurity, thus
within the 0-500 ppm impurity range of ultrapure material.
Application of still larger weights of sample could extend
the analysis to yet lower levels of impurities, and use of
a 5 mg cholesterol sample for thin-layer chromatography has
been advanced for this purpose [179]! In such samples a 50
ng impurity (10 ppm) conceivably might be detectable, but
the streaking of the major sterol spot would limit the matter to impurities that were not incorporated into the
spreading cholesterol zone.

It may be possible to extend the sensitivity of detection of impurities by H_2SO_4 spraying in two other ways, by
viewing under 366 nm light for fluorescent spots or by photodensitometric measurement of spot intensities. Although
systematic examinations have not been made, sterol impurities revealed by H_2SO_4 on thin-layer chromatograms appear
to be more sensitively detected under 366 nm light [501,
617,951,993,1188,1656]. Furthermore, photodensitometric
determination of cholesterol on thin-layer chromatograms

TABLE 31. Detection Limits on Thin Layer Chromatograms

Sterols	Reagent	Detection Limit	Reference
Cholesterol	50% aq. H_2SO_4	50-100 ng	
	20% aq. H_2SO_4	50-100 ng	[2198]
	8% phosphomolybdic acid (2-propanol)	50-100 ng	[2198]
	10% phosphomolybdic acid (95% ethanol)	100 ng	[2293]
	$SbCl_3$ (saturated $CHCl_3$)	100-200 ng	[2198]
	I_2 vapors	200-300 ng	[2198]
Sterol Hydroperoxides	N,N-Dimethyl-p-phenylenediamine	500 ng	[2295]
	N,N,N',N'-Tetramethyl-p-phenylenediamine	50 ng	[2295]
	KI-starch	>1 µg	[2295]
	NH_4SCN-$FeSO_4$	>1 µg	[2295]
3β-Hydroxycholest-5-en-7-one (16)	254 nm light absorption*	25-50 ng	[2303]

* On Silica Gel HF$_{254}$ chromatograms

charred with H_2SO_4 is easily made at the 1-10 μg level [2303, 2464] and by modifications at the 0.05-5 μg level [2070,2406, 2500,2501]. Still more refined microdensitometry procedures extend the method to 20-400 ng cholesterol [770]! In that several cholesterol derivatives, including the 5α-stanol 2, 3β,5α,6β-triol 13, 7-ketones 10 and 16, and 3β, 25-diol 27, give the densitometric responses to charring with H_2SO_4- $Ce(NH_4)_2(SO_4)_3$ [2303], it may be that photodensitometric methods can be adapted to reach into the 50-100 ppm range under favorable circumstances.

Nonetheless, analyses at this extreme would still fail to detect in all likelihood the presence of certain oxidized metabolites of cholesterol present in cholesterol samples as natural matter. Thus, the (24S)-3β,24-diol 25 and 3β,26- diols 29 and/or 31 present at 10.0-18.9 μg/g [2306] and 1.5- 6.0 μg/g [2280,2316] respectively in human brain would es- cape notice. Nonetheless, the power of simple thin-layer chromatography is such as to provide access into the parts- per-million range for all but these most difficult companion sterols.

The inherent sensitivity of the hydrogen flame ioniza- tion detector used in conventional gas chromatography is such as to provide detection of cholesterol at the nanogram level, and we may presume that equal sensitivity obtains for many of the common companion sterols. However, the amount of sterol sample to be analyzed is more limited in gas chromato- graphy than in thin-layer chromatography, and the thermal in- stability of the common cholesterol autoxidation products is such that their gas chromatography for these purposes cannot be assured [2300,2454,2568,2569]. The use of gas chromato- graphy is best retained for analyses of companion sterols that are not altered by the high heat of the chromatograph, thus for the biosynthesis precursors, stanols 2 and 4, oxi- dized metabolites 25 and 29, and C-24 alkyl homologs [992].

The increasingly obvious power of high performance liquid column chromatography for sterol analysis bodes well for future exploitation of this technique for these purposes, but results to date are merely indicative of what may be accomplished. The resolution of cholesterol from its several biosynthesis precursors [956,1106,2481], from C-24 alkyl homologs [505,1919], and from oxidation products [71, 72] has been demonstrated.

Detection of sterols in the effluent from high performance liquid columns is easily accomplished via ultraviolet
light absorption measurements at very short wavelengths, thus
205 nm [1106] or at 212 nm for much of our work [72]. The
relatively low molar absorbances even at this short wavelength for noncarbonyl and nondienoid sterol derivatives
limits the sensitivity with which detection is possible, but
generally light absorption below 220 nm is a sensitive means
of detection. Indeed, detection of 600 ng of cholesterol,
sitosterol, etc. by absorption at a much higher wavelength
254 nm has been achieved [505]! However, some sterols defy
detection by low wavelength light absorption, among these
being the 5,6-epoxides 35, and 36 which must be detected
by the less sensitive differential refractive index measurements [72].

Detection of ketosteroids such as the enones 8 and 16
and dienones 10 and 12 with much greater sensitivity is
possible at wavelengths near their absorption maxima, thus
at 240 and 285 nm respectively. Furthermore, as little as
700 pg of ergosterol (65) has been detected by absorption
measurements at 282 nm [505]. Increased sensitivity of detection of sterols may also be attained by derivitization
with reagents that absorb or fluoresce strongly. Thus,
cholesterol 3β-benzoate with λ_{max} 230 nm (ϵ14,400) is detected at 10-40 ng amounts, cholesterol 3β-nitrobenzoate
with λ_{max} 254 nm (ϵ 10,000) at 1 ng amounts, using 254
light [772]! The 5,6-epoxides 35 and 36 are also readily
detected as their 3β-benzoate esters by 230-240 nm light
absorption [72,1797].

Exploitation of high performance liquid chromatography
for estimation of sterol impurity content of purified cholesterol samples is yet to be had, but the method should be of
great utility.

Optical Methods. Colorimetric and spectrophotometric
measurements have limited use in the assessment of cholesterol purity. As previously mentioned, the preparation of
purified primary standards for clinical assay of serum
cholesterol is of practical importance, and the commonly
used Liebermann-Burchard and H_2SO_4-$FeCl_3$ colorimetric procedures thus become of interest for assessment of the purity of reference samples. Furthermore, colorimetric assay
of the H_2O_2 produced in sterol assays utilizing microbial
oxidases also falls into this category, and such assay

has been used in assessing cholesterol purity for proposed
evaluations of HMG CoA reductase inhibitory activity [179].
However, these colorimetric methods are very much limited
to such general matters, as discrimination among the several
companion sterols and cholesterol is not obtained. Thus
the Liebermann-Burchard reagent fails to measure congeners
56 or 3β,7-diols 14 and 15 in the presence of cholesterol
[358,515,2319]. Moreover, the microbial cholesterol oxidase assay procedures also do not differentiate among
3β-hydroxysteroids effectively [1129,2275].

Spectrophotometric assay of purified cholesterol is
limited to detection of impurities with selective absorption, thus to dienols such as 56 after the fashion of the
pioneering work in search of provitamin in cholesterol and
to ketosteroids. Spectrophotometric measurement of the
7-ketone 16 in purified cholesterol has been used in cal-
culating the apparent purity of the major component cholesterol, (cf. TABLE 30). In that chromatographic methods are
essential for the analysis of most other sterol components
likely to be found in cholesterol and these methods also
are effective for sterols that may be measured by spectro-
photometry, there is but a limited interest in such methodology.

Other Methods. Several other physical methods have
been suggested for analysis of pure cholesterol. Surface
area measurements of monomolecular layers of sterol have
been suggested as a means of estimation of cholesterol,
the surface pressure-surface area effects of oxidized
cholesterol derivatives potentially measuring their presence [401]. Furthermore, simple counter-current distribution has been used for analysis of cholesterol [5], but
the method was never developed. These and related
odd methods do not appear to meet the needs of modern
sterol analysis, and the vastly more superior thin-layer
and high performance liquid chromatographic methods have
in essence displaced most such techniques for practical
analyses.

Purity measurement of steroids has also been approached
by phase solubility analysis [1821,2448]. Application of
this simple physical method to high purity cholesterol has
been reported with a purity of 99.96% being indicated for
the National Bureau of Standards standard reference material [501]!

Differential scanning calorimetry may be used to assess the purity of divers organic compounds in the range of 98.0-99.95 mole % purities, and the method has been applied to the cholesterol purity problem to advantage [441,501,1208, 2198]. Indeed, analysis of the National Bureau of Standards reference sample gave a 99.84 mole % purity in good agreement with that obtained by phase solubility measurements. Differential scanning calorimetry necessarily must be conducted in sealed containers in the absence of air lest air oxidation of the pure sample interfere with the analysis [501].

In that phase solubility analysis and differential scanning calorimetry may be operating at the extreme limit of their sensitivity for analysis of such high purity cholesterol samples, other more powerful methods must be examined for their utility in such matters. Radioisotope dilution methods should prove suitable to extend specific analysis of sterol impurities in cholesterol to the part-per-million range, but these methods have not been directly applied for such purposes. However, radioisotope dilution analyses using specially labeled analytes have been conducted for hydroxysterols in adrenal cortex tissues, the (22R)-3β,22-diol 24 being measured at 1.5 μg/g tissue [619], the (20S)-3β,20-diol 21 at 37 ng/g [1963]! In like fashion, radioisotope derivitization by acetylation with [^{3}H] acetic anhydride provided means of measurement of cholesterol in the presence of its autoxidation products 14-16 [1758] and of the 3β,7-diol 14 [2215], both at approximately 40 ng levels.

Perhaps the most sensitive means of specific detection and measurement of traces of sterol impurities in pure cholesterol is via mass spectrometric selected ion monitoring. When coupled with gas chromatography for introduction of the sample, mass spectrometric detection of an ion unique to the sterol analyte provides a versatile and sensitive procedure, one that has been applied to human plasma and urine cholesterol measurements already. Silyl ether derivitization is recommended, the ions m/z 458 (M)$^+$, 443 (M-CH$_3$)$^+$, 368 (M-C$_3$H$_9$SiOH)$^+$, or 329 (M-129)$^+$ being monitored for the 3β-trimethylsilyl ether [218,1768,2074,2194, 2250], the ion m/z 433 (M-C$_3$H$_9$SiOH)$^+$ for the 3β-*tert*. butyl dimethylsilyl ether [218]. Using isotopically labeled internal reference sterols the technique detects cholesterol in the 4-10 ng range [218,252,1768].

Selected ion monitoring has also been applied to detection and measurement of companion sterols. Monitoring ions
m/z 370 (M)$^+$ of the 5α-stanol 2 or m/z 306 (M-154)$^+$ of 2
3β-trimethylsilyl ether affords an assay sensitive to 1-10
ng levels [2194,2488]. The ion m/z 384 (M)$^+$ likewise is the
basis for assay of the enone 8 sensitive to 2 ng [2488].
Similarly, monitoring of the ion m/z 456 (M-C$_3$H$_9$SiOH)$^+$ of
the 3β,7α-diol 14 3β,7α-ditrimethylsilyl ether [253,254,
256], 3β,25-diol 27 3β,25-ditrimethylsilyl ether [928], and
3β,26-diol 29/31 3β,26-ditrimethylsilyl ether [928] provides a highly sensitive assay for these oxidized sterols
in tissues. Furthermore, the molecular ions m/z 546 for the
ditrimethylsilyl ethers of 14,15, and 27 and m/z 472 for
the 3β-trimethylsilyl ether of the 7-ketone 16 have been
utilized for measurement of tissue levels [2074]. The
methodology should be directly applicable to analysis of
pure cholesterol samples for these oxidized sterols as well.

Moreover, the sensitivity of selected ion monitoring
can be increased using chemical ionization mass spectrometry
instead of electron impact ionization, witness detection of
48 pg cholesterol using CH$_4$ chemical ionization and monitoring the ion m/z 443.5 (M-CH$_3$)$^+$ of cholesterol 3β-trimethyl-
silyl ether [847]! These methods obviously deserve further
exploitation.

Specific Analyses

The removal of detectable amounts of recognized companion sterols from cholesterol purified via the bromination-debromination procedure of Fieser is so well accepted
that improved protocols for the analysis of such cholesterol
preparations for the individuals companion sterols at very
low levels have not been developed. Nonetheless, it is
important to give consideration to the prospects for detection and measurement of individual congeners. Although few
ancillary techniques and specializied tests are applicable,
the general approach to the analysis of traces of companion
sterols is via chromatography. However, no methods now
described are suitable for work below about 500 ppm of
impurities, but work within the 0.05-0.5% impurities range
appears accessible.

Specific methods for three biosynthesis precursors 56, 57, and 78, two reduced metabolites 2 and 4, four oxidized metabolites 21, 25, 29, and 31, typical cholesterol fatty acyl esters, typical C-24 homologs such as sitosterol and campesterol, a dozen common autoxidation products 8, 10, 12-16, 27, 35, 46, and 47, and four process artifacts cholesta-2,4-diene (284), cholesta-3,5-diene (11), cholesteryl methyl ether (17), and dicholesteryl ether (18) deserve attention.

Biosynthesis Precursors. Although cholesterol biosynthesis precursors include many other sterols, such as lanosterol (67), 5α-lanost-8-en-3β-ol (68), divers 4-methylsterols, etc., its is the 5,7-dienol 56, 5,24-dienol 78, and 7-stenol 57 late precursors that generally attract interest, as these sterols may be present in cholesterol to the extent of several percent. Their detection in pure cholesterol samples is not easily accomplished by direct absorption chromatography. Precursors 56 or 57 may be resolved from cholesterol by adsorption thin-layer chromatography in some cases, but resolution of cholesterol from desmosterol is not readily achieved thereby [111,183]. Argentation thin-layer chromatography resolves the three sterols 56, 57, and 78 from one another and from cholesterol, thereby permitting an assay of pure cholesterol for these potential contaminants [522,617,656,1220,1687,2505, 2682]. Moreover, argentation thin-layer and column chromatography of the acetate or propionate esters of these sterols also resolves the derivatives [111,491,841,842, 1116,1339,2619], and reversed phase partition thin-layer chromatography resolves the sterols and esters as well [521,522,606]. Resolution of cholesterol from desmosterol using $PdCl_2$-impregnated silicic acid comumns also appears to be effective, desmosterol being selectively retained on the column via Pd^{++} complexing [2166].

The resolution of the important biosynthesis precursors is also achieved by high performance liquid column chromatography by direct and reversed phase partition chromatography of the sterols [956,1106] and by adsorption chromatography of the acetates [2481].

The dienol 56 may be specifically detected by its selective ultraviolet light absorption. The 7-stenol 57 may be sought with the SeO_2 micro test of Fieser wherein 57 is rapidly oxidized, yielding elemental Se as a yellow color

or red precipitate [740]. No sensitivity of detection of
57 in bulk cholesterol has been established, but the method
has been applied to the National Bureau of Standards reference cholesterol preparations [501].

    Reduced Metabolites. Specific detection of the stanols
2 and 4 is also best achieved by chromatography. Adsorption
thin-layer chromatography is capable of resolution of cholesterol from 2 [183,2678] and from 4 [2069], but argentation
thin-layer chromatography is more effective in resolution
of 2 [522,617,1220,1339,2505,2682] and of 2 3β-acetate
and 3β-propionate esters [491,1116,2269,2619], as is also
reversed phase partition thin-layer chromatography [521,606].
The 5α-stanol 2 is also resolved from cholesterol biosynthesis precursors by high performance liquid chromatography
[2481].

    A special analysis of cholesterol samples for the pre-
sence of the stanols 2 and 4 may also be conducted following
chemical alteration of cholesterol by bromination. The
sterol dibromides formed have chromatographic mobilities
different from the stanols, thereby permitting ready detection of 2 or 4 following such treatment [442,522,1126,1624].
Similarly, oxidations of cholesterol to the 5,6-epoxides
35 and 36, with or without further hydrolysis to the 3β,
5α,6β-triol 13 or acetylation has also provided a means
of determination of the 5α-stanol 2 [116,1408,1689,2008].

    Oxidized Metabolites. Specific detection of the cho-
lesterol metabolites oxidized in the side-chain (the (20S)-
3β,20-diol 21, (24S)-3β,24-diol 25, 3β,25-diol 27, and isomeric 3β,26-diols 29 and 31) and in the nucleus (the 3β, 7α-
diol 14) involves the same chromatographic operations as
does detection of the autoxidation products of cholesterol
from which they are all readily resolved by paper, thin-
layer, gas, and column chromatography, including high performance liquid chromatography. Their analysis in purified
cholesterol samples is thus not a matter of resolution but
of sensitivity of detection, as the oxidized metabolites
are generally present only at very low levels. The (20S)-
3β,20-diol 21, 3β,25-diol 27, and 3β,7α-diol 14 being both
metabolites and common autoxidation products may be present
in greater amounts than the (24S)-3β,24-diol 25 and isomeric 3β,26-diols 29 and 31.

    Moreover, with the exception of the isomeric 3β,26-

diols 29 and 31 that remain unresolved, the several monohy-
droxy-cholesterol derivatives are readily resolved chromatographically from one another. Resolution of the 3β,26-
diols 29 and 31 is accomplished after acetylation by high
performance liquid chromatography operated in a recycling
mode, the (25R)-3β,26-diol 29 3β,26-diacetate being the
more polar [1915]. In that the terminal C-26 and C-27
methyl groups of cholesterol are distinguished from one
another by ^{13}C nuclear magnetic resonance data [308,1362,
1921,2177,2318,2519,2723] as are also the C-22 resonances
of the epimeric 3β-acetoxycholest-5-en-22-ols [1455] and
C-24 resonances of the (24S)-3β,24-diol 25 and the epimeric
(24R)-3β,24-diol 107 [308,1362], so also ^{13}C spectra might
also distinguish the isomeric 3β,26-diols 29 and 31.

The analysis of tissues for several of these stenediols by radioisotope dilution and derivitization techniques
and by gas chromatography coupled with selected ion mass
spectrometry has already been mentioned. These methods
should be applicable to the analysis of pure cholesterol
for these individual sterols equally well.

Cholesterol Esters. The analysis of pure cholesterol
for traces of steryl fatty acyl esters is best approached
by chromatography, simple adsorption thin-layer chromatography serving to resolve cholesterol from its esters readily [992]. Simple inspection of such chromatograms should
accord an assay for cholesterol esters in pure cholesterol
at the 0.1% level. However, gas chromatography [992] and
high performance liquid chromatography [638] also are excellent means for conducting such assay, these methods
being suited to resolution of the individual fatty acyl
esters as well. None of these approaches have been systematically applied to the analysis of purified cholesterol,
but the presence of steryl esters in USP cholesterol and
in other samples has been randomly noted in a few cases.

Spectral methods are also of use in detection of steryl
ester impurities in pure cholesterol. Strong infrared absorption near 1730 cm^{-1} and in the 1175-1250 cm^{-1} region in
steryl ester spectra not observed in the spectrum of cholesterol clearly distinguished the two classes [1414] and
could be the basis for an assay. Furthermore, the fatty
acyl moiety should be uniquely detected by ^{1}H or ^{13}C magnetic resonance spectra. Finally, chemical ionization mass
spectrometry offers another direct approach to steryl ester

analysis, as ions characteristic of the fatty acyl moiety
are derived from scission of the ester bond upon chemical
ionization. Protonated molecular ions (M+H)$^+$ for the fatty
acid are formed in positive ion spectra, hydride abstraction ions (M-H)$^-$ in negative ion spectra, these ions derived
from the fatty acid being in a spectral region not burdened
by fragmentation ions from cholesterol. Thus, cholesterol
3β-palmitate is characterized by ions m/z 257 (M+H)$^+$ and 255
(M-H)$^-$ derived from palmitic acid (M=256) in positive and
negative ion modes respectively [1698,2033]; cholesterol
3β-oleate gives rise similarly to ions m/z 283 (M+H)$^+$ and
281 (M-H)$^-$ from oleic acid (M=282) [1514,2033]. Furthermore, cholesterol 3β-palmitate may be measured in the nanogram range in the presence of cholesterol by chemical ionization mass spectrometric selected ion monitoring of the ion
m/z 211 derived from palmitic acid by decarboxylation [698].
Although none of these spectral approaches have been applied
to the analysis of pure cholesterol, all should provide access to this easy access for measure of ester contaminants.

C-24 Homologs. Analysis of traces of the C-24 alkylated homologs of cholesterol in pure cholesterol preparations is best approached via chromatography using partition
systems. Although adsorption chromatography may resolve cholesterol and sitosterol in some casses [992,1339], such systems, including argentation chromatography, generally do not
provide the resolution needed, whether for the free sterols
or for their acetates [505,1116,2619]. Reversed phase thin-
layer chromatography resolves cholesterol, campesterol and
sitosterol as well as their 3β-acetates [521,522,606] as
does also both reversed phase and direct high performance
liquid partition column chromatography [505,1106,1293].
Gas chromatography is also very effective for resolution
of the sterols, their esters, and trimethylsilyl ethers
[366,367]. Gas chromatography may be the most suitable
means of analysis of cholesterol for C-24 homolog composition.

Although the presence of C-24 homologs sitosterol,
campesterol, and others is not of major concern for cholesterol samples of mammalian origins, thus USP cholesterol,
sitosterol and other C-24 homologs have been detected in
mammalian tissues, so that matter is not irrelevant. Moreover, the presences of other exotic C-24 alkyl substituted
sterols in cholesterol preparations from marine invertebrates, etc., pose this question as a general one.

As naturally occurring cholesterol is but the one
stereoisomer, there is not need to consider analyses
for stereoisomeric forms. Those stereoisomers that are
known (20S)-cholest-5-en-3β-ol [2441], 10α-cholest-5-en-
3β-ol [658], and 14β-cholest-5-en-3β-ol [54] may be resolved from cholesterol chromatographically.

Autoxidation Products. The analysis of cholesterol
samples for the common autoxidation products has been discussed throughout this monograph, the obvious conclusion
being that chromatography be the analytical method of
choice in all matters. Resolution of the common autoxidation products in USP and other pure cholesterol samples has
been approached by paper [1175,1177,1396,1579,2288], partition column [1690], thin-layer [94,492,494-496,1072,1744,
2303,2553], gas [302,327,363,364,366,367,369,479,480,492,495,
496,765,856,914,1066,2568], Sephadex LH-20[2570] and modified
Sephadex LH-20 [94], and high performance liquid column chromatography [71,72,1105,2506] all of which serve the purpose
well. In order to avoid thermal destruction of the 3β,7-
diols 14 and 15 and also as a general matter, gas chromatography is conducted on the trimethysilyl ether derivatives.

Qualitative identification of the several common autoxidation products 10,13-16,27,35,36,46, and 47 may readily be achieved using combinations of thin-layer, gas, and
high performance liquid chromatography and authentic reference sterols for comparison of color responses and chromatographic mobilities. Quantitative analyses for individual
cholesterol autoxidation products have not been developed
except in those cases where the autoxidation product is
also a cholesterol metabolite of interest. In those cases
where highly sensitive specific analyses have been described,
such as for the 3β,7α-diol 14, (20S)-3β, 20-diol 21, 3β, 25-
diol 27, and 3β,26-diols 29/31, the methods have been applied only to tissues and not to pure cholesterol preparations.

Besides the mass spectrometric selected ion monitoring
assays for trimethylsilyl ethers of the 3β,7-diols 14 and
15, 7-ketone 16, and 3β,25-diol 27 already mentioned which
are sensitive to the nanogram range [254,256,928,2074], a
gas chromatographic analysis of tissues for the 3β,7α-diol
14 as the 3β,7α-ditrimethylsilyl ether has been described
sensitive to 20 ng [1560]. Photodensitometric evaluation
of thin-layer chromatograms bearing the common autoxidation

products provides a workable assay procedure but at a less
sensitive (1-10 µg) range [2303].

        Quantitation of autoxidation products from peak measure-
ments on elution curves from high performance liquid column
chromatography will probably provide the needed ready means
of quantitation for the common cholesterol autoxidation products, as pen excursions appear to be linear for components
detected at 212 nm in our general work and at 238 nm for 7-
ketone 16, 280 nm for dienone 12, etc. in a method also applicable to enone 8 and dienone 10 [72,1105].

        The analysis of cholesterol preparations for autoxida-
tion products is now unlikely to be conducted by isolation
of individual oxidized sterols, but preparations of individual autoxidation products to serve as reference material for
chromatographic and spectral analyses are of importance.
Moreover, preparations of pure autoxidation products are
needed for evaluation of their biological properties. The
ten oxidized sterols most closely linked to cholesterol autoxidation are the epimeric 7-hydroperoxides 46 and 47, 3β,
7-diols 14 and 15, 7-ketones 10 and 16, 5,6-epoxides 35 and
36, and the 3β,5α,6β-triol 13, some properties of which are
summarized in TABLE 32.

        Access to the several common cholesterol autoxidation
products has been gained by controlled chemical syntheses
in most cases, so that pure reference samples of 10, 13-
16, 27, 35, 36, 46, and 47 may be had without recourse to
the tedious and demanding isolation processes from autoxidized cholesterol. The 7α-hydroperoxide 46 is best prepared
by solvent isomerization of the 5α-hydroperoxide 51 in turn
readily prepared from cholesterol by photosensitized oxygenation [1544,1900,2103,2104]. The epimeric 7β-hydroperoxide
47 is then available from the 7α-hydroperoxide 46 by epimerization [2454,2455], but radiation-induced oxygenation of cholesterol also affords access to the 7β-hydroperoxide [2313].

        The 3β,7-diols 14 and 15 have been most commonly pre-
pared by reduction of the corresponding 7-ketone 16, the
quasiequatorial 3β,7β-diol 15 predominating [480,540,2038,
2242]. Borohydride reduction of the parent 7-hydroperoxides
46 and 47 also yields the 3β,7-diols 14 and 15 [540,2300,
2455,2576] but only the 3β,7α-diol 14 is suitably prepared
by this approach as the 7β-hydroperoxide 47 is too difficult
to obtain. Rearrangement of the 3β,5α-diol 50 available

TABLE 32. Physical Properties of Ten Common Cholesterol Autoxidation Products

Autoxidation Product	M.p., °C	$[\alpha]_D$, ° (CHCl$_3$)	References
Cholesterol 7α-Hydroperoxide (46)	150-158 d	-133 to -139	[447,1544,1900,2103,2104,2300,2576]
Cholesterol 7β-Hydroperoxide (47)	147-150 d	+40.2	[2300,2455]
3β-Monoacetate	80-82	+91.1	[2455]
Cholest-5-ene-3β,7α-diol (14)	180-189	-79 to -94	[77,141,419,480,540,543,575,753,796, 919,959,960,1007,1008,1194,1290, 2116,2242,2300,2336,2455,2713,2174]
3β,7α-Diacetate	150-162*	-90 to -96.8	[750,1518,2104,2336,2713,2714]
	167-178.5**	-87 to -87.3	[77,2320,2714]
3β,7α-Dibenzoate	121-124	-174.6 to -177	[141,1007,1008,1754,2051,2242]
Cholest-5-ene-3β,7β-diol (15)	150-156.5	-105 to -112.5	[141,419,960,1008,2116,2713,2714]
	170-180	0 to +13	[419,480,540,543,753,959,1008,1194, 1722,1729,1758,2038,2242,2300,2455, 2710,2714]
3β,7β-Diacetate	165-170*	-	[1544,1888]
	105-110	+52 to +55	[1008,2038,2051,2242,2714]
	98-100*	-	[1007,2710]
3β,7β-Dibenzoate	169.5-5-176	+93 to +97	[197,201,419,750,971,972,1007,1008, 1022,1888,2038,2051,2242,2701,2703, 2710,2711,2714]
3β-Hydroxycholest-5-en-7-one (16)	169-173	-100 to -107	[197,201,480,529,543,754,1160,1900, 2038,2104,2153,2242,2300,2567,2711]

(Continued)

TABLE 32 (Continued)

3β-Acetate	153-164*		[1595,1790,2703]
	155-160	-95 to -110	[197,201,540,750,1007,1194,1633,2038,2567]
3β-Benzoate	158-159.5	-103.2	[529,1194]
Cholesta-3,5-dien-7-one (10)	109-114.5	-258 to -314	[138,197,201,509,529,1160,2050]
5,6α-Epoxy-5α-cholestan-3β-ol (35)	139-148	-40 to -48.5	[72,331,479,757,821,945,979,1578,1871,2332,2660]
3β-Acetate	97-99	-46.0 to -46.2	[331,503,979,1871,2660]
3β-Benzoate	167-173	-28 to -31	[72,154,165,330,331,503]
5,6β-Epoxy-5β-cholestan-3β-ol (36)	130-136	+8 to +11.5	[72,165,330,479,945,979,1171,1578,1871,2032]
3β-Acetate	107-114	+0.5 to -1.7	[165,331,1171,1871,2032]
3β-Benzoate	167-178	+13 to +20.3	[72,154,165,331,503,2332]
5α-Cholestane-3β,5,6β-triol (13)	236-244	+3°	[331,757,1633,1864,1954,2660,2717]
3β,6β-Diacetate	163-169	-40 to -47.5	[26,503,746,1275,1633,1841,1870,2050,2660,2717]
3β,5α,6β-Triacetate	148-151.8	-32 to -34.6	[570,2032,2447,2717]
3β,6β-Dibenzoate	110-111	-	[72]
Cholest-5-ene-3β,25-diol (27)	172-183	-38 to -41.0	[169,170,191,438,565,754,1127,1199,1719,1788,1823,2052,2188,2567,2576,2580,2585,2672]
3β-Monoacetate	138.5-142.8	-39.7 to 42.1	[169,170,191,565,754,1199,1678,1823,2052,2567,2576,2585]

(Continued)

TABLE 32. Continued

3β-Monobenzoate	176–178	–13	[170]
3β,25-Diacetate	119–120.5	–35 to –35.5	[170,565]
3β,25-Dibenzoate	100–102	–10	[170]

*A second, lower melting crystalline form
**Mixed crystalline forms

(via the 5α-hydroperoxide 51) from photoxygenation of cholesterol also yields the 3β,7α-diol 14 [2455]. Other chemical syntheses of the 3β,7α-diol 14 have been described for
special purposes [429,919,1194,2336].

Chromatography is required for resolution of the 3β,
7-diols 14 and 15 one from the other, and purification of
either diol present some difficulties in crystallization
[480]. The 3β,7α-diol 14 is notorious for its poor melting
behavior. As evinced in TABLE 32, a solvated form melting
in the range 180-189°C is recovered from methanol; an unsol-
vated form of m.p. 150-162°C is obtained from other solvents.
However, mixed crystals also form, apparently with intermediate melting points within the range 167-178.5°C. Furthermore, in one case a sample of 14 of m.p. 206-216°C has been
reported [1162].

The 3β,7β-diol 15 also exhibits irregularities in melting behavior, including double melting points 172-176°C and
180-181°C [2242]. Recrystallization of pure 15 may yield
material of broader and lower melting point [2714], and
other samples of 15 with low melting points in the range
165-170°C have been recorded [1162,1888]. Variable melting
behavior is also recorded for the 3β,7β-diol 15 3β,7β-
diacetate [1007,2710].

The 3β,7-diols 14 and 15 are also characterized spec-
trally. Infrared absorption bands at 3400-3600 cm^{-1} arising from hydroxylic H-0 stretching characterize both epimers,
but the 3β,7-diols may be distinguished from one another by
absorption bands in the fingerprint regions [480,540,1194,
2242]. Proton nuclear magnetic resonance spectra of the
epimers are also distinct, the C-7 proton signal of the
3β,7α-diol 14 being a doublet of doublets (J = 5.5 and 1.5
Hz) centered at 3.84-3.85 ppm, that of the epimeric 3β,7β-
diol 15 being a doublet of doublets (J=7 and 1.5 Hz) centered at 3.80-3.86 ppm [540,2242,2455]. Moreover, the C-6
vinyl proton signal appears as a doublet (J = 6 Hz) at
5.58-5.60 ppm for the 3β,7α-diol 14, as a doublet (J = 1.5
Hz) at 5.26-5.30 ppm for the 3β,7β-diol 15 [540,1194,2242,
2455]. Proton spectra of the corresping 3β,7-diacetates
are likewise distinguished [1669].

Electron impact mass spectra of 3β,7-diols 14 and 15
are alike, as elimination processes occur equally for both,
yielding ions m/z 402 (M)$^{+}$, 384 (M-H_2O)$^{+}$, 366 (M-$2H_2O$)$^{+}$,

and other fragments ions [103,476]. Spectra of the 3β,7-diacetates do not exhibit a molecular ion but include ions m/z 426 (M-CH$_3$CO$_2$H)$^+$, 384 (M-CH$_3$CO-CH$_3$CO$_2$H)$^+$, and 366 (M-2CH$_3$CO$_2$H)$^+$ [2109]. Spectra of 14/15 3β,7-ditrimethylsilyl ethers contain ions m/z 546 (M)$^+$ and 456 (M-C$_3$H$_9$SiOH)$^+$, as well as fragment ions m/z 233,208,159,145,153,129, and 119 [103,366]. Chemical ionization mass spectra of 14 and 15 are much simpler, including m/z 420 (M+NH$_4$)$^+$, 402 (M-OH+NH$_3$)$^+$, 385 (M-OH)$^+$, and 367 (M-H$_2$O-OH)$^+$ using NH$_3$ as reagent gas, ions m/z 403 (M+H)$^+$, 402 (M)$^+$, 401 (M-H)$^+$, 385, and 367 using CH$_4$ or isobutane as reagent gas [1514]. Negative ion chemical ionization mass spectra exhibit ions m/z 402 (M)$^-$, 401 (M-H)$^-$, 399 (M-H-H$_2$)$^-$, 383 (M-H-H$_2$O)$^-$, 381 (M-H-H$_2$O-H$_2$)$^-$, and 365 (M-H-2H$_2$O)$^-$ [2033].

The 7-ketone 16 has been synthesized by the direct oxidation of cholesterol 3β-acetate (or 3β-benzoate) by CrO$_3$ in divers solvents [1194,1564,1595,2065,2703], MnO$_2$ [2038], or tert.-butyl chromate [2038]. The 7-ketone 16 is also readily prepared from cholesterol by photosensitized oxygenation to the 5α-hydroperoxide 51 and subsequent rearrangement and dehydration with Cu(II) salts in pyridine [2104]. This method has also been used in chemical synthesis of antheridiol (97) and related sterols from 7-deoxygenated 5-stenol analogs [659,660,916,1609,1879,2286, 2649].

The 7-ketone 16 is a well characterized sterol, but double melting behavior has been observed [2242]. Strong ultraviolet light absorption at 236-238 nm (ε12,500-15,500) also characterizes the sterol [197,201,529,1633,1900,2104, 2153,2242,2300,2567,2711] as does also strong carbonyl absorption near 1640-1690 cm^{-1} and hydroxyl absorption at 3500 cm^{-1} [509,889,2300]. The 7-ketone 16 gives characteristic optical rotatory dispersion [917] and circular dichromism [916] spectra.

Proton nuclear magnetic resonance spectra of 16 3β-acetate include the signals 0.68 (C-18), 0.85 (d,J=5 Hz, C-26, C-27), 0.92 (d,J=5 Hz, C-21), 1.21 (C-19), 2.05 (acetyl CH$_3$), 4.70 (3α-proton), and 5.70 ppm (C-6 vinyl proton) [2455]. Furthermore, 24 of the 29 carbon atoms of 3β-acetate are resolved in ^{13}C spectra [308,1921]. Mass spectra of 16 3β-acetate include ions m/z 382 (M-CH$_3$CO$_2$H)$^+$, 367,269,187,174,etc. [2109], whereas spectra of 16 3β-trimethylsilyl ether show

ions m/z 472 (M)$^+$, 457 (M-CH$_3$)$^+$, 382 (M-C$_3$H$_9$SiOH)$^+$, 367,187,
174,161,142,129, etc [367]. Chemical ionization mass spec-
tra of 16 are simplified, showing ions m/z 418 (M+NH$_4$)$^+$ and
401 (M+H)$^+$ using NH$_3$ as reagent gas, ions m/z 401 (M+H)$^+$,
400 (M)$^+$, 399 (M-H)$^+$, and 383 with CH$_4$ [1514]. Negative
ion spectra include ions m/z 400 (M)$^-$, 399 (M-H)$^-$, and
381 (M-H-H$_2$O)$^-$ [2033].

The dienone 10 is an elimination product of the 7-ke-
tone 16 or its fatty acyl esters, derived by treatment of
these derivatives with acid or with alkali. Besides the
physical data of TABLE 32, the 7-ketone 10 is characterized
by strong ultraviolet light absorption near 277-280 nm
(ε23,000-24,000) in alcohol [197,201,529,2050,2300,2567],
268-259 nm (ε25,000) in hydrocarbons [509,1242]. Infrared
absorption bands for the carbonyl group at 1640-1661 cm^{-1}
and for the diene feature at 1610-1629 cm^{-1} also characterize the 7-ketone 10 [509,2300,2567]. The proton nuclear
magnetic resonance spectrum of 10 includes distinctive vinyl
proton signals at 5.59 ppm (C-6) and 6.12 (C-3,C-4) [509].
Moreover, signals from most of the carbon atoms of 10 are
resolved in its ^{13}C nuclear magnetic resonance spectrum
[308,1921]. Chemical ionization mass spectra of the dienone 10 include ions m/z 400 (M+NH$_4$)$^+$ and 383 (M+H)$^+$ with
NH$_3$, ions m/z 383 (M+H)$^+$, 382 (M)$^+$, (M-H)$^+$, and 367 with
CH$_4$, and ion 383 (M+H)$^+$ with isobutane as reagent gases
[1514]. Negative ion spectra include ions m/z 382 (M)$^-$,
381 (M-H)$^-$, 379 (M-H-H$_2$)$^-$, 377, and an adduct ion m/z 397
(M+N$_2$O-H-N$_2$)$^-$ derived by reaction with the reagent gas
N$_2$O [2033].

The 5α,6α-epoxide 35 is prepared from cholesterol by
the attack of peracids. Some 5β,6β-epoxide 36 in also
formed, but fairly pure 5α,6α-epoxide 35 can be recovered
in high yield from such oxidations [330,757,1871,2049,2189,
2660,2692]. Commercial 5α,6α-epoxide 35 prepared by peracid
oxidations contains approximately 5% 5β,6β-epoxide 36
[2509], necessitating high performance liquid column chromatography for ultimate removal of the isomer for careful
work. The 5β,6β-epoxide 36 cannot be recovered pure from
peracid oxidations of cholesterol but is best prepared by
treatment of the 3β,5α,6β-triol 13 3β,5α,6β-triacetate
with alkali [479,2032]. Other syntheses involving epoxidations give various proportions of 5α,6α-epoxide 35 and
5β,6β-epoxide 36, from 8:1 to 1:4 [33,2489,2492-2494,2513].

Spectral characterizations of the isomeric 5,6-epoxides 35 and 36 include infrared spectra, both isomers displaying hydroxyl absorption at 3600 cm^{-1}. However, the isomers may be distinguished by absorption at 1030 cm^{-1} for the 5α,6β-epoxide 35, at 1060 cm^{-1} for the 5β,6β-epoxide 36 [479]. Furthermore, proton spectra distinguish the isomers by chemical shift and coupling of the C-6 hydrogen signal, these differences being useful in the analysis of mixtures of the 5,6-epoxides. One may also project that ^{13}C spectra of the isomeric 5,6-epoxides be distinguishable [1061].

The analysis of the 5,6-epoxides 35 and 36 poses a problem not faced or overcome in some investigations. Although the epoxides are readily resolved from other cholesterol autoxidation products by simple chromatography, differentiation between the isomers requires additional work lest misidentifications such as those potentially made in biological [276,277,279,281,1523] and chemical [765,1303] systems be had. Reliance on infrared absorption spectra [479,481], colorimetry [232,245], or melting points [232] is unwarranted.

Other than product isolation which is not feasible with small samples, three analysis methods (chromatography, chemical alterations, proton spectra) have been advanced for identification of the 5,6-epoxides 35 and 36 found in biological and chemical systems. Conventional chromatography does not resolve the epoxides [93,327,479,1578,2297], although the 5β,6β-epoxide 36 is slightly more mobile than 35 by absorption chromatography [93,327], gas chromatography on 3% SE-30 [92,2297] and on 1% QF-2 [479], and modified Sephadex LH-20 chromatography [93]. The order of elution is reversed in gas chromatography on 1% NGS [479], but none of these systems is satisfactory for general analyses of the 5,6-epoxides in admixture.

The 5,6-epoxides 35 and 36 may be analyzed chromatographically as their 3β-acetate or 3β-benzoate esters or as their 3β-trimethylsilyl ethers. Conventional adsorption chromatography and gas chromatography resolve the 3β-acetates [93,226,327] and the 3β-trimethylsilyl ethers [93, 479,2633] of 35 and 36, and thin-layer and high performance liquid column chromatography resolve the 3β-benzoates [72, 74,1797]. High performance reverse phase partition column chromatography likewise resolves the 3β-benzoates, the 5β,

6β-epoxide 36 3β-benzoate being the more mobile, thus the same elution order as found for the free sterols 35 and 36 discussed next.

It is not necessary to transform the 5,6-epoxides 35 and 36 to ester or ether derivatives for their resolution, as high performance liquid column chromatography using microparticulate adsorption (μPorasil) or reversed phase partition (μBondapak C_{18}) columns are satisfactory for analyses of 5,6-epoxide mixtures [72,74,2507,2509]. High performance adsorption chromatography eluted the 5α,6α-epoxide 35 first, whereas high performance reverse phase operations eluted the 5β,6β-epoxide 36 first, as expected.

In this series care must be exercised in drawing conclusions from chromatographic mobility as to which 5,6-epoxide or derivative be present, as reversals of mobility abound. The 5β,6β-epoxide 36 and its derivatives are the more mobile except for gas chromatography on 1% NGS [479], thin-layer chromatography of the 3β-trimethylsilyl ethers [93], and high performance liquid chromatography of the 3β-benzoates [72,1797]. However, the greater mobility of the 5β,6β-epoxide 36 3β-benzoate over that of 35 3β-benzoate on high performance reverse phase chromatography, thus the same as for high performance adsorption chromatography [72], must be viewed as a reversion of normal elution order.

Detection of the 5,6-epoxides 35 and 36 in effluents from such column chromatographic analyses is not readily achieved by ultraviolet light absorption methods, even at wavelengths as low as 210 nm, for in distinction to many other sterol derivatives, the epoxides have very low absorption properties. The resolutions obtained were recorded using differential refractive index measurements on column effluents [72].

Epoxide mixtures may also be conveniently analyzed following LiAlH₄ reduction to the corresponding alcohols which are readily resolved chromatographically. Catalytic reduction of the 3β-acetates of the isomeric 5,6-epoxides 35 and 36 gave readily resolved product alcohols, 5α-cholestane-3β,5-diol (428) from 35 3β-acetate, 5α-cholestane-3β, 6β-diol (264) from 36 3β-acetate [1870,1871], whereas the more readily adapted reduction using LiAlH₄ gave the 3β,5α-diol 428 from 35 3β-acetate but 3β,6β-diol 264 and 5β-cholestane-3β,5-diol (265) (as minor product) from 36 3β-acetate [1869].

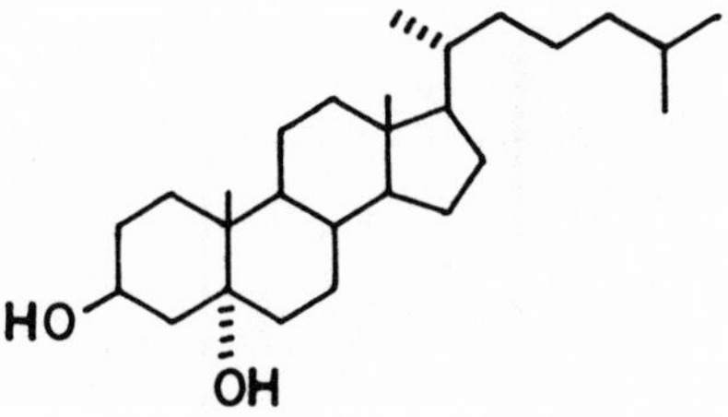

428

We confirm that LiAlH$_4$ reduction of the free sterols
35 and 36 give the same products, 428 from 35, 264 and 265
from 36. This reduction method has been used for such
purposes in epoxides analyses in biological [914] and
chemical [2297-2299] systems. An alternative chemical
analysis scheme involving chromatography of the different
methanolysis products formed from the 5,6-epoxides 35 and
36 with BF$_3$-methanol has also been applied successfully
[2192].

Mixtures of 5,6-epoxides 35 and 36 and their esters
are also conveniently analyzed by proton spectra, the C-6
hydrogen signal appearing as a doublet in both spectra but
with distinctive coupling constants and chemical shifts.
Thus, the 6β-hydrogen of the 5α,6α-epoxide 35 is a doublet
(J = 3.3-4.1 Hz) near 2.8 ppm, whereas the 6α-hydrogen of
the isomeric 5β,6β-epoxide 36 is also a doublet (J = 2.1-
3 Hz) near 3.0 ppm [33,103,226,537,1301,2483,2494,2499].
Our analysis of the 5,6-epoxides formed by the attack of
HO· on cholesterol is shown in FIGURE 29, the resolved
doublet (J = 4 Hz) at 2.91 ppm and doublet (J = 2 Hz) at
3.06 ppm representing a 3.5:1 proportion of 5α,6α-epoxide
35 and 5β,6β-epoxide 36 [73,2291].

Electron impact mass spectra of the isomeric 5,6-epox-
ides 35 and 36 include ions m/z 402 (M)$^+$, 384 (M-H$_2$O)$^+$, 369
(M-H$_2$O-CH$_3$)$^+$,366 (M-2H$_2$O)$^+$, 356, etc. [479,914]. Spectra
of the 3β-trimethylsilyl ethers of either 5,6-epoxide 35
or 36 display ions m/z 474 (M)$^+$, 456 (M-H$_2$O)$^+$, 384 (M-
C$_3$H$_9$SiOH)$^+$, and 366 (M-C$_3$H$_9$SiOH-H$_2$O)$^+$ [93,914]. Chemical
ionization mass spectra of 5,6-epoxides 35 and 36 include
ions m/z 420 (M+NH$_4$)$^+$, 403 (M+H)$^+$, 402 (M)$^+$, 385 (M+H-H$_2$O)$^+$
using NH$_3$ as reagent gas, ions m/z 403,402,401,385,383,

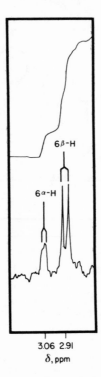

FIGURE 29.   Proton spectra of a mixture of 5,6 epoxides <u>35</u>
              and <u>36</u>.

and 367 with CH_4 or isobutane [1514]. Negative ion spectra show ions m/z 402 (M)⁻, 401 (M-H)⁻, 383 (M-H-H_2O)⁻, and
367 (M-H-2H_2O)⁻ [2033].

        One of the most readily accessible cholesterol autoxi-
dation products is the 3β,5α,6β-triol <u>13</u> formed from cholesterol by epoxidation and subsequent epoxide hydration.   The
oxidation may be accomplished in one reaction from cholesterol with concentrate H_2O_2 and acetic or formic acids
[757,1857,1864,2717] or with HIO_4 [911] or from either 5,6-
epoxide <u>35</u> or <u>36</u> with a variety of hydration conditions
[821,1273,1335,2692].

        Besides data of TABLE 32, the 3β,5α,6β-triol <u>13</u> is
characterized by infrared absorption at 3440 and 1048 cm⁻¹
[1335]and ¹³C nuclear resonance spectra characterize the
triol <u>13</u> and <u>13</u> 3β,6β-diacetate, most carbons being differ-
entiated suitably [307,308,1368].   Electron impact mass spec-

tra of 13 include ions m/z 420 (M)$^+$, 402 (M-H$_2$O)$^+$, 387 (M-H$_2$O-CH$_3$)$^+$, 384 (M-2H$_2$O)$^+$, 369 (M-2H$_2$O-CH$_3$)$^+$, etc. [1335]; of 13 3β,6β-ditrimethylsilyl ether include ions m/z 564 (M)$^+$, 546 (M-H$_2$O)$^+$, 459 (M-CH$_3$-C$_3$H$_9$SiOH)$^+$, 456 (M-H$_2$O-C$_3$H$_9$SiOH)$^+$, 403, 367, 321, 129, etc.; of 13 3β,5α,6β-tritrimethylsilyl ether ions m/z 636 (M)$^+$, 546, 456, 403, 367, 321, 129, etc. [93, 366]. Positive ion chemical ionization mass spectra of 13 include ions m/z 438 (M+NH$_4$)$^+$, 420 (M-OH+NH$_3$)$^+$, 403 (M-OH)$^+$, and 385 (M-OH-H$_2$O)$^+$ with NH$_3$, ions m/z 421 (M+H)$^+$, 420 (M)$^+$, 419 (M-H)$^+$, 403, 401, 385, and 367 with CH$_4$ or isobutane as reagent gases [1514]. Negative ion spectra contain ions m/z 420 (M)$^-$, 419 (M-H)$^-$, 417 (M-H-H$_2$)$^-$, 401 (M-H-H$_2$O)$^-$, 399 (M-H-H$_2$O-H$_2$)$^-$, 383 (M-H-2H$_2$O)$^-$, 381 (M-H-2H$_2$O-H$_2$)$^-$ and 365 (M-H-3H$_2$O)$^-$ [2033].

The chemical synthesis of the 3β,25-diol 27 has attracted much more attention because of the crucial relationship of 27 to proposed syntheses of the cholecalciferol 25-hydroxyderivative 93 of current commercial interest. Syntheses of 27 3β-acetate directly from cholesterol 3β-acetate with suitable protection of the double bond have been accomplished by CrO$_3$ oxidation [170] and photochemically [1273, 2028]. Syntheses from desmosterol 3β-acetate involving oxidative attack on the Δ^{24}-double bond (by epoxidation, photo-oxygenation, or oxymercuration) with subsequent reduction of intermediate oxidation products [1678] also provide the 3β,25-diol 27 3β-acetate directly. Other syntheses require C$_{19}$-C$_{26}$ sterol starting materials variously derived from degradations of common sterols, with reconstruction of the 25-hydroxylated side-chain. Thus syntheses from the 27-nor-25-ketone 299 3β-acetate [191,565,2052,2671,2672], the C$_{24}$-acid 99 3β-acetate [1127,1199], bile acid derivatives [1788], pregnenolone (23) [1719], and degradation products of stigmasterol [1823,2066,2188] have been described.

Besides the physical properties of the 3β,25-diol 27 in TABLE 32, the diol is characterized by spectral data, including 3325-3330 cm^{-1} representing hydroxyl O-H stretching [2567,2576,2578,2585]. Proton nuclear magnetic resonance spectra 27 include signals at 0.68-0.71 (C-18), 0.93-0.95 (d, J = 6-7 Hz, C-21), 1.00-1.02 (C-19), 1.21-1.22 (s, C-26/27), 3.50 (3α-proton), and 5.30-5.49 ppm (C-6 proton) [438,2585,2672].

Electron impact mass spectra of the 3β,25-diol 27, include ions m/z 402 (M)$^+$, 384 (M-H$_2$O)$^+$, 369 (M-H$_2$O-CH$_3$)$^+$,

and 351 (M-2H$_2$O-CH$_3$)$^+$ [2455,2585,2672],of $\underline{27}$ 3β,25-ditri-
methylsilyl ether ions m/z 546 (M)$^+$, 456 $\overline{(M}$-C$_3$H$_9$SiOH)$^+$, 366
(M-2C$_3$H$_9$SiOH)$^+$, 327, 271, 255, and 131 (C$_3$H$_9$SiOC(CH$_3$)$_2$)
(the principal ion)   [268,367,856]. Chemical ionization
spectra of $\underline{27}$ show ions m/z 420 (M+NH$_4$)$^+$, 402 (M-OH+NH$_3$)$^+$,
and 385 (M-$\overline{OH}$)$^+$ using NH$_3$, ions m/z 403 (M+H)$^+$, 402 (M)$^+$,
401 (M-H)$^+$, 385, and 367 using CH$_4$ or isobutane as reagent
gases. Chemical ionization mass spectra of $\underline{27}$ 3β-ditri-
methylsilyl ether include ions m/z 547 (M+H)$^{+}$, 546 (M)$^+$,
545 (M-H)$^+$, 457 (M-C$_3$H$_9$SiO)$^+$, 385, 383, 367, and 255.
Negative ion mass spectra of the 3β,25-diol $\underline{27}$ show ions
m/z 402 (M)$^-$, 401 (M-H)$^-$, 399 (M-H-H$_2$)$^-$, 397, 383 (M-H-
H$_2$O)$^-$, 381 (M-H-H$_2$O-H$_2$)$^-$, and 365 (M-H-2H$_2$O)$^-$[2033].

Process Artifacts. Other artifacts of manipulation
likely to be found in various preparations of cholesterol
include several steroid olefins cholest-2-ene, cholest-3-
ene (344), cholesta-2,4-diene (284), cholesta-3,5-diene (11),
cholesta-2,4,6-triene (37), cholesta-3,5,7-triene (38), and
dicholesteryl ether (18) variously formed as elimination
products. The diene and triene contaminants absorb selectively in the ultraviolet region [192,904,1464,1738], and
may be detected thereby; however, no specific assay for
such steradienes and steratrienes in cholesterol has been
recorded.

These steroid olefins are chromatographically less polar than cholesterol,and resolution of most from one another
is to be had by argentation thin-layer chromatography [370]
and gas chromatography [370,684,2300,2454,2568]. Detection
of 0.5 μg of steradiene impurity in cholesterol by thin-layer
chromatography is readily achieved [1383].

Dicholesteryl ether (18) also an elimination product
of cholesterol is resolved chromatographically from cholesterol by thin-layer and gas chromatography [2199]. Under
some circumstances in which methanol (or other specific
alcohol) is involved, cholesterol may be transformed to
the 3β-O-methyl ether 17 detected as an artifact by chromatography [1383,1688,2199].

In all studies of contaminants in highly pure cholesterol it must be recalled that the mere act of analysis may
contribute to artifact formation, thus generating a "choles-
sterol uncertainty principle". Depending upon the sensitivity with which such analyses are conducted, the presence of

undefined and unidentified artifacts may be directly observed. The formation of such process artifacts during
thin-layer chromatography on silica gel chromatoplates has
been noted [501,2198], and deterioration of pure cholesterol upon gas chromatography is also known [2569]. Furthermore, although systematic studies of the matter have
not been reported, if cholesterol or other sensitive congener present be applied to thin-layer chromatograms and
unduly exposed to air prior to chromatography, some autoxidation may occur. Indeed, chromatography conducted without
exclusion of air may lead to adventitious autoxidations in
special cases. For example, the air oxidation during chromatography of 3β-acetoxy-23-hydroxy-22-oxo-(23S)-stigmasta-5,
24(28)-dien-29-oic lactone (29→23) via a putative 22-enolate
anion generated by irrigation solvent methanol acting as a
base has been noted [1609]. Thin-layer chromatography artifacts may be recognized simply by conducting the analysis in
two dimensions using the same solvent system, any process
artifacts formed during irrigation falling off a 45° diagonal
line on the two-dimensional chromatogram [578].

CHOLESTEROL STABILITY

At this point it must be accepted that cholesterol is
inherently unstable in contact with air, witness the extensive evidence adduced in this monograph. However, awareness
of cholesterol instability to storage in air discovered
about 1901 [1960,2163,2164] and cholesterol sensitivity to
deterioration upon heating in air noted in 1914 [1420] and
definitively established by 1930 [247,296,1352] has repeatedly been compromised by attribution of the effects to heat
rather than air [1315,1666], by misunderstandings arising
from comparison of cholesterol stability with that of 5,7-
dienols [190,2353], and recent assertion that cholesterol
be stable [684].

The problem of cholesterol stability has passed through
several cycles of discovery and rediscovery as an awareness
phase has lapsed into periods of discounted, overlooked,
or forgotten information, only to be rediscovered once more.
At each cycle, improved methods and understanding have made
the matter all the more obvious, and none of the information
has been kept secret. Nonetheless, an awareness phase may
be in decline, as several senior investigators who should
know of such matters have made comments in my presence at

two international meetings in 1979, one on lipid peroxidation, one on atherosclerosis, that suggest unawareness!

Some of the awareness problem may be attributed to rational differences as to what constitutes stability for an
organic compound in air. As there are no absolute standards
to be applied, conclusions of stability are necessarily relative ones, those suited to the viewpoint of the investigator.
It is within this relative framework that my comments are
given that cholesterol is unstable. For those working daily
with cholesterol in various aspects of biochemistry, biological activities, regulation of endogenous metabolism, etc.
the purity of sterol is very important, and misunderstandings
on the matter can be seriously misleading. The present section of the monograph is directed toward recognization of
stability problems and towards selection of storage means
that limit deterioration.

Criteria of Deterioration

The deterioration of cholesterol upon ordinary storage
in containers not depleted of air may not all be from autoxidation, as the action of trace acid, metal, etc. or of
heat and time conceivably could lead to dehydration reactions, yielding cholestadienes 11 and 284 or dicholesteryl
ether (18). Moreover, the presence of low levels of sterol
congeners may also be likewise affected. However, most of
the deterioration of stored cholesterol samples may be
reasonably ascribed to autoxidation.

Cholesterol deterioration on storage is evinced by
numerous criteria, most of which also apply to cholesterol
that has been subjected to heat or radiation. Indeed, we
may view the autoxidative deterioration of cholesterol under all conditions to undergo the same processes, therefore
to display the same changes in properties, whether natural
aging or accelerated decomposition be the case.

The first sign of cholesterol deterioration likely to
be detected is that of odor, as an opened bottle of aged
cholesterol releases an acrid odor immediately, an odor
variously described in unpleasant terms such as "urine
odor" [740]. As bottles of cholesterol are usually brown

glass, the yellow coloration of aged cholesterol is usually
the second evidence of deterioration observed after opening
the bottle [1897,1960,2163,2164]. Other common properties
that are altered include diminished melting point [754,1352,
1731,1960,2020,2163,2164], specific rotation, generally
less negative [1352,1771,1927], increased ultraviolet
[978,1019,1352,1600,1731,1927,1997] and infrared [567]
light absorption and fluorescence [1731,1974], altered
organic solvent solubilities [1600, 1731] increased water
solubility, positive color tests for oxidation and for
peroxides [2205], increased amounts of glassy or resinous
material upon recrystallization [1997], and decreased
yields of cholesterol purified by recrystallization or
via the dibromide [744]. Many of these physical property
alterations are also noted in cholesterol exposed to ultraviolet light [1988]. Furthermore, in both natural air-
aging and in irradiation experiments more refined analyses
for the presence of autoxidation products confirm these
several indications.

Storage Conditions

It is instructive to examine a few example of deterioration of cholesterol under specific conditions, thereby
to recognize common problems. Numerous descriptions of
deterioration of cholesterol adorn the literature, such as
that of a sudden deterioration over a period of 6 mos after
a stable shelf life of several years [1703], melting point
changes suggesting decomposition during pelleting of cholesterol [232], lower cholesterol levels by assay of samples kept in the dark versus kept in the light, etc. Moreover, loss of cholesterol from plasma stored cold away
from light [337], from alcohol or ether-alcohol solutions
stored [337], and from fresh water stored at 4°C [646] is
indicated.

However, relatively few quantitative studies under
stipulated conditions have been conducted. Fieser compared the progressive deterioration of cholesterol samples
with time, using the isolation of the 3β,25-diol 27 as
measure of deterioration together with amount of glassy material formed. Cholesterol samples 2 mo., 2 or 4 y, and
24 y of age contained 0.0%, 0.14%, and 0.34% of the 3β,25-
diol 27, 0.9%, 2.4%, and 18% of glassy material respectively
[748,754].

Of even greater interest are descriptions of conditions under which stored cholesterol was found to be stable. In general these conditions emphasize the absence of oxygen, but light and heat are also factors to minimize. Data of TABLE 33 summarize representative cases.

The storage of pure cholesterol in ampules sealed under vacuum or under inert atmosphere free from oxygen, at low temperature in the dark, is thus the recommended procedure. However, the mere act of sealing cholesterol into glass ampules appears to be destructive of sample, the purity measured by differential scanning calorimetry being diminished by such treatment [1208]! Moreover, thermal decomposition of high purity cholesterol on gas chromatography is indicated by lowered melting point behavior [2569] and by detection of thermal elimination products in effluent [684]. Obviously great care must be exercised in any manipulation of high purity cholesterol.

Suitable storage conditions for other sterols may be quite different from those for cholesterol. Thus, the notoriously unstable 5,6-dienol 56 is best stored as a paste in methanol [748], whereas desmosterol (78) is more stable in benzene solution than as crystalline material [2346, 2480].

Special Cases

There are several cases of cholesterol stability that merit individual discussion, that of radioactive samples, that of a unique sample of Engel not deteriorated since 1937, and that of very old material from mummified and fossilized remains.

Radioactive Samples. The special case of the stability of pure cholesterol labeled with radioactive isotopes is well know, the radiation produced upon nuclear decay serving to catalyze cholesterol decomposition in air. Both $[^{14}C]$- and $[^{3}H]$-cholesterol preparations are rapidly destroyed by oxidative processes involving autoradiolysis [567, 707,940-942,1969,2490]. The amounts of unaltered cholesterol remaining after room temperature storage of radioactive cholesterol in contact with air are: for $[4-^{14}C]$-cholesterol (3.83 μCi/mg) stored 3 mo, 92%; for $[24-^{14}C]$-cholesterol (0.18 μCi/mg) stored 15 mo, 93%; for $[26-^{14}C]$-cho-

TABLE 33. Safe Cholesterol Storage Conditions

Condition	Time	Comments	Reference
CO_2 atmosphere	2 y	No deterioration	[2164]
N_2 atmosphere	–	No autoxidation	[894]
N_2, sealed ampules, dark, 4°C	9 mo	No autoxidation	[1528]
Evacuated vial, dark	1 y	No autoxidation	[567]
In water, 4°C, dark	3 mo	No autoxidation	[1528]
Over P_2O_5, vacuum dessicator, 4°C, dark	–	No autoxidation	[1528]
In vacuum, 2-4 MeV X-rays	–	Stable	[2491]
In vacuum, ^{60}Co γ-rays	–	Stable	[2491]
[^{14}C]-Cholesterol, in vacuum, dark	1 y	No decomposition	[991]
[4-^{14}C]-Cholesterol, in vacuum	2.5 mo	No deterioration	[567]
1 mM in ethanol, 4°C	1 y	No deterioration	[684]
Screw-cap amber vial, N_2, -30°C	6 mo	No deterioration	[2198]

lesterol (1.5 µCi/mg) stored 2 y, 65-66% [567]. Moreover,
we have heated [1,2-^{3}H]-cholesterol (2.0 µCi/mg) at 100°C
in air, 95% unaltered cholesterol remaining after 10 h,
90% after 100 h. The common cholesterol autoxidation products accounted for the balance of radioactivity. Although
systematic studies have not been recorded, temperature and
amount of ionizing radiation appear to be factors influencing rate.

The decomposition of radioactive cholesterol samples
depends absolutely on the presence of O_2, without which decomposition is not observed. Thus, [^{14}C]-cholesterol stored
in a vacuum [567] and unlabeled cholesterol subjected to
ionizing radiation in vacuum [2491] did not decompose. However cholesterol is affected by ionizing radiation, generation of the C-7 radical 3β-hydroxycholest-5-en-7-yl 262
being reliably demonstrated [661,899,1004]. In the presence of O_2 the process of autoxidation then ensues, but
no systematic examination of possible products of further
free radical reactions of the C-7 radical 262 not involving
O_2 has been attempted. The presence of water-soluble radioactive impurities in commercially available [^{14}C]-choleste-
rol preparations [872,873] may be byproducts of manufacture
but may also be unrecognized radiolysis and oxidation products.

Radioactive cholesterol samples must be stored cold in
vacuum, under a suitable inert atmosphere, or in benzene (or
other) solution free from air such that solvent dilution limits autoradiolysis.

The Engel Sample. The instability of pure cholesterol
towards oxidation by the molecular oxygen of the air is
fully supported by the evidence adduced throughout this
monograph, yet assertions to the contrary have been made.
Engel has declared that "Cholesterol Is Stable" in a paper
describing the special stability properties of a sample of
cholesterol purified by him in 1937 [684], and this interpretation has been incorporated into a major monograph on
sterols without qualification of the unique status of this
sample [76]. Lest this viewpoint prevail, it is essential
to give this one sample added attention.

The special sample had been brominated, rapidly recrystallized, and debrominated by Schoenheimer's method using
NaI/ethanol, the bromination-debromination steps repeated,

and the sample recrystallized from ethanol. Analysis of the
sample 31 years later revealed its freedom from detectable
amounts of autoxidation products, thus its unique stability
over the period 1937-1968 [684]. Our analysis in 1971-1972
of the sample kindly provided by Professor Engel [2295,2313]
confirmed the freedom of the sample from detectable autoxidation products.

     Data of TABLE 34 summarize properties of Engel's special
sample and of a sample of cholesterol purified by me in 1953
via the dibromide twice. In this case, the debromination was
accomplished using Fieser's procedure with Zn/acetic acid
and the sample was recrystallized from methanol. The contrast between the two samples is impressive, as my 1953
sample had autoxidized extensively over the period 1953-1966
[2303], whereas Engel's sample had not! Moreover, Engel's
sample remains but minimally autoxidized at this writing in
1980.

     Engel has suggested that the recrystallization of the
5α,6α-dibromide 426 before debromination be the only difference in the two purification procedures. However, other
issues appear to me to be of greater importance in providing
possible explanations. Rather than conclude that pure cholesterol is stable, it is to me much more probable that the
unrecognized presence of traces of unidentified impurities
contribute to the differential stabilities observed. Either
traces of catalytic transition elements increase the
autoxidation of cholesterol purified by Fieser's process or
traces of protection agents stabilize the sample purified
by the Schoenheimer method. Trace metals analyses on cholesterol samples have not been commonly reported, only the
ash content being required for official USP purposes.

     However, the halogen content of Engel's 1937 sample
and of my 1953 sample has been measured by neutron activation, the intensities of the ^{82}Br and ^{128}I photopeaks at
554.3 and 442.9 keV respectively following neutron irradiation providing the absolute levels recorded in TABLE 34.
Clearly the Engel sample retains a 3.5-fold increased level
of Br over the sample debrominated witn Zn/acetic acid
and also has a measureable I content. It is not known
whether the halogens found are present as inorganic salts
or as halogenated sterols.

     In accelerated stability studies we have irradiated

TABLE 34. Stability of Two Naturally Aged Pure Cholesterol Samples

Sample	Date	M.p. (Kofler), C°	Chromatographic Examinations*	Halogen Content, ppm Br	I
Engel	1937	148-149**	-	-	-
	1968	147.5-149	No autoxidation	-	-
	1972	-	No autoxidation	-	-
	1980	150.0-150.5	Trace autoxidation	771.7+115.7	10.7+2.0
Smith	1953	149.0-150.0	No autoxidation	-	-
	1966	138.0-140.0	Autoxidation	-	-
	1971	-	Autoxidation	-	-
	1980	138-140	Autoxidation	215.6+32.3	<7.2***

*From thin-layer, paper, gas, and high performance liquid chromatography
**Capillary m.p. from [684]
***Upper limit representing 6σ deviation of the base area in the I region.
 No specific I signas observed

cholesterol purified via Fieser's method or by Schoenheimer's method with ^{60}Co γ-radiation and sought cholesterol 7-hydroperoxide products. Cholesterol purified by Fieser's method contained cholesterol 7-hydroperoxides after 1 min, that purified via the Schoenheimer process only after 5 min. Cholesterol purified by Fieser's process and then treated with 1-10 ppm levels of NaI (together with Na_2SO_4 used in the Schoenheimer process to reduce I_2 formed) contained sterol hydroperoxides following ^{60}Co γ-radiation for 2 min at 1 ppm, after 1 min at 10 ppm levels. These experiments suggest that radiation-induced autoxidation of cholesterol may be retarded by these added agents, but the matter is obviously not settled.

Yet another factor may influence the matter. From its crystal form Engel's sample is cholesterol monohydrate, recovered from ethanol (95%), whereas our highly purified samples have uniformly been recrystallized from methanol and are anhydrous crystals. We have not conducted comparative stability studies of anhydrous cholesterol and cholesterol monohydrate and cannot comment further on the influence of the state of hydration or crystallinity on the autoxidation of cholesterol.

We have compared the stability to natural aging of two cholesterol samples purified via the dibromide, one debrominated by Fieser's method, the other by Schoenheimer's method. Both samples recrystallized from methanol are anhydrous crystals. Data of TABLE 35 show that both samples autoxidized upon shelf storage in a brown bottle, but the sample prepared by the Schoenheimer process was the more stable over a two-year period. Moreover, high performance liquid chromatography of equal fractions (5%) of the autoxidation products obtained from each sample aged for 1 y showed somewhat higher levels of the common cholesterol autoxidation products, including the epimeric 7-hydroperoxides 46 and 47 and secondary 14-16 in the sample purified via the Fieser procedure, cf. FIGURE 30.

More significantly, the elution region occupied by sterols oxidized in the side-chain (12-22 min) discloses the presence of at least ten components, most of which are present in both samples. However, the elution profile is not the same for both samples, and one component (No.7) present in the sample prepared by Fieser's method is absent from that prepared by Schoenheimer's method. Similar chro-

TABLE 35. Controlled Comparison of Stabilities

Storage Time, y	Odor	M.p. (Kofler), C°	Chromatographic Examination*	Recrystallization Recovery, mg** Cholesterol	Autoxidation Products
Fieser Method					
0	None	150-151	No autoxidation	–	–
0.5	Odor present	148-149	Traces	–	–
1.0	Odor present	149-150	Full spate	951	22
2.0	Odor present	149-150	Full spate	968	22
Schoenheimer Method					
0	None	149-150	No autoxidation	–	–
0.5	Weak odor	149-150	No autoxidation	–	–
1.0	Odor present	149-150	Full spate	940	17
2.0	Odor present	150-151	Full spate	981	17

*Thin-layer and high performance liquid column chromatography
**Recrystallization of 1 g samples from methanol

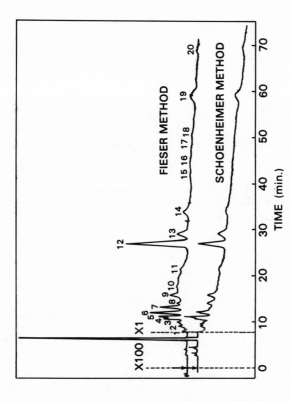

FIGURE 30: Cholesterol purified via the dibromide and stored for 1 y at ambient temperature. Component identities: No. 4, 21; No. 7, 27; No. 12, 16; No. 13, 47; No. 14, 46; No. 19, 15; No. 20, 14.

matograms of samples aged for 2 y showed the same distributions of the common autoxidation products 14-16, 46, and 47,
again with somewhat more oxidized products being present in
the sample prepared by Fieser's procedure. However, component No. 4 diminished in both samples, and in the sample
prepared by the Schoenheimer process, component No. 5 diminished but component No. 7 appeared after 2 y. The sterols
responsible for these dynamic features have not been identified, but is clear that autoxidation processes occur at
different rates in the two samples.

Careful high performance liquid column chromatography
of 900 μg of Engel's sample on μPorasil irrigated with
hexane/2-propanol (24:1) and detection of sterols by their
absorption at 212 nm established that traces of the 3β,7β-
diol 15 and the 7-ketone 16 were present in 1980 but to a
very limited extent in comparison with my own 1953 sample
of cholesterol, which was contaminated with full spate of
cholesterol autoxidation products resolved in this system,
including side-chain oxidation products.

We may conclude that cholesterol purified via Fieser's
process is more susceptible to autoxidation than is cholesterol purified via the Schoenheimer method but that both
processes provide pure cholesterol that is subject to autoxidation.

Very Old Samples. The stability of cholesterol in geologic time is obviously not indicated, as diagenesis processes altering deposits containing tissues bearing cholesterol
lead ultimately to reduction of both the Δ^5-olefin and the
3β-hydroxyl group, as previously mentioned. Under these
circumstances the anaerobic state of geologic sediments
mitigates against an autoxidative disposition of cholesterol.

There remains the question of the long term stability
of cholesterol in the presence of air. However, tissue
samples more than a few thousand years old have not survived
or have not been examined for sterol content. For what answer can be provided, one must turn to tissue samples of established antiquity that contain cholesterol. Egyptian
mummy brain as old as 6000 y B.P. has yielded identifiable
cholesterol but also cholesterol esters [1,1305,1399,1428,
1555,2113], thus evincing the stability of cholesterol and
its esters for this period. As human brain contains very
little esterified cholesterol, the presence of cholesterol

esters in mummy brain suggests a *post mortem* esterification
of the sterol, one which might well stabilize the material
against further alteration. Some autoxidation of material
from more recent unembalmed Egyptian Coptic mummy brain may
have occurred [1305], but autoxidation of cholesterol was
not in evidence in the sterols from 2000 y old American
Indian coproliths which contained identifiable cholesterol,
stanols, and bile acids [1509].

In view of the demonstrated survival of cholesterol in
these mummified and fossilized samples versus the demonstrated deterioration of pure cholesterol in contact with air,
it must be posited that the human artifacts retained adequate antimicrobial and antioxidant protection or else that
a barrier impervious to O_2 diffusion existed in these materials.

REFERENCES

1. E. Abderhalden, Z. Physiol. Chem., 74, 392 (1911).
2. Y.H. Abdulla, C.W.M. Adams, and R.S. Morgan, J. Path. Bact., 94, 63 (1967).
3. J.J. Abel, Monatsh., 11, 61 (1890); Sitzber. Akad. Wiss. Wien, [IIb], 99, 77 (1890).
4. J. Abele and H. Khayam-Bashi, Clin. Chem., 25, 132 (1979).
5. L.L. Abell, B.B. Levy, B.B. Brodie, and F.E. Kendall, J. Biol. Chem., 195, 357 (1952).
6. R.A. Abramovitch and R.G. Micetich, Can. J. Chem., 40, 2017 (1962).
7. H.F. Acevedo and E.A. Campbell, Steroids, 16, 569 (1970).
8. H.F. Acevedo, E.A. Campbell, D:W. Hayeslip, J. Gilmore, L.P. Merkow, J.C. Frich, and R.C. Grauer, Obst. Gynecol., 37, 425 (1971).
9. H.F. Acevedo, E.A. Campbell, E.L. Saier, J.C. Frich, L.P. Merkow, D.W. Hayeslip, S.P. Bartok, R.C. Grauer, and J.L. Hamilton, Cancer, 32, 196 (1973).
10. B. Achari, P.C. Majumdar, and S.C. Pakrashi, Indian J. Chem., 17B, 215 (1979).
11. T. Achtermann, Z. Physiol. Chem., 217, 281 (1933).
12. L. Acker and H. Greve, Fette Seif. Anstrich., 65, 1009 (1963).
13. W.E. Acree and G.L. Bertrand, Nature, 269, 450 (1977).
14. A. Adachi and T. Kobayashi, J. Vitaminology, 16, 103 (1970).
15. C.W.M. Adam and O.B. Bayliss, Abstracts, 5th Int. Symp. Atherosclerosis, Houston, Texas, Nov. 6-9, 1979, Abstract No. 427.
16. H.K. Adam, I.M. Campbell, and N.J. McCorkindale, Nature, 216, 397 (1967).
17. N.K. Adam, F.A. Askew, and J.F. Danielli, Biochem. J., 29, 1786 (1935).
18. N.K. Adam and G. Jessop, Proc. Royal Soc. London, [A], 120, 473 (1928).
19. N.K. Adam and O. Rosenheim, Proc. Royal Soc. London, [A], 126, 25 (1930).
20. W. Adam and H.J. Eggelte, Angew. Chem. Int. Ed., 16, 713 (1977).
21. A.M. Adamany and J.T. Mills, Federation Proc., 39, 1830 (1980).
22. B.G. Adams and L.W. Parks, Biochem. Biophys. Res.

Commun., <u>28</u>, 490 (1967).

23. B.G. Adams and L.W. Parks, J. Lipid Res., <u>9</u>, 8 (1968).

24. J.B. Adams and K.N. Wynne, Steroidologia, <u>2</u>, 321 (1971).

25. A. Adler, Z. Ges. Exp. Med., <u>46</u>, 371 (1925).

26. H. Aebli, C.A. Grob, and E. Schumacher, Helv. Chim. Acta, <u>41</u>, 774 (1958).

27. A. Afonso, Can. J. Chem., <u>47</u>, 3693 (1969).

28. H. Ageta, K. Shiojima, R. Kamaya, and K. Masuda, Tetrahedron Lett.,899 (1978).

29. E.J. Agnello, R. Pinson, and G.D. Laubach, J. Am. Chem. Soc., <u>78</u>, 4756 (1956).

30. N. Ahmad and D.B. Gower, Biochem. J., 108, 233 (1968).

31. A.A. Akhrem, I.S. Levina, and Yu. A. Titov, EKDIZONY-STEROIDNYE GORMONY NASEKOMYKH, Byelorussian Academy of Sciences, Minsk, 1973.

32. A.A. Akhrem, V.M. Shkumatov, and V.L. Chashchin, Dok. Akad. Nauk S.S.S.R., <u>237</u>, 1509 (1977).

33. A.A. Akhrem, T.A. Timoshchuk, and D.I. Metelitsa, Dok. Akad. Nauk S.S.S.R., <u>219</u>, 891 (1974).

34. A.A. Akhrem and Yu. A. Titov, STEROIDI I MIKROORGANIZMI, Academy of Sciences U.S.S.R., Moscow, 1970, p. 488.

35. M. Akhtar, D.C. Wilton, and K.A. Munday, Biochem. J., <u>101</u>, 23C (1966).

36. Zh. I. Akopyan, N.V. Blazheievich, I.V. Veryovkina, V.Z. Gorkin, O.V. Syomina, and V. B. Spirichev, Int. Z. Vitaminforsch., <u>40</u>, 497 (1970).

37. M. Albericci, M. Collart-Lempereur, J.C. Braekman, D. Daloze, B. Tursch, J.P. Declerq, G. Germain, and M. Van Meerssche, Tetrahedron Lett., 2687 (1979).

38. A. Alcaide, J. Viala, F. Pinte, M. Itoh, T. Nomura, and M. Barbier, Compt. Rend., [C], <u>273</u>, 1386 (1971).

39. G.J. Alexander and E. Schwenk, Arch. Biochem. Biophys., <u>66</u>, 381 (1957).

40. Z.B. Alfassi, L. Feldman, and A.P. Kushelvesky, Radiation Effects, <u>32</u>, 67 (1977).

41. M.J. Allen, E. Boyland, C.E. Dukes, E.S. Horning, and J.G. Watson, Brit. J. Cancer, <u>11</u>, 212 (1957).

42. N.L. Allinger and F. Wu, Tetrahedron, <u>27</u>, 5093 (1971).

43. G.H. Alt and D.H.R. Barton, Chem. Ind., 1103 (1952); J. Chem. Soc., 1356 (1954).

44. R.F.A. Altman, Arch. Geschwulstforsch., <u>41</u>, 107 (1973).

45. R. Altschul, Arch. Pathol., <u>42</u>, 277 (1946).

46. R. Altschul, Arch. Pathol., <u>44</u>, 282 (1947).

47. R. Altschul, Anat. Rec., <u>103</u>, 566 (1949).

48. R. Altschul, SELECTED STUDIES ON ARTERIOSCLEROSIS, C.C. Thomas Publisher, Springfield, Ill., 1950, pp. 135-154.

49. R. Altschul and E.Y. Spencer, Rev. Canad. Biol., 11, 250 (1952).
50. M.D. Altschule, Med. Clin. N. America, 58, 397 (1974).
51. M.M. Amer, A.K.S. Ahmad, and S.P. Varda, Fette Seif. Anstrich., 72, 1040 (1970).
52. C.E. Anagnostopoulos and L.F. Fieser, J. Am. Chem. Soc., 76, 532 (1954).
53. M. Anastasia, G.C. Galli, and P. Allevi, J. Org. Chem., 44, 4983 (1979).
54. M. Anastasia, A. Scala, and G. Galli, J. Org. Chem., 41, 1064 (1976).
55. M. Anchel and R. Schoenheimer, J. Biol. Chem., 125, 23 (1938).
56. D.E. Anders and W.E. Robinson, Geochim. Cosmochim. Acta, 35, 661 (1971).
57. C.A. Anderson, Austr. J. Chem., 17, 949 (1964).
58. C.A. Anderson and G.F. Wood, Nature, 193, 742 (1962).
59. K.E. Anderson, E. Kok, and N.B. Javitt, J. Clin. Invest., 51, 112 (1972).
60. P.C. Anderson, P.M. Gardner, E.V. Whitehead, D.E. Anders, and W.E. Robinson, Geochim. Cosmochim. Acta, 33, 1304 (1969).
61. R.J. Anderson, J. Biol. Chem., 71, 407 (1927).
62. T. Ando, A. Kanazawa, S. Teshima, and H. Miyawaki, Marine Biol., 50, 169 (1979).
63. T. Ando, M. Moueza, and H.J. Ceccaldi, Compt. Rend. Soc. Biol., 170, 149 (1976).
64. F.A.L. Anet and G.C. Levy, Science, 180, 141 (1973).
65. N. Anitschkow, Beitr. Path. Anat. Allg. Pathol., 56, 379 (1913).
66. N. Anitschkow, in ARTERIOSCLEROSIS, A SURVEY OF THE PROBLEM, E.V. Cowdry, Editor, MacMillan Co., New York, N.Y., 1933, pp. 271-322.
67. N. Anitschkov and S. Chalatow, Zentralblat Allg. Pathol. Anat., 24, 1 (1913).
68. Anon., PL Analects, 7, No. 2, 1 (1979).
69. G.A.S. Ansari, V.B. Smart, and L.L. Smith, Pharmacologist, 20, No. 3, 155 (1978).
70. G.A.S. Ansari and L.L. Smith, Chem. Phys. Lipids, 22, 55 (1978).
71. G.A.S. Ansari and L.L. Smith, Abstracts of Papers, 77th Natl. Meeting, American Chemical Society, Honolulu, April 1-6, 1979, Abstract BIOL-86.
72. G.A.S. Ansari and L.L. Smith, J. Chromatog., 175, 307 1979.
73. G.A.S. Ansari and L.L. Smith, Photochem. Photobiol.,

$\underline{30}$, 147 (1979).

74. G.A.S. Ansari and L.L. Smith, Abstracts of Papers, 35th
    Ann. Southwest Regional Meeting, American Chemical Soc.,
    Dec. 5-7, 1979, Austin, TX, p. 21.
75. G.A.S. Ansari, R.D. Walker, and L.L. Smith, Abstracts,
    19th Ann. Meeting, Society of Toxicology, Washington
    City, D.C., March 9-13, 1980, p. A107.
76. M.F. Ansell, Editor, RODD'S CHEMISTRY OF CARBON COM-
    POUNDS, 2nd Ed., Elsevier Scientific Publishing Co.,
    Amsterdam, 1974, Vol. 2, Alicyclic Compounds, Parts C,
    D, E, Supplements, p. 156.
77. D. Apotheker, J.L. Owades, and A.E. Sobel, J. Am. Chem.
    Soc., $\underline{76}$, 3685 (1954).
78. J.W. ApSimon, H. Beierbeck, and J.K. Saunders, Can. J.
    Chem., $\underline{51}$, 3974 (1973).
79. J.W. ApSimon, H. Beierbeck, and J.K. Saunders, Can. J.
    Chem., $\underline{53}$, 338 (1975).
80. J.W. ApSimon, J.A. Buccini, and S. Badripersaud, Can.
    J. Chem., $\underline{51}$, 850 (1973).
81. K. Arai, T. Golab, D.S. Layne, and G. Pincus, Endocrinol.,
    $\underline{71}$, 639 (1962).
82. Y. Aramaki, T. Kobayashi, Y. Imai, S. Kikuchi, T. Matsu-
    kawa, and K. Kanazawa, J. Atherosclerosis Res., $\underline{7}$, 653
    (1967).
83. Yu. A. Arbuzov, Uspekhi Khimii, $\underline{34}$, 558 (1965).
84. J.C. Arcos and M.F. Argus, CHEMICAL INDUCTION OF CANCER,
    STRUCTURAL BASES AND BIOLOGICAL MECHANISMS, Academic
    Press, New York/London, 1974, Vol IIA, pp. 52-54, 70-71,
    82-86.
85. J. Arditti, R.E.M.H. Fisch, and B.H. Flick, J. Chem.
    Soc. Chem. Commun., 1217 (1972).
86. M. Ardon, OXYGEN, ELEMENTARY FORMS AND HYDROGEN PEROX-
    IDE, W.A. Benjamin, New York, 1965, pp. 7-11, 29-35.
87. E. Arffmann, J. Natl. Cancer Inst., $\underline{25}$, 893 (1960).
88. E. Arffmann, Acta Pathol. Microbiol. Scand., $\underline{61}$, 161
    (1964).
89. E. Arffmann and J. Glavind, Acta Pathol. Microbiol.
    Scand., $\underline{70}$, 185 (1967).
90. K. Arima, M. Nagasawa, M. Bae, and G. Tamura, Agr. Biol.
    Chem., $\underline{33}$, 1636 (1969).
91. L. Aringer, J. Lipid Res., $\underline{19}$, 933 (1978).
92. L. Aringer and P. Eneroth, J. Lipid Res., $\underline{14}$, 563
    (1973).
93. L. Aringer and P. Eneroth, J. Lipid Res., $\underline{15}$, 389
    (1974).
94. L. Aringer, P. Eneroth, and L. Nordström, J. Lipid Res.,

17, 263 (1976).

95. L. Aringer, P. Eneroth, and L. Nordström, J. Steroid
    Biochem., 11, 1271 (1979).
96. M.L. Armstrong, W.E. Connor, and J.C. Hoak, Circula-
    tion, 38, VI-1 (1968).
97. G.P. Arsenault, K. Biemann, A.W. Barksdale, and T.C.
    McMorris, J. Am. Chem. Soc., 90, 5635 (1968).
98. J.A. Arthur, H.A.F. Blair, G.S. Boyd, N.G. Hattersley,
    and K.E. Suckling, Biochem. Soc. Trans., 3, 963 (1975).
99. J.A. Arthur, H.A.F. Blair, G.S. Boyd, J.I. Mason, and
    K.E. Suckling, Biochem. J., 158, 47 (1976).
100. N.R. Artman, Adv. Lipid Res., 7, 245 (1969).
101. F.A. Askew, R.B. Bourdillon, H.M. Bruce, R.G.C. Jen-
     kins, and T.A. Webster, Proc. Royal Soc. London, [B],
     107, 76 (1930).
102. F.A. Askew, R.B. Bourdillon, H.M. Bruce, R.G.C. Jen-
     kins, and T.A. Webster, Proc. Royal Soc. London, [B],
     107, 91 (1930).
103. G. Assmann, D.S. Fredrickson, H.R. Sloan, H.M. Fales,
     and R.J. Highet, J. Lipid Res., 16, 28 (1975).
104. M. Astruc, M. Laporte, C. Tabacik, and A. Crastes de
     Paulet, Biochem. Biophys. Res. Commun., 85, 691 (1978).
105. A.M. Atallah and H.J. Nicholas, Phytochem., 10, 3139
     (1971).
106. D. Attaway and P.L. Parker, Science, 169, 674 (1970).
107. J. Attenburrow, J.E. Connett, W. Graham, J.F. Oughton,
     A.C. Ritchie, and P.A. Wilkinson, J. Chem. Soc., 4547
     (1961).
108. D. Autenreith and J.E. van Lier, Abstracts, 9th Int.
     Symp. Chemistry Natural Products, Ottawa, June 24-28,
     1974, p. 22D.
109. W. Autenreith and A. Funk, Münch. Med. Woch., 60, 1243
     (1913).
110. J. Avigan, J. Biol. Chem., 234, 787 (1959).
111. J. Avigan, D.S. Goodman, and D. Steinberg, J. Lipid
     Res., 4, 100 (1963).
112. J. Avigan, C.D. Williams, and J.P. Blass, Biochim.
     Biophys. Acta, 218, 381 (1970).
113. Y. Ayaki and K. Yamasaki, J. Biochem., 68, 341 (1970).
114. Y. Ayaki and K. Yamasaki, J. Biochem., 71, 85 (1972).
115. E. Ayanoglu, C. Djerassi, T.R. Erdman, and P.J.
     Scheuer, Steroids, 31, 815 (1978).
116. D.L. Azarnoff and D.R. Tucker, Biochim. Biophys. Acta,
     70, 589 (1963).
117. P. Back, Clin. Chim. Acta, 44, 199 (1973).
118. P. Back and K. Ross, Z. Physiol. Chem., 354, 83 (1973).

119. P. Back, J. Sjövall, and K. Sjövall, Scand. J. Clin.
     Lab. Invest., 26, Suppl. 126, 17.12 (1972).
120. E.J. Bailey, D.H.R. Barton, J. Elks, and J.F. Temple-
     ton, J. Chem. Soc., 1578 (1962).
121. E.J. Bailey, J. Elks, and D.H.R. Barton, Proc. Chem.
     Soc., 214 (1960).
122. M.B. Baird, H.R. Massie, and M.J. Piekielniak, Chem.-
     Biol. Interactions, 16, 145 (1977).
123. W.S. Baird, H.M. O'Bryan, G. Ogden, and D. Lee, J. Opt.
     Soc. Am., 37, 754 (1947).
124. C.P. Bakker, M.P.I. van der Plank-van Winsen, and H.J.
     van der Molen, Biochim. Biophys. Acta, 584, 94
     (1979).
125. S.A. Bakker, J. Lugtenburg, and E. Havinga, Rec. Trav.
     Chim., 91, 1459 (1972).
126. S. Balasubramaniam and K.A. Mitropoulos, Biochem. J.,
     125, 13P (1971).
127. S. Balasubramaniam, K.A. Mitropoulos, and N.B. Myant,
     Eur. J. Biochem., 34, 77 (1973).
128. S. Balasubramaniam, K.A. Mitropoulos, and N.B. Myant,
     Biochim. Biophys. Acta, 398, 172 (1975).
129. J.E. Baldwin, H.H. Basson, and H. Krauss, Chem. Com-
     mun., 984 (1968).
130. J.A. Ballantine, A. Lavis, and R.J. Morris, Comp. Bio-
     chem. Physiol., 63B, 119 (1979).
131. J.A. Ballantine and J.C. Roberts, Tetrahedron Lett.,
     105 (1975).
132. J.A. Ballantine, J.C. Roberts, and R.J. Morris, Biomed.
     Mass Spectrometry, 3, 14 (1976).
133. J.A. Ballantine and K. Williams, J. Chromatog., 148,
     504 (1978).
134. J.A. Ballantine, K. Williams, and B.A. Burke, Tetra-
     hedron Lett., 1547 (1977).
135. J.A. Ballantine, K. Williams, and R.J. Morris, J.
     Chromatog., 166, 491 (1978).
136. B. Balogh, D.M. Wilson, and A.L. Burlingame, Nature,
     233, 261 (1971).
137. P. Baranger, Atti 10th Congr. Intern. Chimica, 5, 261
     (1938).
138. J. Barnett, B.E. Ryman, and F. Smith, J. Chem. Soc.,
     526 (1946).
139. C. Baron and N. Le Boulch, Bull. Soc. Chim. France,
     300 (1958).
140. C. Baron, N. Le Boulch, and Y. Raoul, Bull. Soc. Chim.
     France, 948 (1955).

141. T. Barr, I.M. Heilbron, E.G. Parry, and F.S. Spring, J. Chem. Soc., 1437 (1936).

142. A.G.M. Barrett, D.H.R. Barton, M.H. Pendlebury, L. Phillips, R.A. Russell, D.A. Widdowson, C.H. Carlisle, and P.F. Lindley, J. Chem. Soc. Chem. Commun., 101 (1975).

143. A.G.M. Barrett, D.H.R. Barton, R.A. Russell, and D.A. Widdowson, J. Chem. Soc. Chem. Commun., 102 (1975).

144. G. Barry, J.W. Cook, G.A.D. Haslewood, C.L. Hewett, I. Hieger, and E.L. Kennaway, Proc. Royal Soc. London, [B], 117, 318 (1935).

145. N. Bartlett and D.H. Lohmann, Proc. Chem. Soc., 115 (1962).

146. D.H.R. Barton, J. Chem. Soc., 2174 (1949).

147. D.H.R. Barton and R.K. Haynes, J. Chem. Soc., Perkin Trans. I, 2065 (1975).

148. D.H.R. Barton, R.K. Haynes, G. Leclerc, P.D. Magnus, and I.D. Menzies, J. Chem. Soc., Perkin Trans. I, 2055 (1975).

149. D.H.R. Barton, R.K. Haynes, P.D. Magnus, and I.D. Menzies, J. Chem. Soc. Chem. Commun., 511 (1974).

150. D.H.R. Barton, R.H. Hesse, M.M. Pechet, and E. Rizzardo, J. Am. Chem. Soc., 95, 2748 (1973).

151. D.H.R. Barton, R.H. Hesse, M.M. Pechet, and L.C. Smith, J. Chem. Soc. Chem. Commun. 754 (1977).

152. D.H.R. Barton and G.F. Laws, J. Chem. Soc., 52 (1954).

153. D.H.R. Barton, G. Leclerc, P.D. Magnus, and I.D. Menzies, J. Chem. Soc. Chem. Commun., 447 (1972).

154. D.H.R. Barton and E. Miller, J. Am. Chem. Soc., 72, 370 (1950).

155. D.H.R. Barton and E. Miller, J. Am. Chem. Soc., 72, 1066 (1950).

156. P.G. Barton and J. Glover, Biochem. J., 84, 53P (1962).

157. J.T. Bashour and L. Bauman, J. Biol. Chem., 121, 1 (1937).

158. S.K. Basu, J.L. Goldstein, R.G.W. Anderson, and M.S. Brown, Proc. Natl. Acad. Sci., 73, 3178 (1978).

159. S.K. Basu, J.L. Goldstein, and M.S. Brown, J. Biol. Chem., 253, 3852 (1978).

160. M.L. Bates, W.W. Reid, and J.D. White, J. Chem. Soc. Chem. Commun., 44 (1976).

161. E.-E. Baulieu, J. Clin. Endocrinol. Metab., 22, 501 (1962).

162. E.E. Baulieu, R. Emiliozzi, and C. Corpechot, Experientia, 17, 110 (1961).

163. C.A. Baumann, H.P. Rusch, B.E. Kline, and H.P. Jacobi,

Am. J. Cancer, 38, 76 (1940).

164. G. Bauslaugh, G. Just, and F. Blank, Nature, 202, 1218 (1964).

165. R.A. Baxter and F.S. Spring, J. Chem. Soc., 613 (1943).

166. K.D. Bayes, in OZONE AND OTHER PHOTOCHEMICAL OXIDANTS, National Academy of Sciences, Washington City, D.C., 1977, pp. 13-44.

167. V.I. Bayunova and G.S. Grinenko, Khim. Prirod. Soed., 179 (1973).

168. S. Beck, A.H.M. Kirby, and P.R. Peacock, Cancer Res., 5, 135 (1945).

169. A.L.J. Beckwith, Proc. Chem. Soc., 194 (1958).

170. A.L.J. Beckwith, J. Chem. Soc., 3162 (1961).

171. M.S. Bedour, D. El-Munajjed, M.B.E. Fayez, and A.N. Girgis, J. Pharm. Sci., 53, 1276 (1964).

172. Z.H. Beg, J.A. Stonik, and H.B. Brewer, Proc. Natl. Acad. Sci., 75, 3678 (1978).

173. D. Behar, G. Czapski, J. Rabani, L.M. Dorfman, and H.A. Schwarz, J. Phys. Chem., 74, 3209 (1970).

174. W.T. Beher and W.L. Anthony, J. Nutrition, 52, 519 (1954).

175. W.T. Beher, J. Parsons, and G.D. Baker, Anal. Chem., 29, 1147 (1957).

176. O.R. Beirne, R. Heller, and J.A. Watson, J. Biol. Chem., 252, 950 (1977).

177. J.A. Beisler and Y. Sato, J. Org. Chem., 36, 3946 (1971).

178. J.J. Bell, S.C. Cheng, and B.W. Harding, Ann. N.Y. Acad. Sci., 212, 290 (1973).

179. J.J. Bell, T.E. Sargeant, and J.A. Watson, J. Biol. Chem., 251, 1745 (1976).

180. E.P. Benditt, Circulation, 50, 650 (1974).

181. E.P. Benditt, Am. J. Pathol., 86, 693 (1977).

182. J.G. Bendoraitis, in ADVANCES IN ORGANIC GEOCHEMISTRY, 1973, B. Tissot and F. Bienner, Editors, Éditions Technip, Paris, 1974, pp. 209-224.

183. R.D. Bennett and E. Heftmann, J. Chromatog., 9, 359 (1962).

184. R.D. Bennett, E. Heftmann, and R.A. Joly, Phytochem., 9, 349 (1970).

185. S.W. Benson and R. Shaw, in ORGANIC PEROXIDES, D. Swern, Editor, Wiley-Interscience, New York, 1970, Vol. 1, pp. 105-139.

186. C.B. Berde, R.A. Heller, and R.D. Simoni, Biochim. Biophys. Acta, 488, 112 (1977).

187. M. Berenstein, A. Georg, and E. Briner, Helv. Chim.

Acta, 29, 258 (1946).

188. M. Berenstein, H. Paillard, and E. Briner, Helv. Chim. Acta, 29, 271 (1946).

189. W. Bergmann, Z. Krebsforsch., 48, 546 (1939).

190. W. Bergmann, in COMPARATIVE BIOCHEMISTRY, A COMPREHENSIVE TREATISE, M. Florkin and H.S. Mason, Editors, Academic Press, New York City, NY, Vol. IIIA, 1962, pp. 103-162 (cf. p. 105).

191. W. Bergmann and J.P. Dusza, J. Org. Chem., 23, 459 (1958).

192. W. Bergmann and F. Hirschmann, J. Org. Chem., 4, 40 (1939).

193. W. Bergmann, F. Hirschmann, and E.L. Skau, J. Org. Chem., 4, 29 (1939).

194. W. Bergmann and M.J. McLean, Chem. Rev., 28, 367 (1941).

195. W. Bergmann and M.B. Meyers, Chem. Ind., 655 (1958); Liebigs Ann., 620, 46 (1959).

196. W. Bergmann, H.E. Stavely, L.C. Strong, and G.M. Smith, Am. J. Cancer, 38, 81 (1940).

197. S. Bergström, Arkiv Kemi, Mineral. Geol., 16A, No. 10, 1-72 (1942).

198. S. Bergström, Naturwiss., 30, 684 (1942).

199. S. Bergström, S. Lindstredt, B. Samuelson, E.J. Corey, and G.A. Gregoriou, J. Am. Chem. Soc., 80, 2337 (1958).

200. S. Bergström and B. Samuelsson, in AUTOXIDATION AND ANTIOXIDANTS, W.O. Lundberg, Editor, Interscience Publishers, New York, 1961, Vol. 1, pp. 233-248.

201. S. Bergström and O. Wintersteiner, J. Biol. Chem., 141, 597 (1941).

202. S. Bergström and O. Wintersteiner, J. Biol. Chem., 143, 503 (1942).

203. S. Bergström and O. Wintersteiner, J. Biol. Chem., 145, 309 (1942).

204. S. Bergström and O. Wintersteiner, J. Biol. Chem., 145, 327 (1942).

205. J.D. Bernal, D. Crowfoot, and I. Fankuchen, Philos. Trans. Royal Soc., [A], 239, 135 (1940).

206. A.W. Bernheimer and L.S. Avigad, Biochim. Biophys. Acta, 541, 96 (1978).

207. O. Berséus, Acta Chem. Scand., 19, 325 (1965).

208. O. Berséus, H. Danielsson, and K. Einarsson, J. Biol. Chem., 242, 1211 (1967).

209. O. Berséus and K. Einarsson, Acta Chem. Scand., 21, 1105 (1967).

210. M.P.E. Berthelot, Ann. Chim. Phys., [3], 56, 51 (1859).

211. H. Beumer, Klin. Wochenschr., 5, 1962 (1926).

212. N.S. Bhacca and D.H. Williams, APPLICATIONS OF NMR
     SPECTROSCOPY IN ORGANIC CHEMISTRY. ILLUSTRATIONS FROM
     THE STEROID FIELD, Holden-Day Inc., San Francisco/
     London/Amsterdam, 1964, p.5.
213. B.R. Bhavnani and C.A. Woolever, in STEROID BIOCHEM-
     ISTRY, R. Hobkirk, Editor, CRC Press Inc., Boca Raton,
     Fla., Vol. II, 1979, pp. 1-50.
214. A.E. Bide, H.B. Henbest, E.R.H. Jones, and P.A. Wilkin-
     son, J. Chem. Soc., 1788 (1948).
215. J.-F. Biellmann and M. Rajić, Bull. Soc. Chim. France,
     441 (1962).
216. B.H. Bielski and A.O. Allen, J. Phys. Chem., 81, 1048
     (1977).
217. B.H.J. Bielski and J.M. Gebicki, in FREE RADICALS IN
     BIOLOGY, W.A. Pryor, Editor, Academic Press, New York,
     Vol. 3, 1977, pp. 1-51.
218. D.A. Bigham, R.K. Farrant, G. Phillipou, and R.F.
     Seamark, Clin. Chim. Acta, 69, 537 (1976).
219. M. Bihari-Varga and S. Gero, Artery, 2, 374 (1976).
220. J.H. Bill, Am. J. Med. Sci., 44, 365 (1862).
221. C.E. Bills, Physiol. Rev., 15, 1 (1935).
222. C.E. Bills and E.M. Honeywell, J. Biol. Chem., 80, 15
     (1928).
223. C.E. Bills, E.M. Honeywell, and W.A. MacNair, J. Biol.
     Chem., 76, 251 (1928).
224. R.J. Bing and J.S.M. Sarma, Biochem. Biophys. Res.
     Commun., 62, 711 (1975).
225. R.J. Bing, J.S.M. Sarma, and S.I. Chan, Artery, 5, 14
     (1979).
226. K.D. Bingham, T.M. Blaiklock, R.C.B. Coleman, and G.D.
     Meakins, J. Chem. Soc.,(C), 2330 (1970).
227. J. Binkert, E. Angliker, and A. von Wartburg, Helv.
     Chim. Acta, 45, 2122 (1962).
228. M.K. Birmingham, H. Traikov, and P.J. Ward, Steroids,
     1, 463 (1963).
229. F. Bischoff, Federation Proc., 16, 155 (1957).
230. F. Bischoff, J. Natl. Cancer Inst., 19, 977 (1957).
231. F. Bischoff, Progr. Exp. Tumor Res., 3, 412 (1963).
232. F. Bischoff, Advances Lipid Res., 7, 165 (1969).
233. F. Bischoff and G. Bryson, Abstracts of Papers, 138th
     Natl. Meeting, American Chemical Soc., New York City,
     N.Y., Sept. 11-16, 1960, p. 62C.
234. F. Bischoff and G. Bryson, Federation Proc., 20, 282
     (1961).
235. F. Bischoff and G. Bryson, Progr. Exp. Tumor Res., 5,
     85 (1964).

236. F. Bischoff and G. Bryson, Federation Proc., 29, 860Abs
     (1970).
237. F. Bischoff and G. Bryson, Proc. Amer. Assoc. Cancer
     Res., 15, 6 (1974).
238. F. Bischoff and G. Bryson, Adv. Lipid Res., 15, 61
     (1977).
239. F. Bischoff and R.E. Katherman, Am. J. Physiol., 152,
     189 (1948).
240. F. Bischoff, G. Lopez, and J.J. Rupp, Abstracts of
     Papers, 125th Natl. Meeting, American Chemical Soc.,
     Kansas City, Mo., Mar. 24-Apr. 1, 1954, p. 3C.
241. F. Bischoff, G. Lopez, J.J. Rupp, and C.L. Gray,
     Federation Proc., 14, 183 (1955).
242. F. Bischoff and J.J. Rupp, Cancer Res., 6, 403 (1946).
243. F. Bischoff, J.J. Rupp, J.G. Turner, and G. Bryson,
     Proc. Am. Assoc. Cancer Res., 2, 281 (1958).
244. F. Bischoff, R.D. Stauffer, and G. Bryson, Federation
     Proc., 32, 677Abs (1973).
245. F. Bischoff and J.G. Turner, Clin. Chem., 4, 300 (1958).
246. F. Bischoff, J.G. Turner, and G. Bryson, Abstracts,
     134th Natl. Meeting, American Chemical Soc., Chicago,
     Ill., Sept. 7-12, 1958, p. 54C.
247. G. Bischoff, Z. Exp. Med., 70, 83 (1930).
248. R.H. Bishara and P.L. Schiff, Lloydia, 33, 477 (1970).
249. S. Biswas, J.D.B. MacDougall, and R.P. Cook. Brit. J.
     Exptl. Pathol., 45, 13 (1964).
250. I. Björkhem, Eur. J. Biochem., 8, 337 (1969).
251. I. Björkhem, Eur. J. Biochem., 18, 299 (1971).
252. I. Björkhem, R. Blomstrand, and L. Svensson, Clin.
     Chim. Acta, 54, 185 (1974).
253. I. Björkhem, R. Blomstrand, and L. Svensson, J. Lipid
     Res., 19, 359 (1978).
254. I. Björkhem and H. Danielsson, Anal. Biochem., 59,
     508 (1974).
255. I. Björkhem and H. Danielsson, Molecular Cell Biochem.,
     4, 79 (1974).
256. I. Björkhem and H. Danielsson, Eur. J. Biochem., 53,
     63 (1975).
257. I. Björkhem, H. Danielsson, and K. Einarsson, Eur. J.
     Biochem., 2, 294 (1967).
258. I. Björkhem, H. Danielsson, and K. Einarsson, Eur. J.
     Biochem., 4, 458 (1968).
259. I. Björkhem, H. Danielsson, K. Einarsson, and G.
     Johansson, J. Clin. Invest., 47, 1573 (1968).
260. I. Björkhem, H. Danielsson, and J. Gustafsson, FEBS
     Lett., 31, 20 (1973).

261. I. Björkhem, H. Danielsson, and K. Wikvall, Eur. J.
     Biochem., 36, 8 (1973).
262. I. Björkhem, H. Danielsson, and K. Wikvall, Biochem.
     Biophys. Res. Commun., 61, 934 (1974).
263. I. Björkhem, H. Danielsson, and K. Wikvall, Biochem.
     Soc. Trans., 3, 825 (1975).
264. I. Björkhem, K. Einarsson, J.-Å. Gustafsson, and A.
     Somell, Acta Endocrinol., 71, 569 (1972).
265. I. Björkhem, K. Einarsson, and G. Johansson, Acta
     Chem. Scand., 22, 1595 (1968).
266. I. Björkhem and J. Gustafsson, Eur. J. Biochem., 18,
     207 (1971).
267. I. Björkhem and J. Gustafsson, Eur. J. Biochem., 36,
     201 (1973).
268. I. Björkhem and J. Gustafsson, J. Biol. Chem., 249,
     2528 (1974).
269. I. Björkhem, J. Gustafsson, G. Johansson, and B. Pers-
     son, J. Clin. Invest., 55, 478 (1975).
270. I. Björkhem, J.-Å. Gustafsson, and O. Wrange, Eur. J.
     Biochem., 37, 143 (1973).
271. I. Björkhem and I. Holmberg, J. Biol. Chem., 253, 842
     (1978).
272. I. Björkhem and I. Holmberg, J. Biol. Chem., 254, 9518
     (1979).
273. I. Björkhem and K.-E. Karlmar, Biochim. Biophys. Acta,
     337, 129 (1974).
274. H.S. Black, J. Am. Oil Chem. Soc., 57, 133A (1980).
275. H.S. Black and J.T. Chan, Oncology, 33, 119 (1976).
276. H.S. Black and D.R. Douglas, Cancer Res., 32, 2630
     (1972).
277. H.S. Black and D.R. Douglas, Cancer Res., 33, 2094
     (1973).
278. H.S. Black, S.V. Henderson, C.M. Kleinhans, A.W.
     Phelps, and J.I. Thornby, Cancer Res., 39, 5022 (1979).
279. H.S. Black and J.L. Laseter, Comp. Biochem. Physiol.,
     53C, 29 (1976).
280. H.S. Black and W.A. Lenger, Anal. Biochem., 94, 383
     (1979).
281. H.S. Black and W.-B. Lo, Nature, 234, 306 (1971).
282. H.S. Black, Y. Tsurumaru, and W.-B. Lo, Res. Commun.
     Chem. Pathol. Pharmacol., 10, 177 (1975).
283. G.M. Blackburn, A. Rashid, and M.H. Thompson, J. Chem.
     Soc. Chem. Commun., 420 (1979).
284. P. Bladon, J. Chem. Soc., 2176 (1955).
285. P. Bladon, in CHOLESTEROL CHEMISTRY, BIOCHEMISTRY, AND
     PATHOLOGY, R.P. Cook, Editor, Academic Press Inc.,

New York City, 1958, pp. 15-115.

286. P. Bladon, R.B. Clayton, C.W. Greenhalgh, H.B. Henbest, E.R.H. Jones, B.J. Lovell, G. Silverstone, G.W. Wood, and G.F. Woods, J. Chem. Soc., 4883 (1952).

287. P. Bladon, J.M. Fabian, H.B. Henbest, H.P. Koch, and G.W. Wood, J. Chem. Soc., 2402 (1951).

288. P. Bladon, H.B. Henbest, E.R.H. Jones, B.J. Lovell, and G.F. Woods, J. Chem. Soc., 125 (1954).

289. P. Bladon, H.B. Henbest, E.R.H. Jones, G.W. Wood, and G.F. Woods, J. Chem. Soc., 4890 (1952).

290. P. Bladon, H.B. Henbest, and G.W. Wood, Chem. Ind., 866 (1951); J. Chem. Soc., 2737 (1952).

291. P. Bladon and T. Sleigh, J. Chem. Soc., 6991 (1965).

292. J. Bland and B. Craney, Tetrahedron Lett., 4041 (1974).

293. G. Bleau, F.H. Bodley, J. Longpré, A. Chapdelaine, and K.D. Roberts, Biochim. Biophys. Acta, $\underline{352}$, 1 (1974).

294. G. Bleau, G. Lalumière, A. Chapdelaine, and K.D. Roberts, Biochim. Biophys. Acta, $\underline{375}$, 220 (1975).

295. G. Blix, J. Biol. Chem., $\underline{137}$, 495 (1941).

296. G. Blix and G. Löwenhielm, Biochem. J., $\underline{22}$, 1313 (1928).

297. S.H. Blobstein, D. Grunberger, I.B. Weinstein, and K. Nakanishi, Biochemistry $\underline{12}$, 188 (1973).

298. J.H. Block, Steroids, $\underline{23}$, 421 (1974).

299. R. Blomstrand and J. Gürtler, Arkiv Kemi, $\underline{30}$, 233 (1968).

300. G.A. Blondin, B.D. Kulkarni, J.P. John, R.T. van Aller, P.T. Russell, and W.R. Nes, Anal. Chem., $\underline{39}$, 36 (1967).

301. G.A. Blondin, B.D. Kulkarni, and W.R. Nes, J. Am. Chem. Soc., $\underline{86}$, 2528 (1964).

302. D.K. Bloomfield, Anal. Chem., $\underline{34}$, 737 (1962).

303. E.R. Blout, M. Parrish, G.R. Bird, and M.J. Abbate, J. Opt. Soc. Am., $\underline{42}$, 966 (1952).

304. H.F. Blum, J. Natl. Cancer Inst., $\underline{1}$, 397 (1940).

305. G. Blunden, R. Hardman, and J.C. Morrison, J. Pharm. Sci., $\underline{56}$, 948 (1967).

306. G. Blunden and C.T. Rhodes, J. Pharm. Sci., $\underline{57}$, 602 (1968).

307. J.W. Blunt, Austr. J. Chem., $\underline{28}$, 1017 (1975).

308. J.W. Blunt and J.B. Strothers, Org. Magnetic Resonance, $\underline{9}$, 439 (1977).

309. R.B. Boar, D.A. Lewis, and J.F. McGhie, J. Chem. Soc., Perkin Trans. I, 2331 (1972).

310. K. Bodendorf, Arch. Pharm., $\underline{271}$, 1 (1933).

311. H. Bogren, Acta Radiol., Suppl. 226, 1 (1964).

312. H. Bogren and K. Larsson, Biochim. Biophys. Acta, 75, 65 (1963).

313. P. Boldrini, Physiol. Chem. Physics, 10, 383 (1978).

314. A. Bömer, Z. Untersuch. Nahr. Genuss., 1, 21 (1898).

315. A. Bömer, Z. Untersuch. Nahr. Genuss., 1, 81 (1898).

316. St. Bondzyński and V. Humnicki, Z. Physiol. Chem., 22, 408 (1896/1897).

317. J.J. Bonet-Surgrañes, J. Boix, and S. Aguilá, Afinidad, 26, 371 (1969); via Chem. Abstr., 71, 124778v (1969).

318. A.M. Bongiovanni, J. Clin. Endocrinol. Metab., 26, 1240 (1966).

319. F. Boomsma, H.J.C. Jacobs, E. Havinga, and A. van der Gen, Rec. Trav. Chim., 92, 1361 (1973).

320. F. Boomsma, H.J.C. Jacobs, E. Havinga, and A. van der Gen, Rec. Trav. Chim., 96, 104 (1977).

321. J.J. Boon, J.W. de Leeuw, and A.L. Burlingame, Geochim. Cosmochim. Acta, 42, 631 (1978).

322. J.J. Boon, W.I.C. Rijpstra, F. De Lang, J.W. De Leeuw, M. Yoshioka, and Y. Shimizu, Nature, 277, 125 (1979).

323. W. Bors, M. Saran, E. Lengfelder, R. Spöttl, and C. Michel, Curr. Topics Radiation Res. Quart., 9, 247(1974).

324. M. Bortolotto, J.C. Braekman, D. Daloze, D. Losman, and B. Tursch, Steroids, 28, 461 (1976).

325. M. Bortolotto, J.C. Braekman, D. Daloze, and B. Tursch, Bull. Soc. Chim. Belge, 85, 27 (1976); via Chem. Abstr., 84, 161991y (1976).

326. M. Bortolotto, J.C. Braekman, D. Daloze, B. Tursch, and R. Karlsson, Steroids, 30, 159 (1977).

327. E. Bosisio, G. Galli, S. Nicosia, and M. Galli Kienle, Eur. J. Biochem., 63, 491 (1976).

328. K.M. Botham, G.J. Beckett, I.W. Percy-Robb, and G.S. Boyd, Eur. J. Biochem., 103, 299 (1980).

329. J. Bourdon and B. Schnuriger, in PHYSICS AND CHEMISTRY OF THE ORGANIC SOLID STATE, D. Fox, M.M. Labes, and A. Weissberger, Editors, Interscience Publishers, New York City, N.Y., Vol. 3, 1967, p. 96.

330. R. Bourdon, Bull. Soc. Chim. France, 1117 (1958).

331. R. Bourdon and S. Ranisteano, Bull. Soc. Chim. France, 1982 (1960).

332. J.-L. Boutry, A. Alcaide, and M. Barbier, Compt. Rend., [D], 272, 1022 (1971).

333. J.-L. Boutry and C. Baron, Bull. Soc. Chim. Biol., 49, 157 (1967).

334. J.L. Boutry, A. Saliot, and M. Barbier, Experientia, 35, 1541 (1979).

335. J.P. Bowden, G.M. Muschik, and J.C. Kawalek, Lipids,

$\underline{14}$, 623 (1979).

336. D.V. Bowen and F.H. Field, Org. Mass Spectrometry, $\underline{9}$, 195 (1974).

337. E.M. Boyd, J. Biol. Chem., $\underline{121}$, 485 (1937).

338. G.S. Boyd, Federation Proc., $\underline{21}$, Suppl. II, 86 (1962).

339. G.S. Boyd, J.R. Arthur, G.J. Beckett, J.I. Mason, and W.H. Trzeciak, J. Steroid Biochem., $\underline{6}$, 427 (1975).

340. G.S. Boyd, M.J.G. Brown, N.G. Hattersley, and K.E. Suckling, Biochim. Biophys. Acta, $\underline{337}$, 132 (1974).

341. G.S. Boyd, A.C. Brownie, C.R. Jefcoate, and E.R. Simpson, Biochem. J., $\underline{125}$, 1P (1971).

342. G.S. Boyd, A.M. Grimwade, and M.E. Lawson, Eur. J. Biochem., $\underline{37}$, 334 (1973).

343. G.S. Boyd and E.B. Mawer, Biochem. J., $\underline{81}$, 11P (1961).

344. K.R. Brain and R. Hardman, J. Chromatog., $\underline{38}$, 355 (1968).

345. E. Brand, E. Washburn, B.F. Erlanger, E. Ellenbogen, J. Daniel, F. Lippmann, and M. Scheu, J. Am. Chem. Soc., $\underline{76}$, 5037 (1954).

346. K. Brandau and U. Keup, Arzneimittel Forsch., $\underline{26}$, 1837 (1976).

347. R.D. Brandt, G. Ourisson, and R.J. Pryce, Biochem. Biophys. Res. Commun., $\underline{37}$, 399 (1969).

348. W. Brandt, Biochem. Z., $\underline{288}$, 257 (1936).

349. I.A. Breger, J. Phys. Colloid Chem., $\underline{52}$, 551 (1948).

350. G. Brenner, F.E. Roberts, A. Hoinowski, J. Budavari, B. Powell, D. Hinkley, and E. Schoenewaldt, Angew. Chem., Int. Ed. $\underline{8}$, 975 (1969).

351. J.L. Breslow, D.A. Lothrop, D.R. Spaulding, and A.A. Kandutsch, Biochim. Biophys. Acta, $\underline{398}$, 10 (1975).

352. H. Bretschneider and M. Ajtai, Sitzber. Akad. Wiss. Wien, [IIb], $\underline{150}$, 131 (1941); Monatsh., $\underline{74}$, 57 (1944).

353. F.L. Breusch, Biochem. Z., $\underline{293}$, 280 (1937).

354. J.H. Brewster, J. Am. Chem. Soc., $\underline{81}$, 5493 (1959).

355. C.H. Brieskorn and G. Dertinger, Tetrahedron Lett. 6237 (1968).

356. C.H. Brieskorn and G. Dertinger, Arch. Pharm., $\underline{303}$, 960 (1970).

357. C.H. Brieskorn and G. Dertinger, Arch. Pharm., $\underline{303}$, 968 (1970).

358. C.H. Brieskorn and H. Hofmann, Arch. Pharm., $\underline{297}$, 577 (1964).

359. British Patent No. 607,309, Aug. 27, 1948; via Chem. Abstr., $\underline{43}$, 1158e (1949).

360. THE BRITISH PHARMACOPOEIA 1973, Her Majesty's Stationary Office, London, 1973, p. A17.

361. C.J.W. Brooks, in RODD'S CHEMISTRY OF CARBON COMPOUNDS,
     A MODERN COMPREHENSIVE TREATISE, S. Coffey, Editor,
     2nd Ed., Elsevier Publishing Co., Amsterdam, 1970,
     Vol. II, Part D, p. 96, pp. 164-166.
362. C.J.W. Brooks, J.D. Gilbert, and W.A. Harland, Protides
     Biol. Fluids, 19, 101 (1972).
363. C.J. W. Brooks and L. Hanainah, Biochem. J., 87, 151
     (1963).
364. C.J.W. Brooks, W.A. Harland, and G. Steel, Biochim.
     Biophys. Acta, 125, 620 (1966).
365. C.J.W. Brooks, W.A. Harland, G. Steel, and J.D. Gilbert,
     Biochim. Biophys. Acta, 202, 563 (1970).
366. C.J.W. Brooks, W. Henderson, and G. Steel, Biochim.
     Biophys. Acta, 296, 431 (1973).
367. C.J.W. Brooks, E.C. Horning, and J.S. Young, Lipids,
     3, 391 (1968).
368. C.J.W. Brooks and A.G. Smith, J. Chromatog., 112, 499
     (1975).
369. C.J.W. Brooks, G. Steel, J.D. Gilbert, and W.A. Har-
     land, Atherosclerosis, 13, 223 (1971).
370. C.J.W. Brooks, G. Steel, and W.A. Harland, Lipids, 5,
     818 (1970).
371. B.W.L. Brooksbank and D.A. Wilson, Steroidologia, 1,
     113 (1970).
372. B.W.L. Brooksbank, D.A.A. Wilson, and J.-Å. Gustafsson,
     Steroids Lipids Res., 3, 263 (1972).
373. P.J. Brophy and D.B. Gower, Biochem. J., 128, 945
     (1972).
374. H.H. Brown, A. Zlatkis, B. Zak, and A.J. Boyle, Anal.
     Chem., 26, 397 (1954).
375. J.J. Brown and S. Bernstein, Steroids, 8, 87 (1966).
376. M.J.G. Brown and G.S. Boyd, Eur. J. Biochem., 44, 37
     (1974).
377. M.S. Brown, P.G. Brannan, H.A. Bohmfalk, G.Y. Bruns-
     chede, S.E. Dana, J. Helgeson, and J.L. Goldstein, J.
     Cell Physiol., 85, 425 (1975).
378. M.S. Brown, S.E. Dana, and J.L. Goldstein, J. Biol.
     Chem., 250, 4025 (1975).
379. M.S. Brown, J.R. Faust, and J.L. Goldstein, J. Clin.
     Invest., 55, 783 (1975).
380. M.S. Brown, J.R. Faust, J.L. Goldstein, I. Kaneko, and
     A. Endo, J. Biol. Chem., 253, 1121 (1978).
381. M.S. Brown and J.L. Goldstein, J. Biol. Chem., 249,
     7306 (1974).
382. M.S. Brown and J.L. Goldstein, Cell, 6,307 (1975).
383. R.L. Brown and G.E. Peterson, J. Gen. Microbiol., 45,

441 (1966).

384. R.R. Brown, J.A. Miller, and E.C. Miller, J. Biol.
     Chem., 209, 211 (1954).
385. K.R. Bruckdorfer, R.A. Demel, J. de Gier, and L.L. van
     Deenen, Biochim. Biophys. Acta, 183, 334 (1969).
386. K.R. Bruckdorfer, J.M. Graham, and C. Green, Eur. J.
     Biochem., 4, 512 (1968).
387. T. Brunn and A.-M. Motzfeldt, Acta Chem. Scand., 29B,
     274 (1975).
388. G. Bruns, K. Schubert, W. Zschiesche, and G. Rose,
     Arch. Geschwulstforsch., 22, 52 (1963).
389. G.T. Bryan, R.R. Brown, and J.M. Price, Ann. N.Y.
     Acad. Sci., 108, 924 (1963).
390. G.T. Bryan, R.R. Brown, and J.M. Price, Cancer Res.,
     24, 596 (1964).
391. M.J. Bryson and M.L. Sweat, J. Biol. Chem., 243, 2799
     (1968).
392. J. Buckingham and R.D. Guthrie, J. Chem. Soc.,(C),
     1445 (1968).
393. B.C. Buckland, P. Dunnill, and M.D. Lilly, Biotech.
     Bioengineering, 17, 815 (1975).
394. B.C. Buckland, M.D. Lilly, and P. Dunnill, Biotech.
     Bioengineering, 18, 601 (1976).
395. J.A. Buege and S.D. Aust, Biochim. Biophys. Acta, 444,
     192 (1976).
396. J.F. Bunnett and L.A. Retallick, J. Am. Chem. Soc.,
     89, 423 (1967).
397. H. Burchard, Inaugural Diss., Rostock, 1890; via Chem.
     Zentrl., 61, 25 (1890).
398. J.J. Buresh and T.J. Urban, J. Dental Res., 43, 548
     (1964).
399. J.J. Buresh and T.J. Urban, Arch. Oral Biol., 12, 1221
     (1967).
400. A.W. Burgstahler, J. Am. Chem. Soc., 79, 6047 (1957).
401. L.I. Burke, G.S. Patil, R.V. Panganamala, J.C. Geer,
     and D.G. Cornwell, J. Lipid Res., 14, 9 (1973).
402. A.L. Burlingame, P. Haug, T. Belsky, and M. Calvin,
     Proc. Nat. Acad. Sci., U.S., 54, 1406 (1965).
403. E.P. Burrows, G.M. Hornby, and E. Caspi, J. Org. Chem.,
     34, 103 (1969).
404. H. Burrows and W.V. Mayneord, Am. J. Cancer, 31, 484
     (1937).
405. S. Burstein, C.Y. Byon, H.L. Kimball, and M. Gut,
     Steroids, 27, 691 (1976).
406. S. Burstein, N. Co, M. Gut, H. Schleyer, D.Y. Cooper,
     and O. Rosenthal, Biochemistry, 11, 573 (1972).

407. S. Burstein, J. Dinh, N. Co, M. Gut, H. Schleyer, D.Y.
     Cooper, and O. Rosenthal, Biochemistry, 11, 2883,4012
     (1972).
408. S. Burstein and M. Gut, Steroids, 14, 207 (1969).
409. S. Burstein and M. Gut, Rec. Progr. Hormone Res., 27,
     303 (1971).
410. S. Burstein and M. Gut, Ann. N.Y. Acad. Sci., 212, 262
     (1973).
411. S. Burstein and M. Gut, Steroids, 28, 115 (1976).
412. S. Burstein, M. Gut, and N. Co, Federation Proc., 30,
     1091Abs (1971).
413. S. Burstein, H.L. Kimball, and M. Gut, Steroids, 15,
     809 (1970).
414. S. Burstein, B.S. Middleditch, and M. Gut, Biochem.
     Biophys. Res. Commun., 61, 692 (1974).
415. S. Burstein, B.S. Middleditch, and M. Gut, J. Biol.
     Chem., 250, 9028 (1975).
416. S. Burstein, H. Zamoscianyk, H.L. Kimball, N.K.
     Chaudhuri, and M. Gut, Steroids, 15, 13 (1970).
417. S.H. Burstein, F.G. Péron, and E. Williamson, Steroids,
     13, 399 (1969).
418. J.S. Bus, S.D. Aust, and J.E. Gibson, Biochem. Biophys.
     Res. Commun., 58, 749 (1974).
419. W. Buser, Helv. Chim. Acta, 30, 1379 (1947).
420. I.E. Bush, THE CHROMATOGRAPHY OF STEROIDS, Pergamon
     Press, Ltd., Oxford, 1961, pp. 91-92, p. 379.
421. P. Busse, Z. Physiol. Chem., 214, 211 (1933).
422. A. Butenandt and H. Dannenbaum, Z. Physiol. Chem.,
     248, 151 (1937).
423. A. Butenandt and H. Dannenberg, Naturwiss., 30, 52
     (1942).
424. A. Butenandt and E. Hausmann, Ber., 70, 1154 (1937).
425. A. Butenandt and H. Kudssus, Z. Physiol. Chem., 253, I
     (1938).
426. A. Butenandt and J. Paland, Ber., 72, 424 (1939).
427. K.W. Butler, I.C.P. Smith, and H. Schneider, Biochim.
     Biophys. Acta, 219, 514 (1970).
428. C.-Y. Byon, G. Büyüktür, P. Choay, and M. Gut, J. Org.
     Chem., 42, 3619 (1977). |
429. C.-Y. Byon, H.L. Kimball, and M. Gut, Steroids, 30,
     419 (1977).
430. E. Cabre, J. de Bolas, and S. Garcia, J. Chim. Phys.,
     76, 983 (1979).
431. J.C. Cain and R.A. Morton, Biochem. J., 60, 274
     (1955).
432. R.K. Callow, Biochem. J., 25, 79 (1931).

433. R.K. Callow and F.G. Young, Proc. Royal Soc. London, [A], 157, 194 (1936).
434. B. Camarino, B. Patelli, and R. Sciaky, Tetrahedron Lett., 554 (1961).
435. B. Camarino, B. Patelli, and R. Sciaky, Gazz. Chim. Ital., 92, 676 (1962).
436. B. Camarino, B. Patelli, and R. Sciaky, Gazz. Chim. Ital., 92, 693 (1962).
437. A.C. Campbell, J. McLean, and W. Lawrie, Tetrahedron Lett., 483 (1969).
438. J.A. Campbell, D.M. Squires, and J.C. Babcock, Steroids, 13, 567 (1969).
439. S. Cannistraro and A. Van de Vorst, Biochem. Biophys. Res. Commun., 74, 1177 (1977).
440. L. Canonica, N. Pacini, C. Scolastico, and U. Valcavi, Gazz. Chim. Ital., 93, 787 (1963).
441. M.C. Carey and D.M. Small, J. Clin. Invest., 61, 998 (1978).
442. D.I. Cargill, Analyst, 87, 865 (1962).
443. D.I. Cargill and R.P. Cook, Biochem. J., 93, 504 (1964).
444. R.M.K. Carlson, S. Popov, I. Massey, C. Delseth, E. Ayanoglu, T.H. Varkony, and C. Djerassi, Bioorganic Chem., 7, 453 (1978).
445. M.G. Caron, S. Goldstein, K. Savard, and J.M. Marsh, J. Biol. Chem., 250, 5137 (1975).
446. E. Caspi, M. Galli Kienle, K.R. Varma, and L.J. Mulheirn, J. Am. Chem. Soc., 92, 2161 (1970).
447. E. Caspi, J.B. Greig, P.J. Ramm, and K.R. Varma, Tetrahedron Lett., 3829 (1968).
448. W.K. Cavenee, G.F. Gibbons, H.W. Chen, and A.A. Kandutsch, Biochim. Biophys. Acta, 575, 255 (1979).
449. W.K. Cavenee, D. Johnston, and G. Melnykovych, Proc. Natl. Acad. Sci., 75, 2103 (1978).
450. P. Ceccherelli, R. Fringuelli, and G.F. Madruzza, Phytochem., 14, 1434 (1975).
451. V. Černý, A. Kasal, and F. Šorm, Coll. Czech. Chem. Commun., 35, 1235 (1970).
452. V. Černý, A. Trka, J. Kohoutová, J. Smolíková, and M. Budešínský, Coll. Czech. Chem. Commun., 41, 2788 (1976).
453. M.S. Chadha, N.K. Joshi, V.R. Mamdapur, and A.T. Sipahimalani, Tetrahedron, 26, 2061 (1970).
454. A. Chait, E.L. Bierman, and J.J. Albers, J. Clin. Invest., 64, 1309 (1979).
455. J.T. Chan and H.S. Black, Science, 186, 1216 (1974).

456. J.T. Chan and H.S. Black, Experientia, 34, 110 (1978).
457. W. Charney and H.L. Herzog, MICROBIAL TRANSFORMATIONS
     OF STEROIDS. A HANDBOOK, Academic Press, New York/
     London, 1967, pp. 19-22.
458. A.C. Chaudhuri, Y. Harada, K. Shimizu, M. Gut, and
     R.I. Dorfman, J. Biol. Chem., 237, 703 (1962).
459. N.K. Chaudhuri, R.C. Nickolson, and M. Gut, Steroids,
     16, 495 (1970).
460. R. Chaudhri, G. Cooley, and W.F. Coulson, Steroids,
     31, 495 (1978).
461. A.S. Chawla, V.K. Kapoor, and P.K. Sangal, Planta Med.,
     34, 109 (1978).
462. C. Chen, Abstracts, 9th Int. Congress Biochem., Stock-
     holm, July 1-7, 1973, p. 342.
463. C. Chen and C.-C. Lin, Biochim. Biophys. Acta, 170,
     366 (1968).
464. C. Chen and C.-C. Lin, Biochim. Biophys. Acta, 184,
     634 (1969).
465. C. Chen and Y.-S. Shaw, Biochem. J., 144, 497 (1974).
466. C. Chen and M.-H. Tu, Biochem. J., 160, 805 (1976).
467. H.W. Chen, W.K. Cavanee, and A.A. Kandutsch, J. Biol.
     Chem., 254, 715 (1979).
468. H.W. Chen, H.-J. Heiniger, and A.A. Kandutsch, Proc.
     Natl. Acad. Sci., 72, 1950 (1975).
469. H.W. Chen, H.-J. Heiniger, and A.A. Kandutsch, J. Biol.
     Chem., 253, 3180 (1978).
470. H.W. Chen and A.A. Kandutsch, Artery, 2, 287 (1976).
471. H.W. Chen and A.A. Kandutsch, in ATHEROSCLEROSIS DRUG
     DISCOVERY (1976), C.E. Day, Editor, Plenum Publishing
     Corp., New York City, N.Y., 1976, pp. 405-417.
472. H.W. Chen, A.A. Kandutsch, and H.J. Heiniger, Progress
     Exptl. Tumor Res., 22, 275 (1978).
473. H.W. Chen, A.A. Kandutsch, and C. Waymouth, Nature
     251, 419 (1974).
474. P.S. Chen, A.R. Terepka, K. Lane, and A. Marsh, Anal.
     Biochem., 10, 421 (1965).
475. K.-P. Cheng, Luu Bang, G. Ourisson, and J.-P. Beck,
     J. Chem. Research, (S), 84(1979); (M), 1101(1979).
476. K.P. Cheng, H. Nagano, Luu Bang, G. Ourisson, and
     J.-P. Beck, J. Chem. Research (S),217 (1977), (M),
     2501 (1977).
477. M.E. Chevreul, Ann. Chimie, [1], 95, 5 (1815).
478. M.E. Chevreul, Ann. Chim. Phys., [2], 2, 339 (1816).
479. E. Chicoye, W.D. Powrie, and O. Fennema, Lipids, 3,
     335 (1968).
480. E. Chicoye, W.D. Powrie, and O. Fennema, Lipids, 3,

551 (1968).

481. E. Chicoye, W.D. Powrie, and O. Fennema, J. Food Sci.,
     33, 581 (1968).

482. L.J. Chinn, J. Org. Chem., 29, 3304 (1964).

483. H.E. Christensen, T.T. Luginbyhl, and B.S. Carroll,
     SUSPECTED CARCINOGENS, A SUBFILE OF THE NIOSH TOXIC
     SUBSTANCES LIST, U.S. Public Health Service, Washington
     City, D.C., 1975, pp. 97-98.

484. T. Chulski and A.A. Forist, J. Am. Pharm. Assoc., Sci.
     Ed., 47, 553 (1958).

485. P.F. Churchill and T. Kimura, J. Biol. Chem., 254,
     10443 (1979).

486. J. Churý, Chem. Obzor, 21, 37 (1946); via Chem. Abstr.,
     42, 7405^i (1948).

487. A. Cier, C. Nofre, and A. Revol, Compt. Rend., 247,
     542 (1958).

488. G. Cimino, B. Desiderio, S. De Stefano, and G. Sodano,
     Experientia, 35, 298 (1979).

489. R. Clarenburg, I.L. Chaikoff, and M.D. Morris, J.
     Neurochem., 10, 135 (1963).

490. S.M. Clarke and M. McKenzie, Nature, 213, 504 (1967).

491. J.R. Claude, J. Chromatog., 17, 596 (1965).

492. J.R. Claude, J. Chromatog., 23, 267 (1966).

493. J.R. Claude, Clin. Chim. Acta, 17, 371 (1967).

494. J.R. Claude and J.L. Beaumont, Ann. Biol. Clin., 22,
     815 (1964).

495. J.R. Claude and J.L. Beaumont, Compt. Rend., 260, 3204
     (1965).

496. J.R. Claude and J.L. Beaumont, J. Chromatog., 21, 189
     (1966).

497. D.B. Clayson, J.W. Jull, and G.M. Bonser, Brit. J.
     Cancer, 12, 222 (1958).

498. R.B. Clayton, A. Crawshaw, H.B. Henbest, E.R.H. Jones,
     B.J. Lovell, and G.W. Wood, J. Chem. Soc., 2009 (1953).

499. R.B. Clayton, H.B. Henbest, and E.R.H. Jones, J. Chem.
     Soc., 2015 (1953).

500. G.R. Clemo, M. Keller, and J. Weiss, J. Chem. Soc.,
     3470 (1950).

501. A. Cohen, NATIONAL BUREAU OF STANDARDS TECHNICAL NOTE
     457, R. Schaffer, Editor, U.S. Dept. Commerce, Washing-
     ton City, D.C., 1968, pp. 73-90.

502. B.I. Cohen, T. Kuramoto, M.A. Rothschild, and E.H.
     Mosbach, J. Biol. Chem., 251, 2709 (1976).

503. B. Coleby, M. Keller, and J. Weiss, J. Chem. Soc.,
     66 (1954).

504. D.L. Coleman and C.A. Baumann, Arch. Biochem. Biophys.,

66, 226 (1957).

505. H. Colin, G. Guiochon, and A. Siouffi, Anal. Chem.,
     51, 1661 (1979).
506. R.J. Conca and W. Bergmann, J. Org. Chem., 18, 1104
     (1953).
507. R.L. Conner, J.R. Landrey, J.M. Joseph, and W.R. Nes,
     Lipids, 13, 692 (1978).
508. J.P. Connolly, P.J. Flanagan, R.O. Dorchaí, and J.B.
     Thompson, J. Chromatog., 15, 105 (1964).
509. J.P. Connolly, A.R. Manning, and J.B. Thomson, J. Chem.
     Soc., (C), 1773 (1967).
510. B.G.F. Conradi, Dissertation, Jena, 1775; via
     THESAURUS DISSERTATIONIUM MEDICARUM RARIORUM, Heidel-
     berg, 1784, Vol. 1, pp. 191-211.
511. G. Constantopoulos and T.T. Tchen, Biochem. Biophys.
     Res. Commun., 4, 460 (1961).
512. A.H. Cook, J. Chem. Soc., 1774 (1938).
513. J.W. Cook and G.A.D. Haslewood, Chem. Ind., 11, 758
     (1933); J. Chem. Soc., 428 (1934).
514. R.P. Cook, CHOLESTEROL, CHEMISTRY, BIOCHEMISTRY, AND
     PATHOLOGY, Academic Press Inc., New York, 1958, p. 494.
515. R.P. Cook, Analyst, 86, 373 (1961).
516. R.P. Cook, A. Kliman, and L.F. Fieser, Arch. Biochem.
     Biophys., 52, 439 (1954).
517. R.P. Cook and J.D.B. MacDougall, Brit. J. Exptl.
     Pathol., 49, 265 (1968).
518. G.M. Cooke and D.B. Gower, Biochim. Biophys. Acta,
     498, 265 (1977).
519. R.G. Cooks, R.D. Daftary, and Y. Pomeranz, Ag. Food
     Chem., 18, 620 (1970).
520. G.R. Cooper, J.S. Hazlehurst, and P.H. Duncan, Clin.
     Chem., 23, 1153 (1977).
521. J.W. Copius-Peereboom and H.W. Beekes, J. Chromatog.,
     9, 316 (1962).
522. J.W. Copius-Peereboom and H.W. Beekes, J. Chromatog.,
     17, 99 (1965).
523. R. Cornell, G.L. Grove, G.H. Rothblat, and A.F.
     Horwitz, Exptl. Cell Res., 109, 299 (1977).
524. J.W. Cornforth, G.D. Hunter, and G. Popják, Biochem.
     J., 54, 590 (1953).
525. R.A.F. Couch, S.J.M. Skinner, C.J.P. Tobler, and T.W.
     Doouss, Steroids, 26, 1 (1975).
526. R.E. Counsell, M.C. Lu, S. El Masry, and P.A. Weinhold,
     Biochem. Pharmacol., 20, 2912 (1971).
527. J.L. Cowell and A.W. Bernheimer, Arch. Biochem. Biop-
     phys., 190, 603 (1978).

528. A.J. Cox, J. Org. Chem., 30, 2052 (1965).
529. R.H. Cox and E.Y. Spencer, Can. J. Chem., 29, 398
     (1951).
530. T.A. Crabb, P.J. Dawson, and R.O. Williams, Tetra-
     hedron Lett., 3623 (1975).
531. P. Crabbé and K. Mislow, J. Chem. Soc. Chem. Commun.,
     657 (1968).
532. I.F. Craig, G.S. Boyd, R.J. McLeod, and K.E. Suckling,
     Biochem. Soc. Trans., 7, 967 (1979).
533. I.F. Craig, G.S. Boyd, and K.E. Suckling, Biochim.
     Biophys. Acta, 508, 418 (1978).
534. P.A. Cranwell, Geochim. Cosmochim. Acta, 42, 1523
     (1978).
535. B.M. Craven, Nature, 260, 727 (1976).
536. R. Criegee, B. Marchand, and H. Wannowius, Liebigs Ann.,
     550, 99 (1942).
537. A.D. Cross, J. Am. Chem. Soc., 84, 3206 (1962).
538. W.D. Crow, W. Nicholls, and M. Sterns, Tetrahedron
     Lett., 1353 (1971).
539. J.P. Cruse, M.R. Lewin, G.P. Ferulano, and C.G. Clark,
     Nature, 276, 822 (1978).
540. I.M. Cunningham and K.H. Overton, J. Chem. Soc.,
     Perkin Trans. I, 2458 (1974).
541. F. Dalton and G.D. Meakins, J. Chem. Soc., 1880 (1961).
542. D. Daniel, E. Lederer, and L. Velluz, Bull. Soc. Chim.
     Biol., 27, 218 (1945).
543. H. Danielsson, Acta Chem. Scand., 14, 846 (1960).
544. H. Danielsson, Arkiv Kemi, 17, 363 (1961).
545. H. Danielsson, Arkiv Kemi, 17, 373 (1961).
546. H. Danielsson, Arkiv Kemi, 17, 381 (1961).
547. H. Danielsson, Steroids, 20, 63 (1972).
548. H. Danielsson and K. Einarsson, Acta Chem. Scand., 18,
     831 (1964).
549. H. Danielsson and K. Einarsson, J. Biol. Chem., 241,
     1449 (1966).
550. H. Danielsson and G. Johansson, Acta Chem. Scand., 18,
     788 (1964).
551. H. Danielsson and G. Johansson, FEBS Lett., 25, 329
     (1972).
552. H. Danielsson and K. Wikvall, FEBS Lett., 66, 299
     (1976).
553. B. Danilewsky, Arch. Gesam. Physiol., 120, 181 (1907).
554. H. Dannenberg, Deutsch. Med. Woch., 83, 1726 (1958).
555. H. Dannenberg, Z. Krebsforsch., 63, 523 (1960).
556. H. Dannenberg and K.-F. Hebenbrock, Z. Physiol. Chem.,
     342, 199 (1965).

557. H. Dannenberg, H.-G. Neumann, and D. Dannenberg-von
     Dresler, Liebigs Ann., 674, 152 (1964).
558. L. Darmstaedter and I. Lifschütz, Ber., 28, 3133 (1895).
559. L. Darmstaedter and I. Lifschütz, Ber., 29, 1474 (1896).
560. L. Darmstaedter and I. Lifschütz, Ber., 29, 2890 (1896).
561. L. Darmstaedter and I. Lifschütz, Ber., 31, 1122 (1898).
562. S.K. Dasgupta, D.R. Crump, and M. Gut, J. Org. Chem.,
     39, 1658 (1974).
563. M. Dastillung and P. Albrecht, Nature, 269, 678 (1977).
564. W.G. Dauben, G.A. Boswell, and W. Templeton, J. Org.
     Chem., 25, 1853 (1960).
565. W.G. Dauben and H.L. Bradlow, J. Am. Chem. Soc., 72,
     4248 (1950).
566. W.G. Dauben and G.J. Fonken, J. Am. Chem. Soc., 81,
     4060 (1959).
567. W.G. Dauben and P.H. Payot, J. Am. Chem. Soc., 78,
     5657 (1956).
568. A.G. Davies, ORGANIC PEROXIDES, Butterworths, London,
     1961, pp. 11-24.
569. J.R. Davis and G.E. Peterson, Texas Rpts. Biol. Med.,
     20, 145 (1962).
570. M. Davis and V. Petrow, J. Chem. Soc., 2536 (1949).
571. A.N. Davison, J. Dobbing, R.S. Morgan, and G.P. Wright,
     J. Neurochem., 3, 89 (1958).
572. A.J. Day, N.H. Fidge, and G.N. Wilkinson, J. Lipid
     Res., 7, 132 (1966).
573. E.A. Day, G.T. Malcom, and M.F. Beeler, Metabolism, 18,
     646 (1969).
574. S. Dayton, E.H. Mosbach, and F.E. Kendall, Proc. Soc.
     Exptl. Biol. Med., 84, 608 (1954).
575. P.D.G. Dean and M.W. Whitehouse, Biochem. J., 98, 410
     (1966).
576. P.D.G. Dean and M.W. Whitehouse, Biochim. Biophys.
     Acta, 137, 328 (1967).
577. M. Debono and R.M. Molloy, Steroids, 14, 219 (1969).
578. P. Decker, Naturwissenschaften, 44, 305 (1957).
579. R. de Fazi, Gazz. Chim. Ital., 61, 630 (1931).
580. R. de Fazi and A. Banchetti, Chem. Zentrl., 112, I,
     1039 (1941).
581. R. de Fazi, L. de Fazi, and F. Pirrone, Atti X° Congr.
     Intern. Chim., 5, 43 (1938).
582. H.J. Degenhart, R.J. Kraaipoel, H.E. Falke, J.G. van
     Beek, H. de Leeuw-Boon, G. Abelin, and H.K.A. Visser,
     J. Steroid Biochem., 5, 308 (1974).
583. H.J. Degenhart, H.K.A. Visser, H. Boon, and N.J.
     O'Doherty, Acta Endocrinol., 71, 512 (1972).

584. A.F.P.M. De Goeij and J. Van Stevenick, Clin. Chim.
     Acta, 68, 115 (1976).
585. R. Degkwitz, G. Cadenbach, and H. Lapp, Kolloid-Zeit-
     schr., 78, 311 (1937).
586. W.M. Dehn, J. Am. Chem. Soc., 39, 1399 (1917).
587. D.H. de Kock and J.H. Barnardt, S. Afr. Med. J., 44,
     1274 (1970).
588. P.B.D. de la Mare and R.D. Wilson, J. Chem. Soc.,
     Perkin Trans. II, 157 (1977).
589. G. Délèze, G. Karlaganis, W. Giger, M. Reinhard, D.
     Sideropoulos, and G. Paumgartner, in BILE ACID METABO-
     LISM IN HEALTH AND DISEASE, G. Paumgartner and A.
     Stiehl, Editors, University Park Press, Baltimore, Md.,
     1977, pp. 59-62.
590. G. Délèze and G. Paumgartner, Helv. Paediatr. Acta 32,
     29 (1977).
591. G. Délèze, G. Paumgartner, G. Karlaganis, W. Giger,
     M. Reinhard, and D. Sidiropoulos, Eur. J. Clin. Invest.,
     8, 41 (1978).
592. C. Delseth, R.M.K. Carlson, C. Djerassi, T.R. Erdman,
     and P.J. Scheuer, Helv. Chim. Acta, 61, 1470 (1978).
593. C. Delseth, Y. Kashman, and C. Djerassi, Helv. Chim.
     Acta, 62, 2037 (1979).
594. C. Delseth, L. Tolela, P.J. Scheuer, R.J. Wells, and
     C. Djerassi, Helv. Chim. Acta, 62, 101 (1979).
595. R.A. Demel, K.R. Bruckdorfer, and L.L.M. van Deenen,
     Biochim. Biophys. Acta, 255, 311 (1972).
596. R.A. Demel, K.R. Bruckdorfer, and L.L.M. van Deenen,
     Biochim. Biophys. Acta, 255, 321 (1972).
597. R.A. Demel and B. de Kruyff, Biochim. Biophys. Acta,
     457, 109 (1976).
598. M.E. Dempsey, J. Biol. Chem., 240, 4176 (1965).
599. M.E. Dempsey, M.C. Ritter, D.T. Witiak, and R.A.
     Parker, in ATHEROSCLEROSIS, PROCEEDINGS OF THE SECOND
     INTERNATIONAL SYMPOSIUM, R.J. Jones, Ed., Springer-
     Verlag, New York/Heidelberg/Berlin, 1970, pp. 290-295.
600. M.E. Dempsey, J.D. Seaton, G.J. Schroepfer, and R.W.
     Trockman, J. Biol. Chem., 239, 1381 (1964).
601. C. Denigès, Compt. Rend., 196, 1504 (1933).
602. R.W. Denny and A. Nickon, Org. Reactions, 20, 133
     (1970).
603. D. de Reinach-Hirtzbach and G. Ourisson, Tetrahedron,
     28, 2259 (1972).
604. M. De Rosa, L. Minale, and G. Sodano, Comp. Biochem.
     Physiol., 46B, 823 (1973).
605. S.S. Deshmane and S. Dev, Tetrahedron, 27, 1109 (1971).

606. N.J. De Souza and W.R. Nes, J. Lipid Res., 10, 240
     (1969).
607. A.K. Dhar, J.I. Teng, and L.L. Smith, J. Neurochem.,
     21, 51 (1973).
608. F. Dickens, Brit. Med. Bull., 4, 348 (1947).
609. J. Diekman and C. Djerassi, J. Org. Chem., 32, 1005
     (1967).
610. O. Diels, Ber., 41, 2596 (1908).
611. O. Diels and K. Linn, Ber., 41, 260 (1908).
612. R. Dietz, A.E.J. Forno, B.E. Larcombe, and M.E. Peover,
     J. Chem. Soc., (B), 816 (1970).
613. S. Di Frisco, P. de Ruggieri, and A. Ercoli, Boll. Soc.
     Ital. Biol. Sper., 29, 1351 (1953).
614. G.E. Dingchyan, R.L. Vardanyan, M.G. Vardanyan, and
     B.B. Kanukaev, Arm. Khim. Zh., 30, 295 (1977); via
     Chem. Abstr., 87, 136142b (1977); 89, 24612q (1978).
615. A. Dinner and K.Z. Farid, Lloydia, 39, 144 (1976).
616. W. Dirscherl and H. Breuer, Acta Endocrinol., 19, 37
     (1955).
617. N.W. Ditullio, C.S. Jacobs, and W.L. Holmes, J. Chroma-
     tog., 20, 354 (1965).
618. G.D. Dixon and L.C. Scala, Mol. Cryst. Liq. Cryst., 10,
     317 (1970).
619. R. Dixon, T. Furutachi, and S. Lieberman, Biochem.
     Biophys. Res. Commun., 40, 161 (1970).
620. C. Djerassi, R.M.K. Carlson, S. Popov, and T.H. Varkony,
     in MARINE NATURAL PRODUCTS CHEMISTRY, D.J. Faulkner and
     W.H. Fenical, Editors, Plenum Press, New York/London,
     1977, pp. 111-123.
621. C. Djerassi, R.R. Engle, and A. Bowers, J. Org. Chem.,
     21, 1547 (1956).
622. C. Djerassi, J.C. Knight, and H. Brockmann, Chem. Ber.,
     97, 3118 (1964).
623. C. Djerassi and R. McCrindle, J. Chem. Soc., 4034
     (1962).
624. J. Dobbing, J. Neurochem., 10, 739 (1963).
625. R.M. Dodson, R.T. Nicholson, and R.D. Muir, J. Am.
     Chem. Soc., 81, 6295 (1959).
626. F.H. Doleiden, S.R. Fahrenholtz, A.A. Lamola, and A.M.
     Trozzolo, Photochem. Photobiol., 20, 519 (1974).
627. M. Donbrow, E. Azaz, and A. Pillersdorf, J. Pharm.
     Sci., 67, 1676 (1978).
628. C. Dorée, J. Chem. Soc., 95, 638 (1909).
629. C. Dorée and J.A. Gardner, J. Chem. Soc., 93, 1328
     (1908).
630. C. Dorée and L. Orange, J. Chem. Soc., 109, 46 (1916).

631. L.M. Dorfman and G.E. Adams, National Bureau of Stand-
     ards Report No. NSRDS-NBS-46, U.S. Government Printing
     Office, Washington City, D.C., 1973.
632. R.W. Doskotch, F.S. El-Feraly, E.H. Fairchild, and
     C.-T. Huang, J. Chem. Soc. Chem. Commun, 402 (1976).
633. R.W. Doskotch, F.S. El-Feraly, E.H. Fairchild, and
     C.-T. Huang, J. Org. Chem., $\underline{42}$, 3614 (1977).
634. Drogas, Vacunas, y Sueros, S.A., Spanish Patent No. 217,
     089, Sept. 24, 1954; via Chem. Abstr., $\underline{50}$,6517c (1956).
635. D.J. Duchamp, C.G. Chidester, J.A.F. Wickramasinghe,
     E. Caspi, and B. Yagen, J. Am. Chem. Soc., $\underline{93}$, 6283
     (1971).
636. G.L. Duff, Arch. Pathol., $\underline{20}$, 81 (1935).
637. G.L. Duff, Arch. Pathol., $\underline{20}$, 259 (1935).
638. E.W. Duncan, P.H. Culbreth, and C.A. Burtis, J. Chroma-
     tog., $\underline{162}$, 281 (1979).
639. J.L. Duncan and L. Buckingham, Infection & Immunity,
     $\underline{22}$, 94 (1978).
640. J.A. Dunn, Brit. J. Cancer, $\underline{19}$, 496 (1965).
641. J.L. Dunn, I.M. Heilbron, R.F. Phipers, K.M. Samant,
     and F.S. Spring, J. Chem. Soc., 1576 (1934).
642. C. Duque, M. Morisaki, and N. Ikekawa, Biochem. Bio-
     phys. Res. Commun., $\underline{82}$, 179 (1978).
643. C. Duque, M. Morisaki, N. Ikekawa, and M. Shikita,
     Tetrahedron Lett., 4479 (1979).
644. Dutch Patent No. 58,771, Jan. 15, 1947; via Chem.
     Abstr., $\underline{41}$, 5154b (1947).
645. Dutch Patent No. 60,735, Mar. 15, 1948; via Chem.
     Abstr., $\underline{42}$, 4206d (1948).
646. B.J. Dutka, A.S.Y. Chau, and J. Coburn, Water Res., $\underline{8}$,
     1047 (1974).
647. B.J. Dutka and A. El-Shaarawi, Can. J. Microbiol., $\underline{21}$,
     1386 (1975).
648. D. Dvornik, M. Kraml, and J.F. Bagli, J. Am. Chem. Soc.,
     $\underline{86}$, 2739 (1964).
649. D. Dvornik, M. Kraml, and J.F. Bagli, Biochemistry, $\underline{5}$,
     1060 (1966).
650. D. Dvornik, M. Kraml, J. Dubuc, M. Givner, and R. Gau-
     dry, J. Am. Chem. Soc., $\underline{85}$, 3309 (1963).
651. J.H. Dygos and B.M. Desai, J. Org. Chem., $\underline{44}$, 1590
     (1979).
652. U.M. Dzhemilev, V.P. Yur'ev, and G.A. Tolstikov, Zh.
     Obshcheii Khim., $\underline{40}$, 2518 (1970); Zh. Obshcheii Khim.,
     (English Trans.), $\underline{40}$, 2505 (1970).
653. H. Eagle, J. Exp. Med., $\underline{52}$, 747 (1930).
654. W.R. Eberlein and A.A. Patti, J. Clin. Endocrinol.

Metab., 25, 1101 (1965).

655. E.A.V. Ebsworth, J.A. Connor, and J.J. Turner, in COMPREHENSIVE INORGANIC CHEMISTRY, J.C. Bailar, H.J. Emeleus, R. Nyholm, and A.F. Trotman-Dickenson, Editors, Pergamon Press, Oxford, 1975, pp. 685-794.

656. D. Edgar, J. Chromatog., 43, 271 (1969).

657. C.G. Edmonds, A.G. Smith, and C.J.W. Brooks. J. Chromatog., 133, 372 (1977).

658. J.T. Edward and N.E. Lawson, J. Org. Chem., 35, 1426 (1970).

659. J.A. Edwards, J.S. Mills, J. Sundeen, and J.H. Fried, J. Am. Chem. Soc., 91, 1248 (1969).

660. J.A. Edwards, J. Sundeen, W. Salmond, T. Iwadare, and J.H. Fried, Tetrahedron Lett. 791 (1972).

661. A. Ehrenberg, L. Ehrenberg, and G. Löfroth, Risö Report No. 16, 21 (1960).

662. N. Eickman, J. Clardy, R.J. Cole, and J.W. Kirksey, Tetrahedron Lett., 1051 (1975).

663. K. Einarsson, Eur. J. Biochem., 5, 101 (1968).

664. K. Einarsson, Eur. J. Biochem., 6, 299 (1968).

665. W. Eisfeld, Dissertation, Göttingen, 1965.

666. P. Ekwall and H. Baltscheffsky, Acta Chem. Scand., 15, 1198 (1961).

667. P. Ekwall, H. Baltscheffsky, and L. Mandell, Acta. Chem. Scand., 15, 1195 (1961).

668. P. Ekwall and L. Mandell, Acta Chem. Scand., 15, 1403 (1961).

669. P. Ekwall and L. Mandell, Acta Chem. Scand., 15, 1404 (1961).

670. F.S. El-Feraly, Y.-M. Chan, G.A. Capiton, R.W. Doskotch, and E.H. Fairchild, J. Org. Chem., 44, 3952 (1979).

671. F.S. El-Feraly, Y.-M. Chan, E.H. Fairchild, and R.W. Doskotch, Tetrahedron Lett., 1973 (1977).

672. R. Eliasson, Biochem. J., 98, 242 (1966).

673. C.G. Elliott, M.R. Hendrie, and B.A. Knights, J. Gen. Microbiol., 42, 425 (1966).

674. M.T. Ellis, Biochem. J., 12, 173 (1918).

675. S.E. El Masry, J.A. Fee, A.H. El Masry, and R.E. Counsell, Biochem. Pharmacol., 26, 1109 (1977).

676. G.B. Elyakov, T.A. Kuznetsova, A.K. Dzizenko, and Yu. N. Elkin, Tetrahedron Lett.,1151 (1969).

677. H. Elze, H. Pilgrim, and E. Teuscher, Pharmazie, 29, 727 (1974).

678. E. Enderlin, S.J. Thannhauser, and M. Jenke, Arch. Exptl. Pathol. Pharmakol., 130, 292 (1928).

679. M. Endo, M. Kajiwara, and K. Nakanishi, J. Chem. Soc. Chem. Commun., 309 (1970).

680. P. Eneroth and J.-Å. Gustafsson, FEBS Lett., 3 129 (1969).

681. P. Eneroth, K. Hellström, and R. Ryhage, J. Lipid Res., 5, 245 (1964).

682. P. Eneroth and E. Nyström, Biochim. Biophys. Acta, 144, 149 (1967).

683. L.F. Eng, Y.L. Lee, R.B. Hayman, and B. Gerstl, J. Lipid Res., 5, 128 (1964).

684. L.L. Engel and P. Brooks, Steroids, 17, 531 (1971).

685. J.P. Engelbrecht, B. Tursch, and C. Djerassi, Steroids, 20, 121 (1972).

686. P.R. Enslin, Tetrahedron, 27, 1909 (1971).

687. P.R. Enslin, C.W. Holzapfel, and K.B. Norton, J. Chem. Soc.,(C), 964 (1967).

688. A. Ensminger, G. Joly, and P. Albrecht, Tetrahedron Lett., 1575 (1978).

689. N. Entwistle and A.D. Pratt, Tetrahedron, 24, 3949 (1968).

690. N. Entwistle and A.D. Pratt, Tetrahedron, 25, 1449 (1969).

691. N. Entwistle and A.D. Pratt, J. Chem. Soc., Perkin Trans. I, 1235 (1973).

692. V. Erä and K. Noronen, J. Am. Oil Chem. Soc., 56, 992 (1979).

693. A. Ercoli, S. Di Frisco, and P. de Ruggieri, Boll. Soc. Ital. Biol. Sper., 29, 494 (1953).

694. A. Ercoli, S. Di Frisco, and P. de Ruggieri, Gazz. Chim. Ital., 83, 78 (1953).

695. A. Ercoli and P. de Ruggieri, Gazz. Chim. Ital., 83, 720 (1953).

696. A. Ercoli and P. de Ruggieri, Gazz. Chim. Ital., 83, 738 (1953).

697. A. Ercoli and P. de Ruggieri, J. Am. Chem. Soc., 75, 3284 (1953).

698. W.L. Erdahl and O.S. Privett, Lipids, 12, 797 (1977).

699. T.R. Erdman and P.J. Scheuer, Lloydia, 38, 359 (1975).

700. T.R. Erdman and R.H. Thomson, Tetrahedron, 28, 5163 (1972).

701. S.K. Erickson, A.D. Cooper, S.M. Matsui, and R.G. Gould, J. Biol. Chem., 252, 5186 (1977).

702. S.K. Erickson, S. Matsui, A.D. Cooper, and R.G. Gould, Abstracts, 10th Int. Congress Biochem., Hamburg, July 25-31, 1976, p. 382.

703. S.K. Erickson, S.M. Matsui, M.A. Shrewsbury, A.D.

Cooper, and R.G. Gould, J. Biol. Chem., $\underline{253}$, 4159
(1978).

704. S.K. Erickson, D.J. Meyer, and R.G. Gould, J. Biol.
Chem., $\underline{253}$, 1817 (1978).

705. J. Eriksen, C.S. Foote, and T.L. Parker, J. Am. Chem.
Soc., $\underline{99}$, 6455 (1977).

706. R. Estripeaut, Bol. Inst. Med. Exp. Estud. Tratamiento
Cáncer, $\underline{14}$, 159 (1937); via Chem. Zentr., II, 1788
(1938).

707. E.A. Evans and F.G. Stanford, Nature, $\underline{197}$, 551 (1963).

708. J. Everall and E.V. Truter, J. Invest. Dermatol., $\underline{22}$,
493 (1954).

709. S.R. Fahrenholtz, F.H. Doleiden, A.M. Trozzolo, and
A.A. Lamola, Photochem. Photobiol., $\underline{20}$, 505 (1974).

710. R. Faiman, Chem. Phys. Lipids, $\underline{18}$, 84 (1977).

711. P. Falardeau and L. Tan, Proc. Can. Fed. Biol. Soc.,
$\underline{14}$, 373 (1971).

712. P. Falardeau and L. Tan, J. Labelled Compounds, $\underline{10}$,
239 (1974).

713. H.L. Falk, S. Goldfein, and P.E. Steiner, Cancer Res.,
$\underline{9}$, 438 (1949).

714. H.E. Falke, H.J. Degenhart, G.J.A. Abeln, and H.K.A.
Visser, Molec. Cell. Endocrinol., $\underline{3}$, 375 (1975).

715. H.E. Falke, H.J. Degenhart, G.J.A. Abeln, and H.K.A.
Visser, Molec. Cell. Endocrinol., $\underline{4}$, 107 (1976).

716. I. Faredin, A.G. Fazekas, K. Tóth, K. Kókai, and M.
Julesz, J. Invest. Dermatol., $\underline{52}$, 357 (1969).

717. E. Fattorusso, S. Magno, C. Santacroce, and D. Sica,
Gazz. Chim. Ital., $\underline{104}$, 409 (1974).

718. E. Fattorusso, S. Magno, C. Santacroce, D. Sica, G.
Impellizzeri, S. Mangiafico, G. Oriente, M. Piattelli,
and S. Sciuto, Phytochem., $\underline{14}$, 1579 (1975).

719. D.J. Faulkner and M.D. Higgs, Abstracts of Papers,
175th Natl. Meeting American Chemical Soc., Anaheim,
Calif., March 13-17, 1978, No. AGFD-27.

720. E.S. Faust, Arch. Exp. Pathol. Pharmakol., $\underline{47}$, 278
(1902).

721. E.S. Faust, Arch. Exp. Pathol. Pharmakol., $\underline{56}$, 236
(1907).

722. E.S. Faust, Arch. Exp. Pathol. Pharmakol., $\underline{64}$, 244
(1911).

723. J.R. Faust, J.L. Goldstein, and M.S. Brown, J. Biol.
Chem., $\underline{252}$, 4861 (1977).

724. J.R. Faust, J.L. Goldstein, and M.S. Brown, Arch.
Biochem. Biophys., $\underline{192}$, 86 (1979).

725. J.R. Faust, J.L. Goldstein, and M.S. Brown, Proc. Natl.

Acad. Sci., 76, 5018 (1979).

726. F. Fawaz, M. Chaigneau, and F. Puisieux, Ann. Pharm. Franç., 32, 301 (1974).

727. J. Fayos, D. Lokensgard, J. Clardy, R.J. Cole, and J.W. Kirksey, J. Am. Chem. Soc., 96, 6785 (1974).

728. J.J. Feher, L.D. Wright, and D.B. McCormick, J. Phys. Chem., 78, 250 (1974).

729. A.M. Feinberg, K. Nakanishi, J. Barciszewski, A.J. Rafalski, H. Augustyniak, and M. Wiewiórowski, J. Am. Chem. Soc., 96, 7797 (1974).

730. L. Feldman, A. Kushelevsky, and Z.B. Alfassi, Radiation Effects, 40, 151 (1979).

731. W. Fenical, J. Am. Chem. Soc., 96, 5580 (1974).

732. J.P. Ferezou, M. Devys, J.P. Allais, and M. Barbier, Phytochem., 13, 593 (1974).

733. J.P. Ferezou, M. Devys, and M. Barbier, Experientia, 28, 153, 407, (1972).

734. H. Fernandez Ferrufino, Bol. Inst. Med. Exp. Estud. Cancer, 20, 233 (1943); via Chem. Abstr., 38, 1276[8] (1944).

735. G.N. Festenstein and R.A. Morton, Biochem. J., 52, 168 (1952).

736. W. Fiehn and D. Seiler, Experientia, 31, 773 (1975).

737. L.F. Fieser, THE CHEMISTRY OF NATURAL PRODUCTS RELATED TO PHENANTHRENE, ACS Monograph No. 70, Reinhold Publishing Corp., New York, 1936, p. 117, p. 174.

738. L.F. Fieser, THE CHEMISTRY OF NATURAL PRODUCTS RELATED TO PHENANTHRENE, ACS Monograph No. 70, Reinhold Publishing Corp., New York, 2nd Ed., 1937, p. 117.

739. L.F. Fieser, J. Am. Chem. Soc., 73, 5007 (1951).

740. L.F. Fieser, J. Am. Chem. Soc., 75, 4395 (1953).

741. L.F. Fieser, J. Am. Chem. Soc., 75, 5421 (1953).

742. L. Fieser, Bull. Soc. Chim. France, 21, 541 (1954).

743. L.F. Fieser, Science, 119, 710 (1954).

744. L.F. Fieser, ORGANIC SYNTHESES, 35, 43 (1955); Collected Volumes, 4, 195 (1963).

745. L.F. Fieser, in FESTSCHRIFT PROF. DR. ARTHUR STOLL, Burkhäuser AG, Basel, 1957, pp. 489-497.

746. L.F. Fieser and B.K. Bhattacharyya, J. Am. Chem. Soc., 75, 4418 (1953).

747. L.F. Fieser and M. Fieser, NATURAL PRODUCTS RELATED TO PHENANTHRENE, 3rd Edition, Reinhold Publishing Corp., New York City, 1949, pp. 233-234.

748. L.F. Fieser and M. Fieser, STEROIDS, Reinhold Publishing Corp., New York City, 1959, p. 158, pp. 233-237.

749. L.F. Fieser and M. Fieser, REAGENTS FOR ORGANIC

SYNTHESIS, John Wiley & Sons Inc., New York/London/
Sydney, Vol. 1, 1967, p. 141.

750. L.F. Fieser, M. Fieser, and R.N. Chakravarti, J. Am.
     Chem. Soc., 71, 2226 (1949).

751. L.F. Fieser, T. Goto, and B.K. Bhattacharyya, J. Am.
     Chem. Soc., 82, 1700 (1960).

752. L.F. Fieser, T.W. Greene, F. Bischoff, G. Lopez, and
     J.J. Rupp, J. Am. Chem. Soc., 77, 3928 (1955).

753. L.F. Fieser, J.E. Herz, M.W. Klohs, M.A. Romero, and
     T. Utne, J. Am. Chem. Soc., 74, 3309 (1952).

754. L.F. Fieser, W.-Y. Huang, and B.K. Bhattacharyya, J.
     Org. Chem., 22, 1380 (1957).

755. L.F. Fieser, W.-Y. Huang, and T. Goto, J. Am. Chem.
     Soc., 82, 1688 (1960).

756. L.F. Fieser and M.S. Newman, J. Am. Chem. Soc., 57,
     961 (1935).

757. L.F. Fieser and S. Rajagopalan, J. Am. Chem. Soc., 71,
     3938 (1949).

758. L.F. Fieser and W.P. Schneider, J. Am. Chem. Soc., 74,
     2254 (1952).

759. L.F. Fieser and R. Stevenson, J. Am. Chem. Soc., 76,
     1728 (1954).

760. B. Fillmann and H. Breuer, Z. Physiol. Chem., 351, 1100
     (1970).

761. B.S. Finkle, R.L. Foltz, and D.M. Taylor, J. Chromatog.
     Sci., 12, 304 (1974).

762. B.J. Finlayson and J.N. Pitts, Science, 192, 111
     (1976).

763. J.A. Fioriti, N. Buide, and R.J. Sims, Lipids, 4, 142
     (1969).

764. J.A. Fioriti, M.J. Kanuk, M. George, and R.J. Sims,
     Lipids, 5, 71 (1970).

765. J.A. Fioriti and R.J. Sims, J. Am. Oil Chem. Soc., 44,
     221 (1967).

766. M.H. Fisch, R. Ernst, B.H. Flick, J. Arditti, D.H.R.
     Barton, P.D. Magnus, and I.D. Menzies, J. Chem. Soc.
     Chem. Commun., 530 (1973).

767. F.G. Fischer and H. Mägerlein, Liebigs Ann., 636 88
     (1960).

768. H. Fischer and A. Treibs, Liebigs Ann., 446, 241
     (1926).

769. A. Fish, in ORGANIC PEROXIDES, D. Swern, Editor, Wiley-
     Interscience, New York, 1970, Vol. 1, pp. 146-151.

770. K.A. Fisher, Proc. Natl. Acad. Sci., 73, 173 (1976).

771. H.J.M. Fitches, Adv. Mass Spectrometry, 2, 428 (1962).

772. F.A. Fitzpatrick and S. Siggia, Anal. Chem., 45, 2310

(1973).

773. P.J. Flanagan and J.B. Thomson, Tetrahedron Lett.,
     1671 (1965).
774. P.J. Flanagan and J.B. Thomson, J. Chem. Soc.,(C), 833
     (1968).
775. V.P. Flanagan and A. Ferretti, Lipids, 9, 471 (1974).
776. V.P. Flanagan, A. Ferretti, D.P. Schwartz, and J.M.
     Ruth, J. Lipid Res., 16, 97 (1975).
777. H.M. Flegg, Ann. Clin. Biochem., 10, 79 (1973); via
     Chem. Abstr., 79, 112935w (1973).
778. J.H. Fleisher, L.J. Marton, N.R. Bachur, and R.S.
     Mann-Kaplan, Life Sci., 13, 1517 (1973).
779. A. Flint, Am. J. Med. Sci., 44, 305 (1862).
780. F. Flury, Arch. Exp. Pathol. Pharmakol., 66, 221
     (1911).
781. G.L. Flynn, Y. Shah, S. Prakongpan, K.H. Kwan, W.I.
     Higuchi, and A.F. Hofmann, J. Pharm. Sci., 68, 1090
     (1979).
782. K.-L. Fong, P.B. McCay, J.L. Poyer, B.B. Keele, and
     H. Misra, J. Biol. Chem., 248, 7792 (1973).
783. C.S. Foote, Abstracts, Internatl. Conf. Chemi- and
     Bioenergized Processes, Guarujá, Brazil, Aug. 8-10,
     1978, p. 25.
784. C.S. Foote, in REPORT, CNPQ-NSF JOINT SEMINAR ON CHEMI-
     AND BIO-ENERGIZED PROCESSES, Sao Paulo, Brasil, August
     2-6, 1978, pp. 42-44.
785. C.S. Foote, F.C. Shook and R.A. Abakerli, J. Am. Chem.
     Soc., 102, 2503 (1980).
786. R.J. Fosbinder, F. Daniels, and H. Steenbock, J. Am.
     Chem. Soc., 50, 923 (1928).
787. A.F. de Fourcroy, Ann. Chim., 3, 120 (1789).
788. A.F. de Fourcroy, Ann. Chim., 3, 242 (1789).
789. A.F. de Fourcroy, Ann. Chim., 5, 154 (1790).
790. A.F. de Fourcroy, Ann. Chim., 8, 17 (1791).
791. D.L. Fox and C.H. Oppenheimer, Arch. Biochem. Biophys.,
     51, 323 (1954).
792. J.E. Fox, A.I. Scott, and D.W. Young, Chem. Commun.,
     1105 (1967).
793. J.E. Fox, A.I. Scott, and D.W. Young, J. Chem. Soc.,
     Perkin Trans. I, 799 (1972).
794. C. Francisco, G. Combaut, J. Teste, C. Tarchini, and
     C. Djerassi, Steroids, 34, 163 (1979).
795. E. Fränkel and G. Löhr, Z. Physiol. Chem., 218, 218
     (1933).
796. E. Frederickson and S. Liisberg, Chem. Ber., 88, 684
     (1955).

797. D.S. Fredrickson, J. Biol. Chem., 222, 109 (1956).
798. D.S. Fredrickson and K. Ono, Biochim. Biophys. Acta, 22, 183 (1956).
799. N.K. Freeman, F.T. Lindgren, Y.C. Ng, and A.V. Nichols, J. Biol. Chem., 203, 293 (1953).
800. N.K. Freeman, F.T. Lindgren, Y.C. Ng, and A.V. Nichols, J. Biol. Chem., 227, 449 (1957).
801. R. Freire Barreira, A. González González, J.A. Salazar Rocío, and E. Suárez López, Ann. Ouím., 66, 415 (1970).
802. F. Frick, German Patent No. 485,198, Dec. 25, 1926; via Chem. Abstr., 24, 983 (1930).
803. I. Fridovich, Accts. Chem. Res., 5, 321 (1972).
804. I. Fridovich, in FREE RADICALS IN BIOLOGY, W.A. Pryor, Editor, Academic Press, New York, Vol. 1, 1976, pp. 239-277.
805. S.S. Friedland, G.H. Lane, R.T. Longman, K.E. Train, and M.J. O'Neal, Anal. Chem., 31, 169 (1959).
806. N. Friedman, M. Gorodetsky, and Y. Mazur, J. Chem. Soc. Chem. Commun., 874 (1971).
807. A.A. Frimer, Chem. Rev., 79, 359 (1979).
808. S. Fritsch, Arch. Geschwulstforsch., 25, 265 (1965).
809. M. Fryberg, A.C. Oehlschlager, and A.M. Umrau, J. Chem. Soc. Chem. Commun., 1194 (1971).
810. L. Fuchs and J. van Niekerk, Biochem. Z., 277, 32 (1935).
811. H. Fühner, Arch. Exp. Pathol. Pharmakol., 63, 374 (1910).
812. G.I. Fujimoto and A.E. Jacobson, J. Org. Chem., 29, 3377 (1964).
813. H. Fukuda, Y. Kawakami, and S. Nakamura, Chem. Pharm. Bull., 21, 2057 (1973).
814. D.K. Fukushima, A.D. Kemp, R. Schneider, M.B. Stokem, and T.F. Gallagher, J. Biol. Chem., 210, 129 (1954).
815. M. Fukuyama and Y. Miyake, J. Biochem., 85, 1183 (1979).
816. R. Fumagalli, G. Galli, and G. Urna, Life Sci., 10, Part II, 25 (1971).
817. H. Funasaki and J.R. Gilbertson, J. Lipid Res., 9, 766 (1968).
818. C.H. Fung and A.K. Khachadurian, Federation Proc., 38, 511 (1979).
819. C.H. Fung and A.K. Khachadurian, J. Biol. Chem., 255, 676 (1980).
820. B.J.A. Furr and G.S. Pope, Steroids, 16, 471 (1970).
821. A. Fürst and F. Koller, Helv. Chim. Acta, 30, 1454 (1947).

822. A. Fürst, L. Labler, W. Meier, and K.-H. Pfoertner, Helv. Chim. Acta, 56, 1708 (1973).
823. W. Fürst, Arch. Pharm., 298, 795 (1965).
824. W. Fürst, Sci. Pharm., Proc. 25th, 2, 229 (1965); via Chem. Abstr., 69, 95099a (1968).
825. W. Fürst, Arch. Pharm., 299, 991 (1966).
826. W. Fürst, Arch. Pharm., 300, 141 (1967).
827. W. Fürst, Arch. Pharm., 300, 144 (1967).
828. W. Fürst, Arch. Pharm., 300, 353 (1967).
829. W. Fürst, Arch. Pharm., 300, 359 (1967).
830. W. Fürst and C. Diethmann, Pharm. Zentralh., 107, 274 (1968); via Chem. Abstr., 69, 21875d (1968).
831. N. Furutachi, Y. Nakadaira, and K. Nakanishi, Chem. Commun., 1625 (1968).
832. R.B. Gagosian, Geochim. Cosmochim. Acta, 39, 1443 (1975).
833. R.B. Gagosian, Limnol. Oceanography, 21, 702 (1976).
834. R.B. Gagosian and J.W. Farrington, Geochim. Cosmochim. Acta, 42, 1091 (1978).
835. R.B. Gagosian and F. Heinzer, Geochim. Cosmochim. Acta, 43, 471 (1979).
836. R.B. Gagosian and S.O. Smith, Nature 277, 287 (1979).
837. T.F. Gallagher, Cancer Res., 17, 520 (1957).
838. T.F. Gallagher, J. Clin. Endocrinol. Metab., 19, 937 (1958).
839. E.J. Gallegos, Anal. Chem., 43, 1151 (1971).
840. E.J. Gallegos, Anal. Chem., 47, 1524 (1975); 47, 2486 (1975).
841. G. Galli and E. Grossi Paoletti, Lipids, 2, 72 (1967).
842. G. Galli and E. Grossi Paoletti, Lipids, 2, 84 (1967).
843. G. Galli and S. Maroni, Steroids, 10, 189 (1967).
844. G. Galli, E. Grossi Paoletti, and J.F. Weiss, Science, 162, 1495 (1968).
845. M. Galli Kienle, G. Cighetti, M. Anastasia, and C.R. Sirtori, J. Steroid Biochem., 9, 127 (1978).
846. M. Galli Kienle, R.K. Varma, L.J. Mulheirn, B. Yagen, and E. Caspi, J. Am. Chem. Soc., 95, 1996 (1973).
847. P. Gambert, C. Lallemant, A. Archambault, B.F. Maume, and P. Padieu, J. Chromatog., 162, 1 (1979).
848. R. Gardi and A. Lusignani, J. Org. Chem., 32, 2647 (1967).
849. J.A. Gardner and H. Gainsborough, Biochem. J., 28, 1631 (1934).
850. J.N. Gardner, F.E. Carlon, and O. Gnoj, J. Org. Chem., 33, 1566 (1968).
851. J.N. Gardner, F.E. Carlon, C.H. Robinson, and E.P.

Oliveto, Steroids, 7, 234 (1966).

852. N. Garti, L. Karpuj, and S. Sarig, Thermochim. Acta, 35, 343 (1980).

853. S.J. Gaskell and G. Eglinton, in ADVANCES IN ORGANIC GEOCHEMISTRY 1973, B. Tissot and F. Bienner, Editors, Éditions Technip, Paris, 1974, pp. 963-976.

854. S.J. Gaskell and G. Eglinton, Nature, 254, 209 (1975).

855. S.J. Gaskell and G. Eglinton, Geochim. Cosmochim. Acta, 40, 1221 (1976).

856. S.J. Gaskell, A.G. Smith, and C.J.W. Brooks, Biomed. Mass Spectrometry, 2, 148 (1975).

857. J.L. Gaylor and C.V. Delwiche, J. Biol. Chem., 251, 6638 (1976).

858. R.L. Gebhard and A.D. Cooper, J. Biol. Chem., 253, 2790 (1978).

859. T.A. Geissman, Phytochem., 9, 2377 (1970).

860. A. Gemant, Life Sci., 1, 233 (1962).

861. General Foods Corp., Oxidation Products of Fats, Cholesterol and Cholesterol Epoxide, Final Report, NIH Contract PH 43-63-1165, White Plains, N.Y., June 4, 1971.

862. W.L.G. Gent, J. Sci. Instruments, [2], 2, 69 (1969).

863. A. George, ELSEVIER'S ENCYCLOPEDIA OF ORGANIC CHEMISTRY. SERIES III. CARBOISOCYCLIC CONDENSED COMPOUNDS. STEROIDS, F. Radt, Editor, Elsevier Publishing Co., Amsterdam, 1954, Vol. 14, Supplement, pp. 1568S-1620S.

864. P. George, Discussions Faraday Soc., 2, 196 (1947).

865. D. Ghosh, R.L. Lyman, and J. Tinoco, Lipids, 8, 61 (1973).

866. G.F. Gibbons, C.R. Pullinger, H.W. Chen, W.K. Cavanee, and A.A. Kandutsch, J. Biol. Chem., 255, 395 (1980).

867. J. Gielen, J. Van Cantfort, and J. Renson, Arch. Int. Physiol. Biochem., 76, 581 (1968).

868. J. Gielen, J. Van Cantfort, and J. Renson, Arch. Int. Physiol. Biochem., 76, 930 (1968).

869. J. Gielen, J. Van Cantfort, B. Robaye, and J. Renson, Compt. Rend., [D], 269, 731 (1969).

870. J. Gielen, J. Van Cantfort, B. Robaye, and J. Renson, Eur. J. Biochem., 55, 41 (1975).

871. J. Gielen, B. Robaye, J. Van Cantfort, and J. Renson, Arch. Int. Pharmacodynam., 183, 403 (1970).

872. D.B. Gilbert and J.A. Reynolds, Biochemistry, 15, 71 (1976).

873. D.B. Gilbert, C. Tanford, and J.A. Reynolds, Biochemistry, 14, 444 (1975).

874. J.D. Gilbert, C.J.W. Brooks, and W.A. Harland, Biochim. Biophys. Acta, 270, 149 (1972).

875. J.D. Gilbert, W.A. Harland, G. Steel, and C.J.W. Brooks, Biochim. Biophys. Acta, 187, 453 (1969).

876. M.W. Gilgan, R.K. Pike, and J.W. ApSimon, Comp. Biochem. Physiol., 54B, 561 (1976).

877. D. Ginsburg, J. Am. Chem. Soc., 75, 5489 (1953).

878. L.J. Goad, I. Rubinstein, and A.G. Smith, Proc. Royal Soc. London, [B], 180, 223 (1972).

879. N.-T. Gobley, J. Pharm. Chim., [3], 9, 81 (1846).

880. J. Godeaux, Bull. Soc. Roy. Sci. Liége, 12, 579, (1943); via Chem. Abstr., 39, 21343 (1945).

881. M.W. Goldblatt, Biochem. J., 29, 1346 (1935).

882. J.L. Goldstein and M.S. Brown, Am. J. Med., 58, 147 (1975).

883. J.L. Goldstein and M.S. Brown, Current Topics Cell. Regulation, 11, 147 (1976).

884. J.L. Goldstein, J.R. Faust, G.Y. Brunschede, and M.S. Brown, in LIPIDS, LIPOPROTEINS, AND DRUGS, D. Kritchevsky, R. Paoletti, and W.L. Holmes, Editors, Plenum Press, New York/London, 1975, pp. 77-84.

885. J.L. Goldstein, J.R. Faust, J.H. Dygos, R.J. Chorvat, and M.S. Brown, Proc. Natl. Acad. Sci., 75, 1877 (1978).

886. J.L. Goldstein, Y.K. Ho, S.K. Basu, and M.S. Brown, Proc. Natl. Acad. Sci., 76, 333 (1979).

887. K. Gollnick and G.O. Schenck, Pure Applied Chem., 9, 507 (1964).

888. N.P. Gontscharow, K. Wehrberger, and K. Schubert, J. Steroid Biochem., 2, 389 (1971).

889. N.P. Gontscharow, K. Wehrberger, K. Schubert, and S.W. Schewzowa, Acta Biol. Med. Germ., 23, 713 (1969).

890. A. González González, J. Delgado Martín, and M. Melián Martín, An. Quim., 69, 807 (1973); via Chem. Abstr., 79, 134328c (1973).

891. A.G. González, C.G. Francisco, R. Freire, R. Hernández, J.A. Salazar, and E. Suárez, J. Chem. Soc. Chem. Commun., 905 (1974).

892. A.G. González, R. Freire, J.A. Salazar, and E. Suárez, Phytochem., 10, 1339 (1971).

893. E.L. Gonzalez, Nature, 219, 274 (1968).

894. R. Gonzalez, J.P. Carlson, and M.E. Dempsey, Arch. Biochem. Biophys., 196, 574 (1979).

895. G.S. Gordan, T.J. Cantino, L. Erhardt, J. Hansen, and W. Lubich, Science, 151, 1226 (1966).

896. G.S. Gordan, M.E. Fitzpatrick, and W.P. Lubich, Trans.

Assoc. Am. Physicians, 80, 183 (1967).

897. L.I. Gordon, J. Bass, and S. Yachnin, Clin. Res., 27, 688A (1979).

898. L.I. Gordon, J. Bass, and S. Yachnin, Proc. Natl. Acad. Sci., 77, 4313 (1980).

899. W. Gordy, Radiation Res., Suppl. 1, 491 (1959).

900. P.D. Gorizontov, Biokhimiya, 5, 102 (1940); via Chem. Abstr., 36, 2321³ (1942).

901. M.B. Gorovitz, A.D. Koveshnikov, and N.K. Abubakirov, Khim. Prirod. Soed., 125 (1969).

902. M.H. Gottlieb, Biochim. Biophys. Acta, 466, 422 (1977).

903. R.G. Gould, in CHOLESTEROL METABOLISM AND LIPOLYTIC ENZYMES, J. Polonovski, Editor, Masson Publishing U.S.A. Inc., New York, 1977, pp. 13-38.

904. D.H. Gould, K.H. Schaaf, and W.L. Ruigh, J. Am. Chem. Soc., 73, 1263 (1951).

905. P.A. Govinda Malya, J.R. Wright, and W.R. Nes, J. Chromatog. Sci., 9, 700 (1971).

906. D.B. Gower, J. Steroid Biochem., 3, 45 (1972).

907. D.B. Gower and N. Ahmad, Biochem. J., 104, 550 (1967).

908. D.B. Gower, J.R. Daly, G.J.A.I. Snodgrass and M.I. Stern, Acta Endocrinol., 63, 562 (1970).

909. D.B. Gower and K.H. Loke, Biochim. Biophys. Acta, 250, 614 (1971).

910. D.B. Gower and M.I. Stern, Acta Endocrinol., 60, 265 (1969).

911. R.P. Graber, C.S. Snoddy, H.B. Arnold, and N.L. Wendler, J. Org. Chem., 21, 1517 (1956).

912. U. Graff, Biochem. Z., 298, 179 (1938).

913. J.K. Grant, Biochem. J., 64, 559 (1956).

914. M.F. Gray, T.D.V. Lawrie, and C.J.W. Brooks, Lipids, 6, 836 (1971).

915. M.F. Gray, A. Morrison, E. Farish, T.D.V. Lawrie, and C.J.W. Brooks, Biochim. Biophys. Acta, 187, 163 (1969).

916. D.M. Green, J.A. Edwards, A.W. Barksdale, and T.C. McMorris, Tetrahedron, 27, 1199 (1971).

917. J.R. Green, P.A. Edwards, and C. Green, Biochem. J., 135, 63 (1973).

918. C.W. Greenhalgh, H.B. Henbest, and E.R.H. Jones, J. Chem. Soc., 2375 (1952).

919. C.W. Greenhalgh, H.B. Henbest, and E.R.H. Jones, J. Chem. Soc., 2380 (1952).

920. A.M. Grimwade, M.E. Lawson, and G.S. Boyd, Biochem. J., 125, 14P (1971).

921. G.S. Grinenko and V.I. Bayunova, Khim.-Farm. Zhur., 6,

No. 2, 25 (1972).

922. H. Grossmann, Z. Ges. Exp. Med., 42, 496 (1924).

923. M.D. Grove, G.F. Spencer, W.K. Rohwedder, N. Mandava, J.F. Worley, J.D. Warthen, G.L. Steffens, J.L. Flippen-Anderson, and J.C. Cook, Nature, 281, 216 (1979).

924. C. Grunwald, D.L. Davis, and L.P. Bush, J. Ag. Food Chem., 19, 138 (1971).

925. K.C. Gupta, R.L. Miller, J.R. Williams, R.B. Gagosian, and F. Heinzer, J. Nat. Products, 42, 305 (1979).

926. A.R. Guseva and V.A. Paseshnichenko, Biokhimiya, 27, 853 (1962); via Chem. Abstr., 58, 8240g (1963).

927. J. Gustafsson, J. Lipid Res., 17, 366 (1976).

928. J. Gustafsson, J. Lipid Res., 19, 237 (1978).

929. J.-Å. Gustafsson, E.G. Hrycay, and L. Ernster, Arch. Biochem. Biophys., 174, 440 (1976).

930. J. Gustafsson and S. Sjöstedt, J. Biol. Chem., 253, 199 (1978).

931. J.-Å. Gustafsson and J. Sjövall, Eur. J. Biochem., 8, 467 (1969).

932. M. Gut, Y. Letourneux, J.A. Story, S.A. Tepper, and D. Kritchevsky, Experientia, 30, 1325 (1974).

933. D.E. Guttman and P.D. Meister, J. Am. Pharm. Assoc., Sci. Ed., 47, 773 (1958).

934. J. Güttner, G. Bruns, W. Zschiesche, and M. Horn, Arch. Geschwulstforsch., 38, 10 (1971).

935. F. Haber and J. Weiss, Proc. Royal Soc. London, [A], 147, 332 (1934).

936. F. Haber and R. Willstätter, Ber., 64, 2844 (1931).

937. M.E. Haberland and J.A. Reynolds, Proc. Natl. Acad. Sci., 70, 2313 (1973).

938. J.T. Hackmann, Rec. Trav. Chim., 69, 433 (1950).

939. J.G. Haddad, S.J. Couranz, and L.V. Avioli, J. Clin. Endocrinol. Metabolism, 30, 174 (1970).

940. I.M. Hais, J. Chromatog., 33, 334 (1968).

941. I.M. Hais and N.B. Myant, Biochem. J., 87, 8P (1963).

942. I.M. Hais and N.B. Myant, Biochem. J., 94, 85 (1965).

943. P.F. Hall and S.B. Koritz, Biochemistry, 3, 129 (1964).

944. P.F. Hall and S.B. Koritz, Biochim. Biophys. Acta, 93, 441 (1964).

945. A.S. Hallsworth and H.B. Henbest, J. Chem. Soc., 3571 (1960).

946. S. Hamano, Biochem. Z., 163, 438 (1925).

947. S. Hamano, Biochem. Z., 169, 432 (1926).

948. G.A. Hamilton, J. Am. Chem. Soc., 86, 3391 (1964).

949. G.A. Hamilton, in MOLECULAR MECHANISMS OF OXYGEN ACTIVATION. O. Hayaishi, Editor, Academic Press, New York,

1974, pp. 405-451.

950. J.G. Hamilton and R.N. Castrejon, Federation Proc., 25, 221 (1966).

951. B.L. Hamman and M.M. Martin, Steroids, 10, 169 (1967).

952. R. Hampl and L. Stárka, J. Steroid Biochem., 1, 47 (1969).

953. K.H. Hanewald, F.J. Mulder, and K.J. Keuning, J. Pharm. Sci., 57, 1308 (1968).

954. R. Hanna and G. Ourisson, Bull. Soc. Chim. France, 1945 (1961).

955. R. Hanna and G. Ourisson, Bull. Soc. Chim. France, 3742 (1967).

956. E. Hansbury and T.J. Scallen, J. Lipid Res., 19, 742 (1978).

957. R. Hansson and K. Wikvall, Eur. J. Biochem., 93, 419 (1979).

958. K. Harano, T. Harano, K. Yamasaki, and D. Yoshioka, Proc. Japan Acad., 52, 453 (1976).

959. T. Harano and K. Harano, Kawasaki Med. J., 2, 175 (1976).

960. E. Hardegger, L. Ruzicka, and E. Tagmann, Helv. Chim. Acta, 26, 2205 (1943).

961. W.A. Harland, J.D. Gilbert, and C.J.W. Brooks, Biochim. Biophys. Acta, 316, 378 (1973).

962. W.A. Harland, J.D. Gilbert, G. Steel, and C.J.W. Brooks, Atherosclerosis, 13, 239 (1971).

963. F.M. Harold, S. Abraham, and I.L. Chaikoff, J. Biol. Chem., 221, 435 (1956).

964. C. Harries, Ber., 45, 936 (1912).

965. H. Hart and P.B. Lavrik, J. Org. Chem., 39, 1793 (1974).

966. J. Hartmann, Bull. Acad. Med. (Paris), 111, 26 (1934).

967. J. Hartmann, Bull. Acad. Med. (Paris), 112, 835 (1935).

968. J.L. Hartwell, SURVEY OF COMPOUNDS WHICH HAVE BEEN TESTED FOR CARCINOGENIC ACTIVITY, U.S. Public Health Service Publ. No. 149, Washington City, D.C., 1951, pp. 382-387.

969. J.L. Hartwell and B.J. Abbott, Adv. Pharmacol. Chemotherapy, 7, 117 (1969).

970. R.J. Harwood and E.M. Cohen, J. Soc. Cosmet. Chem., 28, 79 (1977).

971. G.A.D. Haslewood, Biochem. J., 33, 709 (1939).

972. G.A.D. Haslewood, Biochem. J., 35, 708 (1941).

973. G.A.D. Haslewood, Chem. Ind., 60, 220 (1941).

974. G.A.D. Haslewood, Biochem. J., 36, 389 (1942).

975. G.A.D. Haslewood, Nature, 154, 29 (1944).

976. T. Hasselstrom and B.L. Hampton, J. Am. Chem. Soc., $\underline{63}$ 111 (1941).

977. P.G. Hatcher, L.E. Keister, and P.A. McGillivary, Bull. Environ. Contamination Toxicol., $\underline{17}$, 491 (1977).

978. M.L. Hathaway and F.C. Koch, J. Biol. Chem., $\underline{108}$, 773 (1935).

979. J. Hattori, J. Pharm. Soc. Japan, $\underline{60}$, 125 (1940); Yakugaku Zasshi, $\underline{60}$, 334 (1940).

980. C. Havel, E. Hansbury, T.J. Scallen, and J.A. Watson, J. Biol. Chem., $\underline{254}$, 9573 (1979).

981. F.L. Haven and W.R. Bloor, Adv. Cancer Res., $\underline{4}$, 237 (1956).

982. E. Havinga, Experientia, $\underline{29}$, 1181 (1973).

983. E.G.E. Hawkins, ORGANIC PEROXIDES THEIR FORMATION AND REACTIONS, D. Van Nostrand Co. Inc., New York, 1961, pp. ix-xiv.

984. S. Hayakawa and Y. Sato, J. Org. Chem., $\underline{27}$, 704 (1962).

985. S. Hayakawa and Y. Sato, J. Org. Chem., $\underline{28}$, 2742 (1963).

986. M. Hayano, in OXYGENASES, O. Hayaishi, Editor, Academic Press, New York, 1962, pp. 181-240.

987. S.D. Hayes and L. Saunders, Biochim. Biophys. Acta, $\underline{116}$, 184 (1966).

988. R.K. Haynes, Austr. J. Chem., $\underline{31}$, 121 (1978).

989. R.K. Haynes, Austr. J. Chem., $\underline{31}$, 131 (1978).

990. R.B. Head, Nature, $\underline{191}$, 1058 (1961).

991. R.D.H. Heard, E. Schwenk, and T.F. Gallagher, Rec. Prog. Hormone Res., $\underline{9}$, 451 (1954).

992. E. Heftmann, CHROMATOGRAPHY OF STEROIDS, Elsevier Scientific Publishing Co., Amsterdam/Oxford/New York, 1976, p. 24, pp. 55-69.

993. E. Heftmann, S.-T. Ko, and R.D. Bennett, J. Chromatog., $\underline{21}$, 490 (1966).

994. E. Heftmann and M.L. Weaver, Phytochem., $\underline{13}$, 1801 (1974).

995. I.M. Heilbron, H. Jackson, E.R.H. Jones, and F.S. Spring, J. Chem. Soc., 102 (1938).

996. I.M. Heilbron, R.A. Morton, and W.A. Sexton, J. Chem. Soc., 47 (1928).

997. I.M. Heilbron and W.A. Sexton, J. Chem. Soc., 347 (1928).

998. I.M. Heilbron, J.C.E. Simpson, and F.S. Spring, J. Chem. Soc., 626 (1933).

999. H.-J. Heiniger, K.T. Brunner, and J.-C. Cerottini, Proc. Natl. Acad. Sci., $\underline{75}$, 5683 (1978).

1000. H.-J. Heiniger, A.A. Kandutsch, and H.W. Chen, Nature,

263, 515 (1976).

1001. W.L. Heinrichs, R.L. Musher, and A. Colás, Steroids, 9, 23 (1966).

1002. W. Heintz, Ann. Physik., 79, 524 (1850).

1003. R.A. Heller, R. Klotzbücher, and W. Stoffel, Proc. Natl. Acad. Sci., 76, 1721 (1979).

1004. O. Hellinger, Biophysik, 6, 63 (1969).

1005. O. Hellinger, H. Heusinger, and O. Hug, Biophysik, 6, 193 (1970).

1006. H.B. Henbest and E.R.H. Jones, Nature, 158, 950 (1946).

1007. H.B. Henbest and E.R.H. Jones, J. Chem. Soc., 1792 (1948).

1008. H.B. Henbest and E.R.H. Jones, J. Chem. Soc., 1798 (1948).

1009. H.B. Henbest, E.R.H. Jones, G.W. Wood, and G.F. Woods, J. Chem. Soc., 4894 (1952).

1010. H.B. Henbest and R.A.L. Williams, Chem. Ind., 86 (1956).

1011. A.E. Henderson, J. Histochem. Cytochem., 4, 153 (1956).

1012. A.E. Henderson and J.D.B. MacDougall, Biochem. J., 57, xxi (1954).

1013. W. Henderson, W.E. Reed, G. Steel, and M. Calvin, Nature, 231, 308 (1971).

1014. W. Henderson and G. Steel, J. Chem. Soc. Chem. Commun., 1331 (1971).

1015. W. Henderson, V. Wollrab, and G. Eglinton, Chem. Commun., 710 (1968).

1016. W. Henderson, V. Woolrab, and G. Eglinton, in ADVANCES IN ORGANIC GEOCHEMISTRY 1968, P.A. Schenck and I. Havenaar, Editors, Pergamon Press, Oxford, 1969, pp. 181-207.

1017. J.E. Herz, J. Lucero, Y. Santoyo, and E.S. Waight, Can. J. Chem., 49, 2418 (1971).

1018. A.F. Hess and M. Weinstock, J. Biol. Chem., 64, 181 (1925).

1019. A.F. Hess and M. Weinstock, J. Biol. Chem., 64, 193 (1925).

1020. G. Hesse, W. Geiger, and G. Lettenbauer, Liebigs Ann., 625, 167 (1959).

1021. O. Hesse, Liebigs Ann., 192, 175 (1870).

1022. H. Heymann and L.F. Fieser, Helv. Chim. Acta, 35, 631 (1952).

1023. R. Hiatt, in OXIDATION, R.L. Augustine and D.J. Trecker, Editors, Marcel Dekker Inc., New York, Vol. 2, 1971, pp. 113-140.

1024. R. Hiatt, in ORGANIC PEROXIDES, D. Swern, Editor,
      Wiley-Interscience, New York, Vol. 2, 1971, pp. 1-151.
1025. I. Hieger, Biochem. J., 24, 505 (1930).
1026. I. Hieger, Am. J. Cancer, 39, 496 (1940).
1027. I. Hieger, Science, 93, 262 (1941).
1028. I. Hieger, Cancer Res., 6, 657 (1946).
1029. I. Hieger, Brit. Med. Bull., 4, 360 (1947).
1030. I. Hieger, Nature, 160, 270 (1947).
1031. I. Hieger, Brit. J. Cancer, 3, 123 (1949).
1032. I. Hieger, Proc. Royal Soc. London, [B], 147, 84 (1957).
1033. I. Hieger, Brit. Med. Bull., 14, 159 (1958).
1034. I. Hieger, Acta Unio Int. Contra Cancrum, 15, 603
      (1959).
1035. I. Hieger, Brit. J. Cancer, 13, 439 (1959).
1036. I. Hieger, Brit. J. Cancer, 16, 716 (1962).
1037. I. Hieger, and S.F.D. Orr, Brit. J. Cancer, 8, 274
      (1954).
1038. E. Hietanen, O. Hänninen, M. Laitinen, and M. Lang,
      Pharmacology, 17, 163 (1978).
1039. E. Hietanen and M. Laitinen, Biochem. Pharmacol., 27,
      1095 (1978).
1040. J. Higginson, J.A. Dunn, and D.A. Sutton, Acta Unio
      Int. Contra Cancrum, 15, 607 (1959).
1041. J. Higginson, J.A. Dunn, and D.A. Sutton, Exp. Molec.
      Pathol., 3, 297 (1964).
1042. M.D. Higgs and D.J. Faulkner, Steroids, 30, 379 (1977).
1043. M.D. Higgs and D.J. Faulkner, J. Org. Chem., 43, 3454
      (1978).
1044. H. Hikino and Y. Hikino, Fortschr. Chem. Org. Natur-
      stoffe, 28, 256 (1970).
1045. M.J. Hill, Digestion, 11, 289 (1974).
1046. M.J. Hill, CRC Crit. Rev. Toxicol., 4, 32 (1975).
1047. M.J. Hill, J.S. Crowther, B.S. Drasar, G. Hawksworh,
      V. Aries, and R.E.O. Williams, Lancet, 1, 95 (1971).
1048. I.R. Hills, G.W. Smith, and E.V. Whitehead, J. Inst.
      Petroleum, 56, 127 (1970).
1049. T. Hirayama, F. Fujikawa, I. Yosioka, and I. Kitagawa,
      Chem. Pharm. Bull., 23, 693 (1975).
1050. H. Hirschmann, J. Am. Chem. Soc., 74, 5357 (1952).
1051. C.K. Ho and L. Tan, IRCS J. Med. Sci., 1, 14 (1973).
1052. Y.K. Ho, M.S. Brown, D.W. Bilheimer, and J.L. Gold-
      stein, J. Clin. Invest., 58, 1465 (1976).
1053. R.B. Hochberg, S. Ladany, and S. Lieberman, J. Biol.
      Chem., 251, 3320 (1976).
1054. R.B. Hochberg, P.D. McDonald, M. Feldman, and S. Lie-
      berman, J. Biol. Chem., 249, 1277 (1974).

1055. R.B. Hochberg, P.D. McDonald, M. Feldman, and S. Lie-
      berman, J. Biol. Chem., 251, 2087 (1976).
1056. R.B. Hochberg, P.D. McDonald, S. Ladany, and S. Lie-
      berman, J. Steroid Biochem., 6, 323 (1975).
1057. R.B. Hochberg, P.D. McDonald, L. Ponticorvo, and S.
      Lieberman, J. Biol. Chem., 251, 7336 (1976).
1058. P.C. Hoffman, C.M. Richman, R.C. Hsu, and S. Yachnin,
      Clin. Res., 27, 297A (1979).
1059. T.R. Hogness, A.E. Sidwell, and F.P. Zscheile, J.
      Biol. Chem., 120, 239 (1937).
1060. H.L. Holland and P.R.P. Diakow, Can. J. Chem., 57,
      436 (1979).
1061. H.L. Holland, P.R.P. Diakow, and G.J. Taylor, Can. J.
      Chem., 56, 3121 (1978).
1062. A. Hollerbach, Naturwissenschaften, 65, 202 (1978).
1063. R.T. Holman and P.R. Edmondson, Anal. Chem., 28, 1533
      (1956).
1064. E. Homberg, Fette Seif. Anstrich., 77, 8 (1975).
1065. E. Homberg, Fette Seif. Anstrich., 81, 148 (1979).
1066. E. Homberg and B. Bielefeld, J. Chromatog., 180, 83
      (1979).
1067. E. Homberg and A. Seher, Z. Lebensmitt. Untersuch.
      Forsch., 148, 133 (1972).
1068. E.M. Honeywell and C.E. Bills, J. Biol. Chem., 99,
      71 (1932-1933).
1069. D.H.S. Horn, J. Sci. Food Agric., 9, 632 (1958).
1070. D.H.S. Horn and D. Ilse, Chem. Ind., 524 (1956).
1071. D.H.S. Horn and D. Ilse, J. Chem. Soc., 2280 (1957).
1072. C. Horvath, J. Chromatog., 22, 52 (1966).
1073. T. Hoshita, J.Biochem., 61, 440 (1967).
1074. A. HovanHap, J.A. Omelon, G.R. Duncan, and G.C. Walker,
      Can. J. Microbiol., 11, 869 (1965).
1075. B.M. Howard, W. Fenical, J. Finer, K. Hirotsu, and J.
      Clardy, J. Am. Chem. Soc., 99, 6440 (1977).
1076. R.M. Hoyte and R.B. Hochberg, J. Biol. Chem., 254,
      2278 (1979).
1077. J. Hradec, Biochim. Biophys. Acta, 47, 149 (1961).
1078. J. Hradec, Acta Unio Int. Contra Cancrum, 20, 926
      (1964).
1079. J. Hradec, in TUMOR LIPIDS:BIOCHEMISTRY AND METABO-
      LISM, R. Wood, Editor, American Oil Chemists Society
      Press, Champaign, Ill., 1973, pp. 54-65.
1080. J. Hradec and L. Dolejs, Biochem. J., 105, 129 (1968).
1081. J. Hradec and J. Kruml, Nature, 185, 55 (1960).
1082. J. Hradec and J. Sommerau, J. Chromatog., 32, 230
      (1968).

1083. J. Hradec and A. Štroufová, Biochim. Biophys. Acta, 40, 32 (1960).

1084. E.G. Hrycay, J.-Å. Gustafsson, M. Ingelman-Sundberg, and L. Ernster, Biochem. Biophys. Res. Commun., 66, 209 (1975).

1085. E.G. Hrycay, J.-Å. Gustafsson, M. Ingelman-Sundberg, and L. Ernster, FEBS Lett., 56, 161 (1975).

1086. E.G. Hrycay, J.-Å. Gustafsson, M. Ingelman-Sundberg, and L. Ernster, Eur. J. Biochem., 61, 43 (1976).

1087. E.G. Hrycay and P.J. O'Brien, Biochem. J., 125, 12P (1971).

1088. E.G. Hrycay and P.J. O'Brien, Arch. Biochem. Biophys., 153, 480 (1972).

1089. E.G. Hrycay and P.J. O'Brien, Arch. Biochem. Biophys., 157, 7 (1973).

1090. E.G. Hrycay and P.J. O'Brien, Arch. Biochem. Biophys., 160, 230 (1974).

1091. E.G. Hrycay, P.J. O'Brien, J.E. van Lier, and G. Kan, Arch. Biochem. Biophys., 153, 495 (1972).

1092. J.C. Hsia, R.A. Long, F.E. Hruska, and H.D. Gesser, Biochim. Biophys. Acta, 290, 22 (1972).

1093. R.C. Hsu, J.R. Kanofsky, and S. Yachnin, Blood, 56, 109 (1980).

1094. W.-Y. Huang and W.G. Meinschein, Geochim. Cosmochim. Acta, 40, 323 (1976).

1095. W.-Y. Huang and W.G. Meinschein, Geochim. Cosmochim. Acta, 42, 1391 (1978).

1096. W.-Y. Huang and W.G. Meinschein, Geochim. Cosmochim. Acta, 43, 739 (1979).

1097. J.E. Huber, Tetrahedron Lett., 3271 (1968).

1098. W. Huber and O.W. Barlow, J. Biol. Chem., 149, 125 (1943).

1099. J.W. Huffman, J. Org. Chem., 41, 3847 (1976).

1100. L. Hugounenq and E. Couture, Bull. Soc. Chim. Biol., 11, 956 (1929).

1101. L. Hugounenq and E. Couture, Compt. Rend., 188, 349 (1929).

1102. L. Hugounenq and E. Couture, Compt. Rend., 188, 742 (1929).

1103. J.G. Huijmans, H.J. Degenhart, and D.J. Kortleve, J. Endocrinol., 72, 3P (1977).

1104. G.M.K. Humphries and H.M. McConnell, J. Immunol., 122, 121 (1979).

1105. I.R. Hunter, M.K. Walden, G.F. Bailey, and E. Heftmann, Lipids, 14, 687 (1979).

1106. I.R. Hunter, M.K. Walden, and E. Heftmann, J. Chromatog., 153, 57 (1978).

1107. H.R.B. Hutton and G.S. Boyd, Biochim. Biophys. Acta, 116, 336 (1966).

1108. H.R.B. Hutton and G.S. Boyd, Biochim. Biophys. Acta, 116, 362 (1966).

1109. K.-K. Hwang and M.I. Kelsey, Cancer Biochem. Biophys., 3, 31 (1978).

1110. IARC Monographs on the Evaluation of Carcinogenic Risk of Chemicals to Man. Some Naturally Occurring Substances, Vol. 10, Int. Agency for Research on Cancer, Lyon, 1976, pp. 99-111.

1111. S. Ichii, E. Forchielli, and R.I. Dorfman, Steroids, 2, 631 (1963).

1112. S. Ichii, S. Omata, and S. Kobayashi, Biochim. Biophys. Acta, 139, 308 (1967).

1113. S. Ichii, N. Yago, S. Kobayashi, and S. Omata, J. Biochem., 62, 740 (1967).

1114. D.R. Idler, M.W. Khalil, C.J.W. Brooks, C.G. Edmonds, and J.D. Gilbert, Comp. Biochem. Physiol., 59B, 163 (1978).

1115. D.R. Idler, M.W. Khalil, J.D. Gilbert, and C.J.W. Brooks, Steroids, 27, 155 (1976).

1116. D.R. Idler and L.M. Safe, Steroids, 19, 315 (1972).

1117. D.R. Idler, L.M. Safe, and S.H. Safe, Steroids, 16, 251 (1970).

1118. D.R. Idler, T. Tamura, and T. Wainai, J. Fish. Res. Bd. Canada, 21, 1035 (1964).

1119. D.R. Idler and P. Wiseman, Comp. Biochem. Physiol., 35, 679 (1970).

1120. D.R. Idler and P. Wiseman, Comp. Biochem. Physiol., 38A, 581 (1971).

1121. D.R. Idler and P. Wiseman, Int. J. Biochem., 2, 516 (1971).

1122. D.R. Idler, P.M. Wiseman, and L.M. Safe, Steroids, 16, 451 (1970).

1123. T. Iida, M. Kikuchi, T. Tamura, and T. Matsumoto, J. Lipid Res., 20, 279 (1979).

1124. R. Ikan, M.J. Baedecker, and I.R. Kaplan, Geochim. Cosmochim. Acta, 39, 195 (1975).

1125. R. Ikan and A. Bortinger, Israel J. Chem., 9, 679 (1971).

1126. R. Ikan, S. Harel, J. Kashman, and E.D. Bergmann, J. Chromatog., 14, 504 (1964).

1127. R. Ikan, A. Markus, and Z. Goldschmidt, J. Chem. Soc., Perkin Trans. I, 2423 (1972).

1128. S. Ikawa, J. Biochem., 82, 1093 (1977).

1129. S. Ikawa, M. Takita, and M. Ogura, J. Biochem., 85, 1147 (1979).
1130. S. Ikawa and K. Yamasaki, Yonago Acta Med., 15, 21 (1971).
1131. M. Ikeda and K. Yamasaki, J. Biochem., 83, 799 (1978).
1132. S. Ikegami, Y. Kamiya, and S. Tamura, Agr. Biol. Chem., 36, 1777 (1972).
1133. S. Ikegami, Y. Kamiya, and S. Tamura, Tetrahedron Lett., 1601 (1972).
1134. S. Ikegami, Y. Kamiya, and S. Tamura, Tetrahedron Lett., 3725 (1972).
1135. S. Ikegami, Y. Kamiya, and S. Tamura, Agr. Biol. Chem., 37, 367 (1973).
1136. S. Ikegami, Y. Kamiya, and S. Tamura, Tetrahedron Lett., 731 (1973).
1137. S. Ikegami, Y. Kamiya, and S. Tamura, Tetrahedron, 29, 1807 (1973).
1138. N. Ikekawa, Y. Fujimoto, M. Ishiguro, S. Suwa, Y. Hirayama, and H. Mizunuma, Science, 204, 1223 (1979).
1139. N. Ikekawa, M. Morisaki, and K. Hirayama, Phytochem., 11, 2317 (1972).
1140. N. Ikekawa, N. Morisaki, K. Tsuda, and T. Yoshida, Steroids, 12, 41 (1968).
1141. N. Ikekawa, K. Tsuda, and N. Morisaki, Chem. Ind., 1179 (1966).
1142. H. Imai, Federation Proc., 33, 623 (1974).
1143. H. Imai, Federation Proc., 38, 1457 (1979).
1144. H. Imai, Abstracts, Autoxidation Processes in Food and Related Biological Systems Workshop, U.S. Army Res. Dev. Command, Natick, Mass., Oct. 29-31, 1979.
1145. H. Imai, M. Kanisawa, K. Kojima, and N.T. Werthessen, Federation Proc., 36, 392 (1977).
1146. H. Imai, K.T. Lee, C.B. Taylor, and N.T. Werthessen, Circulation, 47/48, Suppl. IV, 42 (1973).
1147. H. Imai and H. Nakamura, Federation Proc., 39, 772 (1980).
1148. H. Imai, V. Subramanyam, P.W. LeQuesne, and N.T. Werthessen, Federation Proc., 37, 934 (1978).
1149. H. Imai, V. Subramanyam, N.T. Werthessen, and A.H. Soloway, 10th International Congress Angiology, Tokyo, Aug. 30-Sept. 3, 1976.
1150. H. Imai, N.T. Werthessen, V. Subramanyam, P.W. LeQuesne, A.H. Soloway, and M. Kanisawa, Science, 207, 651 (1980).
1151. H. Imai, N.T. Werthessen, C.B. Taylor, and K.T. Lee, Arch. Pathol. Lab Med., 100, 565 (1976).

1152. Y. Imai, S. Kikuchi, T. Matsuo, Suzuoki-Z., and
      K. Nishikawa, J. Atherosclerosis Res., 7, 671 (1967).
1153. A.A. Imshenetskiĭ, E.F. Efimochkina, L.E. Nikitin,
      and T.S. Nazarova, Dok. Akad. Nauk S.S.S.R., 170,
      960 (1966).
1154. A.A. Imshenetskiĭ, E.F. Efimochkina, V.A. Zanin, and
      L.E. Nikitin, Mikrobiologiya, 37, 31 (1968).
1155. A.A. Imshenetskii and L.A. Mavrina, Mikrobiologiya,
      37, 620 (1968).
1156. H. Inano and B.-I. Takaoki, Biochemistry, 10, 1503
      (1971).
1157. K.U. Ingold, Accts. Chem. Res., 2, 1 (1969).
1158. H.H. Inhoffen, Liebigs Ann., 497, 130 (1932).
1159. H.H. Inhoffen and H. Hauptmann, Z. Physiol. Chem.,
      207, 259 (1932).
1160. R. Irie, M. Iwanaga, and T. Yamakawa, J. Biochem.,
      50, 122 (1961).
1161. S. Isaka and P.F. Hall, Biochem. Biophys. Res. Com-
      mun., 43, 747 (1971).
1162. E. Iseli, M. Kotake, E. Weiss, and T. Reichstein,
      Helv. Chim. Acta, 48, 1093 (1965).
1163. R. Ishiwatari and T. Hanya, in ADVANCES IN ORGANIC
      GEOCHEMISTRY 1973, B. Tissot and F. Bienner, Editors,
      Éditions Technip, Paris, 1974, pp. 1051-1065.
1164. M. Ito, W.E. Connor, E.J. Blanchette, C.R. Treadwell,
      and G.V. Vahouny, J. Lipid Res., 10, 694 (1969).
1165. H. Itokawa, M. Akasu, and M. Fujita, Chem. Pharm.
      Bull., 21, 1386 (1973).
1166. H. Itokawa, Y. Tachi, Y. Kamano, and Y. Iitaka, Chem.
      Pharm. Bull., 26, 331 (1978).
1167. S. Iwasaki and K. Tsuda, Chem. Pharm. Bull., 11, 1034
      (1963).
1168. H.J.C. Jacobs, F. Boomsma, E. Havinga, and A. van der
      Gen, Rec. Trav. Chim., 96, 113 (1977).
1169. G.M. Jacobsohn and R.J. Montiel, Lipids, 4, 380 (1969).
1170. J. Jacques, H. Kagan, and G. Ourisson, CONSTANTES
      SÉLECTIONNÉES. POUVOIR ROTATOIRE NATUREL. Ia.
      STÉROÏDES, Pergamon Press, Oxford, 1965, p. 472.
1171. D.R. James and C.W. Shoppee, J. Chem. Soc., 4224
      (1954).
1172. M.J. James and A.A. Kandutsch, J. Biol. Chem., 254,
      8442 (1979).
1173. R.J. Jandacek, M.R. Webb, and F.H. Mattson, J. Lipid
      Res., 18, 203 (1977).
1174. H. Janecke and W. Fülberth, Arch. Pharm., 297, 193
      (1964).

1175. H. Janecke and G. Senft, Naturwissenschaften, 43, 502 (1956).

1176. H. Janecke and G. Senft, Pharmazie, 12, 555 (1957).

1177. H. Janecke and G. Senft, Pharmazie, 12, 673 (1957); via Chem. Abstr., 52, 10609h (1958)

1178. H. Janecke and H. Vögele, Arzneimittel Forsch., 15, 873 (1965).

1179. H. Janistyn, Fette Seifen, 47, 351 (1940).

1180. H. Janistyn, Fette Seifen, 47, 405 (1940).

1181. O. Jänne and R. Vihko, J. Steroid Biochem., 1, 177 (1970).

1182. C.R. Jefcoate, J. Biol. Chem., 250, 4663 (1975).

1183. C.R. Jefcoate, J. Biol. Chem., 252, 8788 (1977).

1184. C.R. Jefcoate, R. Hume, and G.S. Boyd, FEBS Lett., 9, 41 (1970).

1185. C.R. Jefcoate and W.H. Orme-Johnson, J. Biol. Chem., 250, 4671 (1975).

1186. C.R. Jefcoate, E.R. Simpson, and G.S. Boyd, Eur. J. Biochem., 42, 539 (1974).

1187. C.R. Jefcoate, E.R. Simpson, G.S. Boyd, A.C. Brownie, and W.H. Orme-Johnson, Ann. N.Y. Acad. Sci., 212, 243 (1973).

1188. J. Jeffery, J. Chromatog., 59, 216 (1971).

1189. L.M. Jeffrey, J. Am. Oil Chem. Soc., 43, 211 (1966).

1190. L.M. Jeffrey, B.F. Pasby, B. Stevenson, and D.W. Hood, in ADVANCES IN ORGANIC GEOCHEMISTRY, U. Colombo and G.D. Hobson, Editors, Pergamon Press-MacMillan Co., Oxford/New York City, 1964, pp. 175-197.

1191. C. Jett and J.E. Miller, Abstracts, 5th Int. Symp. Atherosclerosis, Houston, Texas, Nov. 6-9, 1979, Abstract No. 25.

1192. G. Johansson, Eur. J. Biochem., 21, 68 (1971).

1193. W.F. Johns, J. Org. Chem., 36, 2391 (1971).

1194. D.B. Johnson and L. Lack, J. Lipid Res., 17, 91 (1976).

1195. D.B. Johnson, M.P. Tyor, and L. Lack, J. Lipid Res., 17, 353 (1976).

1196. D.R. Johnson, D.R. Idler, V.W. Meloche, and C.A. Baumann, J. Am. Chem. Soc., 75, 52 (1953).

1197. W.S. Johnson, J.A. Marshall, J.F.W. Keana, R.W. Franck, D.G. Martin, and V.J. Bauer, Tetrahedron, Suppl. 8, II, 541 (1966).

1198. G.W. Johnston, and C.O. Lee, J. Am. Pharm. Assoc., Sci. Ed., 29, 236 (1940).

1199. R.A. Joly, H.H. Sauer, and J. Bonner, J. Labelled Compounds, 5, 80 (1969).

1200. R. Joly and C. Tamm, Tetrahedron Lett., 3535 (1967).
1201. R. Joly, J. Warnant, J. Jolly, and J. Mathieu, Compt.
      Rend., 258, 5669 (1964).
1202. H.I. Jones, U.S. Patent No. 2,048,008, July 21, 1936;
      via Chem. Abstr., 30, 6098^3 (1936).
1203. J.B. Jones and K.D. Gordon, Can. J. Chem., 50, 2712
      (1972).
1204. R.N. Jones, P. Humphries, E. Packard, and K. Dobriner,
      J. Am. Chem. Soc., 72, 86 (1950).
1205. E. Joseph and F. Radt, Editors, ELSEVIER'S ENCYCLO-
      PEDIA OF ORGANIC CHEMISTRY. VOL. 14.   TETRACYCLIC
      AND HIGHER CYCLIC COMPOUNDS. SERIES III. CARBOISO-
      CYCLIC CONDENSED COMPOUNDS, Elsevier Publishing Co.
      Inc., New York/Amsterdam, 1940, pp. 38-49.
1206. P. Joseph-Nathan, G. Mejía, and D. Abramo-Bruno, J.
      Am. Chem. Soc., 101, 1289 (1979).
1207. J. Joska, Ž. Procházka, J. Fajkoš, and F. Šorm, Coll.
      Czech. Chem. Commun., 38, 1398 (1973).
1208. E.F. Joy, J.D. Bonn, and A.J. Barnard, Thermochim.
      Acta, 2, 57 (1971).
1209. M. Julesz, I. Faredin, and I. Tóth, STEROIDS IN HUMAN
      SKIN, Akadémiai Kiadó, Budapest, 1971, pp. 109-120.
1210. B. Kadis, J. Steroid Biochem., 9, 75 (1978).
1211. F.F. Kadlubar, K.C. Morton, and D.M. Ziegler, Biochem.
      Biophys. Res. Commun., 54, 1255 (1973).
1212. G. Kakis, A. Kuksis, J.J. Myher, Adv. Exptl. Med.
      Biol., 82, 297 (1977).
1213. J. Kalvoda, K. Heusler, H. Ueberwasser, G. Anner, and
      A. Wettstein, Helv. Chim. Acta, 46, 1361 (1963).
1214. T. Kamei, Y. Takiguchi, H. Suzuki, M. Matsuzaki, and
      S. Nakamura, Chem. Pharm. Bull., 26, 2799 (1978).
1215. A.M. Kamel, A. Felmeister, and N.D. Wiener, J. Pharm.
      Sci., 59, 1807 (1970).
1216. A.M. Kamel, A. Felmeister, and N.D. Weiner, J. Lipid
      Res., 12, 155 (1971).
1217. A.M. Kamel, N.D. Weiner, and A. Felmeister, Chem.
      Phys. Lipids, 6, 225 (1971).
1218. A.M. Kamel, N.D. Weiner, and A. Felmeister, J. Colloid
      Interface Sci., 35, 163 (1971).
1219. Y. Kamiya, S. Ikegami, and S. Tamura, Tetrahedron
      Lett., 655 (1974).
1220. R. Kammereck, W.-H. Lee, A. Paliokas, and G.J. Schroep-
      fer, J. Lipid Res., 8, 282 (1967).
1221. A. Kanazawa and S. Teshima, Nippon Suisan Gakkaishi,
      37, 675 (1971); via Chem. Abstr., 75, 106684q (1971).
1222. A. Kanazawa, S.-I. Teshima, and T. Ando, Comp.

Biochem. Physiol., 57B, 317 (1977).

1223. A. Kanazawa, S. Teshima, T. Ando, and S. Tomita, Marine Biol., 34, 53 (1976).

1224. A. Kanazawa, S.-I. Teshima, and S.-I. Hyodo, Comp. Biochem. Physiol., 62B, 521 (1979).

1225. A.A. Kandutsch, J. Lipid Res., 7, 603 (1966).

1226. A.A. Kandutsch, Steroids, 10, 31 (1967).

1227. A.A. Kandutsch, Abstracts, Autoxidation Processes In Food and Related Biological Systems, U.S. Army Res. Dev. Command, Natick, Mass., Oct. 29-31, 1979.

1228. A.A. Kandutsch and H.W. Chen, J. Biol. Chem., 248, 8408 (1973).

1229. A.A. Kandutsch and H.W. Chen, J. Biol. Chem., 249, 6057 (1974).

1230. A.A. Kandutsch and H.W. Chen, J. Cell. Physiol., 85, 415 (1975).

1231. A.A. Kandutsch and H.W. Chen, J. Biol. Chem., 252, 409 (1977).

1232. A.A. Kandutsch and H.K. Chen, Lipids, 13, 704 (1978).

1233. A.A. Kandutsch, H.W. Chen., and H.-J. Heiniger, Science, 201, 498 (1978).

1234. A.A. Kandutsch, H.W. Chen, and E.P. Shown, Proc. Natl. Acad. Sci., 74, 2500 (1977).

1235. A.A. Kandutsch, H.-J. Heiniger, and H.W. Chen, Biochim. Biophys. Acta, 486, 260 (1977).

1236. A.A. Kandutsch and R.M. Packie, Arch. Biochem. Biophys., 140, 122 (1970).

1237. C. Kaneko, S. Yamada, A. Sugimoto, M. Ishikawa, S. Sasaki, and T. Suda, Tetrahedron Lett., 2339 (1973).

1238. C. Kaneko, S. Yamada, A. Sugimoto, M. Ishikawa, T. Suda, M. Suzuki, and S. Sasaki, J. Chem. Soc., Perkin Trans. I, 1104 (1975).

1239. I. Kaneko, Y. Hazama-Shimada, M. Kuroda, and A. Endo, Biochem. Biophys. Res. Commun., 76, 1207 (1977).

1240. K. Kaneko, H. Mitsuhashi, and K. Hirayama, Chem. Pharm. Bull., 17, 2031 (1969).

1241. K. Kaneko, S. Terada, N. Yoshida, and H. Mitsuhashi, Phytochem., 16, 791 (1977).

1242. N.L. Kantiengar and R.A. Morton, Biochem. J., 60, 25 (1955).

1243. N.L. Kantiengar and R.A. Morton, Biochem. J., 60, 28 (1955).

1244. N.L. Kantiengar and R.A. Morton, Biochem. J., 60, 30 (1955).

1245. N.L. Kantiengar and R.A. Morton, Biochem. J., 60, 34 (1955).

1246. G. Kapke and J. Baron, Biochem. Biophys. Res. Commun., 70, 1097 (1976).

1247. G.F. Kapke and J. Baron, Biochem. Pharmacology, 29, 845 (1980).

1248. G.F. Kapke, J.A. Redick, and J. Baron, J. Biol. Chem., 253, 8604 (1978).

1249. P. Karrer and A.R. Naik, Helv. Chim. Acta, 31, 1617, 2244 (1948).

1250. R.M. Kaschnitz and Y. Hatefi, Arch. Biochem. Biophys., 171, 292 (1975).

1251. M. Kasha and A.U. Khan, Ann. N.Y. Acad. Sci., 171, 5 (1970).

1252. Y. Kashio and J.A. Gardner, Biochem. J., 24, 1047 (1930).

1253. Y. Kashman and M. Rotem, Tetrahedron Lett., 1707 (1979).

1254. K. Katayama, Yonago Acta Med., 12, 93 (1968).

1255. K. Katayama and K. Yamasaki, Yonago Acta Med., 12, 103 (1968).

1256. T. Katkov and D.B. Gower, Biochem. J., 115, 16P (1969); 117, 533 (1970).

1257. N. Katsui, H. Matsue, T. Hirata, and T. Masamune, Bull. Chem. Soc. Japan, 45, 223 (1972).

1258. S.S. Katz, G.G. Shipley, and D.M. Small, J. Clin. Invest., 58, 200 (1976).

1259. H.P. Kaufmann and Y. Hamza, Fette Seif. Anstrich., 72, 432 (1970).

1260. H.P. Kaufmann, E. Vennekel, and Y. Hamza, Fette Seif. Anstrich., 72, 242 (1970).

1261. G.J. Kautsky, C.J. Bouboulis, R.R. Becker, and C.G. King, J. Biol. Chem., 233, 1340 (1958).

1262. J.C. Kawalek and A.W. Andrews, Federation Proc., 36, 844 (1977).

1263. N. Kawamura and T. Taketomi, Jap. J. Exp. Med., 43, 157 (1973).

1264. M. Kawata, M. Tohma, T. Sawaya, and M. Kimura, Chem. Pharm. Bull., 24, 3109 (1976).

1265. F. Kayser and R. Balat, Compt. Rend. Soc. Biol., 146, 1598 (1952).

1266. K. Kaziwara, H. Yokotani, Y. Imai, S. Ogiwara, and Y. Aramaki, Jap. J. Pharmacol., 17, 340 (1967).

1267. J.F.W. Keana and W.S. Johnson, Steroids, 4, 457 (1964).

1268. D.R. Kearns, Chem. Rev., 71, 395 (1971).

1269. D.R. Kearns, R.A. Hollins, A.U. Khan, R.W. Chambers, and P. Radlick, J. Am. Chem. Soc., 89, 5455 (1967).

1270. D.R. Kearns, R.A. Hollins, A.U. Khan, and P. Radlick,

J. Am. Chem. Soc., 89, 5456 (1967).

1271. E. Keeser, Biochem. Z., 154, 321 (1924).

1272. E. Keeser, Biochem. Z., 157, 166 (1925).

1273. E. Keinan and Y. Mazur, J. Org. Chem., 43, 1020 (1978).

1274. M. Keller, J. Clin. Endocrinol. Metab., 16, 1075 (1956).

1275. M. Keller and J. Weiss, J. Chem. Soc., 2709 (1950).

1276. E.W. Kellogg and I. Fridovich, J. Biol. Chem., 252, 6721 (1977).

1277. E.W. Kellogg and I. Fridovich, J. Biol. Chem., 250, 8812 (1975).

1278. M.I. Kelsey and R.J. Pienta, Cancer Lett., 6, 143 (1979).

1279. F.E. Kendall, W. Meyer, and M. Bevans, Federation Proc., 7, 273 (1948).

1280. F.E. Kendall, E.H. Mosbach, L.L. Abell, and W. Meyer, Federation Proc., 12, 449 (1953).

1281. E.L. Kennaway, Brit. Med. J., 2, 1 (1925).

1282. E.L. Kennaway, Biochem. J., 24, 497 (1930).

1283. E.L. Kennaway, in CANCER, R.W. Raven, Editor, Butterworth & Co. Publishers Ltd., London, 1957, Vol. 1, pp. 24-31.

1284. E.L. Kennaway and J.W. Cook, Chem. Ind., 10, 521 (1932).

1285. E.L. Kennaway and B. Sampson, J. Path. Bact., 31, 609 (1928).

1286. W.C. Kenney, T.P. Singer, M. Fukuyama, and Y. Miyake, J. Biol. Chem., 254, 4689 (1979).

1287. G. Kerényi, A. Szentirmai, and M. Natonek, Acta Microbiol. Acad. Sci. Hung., 22, 487 (1975).

1288. W.O. Kermack and P. MacCallum, Proc. Royal Soc. Edinburgh, 45, 71 (1924/1925).

1289. A.S. Keston and R. Carsiotis, Arch. Biochem. Biophys., 52, 282 (1954).

1290. J.A. Keverling Buisman, W. Stevens, and J.v.d. Vliet, Rec. Trav. Chim., 66, 83 (1947).

1291. B.B. Khanukaev, R.L. Vardanyan, and N.S. Khanukaeva, Khim. Kinet. Katal., 43 (1979); via Chem. Abstr., 92, 65316b (1980).

1292. T. Kido, M. Arakawa, and T. Kimura, J. Biol. Chem., 254, 8377 (1979).

1293. N. Kikuchi and T. Miki, Mikrochim. Acta, 89 (1978).

1294. S. Kikuchi, Y. Imai, Suzuoki-Ziro, T. Matsuo, and S. Noguchi, J. Pharmacol. Exptl. Therap., 159, 399 (1968).

1295. T. Kikuchi, T. Toyoda, M. Arimoto, M. Takayama, and
      M. Yamano, Yakugaku Zasshi, 89, 1358 (1969).
1296. M. Kimura, A. Fujino, K. Yamazaki, and T. Sawaya, Chem.
      Pharm. Bull., 24, 2301 (1976).
1297. M. Kimura, Y. Jin, and T. Sawaya, Chem. Pharm. Bull.,
      27, 710 (1979).
1298. M. Kimura, M. Kawata, and T. Sawaya, Chem. Pharm.
      Bull., 24, 2258 (1976).
1299. M. Kimura, M. Kawata, M. Tohma, A. Fujino, and K.
      Yamasaki, Tetrahedron Lett., 2021 (1970).
1300. M. Kimura, M. Kawata, M. Tohma, A. Fujino, K. Yamazaki,
      and T. Sawaya, Chem. Pharm. Bull., 20, 1883 (1972).
1301. M. Kimura and T. Muto, Chem. Pharm. Bull., 27, 109
      (1979).
1302. M. Kimura and T. Sawaya, Chem. Pharm. Bull., 25, 1073
      (1977).
1303. M. Kimura, M. Tohma, and T. Tomita, Chem. Pharm. Bull.,
      20, 2185 (1972).
1304. M. Kimura, M. Tohma, and T. Tomita, Chem. Pharm. Bull.,
      21, 2521 (1973).
1305. H. King, O. Rosenheim, and T.A. Webster, Biochem. J.,
      23, 166 (1929).
1306. M.M. King, E.K. Lai, and P.B. McCay, J. Biol. Chem.,
      250, 6496 (1975).
1307. J.F. Kingston, E. Benson, B. Gregory, and A.G. Fallis,
      J. Nat. Products, 42, 528 (1979).
1308. J.F. Kingston, B. Gregory, and A.G. Fallis, Tetrahed-
      ron Lett., 4261 (1977).
1309. J.F. Kingston, B. Gregory, and A.G. Fallis, J. Chem.
      Soc., Perkin Trans. I, 2064 (1979).
1310. R.A. Kinsella, F.E. Francis, R.T. Aexel, and P.R. Mohr,
      J. Clin. Endocrinol. Metab., 32, 801 (1971).
1311. A.H.M. Kirby, Cancer Res., 3, 519 (1943).
1312. A.H.M. Kirby, Cancer Res., 4, 94 (1944).
1313. A.H.M. Kirby, Cancer Res., 5, 129 (1945).
1314. A.H.M. Kirby, Brit. J. Cancer, 2, 70 (1948).
1315. R.Y. Kirdani and D.S. Layne, Biochemistry, 4, 330
      (1965).
1316. J. Kirimura, M. Saito, and M. Kobayashi, Nature, 195,
      729 (1962).
1317. A. Kisic, D. Menger, E.J. Parish, S. Satterfield,
      D.L. Raulston, and G.J. Schroepfer, Artery, 3, 421
      (1977).
1318. I. Kitagawa and M. Kobayashi, Tetrahedron Lett., 859
      (1977).
1319. I. Kitagawa and M. Kobayashi, Chem. Pharm. Bull., 26,

1864 (1978).

1320. I. Kitagawa, M. Kobayashi, and T. Sugiwara, Chem. Pharm. Bull., 26, 1852 (1978).

1321. I. Kitagawa, M. Kobayashi, T. Sugiwara, and I. Yosioka, Tetrahedron Lett., 967 (1975).

1322. I. Kitagawa, T. Sugawara, and I. Yosioka, Tetrahedron Lett., 4111 (1974).

1323. I. Kitagawa, T. Sugawara, I. Yosioka, and K. Kuriyama, Tetrahedron Lett., 963 (1975).

1324. M. Kitagiri, O. Takikawa, H. Sato, and K. Suhara, Biochem. Biophys. Res. Commun., 77, 804 (1977).

1325. F. Kitame, H. Saito, and N. Ishida, Jpn. Kokai Tokkyo 78,104,735, 12 September 1978; via Chem. Abstr., 90, 127552v (1979).

1326. E. Klappauf and D. Schubert, Z. Physiol. Chem., 360, 1225 (1979).

1327. A.J. Klein and W.L. Palmer, Arch. Pathol., 29, 814 (1940): J. Natl. Cancer Inst., 1, 559 (1941).

1328. B. Klein, N.B. Kleinman, and J.A. Foreman, Clin. Chem., 20, 482 (1974).

1329. P.D. Klein, J.C. Knight, and P.A. Szczepanik, J. Am. Oil Chem. Soc., 43, 275 (1966).

1330. P.D. Klein and P.A. Szczepanik, J. Lipid Res., 3, 460 (1962).

1331. F. Klötzer, Z. Kristallograph., 95A, 338 (1936).

1332. W. Klyne, Helv. Chim. Acta, 35, 1224 (1952).

1333. W. Klyne, THE CHEMISTRY OF THE STEROIDS, Methuen & Co., Ltd., London, 1957, pp. 107-112.

1334. F.F. Knapp and G.J. Schroepfer, Chem. Phys. Lipids, 17, 466 (1976).

1335. F.F. Knapp, M.S. Wilson, and G.J. Schroepfer, Chem. Phys. Lipids, 16, 31 (1976).

1336. J.C. Knight and G.R. Petit, Phytochem., 8, 477 (1969).

1337. B.A. Knights, J. Gas Chromatog., 5, 273 (1967).

1338. B.A. Knights, Phytochem., 9, 903 (1970).

1339. B.A. Knights, Anal. Biochem., 80, 324 (1977).

1340. M. Kobayashi, T. Hayashi, F. Nakajima, and H. Mitsuhashi, Steroids, 34, 285 (1979).

1341. M. Kobayashi and H. Mitsuhashi, Steroids, 24, 399 (1974).

1342. M. Kobayashi and H. Mitsuhashi, Tetrahedron, 30, 2147 (1974).

1343. M. Kobayashi and H. Mitsuhashi, Steroids, 26, 605 (1975).

1344. M. Kobayashi, M. Nishizawa, K. Todo, and H. Mitsuhashi, Chem. Pharm. Bull., 21, 323 (1973).

1345. M. Kobayashi, K. Todo, and H. Mitsuhashi, Chem. Pharm. Bull., 22, 236 (1974).
1346. M. Kobayashi, R. Tsuru, K. Todo, and H. Mitsuhashi, Tetrahedron Lett., 2935 (1972).
1347. K. Kobayashi, R. Tsuru, K. Todo, and H. Mitsuhashi, Tetrahedron, 29, 1193 (1973).
1348. S. Kobayashi and S. Ichi, J. Biochem., 66, 51 (1969).
1349. T. Kobayashi, J. Vitaminology, 13, 268 (1967).
1350. E.M. Koch and F.C. Koch, J. Biol. Chem., 116, 757 (1936).
1351. E.M. Koch, F.C. Koch, and H.B. Lemon, J. Biol. Chem., 85, 159 (1929/1930).
1352. F.C. Koch, E.M. Koch, and I.K. Ragins, J. Biol. Chem., 85, 141 (1929).
1353. R. Koch, Naturwissenschaften, 45, 134 (1958).
1354. R. Koch, Arzneimittel Forsch., 13, 1116 (1963).
1355. R. Koch and G.O. Schenck, Strahlentherapie, 116, 364 (1961).
1356. R. Koch and G.O. Schenck, Naturwissenschaften, 54, 172 (1967).
1357. R. Koch, G.O. Schenck, and O.-A. Neumüller, 124, Strahlentherapie, 124, 626 (1964).
1358. M. Kocor and A. Schmidt-Szalowska, Bull. Acad. Pol. Sci., Ser. Sci. Chim., 20, 515 (1972); via Chem. Abstr., 77, 137340v (1972).
1359. L. Kofler and A. Kofler, Z. Physiol. Chem., 204, 211 (1932).
1360. L.M. Kogan, E.A. Elin, A.S. Barmenkov, and I.V. Torgov, Izv. Akad. Nauk S.S.S.R., Ser. Khim., 2016 (1964).
1361. L. Kohout and J. Fajkoš, Coll. Czech. Chem. Commun., 43, 1134 (1978).
1362. N. Koizumi, Y. Fujimoto, T. Takeshita, and N. Ikekawa, Chem. Pharm. Bull., 27, 38 (1979).
1363. N. Koizumi, M. Morisake, N. Ikekawa, A. Suzuki, and T. Takeshita, Tetrahedron Lett., 2203 (1975).
1364. S. Kominami, H. Ochi, and S. Takemori, Biochim. Biophys. Acta, 577, 170 (1979).
1365. S.H. Kon, Med. Hypotheses, 4, 559 (1978).
1366. S.-K. Kon, F. Daniels, and H. Steenbock, J. Am. Chem. Soc., 50, 2573 (1928).
1367. Y. Kondo, J.A. Waters, B. Witkop, D. Guenard, and R. Beugelmans, Tetrahedron, 28, 797 (1972).
1368. C. Konno and H. Hikino, Tetrahedron, 32, 325 (1976).
1369. W.H. Koppenol and J. Butler, FEBS Lett., 83, 1 (1977).
1370. M. Koreeda and K. Koizumi, Tetrahedron Lett., 1641 (1978).

1371. S.B. Koritz and P.F. Hall, Biochemistry, 3, 1298 (1964).

1372. S.B. Koritz and P.F. Hall, Biochim. Biophys. Acta, 93, 215 (1964).

1373. L. Kornel and K. Motohashi, Steroids, 6, 9 (1965).

1374. T. Koschinsky, T.E. Carew, and D. Steinberg, J. Lipid Res., 18, 451 (1977).

1375. P.T. Kovanen, J.R. Faust, M.S. Brown, and J.L. Goldstein, Endocrinology, 104, 599 (1979).

1376. Y. Kōzaki, M. Makiguchi, Y. Nagase, and S. Baba, J. Pharm. Soc. Japan, 81, 93 (1961).

1377. Z.P. Kozhina, N.A. Bogosklovskii, V.B. Spirichev, and E.V. Kabanova, Khim. Priorod. Soed., 7, 480 (1971).

1378. R.J. Kraaipoel, H.J. Degenhart, and J.G. Leferink, FEBS Lett., 57, 294 (1975).

1379. R.J. Kraaipoel, H.J. Degenhart, J.G. Leferink, V. van Beek, H. de Leeuw-Boon, and H.K.A. Visser, J. Steroid Biochem., 5, 308 (1974).

1380. R.J. Kraaipoel, H.J. Degenhart, J.G. Leferink, V. van Beek, H. de Leeuw-Boon, and H.K.A. Visser, FEBS Lett., 50, 204 (1975).

1381. R.J. Kraaipoel, H.J. Degenhart, V. van Beek, H. de Leeuw-Boon, G. Abeln, H.K.A. Visser, and J.G. Leferink, FEBS Lett., 54, 172 (1975).

1382. B. Kramer, M.J. Shear, and D.H. Shelling, J. Biol. Chem., 71, 221 (1926).

1383. J.K.G. Kramer and H.W. Hulan, J. Lipid Res., 17, 674 (1976).

1384. A. Krámli and J. Horváth, Nature, 162, 619 (1948).

1385. A. Krámli and J. Horváth, Nature, 163, 219 (1949).

1386. M. Krieger, M.S. Brown, J.R. Faust, and J.L. Goldstein, J. Biol. Chem., 253, 4093 (1978).

1387. M. Krieger, J.L. Goldstein, and M.S. Brown, Proc. Natl. Acad. Sci., 75, 5052 (1978).

1388. D. Kritchevsky, CHOLESTEROL, John Wiley and Sons, New York, 1958, p. 24, p. 46.

1389. D. Kritchevsky and S.A. Tepper, Proc. Soc. Exptl. Biol. Med., 116, 104 (1964).

1390. D. Kritchevsky and S.A. Tepper, Experientia, 22, 84 (1966).

1391. D. Kritchevsky, S.A. Tepper, D.M. Klurfeld, and J.A. Story, Nutrition Rpts. Int., 20, 663 (1979).

1392. A.M. Krubiner, G. Saucy, and E.P. Oliveto, J. Org. Chem., 33, 3548 (1968).

1393. T. Kubota and Y. Shimazu, J. Inst. Polytech. Osaka City Univ., [C], 3, 83 (1952); via Chem. Abstr., 48,

10760c (1954).

1394. T. Kubota and Y. Shimazu, J. Inst. Polytech. Osaka
City Univ., [C], 3, 91 (1952); via Chem. Abstr., 48,
10760f (1954).

1395. T. Kubota and Y. Shimazu, Proc. Japan Acad., 28, 80
(1952); via Chem. Abstr., 47, 2763d (1953).

1396. J. Kučera, Ž. Procházka, and K. Vereš, Coll. Czech.
Chem. Commun., 22, 1185 (1957); Chem. Listy, 51, 97
(1957).

1397. I. Kühl and K. Schubert, Experientia, 16, 549 (1960).

1398. R. Kuhn and K. Meyer, Z. Physiol. Chem., 185, 193
(1929).

1399. A. Kuksis, P. Child, J.J. Myher, L. Marai, I.M. Yousef,
and P.K. Lewin, Can. J. Biochem., 56, 1141 (1978).

1400. A. Kuksis, L. Marai, J.J. Myher, and K. Geher, Lipids,
11, 581 (1976).

1401. M.J. Kulig and L.L. Smith, J. Org. Chem., 38, 3639
(1973).

1402. M.J. Kulig and L.L. Smith, J. Org. Chem., 39, 3398
(1974).

1403. M.J. Kulig and L.L. Smith, J. Steroid Biochem., 5,
485 (1974).

1404. M.J. Kulig, J.I. Teng, and L.L. Smith, Lipids, 10, 93
(1975).

1405. F.A. Kummerow, Abstracts, Autoxidation Processes in
Food and Related Biological Systems Workshop, U.S.
Army Res. Dev. Command, Natick, Mass., Oct. 29-31,
1979.

1406. S.M. Kupchan, C.W. Sigel, L.J. Guttman, R.J. Restivo,
and R.F. Bryan, J. Am. Chem. Soc., 95, 1353 (1972).

1407. S.M. Kupchan, C.J. Sih, S. Kubota, and A.M. Rahim,
Tetrahedron Lett., 1767 (1963).

1408. M. Kuroda, H. Werbin, and I.L. Chaikoff, Anal. Biochem., 9, 75 (1964).

1409. Y. Kurosawa, Nippon Nõgei Kagaku Kaishi, 32, 415
(1958); via Chem. Abstr., 53, 11510a (1959).

1410. G. Kusano, T. Takemoto, J.A. Beisler, and D.F. Johnson, Phytochem., 14, 1679 (1975).

1411. G. Kusano, T. Takemoto, J.A. Beisler, and Y. Sato,
Phytochem., 14, 529 (1975).

1412. W. Küster and K. Kimmich, Z. Physiol. Chem., 172, 199
(1927).

1413. C.N. Kwong, R.E. Heikkila, and D.G. Cornwell, J. Lipid
Res., 12, 31 (1971).

1414. J.A. Labarrère, J.R. Chipault, and W.O. Lundberg, Anal.
Chem., 30, 1466 (1958).

1415. L.C. Labowitz, Thermochim. Acta, 3, 419 (1972).
1416. A. Lacassagne, N.P. Buu-Hoï, and F. Zajdela, Nature, 209, 1026 (1966).
1417. R.E. Lack and A.B. Ridley, J. Chem. Soc., (C), 3017 (1968).
1418. C.-S. Lai and L.H. Piette, Biochem. Biophys. Res. Commun., 78, 51 (1977).
1419. M. Laitinen, Acta Pharmacol. Toxicol., 39, 241 (1976).
1420. F.W. Lamb, J. Physiol., 48, lvi (1914).
1421. A.A. Lamola, T. Yamane, and A.M. Trozzolo, Science, 179, 1131 (1973).
1422. M. Lang, M. Laitinen, E. Hietanen, and H. Vainio, Acta Pharmacol. Toxicol., 39, 273 (1976).
1423. R. Langdon, S. El-Masry, and R.E. Counsell, J. Lipid Res., 18, 24 (1977).
1424. W.E. Lange and M.E. Amundson, J. Pharm. Sci., 51, 1102 (1962).
1425. R.J. Langenbach and H.W. Knoche, Steroids, 11, 123 (1968).
1426. K. Langheld, Ber., 41, 1023 (1908).
1427. R. Langlois and J.E. van Lier, Proc. Can. Fed. Biol. Soc., 15, 44 (1972).
1428. A. Lapworth and F.A. Royle, J. Pathol. Bacteriol., 19, 474 (1915).
1429. C.D. Larsen and M.K. Barrett, J. Natl. Cancer Inst., 4, 587 (1944).
1430. S. Lassen and E. Geiger, Proc. Soc. Exptl. Biol. Med., 53, 181 (1943).
1431. G.D. Laubach, E.C. Schreiber, E.J. Agnello, and K.J. Brunings, J. Am. Chem. Soc., 78, 4746 (1956).
1432. G.D. Laubach, E.C. Schreiber, E.J. Agnello, E.N. Lightfoot, and K.J. Brunings, J. Am. Chem. Soc., 75, 1514 (1953).
1433. D. Lavie and E. Glotter, Fortschr. Chem. Org. Naturstoffe, 29, 307 (1971).
1434. D. Lavie and I.A. Kaye, J. Chem. Soc., 5001 (1963).
1435. U. Lavy, S. Burstein, M. Gut, and N.B. Javitt, J. Lipid Res., 18, 232 (1977).
1436. A.S.C. Lawrence, LIQUID CRYSTALS AND ORDERED FLUIDS, J.F. Johnson and R.S. Porter, Editors, Pergamon Press, New York/London, 1970, pp. 289-291.
1437. G.F. Laws, J. Chem. Soc., 4185 (1953).
1438. D.S. Layne, T. Golab, K. Arai, and G. Pincus, Biochem. Pharmacol., 12, 905 (1963).
1439. C. Le Borgne de Kaouël, J. Duron, C. Aubert, and P. Juret, J. Chromatog., 27, 282 (1967).

1440. E. Lederer, F. Marx, D. Mercier, and G. Pérot, Helv. Chim. Acta, 29, 1354 (1946).

1441. A. Lee, H.M. Dyer, M.S. Murray, F.J. Letkiewicz, and M.R. Smith, SURVEY OF COMPOUNDS WHICH HAVE BEEN TESTED FOR CARCINOGENIC ACTIVITY, 1972-1973 VOLUME, U.S. Public Health Service, Washington City, D.C., no date, pp. 530-531, 934-936, 974.

1442. C. Lee, R.B. Gagosian, and J.W. Farrington, Geochim. Cosmochim. Acta, 41, 985 (1977).

1443. K.T. Lee, H. Imai, N.T. Werthessen, and C.B. Taylor, in ATHEROSCLEROSIS III, PROCEEDINGS OF THE THIRD INTERNATIONAL SYMPOSIUM, G. Schettler and A. Weizel, Editors, Springer-Verlag, Berlin, 1974, pp. 344-347.

1444. P.K. Lee, D.P. Carew, and J. Rosazza, Lloydia, 35, 150 (1972).

1445. S.-G. Lee and C. Chen, Federation Proc., 31, 879 Abs (1972).

1446. E. Lee-Ruff, Chem. Soc. Rev., 6, 195 (1977).

1447. A. Lefèvre, A.-M. Morera, and J.M. Saez, FEBS Lett., 89, 287 (1978).

1448. A. Lefèvre and J.M. Saez, Compt. Rend., [D], 284, 561 (1977).

1449. T. Legatt and E.L. Shapiro, U.S. Patent No. 3,186,986, June 1, 1965; via Chem. Abstr., 63, 14941d (1965).

1450. T. Legatt and E.L. Shapiro, U.S. Patent No. 3,280,157, Oct. 18, 1966; via Chem. Abstr., 66, P29000y (1967).

1451. M. Legrand and R. Viennet, Compt. Rend., [C], 262, 1290 (1966).

1452. T. Lehtinen, E. Elomaa, J. Alhojärvi, and H.H. Bruun, Suomen Kemistilehti, 38, 127 (1965).

1453. R.A. LeMahieu, A. Boris, M. Carson, and R.W. Kierstead, J. Med. Chem., 14, 629 (1971).

1454. J.R. Lenton, L.J. Goad, and T.W. Goodwin, Phytochem., 12, 1135 (1973).

1455. Y. Letourneaux, Q. K.-Huu, M. Gut, and G. Lukacs, J. Org. Chem., 40, 1674 (1975).

1456. H. Lettré, Progr. Exp. Tumor Res., 1, 329 (1960).

1457. H. Lettré and O. Flandre, Rev. Franç. Etud. Clin. Biol., 4, 335 (1959).

1458. H. Lettré and D. Hotz, Liebigs Ann., 620, 63 (1959).

1459. H. Lettré and H.H. Inhoffen, ÜBER STERINE, GALLENSÄUREN UND VERWANDTE NATURSTOFFE, HERZGIFTE, HORMONE, SAPONINE, UND VITAMIN D, F. Enke, Stuggart, 1936, 320 pp.

1460. H.Lettré, H.H. Inhoffen, and R. Tschesche, ÜBER STERINE, GALLENSÄUREN UND VERWANDTE NATURSTOFFE, F. Enke, Stuttgart, 1954, 2 vols.

1461. H. Lettré and A. Jahn, Agnew. Chem., <u>69</u>, 266 (1957).

1462. H. Lettré and A. Jahn, Liebigs Ann., <u>608</u>, 43 (1957).

1463. H. Lettré, L. Knof, and A. Egle, Liebigs Ann., <u>640</u>, 168 (1961).

1464. D. Levy and R. Stevenson, J. Org. Chem., <u>30</u>, 3469 (1965).

1465. M.L. Lewbart, Nature, <u>222</u>, 663 (1969).

1466. G.N. Lewis, Chem. Rev., <u>1</u>, 243 (1924).

1467. H.L. Lewis and G.E. Peterson, Texas Rpts. Biol. Med., <u>20</u>, 144 (1962).

1468. K. Li, T. Foo, and J.B. Adams, Steroids, <u>31</u>, 113 (1978).

1469. A.H. Lichtenstein, P. Brecher, and A.V. Chobanian, Federation Proc., <u>39</u>, 1039 (1980).

1470. S. Lieberman, L. Bandy, V. Lippman, and K.D. Roberts, Biochem. Biophys. Res. Commun., <u>34</u>, 367 (1969).

1471. S. Lieberman, K. Dobriner, B.R. Hill, L.F. Fieser, C.P. Rhoads, and L.B. Hariton, J. Biol. Chem., <u>172</u>, 263 (1948).

1472. C. Liebermann, Ber., <u>18</u>, 1803 (1885).

1473. I. Lifschütz, Deutsch. Med. Woch., Therapeut. Beilage, <u>23</u>, No. 6, 44 (1897).

1474. I. Lifschütz, Pharm. Ztg., <u>43</u>, 230 (1898).

1475. I. Lifschütz, Z. Physiol. Chem., <u>50</u>, 436 (1907).

1476. I. Lifschütz, Z. Physiol. Chem., <u>53</u>, 140 (1907).

1477. I. Lifschütz, Ber., <u>41</u>, 252 (1908).

1478. I. Lifschütz, Z. Physiol. Chem., <u>55</u>, 1 (1908).

1479. I. Lifschütz, Z. Physiol. Chem., <u>58</u>, 175 (1908).

1480. I. Lifschütz, Z. Physiol. Chem., <u>63</u>, 222 (1909).

1481. I. Lifschütz, Biochem. Z., <u>48</u>, 373 (1913).

1482. I. Lifschütz, Biochem. Z., <u>52</u>, 208 (1913).

1483. I. Lifschütz, Biochem. Z., <u>54</u>, 212 (1913).

1484. I. Lifschütz, Ber., <u>47</u>, 1459 (1914).

1485. I. Lifschütz, Biochem. Z., <u>62</u>, 219 (1914).

1486. I. Lifschütz, Z. Physiol. Chem., <u>91</u>, 309 (1914).

1487. I. Lifschütz, Z. Physiol. Chem., <u>92</u>, 383 (1914).

1488. I. Lifschütz, Z. Physiol. Chem., <u>93</u>, 209 (1914).

1489. I. Lifschütz, Z. Physiol. Chem., <u>96</u>, 342 (1916).

1490. I. Lifschütz, Z. Physiol. Chem., <u>101</u>, 89 (1916).

1491. I. Lifschütz, Biochem. Z., <u>83</u>, 18 (1917).

1492. I. Lifschütz, Z. Physiol. Chem., <u>106</u>, 271 (1919).

1493. I. Lifschütz, Z. Physiol. Chem., <u>110</u>, 29 (1920).

1494. I. Lifschütz, Z. Physiol. Chem., <u>114</u>, 108 (1921).

1495. I. Lifschütz, Z. Physiol. Chem., <u>117</u>, 201 (1921).

1496. I. Lifschütz, Z. Physiol. Chem., <u>117</u>, 212 (1921).

1497. I. Lifschütz, Biochem. Z., <u>129</u>, 115 (1922).

1498. I. Lifschütz, Z. Physiol. Chem., <u>141</u>, 146 (1924).

1499. I. Lifschütz, Arch. Pharm., <u>265</u>, 450 (1927).

1500. I. Lifschütz, Arch. Pharm., <u>266</u>, 518 (1928).

1501. I. Lufschütz, Chem.-Ztg., <u>52</u>, 609 (1928).

1502. I. Lifschütz, Arch. Pharm., <u>268</u>, 166 (1930).

1503. I. Lifschütz, Arch. Pharm., <u>270</u>, 205 (1932).

1504. I. Lifschütz, Arch. Pharm., <u>270</u>, 253 (1932).

1505. I. Lifschütz, Biochem. Z., <u>280</u>, 65 (1935).

1506. I. Lifschütz and T. Grethe, Ber., <u>47</u>, 1453 (1914).

1507. D.A. Lightner and R.D. Norris, New England J. Med.,
      <u>290</u>, 1260 (1974).

1508. C.-C. Lin and C. Chen, Biochim. Biophys. Acta, <u>192</u>,
      133 (1969).

1509. D.S. Lin, W.E. Connor, L.K. Napton, and R.F. Heizer,
      J. Lipid Res., <u>19</u>, 215 (1978).

1510. R.C. Lin and P.J. Snodgrass, FEBS Lett., <u>83</u>, 89 (1977).

1511. Y.Y. Lin and L.L. Smith, Biochim. Biophys. Acta, <u>384</u>,
      189 (1974).

1512. Y.Y. Lin and L.L. Smith, J. Labelled Compounds, <u>10</u>,
      541 (1974).

1513. Y.Y. Lin and L.L. Smith, J. Neurochem., <u>25</u>, 659 (1975).

1514. Y.Y. Lin and L.L. Smith, Biomed. Mass Spectrometry, <u>5</u>,
      604 (1978).

1515. Y.Y. Lin and L.L. Smith, Biomed. Mass Spectrometry, <u>6</u>,
      15 (1979).

1516. O. Lindenmeyer, J. Prakt. Chem., [1], <u>90</u>, 321 (1863).

1517. O. Lindenmeyer, J. Prakt. Chem., [1], <u>90</u>, 331 (1863).

1518. S. Lindstedt, Acta Chem. Scand., <u>11</u>, 417 (1957).

1519. K. Lippel, S. Ahmed, J.J. Albers, P. Bachorik, G.
      Cooper, R. Helms, and J. Williams, Clin. Chem., <u>23</u>,
      1744 (1977).

1520. A.E. Lippman, E.W. Foltz, and C. Djerassi, J. Am.
      Chem. Soc., <u>77</u>, 4364 (1955).

1521. C. Little, Biochim. Biophys. Acta, <u>284</u>, 375 (1972).

1522. W.-B. Lo and H.S. Black, Experientia, <u>27</u>, 1397 (1971).

1523. W.-B. Lo and H.S. Black, J. Invest. Dermatol., <u>58</u>,
      278 (1972).

1524. W.-B. Lo and H.S. Black, Nature, <u>246</u>, 489 (1973).

1525. W.-B. Lo, H.S. Black, and J.M. Knox, Clin. Res., <u>22</u>,
      618A (1974).

1526. K.H. Loke and D.B. Gower, Biochem. J., <u>127</u>, 545 (1972).

1527. F.J. Loomeijer, Biochim. Biophys. Acta, <u>29</u>, 168 (1958).

1528. C.R. Loomis, G.G. Shipley, and D.M. Small, J. Lipid
      Res., <u>20</u>, 525 (1979).

1529. N.G. Lordi, A. Felmeister, and N.D. Weiner, J. Pharm.
      Sci., <u>60</u>, 933 (1971).

1530. E. Lorenz, M.B. Shimkin, and H.L. Stewart, J. Natl. Cancer Inst., 1, 355 (1940).

1531. O. Lossen, R. Brennecke, and D. Schubert, Biochim. Biophys. Acta, 330, 132 (1973).

1532. E.S. Lower, Soap Perfumery Cosmetics, 18, 125 (1945).

1533. E.S. Lower, Soap Perfumery Cosmetics, 18, 219 (1945).

1534. E.S. Lower, Soap Perfumery Cosmetics, 18, 299 (1945).

1535. E.S. Lower, Drug Cosmetic Ind., No. 2, 54 (1975).

1536. E.S. Lower, Drug Cosmetic Ind., No. 3, 48 (1975).

1537. E.S. Lower, Drug Cosmetic Ind., No. 4, 57 (1975).

1538. R.R. Lowry, J. Lipid Res., 9, 397 (1968).

1539. M.C. Lu, P. Afiatpour, C.B. Sullivan, and R.E. Counsell, J. Med. Chem., 15, 1284 (1972).

1540. N.S. Lucas, Biochem. J., 20, 23 (1926).

1541. H. Ludwig, J. Reiner, and G. Spiteller, Chem. Ber., 110, 217 (1977).

1542. W.O. Lundberg, Editor, AUTOXIDATION AND ANTIOXIDANTS IN TWO VOLUMES, Interscience Publishers, New York, 1961, 450 pp.

1543. B. Luttrell, R.B. Hochberg, W.R. Dixon, P.D. McDonald, and S. Lieberman, J. Biol. Chem., 247, 1462 (1972).

1544. B. Lythgoe and S. Trippett, J. Chem. Soc., 471 (1959).

1545. I.A. Macdonald, G. Singh, D.E. Mahony, and C.E. Meier, Steroids, 32, 245 (1978).

1546. J.D.B. MacDougall, S. Biswas, and R.P. Cook, Brit. J. Exptl. Pathol., 46, 549 (1965).

1547. J.D.B. MacDougall, S. Biswas, and R.P. Cook, J. Anatomy, 99, 211 (1965).

1548. D.I. Macht, J. Pharmacol. Exptl. Therap., 24, 213 (1924/1925).

1549. R.M.B. MacKenna, V.R. Wheatley, and A. Wormall, Biochem. J., 52, 161 (1952).

1550. A.M. Mackie and A.B. Turner, Biochem. J., 117, 543 (1970).

1551. H.B. MacPhillamy, J. Am. Chem. Soc., 62, 3518 (1940).

1552. D.K. Madan and D.E. Cadwallader, J. Pharm. Sci., 59, 1362 (1970).

1553. O.L. Mageli and C.S. Sheppard, in ORGANIC PEROXIDES, D. Swern, Editor, Wiley-Interscience, New York, 1970, Vol. 1, pp. 14-24.

1554. H.R. Mahler, G. Neiss, P.P. Slonimski, and B. Mackler, Biochemistry, 3, 893 (1964).

1555. H. Mair, J. Pathol. Bacteriol., 18, 179 (1914).

1556. R.F. Majewski, J.M. Berdahl, L.D. Jost, T.A. Martin, J.C. Simms, J.G. Schmidt, and J.R. Corrigan, Steroids, 16, 15 (1970).

1557. I. Makino, J. Sjövall, A. Norman, and B. Strandvik, FEBS Lett., 15, 161 (1971).

1558. G.G. Malanina, L.I. Klimova, L.M. Morozovskaya, O.S. Anisimova, A.G. Arsentev, and G.S. Grinenko, Khim.-Farm. Zhur., 12, 101 (1978).

1559. G.G. Malanina, L.I. Klimova, L.M. Morozovskaya, and G.S. Grinenko, Khim.-Farm. Zhur., No. 8, 98 (1976).

1560. M.R. Malinov, P. McLaughlin, L. Papworth, and G.W. Kittinger, Steroids, 25, 663 (1975).

1561. G.E. Mallet and D.S. Fukuda, Bacteriol. Proc., 26 (1962).

1562. H.H. Mantsch and I.C.P. Smith, Can. J. Chem., 51, 1384 (1973).

1563. F.N. Manzur and A.V. Chobanian, Circulation, 41/42, Suppl. III, 5 (1970).

1564. E. Mappus and C.-Y. Cuilleron, J. Chem. Res., (S), 42 (1979); (M), 501 (1979).

1565. P. Marfey and H. Chessin, Biochim. Biophys. Acta, 337, 136 (1974).

1566. S.P. Marfey and A.R. Schultz, Artery, 1, 298 (1975).

1567. E.A. Markaryan, Biokhimiya, 6, 229 (1941); via Chem. Abstr., 35, 7655[8] (1941).

1568. R.E. Marker, J. Am. Chem. Soc., 61, 1287 (1939).

1569. R.E. Marker, E. Rohrmann, E.J. Lawson, and E.L. Wittle, J. Am. Chem. Soc., 60, 1901 (1938).

1570. S. Marklund, J. Biol. Chem., 251, 7504 (1976).

1571. L.J. Marnett, P. Wlodawer, and B. Samuelsson, J. Biol. Chem., 250, 8510 (1975).

1572. C.W. Marshall, R.E. Ray, I. Laos, and B. Riegel, J. Am. Chem. Soc., 79, 6303 (1957).

1573. C.W. Marshall, R.E. Ray, I. Laos, and B. Riegel, J. Am. Chem. Soc., 79, 6308 (1957).

1574. J.G. Marshall and E.J. Staba, Lloydia, 39, 84 (1976).

1575. W.J. Marshall and A.E.M. McLean, Biochem. J., 122, 569 (1971).

1576. H.R. Marston, Austr. J. Exptl. Biol. Med. Sci., 1, 53 (1924).

1577. C.K.A. Martin, Adv. Applied Microbiology, 22, 29 (1977).

1578. C.M. Martin and H.J. Nicholas, J. Lipid Res., 14, 618 (1973).

1579. R.P. Martin, Biochim. Biophys. Acta, 25, 408 (1957).

1580. A.F. Marx, H.C. Beck, W.F. van der Waard, and J. de Flines, Steroids, 8, 421 (1966).

1581. S.J. Masiak and P.G. LeFevre, Arch. Biochem. Biophys., 162, 442 (1974).

1582. J.I. Mason, J.R. Arthur, and G.S. Boyd, Biochem. J., 173, 1045 (1978).

1583. J.I. Mason, R.J. Park, and G.S. Boyd, Biochem. Soc. Trans., 7, 641 (1979).

1584. B. Matkovics, P. Pénzes, and G. Göndös, Steroids, 5, 451 (1965).

1585. T. Matsuno, S. Nagata, and K. Hashimoto, Bull. Japan Soc. Sci. Fisheries, 38, 1261 (1972).

1586. T.Matsuura, Tetrahedron, 33, 2869 (1977).

1587. G. Mattern, P. Albrecht, and G. Ourisson, J. Chem. Soc. Chem. Commun., 1570 (1970).

1588. W.S. Matthews and L.L. Smith, Texas Reports Biol. Med., 24, 515 (1966).

1589. W.S. Matthews and L.L. Smith, Lipids, 3, 239 (1968).

1590. M. Maumy, Bull. Soc. Chim. France, 2895 (1974).

1591. M. Maumy and J. Rigaudy, Bull. Soc. Chim. France, 1487 (1974).

1592. M. Maumy and J. Rigaudy, Bull. Soc. Chim. France, 1879 (1975).

1593. M. Maumy and J. Rigaudy, Bull. Soc. Chim. France, 2021 (1976).

1594. J. Mauthner and W. Suida, Monatsh., 15, 85 (1894).

1595. J. Mauthner and W. Suida, Monatsh., 17, 579 (1896).

1596. J. Mauthner and W. Suida, Sitzber. Akad. Wiss. Wien, [IIb], 112, 482 (1903); Monatsh., 24, 648 (1903).

1597. A. Mayer, C.W. Picard, and F. Wolkes, Pharm. Acta Helv., 33, 603 (1958).

1598. D. Mayer, F.-W. Koss, and A. Glasenapp, Z. Physiol. Chem., 353, 921 (1972).

1599. D. Mayer and A. Voges, Z. Physiol. Chem., 353, 1187 (1972).

1600. W.V. Mayneord and E.M.F. Roe, Am. J. Cancer, 31, 476 (1937).

1601. C.S. McArthur, Biochem. J., 36, 559 (1942).

1602. P.J. McCabe and C. Green, Chem. Phys. Lipids, 20, 319 (1977).

1603. C.J. McClure and J.A. Fee, FEBS Lett., 67, 294 (1976).

1604. R. McCrindle and C. Djerassi, Chem. Ind., 1311 (1961).

1605. E.N. McIntosh, F. Mitani, V.I. Uzgiris, C. Alonso, and H.A. Salhanick, Ann. N.Y. Acad. Sci., 212, 392 (1973).

1606. H. McKennis, J. Biol. Chem., 167, 645 (1947).

1607. E. McKeown and W.A. Waters, J. Chem. Soc., (B), 1040 (1966).

1608. T.C. McMorris, S.R. Schow, and G.R. Weihe, Tetrahedron Lett., 335 (1978).

1609. T.C. McMorris, R. Seshadri, and T. Arunachalam, J.
      Org. Chem., 39, 669 (1974).
1610. T.C. McMorris, R. Seshadri, G.R. Weihe, G.P. Arsen-
      ault, and A.W. Barksdale, J. Am. Chem. Soc., 97, 2544
      (1975).
1611. H. Meffert and E. Schnarrer, Acta Biol. Med. Germ.,
      32, 263 (1974).
1612. J.W.A. Meijer and C.J.F. Böttcher, Rec. Trav. Chim.,
      78, 622 (1959).
1613. W.G. Meinshein and G.S. Kenny, Anal. Chem., 29, 1153
      (1957).
1614. M.J. Mellies, T.T. Ishikawa, C.J. Glueck, and J.D.
      Crissman, Cancer Res., 37, 3034 (1977).
1615. D. Mendelsohn and L. Mendelsohn, Biochemistry 7, 4167
      (1968).
1616. D. Mendelsohn, L. Mendelsohn, and E. Staple, Biochim.
      Biophys. Acta, 97, 379 (1965).
1617. D. Mendelsohn, L. Mendelsohn, and E. Staple, Biochem-
      istry, 5, 1286 (1966).
1618. E. Menini and J.K. Norymberski, Biochem. J., 84,
      195 (1962).
1619. C. Merritt, Report on Workshop on Study and Review of
      Angiotoxic and Carcinogenic Sterols in Processed
      Foods, Boston, Mass., Oct. 12-14, 1977, pp. 79-82.
1620. C. Merritt, P. Angelini, and D.J. McAdoo, in RADIATION
      PRESERVATION OF FOODS, Advances in Chemistry Series
      No. 65, E.S. Josephson and J.H. Frankfort, Editors,
      American Chemical Society, Washington City, D.C.,
      1967, pp. 26-34.
1621. K. Meyer, J. Biol. Chem., 103, 607 (1933).
1622. Č. Michalec,in HANDBUCH DER PAPIERCHROMATOGRAPHIE, I.
      M. Hais and K. Macek, Editors, Gustav Fischer Verlag,
      Jena, Vol. 1, 1958, p. 373.
1623. Č. Michalec, in PAPER CHROMATOGRAPHY, A COMPREHENSIVE
      TREATISE, I.M. Hais and K. Macek, Editors, Academic
      Press Inc., New York, 1963, p. 393.
1624. Č. Michalec, J. Lipid Res., 4, 110 (1963).
1625. Č. Michalec and C. Sobĕslavský, Chem. Listy, 53, 1170
      (1959).
1626. R.A. Micheli and T.H. Applewhite, J. Org. Chem., 27,
      345 (1962).
1627. J. Michnowicz and B. Munson, Org. Mass Spectrometry,
      6, 765 (1972).
1628. K. Miescher and H. Kägi, Helv. Chim. Acta, 24, 986
      (1941).
1629. M.Lj. Mihailović, Lj. Lorenc, V. Pavlović, and J.

Kalvoda, Tetrahedron, 33, 441 (1977).

1630. A. Mijares, D.I. Cargill, J.A. Glasel, and S. Lieberman, J. Org. Chem., 32, 810 (1967).

1631. A.H. Milburn and E.V. Truter, J. Chem. Soc., 1736 (1956).

1632. A.H. Milburn and E.V. Truter, J. Applied Chem., 7, 491 (1957).

1633. A.H. Milburn, E.V. Truter, and F.P. Woodford, J. Chem. Soc., 1740 (1956).

1634. L.R. Miller, A. Izumi, and F.D. Pinkerton, Federation Proc., 39, 1907 (1980).

1635. J.T. Mills and A.M. Adamany, J. Biol. Chem., 253, 5270 (1978).

1636. Y.W. Mirhom and F.E. Szontágh, J. Endocrinol., 50, 301 (1971).

1637. F. Mitani, Mol. Cell. Biochem., 24, 21 (1979).

1638. F. Mitani and S. Horie, J. Biochem., 65, 269 (1969).

1639. K.A. Mitropoulos, M.D. Avery, N.B. Myant, and G.F. Gibbons, Biochem. J., 130, 363 (1972).

1640. K.A. Mitropoulos and S. Balasubramaniam, Biochem. J., 128, 1 (1972).

1641. K.A. Mitropoulos, S. Balasubramaniam, G.F. Gibbons, and B.E.A. Reeves, FEBS Lett., 27, 203 (1972).

1642. K.A. Mitropoulos, P.D.G. Dean, M.W. Whitehouse, and N.B. Myant, Biochem. J., 105, 31P (1967).

1643. K.A. Mitropoulos and N.B. Myant, Biochem. J., 97, 26c (1965).

1644. K.A. Mitropoulos and N.B. Myant, Biochem. J., 101, 38P (1966).

1645. K.A. Mitropoulos and N.B. Myant, Biochem. J., 103, 472 (1967).

1646. K.A. Mitropoulos and N.B. Myant, Biochim. Biophys. Acta, 144, 430 (1967).

1647. H. Mitsuhashi, T. Nomura, and M. Hirano, Chem. Pharm. Bull., 14, 717 (1966).

1648. H. Mitsuhashi and Y. Shimizu, Chem. Pharm. Bull., 8, 313 (1960).

1649. H. Mitsuhashi and Y. Shimizu, Chem. Pharm. Bull., 8, 318 (1960).

1650. H. Mitsuhashi and Y. Shimizu, Chem. Pharm. Bull., 8, 565 (1960).

1651. H. Mitsuhashi and Y. Shimizu, Chem. Pharm. Bull., 10, 433 (1962).

1652. T. Mitsui, J. Agr. Chem. Soc. Japan, 15, 526 (1939); via Chem. Abstr., 34, 383[7] (1940).

1653. J.R. Mitton and G.S. Boyd, Biochem. J., 96, 60P(1965).

1654. J.R. Mitton and G.S. Boyd, Biochem. J., 103, 17P (1967).
1655. J.R. Mitton, N.A. Scholan, and G.S. Boyd, Eur. J.
      Biochem., 20, 569 (1971).
1656. W. Mlekusch, W. Truppe, and B. Paletta, J. Chromatog.,
      78, 438 (1973).
1657. D. Mohr, Dtsch. Z. Verdauungs- u. Stoffwechselkrankh.,
      6, 252 (1943); via Chem. Zentralblat, I, 2101 (1943).
1658. J.M. Moldowan, W.L. Tan, and C. Djerassi, Steroids,
      26, 107 (1975).
1659. J.M. Moldowan, B.M. Tursch, and C. Djerassi, Steroids,
      24, 387 (1974).
1660. E. Molinari and P. Fenaroli, Ber., 41, 2785 (1908).
1661. C. Monder, Endocrinology, 82, 318 (1968).
1662. E. Montignie, Bull. Soc. Chim. France, 47, 1323 (1930).
1663. E. Montignie, Bull. Soc. Chim. France, 53, 1411 (1933).
1664. E. Montignie, Bull. Soc. Chim. France, 53, 1412 (1933).
1665. K.H. Moon and R.G. Bunge, Fertility Sterility, 21, 80
      (1970).
1666. T. Moore and S.G. Willimott, Biochem. J., 21, 585
      (1927).
1667. P. Morand and A. Van Tongerloo, J. Chem. Soc. Chem.
      Commun., 7 (1972).
1668. R. Morand and A. Van Tongerloo, Steroids, 21, 47
      (1973).
1669. P. Morand and A. Van Tongerloo, Steroids, 21, 65
      (1973).
1670. H. Mori, K. Shibata, K. Tsuneda, M. Sawai, and K.
      Tsuda, Chem. Pharm. Bull., 16, 1407 (1968).
1671. R.M. Moriarty and T.D.J. D'Silva, J. Org. Chem., 28,
      2445 (1963); Tetrahedron, 21, 547 (1965).
1672. H. Morimoto, I. Imada, T. Murata, and N. Matsumoto,
      Liebigs Ann., 708, 230 (1967).
1673. M. Morisaki, K. Bannai, and N. Ikekawa, Chem. Pharm.
      Bull., 21, 1853 (1973).
1674. M. Morisaki, K. Bannai, and N. Ikekawa, Chem. Pharm.
      Bull., 24, 1948 (1976).
1675. M. Morisaki, K. Bannai, N. Ikekawa, and M. Shikita,
      Biochem. Biophys. Res. Commun., 69, 481 (1976).
1676. M. Morisaki, S. Kidooka, and N. Ikekawa, Chem. Pharm.
      Bull., 24, 3214 (1976).
1677. M. Morisaki, H. Ohtaka, M. Okubayashi, N. Ikekawa,
      Y. Horie, and S. Nakasone, J. Chem. Soc. Chem. Commun.,
      1275 (1972).
1678. M. Morisaki, J. Rubio-Lightbourn, and N. Ikekawa,
      Chem. Pharm. Bull., 21, 457 (1973).
1679. M. Morisaki, A. Saika, K. Bannai, M. Sawamura, J.

Rubio-Lightbourn, and N. Ikekawa, Chem. Pharm. Bull., 23, 3272 (1975).

1680. M. Morisaki, S. Sato, N. Ikekawa, and M. Shikita, FEBS Lett., 72, 337 (1976).

1681. S. Moriuchi, F. Tsuruki, Y. Otawara, N. Hosoya, S. Yamada, K. Nakayama, and H. Takayama, J. Nutr. Sci. Vitaminol., 25, 455 (1979).

1682. R.P. Morozova, Ukr. Biokhim. Zh., 41, 428 (1969); via Chem. Abstr., 72, 1281r (1970).

1683. R.P. Morozova, Ukr. Biokhim. Zh., 41, 480 (1969); via Chem. Abstr., 72, 88598w (1970).

1684. R.P. Morozova, Ukr. Biokhim. Zh., 41, 702 (1969); via Chem. Abstr., 72, 130954m (1970).

1685. R.P. Morozova and I.A. Nikolenko, Vitaminy, 6, 101 (1971); via Chem. Abstr., 77, 56602e (1972).

1686. R.P. Morozova and V.P. Vendt, Vitaminy, 6, 92 (1971); via Chem. Abstr., 77, 56251q (1972).

1687. L.J. Morris, J. Lipid Res., 7, 717 (1966).

1688. W.R. Morrison and L.M. Smith, J. Lipid Res., 5, 600 (1964).

1689. E.H. Mosbach, J. Blum, E. Arroyo, and S. Milch, Anal. Biochem., 5, 158 (1963).

1690. E.H. Mosbach, M. Nierenberg, and F.E. Kendall, J. Am. Chem. Soc., 75, 2358 (1953).

1691. E. Mosettig and I. Scheer, J. Org. Chem., 17, 764 (1952).

1692. A.-M. Motzfeldt, Acta Chem. Scand., 24, 1846 (1970).

1693. L.S. Moyer, Biochem. Z., 273, 122 (1934).

1694. D. Mufson, J.E. Zarembo, L.J. Ravin, and K. Meksuwan, Thermochim. Acta, 5, 221 (1972).

1695. L.J. Mulheirn and G. Ryback, Nature, 256, 301 (1975).

1696. G. Müller, A. Kanazawa, and S.-I. Teshima, Naturwissenschaften, 66, 520 (1979).

1697. M. Müller, Z. Physiol. Chem., 231, 75 (1935).

1698. T. Murata, S. Takahashi, and T. Takeda, Anal. Chem., 47, 577 (1975).

1699. M.T.J. Murphy, A. McCormick, and G. Eglinton, Science, 157, 1040 (1967).

1700. J. Murtaugh and R.L. Bunch, J. Water Pollution Control Fed., 39, 404 (1967).

1701. N.B. Myant and K.A. Mitropoulos, J. Lipid Res., 18, 135 (1977).

1702. H.R. Nace and M. Inaba, J. Org. Chem., 27, 4024 (1962).

1703. L. Naftalin and A. Stephens, Life Sci., 639 (1962).

1704. N. Nagano, J.P. Poyser, K.-P. Cheng, Luu Bang, and G. Ourisson, J. Chem. Research (S), 218 (1977); (M)

2522 (1977).

1705. M. Nagasawa, M. Bae, G. Tamura, and K. Arima, Agr.
      Biol. Chem., 33, 1644 (1969).

1706. M. Nagasawa, H. Hashiba, N. Watanabe, M. Bae, G. Tam-
      ura, and K. Arima, Agr. Biol. Chem., 34, 801 (1970).

1707. F. Nakada, R. Oshio, S. Sasaki, H. Yamasaki, N. Yamaga,
      and K. Yamasaki, J. Biochem., 64, 495 (1968).

1708. S. Nakajima, R. Konaka, and K. Takeda, Yakugaku
      Zasshi, 96, 764 (1976).

1709. S. Nakajima, R. Konaka, and K. Takeda, Yakugaku
      Zasshi, 96, 863 (1976).

1710. S. Nakajima and K. Takeda, Chem. Pharm. Bull., 12,
      1530 (1964).

1711. T. Nakamura, M. Nishikawa, K. Inoue, S. Nojima, T.
      Akiyama, and U. Sankawa, Chem. Phys. Lipids, 26, 101
      (1980).

1712. K. Nakanishi, B.K. Bhattacharyya, and L.F. Fieser,
      J. Am. Chem. Soc., 75, 4415 (1953).

1713. S.H.M. Naqvi, Lipids, 8, 766 (1973).

1714. S.H.M. Naqvi, B.L. Herndon, M.T. Kelley, V. Bleisch,
      R.T. Aexel, and H.J. Nicholas, J. Lipid Res., 10, 115
      (1969).

1715. C.R. Narayanan and K.N. Iyer, J. Org. Chem., 30, 1734
      (1965).

1716. C.R. Narayanan and A.K. Lala, Org. Mass Spectrometry,
      6, 119 (1972).

1717. C.R. Narayanan, M.S. Parkar, and P.S. Ramaswamy, Chem.
      Ind., 208 (1974).

1718. C.R. Narayanan, M.S. Parker, and M.S. Wadia, Tetra-
      hedron Lett., 4703 (1970).

1719. T.A. Narwid, K.E. Cooney, and M.R. Uskoković, Helv.
      Chim. Acta, 57, 771 (1974).

1720. S.V. Nedzvetskiĭ, Biokhimiya, 6, 425 (1941); via Chem.
      Abstr., 36, 6551[9] (1942).

1721. S.V. Nedzvetskiĭ and G.A. Gaukhman, Biokhimiya, 13,
      234 (1948).

1722. S.V. Nedzvetskiĭ, A.N. Panyukov, and T.A. Shpats, Bio-
      khimiya, 18, 315 (1953).

1723. G. Neef, U. Eder, A. Seeger, and R. Wiechert, Chem.
      Ber., 113, 1184 (1980).

1724. R. Neher, J. Chromatog., 1, 122 (1958).

1725. R. Neher, J. Chromatog., 1, 205 (1958).

1726. R. Neher, CHROMATOGRAPHIE VON STERINEN, STEROIDEN UND
      VERWANDTEN VERBINDUNGEN, Elsevier Publishing Co.,
      Amsterdam, 1958, p. 38, p. 59, p. 69.

1727. R. Neher, Chromatog. Rev., 1, 99 (1959).

1728. R. Neher, STEROID CHROMATOGRAPHY, Elsevier Publishing
      Co., Amsterdam, 1964, p. 73, p. 136, p. 264.
1729. R. Neher and A. Wettstein, Helv. Chim. Acta, 43, 1628
      (1960).
1730. D.H. Neiderhiser and H.P. Roth, Proc. Soc. Exptl.
      Biol. Med., 128, 221 (1968).
1731. S.A. Neifakh, Biokhimiya, 5, 348 (1940); via Chem.
      Abstr., 35, 4789^2 (1941).
1732. E.K. Nelson, J. Am. Chem. Soc., 33, 1404 (1911).
1733. E.K. Nelson, J. Am. Chem. Soc., 35, 84 (1913).
1734. J.A. Nelson, M.R. Czarny, T.A. Spencer, J.S. Limanek,
      K.R. McCrae, and T.Y. Chang, J. Am. Chem. Soc., 100,
      4900 (1978).
1735. W.R. Nes, J. Am. Chem. Soc., 78, 193 (1956).
1736. W.R. Nes, Lipids, 9, 596 (1974).
1737. W.R. Nes and D.L. Ford, J. Am. Chem. Soc., 85, 2137
      (1963).
1738. W.R. Nes, R.B. Kostic, and E. Mosettig, J. Am. Chem.
      Soc., 78, 436 (1956).
1739. W.R. Nes and M.L. McKean, BIOCHEMISTRY OF STEROIDS AND
      OTHER ISOPENTENOIDS, University Park Press, Baltimore/
      London/Tokyo, 1977, pp. 352-354.
1740. W.R. Nes and E. Mosettig, J. Am. Chem. Soc., 76, 3182
      (1954).
1741. W.D. Nes, G.W. Patterson, and G.A. Bean, Lipids, 14,
      458 (1979).
1742. W. Neudert and H. Röpke, ATLAS OF STEROID SPECTRA,
      Springer-Verlag New York Inc., 1965, Spectrum No. 843.
1743. S. Neukomm, J. Bonnet, T. Baer, and M. de Trey, Oncol-
      ogia, 13, 279 (1960).
1744. F. Neuwald and K.-E. Fetting, Pharm. Zeit., 108, 1490
      (1963).
1745. A.M. Neville and J.L. Webb, Steroids, 6, 421 (1965).
1746. J.D. Newburger, J.J. Uebel, M. Ikawa, K.K. Andersen,
      and R.B. Gagosian, Phytochem., 18, 2042 (1979).
1747. C.W. Nichols, S. Lindsay, and I.L. Chaikoff, Proc.
      Soc. Exptl. Biol. Med., 89, 609 (1955).
1748. C.W. Nichols, M.D. Siperstein, and I.L. Chaikoff,
      Proc. Soc. Exptl. Biol. Med., 83, 756 (1953).
1749. N. Nicholson, L. Flanders, B.N. Desai, and L. Chinn,
      J. Steroid Biochem., 10, 709 (1979).
1750. S.H. Nicholson and A.B. Turner, J. Chem. Soc., Perkin
      Trans. I, 1357 (1976).
1751. A. Nickon and J.F. Bagli, J. Am. Chem. Soc., 81, 6330
      (1959); 83, 1498 (1961).
1752. A. Nickon, V.T. Chuang, P.J.L. Daniels, R.W. Denny,

J.B. DiGiorgio, J. Tsunetsugu, H.G. Vilhuber, and E. Werstiuk, J. Am. Chem. Soc., 94, 5517 (1972).

1753. A. Nickon, J.B. DiGiorgio, and P.J.L. Daniels, J. Org. Chem., 38, 533 (1973).

1754. A. Nickon and W.L. Mendelson, J. Am. Chem. Soc., 85, 1894 (1963); 87, 3921 (1965).

1755. A. Nickon and W.L. Mendelson, Can. J. Chem., 43, 1419 (1965).

1756. A. Nickon and W.L. Mendelson, J. Org. Chem., 30, 2087 (1965).

1757. A. Nickon, N. Schwartz, J.B. DiGiorgio, and D.A. Widdowson, J. Org. Chem., 30, 1711 (1965).

1758. G. Nicolau, S. Shefer, and E.H. Mosbach, Anal. Biochem., 68, 255 (1975).

1759. G. Nicolau, S. Shefer, G. Salen, and E.H. Mosbach, J. Lipid Res., 15, 146 (1974).

1760. H. Niewiadomski, Nahrung, 19, 525 (1975); via Chem. Abstr., 83, 94944z (1975).

1761. H. Niewiadomski and M. Budny, Rocz. Technol. Chem. Zywn., 13, 23 (1967); via Chem. Abstr., 68, 21023d (1968).

1762. H. Niewiadomski and J. Sawicki, Grasas Aceites, 9, 306 (1958).

1763. H. Niewiadomski and J. Sawicki, Fette Seif. Anstrich., 66, 930 (1964).

1764. H. Niewiadomski and J. Sawicki, Zesz. Probl. Postepow Nauk Rolniczych, 53, 175 (1965); via Chem. Abstr., 63, 13588e (1965).

1765. H. Niewiadomski and J. Sawicki, Fette Seif. Anstrich., 68, 641 (1966).

1766. H. Niewiadomski, E. Szukalaka, and E. Rakowska, Zesz. Probl. Postepow Nauk Rolniczych, No. 136, 171, (1972); via Chem. Abstr., 79, 3982p (1973).

1767. I.A. Nikolenko and R.P. Morozova, Vitaminy, 6, 110 (1971); via Chem. Abstr., 77, 56603f (1972).

1768. Y. Nishikawa, K. Yamashita, M. Ishibashi, and H. Miyazaki, Chem. Pharm. Bull., 26, 2922 (1978).

1769. M. Nishimura, Nature, 270, 711 (1977).

1770. M. Nishimura and T. Kyoama, Chem. Geology, 17, 229 (1976).

1771. M. Nishimura and T. Koyama, Geochim. Cosmochim. Acta, 41, 379 (1977).

1772. C. Nofre, A. Revol, and A. Cier, Compt. Rend., 250, 2638 (1960).

1773. A. Noma and K. Nakayama, Clin. Chim. Acta, 73, 487 (1976).

1774. T. Nomura, M. Itoh, A. Alcaide, and M. Barbier, Bull. Japan Soc. Sci. Fisheries, 38, 1365 (1972).

1775. T. Nomura, S. Yamada, and H. Mitsuhashi, Chem. Pharm. Bull., 27, 508 (1979).

1776. L.N. Norcia, J. Am. Oil Chem. Soc., 38, 238 (1961).

1777. L.N. Norcia and W.F. Janusz, J. Am. Oil Chem. Soc., 42, 847 (1965).

1778. L.N. Norcia and V. Mahadevan, Lipids, 8, 17 (1973).

1779. F.F. Nord, Chem.-Zeit., 58, 347 (1934).

1780. F.F. Nord and G. Weiss, Biochem. Z., 263 353 (1933).

1781. T. Norii, N. Yamaga, and K. Yamasaki, Steroids, 15, 303 (1970).

1782. A.W. Norman, VITAMIN D, THE CALCIUM HOMEOSTATIC STEROID HORMONE, Academic Press, New York/San Francisco/London, 1979, p. 50.

1783. B.E. North, S.S. Katz, and D.M. Small, Atherosclerosis, 30, 211 (1978).

1784. T.J. Novak, E.J. Poziomek, and R.A. Mackay, Mol. Cryst. Liq. Cryst., 20, 203 (1973).

1785. E.C. Noyons and M.K. Polano, Biochem. Z., 303, 415 (1940).

1786. W.H.R. Nye, R. Terry, and D.L. Rosenbaum, Am. J. Clin. Pathol., 49, 718 (1968).

1787. K. Obermüller, Z. Physiol. Chem., 15, 37 (1891).

1788. K. Ochi, I. Matsunaga, M. Shindo, and C. Kaneko, Chem. Pharm. Bull., 27, 252 (1979).

1789. K. Oette, J. Lipid Res., 6, 449 (1965).

1790. A. Ogata and I. Kawakami, J. Pharm. Soc. Japan, 58, 738 (1938); J. Pharm. Soc. Japan Transactions, 58, 18 (1938).

1791. K. Ogura and T. Hanya, Proc. Japan Acad., 49, 201 (1973).

1792. M. Ogura, J. Shiga, and K. Yamasaki, J. Biochem., 70, 967 (1971).

1793. M. Ogura and K. Yamasaki, J. Biochem., 67, 643 (1970).

1794. M. Okada, D.K. Fukushima, and T.F. Gallagher, J. Biol. Chem., 234, 1688 (1959).

1795. M. Okada and Y. Saito, Steroids, 6, 651 (1965).

1796. R.A. Okerholm, P.I. Brecher, and H.H. Wotiz, Steroids, 12, 435 (1968).

1797. T. Okumura, Bunseki Kagaku, 26, 396 (1977).

1798. M. Onda and A. Azuma, Chem. Pharm. Bull., 19, 859 (1971).

1799. M. Onda and A. Azuma, Chem. Pharm. Bull., 20, 1467 (1972).

1800. M. Onda, Y. Konda, and R. Yabuki, Chem. Pharm. Bull.,

$\underline{23}$, 611 (1975).

1801. N.R. Orme-Johnson, D.R. Light, R.W. White-Stevens, and W.H. Orme-Johnson, J. Biol. Chem., $\underline{254}$, 2103 (1979).

1802. P.R. Ortiz de Montellano, J.P. Beck, and G. Ourisson, Biochem. Biophys. Res. Commun., $\underline{90}$, 897 (1979).

1803. A. Osol and R. Pratt, THE UNITED STATES DISPENSATORY, J.B. Lippincott Co., Philadelphia/Toronto, 27th Ed., 1973, pp. 312-313.

1804. P.K. Paasonen, Suomen Kemistilehti, $\underline{B40}$, 277 (1967).

1805. R.M. Packie and A.A. Kandutsch, Biochem. Genetics, $\underline{4}$, 203 (1970).

1806. I.H. Page, J. Biol. Chem., $\underline{57}$, 471 (1923).

1807. I.H. Page and W. Menschick, Naturwissenschaften, $\underline{18}$, 585 (1930).

1808. H. Paget, J. Chem. Soc., 829 (1938).

1809. H. Paillard, M. Berenstein, and E. Briner, Compt. Rend. Soc. Physique Hist. Nat., $\underline{61}$, 67 (1944).

1810. S.C. Pakrashi and B. Achari, Tetrahedron Lett., 365 (1971).

1811. S.C. Pakrashi, B. Achari, and P.C. Majumdar, Indian J. Chem., $\underline{13}$, 755 (1975).

1812. A. Pakrashi and B. Basak, J. Reprod. Fert., $\underline{46}$, 461 (1976).

1813. A.M. Paliokas and G.J. Schroepfer, Biochim. Biophys. Acta, $\underline{144}$, 167 (1967).

1814. T. Panalaks, Int. Z. Vitaminforsch., $\underline{39}$, 426 (1969).

1815. P. Paoletti, F.A. Vandenheuvel, R. Fumagalli, and R. Paoletti, Neurology, $\underline{19}$, 190 (1969).

1816. E.J. Parish, M. Tsuda, and G.J. Schroepfer, Chem. Phys. Lipids, $\underline{24}$, 209 (1979).

1817. F.S. Parker and K.R. Bhaskar, Biochemistry, $\underline{7}$, 1286 (1968).

1818. O.W. Parks, D.P. Schwartz, M. Keeney, and J.N. Damico, Nature, $\underline{210}$, 417 (1966).

1819. G. Parmentier and H. Eyssen, Biochim. Biophys. Acta, $\underline{348}$, 279 (1974).

1820. P.G. Parsons and P. Goss, Aust. J. Exp. Biol. Med. Sci., $\underline{56}$, 287 (1978).

1821. W.H. Parsons, A.R. Gennaro, and A. Osol, Am. J. Pharmacy, $\underline{133}$, 351 (1961).

1822. J.R. Partington, J. Chem. Soc., $\underline{99}$, 313 (1911).

1823. J.J. Partridge, S. Faber, and M.R. Uskoković, Helv. Chim. Acta, $\underline{57}$, 764 (1974).

1824. E.L. Patterson, W.W. Andres, and R.E. Hartman, Experientia, $\underline{20}$, 256 (1964).

1825. G.W. Patterson, Comp. Biochem. Physiol., $\underline{24}$, 501 (1968).
1826. E.J. Patzer, R.R. Wagner, and Y. Barenholz, Nature, $\underline{274}$, 394 (1978).
1827. L. Pauling, J. Am. Chem. Soc., $\underline{53}$, 3225 (1931).
1828. L. Pauling, THE NATURE OF THE CHEMICAL BOND AND THE STRUCTURE OF MOLECULES AND CRYSTALS, Cornell University Press, Itaca, N.Y., 1945, pp. 271-274.
1829. G. Paumgartner, G. Délèze, G. Karlaganis, W. Giger, M. Reinhard, and D. Sidiropoulos, Schweiz. Med. Wschr., $\underline{107}$, 529 (1977).
1830. G. Paumgartner, G. Délèze, G. Karlaganis, and D. Sidiropoulos, Gasteroenterology, $\underline{71}$, 923 (1976).
1831. P.R. Peacock, Brit. Med. Bull., $\underline{4}$, 364 (1946/1947).
1832. P.R. Peacock, Brit. J. Nutrition, $\underline{2}$, 201 (1948).
1833. W.J. Peal, Chem. Ind., 1451 (1957).
1834. J.I. Pedersen, FEBS Lett., $\underline{85}$, 35 (1978).
1835. J.I. Pedersen, I. Björkhem, and J. Gustafsson, J. Biol. Chem., $\underline{254}$, 6464 (1979).
1836. J.I. Pedersen, H. Oftbro, and T. Vänngård, Biochem. Biophys. Res. Commun., $\underline{76}$, 666 (1977).
1837. J.I. Pedersen and K. Saarem, J. Steroid Biochem., $\underline{9}$, 1165 (1978).
1838. T.C. Pederson and S.D. Aust, Biochem. Biophys. Res. Commun., $\underline{48}$, 789 (1972).
1839. T.C. Pederson and S.D. Aust, Biochem. Biophys. Res. Commun., $\underline{52}$, 1071 (1973).
1840. T.C. Pederson and S.D. Aust, Biochim. Biophys. Acta, $\underline{385}$, 232 (1975).
1841. B. Pelc, Coll. Czech Chem. Commun., $\underline{22}$, 1457 (1957).
1842. B. Pelc and E. Kodicek, J. Chem. Soc., (C), 1624 (1970).
1843. B. Pelc and D.H. Marshall, Steroids, $\underline{31}$, 23 (1978).
1844. N. Pelick and J.W. Shigley, J. Am. Oil Chem. Soc., $\underline{44}$, 121 (1967).
1845. P. Pelletier and J.B. Caventou, Ann. Chim. Phys., [2], $\underline{6}$, 401 (1817).
1846. S.K. Peng, C.B. Taylor, P. Tham, and B. Mikkelson, Federation Proc., $\underline{36}$, 392 (1977).
1847. S.-K. Peng, C.B. Taylor, P. Tham, N.T. Werthessen, and B. Mikkelson, Arch. Path. Lab. Med., $\underline{102}$, 57 (1978).
1848. S.-K. Peng, P. Tham, C.B. Taylor, and B. Mikkelson, Am. J. Clin. Nutrition, $\underline{32}$, 1033 (1979).
1849. S.-K. Peng, P. Tham. C.B. Taylor, and B. Mikkelson, Federation Proc., $\underline{38}$, 1243 (1979).

1850. J.F. Pennock, G. Neiss, and H.R. Mahler, Biochem. J., 85, 530 (1962).

1851. J.A.M. Peters, N.P. van Vliet, and F.J. Zeelen, Rec. Trav. Chim. 98, 459 (1979).

1852. J.W. Peters and C.S. Foote, J. Am. Chem. Soc., 98, 873 (1976).

1853. R. Peters, J. Royal Soc. Med., 71, 459 (1978).

1854. G.E. Peterson and J.R. Davis, Steroids, 4, 677 (1964).

1855. G.E. Peterson, W.J. Mandy, H. Futch, and D. Luckey, Can. J. Microbiol., 8, 193 (1962).

1856. N.N. Petropavlov and N.F. Kostin, Sov. Phys. Crystallogr., 21, 88 (1976).

1857. V.A. Petrow, J. Chem. Soc., 1077 (1937).

1858. K. Petzoldt and K. Kieslich, Liebigs Ann., 724, 194 (1969).

1859. E.H. Pfeiffer, Naturwissenschaften, 60, 525 (1973).

1860. J.R. Philippot, A.G. Cooper, and D.F.H. Wallach, Biochim. Biophys. Acta, 406, 161 (1975).

1861. J.R. Philippot, A.G. Cooper, and D.F.H. Wallach, Biochem. Biophys. Res. Commun., 72, 1035 (1976).

1862. J.R. Philippot, A.G. Cooper, and D.F.H. Wallach, Proc. Natl. Acad. Sci., 74, 956 (1977).

1863. C.W. Picard and D.E. Seymour, J. Soc. Chem. Ind. Trans., 37, 304 (1945).

1864. R.H. Pickard and J. Yates, J. Chem. Soc., 93, 1678 (1908).

1865. C.G.P. Pillai and J.D. Weete, Phytochem., 14, 2347 (1975).

1866. J.N. Pitts and B.J. Finlayson, Angew. Chem. Int. Ed., 14, 1 (1975).

1867. O. Pizzolato and H.H. Beard, Exp. Med. Surgery, 3, 95 (1945).

1868. P.A. Plattner and H. Heusser, Helv. Chim. Acta, 27, 748 (1944).

1869. P.A. Plattner, H. Heusser, and M. Feurer, Helv. Chim. Acta, 32, 587 (1949).

1870. P.A. Plattner and W. Lang, Helv. Chim. Acta., 27, 1872 (1944).

1871. P.A. Plattner, T. Petrzilka, and W. Lang, Helv. Chim. Acta, 27, 513 (1944).

1872. B. Polettini, Boll. Soc. Ital. Biol. Sper., 11, 951 (1936).

1873. N.M. Polyakova, Dok. Akad. Nauk S.S.S.R., 93, 321 (1953); via Chem. Abstr., 48, 2858d (1954).

1874. N.M. Polyakova and V.P. Vendt, Ukraïn. Biokhim. Zhur., 25, 419 (1953).

1875. G. Ponsinet and G. Ourisson, Phytochem., 6, 1235
      (1967).
1876. G. Popják, J. Edmond, F.A.L. Anet, and N.R. Easton,
      J. Am. Chem. Soc., 99, 931 (1977).
1877. S. Popov, R.M.K. Carlson, A. Wegmann, and C. Djerassi,
      Steroids, 28, 699 (1976).
1878. S. Popov, R.M.K. Carlson, A.-M. Wegmann, and C. Djer-
      assi, Tetrahedron Lett., 3491 (1976).
1879. C.R. Popplestone and A.M. Unrau, Can. J. Chem., 51,
      1223 (1973).
1880. O. Porges and E. Neubauer, Biochem. Z., 7, 152 (1908).
1881. J.K. Porter, C.W. Bacon, J.D. Robbins, and H.C. Higman,
      J. Ag. Food Chem., 23, 771 (1975).
1882. C.J. Pouchert, THE ALDRICH LIBRARY OF INFRARED SPECTRA,
      Aldrich Chemical Co., Milwaukee, Wis., 1970, Spectrum
      1069F.
1883. J.P. Poyser, F. de Reinach Hirtzbach, and G. Ourisson,
      Tetrahedron, 30, 977 (1974).
1884. J.P. Poyser and G. Ourisson, J. Chem. Soc., Perkin
      Trans. I, 2061 (1974).
1885. H.P.M. Pratt, P.A. Fitzgerald, and A. Saxon, Cellular
      Immunol., 32, 160 (1977).
1886. V. Prelog and H.C. Beyerman, Experientia, 1, 64 (1945).
1887. V. Prelog and L. Ruzicka, Helv. Chim. Acta, 27, 61
      (1944).
1888. V. Prelog, L. Ruzicka, and P. Stein, Helv. Chim. Acta,
      26, 2222 (1943).
1889. V. Prelog, E. Tagmann, S. Lieberman, and L. Ruzicka,
      Helv. Chim. Acta, 30, 1080 (1947).
1890. E. Premuzic, Fortschr. Chem. Org. Naturstoffe, 29,
      417 (1971).
1891. M.J. Price and G.K. Worth, Aust. J. Chem., 27, 2505
      (1974).
1892. Ž. Procházka and V. Šašek, Coll. Czech. Chem. Commun.,
      32, 610 (1967).
1893. G.J. Proksch and D.P. Bonderman, Clin. Chem., 24, 1924
      (1978).
1894. M. Prost, B.F. Maume, and P. Padieu, Biochim. Biophys.
      Acta, 360, 230 (1974).
1895. W.A. Pryor, in FREE RADICALS IN BIOLOGY, W.A. Pryor,
      Editor, Academic Press, New York, Vol. 1, 1976, pp.
      1-49.
1896. J. Pusset, D. Guénard, and R. Beugelmans, Tetrahedron,
      27, 2939 (1971).
1897. N. Radin and A.L. Gramza, Clin. Chem., 9, 121 (1963).
1898. P.R. Raggatt and M.W. Whitehouse, Biochem. J., 101,

819 (1966).

1899. C.K. Ramachandran, S.L. Gray, and G. Melnykovych, Federation Proc., 38, 632 (1979).

1900. P.J. Ramm and E. Caspi, J. Biol. Chem., 244, 6064 (1969).

1901. J. Ramseyer and B.W. Harding, Biochim. Biophys. Acta, 315, 306 (1973).

1902. F. Ransom, Deutsch. Med. Wochenschr., 27, 194 (1901).

1903. Y. Raoul and N. Le Boulch, Bull. Soc. Chim. Biol., 45, 145 (1963).

1904. Y. Raoul, N. Le Boulch, C. Baron, R. Baizer, and A. Gueriollot-Vinet, Bull. Soc. Chim. Biol., 38, 495 (1956).

1905. Y. Raoul, N. Le Boulch, C. Baron, R. Bazier, and A. Guerillot-Vinet, Bull. Soc. Chem. Biol., 38, 885 (1956).

1906. Y. Raoul, N. Le Boulch, A. Guerillot-Vinet, R. Dulou, and C. Baron, Compt. Rend., 241, 1882 (1955).

1907. D.L. Raulston, C.O. Mishaw, E.J. Parish, and G.J. Schroepfer, Biochem. Biophys. Res. Commun., 71, 984 (1976).

1908. B.N. Ravi, R.W. Armstrong, and D.J. Faulkner, J. Org. Chem., 44, 3109 (1979).

1909. J.P.D. Reckless, D.B. Weinstein, and D. Steinberg, Biochim. Biophys. Acta, 529, 475 (1978).

1910. B.S. Reddy, S. Mangat, A. Sheinfil, J.H. Weisburger, and E.L. Wynder, Cancer Res., 37, 2132 (1977).

1911. B.S. Reddy, C.W. Martin, and E.L. Wynder, Cancer Res., 37, 1697 (1977).

1912. B.S. Reddy and K. Watanabe, Cancer Res., 39, 1521 (1979).

1913. B.S. Reddy and E.L. Wynder, Cancer, 39, 2533 (1977).

1914. J. Redel, J. Chromatog., 168, 273 (1979).

1915. J. Redel and J. Capillon, J. Chromatog., 151, 418 (1978).

1916. E.H. Reerink and A. van Wijk, Biochem. J., 23, 1294 (1929).

1917. E.H. Reerink and A. van Wijk, Biochem. J., 25, 1001 (1931).

1918. E.H. Reerink and A. van Wijk, Strahlentherapie, 40, 728 (1931).

1919. H.H. Rees, P.L. Donnahey, and T.W. Goodwin, J. Chromatog., 116, 281 (1976).

1920. H. Reich, F.E. Walker, and R.W. Collins, J. Org. Chem., 16, 1753 (1951).

1921. H.J. Reich, M. Jautelat, M.T. Messe, F.J. Weigert, and

J.D. Roberts, J. Am. Chem. Soc., 91, 7443 (1969).

1922. D. Reichl, N.B. Myant, M.S. Brown, and J.L. Gold-stein, J. Clin. Invest., 61, 64 (1978).

1923. P. Reichstein, H. Kaufmann, W. Stöcklin, and T. Reichstein, Helv. Chim. Acta, 50, 2114 (1967).

1924. T. Reichstein and H. Reich, Ann. Rev. Biochem., 15, 155 (1946).

1925. A.E. Reif, R.R. Brown, V.R. Potter, E.C. Miller, and J.A. Miller, J. Biol. Chem., 209, 223 (1954).

1926. F. Reindel and A. Detzel, Liebigs Ann., 475, 78 (1929).

1927. M.C. Reinhard and K.W. Buchwald, J. Biol. Chem., 73, 383 (1927).

1928. F. Reinitzer, Monatsh., 9, 421 (1888).

1929. I. Remezov, Biochem. Z., 218, 86 (1930).

1930. I. Remesov, Biochem. Z., 218, 134 (1930).

1931. I. Remesov, Biochem. Z., 218, 157 (1930).

1932. I. Remesov, Biochem. Z., 218, 173 (1930).

1933. I. Remesov, Biochem. Z., 246, 431 (1932).

1934. I. Remesov, Biochem. Z., 248, 256 (1932).

1935. I. Remesov, Ber., 67B, 134 (1934).

1936. I. Remesov, Biochem. Z., 269, 63 (1934).

1937. I. Remesov, XI Congr. Intern. Quim. Pura Aplicada (Madrid), 5, 325 (1934); via Chem. Abstr., 31, 5243[4] (1937).

1938. I. Remesov, Verhandl. Deut. Path. Ges., 142 (1934); via Chem. Abstr., 29, 8029[2] (1935).

1939. I. Remesov, Biochem. Z., 288, 429 (1936).

1940. I.A. Remesov and M.I. Karlina, Biokhimiya, 2, 337 (1937); via Chem. Abstr., 31, 7914[2] (1937).

1941. I. Remesov and O. Sepalowa, Biochem. Z., 266, 330 (1933).

1942. I. Remesov and O. Sepalowa, J. Biochem., 22, 71 (1935).

1943. I. Remesov and O. Sepalowa, Biochem. Z., 287, 345 (1936).

1944. I. Remesov and J. Sosi, Biochem. Z., 287, 358 (1936).

1945. I. Remesov and N. Tavaststyerna, Biochem. Z., 218, 147 (1930).

1946. R. Repke, U. Kubasch, and M. Čarman-Kržan, Arzneimit-tel Forsch., 16, 1469 (1966).

1947. A. Revol, C. Nofre, and A. Cier, Compt. Rend., 247, 2486 (1958).

1948. H.N. Rexroad and W. Gordy, Proc. Natl. Acad. Sci., 45, 256 (1959).

1949. B. Řežábová, J. Hora, V. Landa, C. Černý, and F. Šorm, Steroids, 11, 475 (1968).

1950. M.M. Rhead, G. Eglinton, and G.H. Draffan, Chem. Geol.,

$\underline{8}$, 277 (1971).

1951. W. Richmond, Clin. Chem., $\underline{19}$, 1350 (1973).

1952. R. Richter and H. Dannenberg, Z. Physiol. Chem., $\underline{350}$, 1213 (1969).

1953. C. Riddell and R.P. Cook, Biochem. J., $\underline{61}$, 657 (1955).

1954. B. Riegel and I.A. Kaye, J. Am. Chem. Soc., $\underline{66}$, 723 (1944).

1955. A. Rigo, R. Stevanato, A. Finazzi-Agrò, and G. Rotilio, FEBS Lett. $\underline{80}$, 130 (1977).

1956. C. Riley, Biochem. J., $\underline{87}$, 500 (1963).

1957. H.J. Ringold and S.K. Malhotra, Tetrahedron Lett., 669 (1962).

1958. D.A. Rintoul and R.D. Simoni, Biochim. Biophys. Acta, $\underline{531}$, 322 (1978).

1959. H. Ripperger, Pharmazie, $\underline{33}$, 82 (1978).

1960. E. Ritter, Z. Physiol. Chem., $\underline{34}$, 461 (1901/1902).

1961. F.J. Ritter, Nature, $\underline{202}$, 694 (1964).

1962. J. Robberecht, Bull. Soc. Chim. Belge, $\underline{47}$, 597 (1938).

1963. K.D. Roberts, L. Bandy, and S. Lieberman, Biochemistry, $\underline{8}$, 1259 (1969).

1964. D.M. Robertson, Biochem. J., $\underline{61}$, 681 (1955).

1965. T.B. Robertson, Austr. J. Exptl. Biol. Med. Sci., $\underline{2}$, 83 (1925).

1966. T.B. Robertson and T.C. Burnett, J. Exptl. Med., $\underline{17}$, 344 (1913).

1967. J. Robeson, B. Foster, E.F. Crawford, N. Tagata, and E.T. Adams, Federation Proc., $\underline{39}$, 1721 (1980).

1968. R.L. Robinson and A. Strickholm, Biochim. Biophys. Acta, $\underline{509}$, 9 (1978).

1969. P. Rochlin, Chem. Rev., $\underline{65}$, 685 (1965).

1970. H. Roderbourg and S. Kuzdzal-Savoie, J. Am. Oil Chem. Soc., $\underline{56}$, 485 (1979).

1971. A.W. Rodwell, J. Gen. Microbiol., $\underline{32}$, 91 (1963).

1972. A.E. Roffo, Bol. Inst. Med. Exp. Estud. Cáncer, No. $\underline{36}$, 518 (1934); via Chem. Abstr., $\underline{31}$, 4347[7] (1937).

1973. A.E. Roffo, Bol. Inst. Med. Exp. Estud. Cáncer, No. $\underline{41}$, 41 (1936); via Chem. Abstr., $\underline{31}$, 4347[9] (1937).

1974. A.E. Roffo, Bol. Inst. Med. Exp. Estud. Cáncer, $\underline{13}$, 369 (1936); Anales Asoc. Quím. Argentina, $\underline{25}$, 18B (1937); via Chem. Abstr., $\underline{32}$, 2159[6] (1938).

1975. A.E. Roffo, Bol. Inst. Med. Exp. Estud. Cáncer, $\underline{14}$, 107 (1937); via Chem. Abstr., $\underline{31}$, 7761[3] (1937).

1976. A.E. Roffo, Bol. Inst. Med. Exp. Estud. Cáncer, $\underline{14}$, 447 (1937); via Chem. Abstr., $\underline{32}$, 3812[5] (1938).

1977. A.H. Roffo, Bol. Inst. Med. Exp. Estud. Cáncer, $\underline{7}$, 555 (1930); via Am. J. Cancer, $\underline{15}$, 915 (1930).

1978. A.H. Roffo, Bol. Inst. Med. Exp. Estud. Cáncer, 7,
1216 (1930); via Am. J. Cancer, 15, 1628 (1931).

1979. A.H. Roffo, Am. J. Cancer, 17, 42 (1933).

1980. A.H. Roffo, Bol. Inst. Med. Exp. Estud. Cáncer, No. 39,
390 (1935); via Chem. Abstr., 31, 4347[8] (1937).

1981. A.H. Roffo, Bol. Inst. Med. Exp. Estud. Cáncer, 14, 5
(1937); via Chem. Zentr., II, 1788 (1938).

1982. A.H. Roffo, Bol. Inst. Med. Exp. Estud. Cáncer, 14,
19 (1937); via Chem. Zentr., II, 1788 (1938).

1983. A.H. Roffo, Bol. Inst. Med. Exp. Estud. Cáncer, 14, 46,
589 (1938); via Chem. Abstr., 33, 1030[9] (1939).

1984. A.H. Roffo, Bol. Inst. Med. Exp. Estud. Cáncer, 15,
837 (1938); via Chem. Abstr., 33, 8277[7] (1939).

1985. A.H. Roffo, Z. Krebsforsch., 47, 473 (1938).

1986. A.H. Roffo, Bull. Assoc. Franc. Etude Cancer, 28,
556 (1939).

1987. A.H. Roffo, Z. Krebsforsch., 49, 341 (1939/1940).

1988. A.H. Roffo, Österr. Chem.-Ztng., 43, 87 (1940).

1989. A.H. Roffo, Bol. Inst. Med. Exp. Estud. Cáncer, 19,
503 (1942); via Chem. Abstr., 37, 6335[8] (1943).

1990. A.H. Roffo, Bol. Inst. Med. Exp. Estud. Cáncer, 20, 51
(1943); via Chem. Abstr., 38, 1275[9] (1944).

1991. A.H. Roffo, Bol. Inst. Med. Exp. Estud. Cáncer, 20, 65
(1943); via Chem. Abstr., 38, 1281[1] (1944).

1992. A.H. Roffo, Bol. Inst. Med. Exp. Estud. Cáncer, 20,
123 (1943); via Chem. Abstr., 38, 1276[1] (1944).

1993. A.H. Roffo, Bol. Inst. Med. Exp. Estud. Cáncer, 20,
471 (1943); via Chem. Abstr., 38, 2730[1] (1944).

1994. A.H. Roffo, Bol. Inst. Med. Exp. Estud. Cáncer, 20,
515 (1943); via Chem. Abstr., 38, 2753[7] (1944).

1995. A.H. Roffo and L.M. Correa, Bol. Inst. Med. Exp. Estud.
Cáncer, 15, 847 (1939); via Chem. Abstr., 33, 8209[3]
(1939).

1996. A.H. Roffo and L.M. Correa, Prensa Med. Argentina, 26,
955 (1939); via Chem. Abstr., 36, 5838[2] (1942).

1997. A.H. Roffo and L.M. Correa Urquiza, Anal. Asoc. Quím.
Argentina, 30, 177 (1942); via Chem. Abstr., 37, 1932[4]
(1943).

1998. A.H. Roffo and L.M. Correa Urquiza, Bol. Inst. Med.
Exp. Estud. Cáncer, 19, 609 (1942); via Chem. Abstr.,
37, 6336[2] (1943).

1999. A.H. Roffo and H. De Giorgi, Bol. Inst. Med. Exp.
Estud. Cáncer, 7, 970 (1930); via Am. J. Cancer, 15,
1628 (1931).

2000. A.H. Roffo and A.E. Roffo, Bol. Inst. Med. Exp.
Estud. Cáncer, 20, 143 (1943); via Chem. Abstr., 38,

12763 (1944).

2001. P. Roller, C. Djerassi, R. Cloetens, and B. Tursch, J. Am. Chem. Soc., 91, 4918 (1969).

2002. P. Rona and W. Deutsch, Biochem. Z., 171, 89 (1926).

2003. F. Ronchetti and G. Russo, J. Chem. Soc. Chem. Commun., 184 (1973).

2004. F. Ronchetti and G. Russo, J. Chem. Soc. Chem. Commun., 785 (1974).

2005. F. Ronchetti and G. Russo, Tetrahedron Lett., 85 (1975).

2006. H.G. Roscoe and M.J. Fahrenbach, J. Lipid Res., 12, 17 (1971).

2007. H.G. Roscoe, R. Goldstein, and M.J. Fahrenbach, Biochem. Pharmacol., 17, 1189 (1968).

2008. R.S. Rosenfeld, Anal. Biochem., 12, 483 (1965).

2009. R.S. Rosenfeld and L. Hellman, J. Biol. Chem., 233, 1089 (1958).

2010. R.S. Rosenfeld, B. Zumoff, and L. Hellman, Arch. Biochem. Biophys., 96, 84 (1962).

2011. R.S. Rosenfeld, B. Zumoff, and L. Hellman, J. Lipid Res., 8, 16 (1967).

2012. M.C. Rosenheim, Biochem. J., 8, 74 (1914).

2013. M.C. Rosenheim, Biochem. J., 8, 82 (1914).

2014. M.C. Rosenheim, Biochem. J., 10, 176 (1916).

2015. O. Rosenheim, Biochem. J. 21, 386 (1927).

2016. O. Rosenheim and R.K. Callow, Biochem. J., 25, 74 (1931).

2017. O. Rosenheim and H. King, Ann. Rev. Biochem., 3, 87 (1934).

2018. O. Rosenheim and W.W. Starling, Chem. Ind., 52, 1056 (1933).

2019. O. Rosenheim and W.W. Starling, J. Chem. Soc., 377 (1937).

2020. O. Rosenheim and T.A. Webster, Biochem. J., 20, 537 (1926).

2021. O. Rosenheim and T.A. Webster, Biochem. J., 27, 389 (1927).

2022. O. Rosenheim and T.A. Webster, Nature, 136, 474 (1935).

2023. O. Rosenheim and T.A. Webster, Biochem. J., 37, 513 (1943).

2024. H. Rosenkrantz, A.T. Milhorat, and M. Farber, J. Biol. Chem., 195, 509 (1952).

2025. R.A. Ross and P.J. Scheuer, Tetrahedron Lett., 4701 (1979).

2026. W. Rossner, Z. Physiol. Chem., 249, 267 (1937).

2027. G.H. Rothblat and M.K. Buchko, J. Lipid Res., 12, 647 (1971).

2028. A. Rotman and Y. Mazur, J. Chem. Soc. Chem. Commun., 15 (1974).

2029. E. Rousseau, Compt. Rend. Soc. Biol., 135, 567, 569 (1941).

2030. J.W. Rowe, Phytochem., 4, 1 (1965).

2031. J.W. Rowe and H.H. Scroggins, J. Org. Chem., 29, 1554 (1964).

2032. A.T. Rowland and H.R. Nace, J. Am. Chem. Soc., 82, 2833 (1960).

2033. T.A. Roy, F.H. Field, Y.Y. Lin, and L.L. Smith, Anal. Chem., 51, 272 (1979).

2034. I. Rubinstein and P. Albrecht, J. Chem. Soc. Chem. Commun., 957 (1975).

2035. I. Rubinstein, L.J. Goad, A.D.H. Clague, and L.J. Mulheirn, Phytochem., 15, 195 (1976).

2036. I. Rubinstein, O. Sieskind, and P. Albrecht, J. Chem. Soc., Perkin Trans. I, 1833 (1975).

2037. W.L. Ruigh, Ann. Rev. Biochemistry, 14, 224 (1945).

2038. H.I. Ruiz, Anal. Inst. Farmacol. Española, 5, 323 (1956).

2039. R. Ruiz and P. Dea, J. Chromatog., 146, 321 (1978).

2040. T.-I. Ruo, S.-S. Loo, and C. Chen, Federation Proc., 36, 663 (1977).

2041. A. Ruokonen, Biochim. Biophys. Acta, 316, 251 (1973).

2042. E. Ruppol, Bull. Soc. Chim. Biol., 24, 324 (1942).

2043. E. Ruppol, J. Pharm. Belg., 2, 96 (1943); via Chem. Zentr., II, 1610 (1945).

2044. G.A. Russell, Science, 161, 423 (1968).

2045. G.A. Russell and E.R. Talaty, J. Am. Chem. Soc., 86, 5345 (1964).

2046. G.A. Russell and E.R. Talaty, Science, 148, 1217 (1965).

2047. G.A. Russell, E.R. Talaty, and R.H. Harrocks, J. Org. Chem., 32, 353 (1967).

2048. W.J. Russell, Proc. Royal Soc. London, 63, 102 (1898); 64, 409 (1899).

2049. L. Ruzicka and W. Bosshard, Helv. Chim. Acta, 20, 244 (1937).

2050. L. Ruzicka and V. Prelog, Helv. Chim. Acta, 26, 975 (1943).

2051. L. Ruzicka, V. Prelog, and E. Tagmann, Helv. Chim. Acta, 27, 1149 (1944).

2052. A.I. Ryer, W.H. Gebert, and N.M. Murrill, J. Am. Chem. Soc., 72, 4247 (1950).

2053. R. Ryhage and E. Stenhagen, J. Lipid Res., 1, 361 (1960).

2054. H.Y. Saad and W.I. Higuchi, J. Pharm. Sci., 54, 1205 (1965).

2055. J.R. Sabine, CHOLESTEROL, Marcel Dekker Inc., New York/ Basel, 1977, p. 33.

2056. J.M. Saez, B. Loras, A.M. Morera, and J. Bèrtrand, J. Steroid Biochem., 1, 355 (1970).

2057. L.M. Safe, C.J. Wong, and R.F. Chandler, J. Pharm. Sci., 63, 464 (1974).

2058. B. Saha, D.B. Naskar, D.R. Misra, B.P. Pradhan, and H.N. Khastgir, Tetrahedron Lett., 3095 (1977).

2059. A. Sakamoto, Yonago Acta Med., 12, 81 (1968).

2060. S. Sakurai, N. Ikekawa, T. Ohtaki, and H. Chino, Science, 198, 627 (1977).

2061. A. Saliot and M. Barbier, in ADVANCES IN ORGANIC GEO- CHEMISTRY 1973, B. Tissot and F. Bienner, Editors, Éditions Technip, Paris, 1974, pp. 607-617.

2062. A. Saliot and M. Barbier, Deep Sea Res., 20, 1077 (1973).

2063. E. Salkowski, Z. Anal. Chem., 26, 557 (1887).

2064. W.G. Salmond, M.A. Barta, A.M. Cain, and M.C. Sobala, Tetrahedron Lett., 1683 (1977).

2065. W.G. Salmond, M.A. Barta, and J.L. Havens, J. Org. Chem., 43, 2057 (1978).

2066. W.G. Salmond and M.C. Sobala, Tetrahedron Lett., 1695 (1977).

2067. D. Samuel, in BIOCHEMIE DES SAUERSTOFFS, B. Hess and Hj. Staudinger, Editors, Springer-Verlag, Heidelberg, 1968, pp. 6-25.

2068. D. Samuel, in MOLECULAR OXYGEN IN BIOLOGY, TOPICS IN MOLECULAR OXYGEN RESEARCH, O. Hayaishi, Editor, North Holland Publ. Co., Amsterdam, 1974, pp. 1-32.

2069. P. Samuel, M. Urivetsky, and G. Kaley, J. Chromatog., 14, 508 (1964).

2070. S. Samuels and C. Fisher, J. Chromatog., 71, 297 (1972).

2071. L. Sanche and J.E. van Lier, Chem. Phys. Lipids, 16, 225 (1976).

2072. L. Sanche and J.E. van Lier, Nature, 263, 79 (1976).

2073. R. Sandmeier and C. Tamm, Helv. Chim. Acta, 56, 2238 (1973).

2074. A. Sanghvi, M. Galli Kienle, and G. Galli, Anal. Biochem., 85, 430 (1978).

2075. G. Santillan, J.S.M. Sarma, G. Pawlik, A. Rackl, A. Grenier, and R.J. Bing, Atherosclerosis, 35, 1 (1980).

2076. J.S.M. Sarma and R.J. Bing, J. Mol. Cell. Cardiol., 10, 197 (1978).

2077. J.S.M. Sarma, R.J. Bing, S. Ikeda, and R. Fischer, Artery, 2, 153 (1976).

2078. J.S.M. Sarma, R. Fischer, S. Ideda, and R.J. Bing, Proc. Soc. Exptl. Biol. Med., 151, 303 (1976).

2079. T. Sato, H.Yamauchi, Y. Ogata, M. Tsujii, T. Kunii, K. Kagei, S. Toyoshima, and T. Kobayashi, Chem. Pharm. Bull., 26, 2933 (1978).

2080. Y. Sato and S. Hayakawa, J. Org. Chem., 26, 4181 (1961).

2081. Y. Sato and S. Hayakawa, J. Org. Chem., 28, 2739 (1963).

2082. Y. Sato, J.A. Waters, and H. Kaneko, J. Org. Chem., 29, 3732 (1964).

2083. S.E. Saucier and A.A. Kandutsch, Biochim. Biophys. Acta, 572, 541 (1979).

2084. L. Saunders, J. Perrin, and D. Gammack, J. Pharm. Pharmacol., 14, 567 (1962).

2085. K. Savard, H.W. Wotiz, P. Marcus, and H.M. Lemon, J. Am. Chem. Soc., 75, 6327 (1953).

2086. D.T. Sawyer, M.J. Gibian, M.M. Morrison, and E.T. Seo, J. Am. Chem. Soc., 100, 627 (1978).

2087. L.C. Scala and G.D. Dixon, Mol. Cryst. Liq. Cryst., 7, 443 (1969).

2088. T.J. Scallen, A.K. Dhar, and E.D. Loughran, J. Biol. Chem., 246, 3168 (1971).

2089. T.J. Scallen and W. Krueger, J. Lipid Res., 9, 120 (1968).

2090. A.P. Schaap, SINGLET MOLECULAR OXYGEN, Dowden, Hutchinson & Ross, Inc., Stroudsburg, Pa., 1976.

2091. A.P. Schaap, K.A. Zaklika, B. Kashir, and L.W.-M. Fung, J. Am. Chem. Soc., 102, 389 (1980).

2092. J.C. Schabort and H.L. Teijema, Phytochem., 7, 2107 (1968).

2093. J. Schaefle, B. Ludwig, P. Albrecht, and G. Ourisson, Tetrahedron Lett., 4163 (1978).

2094. H. Schaltegger and F.X. Müllner, Helv. Chim. Acta, 34, 1096 (1951).

2095. I. Scheer, M.J. Thompson, and E. Mosettig, J. Am. Chem. Soc., 78, 4733 (1956).

2096. F. Schenck, K. Bucholz, and O. Wiese, Ber., 69, 2696 (1936).

2097. G.O. Schenck, Naturwissenschaften, 43, 71 (1956).

2098. G.O. Schenck, Angew. Chem., 69, 579 (1957).

2099. G.O. Schenck, Ind. Eng. Chem., 55, No. 6, 40 (1963).

2100. G.O. Schenck, W. Eisfeld, and O.-A. Neumüller, Liebigs Ann., 701 (1975).

2101. G.O. Schenck, K. Gollnick, and O.-A. Neumüller,

      Liebigs Ann., 603, 46 (1957).
2102. G.O. Schenck and O.-A. Neumüller, Liebigs Ann., 618,
      194 (1958).
2103. G.O. Schenck, O.-A. Neumüller, and W. Eisfeld, Angew.
      Chem., 70, 595 (1958).
2104. G.O. Schenck, O.-A. Neumüller, and W. Eisfeld, Liebigs
      Ann., 618, 202 (1958).
2105. G.O. Schenck and K. Ziegler, Naturwissenschaften, 32,
      157 (1944).
2106. H. Schildknecht and D. Hotz, Angew. Chem., 79, 902
      (1967); Angew. Chem., Int. Ed., 6, 881 (1967).
2107. H. Schildknecht and W. Körnig, Angew. Chem., 80, 45
      (1968); Angew. Chem., Int. Ed., 7, 62 (1968).
2108. H. Schildknecht, R. Siewerdt, and U. Maschwitz, Lie-
      bigs Ann., 703, 182 (1967).
2109. H. Schiller, Fette Seif. Anstrich., 75, 145 (1973).
2110. Schimmel and Co., Geschäftsbericht, April 1908; via
      Chem. Zentr., I, 1839 (1908).
2111. F.W. Schlutz and M.R. Ziegler, J. Biol. Chem., 69,
      415 (1926).
2112. P. Schmid and E. Hunter, Physiol. Chem. Phys., 3, 98
      (1971).
2113. W.A. Schmidt, Z. Allgem. Physiol., 7, 369 (1907).
2114. F.J. Schmitz, in MARINE NATURAL PRODUCTS, CHEMICAL
      AND BIOLOGICAL PERSPECTIVES, P.J. Scheuer, Editor,
      Academic Press, New York City, N.Y., Vol. 1, 1978,
      pp. 241-297.
2115. F.J. Schmitz, D.C. Campbell, and I. Kubo, Steroids, 28,
      211 (1976).
2116. J. Schmutz, H. Schaltegger, and M. Sanz, Helv. Chim.
      Acta, 34, 1111 (1951).
2117. J.J. Schneider, Tetrahedron, 28, 2717 (1972).
2118. J.J. Schneider and M.L. Lewbart, Rec. Progr. Hormone
      Res., 15, 201 (1959).
2119. W.J. Schneider, S.K. Basu, M.J. McPhaul, J.L. Gold-
      stein, and M.S. Brown, Proc. Natl. Acad. Sci., 76,
      5577 (1979).
2120. B. Schnuriger and J. Bourdon, Photochem. Photobiol.,
      8, 361 (1968).
2121. R. Schoenheimer, J. Biol. Chem., 110, 461 (1935).
2122. R. Schoenheimer, H. Dam, and K. von Gottberg, J. Biol.
      Chem., 110, 659 (1935).
2123. R. Schoenheimer and E.A. Evans, J. Biol. Chem., 114,
      567 (1936).
2124. R. Schoenheimer, D. Rittenberg, and M. Graff, J.
      Biol. Chem., 111, 183 (1935).

2125. N.A. Scholan and G.S. Boyd, Biochem. J., 108, 27P (1968).

2126. N.A. Scholan and G.S. Boyd, Z. Physiol. Chem., 349, 1628 (1968).

2127. K.H. Schönemann, N.P. van Vliet, and F.J. Zeelen, Rec. Trav. Chim., 99, 91 (1980).

2128. R. Schönheimer, Z. Physiol. Chem., 177, 143 (1928).

2129. R. Schönheimer, Z. Physiol. Chem., 192, 73 (1930).

2130. R. Schönheimer, Z. Physiol. Chem., 192, 77 (1930).

2131. R. Schönheimer, Z. Physiol. Chem., 192, 86 (1930).

2132. R. Schönheimer, Z. Physiol. Chem., 211, 65 (1932).

2133. R. Schönheimer, H. von Behring, and R. Hummel, Z. Physiol. Chem., 192, 93 (1930).

2134. E. Schreiber, Münch. Med. Woch., 60, 2001 (1913).

2135. E. Schreiber and Lénárd, Biochem. Z., 54, 291 (1913).

2136. E. Schreiber and Lénárd, Biochem. Z., 49, 459 (1913).

2137. H. Schriefers and W. Wagner, Experientia, 32, 18 (1976).

2138. G.J. Schroepfer, B.N. Lutsky, J.A. Martin, S. Huntoon, B. Fourcans, W.-H. Lee, and J. Vermillion, Proc. Royal Soc. London, [B], 180, 125 (1972).

2139. G.J. Schroepfer, D. Monger, A.S. Taylor, J.S. Chamberlain, E.J. Parish, A. Kisic, and A.A. Kandutsch, Biochem. Biophys. Res. Commun., 78, 1227 (1977).

2140. G.J. Schroepfer, E.J. Parish, H.W. Chen, and A.A. Kandutsch, Federation Proc., 35, 1697 (1976).

2141. G.J. Schroepfer, E.J. Parish, H.W. Chen, and A.A. Kandutsch, J. Biol. Chem., 252, 8975 (1977).

2142. G.J. Schroepfer, E.J. Parish, G.L. Gilliland, M.E. Newcomer, L.L. Somerville, F.A. Quiocho, and A.A. Kandutsch, Biochem. Biophys. Res. Commun., 84, 823 (1978).

2143. G.J. Schroepfer, E.J. Parish, and A.A. Kandutsch, J. Am. Chem. Soc., 99, 5494 (1977).

2144. G.J. Schroepfer, E.J. Parish, and A.A. Kandutsch, Chem. Phys. Lipids, 25, 265 (1979).

2145. G.J. Schroepfer, E.J. Parish, M. Tsuda, and A.A. Kandutsch, Biochem. Biophys. Res. Commun., 91, 606 (1979).

2146. G.J. Schroepfer, E.J. Parish, M. Tsuda, D.L. Raulston, and A.A. Kandutsch, J. Lipid Res., 20, 994 (1979).

2147. G.J. Schroepfer, R.A. Pascal, and A.A. Kandutsch, Biochem. Pharmacol., 28, 249 (1979).

2148. G.J. Schroepfer, R.A. Pascal, R. Shaw, and A.A. Kandutsch, Biochem. Biophys. Res. Commun., 83, 1024 (1978).

2149. G.J. Schroepfer, D.L. Raulston, and A.A. Kandutsch, Biochem. Biophys. Res. Commun., 79, 406 (1977).

2150. G.J. Schroepfer, V. Walker, E.J. Parish, and A. Kisic,
      Biochem. Biophys. Res. Commun., 93, 813 (1980).
2151. K. Schubert and W. Fischer, Arch. Geschwulstforsch.,
      20, 177 (1963).
2152. K. Schubert, G. Kaufmann, and H. Budzikiewicz, Bio-
      chim. Biophys. Acta, 176, 170 (1969).
2153. K. Schubert, G. Rose, and G. Bacigalupo, Naturwissen-
      schhaften, 47, 497 (1960).
2154. K. Schubert, G. Rose, and M. Bürger, Z. Physiol. Chem.,
      326, 235 (1961).
2155. K. Schubert, K. Wehrberger, and G. Hobe, Endocrinol.
      Exp., 5, 205 (1971); via Chem. Abstr., 76, 124618w
      (1972).
2156. L.A. Schuler, L. Scavo, T.M. Kirsch, G.L. Flickinger,
      and J.F. Strauss, J. Biol. Chem., 254, 8662 (1979).
2157. W.H. Schuller and R.V. Lawrence, J. Med. Chem., 14,
      466 (1971).
2158. K.E. Schulte, G. Rücker, and H. Fachmann, Tetrahedron
      Lett., 4763 (1968).
2159. A. Schultz, Centrlb. Allgem. Pathol., 35, 314 (1924).
2160. A. Schultz and G. Löhr, Centrlb. Allgem. Pathol., 36,
      529 (1925).
2161. E. Schulze, Ber., 5, 1075 (1872).
2162. E. Schulze, J. Prakt. Chem.,N.F., 7, 163 (1873).
2163. E. Schulze and E. Winterstein, Z. Physiol. Chem., 43,
      316 (1904).
2164. E. Schulze and E. Winterstein, Z. Physiol. Chem., 48,
      546 (1906).
2165. W.C. Schumb, C.N. Satterfield, and R.L. Wentworth,
      HYDROGEN PEROXIDE, ACS Monograph Series No. 128, Rein-
      hold Publishing Corp., New York, 1955, p. 456.
2166. D.P. Schwartz, J. Chromatog., 178, 105 (1979).
2167. V. Schwartz and J. Protiva, Folia Microbiol., 19, 156
      (1974).
2168. R.B. Schwendinger and J.G. Erdman, Science, 144, 1575
      (1964).
2169. E. Schwenk, G.J. Alexander, C.A. Fish, and T.H.
      Stoudt, Federation Proc., 14, 752 (1955).
2170. E. Schwenck, D.F. Stevens, and R. Altschul, Proc. Soc.
      Exptl. Biol. Med., 102, 42 (1959).
2171. E. Schwenk and N.T. Werthessen, Arch. Biochem. Bio-
      phys., 40, 334 (1952).
2172. E. Schwenk, N.T. Werthessen, and H. Rosenkrantz, Arch.
      Biochem. Biophys., 37, 247 (1952).
2173. J. Scotney and E.V. Truter, J. Chem. Soc.,(C), 2516
      (1968).

2174. J. Scotney and E.V. Truter, J. Chem. Soc.,(C), 2184 (1968).

2175. J. Scotney and E.V. Truter, J. Chem. Soc.,(C), 1911 (1968).

2176. A.I. Scott and A.D. Wrixon, J. Chem. Soc.,(D), Chem. Commun., 1184 (1969).

2177. K.N. Scott and T.H. Mareci, Can. J. Chem., 57, 27 (1979).

2178. W.W. Scott, J. Urology, 53, 712 (1945).

2179. J.T. Sears and J.W. Sutherland, J. Phys. Chem., 72, 1166 (1968).

2180. H. Seel, Arch. Exp. Pathol. Pharmakol., 117, 282 (1926).

2181. H. Seel, Arch. Exp. Pathol. Pharmakol., 133, 129 (1928).

2182. C. Seelkopf and K. Salfelder, Z. Krebsforsch., 64, 459 (1962).

2183. W.K. Seifert, Fortschr. Chem. Org. Naturstoffe, 32, 1 (1975).

2184. W.K. Seifert, E.J. Gallegos, and R.M. Teeter, Angew. Chem. Int. Ed., 10, 747 (1971).

2185. W.K. Seifert, E.J. Gallegos, and R.M. Teeter, J. Am. Chem. Soc., 94, 5880 (1972).

2186. N. Seiler and H. Mägerlein, Z. Naturforsch., 21B, 78 (1966).

2187. L.M. Seitz and J.V. Paukstelis, J. Ag. Food Chem., 25, 838 (1977).

2188. G.M. Selal and I.V. Torgov, Bioorgan. Khimiya, 5, 1668 (1979).

2189. G.A. Selter and K.D. McMichael, J. Org. Chem., 32, 2546 (1967).

2190. E.P. Serebryakov, A.V. Simolin, V.F. Kucherov, and B.V. Rosynov, Tetrahedron, 26, 5215 (1970).

2191. A. Sevanian, N.M. Elsayed, and A.D. Hacker, Federation Proc., 39, 788 (1980).

2192. A. Sevanian, J.F. Mead, and R.A. Stein, Lipids, 14, 634 (1979).

2193. A. Sevanian, R.A. Stein, and J.F. Mead, Am. Rev. Respiratory Disease, 119, Suppl. No. 4, 358 (1979).

2194. Y. Seyama, K. Ichikawa, and T. Yamakawa, J. Biochem., 80, 223 (1976).

2195. L.M. Shabad, T.S. Kolesnichenko, and L.A. Savluchinskaya, Neoplasma, 20, 347 (1973).

2196. E. Shapiro, L. Finckenor, and H.L. Herzog, J. Org. Chem., 33, 1673 (1968).

2197. E.L. Shapiro, T. Legatt, and E.P. Oliveto, Tetrahedron

Lett., 663 (1964).

2198. I.L. Shapiro, in ULTRAPURITY, METHODS AND TECHNIQUES,
       M. Zief and R. Speights, Editors, Marcel Dekker Inc.,
       New York, 1972, pp. 193-204.

2199. I.L. Shapiro and D. Kritchevsky, J. Chromatog., 18,
       599 (1965).

2200. N.K. Sharma, D.K. Kulshreshtha, J.S. Tandon, D.S.
       Bhakuni, and M.M. Dhar, Phytochem., 13, 2239 (1974).

2201. R.K. Sharma and J.S. Brush, Arch. Biochem. Biophys.,
       156, 560 (1973).

2202. P.E. Shaw, Steroids, 15, 151 (1970).

2203. Y.-S. Shaw and C. Chen, Biochem. J., 128, 1285 (1972).

2204. M.J. Shear and F.W. Ilfeld, Am. J. Pathol., 16, 287
       (1940).

2205. M.J. Shear and B. Kramer, J. Biol. Chem., 71, 213
       (1926).

2206. M.J. Shear and E. Lorenz, Am. J. Cancer, 36, 201
       (1939).

2207. S. Shefer, F.W. Cheng, A.K. Batta, B. Dayal, G.S.
       Tint, and G. Salen, J. Clin. Invest., 62, 539 (1978).

2208. S. Shefer, F.W. Cheng, A.K. Batta, B. Dayal, G.S.
       Tint, G. Salen, and E.H. Mosbach, J. Biol. Chem., 253,
       6386 (1978).

2209. S. Sheffer, F.W. Cheng, B. Dayal, S. Hauser, G.S. Tint,
       G. Salen, and E.H. Mosbach, J. Clin. Invest., 57, 897
       (1976).

2210. S. Shefer, S. Hauser, and E.H. Mosbach, J. Biol. Chem.,
       241, 946 (1966).

2211. S. Shefer, S. Hauser, and E.H. Mosbach, J. Lipid Res.,
       7, 763 (1966).

2212. S. Shefer, S. Hauser, and E.H. Mosbach, J. Lipid Res.,
       9, 328 (1968).

2213. S. Shefer, S. Hauser, and E.H. Mosbach, J. Lipid Res.,
       13, 69 (1972).

2214. S. Shefer, S. Milch, and E.H. Mosbach, J. Biol. Chem.,
       239, 1731 (1964).

2215. S. Shefer, G.Nicolau,   and E.H. Mosbach, J. Lipid
       Res., 16, 92 (1975).

2216. Y.M. Sheikh and C. Djerassi, Tetrahedron Lett., 2927
       (1973).

2217. Y.M. Sheikh and C. Djerassi, Tetrahedron, 30, 4095
       (1974).

2218. Y.M. Sheikh, M. Kaisin, and C. Djerassi, Steroids, 22,
       835 (1973).

2219. Y.M. Sheikh, B.M. Tursch, and C. Djerassi, J. Am. Chem.
       Soc., 94, 3278 (1972).

2220. Y.M. Sheikh, B. Tursch, and C. Djerassi, Tetrahedron Lett., 3721 (1972).

2221. D.H. Shelling, Proc. Soc. Exptl. Biol. Med., 35, 660 (1937).

2222. J.L. Sheumaker and J.K. Guillory, Thermochim. Acta, 5, 355 (1973).

2223. H.S. Shieh, L.G. Hoard, and C.E. Nordman, Nature, 267, 287 (1977).

2224. M. Shikita and P.F. Hall, J. Biol. Chem., 248, 5598 (1973).

2225. M. Shikita and P.F. Hall, J. Biol. Chem., 248, 5605 (1973).

2226. M. Shikita, P.F. Hall, and S. Isaka, Biochem. Biophys. Res. Commun., 50, 289 (1973).

2227. A. Shimasue, Hiroshima J. Med. Sci., 23, 265 (1974).

2228. K. Shimizu, Biochim. Biophys. Acta, 111, 571 (1965).

2229. K. Shimizu, J. Biol. Chem., 240, 1941 (1965).

2230. K. Shimizu, J. Biochem., 59, 430 (1966).

2231. K. Shimizu, R.I. Dorfman, and M. Gut, J. Biol. Chem., 235, PC25 (1960).

2232. K. Shimizu, M. Hayano, M. Gut, and R.I. Dorfman, J. Biol. Chem., 236, 695 (1961).

2233. K. Shimizu and F. Nakada, Biochim. Biophys. Acta, 450, 441 (1976).

2234. Y. Shimizu, J. Am. Chem. Soc., 94, 4051 (1972).

2235. Y. Shimizu and H. Mitsuhashi, Tetrahedron, 24, 4143 (1968).

2236. A.A. Shishkina, V.M. Rzheznikov, and K.K. Pivnitskii, Khim. Prirod. Soed., 138 (1970).

2237. A.N. Shivrina, Prod. Biosin. Vyssh. Gribov Ikh Ispol's. Akad. Nauk S.S.S.R., 49 (1966); via Chem. Abstr., 66, 17271z (1967).

2238. Z.A. Shkiryak, R.P. Morozova, I.A. Nikolenko, and B.N. Mandzyuk, Vitaminy, 6, 122 (1971); via Chem. Abstr., 77, 56604g (1972).

2239. J.N. Shoolery and M.T. Rogers, J. Am. Chem. Soc., 80, 5121 (1958).

2240. C.W. Shoppee, Ann. Reports Progress Chem. For 1946, 214 (1947).

2241. C.W. Shoppee, CHEMISTRY OF THE STEROIDS, Butterworths & Co. Ltd., Washington City, D.C., 2nd Ed., 1964, p. 46, p. 50.

2242. C.W. Shoppee and B.C. Newman, J. Chem. Soc., (C), 981 (1968).

2243. C.W. Shoppee and E. Shoppee, in CHEMISTRY OF CARBON COMPOUNDS, A MODERN COMPREHENSIVE TREATISE, E.H. Rodd,

Editor, Elsevier Publ. Co., Amsterdam/Houston/London/
New York, 1953, Vol. IIB, pp. 765-875.

2244. M.R. Shreve, P.G. Morrissey, and P.J. O'Brien, Biochem. J., 177, 761 (1979).

2245. P. Shubik and J.L. Hartwell, SURVEY OF COMPOUNDS
WHICH HAVE BEEN TESTED FOR CARCINOGENIC ACTIVITY,
Suppl. 1, U.S. Public Health Service Publ. No. 149,
Washington City, D.C., 1957, pp. 244-247.

2246. P. Shubik and J.L. Hartwell, SURVEY OF COMPOUNDS
WHICH HAVE BEEN TESTED FOR CARCINOGENIC ACTIVITY,
Suppl. 2, U.S. Public Health Service, Washington
City, D.C., 1969, pp. 412-421.

2247. J.B. Siddall, G.V. Baddeley, and J.A. Edwards, Chem.
Ind., 25 (1966).

2248. Zh. S. Sidykov, A.P. Yavkin, G.M. Segal, and K.K.
Koshoev, Khim.-Farm. Zh. S.S.S.R., 12. 138 (1978).

2249. C.M. Siegmann and M.D. De Winter, Rec. Trav. Chim.,
89, 442 (1970).

2250. L. Siekmann, K.P. Hüskes, and H. Breuer, Z. Anal.
Chem., 279, 145 (1976).

2251. D. Siele and W. Fiehn, Experientia, 32, 849 (1976).

2252. H. Silberman and S. Silberman-Martyncewa, J. Biol.
Chem., 159, 603 (1945).

2253. S.J. Silverman and A.W. Andrews, J. Natl. Cancer Inst.,
59, 1557 (1977).

2254. B.R. Simoneit, D.H. Smith, G. Eglinton, and A.L. Burlingame, Arch. Environ. Contamination Toxicol., 1,
193 (1973).

2255. E.R. Simpson and G.S. Boyd, Biochem. Biophys. Res.
Commun., 24, 10 (1966).

2256. E.R. Simpson and G.S. Boyd, Biochem. J., 99, 52
(1966).

2257. E.R. Simpson and G.S. Boyd, Biochem. Biophys. Res.
Commun., 28, 945 (1967).

2258. E.R. Simpson and G.S. Boyd, Eur. J. Biochem., 2, 275
(1967).

2259. E.R. Simpson and D.A. Miller, Arch. Biochem. Biophys.,
190, 800 (1978).

2260. M. Sinensky, Biochem. Biophys. Res. Commun., 78, 863
(1977).

2261. M. Sinensky, Proc. Natl. Acad. Sci., 75, 1247 (1978).

2262. M. Sinensky, G. Duwe, and F. Pinkerton, J. Biol.
Chem., 254, 4482 (1979).

2263. A.T. Sipahimalani, V.R. Mamdapur, N.K. Joshi, and
M.S. Chadha, Naturwissenschaften, 57, 40 (1970).

2264. M.D. Siperstein, C.W. Nichols, and I.L. Chaikoff,

Circulation, 7, 37 (1953).

2265. K.M. Sivanandaiah and W.R. Nes, Steroids, 5, 539 (1965).

2266. E.L. Skau and W. Bergmann, J. Am. Chem. Soc., 60, 986 (1938).

2267. E.L. Skau and W. Bergmann, J. Org. Chem., 3, 166 (1938).

2268. S.J.M. Skinner, C.J.P. Tobler, and R.A.F. Couch, Steroids, 30, 315 (1977).

2269. J.K. Sliwowski and E. Caspi, J. Steroid Biochem., 8, 47 (1977).

2270. J.K. Sliwowski and E. Caspi, J. Am. Chem. Soc., 99, 4479 (1977).

2271. G. Slomp and F.A. MacKellar, J. Am. Chem. Soc., 84, 204 (1962).

2272. D.M. Small and G.G. Shipley, Science, 185, 222 (1974).

2273. A.G. Smith and C.J.W. Brooks, J. Chromatog., 101, 373 (1974).

2274. A.G. Smith and C.J.W. Brooks, Biochem. Soc. Trans., 3, 675 (1975).

2275. A.G. Smith and C.J.W. Brooks, Biomed. Mass Spectro-metry, 3, 81 (1976).

2276. A.G. Smith and C.J.W. Brooks, J. Steroid Biochem., 7, 705 (1976).

2277. A.G. Smith and C.J.W. Brooks, Biochem. J., 167, 121 (1977).

2278. A.G. Smith and C.J.W. Brooks, Biochem. Soc. Trans., 5, 1088 (1977).

2279. A.G. Smith, C.J.W. Brooks, and W.A. Harland, Steroids Lipids Res., 5, 150 (1974).

2280. A.G. Smith, J.D. Gilbert, W.A. Harland, and C.J.W. Brooks, Biochem. J., 139, 793 (1974).

2281. A.G. Smith and L.J. Goad, Biochem. J., 142, 421 (1974).

2282. A.G. Smith, W.A. Harland, and C.J.W. Brooks, Steroids Lipids Res., 4, 122 (1973).

2283. A.G. Smith, I. Rubinstein, and L.J. Goad, Biochem. J., 135, 443 (1973).

2284. D.S.H. Smith and A.B. Turner, Tetrahedron Lett., 5263 (1972).

2285. D.S.H. Smith, A.B. Turner, and A.M. Mackie, J. Chem. Soc., Perkin Trans. I, 1745 (1973).

2286. G.A. Smith and D.H. Williams, J. Chem. Soc., Perkin Trans. I, 2811 (1972).

2287. J.C.H. Smith, Federation Proc., 39, 650 (1980).

2288. L.L. Smith, J. Am. Chem. Soc., 76, 3232 (1954).

2289. L.L. Smith, in SPECIALIST PERIODICAL REPORTS.

TERPENOIDS AND STEROIDS, K.H. Overton, Editor, The
Chemical Society, Vol. 5, 1974, pp. 394-530.

2290. L.L. Smith, Abstracts, Autoxidation Processes in Food
and Related Biological Systems Workshop, U.S. Army
Res. Dev. Command, Natick, Mass., Oct. 29-31, 1979.

2291. L.L. Smith and G.A.S. Ansari, Abstracts, Internatl.
Conf. Chemi- and Bioenergized Processes, Guarujá,
Brazil, Aug. 8-10, 1978, p. 26.

2292. L.L. Smith, A.K. Dhar, J.L. Gilchrist, and Y.Y. Lin,
Phytochem., 12, 2727 (1973).

2293. L.L. Smith and T. Foell, J. Chromatog., 9, 339 (1962).

2294. L.L. Smith and R.E. Gouron, Water Res., 3, 141 (1969).

2295. L.L. Smith and F.L. Hill, J. Chromatog., 66, 101
(1972).

2296. L.L. Smith, D.P. Kohler, J.E. Hempel, and J.E. van
Lier, Lipids, 3, 301 (1968).

2297. L.L. Smith and M.J. Kulig, Cancer Biochem. Biophys.,
1, 79 (1975).

2298. L.L. Smith and M.J. Kulig, J. Am. Chem. Soc., 98,
1027 (1976).

2299. L.L. Smith, M.J. Kulig, D.A. Miiller, and G.A.S.
Ansari, J. Am. Chem. Soc., 100, 6206 (1978).

2300. L.L. Smith, M.J. Kulig, and J.I. Teng, Steroids, 22,
627 (1973).

2301. L.L. Smith, M.J. Kulig, and J.I. Teng, Chem. Phys.
Lipids, 20, 211 (1977).

2302. L.L. Smith, W.S. Matthews, R.C. Bachmann, and B. Reynolds, Federation Proc., 25, 770 (1966).

2303. L.L. Smith, W.S. Matthews, J.C. Price, R.C. Bachmann,
and B. Reynolds, J. Chromatog., 27, 187 (1967).

2304. L.L. Smith and N.L. Pandya, Atherosclerosis, 17, 21
(1973).

2305. L.L. Smith and J.C. Price, J. Chromatog., 26, 509
(1967).

2306. L.L. Smith, D.R. Ray, J.A. Moody, J.D. Wells, and
J.E. van Lier, J. Neurochem., 19, 899 (1972).

2307. L.L. Smith, V.B. Smart, and G.A.S. Ansari, Mutation
Res., 68, 23 (1979).

2308. L.L. Smith and S.J. States, Texas Reports Biol. Med.,
12, 543 (1954).

2309. L.L. Smith and J.P. Stroud, Photochem. Photobiol.,
28, 479 (1978).

2310. L.L. Smith and J.I. Teng, J. Steroid Biochem., 5, 309
(1974).

2311. L.L. Smith and J.I. Teng, J. Am. Chem. Soc., 96, 2640
(1974).

2312. L.L. Smith, J.I. Teng, and M.J. Kulig, Abstracts, 10th
      Int. Congress Biochem., Hamburg, July 25-31, 1976,
      p. 666.
2313. L.L. Smith, J.I. Teng, M.J. Kulig, and F.L. Hill, J.
      Org. Chem., 38, 1763 (1973).
2314. L.L. Smith, J.I. Teng, and Y.Y. Lin, Abstracts of
      Papers, 77th Natl. Meeting, American Chemical Society,
      Honolulu, April 1-6, 1979, Abstract BIOL-65.
2315. L.L. Smith and J.E. van Lier, Atherosclerosis, 12, 1
      (1970); 13, 140 (1971).
2316. L.L. Smith, J.D. Wells, and N.L. Pandya, Texas Reports
      Biol. Med., 31, 37 (1973).
2317. L.L. Smith and N.T. Werthessen, Endocrinology, 53,
      506 (1953).
2318. W.B. Smith, D.L. Deavenport, J.A. Swanzy, and G.A.
      Pate, J. Mag. Resonance, 12, 15 (1973).
2319. A.E. Sobel, M. Goldberg, and S.R. Slater, Anal. Chem.,
      25, 629 (1953).
2320. A.E. Sobel, P.S. Owades, and J.L. Owades, J. Am. Chem.
      Soc., 71, 1487 (1949).
2321. H. Sobotka, THE CHEMISTRY OF THE STERIDS, Williams
      and Wilkins Co., Baltimore, 1938, 634 pp.
2322. H.S. Sodhi, B.J. Kudchodkar, and D.T. Mason, CLINICAL
      METHODS IN STUDY OF CHOLESTEROL METABOLISM, S. Karger,
      Basel, 1979, p. 17.
2323. S. Solomon, P. Levitan, and S. Lieberman, Rev. Can.
      Biol., 15, 282 (1956).
2324. F. Sondheimer, C. Amendolla, and G. Rosenkranz, J. Am.
      Chem. Soc., 75, 5932 (1953).
2325. J.T. Spence and J.L. Gaylor, J. Biol. Chem., 252,
      5852 (1977).
2326. H.L. Spier and K.G. Van Senden, Steroids, 6, 871
      (1965).
2327. W.B. Spirichev and N.V. Blazheievich, Int. Z. Vitamin-
      forsch., 39, 30 (1969).
2328. V.B. Spirichev, A.K. Gazdarov, L.V. Barkova, A.N. Sap-
      rin, and V.A. Belyakov, Tr. Mosk. O-va. Ispyt. Prir.,
      52, 92 (1975); via Chem. Abstr., 83, 159509h (1975).
2329. V.B. Spirichev, A.K. Gazdarov, A.N. Saprin, and V.A.
      Belyakov, Biofizika, 19, 692 (1974).
2330. V.B. Spirichev and I. Ya. Kon, Zh. Vses. Khim. Obshch.,
      23, 425 (1978).
2331. V.B. Spirichev, A.G. Miloserdova, T.A. Zinov'eva, N.A.
      Bogoslovskiĭ, and Z.P. Kozhina, Biokhimiya, 36, 489
      (1971).
2332. F.S. Spring and G. Swain, J. Chem. Soc., 1356 (1939).

2333. A. Stabursvik, Acta Chem. Scand., 7, 1220 (1953).

2334. T.C. Stadtman, A. Cherkes, and C.B. Anfinsen, J. Biol.
      Chem., 206, 511 (1954).

2335. L. Stárka, Coll. Czech. Chem. Commun., 26, 2452 (1961).

2336. L. Stárka, Pharmazie, 17, 126 (1962).

2337. L. Stárka, Naturwissenschaften, 52, 499 (1965).

2338. L. Stárka, and H. Breuer, Z. Physiol. Chem., 344, 124
      (1966).

2339. L. Stárka, E. Döllefeld, and H. Breuer, Z. Physiol.
      Chem., 348, 293 (1967).

2340. L. Stárka and R. Hampl, Natuwissenschaften, 51, 164
      (1964).

2341. L. Stárka and R. Hampl, in HORMONAL STEROIDS, V.H.T.
      James and L. Martini, Editors, Excerpta Medica,
      Amsterdam, 1971, pp. 150-157.

2342. L. Stárka and J. Kůtová, Biochim. Biophys. Acta, 56,
      76 (1962).

2343. L. Stárka, J. Šulcová, K. Dahm, E. Döllefeld, and H.
      Breuer, Biochim. Biophys. Acta, 115, 228 (1966).

2344. L. Stárka, J. Šulcová, and K. Šilink, Clin. Chim.
      Acta, 7, 309 (1962).

2345. A.N. Starratt, Phytochem., 5, 1341 (1966).

2346. A.N. Starratt and C. Madhosingh, Can. J. Microbiol.,
      13, 1351 (1967).

2347. R.D. Stauffer and F. Bischoff, Clin. Chem., 12, 206
      (1966).

2348. R.D. Stauffer, G. Bryson, and F. Bischoff, Res. Com-
      mun. Chem. Path. Pharmacol., 11, 515 (1975).

2349. H.E. Stavely and W. Bergmann, Am. J. Cancer, 30, 749
      (1937).

2350. A.E. Steel, Clin. Chem., 23, 2351 (1977).

2351. G. Steel, C.J.W. Brooks, and W.A. Harland, Biochem.
      J., 99, 51P (1966).

2352. G. Steel and W. Henderson, Nature, 238, 148 (1972).

2353. H. Steenbock and A. Black, J. Biol. Chem., 64, 263
      (1925).

2354. M. Stefanović, A. Jokić, Z. Maksimović, Lj. Lorenc,
      and M. Lj. Mihailović, Helv. Chim. Acta, 53, 1895
      (1970).

2355. D. Steinberg and D.S. Fredrickson, Ann. N.Y. Acad.
      Sci., 64, 579 (1956).

2356. D. Steinberg, D.S. Fredrickson, and J. Avigan, Proc.
      Soc. Exptl. Biol. Med., 97, 784 (1958).

2357. E. Steiner and C. Djerassi, Helv. Chim. Acta, 60, 475
      (1977).

2358. P.H. Steiner, R. Steele, and F.C. Koch, Cancer Res.,

3, 100 (1943).

2359. R. Stern, Arch. Exp. Pathol. Pharmacol., 112, 129
      (1926).

2360. R. Stern, Biochem. Z., 187, 315 (1927).

2361. R. Stern, Biochem. Z., 203, 313 (1928).

2362. P.A. Steudler, F.J. Schmitz, and L.S. Ciereszko, Comp.
      Biochem. Physiol., 56B, 385 (1977).

2363. P.J. Stevens and A.B. Turner, J. Chromatog., 43, 282
      (1969).

2364. R.W. Stevens and C. Green, FEBS Lett., 27, 145 (1972).

2365. D.B. Stierle and D.J. Faulkner, J. Org. Chem., 44,
      964 (1979).

2366. W. Stoffel and R. Klotzbücher, Z. Physiol. Chem., 359,
      199 (1978).

2367. S.J. Stohs and M.M. El-Olemy, J. Steroid Biochem., 2,
      293 (1971).

2368. S.J. Stohs and M.M. El-Olemy, Phytochem., 10, 3053
      (1971).

2369. S.J. Stohs, J.J. Sabatka, and H. Rosenberg, Phytochem.,
      13, 2145 (1974).

2370. W.M. Stokes and W. Bergmann, J. Org. Chem., 17, 1194
      (1952).

2371. C. Stora and R. Freymann, Compt. Rend., 209, 752 (1939).

2372. W.H. Strain, in ORGANIC CHEMISTRY, AN ADVANCED TREAT-
      ISE, H. Gilman, Editor, 2nd Edition, John Wiley &
      Sons, New York, 1943, Vol. 2, pp. 1341-1531.

2373. R.A. Streuli, J. Chung, A.M. Scanu, and S. Yachnin,
      Clin. Res., 27, 475A (1979).

2374. R.A. Streuli, J. Chung, A.M. Scanu, and S. Yachnin,
      J. Immunology, 123, 2897 (1979).

2375. L.I. Strigina, Yu. N. Elkin, and G.B. Elyakov, Phyto-
      chem., 10, 2361 (1971).

2376. L.I. Strigina and V.N. Sviridov, Khim. Prir. Soed.,
      551 (1976).

2377. L.I. Strigina and V.N. Sviridov, Phytochem., 17, 327
      (1978).

2378. J. Stríteský, Biochem. Z., 187, 388 (1927).

2379. H.W. Strobel and M.J. Coon, J. Biol. Chem., 246, 7826
      (1971).

2380. R.G. Strobel, H. Quinn, and W. Lange, Can. J. Micro-
      biol., 13, 121 (1967).

2381. C.A. Strott and C.D. Lyons, J. Steroid Biochem., 9,
      721 (1978).

2382. C.A. Strott and C.D. Lyons, J. Steroid Biochem., 13,
      73 (1980).

2383. J.P. Stroud, M.A. Thesis, Univ. Texas Medical Branch,

Galveston, Texas, May 1979.

2384. B. Stuber, Biochem. Z., 53, 493 (1913).

2385. M.T.R. Subbiah, Mayo Clin. Proc., 46, 549 (1971).

2386. K.E. Suckling, H.A.F. Blair, G.S. Boyd, I.F. Craig, and B.R. Malcolm, Biochim. Biophys. Acta, 551, 10 (1979).

2387. K.E. Suckling and G.S. Boyd, Biochim. Biophys. Acta, 436, 295 (1976).

2388. W. Sucrow, Chem. Ber., 99, 2765 (1966).

2389. W. Sucrow and A. Reimerides, Z. Naturforsch., 23B, 42 (1968).

2390. K. Suhara, T. Gomi, H. Sato, E. Itagaki, S. Takemori, and M. Katagiri, Arch. Biochem. Biophys., 190, 290 (1978).

2391. J. Šulcová, A. Čapková, J.E. Jirašek, and L. Stárka, Acta Endocrinol., 59, 1 (1968).

2392. J. Šulcová, J.E. Jirásek, J. Carlstedt-Duke, and L. Stárka, J. Steroid Biochem., 7, 101 (1976).

2393. J. Šulcová and L. Stárka, Experientia, 19, 632 (1963).

2394. J. Šulcová and L. Stárka, Steroids, 12, 113 (1968).

2395. J. Šulcová and L. Stárka, Experientia, 28, 1361 (1972).

2396. J.A. Summerfield, B.H. Billing, and C.H.L. Shackleton, Biochem. J., 154, 507 (1976).

2397. D.J. Sutor and P.J. Gaston, Gut, 13, 64 (1972).

2398. D.J. Sutor and S.E. Wooley, Gut, 12, 55 (1971).

2399. K. Suwa, T. Kimura, and A.P. Schaap, Biochem. Biophys. Res. Commun., 75, 785 (1977).

2400. K. Suwa, T. Kimura, and A.P. Schaap, Photochem. Photobiol., 28, 469 (1978).

2401. J. Suzuki and K. Tsuda, Chem. Pharm. Bull., 11, 1028 (1963).

2402. V.N. Sviridov and L.I. Strigina, Khim. Priorod. Soed., 669 (1976).

2403. J.A. Svoboda and M.J. Thompson, J. Lipid Res., 8, 152 (1967).

2404. J.A. Svoboda, M.J. Thompson, and W.E. Robbins, Steroids, 12, 559 (1968).

2405. J.A. Svoboda, M. Womack, M.J. Thompson, and W.E. Robbins, Comp. Biochem. Physiol., 30, 541 (1969).

2406. J.R. Swartwout, J.W. Dieckert, O.N. Miller, and J.G. Hamilton, J. Lipid Res., 1, 281 (1960).

2407. L. Swell and M.D. Law, Arch. Biochem. Biophys., 112, 115 (1965).

2408. D. Swern, in AUTOXIDATION AND ANTIOXIDANTS IN TWO VOLUMES, W.O. Lundberg, Editor, Interscience Publishers, New York, 1961, Vol. 1, pp. 1-54.

2409. Zh. S. Sydykov, G.M. Segal, and K.K. Koshoev, Khim. Prirod. Soed., 820 (1977).

2410. J. Szepsenwol, Proc. Soc. Exptl. Biol. Med., 121, 168 (1966).

2411. I. Szundi, Chem. Phys. Lipids, 22, 153 (1978).

2412. T. Tabei and W.L. Heinrichs, Endocrinology, 91, 969 (1972).

2413. T. Tabei and W.L. Heinrichs, Endocrinology, 94, 97 (1974).

2414. N. Taboada, Bol. Inst. Med. Exp. Estud. Cáncer, 20 213 (1943); via Chem. Abstr., 38, 1276[7] (1944).

2415. A.D. Tait, Biochem. J., 128, 467 (1972).

2416. A.D. Tait, Steroids, 22, 239 (1973).

2417. A.D. Tait, Steroids, 22, 609 (1973).

2418. K. Tajima and N.L. Gershfeld, J. Colloid Interface Sci., 52, 619 (1975).

2419. T. Takagi, A. Sakai, K. Hayashi, and Y. Itabachi, Lipids, 14, 5 (1979).

2420. R. Takahashi, O. Tanaka, and S. Shibata, Phytochem., 11, 1850 (1972).

2421. T. Takahashi and R. Yamamoto, Yakugaku Zasshi, 89, 909 (1969).

2422. T. Takahashi and R. Yamamoto, Yakugaku Zasshi, 89, 914 (1969).

2423. T. Takahashi and R. Yamamoto, Yakugaku Zasshi, 89, 919 (1969).

2424. T. Takahashi and R. Yamamoto, Yakugaku Zasshi, 89, 925 (1969).

2425. T. Takahashi and R. Yamamoto, Yakugaku Zasshi, 89, 938 (1969).

2426. S. Takemori, H. Sato, T. Gomi, K. Suhara, and M. Kata-giri, Biochem. Biophys. Res. Commun., 67, 1151 (1975).

2427. S. Takemori, K. Suhara, S. Hashimoto, M. Hashimoto, H. Sato, T. Gomi, and M. Katagiri, Biochem. Biophys. Res. Commun., 63, 588 (1975).

2428. C. Takemoto, H. Nakano, H. Saito, and B.-I. Tamaoki, Biochim. Biophys. Acta, 152, 749 (1968).

2429. H. Takiguchi, H. Nagata, and J. Kanno, J. Biochem., 60, 723 (1966).

2430. O. Takikawa, T. Gomi, K. Suhara, E. Itagaki, S. Take-mori, and M. Katagiri, Arch. Biochem. Biophys., 190, 300 (1978).

2431. E.R. Talaty and G.A. Russell, J. Am. Chem. Soc., 87, 4867 (1965).

2432. E.R. Talaty and G.A. Russell, J. Org. Chem., 31, 3455 (1966).

2433. L. Tan, M. Clemence, and J. Gass, J. Chromatog., 53,
      209 (1970).

2434. L. Tan and P. Falardeau, Biochem. Biophys. Res. Com-
      mun., 41, 894 (1970).

2435. L. Tan and P. Falardeau, Steroidologia, 2, 65 (1971).

2436. L. Tan, P. Falardeau, and J. Rousseau, Hormone Res.,
      6, 213 (1975).

2437. L. Tan and J. Rousseau, Biochem. Biophys. Res. Com-
      mun., 65, 1320 (1975).

2438. L. Tan, H.M. Wang, and P. Falardeau, Biochim. Biophys.
      Acta, 260, 731 (1972).

2439. L. Tan, H.M. Wang, and P. Falardeau, Can. J. Biochem.,
      50, 706 (1972).

2440. L. Tan, H.M. Wang, P. Falardeau, and J.-G. Lehoux,
      Steroid Lipids Res., 5, 28 (1974).

2441. M. Tanabe and K. Hayashi, J. Am. Chem. Soc., 102, 862
      (1980).

2442. M. Tanabe and R.A. Walsh, J. Org. Chem., 28, 3232
      (1963).

2443. Y. Tanahashi and T. Takahashi, Bull. Chem. Soc. Japan,
      39, 848 (1966).

2444. R. Tang, H.J. Yue, J.F. Wolf, and F. Mares, J. Am.
      Chem. Soc., 100, 5248 (1978).

2445. C. Tanret, Ann. Chim. Phys., [6], 20, 289 (1890).

2446. C. Tanret, Ann. Chim. Phys., [8], 15, 313 (1908).

2447. E.J. Tarlton, M. Fieser, and L.F. Fieser, J. Am. Chem.
      Soc., 75, 4423 (1953).

2448. W.Tarpley and M. Yudis, Anal. Chem., 25, 121 (1953).

2449. K.A. Tartivita, J.P. Sciarello, and B.C. Rudy, J.
      Pharm. Sci., 65, 1024 (1976).

2450. C.B. Taylor, E.S. Allen, B. Mikkelson, and K.-J. Ho,
      Paroi Artérielle, 3, 175 (1976).

2451. C.B. Taylor, S.K. Peng, J.C. Hill, and B. Mikkelson,
      Federation Proc., 39, 771 (1980).

2452. B.A. Teicher, N. Koizumi, M. Koreeda, M. Shikita, and
      P. Talalay, Eur. J. Biochem., 91, 11 (1978).

2453. B.A. Teicher, M. Shikita, and P. Talalay, Biochem.
      Biophys. Res. Commun., 83, 1436 (1978).

2454. J.I. Teng, M.J. Kulig, and L.L. Smith, J. Chromatog.,
      75, 108 (1973).

2455. J.I. Teng, M.J. Kulig, L.L. Smith, G. Kan, and J.E.
      van Lier, J. Org. Chem., 38, 119 (1973).

2456. J.I. Teng, C.-E. Low, and L.L. Smith, Chem. Phys.
      Lipids, 22, 63 (1978).

2457. J.I. Teng and L.L. Smith, Federation Proc., 31, 913
      Abs (1972.

2458. J.I. Teng and L.L. Smith, J. Am. Chem. Soc., 95, 4060 (1973).

2459. J.I. Teng and L.L. Smith, J. Chromatog., 115, 648 (1975).

2460. J.I. Teng and L.L. Smith, Texas Reports Biol. Med., 33, 293 (1975).

2461. J.I. Teng and L.L. Smith, Bioorganic Chem., 5, 99 (1976).

2462. J.I. Teng and L.L. Smith, J. Steroid Biochem., 7, 577 (1976).

2463. J.I. Teng and L.L. Smith, Abstracts, VIII World Congress Cardiology, September 17-23, 1978, Tokyo, p. 371.

2464. J.I. Teng and L.L. Smith, in DENSITOMETRY IN THIN LAYER CHROMATOGRAPHY. PRACTICE AND APPLICATIONS, J.C. Touchstone and J. Sherma, Editors, John Wiley & Sons, New York, NY, 1979, pp. 661-676.

2465. J.I. Teng, L.L. Smith, and J.E. van Lier, J. Steroid Biochem., 5, 581 (1974).

2466. S.-I. Teshima and A. Kanazawa, Bull. Japan Soc. Sci. Fisheries, 38, 1299 (1972).

2467. S.-I. Teshima and A. Kanazawa, Comp. Biochem. Physiol., 47B, 555 (1974).

2468. S. Teshima and A. Kanazawa, Mem. Fac. Fish. Kagoshima Univ., 27, 41 (1978).

2469. S.-I. Teshima, A. Kanazawa, and T. Ando, Comp. Biochem. Physiol., 41B, 121 (1972).

2470. S. Teshima, A. Kanazawa, S. Hyodo, and T. Ando, Comp. Biochem. Physiol., 64B, 225 (1979).

2471. S.-I. Teshima, A. Kanazawa, and H. Miyawaki, Comp. Biochem. Physiol., 63B, 323 (1979).

2472. M. Tezuka, Y. Ohkatsu, and T. Osa, Bull. Chem. Soc. Japan, 48, 1471 (1975).

2473. P. Tham, S.-K. Peng, C.B. Taylor, and B. Mikkelson, Federation Proc., 37, 474 (1978).

2474. P. Tham, S.K. Peng, C.B. Taylor, and B. Mikkelson, Federation Proc., 38, 894 (1979).

2475. N. Theobald, J.N. Schoolery, C. Djerassi, T.R. Erdman, and P. Scheuer, J. Am. Chem. Soc., 100, 5574 (1978).

2476. A.H.T. Theorell, Biochem. Z., 175, 297 (1926).

2477. H. Theorell, Biochem. Z., 223, 1 (1930).

2478. E.D. Thompson, B.A. Knights, and L.W. Parks, Biochim. Biophys. Acta, 304, 132 (1973).

2479. M.J. Thompson, S.R. Dutky, G.W. Paterson, and E.L. Gooden, Phytochem., 11, 1781 (1972).

2480. M.J. Thompson, J.N. Kaplanis, and H.E. Vroman, Steroids, 5, 551 (1965).

2481. J.R. Thowsen and G.J. Schroepfer, J. Lipid Res., 20, 681 (1979).

2482. H.T. Tien, in THE CHEMISTRY OF BIOSURFACES, M.L. Nair, Editor, Marcel Dekker Inc., New York, 1971, Vol. 1, pp. 233-348.

2483. H.T. Tien, S. Carbone, and E.A. Dawidowicz, Nature, 212, 718 (1966).

2484. B.E. Tilley, M. Watanuki, and P.F. Hall, Biochim. Biophys. Acta, 488, 330 (1977).

2485. B.E. Tilley, M. Watanuki, and P.F. Hall, Biochim. Biophys. Acta, 493, 260 (1977).

2486. M.E. Toaff, J.F. Strauss, G.L. Flickinger, and S.J. Shattil, J. Biol. Chem., 254, 3977 (1979).

2487. P.P. Tobback, in RADIATION CHEMISTRY OF MAJOR FOOD COMPONENTS, ITS RELEVANCE TO THE ASSESSMENT OF THE WHOLESOMENESS OF IRRADIATED FOODS, P.S. Elias and A.S. Cohen, Editors, Elsevier Publ. Co., Amsterdam, 1977, pp. 187-220.

2488. M. Tohma, Y. Nakata, and T. Kurosawa, J. Chromatog., 171, 469 (1979).

2489. M. Tohma, T. Tomita, and M. Kimura, Tetrahedron Lett., 4359 (1973).

2490. B.M. Tolbert, P.T. Adams, E.L. Bennett, A.M. Hughes, M.R. Kirk, R.M. Lemmon, R.M. Noller, R. Ostwald, and M. Calvin, J. Am. Chem. Soc., 75, 1867 (1953); UCRL-2116, February 27, 1953.

2491. B.M. Tolbert and R.M. Lemmon, Radiation Res., 3, 52(1955).

2492. G.A. Tolstikov, U.M. Dzhemilev, and V.P. Yur'ev, Zh. Org. Khim., 8, 1190 (1972); via Chem. Abstr., 77, 101986w (1972).

2493. G.A. Tolstikov, U.M. Dzhemilev, and V.P. Yur'ev, Zh. Org. Khim., 8, 2204 (1972); via Chem. Abstr., 78, 4383c (1973).

2494. G.A. Tolstikov, V.P. Yur'ev, I.A. Gailyunas, and U.M. Dzhemilev, Zh. Obshcheii Khim., 44, 215 (1973).

2495. H. Tomioka, M. Kagawa, and S. Nakamura, J. Biochem., 79, 903 (1973).

2496. G.M. Tomkins, C.W. Nichols, D.D. Chapman, S. Hotta, and I.L. Chaikoff, Science, 125, 936 (1957).

2497. G.M. Tomkins, H. Sheppard, and I.L. Chaikoff, J. Biol. Chem., 203, 781 (1953).

2498. R.W. Topham and J.L. Gaylor, Biochem. Biophys. Res. Commun., 47, 180 (1972).

2499. K. Tori, T. Komeno, and T. Nakagawa, J. Org. Chem., 29, 1136 (1964).

2500. J.C. Touchstone and T. Murawec, in QUANTITATIVE THIN

LAYER CHROMATOGRAPHY, J.C. Touchstone, Editor, John Wiley & Sons, New York, NY, 1973, pp. 131-152.

2501. J.C. Touchstone, T. Murawec, M. Kasparow, and W. Wortmann, J. Chromatog. Sci., 10, 490 (1972).

2502. T.R. Tritton, S.A. Murphee, and A.C. Sartorelli, Biochem. Pharmacol., 26, 2319 (1977).

2503. R. Truhaut, Ann. Pharm. Franç., 5, 619 (1947).

2504. R. Truhaut, Compt. Rend., 225, 544 (1947).

2505. A.S. Truswell and W.D. Mitchell, J. Lipid Res., 6, 438 (1965).

2506. L.S. Tsai and C.A. Hudson, J. Am. Oil Chem. Soc., 56, 188A (1979).

2507. L.S. Tsai, C.A. Hudson, K. Ijichi, and J.J. Meehan, J. Am. Oil Chem. Soc., 56, 185A (1979).

2508. L.S. Tsai, K. Ijichi, C.A. Hudson, and J.J. Meehan, Abstracts, 39th Annual Meeting, Institute of Food Technologists, St. Louis, Mo., June 10-13, 1979.

2509. L.S. Tsai, K. Ijichi, C.A. Hudson, and J.J. Meehan, Lipids, 15, 124 (1980).

2510. Y.-H. Tsay, J.V. Silverton, J.A. Beisler, and Y. Sato, J. Am. Chem. Soc., 92, 7005 (1970).

2511. L. Tschugaeff, Z. Physik. Chem., 76, 469 (1911).

2512. L. Tschugaeff and W. Formin, Liebigs Ann., 375, 288 (1910).

2513. T. Tsuchiya, H. Arai, and H. Igeta, Tetrahedron Lett., 2747 (1969).

2514. K. Tsuda, K. Arima, and R. Hayatsu, J. Am. Chem. Soc., 76, 2933 (1954).

2515. K. Tsuda and R. Hayatsu, J. Am. Chem. Soc., 77, 3089 (1955).

2516. K. Tsuda and R. Hayatsu, J. Am. Chem. Soc., 81, 5987 (1959).

2517. K. Tsuda, J. Suzuki, and S. Iwasaki, Chem. Pharm. Bull., 11, 405 (1963).

2518. M. Tsuda, E.J. Parish, and G.J. Schroepfer, J. Org. Chem., 44, 1282 (1979).

2519. M. Tsuda and G.J. Schroepfer, J. Org. Chem., 44, 1290 (1979).

2520. A. Tsuji, M. Smulowtiz, J.S.C. Liang, and D.K. Fukushima, Steroids, 24, 739 (1974).

2521. C. Tu, W.D. Powrie, and O. Fennema, J. Food Sci., 32, 30 (1967).

2522. C. Tu, W.D. Powrie, and O. Fennema, Lipids, 5, 369 (1969).

2523. P. Tunmann and H.J. Grimm, Arch. Pharm., 307, 891 (1974).

2524. P. Tunmann and H.J. Grimm, Arch. Pharm., 307, 966
      (1974).
2525. G.E. Turfitt, Biochem. J., 37, 115 (1943).
2526. G.E. Turfitt, Biochem. J., 42, 376 (1948).
2527. A.B. Turner, D.S.H. Smith, and A.M. Mackie, Nature,
      233, 209 (1971).
2528. D.L. Turner and R. Freeman, J. Magnetic Resonance, 29,
      587 (1978).
2529. D.W. Turner, J. Chem. Soc., 30 (1959).
2530. B. Tursch, R. Cloetens, and C. Djerassi, Tetrahedron
      Lett., 467 (1970).
2531. B. Tursch, C. Hootelé, M. Kaisin, D. Losman and R.
      Karlsson, Steroids, 27, 137 (1976).
2532. D.D. Tyler, FEBS Lett., 51, 180 (1975).
2533. K. Ubik and J. Vrkoč, Insect Biochem., 4, 281 (1974).
2534. THE PHARMACOPEIA OF THE UNITED STATES OF AMERICA, U.S.
      Pharmacopeial Convention Inc., 13th Revision, 1947,
      pp. 128-129; 14th Revision, 1950, pp. 139-140; 15th
      Revision, 1955, pp. 157-158; 16th Revision, 1960, pp.
      162-163; 17th Revision, 1965, pp. 133-134; 18th Revi-
      sion, 1970, pp. 131-132.
2535. THE UNITED STATES PHARMACOPEIA, U.S. Pharmacopeial
      Convention Inc., Rockville, Md., 19th Revision, 1974,
      p. 92.
2536. V. Ullrich, Z. Physiol. Chem., 350, 357 (1969).
2537. V. Ullrich and H. Staudinger, in MICROSOMES AND DRUG
      OXIDATIONS, J.R. Gillette, A.H. Conney, G.J. Cosmides,
      R.W. Estabrook, J.R. Fouts, and G.J. Mannering, Edi-
      tors, Academic Press, New York/London, 1969, pp. 199-
      223.
2538. P.G. Unna and L. Golodetz, Biochem. Z., 20, 469
      (1909).
2539. N. Uri, in AUTOXIDATION AND ANTIOXIDANTS, W.O. Lundberg,
      Editor, Interscience Publishers, New York, N.Y., Vol.
      1, 1961, p. 61.
2540. T. Usui and K. Yamasaki, J. Biochem., 48, 226 (1960).
2541. T. Uwajima, H. Yagi, S. Nakamura, and O. Terada, Ag.
      Biol. Chem., 37, 2345 (1973).
2542. T. Uwajima, H. Yagi, and O. Terada, Ag. Biol. Chem.,
      38, 1149 (1974).
2543. M.-J. Vacheron and G. Michel, Phytochem., 7, 1645
      (1968).
2544. M. Vajdi, W.W. Nawar, and C. Merritt, J. Am. Oil Chem.
      Soc., 56, 611 (1979).
2545. A. Vallisneri, OPIRE FISICO-MEDICHE STAMPATE E MANO-
      SCRITE, S. Coleti, Venezia, Vol. 3., 1733, pp.594-597.

2546. R.T. van Aller, H. Chikamatsu, N.J. de Souza, J.P.
      John, and W.R. Nes, J. Biol. Chem., 244, 6645 (1969).
2547. J. Van Cantfort, Compt. Rend., [D], 273, 491 (1971).
2548. J. Van Cantfort, Life Sci., 11, Part II, 773 (1972).
2549. J. Van Cantfort, Biochimie, 55, 1171 (1973).
2550. J. Van Cantfort, J. Interdiscipl. Cycle Res., 5, 89
      (1974).
2551. J. Van Cantfort and J. Gielen, Eur. J. Biochem., 55,
      33 (1975).
2552. J. Van Cantfort, J. Renson, and J. Gielen, Eur. J.
      Biochem., 55, 23 (1975).
2553. M.J. van Dam, G.J. de Kleuver, and J.G. de Huis, J.
      Chromatog., 4, 26 (1960).
2554. D.J. Vanderah and C. Djerassi, Tetrahedron Lett., 683
      (1977).
2555. N.J. van Haeringen and E. Glasius, Exp. Eye Res., 20,
      271 (1975).
2556. J.E. van Lier, A.L. Da Costa, and L.L. Smith, Chem.
      Phys. Lipids, 14, 327 (1975).
2557. J.E. van Lier and G. Kan, Biochem. J., 125, 47P (1971).
2558. J.E. van Lier and G. Kan, Federation Proc., 31, 917
      Abs (1972).
2559. J.E.van Lier and G. Kan, J. Org. Chem., 37, 145 (1972).
2560. J.E. van Lier, G. Kan, and H. Buttemer, Proc. Can.
      Fedn. Biol. Soc., 14, 367 (1971).
2561. J.E. van Lier, G. Kan, and R. Langlois, Abstracts,
      Commun. Meet. Fed. Eur. Biochem. Soc., 8, 845 (1972).
2562. J.E. van Lier, G. Kan, and R. Langlois, Steroids, 21,
      521 (1973).
2563. J.E. van Lier, G. Kan, R. Langlois, and L.L. Smith,
      in BIOLOGICAL HYDROXYLATION MECHANISMS, Biochemical
      Society Symposium No. 34, G.S. Boyd and R.M.S. Smellie,
      Editors, Academic Press, London/New York, 1972, pp.
      21-43.
2564. J.E. van Lier and M. Milot, J. Steroid Biochem., 5,
      308 (1974).
2565. J.E. van Lier and J. Rousseau, Abstracts, 10th Int.
      Congress Biochem., Hamburg, July 25-31, 1976, p. 665;
      FEBS Lett., 70, 23 (1976).
2566. J.E. van Lier, J. Rousseau, R. Langlois, and G.J.
      Fisher, Biochim. Biophys. Acta, 487, 395 (1977).
2567. J.E. van Lier and L.L. Smith, Federation Proc., 26,
      342 (1967); Biochemistry, 6, 3269 (1967).
2568. J.E. van Lier and L.L. Smith, Anal. Biochem., 24, 419
      (1968).
2569. J.E. van Lier and L.L. Smith, J. Chromatog., 36,7(1968).

2570. J.E. van Lier and L.L. Smith, J. Chromatog., 41, 37 (1969).
2571. J.E. van Lier and L.L. Smith, Texas Reports Biol. Med., 27, 167 (1969).
2572. J.E. van Lier and L.L. Smith, Biochem. Biophys. Res. Commun., 40, 516 (1970).
2573. J.E. van Lier and L.L. Smith, Biochim. Biophys. Acta, 210, 153 (1970).
2574. J.E. van Lier and L.L. Smith, Biochim. Biophys. Acta, 218, 320 (1970).
2575. J.E. van Lier and L.L. Smith, J. Chromatog., 49, 555 (1970).
2576. J.E. van Lier and L.L. Smith, J. Org. Chem., 35, 2627 (1970).
2577. J.E. van Lier and L.L. Smith, J. Pharm. Sci., 59, 719 (1970).
2578. J.E. van Lier and L.L. Smith, Steroids, 15, 485 (1970).
2579. J.E. van Lier and L.L. Smith, J. Org. Chem., 36, 1007 (1971).
2580. J.E. van Lier and L.L. Smith, Lipids, 6, 85 (1971).
2581. K. Van Putte, W. Skoda, and M. Petroni, Chem. Phys. Lipids, 2, 361 (1968).
2582. V.H. Van Rheenen and M.J. Visser, French Patent No. 1,556,187, January 31, 1969; via Chem. Abstr., 72, 79324b (1969).
2583. R.L. Vardanyan, G.E. Dingchyan, B.B. Khanukaev, and A.S. Vardanyan, Kinet. Katal., 19, 72 (1978); via Chem. Abstr., 89, 24617v (1978).
2584. R.K. Varma, M. Koreeda, B. Yagen, K. Nakanishi, and E. Caspi, J. Org. Chem., 40, 3680 (1975).
2585. R.K. Varma, J.A.F. Wickramasinghe, and E. Caspi, J. Biol. Chem., 244, 3951 (1969).
2586. L. Vaska, Accounts Chem. Res., 9, 175 (1976).
2587. B.A. Vela and H.F. Acevedo, J. Clin. Endocrinol. Metabolism, 29, 1251 (1969).
2588. B.A. Vela and H.F. Acevedo, Steroids, 14, 499 (1969).
2589. H. Veldstra, Nature, 144, 246 (1939).
2590. L. Velluz, B. Goffinet, J. Warnant, and G. Amiard, Bull. Soc. Chim. France, 1289 (1957).
2591. H. Venner, Chem. Ber., 89, 1634 (1956).
2592. S.P. Verma, D.F.H. Wallach, and J. Philippot, Federation Proc., 39, 1835 (1980).
2593. A.J.M. Vermorken, R. de Waal, W.J.M. van de Ven, H. Bloemendal, and P. Th. Henderson, Biochim. Biophys. Acta, 496, 495 (1977).
2594. J. Viala, M. Devys, and M. Barbier, Bull. Soc. Chim.

France, 3626 (1972).

2595. H.S. Vishniac and F.J. Nielsen, Federation Proc., 15, 620 (1956).

2596. Z.R. Vlahcevic, C.C. Schwartz, J. Gustafsson, L.G. Halloran, H. Danielsson, and L. Swell, J. Biol. Chem., 255, 2925 (1980).

2597. J. Vogel, THE PATHOLOGICAL ANATOMY OF THE HUMAN BODY, Lea and Blanchard, Philadelphia, 1847, pp. 350-351, p. 531.

2598. M.J. Vogel, E.M. Barrall, and C.P. Mignosa, in LIQUID CRYSTALS AND ORDERED FLUIDS, J.F. Johnson and R.S. Porter, Editors, Plenum Press, New York/London, 1970, pp. 333-349.

2599. H.C. Volger and W. Brackman, Rec. Trav. Chim., 84, 579 (1965).

2600. H. Vollmer, Biochem. Z., 172, 465 (1926).

2601. J.J. Volpe and S.W. Hennessy, Biochim. Biophys. Acta, 486, 408 (1977).

2602. O. von Fürth and G. Felsenreich, Biochem. Z., 69, 417 (1915).

2603. A. von Wartburg, J. Binkert, and E. Anglicker, Helv. Chim. Acta, 45, 2139 (1962).

2604. P.A. Voogt, Comp. Biochem. Physiol., 31, 37 (1969).

2605. P.A. Voogt, Comp. Biochem. Physiol., 39B, 139 (1971).

2606. P.A. Voogt, Arch. Int. Physiol. Biochim., 81, 871 (1973).

2607. P.A. Voogt, Netherland. J. Zool., 24, 22 (1974).

2608. P.A. Voogt, Netherland. J. Zool., 24, 469 (1974).

2609. P.A. Voogt, Comp. Biochem. Physiol., 50B, 499 (1975).

2610. P.A. Voogt, Comp. Biochem. Physiol., 50B, 505 (1975).

2611. P.A. Voogt, Netherland. J. Zool., 26, 84 (1976).

2612. P.A. Voogt and H.J. Schoenmakers, Comp. Biochem. Physiol., 45B, 509 (1973).

2613. P.A. Voogt, J.M. van de Ruit, and J.W.A. van Rheenen, Comp. Biochem. Physiol., 48B, 47 (1974).

2614. P.A. Voogt and J.W.A. van Rheenen, Experientia, 29, 1070 (1973).

2615. P.A. Voogt and J.W.A. van Rheenen, Arch. Int. Physiol. Biochim., 83, 563 (1975).

2616. P.A. Voogt and J.W.A. van Rheenen, Comp. Biochem. Physiol., 54B, 473 (1976).

2617. P.A. Voogt and J.W.A. van Rheenen, Comp. Biochem. Physiol., 54B, 479 (1976).

2618. J. Vrkoč, M. Buděšínský, and L. Dolejš, Phytochem., 15, 1782 (1976).

2619. H.E. Vroman and C.F. Cohen, J. Lipid Res., 8, 150 (1967).

2620. N. Wachtel, S. Emerman, and N.B. Javitt, J. Biol. Chem., 243, 5207 (1968).
2621. L. Wacker, Z. Physiol. Chem., 80, 383 (1912).
2622. L. Wacker and W. Hueck, Munch. Med. Woch., 60, 2097 (1913).
2623. F. Wada, K. Hirata, K. Nakao, and Y. Sakamoto, J. Biochem., 66, 699 (1969).
2624. K. Wada and T. Ishida, Phytochem., 13, 2755 (1974).
2625. T. Wainai, T. Tamura, B. Truscott, and D.R. Idler, J. Fish. Res. Bd. Canada, 21, 1543 (1964).
2626. O. Wallach, Liebigs Ann., 392, 59 (1912).
2627. C. Walling, Accts. Chem. Res., 8, 125 (1975).
2628. H.-P. Wang and T. Kimura, J. Biol. Chem., 251, 6068 (1976).
2629. A.M.K. Wardroper, J.R. Maxwell, and R.J. Morris, Steroids, 32, 203 (1978).
2630. A.M. Wartman and W.E. Connor, J. Lab. Clin. Med., 82, 793 (1973).
2631. T. Watabe, M. Kanai, M. Isobe, and N. Ozawa, Biochem. Biophys. Res. Commun., 92, 977 (1980).
2632. T. Watabe and T. Sawahata, Biochem. Biophys. Res. Commun., 83, 1396 (1978).
2633. T. Watabe and T. Sawahata, J. Biol. Chem., 254, 3854 (1979).
2634. T. Watabe, T. Sawahata, and J. Horie, Biochem. Biophys. Res. Commun., 87, 469 (1979).
2635. K. Watanabe, T. Narisawa, C.Q. Wong, and J.H. Weisburger, J. Natl. Cancer Inst., 60, 1501 (1978).
2636. N. Waterman, Acta Cancrologica, 2, 375 (1936).
2637. N. Waterman, Nederland. Tijdschr. Geneesk., 81, 1273 (1937).
2638. N. Waterman, Acta Brevia Neerland., 9, 143 (1939).
2639. N. Waterman, Acta Unio Int. Contra Cancrum, 4, 764 (1939).
2640. N. Waterman, Bull. Assoc. Franç. Etude Cancer, 29, 70 (1940).
2641. J.A. Waters and B. Witkop, J. Org. Chem., 34, 3774 (1969).
2642. A.S. Watnick, J. Gibson, M. Vinegra, and S. Tolksdorf, J. Endocrinol., 33, 241 (1965).
2643. J.A. Watson, C. Havel, E. Hansbury, and T.J. Scallen, Federation Proc., 38, 785 (1979).
2644. K.C. Watson and E.J.C. Kerr, Biochem. J., 140, 95 (1974).
2645. B.W. Wattenberg, C.E. Freter, and D.F. Silbert, J. Biol. Chem., 254, 12295 (1979).

2646. N. Weber, Phytochem., 16, 1849 (1977).

2647. J.D. Weete, Phytochem., 12, 1843 (1973).

2648. J. Weichherz and H. Marschik, Biochem. Z., 249, 312 (1932).

2649. G.R. Weihe and T.C. McMorris, J. Org. Chem., 43, 3942 (1978).

2650. N.D. Weiner, W.C. Bruning, and A. Felmeister, J. Pharm. Sci., 62, 1202 (1973).

2651. N.D. Weiner, P. Noomnont, and A. Felmeister, J. Lipid Res., 13, 253 (1972).

2652. J. Weinman, Bull. Soc. Chim. France, 4259 (1967).

2653. J. Weinman and S. Weinman, Steroids, 6, 699 (1965).

2654. B. Weinstein, T.L. Rold, R. Settine, and J.R. Waaland, Abstracts of Papers, 175th Natl. Meeting American Chem. Soc., Anaheim, Calif., March 13-17, 1978, No. ORGN-188.

2655. J. Weiss, Nature, 153, 748 (1944).

2656. J. Weiss and M. Keller, Experientia, 6, 379 (1950).

2657. J. Weiss, J. Rauschohoff, and H.J. Kayden, Neurology, 22, 187 (1972).

2658. R.J. Wells, Tetrahedron Lett., 2637 (1976).

2659. N.T. Werthessen, Adv. Exptl. Med. Biol., 16A, 253 (1971).

2660. T. Westphalen, Ber., 48, 1064 (1915).

2661. S.A. Wharton and C. Green, Biochem. Soc. Trans., 6, 781 (1978).

2662. V.R. Wheatley, Biochem. J., 58, 167 (1954).

2663. O.H. Wheeler and J.L. Mateos, J. Org. Chem., 21, 1110 (1956).

2664. D.R. Whikehart and M.B. Lees, Biochim. Biophys. Acta, 231, 561 (1971).

2665. G.S. Whitby, Biochem. J., 17, 5 (1923).

2666. J.D. White, D.W. Perkins, and S.I. Taylor, Bioorg. Chem., 2, 163 (1973).

2667. J.D. White, and S.I. Taylor, J. Am. Chem. Soc., 92, 5811 (1970).

2668. J.A. Whysner and B.W. Harding, Biochem. Biophys. Res. Commun., 32, 921 (1968).

2669. J.A. Whysner, J. Ramseyer, and B.W. Harding, J. Biol. Chem., 245, 5441 (1970).

2670. J.A. Whysner, J. Ramseyer, G.M. Kazmi, and B.W. Harding, Biochem. Biophys. Res. Commun., 36, 795 (1969).

2671. J. Wicha and K. Bal, J. Chem. Soc. Chem. Commun., 968 (1975).

2672. J. Wicha and K. Bal, J. Chem. Soc., Perkin Trans. I, 1282 (1978).

2673. E.M.P. Widmark, Nature, 143, 984 (1939).

2674. H. Wieland and G. Coutelle, Liebigs Ann., $\underline{548}$, 270 (1941).

2675. H. Wieland and E. Dane, Z. Physiol. Chem., $\underline{219}$, 240 (1933).

2676. H. Wieland and V. Wiedersheim, Z. Physiol. Chem., $\underline{186}$, 229 (1930).

2677. P. Wieland and V. Prelog, Helv. Chim. Acta, $\underline{30}$, 1028 (1947).

2678. W.H.J.M. Wientjens, R.A. de Zeeuw, and J. Wijsbeek, J. Lipid Res., $\underline{11}$, 376 (1970).

2679. B.S. Wildi, U.S. Patent No. 2,698,853, Jan. 4, 1955; via Chem. Abstr., $\underline{49}$, 7008^b (1955).

2680. M. Wilk and W. Taupp, Z. Naturforsch., $\underline{24B}$, 16 (1969).

2681. J.H. Williams, M. Kuchmak, and R.F. Witter, J. Lipid Res., $\underline{6}$, 461 (1965).

2682. J.H. Williams, M. Kuchmak, and R.F. Witter, Clin. Chem., $\underline{16}$, 423 (1970).

2683. L.D. Williams and A.M. Pearson, J. Ag. Food Chem., $\underline{13}$, 573 (1965).

2684. L.D. Wilson, Biochemistry, $\underline{11}$, 3696 (1972).

2685. L.D. Wilson and B.W. Harding, J. Biol. Chem., $\underline{248}$, 9 (1973).

2686. D.C. Wilton, M. Akhtar, and K.A. Munday, Biochem. J., $\underline{98}$, 29C (1966).

2687. A. Windaus, Ber., $\underline{36}$, 3752 (1903).

2688. A. Windaus, Ber., $\underline{39}$, 518 (1906).

2689. A. Windaus, Chem.-Zeit., $\underline{30}$, 1011 (1906).

2690. A. Windaus, Arch. Pharm., $\underline{246}$, 117 (1908).

2691. A. Windaus, Ber., $\underline{42}$, 238 (1909).

2692. A. Windaus, Z. Physiol. Chem., $\underline{117}$, 146 (1921).

2693. A. Windaus, Ann. Rev. Biochem., $\underline{1}$, 109 (1932).

2694. A. Windaus, Z. Physiol. Chem., $\underline{213}$, 147 (1932).

2695. A. Windaus, Chemie & Industrie, $\underline{40}$, 835 (1938).

2696. A. Windaus, Z. Physiol. Chem., $\underline{276}$, 280 (1942).

2697. A. Windaus and E. Auhagen, Z. Physiol. Chem., $\underline{196}$, 108 (1931).

2698. A. Windaus, W. Bergmann, and A. Lüttringhaus, Liebigs Ann., $\underline{472}$, 195 (1929).

2699. A. Windaus and P. Borgeaud, Liebigs Ann., $\underline{460}$, 235 (1928).

2700. A. Windaus and J. Brunken, Liebigs Ann., $\underline{460}$, 225 (1928).

2701. A. Windaus, K. Bursian, and U. Riemann, Z. Physiol. Chem., $\underline{271}$, 177 (1941).

2702. A. Windaus and R. Langer, Liebigs Ann., $\underline{508}$, 105 (1933).

2703. A. Windaus, H. Lettré, and F. Schenck, Liebigs Ann., 520, 98 (1935).

2704. A. Windaus and O. Linsert, Liebigs Ann., 465, 148 (1928).

2705. A. Windaus and H. Lüders, Z. Physiol. Chem., 109, 183 (1920).

2706. A. Windaus and H. Lüders, Z. Physiol. Chem., 115, 257 (1921).

2707. A. Windaus and A. Lüttringhaus, Liebigs Ann., 481, 119 (1930).

2708. A. Windaus and C. Resau, Ber., 48, 851 (1915).

2709. A. Windaus and U. Riemann, Z. Physiol. Chem., 274, 206 (1942).

2710. A. Windaus and F. Schenck, U.S. Patent No. 2,098,985, Nov. 16, 1937; via Chem. Abstr., 32, 196^4 (1938).

2711. O. Wintersteiner and S. Bergström, J. Biol. Chem., 137, 785 (1941).

2712. O. Wintersteiner and M. Moore, J. Am. Chem. Soc., 65, 1503 (1943).

2713. O. Wintersteiner and J.R. Ritzmann, J. Biol. Chem., 136, 697 (1940).

2714. O. Wintersteiner and W.L. Ruigh, J. Am. Chem. Soc., 64, 2453 (1942).

2715. J. Wislicenus and W. Moldenhauer, Liebigs Ann., 146, 175 (1868).

2716. D.T. Witiak, W.E. Connor, D.M. Brahmankar, A. Wartman, and R. Parker, J. Clin. Invest., 47, 104a (1968).

2717. D.T. Witiak, R.A. Parker, D.R. Brann, M.E. Dempsey, M.C. Ritter, W.E. Connor, and D.M. Brahmankar, J. Med. Chem., 14, 216 (1971).

2718. D.T. Witiak, R.A. Parker, M.E. Dempsey, and M.C. Ritter, J. Med. Chem., 14, 684 (1971).

2719. R.F.Witter, M. Kuchmak, J.H. Williams, V.S. Whitner, and C.L. Winn, Clin. Chem., 16, 743 (1970).

2720. G. Wolf, CHEMICAL INDUCTION OF CANCER, Harvard Univ. Press, Cambridge, 1952, pp. 132-152.

2721. A. Wolff, Angew Chem., 52, 516 (1939).

2722. J. Wolinsky, L.L. Baxter, and J. Hamsher, J. Org. Chem., 33, 438 (1968).

2723. J.L.C. Wright, A.G. McInnes, S. Shimizu, D.G. Smith, J.A. Walter, D. Idler, and W. Khalil, Can. J. Chem., 56, 1898 (1978).

2724. L.D. Wright, Proc. Soc. Exptl. Biol. Med., 121, 265 (1966).

2725. G.-S. Wu, R.A. Stein, and J.F. Mead, Lipids, 13, 517 (1978).

2726. W. Wünderlich, Z. Physiol. Chem., 241, 116 (1936).

2727. S.G. Wyllie, B.A. Amos, and L. Tökés, J. Org. Chem., 42, 725 (1977).

2728. S.G. Wyllie and C. Djerassi, J. Org. Chem., 33, 305 (1968).

2729. E.L. Wynder and B.S. Reddy, Cancer, 40, 2565 (1977).

2730. E.L. Wynder, G. Wright, and J. Lam, Cancer, 11, 1140 (1958).

2731. K.N. Wynne and N. Kraft, J. Steroid Biochem., 9, 1189 (1978).

2732. S. Yachnin, Blood, 52, Suppl. 1, 146 (1978).

2733. S. Yachnin and R. Hsu, Cellular Immunol., 51, 42 (1980).

2734. S. Yachnin, R.C. Hsu, J. Chung, and A.M. Scanu, Clin. Res., 27, 514A (1979).

2735. S. Yachnin, R.A. Streuli, L.I. Gordon, and R.C. Hsu, Current Topics Hematology, 2, 245 (1979).

2736. N. Yago, S. Kobayashi, S. Sekiyama, H. Kurokawa, Y. Iwai, I. Suzuki, and S. Ichii, J. Biochem., 68, 775 (1970).

2737. S. Yahara , R. Kasai, and O. Tanaka, Chem. Pharm. Bull., 25, 2041 (1977).

2738. S. Yamada, K. Nakayama, and H. Takayama, Tetrahedron Lett., 4895 (1978).

2739. S. Yamada, K. Nakayama, H. Takayama, A. Itai, and Y. Iitaka, Chem. Pharm. Bull., 27, 1949 (1979).

2740. Y. Yamada, S. Suzuki, K. Iguchi, H. Kikuchi, Y. Tsukitani, H. Horiai, and H. Nakanishi, Chem. Pharm. Bull., 28, 473 (1980).

2741. T. Yamagishi, K. Hayashi, R. Kiyama, and H. Mitsuhashi, Tetrahedron Lett., 4005 (1972).

2742. T. Yamagishi, K. Hayasi, H. Mitsuhashi, M. Imanari, and K. Matsushita, Tetrahedron Lett., 3527 (1973).

2743. T. Yamagishi, K. Hayashi, H. Mitsuhashi, M. Imanari, and K. Matsushita, Tetrahedron Lett., 3531 (1973).

2744. T. Yamagishi, K. Hayashi, H. Mitsuhashi, M. Imanari, and K. Matsushita, Tetrahedron Lett., 4735 (1973).

2745. H. Yamasaki and K. Yamasaki, J. Biochem., 70, 235 (1971).

2746. H. Yamasaki and K. Yamasaki, J. Biochem., 71, 77 (1972).

2747. K. Yamasaki, Kawasaki Med. J., 4, 227 (1978).

2748. K. Yamasaki, Y. Ayaki, and G. Yamasaki, J. Biochem., 71, 927 (1972).

2749. K. Yamasaki, Y. Ayaki, and H. Yamasaki, J. Biochem., 70, 715 (1971).

2750. K. Yamasaki, F. Noda, and K. Shimizu, J. Biochem., 46, 739 (1959).
2751. K. Yamasaki, F. Noda, and K. Shimizu, J. Biochem., 46, 747 (1959).
2752. T. Yamauchi, Chem. Pharm. Bull., 7, 343 (1959).
2753. T. Yamauchi, F. Abe, Y. Ogata, and M. Takahashi, Chem. Pharm. Bull., 22, 1680 (1974).
2754. M. Yamazaki, H. Fujimoto, and T. Kawasaki, Chem. Pharm. Bull., 28, 245 (1980).
2755. M. Yamazaki, K. Susago, and K. Miyaki, J. Chem. Soc. Chem. Commun., 408 (1974).
2756. M. Yamazaki, S. Suzuki, and K. Miyaki, Chem. Pharm. Bull., 19, 1739 (1971).
2757. N.C. Yang and R.A. Finnegan, J. Am. Chem. Soc., 80, 5845 (1958).
2758. N. Yanishlieva, E. Marinova, and H. Schiller, IUPAC 11th Int. Symp. Chemistry of Natural Products, Golden Sands, Bulgaria, Symp. Papers, 2, 141 (1978).
2759. N. Yanishlieva, H. Schiller, and E. Marinova, IUPAC 11th Int. Symp. Chemistry of Natural Products, Golden Sands, Bulgaria, Symp. Papers, 2, 145 (1978).
2760. S. Yasuda, Comp. Biochem. Physiol., 44B, 41 (1973).
2761. S. Yasuda, Comp. Biochem. Physiol., 48B, 225 (1974).
2762. S. Yasuda, Comp. Biochem. Physiol., 49B, 361 (1974).
2763. S. Yasuda, Comp. Biochem. Physiol., 50B, 399 (1975).
2764. L. Yoder, D.R. Sweeney, and L.K. Arnold, Ind. Eng. Chem., 37, 374 (1945).
2765. P.H. Yu and L. Tan, J. Steroid Biochem., 8, 825 (1977).
2766. M. Zander, P. Koch, Luu Bang, G. Ourisson, and J.-P. Beck, J. Chem. Research, (S) 219 (1977); (M) 2572 (1977).
2767. V.A. Zanin, Mikrobiologiya, 37, 919 (1968).
2768. I.I. Zaretskaya, L.M. Kogan, O.B. Tikhomirova, Jh. D. Sis, N.S. Wulfson, V.I. Zaretskii, V.G. Zaikin, G.K. Skryabin, and I.V. Torgov, Tetrahedron, 24, 1595 (1968).
2769. Z.V. Zaretskii, MASS SPECTROMETRY OF STEROIDS, John Wiley & Sons, New York/Toronto, 1976, pp. 96-101.
2770. B.R. Zeitlin, S.F. Biscardi, and M.N. George, Oxidation Products of Fats, Interim Progess Report No. 2, Biological Phase, General Foods Corp., February 18, 1970, NIH Contract PH 43-68-1005.
2771. A. Zlatkis, B. Zak, and A.J. Boyle, J. Lab. Clin. Med., 41, 486 (1953).
2772. W. Zschiesche and G. Bruns, Oncologia, 18, 289 (1964).
2773. J.J.L. Zwikker, Pharm. Weekblad, 54, 101 (1917); via Chem. Abstr., 11, 1559 (1917).